T0190189

Studies in Systems, Decision and Control

Volume 179

Series editor

Janusz Kacprzyk, Polish Academy of Sciences, Warsaw, Poland
e-mail: kacprzyk@ibspan.waw.pl

The series "Studies in Systems, Decision and Control" (SSDC) covers both new developments and advances, as well as the state of the art, in the various areas of broadly perceived systems, decision making and control–quickly, up to date and with a high quality. The intent is to cover the theory, applications, and perspectives on the state of the art and future developments relevant to systems, decision making, control, complex processes and related areas, as embedded in the fields of engineering, computer science, physics, economics, social and life sciences, as well as the paradigms and methodologies behind them. The series contains monographs, textbooks, lecture notes and edited volumes in systems, decision making and control spanning the areas of Cyber-Physical Systems, Autonomous Systems, Sensor Networks, Control Systems, Energy Systems, Automotive Systems, Biological Systems, Vehicular Networking and Connected Vehicles, Aerospace Systems, Automation, Manufacturing, Smart Grids, Nonlinear Systems, Power Systems, Robotics, Social Systems, Economic Systems and other. Of particular value to both the contributors and the readership are the short publication timeframe and the world-wide distribution and exposure which enable both a wide and rapid dissemination of research output.

More information about this series at http://www.springer.com/series/13304

Cristina Flaut · Šárka Hošková-Mayerová
Daniel Flaut
Editors

Models and Theories in Social Systems

 Springer

Editors
Cristina Flaut
Faculty of Mathematics and Computer
  Science
Ovidius University of Constanţa
Constanţa, Romania

Daniel Flaut
Faculty of History and Political Science
Ovidius University of Constanţa
Constanţa, Romania

Šárka Hošková-Mayerová
Department of Mathematics and Physics,
  Faculty of Military Technology
University of Defence
Brno, Czech Republic

ISSN 2198-4182   ISSN 2198-4190 (electronic)
Studies in Systems, Decision and Control
ISBN 978-3-030-13082-4   ISBN 978-3-030-00084-4 (eBook)
https://doi.org/10.1007/978-3-030-00084-4

This Springer imprint is published by the registered company Springer Nature Switzerland AG
The registered company address is: Gewerbestrasse 11, 6330 Cham, Switzerland

Preface

The book *Models and Theories in Social Systems* is part of the series *Studies in Systems, Decision and Control* published by Springer. This is the result of a scientific collaboration in the fields of Mathematics, Statistics, and Social Science, among Prof. Šárka Hošková-Mayerová from the *University of Defence* of Brno (Czech Republic), Profs. Cristina Flaut and Daniel Flaut from the *Ovidius* University of Constanţa (Romania). The different studies included in this volume, selected after a peer-review process, are collected into four parts.

Part I *General Remark* is composed of the chapter of Syamal K. Sen and Ravi P. Agarwal, which is titled "Perusing the Minds Behind Scientific Discoveries." The authors record the essential discoveries of a few scientists and their interaction/reaction with their respective diverse environments. The chapter records the winding path of their scholarship throughout history, and most importantly, the thought process of each individual that resulted in the mastery of their subject. Despite there are many other legendary scientists with their gigantic contributions, the few instances mentioned in this contribution are good enough to get a feel of the trend of the mindset of the scientific community, in general.

The contribution of Cavallo, D'Apuzzo, Di Nola, Squillante, and Vitale opens the Part II of the volume, which is titled *Theories in Social Systems*. This chapter, namely "A General Framework for Individual and Social Choices," considers some models for both individual and social choices. Specifically, the proposed structures allow unifying some previous models analyzed in the literature and also overcoming or interpreting some critical issues or paradoxes. Moreover, relevant properties are illustrated in detail.

Afterward, Maturo, Migliori, and Paolone propose the research paper "Nationality Board Diversity in Organizations: A Brief Review and Future Research Directions." Because of the increasing interest in the literature on the impact of diversity on firms' performance, the authors focus on the issue of nationality board variety in organizations. The chapter highlights that the concept of diversity is multidimensional and can concern different aspects such as gender, nationality, educational, age, ethnicity, and race. However, the authors concentrate on the previous studies limiting their attention to nationality board diversity.

The final purpose of this contribution is to illustrate and discuss the results and limits of these studies, and understand why most of them lead to conflicting results.

Chapter "Mathematical Modeling of Some Physical Phenomena Through Dynamical Systems," by Olivia Ana Florea, considers that differential equations and system of differential equations represent the kernel of the mathematical modeling, offering tools to predict the natural phenomena from science, technics, medicine, biology, etc. The author analyzes the phase portraits of different dynamical systems linear and nonlinear, the Lagrangian formalism of a problem encountered in aerodynamics, and averaging method for nonlinear differential equation.

Carp, Popa, and Serban, in their Chapter "Methods for Improving the Quality of Image Reconstruction in Computerized Tomography," present several classes of methods which can rise from classical projection-based algorithms, such as Kaczmarz- and Cimmino-type algorithms, for algebraic reconstruction of images in computerized tomography.

Pipina Nikolaidou contributes with the Chapter "Questionnaires, Bar and Hyperstructures." She presents an application of the hyperstructure theory in the field of social sciences; specifically, the bar is suggested as a tool to be used in questionnaires instead of Likert Scale. The utility of this instrument is giving the opportunity of obtaining more accurate results, and also automatically saving them on computers.

The following chapter is titled "Micropolar Thermoelasticity with Voids Using Fractional Order Strain" and is presented by Lavinia Codarcea-Munteanu and Marin Marin. This study deals with thermoelasticity of micropolar materials with voids that use the fractional order strain, to determine some equations of this linear thermoelasticity theory, as well as of a reciprocity relation for the mentioned bodies.

Adina Chirila and Marin Marin are the authors of Chapter "Diffusion in Microstretch Thermoelasticity with Microtemperatures and Microconcentrations." They focus on the linear theory of microstretch thermoelasticity for materials whose particles have microelements that are equipped with microtemperatures and microconcentrations. Specifically, they derive the field equations and constitutive equations for isotropic and homogeneous bodies, introduce some dimensionless quantities and establish the continuous dependence of solutions upon initial data and body loads by means of the Gronwall inequality, and finally provide a rigorous mathematical model with various possible applications in materials science, engineering, and even biology.

Chapter "Axial-Symmetric Potential Flows," by Plaksa, consider axial-symmetric stationary flows of the ideal incompressible fluid as an important case of potential solenoid vector fields. He establishes relations between axial-symmetric potential solenoid fields and principal extensions of complex analytic functions into a special topological vector space containing an infinite-dimensional commutative Banach algebra.

Dušan Knežo and Alena Vagaská, in Chapter "Monte Carlo Method Application and Generation of Random Numbers by Usage of Numerical Methods," deal with the Monte Carlo method, which is often used for simulating systems with many coupled degrees of freedom, for simulation of experiments. The chapter presents some methods of generating random numbers by usage of standard numerical methods for various probability distributions types.

Part III of the book is titled *Models in Social Systems*. The first contribution of this part is Chapter "Rolling Circles of Motions: Yesterday and Today," which is authored by Murat Tosun and Soley Ersoy. In this chapter, they give a short historical survey of basic events which happened during the development of models depending on rolling circles around circles. Specifically, the study focuses on elliptic and cycloidal (epicycloid or hypocycloid) motion by use of the complex forms of Bottema's instantaneous invariants characterizing the infinitesimal properties of motion.

Chapter "Some Remarks on Social Life in Romanian Towns and Cities in the 1930s, Based on Statistical Data," by Daniel Flaut and Enache Tuşa, presents some aspects regarding the social life in some Romanian towns and cities, based on statistical data, in the fourth decade of the twentieth century, a period marked by economic crises, social problems, and the imminent outbreak of World War II.

Dan Vătăman proposes the study in Chapter "Developments in Decision-Making Process Within the European Union System." This research deals with developments in decision-making process within the European Union, the way in which the European legislation is adopted, and also the problem of clarifying principles, theories, and technical issues regarding decision-making process within the European Union system.

Eva Kellnerová, Kristýna Binková, and Šárka Mayerová, in their Chapter "Assessment of the Efficiency of Respiratory Protection Devices Against Lead Oxide Nanoparticles," evaluate the current state of health- and safety-related problems of those people exposed to environmental burden due to occupational requirements. Specifically, the authors focus on personal protective equipment used against inhalation of pollutants from the air. Filter efficiency is determined according to standardized methods given by the standardized Czech technical norms. However, such rehearsals are not specifically focused on an ultrafine aerosol with the content of nanoparticles in the range of 7.6–299.6 nm. The study evaluates permeability of one of the most often used protective filter OF-90 against ultrafine aerosol lead oxide with predetermined characteristics.

Chapter "Community Detection in Social Networks," by Fataneh Dabaghi Zarandi, and Marjan Kuchaki Rafsanjani, deals with the use of social, biologic, communication, and the World Wide Web networks. Particularly, the authors introduce several methods for community detection and their comparison.

Lepellere, Cristea, and Gubiani, in Chapter "The E-Learning System for Teaching Bridging Mathematics Course to Applied Degree Studies," present a comparison between the University of Udine (Italy) and the University of Nova Gorica (Slovenia). Specifically, this study focuses on the e-learning system for teaching the math bridge-course to applied degrees studies.

Anata-Flavia Ionescu and Dorin-Mircea Popovici propose the research paper "Applications of Multi-Agent Systems in Social Sciences: Virtual Enterprises as an Example." The study is a review of the most important applications of the multi-agent systems in social sciences, with a particular focus on virtual enterprises.

Following, Fabrizio Maturo, Viviana Ventre and Angelarosa Longo discuss the topic in Chapter "On Consistency and Incoherence in Analytical Hierarchy Process and Intertemporal Choices Models." The authors focus on two different approaches in decision-making processes, i.e., the analytical hierarchy process and intertemporal choices models, highlighting the consistency conditions usually adopted. After a general discussion on consistence and incoherence in the framework of these two different approaches, they show that sometimes it is preferable to weaken or reinforce coherence conditions according to the specific context.

Šárka Hošková-Mayerová and Antonio Maturo, in their Chapter "On Some Applications of Fuzzy Sets for the Management of Teaching and Relationships in Schools," analyze the problem of uncertainty on the result of an aggregation operation and on the degree to which a relation holds. They suggest the use of hyperoperations, which permit considering together many possible results of the interaction of any ordered pair of elements and fuzzy sets that give the possibility to measure the degree of belonging of an element to a set described by a linguistic property or the degree of a relation between individuals. The authors show some possible applications to social science at the aim to give an efficient tool for modeling of social phenomena.

The research article "Resources and Capabilities for Academic Spin-Offs' Development. An Empirical Analysis of the Italian Context," by Migliori and De Luca, investigates which resources can affect more than others the creation and successful development of university spin-offs (USOs). Using a sample of 100 Italian USOs, their analysis shows that spin-offs appear to be quite innovative but they generally need time and probably more funding to protect their innovation through patents issuing. In this sense, established spin-offs suffer for the difficulties in raising funds, high costs of developing ideas, and lack of governmental support.

Part IV of this volume, which is titled *Mathematical Methods in Social Sciences*, starts with Chapter "A Fixed Point Result on the Interesting Abstract Space: Partial Metric Spaces" by Erdal Karapınar. The author investigates the existence of fixed point of certain mappings via simulation functions in the framework of an interesting abstract space, namely partial metric spaces. The main results of this manuscript not only extend, but also generalize, improve, and unify several existing results on the literature of metric fixed point theory.

Dorina Raducanu, in her study in Chapter "Geometric Properties of Mittag-Leffler Functions", concentrates on certain geometric properties for two-parametric Mittag-Leffler function.

Following, in the contribution in Chapter "Special Numbers, Special Quaternions and Special Symbol Elements," Diana Savin proposes a study of some properties of quaternion algebras and symbol algebras and obtains a specific algebraic structure.

Cristina Flaut presents some applications of quaternions and octonions. Specifically, her Chapter "An Algebraic Model for Real Matrix Representations. Remarks Regarding Quaternions and Octonions" illustrates the real matrix representation for complex octonions and some of its properties which can be used in computations, where these elements are involved. Moreover, she gives a set of invertible elements in a split quaternion algebra and in a split octonion algebra.

Chapter "A Theory of Quaternionic G-Monogenic Mappings in E_3," by Kuzmenko and Shpakivskyi, considers a class of so-called quaternionic G-monogenic mappings and proposes a description of all mappings from this class by using four analytic functions of complex variable. For G-monogenic mappings, they generalize some analogues of classical integral theorems of the holomorphic function theory of the complex variable (the surface and the curvilinear Cauchy integral theorems, the Morera theorem), and Taylor's and Laurent's expansions. Moreover, they introduce a new class of quaternionic H-monogenic (differentiable by Hausdorff) mappings and establish the relation between G-monogenic and H-monogenic mappings. Finally, they prove the theorem of equivalence of different definitions of a G-monogenic mapping.

Serpil Halici, in Chapter "On Bicomplex Fibonacci Numbers and Their Generalization," consider bicomplex numbers with coefficients from Fibonacci sequence and give some identities. He demonstrates the accuracy of such identities by taking advantage of idempotent representations of the bicomplex numbers, and then, by this representation, he gives some identities containing these numbers. Then, the author proposes a generalization that includes these new numbers and calls them Horadam bicomplex numbers. Finally, the Binet formula, the generating function of Horadam bicomplex numbers, and two important identities that relate the matrix theory to the second order recurrence relations are obtained.

Caruso, Gattone, Balzanella, and Di Battista, in their Chapter "Cluster Analysis: An Application to a Real Mixed-Type Data Set," stress the importance of clustering mixed data, and propose an application on a real-world mixed-type data set regarding flight delays.

In Chapter "Ordering in the Algebraic Hyperstructure Theory: Some Examples with a Potential for Applications in Social Sciences," several examples of concepts of the algebraic hyperstructure theory, which are all based on the concept of *ordering* are included. The author Michal Novák pointed the fact that in many aspects related to social sciences the population, i.e., the elements of the carrier set, on which the operation or a hyperoperation is constructed, are somehow put in relations. A typical example of this is *family relations*, in which the set of individuals are linked in two ways: by mating operation (or hyperoperation) and in descendant—ancestor relation. It is also shown how these concepts could be linked. The reason why this selection was made is the fact that in social sciences, objects are often linked in two different ways, which can be represented by an operation (or a hyperoperation) and a relation. The algebraic hyperstructure theory is useful in considerations of social sciences because in this theory the result of an interaction of two objects is, generally speaking, a set of objects instead of one particular object.

The closing chapter of this book is titled "Classical and Weakly Prime *L*-Submodules" by Razieh Mahjoob and Shaheen Qiami. Let *L* be a complete lattice, the authors introduce and characterize classical prime and weakly prime *L*-submodules of a unitary module over a commutative ring with identity. Also, they topologize Cl.L-Spec(M), the collection of all classical prime *L*-submodules of M and investigate the properties of this topological space.

In summary, the book *Models and Theories in Social Systems* collects a broad range of models and theories regarding social systems. Because of the wide spectrum of topics that social systems cover, different issues related to Mathematics, Statistics, Teaching, Social Science, and Economics are discussed. Due to the large number of interests of the papers collected in this volume, the latter is addressed, in equal measure, to Mathematicians, Statisticians, Sociologists, Philosophers, and more generally to scholars and specialists of different sciences.

The second editor thanks for the support of her work provided within project maintained by the Ministry of Defence the Czech Republic—Project code: DZRO K-217.

Constanţa, Romania   Cristina Flaut
Brno, Czech Republic   Šárka Hošková-Mayerová
Constanţa, Romania   Daniel Flaut

Contents

About the Editors

Cristina Flaut is a professor in the Department of Mathematics and Computer Science at the Ovidius University of Constanţa, Romania. She is the author and the co-author of more than 60 monographs, chapters of books, and papers in important journals (as for example in Taylor & Francis, in Springer or in the journals: Ann. Mat. Pura Appl., Adv. Appl. Clifford Algebras, Bull. Korean Math. Soc., Adv. Differ. Equ.-NY, J. Intell. Fuzzy Syst., Results Math., Chaos, Solitons & Fractals, Soft Computing, Algebr. Represent. Theor., etc.).

She is Editor-in-Chief of the journal *Analele Ştiinţifice ale Universităţii Ovidius Constanţa-Seria Matematica*, an ISI journal.

In 2016, she was considered the best researcher of the Ovidius University of Constanţa.

Areas of interest: algebra (nonassociative algebras, logical algebras), coding theory and cryptography.

Šárka Hošková-Mayerová is an associate professor in the Department of Mathematics and Physics at the University of Defense in Brno, Czech Republic. She is the co-author of one monograph, co-editor of 4 books published in Springer Publishing house, author or co-author of several chapters of books, and papers in important journals (e.g., Soft Computing, Computers and Mathematics with Appl., An. Şt. Univ. Ovidius Constanţa-Seria Matematica, Quality & Quantity, Iran. J. Fuzzy sets, Advances in Fuzzy Systems, International Journal of Production Research, Ital. J. Pure and Appl. Math., Deturope, J. of Security Sustainability Issues).

She is Editor in Chief of the journal *Ratio Mathematica*, indexed in various Databases, also member of editorial board of various journals, eg. *Ital. J. Pure and Appl. Math.* and *Advances in Military Technology*.

Areas of interest: algebraic hyperstructures and fuzzy structures, mathematical modelling and decision-making process.

Daniel Flaut is a professor in the Department of History and Political Science at Ovidius University of Constanţa, Romania. He is the author and the co-author of more than 50 books, chapters of books, and papers in important journals.

He is the Editor-in-Chief of the journal *Revista Română de Studii Eurasiatice*, indexed in various Databases. He is also member of editorial board of various journals.

He is Director of the Eurasian Studies Center of Faculty of History and Political Science of Ovidius University of Constanţa, Romania.

As the co-author of the book *Arheologie medievală romînă*, he received in 2006 the "George Potra" prize, awarded by the Cultural Foundation "Magazin Istoric", Romania.

Areas of interest: medieval history, auxiliary sciences of history, history of international relations.

Part I
General Remark

Perusing the Minds Behind Scientific Discoveries

Syamal K. Sen and Ravi P. Agarwal

Abstract All knowledge exists in one's mind and not outside. One needs to do knowledge mining within one's own mind to get at the information being sought. A scientist brings to us the information viz. the scientific knowledge. Over millennia the scientists apparently painstakingly have brought forth for humanity their creations. Their creations have contributed to lift the civilization to such a height that we could not have imagined a few decades ago. The mining goes on and will continue as long as the civilization remains alive. We have read the amazing scientific creations of these revered legends. Now we are reading them—their nature, characters, and lastly how their minds had worked in their quest for scientific truth in both adverse and conducive environments. This study, we believe, will immensely influence the young minds having great respect for these memorable selfless beings and emulate their virtues.

Keywords Consciousness · Extraordinary living computers Infinity of pockets of mind · Knowledge mining · Limit of computation Newton of France · Reading scientists · Teenage giant

S. K. Sen (✉)
GVP-Prof. V. Lakshmikantham Institute for Advanced Studies, GVP College of Engineering, Visakhapatnam, India
e-mail: sksenfit@gmail.com

S. K. Sen
Department of Computer Science
and Engineering, GVP College of Engineering, Visakhapatnam, India

R. P. Agarwal
Department of Mathematics Texas, A & M University-Kingsville, Kingsville, TX, USA
e-mail: Ravi.Agarwal@tamuk.edu

© Springer Nature Switzerland AG 2019
C. Flaut et al. (eds.), *Models and Theories in Social Systems*, Studies in Systems, Decision and Control 179, https://doi.org/10.1007/978-3-030-00084-4_1

1   Introduction

The chapter records the essential discoveries of a few scientists (Agarwal and Sen 2014; Dedron and Itard 1973) following the birth of ideas on the basis of prior ideas ad infinitum. The authors document the winding path of their scholarship throughout history, and most importantly, the thought process of each individual that resulted in the mastery of their subject. The chapter implicitly addresses the nature and character of every scientist as the authors attempt to understand their visible actions in both hostile and hospitable environments. The authors hope that this will enable the reader to understand their mode of thinking and perhaps even to emulate their virtues in life.

Although the authors include only a few of the scientists to conserve space, the essence remains valid across the whole body of scientists — mathematical, computational, physical, spiritual, and psychological. The essence involves focussing the mind intensely on the question whose answer is sought during a period of time.

Section 2 deals with the limit of computation by a living as well as that by a non-living computer along with a few examples of extraordinary living computers. Also included in this section is a comparison of the speed of computation and the information storage capacity of each of the two. Section 3 depicts the scientific contributions of a few of the renowned scientists and an attempt to correlate their visible action compatible with their contributions. Section 4 comprises conclusions.

2   Living Versus Non-living Computers: Limits of Computation, Speed, and Storage

The most vital distinction between a living computer e.g. a living human being and a non-living computer viz. a modern digital computer is **the distinction between the infinity and a non-infinity**.

Limit of computation by a living computer. The term 'living computer' implies any living being such as a living human being, a living animal, a living bird, a living insect, and a living bacterium. A living computer, for instance, a living human has 100 billion neurons (on average). Each neuron has 7000 synaptic connections (on average) to other neurons.

Thus there are 700 trillion synapses on average for an adult. If it is assumed that a neuron can hold one bit (one bit 0 or 1 constitutes the building block of any information i.e. a bit is indivisible) of information, then a man will have at least 10^{11} bits of information stored in his memory, which is comparable to information stored in a public library.

If a neuron has to hold an information then it has to be at least one bit of information since anything less than one bit is simply non-existent (i.e. no information). That is the best a computer scientist could assume. It is the fact that a human being has a finite number of neurons.

But it is possibly not the fact that the information content and hence the knowledge content in the mind of a living human being is finite. As a matter of fact the information/knowledge that exists in one's mind is indefinitely large or possibly limitless. There exists no knowledge outside one's own mind.

One may, however, generate an inference/conclusion using the already existing knowledge and store in his mind. This inference is no independent new knowledge as it is an outcome of the information already existing in the mind.

This implies that the knowledge existing in Mr. X's mind is exactly equal to that existing in Mr. Y's mind. The only difference is the speed of mining a specific knowledge from one's own mind. A scientist (materials/spiritual) only do knowledge mining in the ocean of knowledge that is already there in his/her mind. In the absolute sense nobody creates any knowledge not already existing in him. The proof is not mathematical in which fuzziness cannot be ruled out. The proof is through actual deep meditative state of the mind. The spiritual scientists of the world viz. rishis have experienced it.

This first-hand experience is the best possible proof (better than any mathematical proof based on deduction/induction/contradiction/construction or any other method) totally devoid of fuzziness. The biological neural network responsible for communication has a speed based on a state of mind out of possible infinite states of mind, which can be categorized in 4 major divisions viz. super-conscious, conscious, subconscious, and unconscious (consciousness not exactly zero). In each of these categories, there are infinity of states.

The speed of neural communication is extremely fast — possibly faster than the *fastest* available digital computer today (2018) and that in future — in the unconscious state and not so prone to errors/mistakes (Sen and Agarwal 2016; Lakshmikantham and Sen 2005) while it is much slower (than that of an ordinary laptop) in the conscious state of mind and considerably prone to mistakes/errors. The speed and quality (error) of transmission in the subconscious state will be somewhere in between those of conscious and unconscious states. The probability of committing an error in any state of mind of any living being (except the highest state of mind) is never zero. In this regard there is no exception.

This probability depends on a state of mind and varies from one state to another while the probability of committing an error by a non-living modern computer is zero (within its finite domain subject to the rules of the concerned computation – numerical, semi-numerical, and non-numerical). Thus we have the proverb: ***To err is human. Not to err is computer***.

Limit of computation by a non-living computer. Every eighteen months processor speed is doubling. Every twelve months band-width is doubling and every nine months hard disk space is doubling. Behind all these exponential growth is the computational mathematics superimposed on the hardware. In this context, the following story is significant.

A merchant asked his friend (another merchant) to supply zinc for 60 days – 1 gm on Day 1, 2 on Day 2, 4 on Day 3, 8 on Day 4 and so on. For this 60 days' supply, he would pay his friend $ 10,000.

The friend was very glad to readily accept the order as he felt it was indeed a very profitable offer without realizing that the supply grows exponentially and consequently it soon turns out to be beyond his means. In fact, the friend needs to supply

$$2^0 + 2^1 + 2^2 + 2^3 + 2^4 + \cdots + 2^{58} + 2^{59} = 1 + 2 + 4 + 8 + 16 + \cdots +$$
$$+\ 288230376151711744 + 576460752303423488$$
$$=\ \mathbf{1152921504606846975\ gm\ of\ zinc}$$

computed using the Matlab (Mathews and Fink 2014; Sen and Agarwal 2011) commands

>> **format long g; s = 0; for i = 0 : 59, s = s + vpa(2^i), end;**

where 2^{58} and 2^{59} are computed using vpa (variable precision arithmetic) commands viz.

>> **a1 = vpa(2^58, 25), b1 = vpa(2^59, 25)**

a1 = 288230376151711744. b1 = 576460752303423488

If 1 gm of zinc costs 0.5 cent, then 1152921504606846975 gm of zinc would cost \$ 5764607523034240 $\approx$ **576460752303 times larger than** what the merchant had offered!

The question now arises: Is there any limit/barrier to increase the computational power arbitrarily? The answer is 'yes'. It is physics that sets up the limits.

(i) *Speed of light barrier.* Electrical signals (pulses) cannot propagate faster than the speed of light viz. 3×10^{10} cm/s $\approx$ 186000 miles/s. A random access memory used to 10^9 cycles/s (1 GHtz) will deliver information at 0.1 ns (0.1×10^{-9} s) speed if it has a diameter of 3 cm.

(ii) *Thermal efficiency barrier.* Information processing enhances the entropy of the system. Hence the amount of heat that is absorbed is $kT \log_e 2$/bit, where k = Boltzmann constant $= 1.38 \times 10^{-16}$ erg per degree and T is the absolute temperature (taken as room temperature) $= 300$ K. It is not possible to economize any further on this. If we want to process 10^{30} bits/s, the amount of power that we need is $p = 10^{30} \times k \times 300 \times \log_e 2 = 28696293275181728$ erg/s obtained using the Matlab vpa command .$\gg$ **p = vpa(10^30*1.38*10^ − 16*300*log(2),40).** This produces $p = 2869629328$ W $\approx 2.87 \times 10^9$ W, where 1 W $= 10^7$ erg/s.

(iii) *Quantum barrier.* A wave associated with every moving particle is quantified such that the energy of 1 quantum $E = h\nu$, where h = Plank's constant and $\nu =$ frequency of the wave. The maximum frequency $\nu_{max} = mc^2/h$, where m = mass of the system and c = velocity of the light. Hence the frequency band that can be used for signalling is limited to the maximum frequency ν_{max}. From Shannon's information theory, the no. of information that can be processed per sec cannot exceed ν_{max}. The mass of hydrogen atom is 1.67×10^{-24} gm, c = 3×10^{10} cm/s, and $h = 6 \times 10^{-27}$. Thus per mass of hydrogen atom, maximum $1.67 \times 10^{-24} \times 3^2 \times 10^{20}/(6 \times 10^{-27}) = 2.5050 \times 10^{23}$ bits/s can be transmitted.

The no. of atoms in the observable universe is estimated to be around 10^{81}. Thus if the whole universe is dedicated to information processing i.e. if all the atoms are employed to information processing simultaneously, then no more than 2.5050×10^{104} bits/s or $2.5050*10^{\wedge}104*60*60*24*365 = 7.899768e + 111 = 7.8998 \times 10^{111}$ bits/year can be processed. This is the theoretical limit (for massively parallel processing) set by the laws of physics and we are still too far away from it!

A single processor (atom) can process only 2.5×10^{23} bits per sec or 7.9×10^{30} bits per year and no more!! (Question: Assume that every 18 months, a processor speed is doubling. If the CPU clock speed (i.e. clock rate) is 8.31 GHz and if we assume that the bits/s (bit rate) = clock rate, then what will be the processing speed of the CPU? How many years will be needed to arrive at the theoretical limit?)

Extraordinary living computers. All living beings such as humans, animals, and birds do compute consciously, unconsciously, or/and subconsciously, a few of them have extraordinary computing capability. We give below a few examples of living human computers who, unlike a non-living modern computer, are not completely mistake-free (no living being (without any exception) can always be error/mistake-free).

Johann Martin Zacharias Dase (1824–1861), born in Hamburg, Germany, was a calculating prodigy. (A calculating prodigy has extraordinary ability in some area of mental calculation, such as multiplying/dividing large numbers, factoring large numbers, and finding an n-th root of a large number).

Johann attended schools in Hamburg, but he made only a little progress there. He used to spend a lot of time developing his calculating skills; people around Johann found him quite dull. He suffered from epilepsy throughout his life, beginning in his early childhood. At the age of 15 he gave exhibitions in Germany, Austria, and England.

His extraordinary calculating powers were timed by renowned mathematicians including Gauss. He multiplied 79,532,853 by 93,758,479 in 54 s; two 20-digit numbers in 6 min; two 40-digit numbers in 40 min; and two 100-digit numbers in 8 h 45 min. He, in 1840, struck up an acquaintance with L.K. Schulz von Strasznicky (1803–1852), who advised him to apply his powers to scientific purposes.

When Dase was 20, Strasznicky taught him the use of the formula

$$\pi/4 = \tan^{-1}(1/2) + \tan^{-1}(1/5) + \tan^{-1}(1/8)$$

and asked him to calculate π. In two months he carried the approximation to 205 decimal places, of which 200 are correct. He then calculated a 7-digit logarithm table of the first 1,005,000 numbers during his off-time from 1844 to 1847, while he was also occupied by work on the Prussian survey.

His next contribution was the compilation of a hyperbolic table in his spare time, which was published by the Austrian Government in 1857. Next, he offered to make a table of integer factors of all numbers from 7 million to 10 million. On Gauss' recommendation, the Hamburg Academy of Sciences agreed to compensate him, but Dase died shortly thereafter in Hamburg. He had an uncanny sense of quantity;

he could just tell, without counting, how many cattle were in a field or words in a sentence, and so forth, up to quantities of about 30.

Dase may not be aware about how he does the computations in his mind. Possibly besides a conscious state of mind he would be using other non-conscious state also. Several other calculating/mathematical prodigies have been discovered, such as **Jedediah Buxton** (1707–1772) of England, **Thomas Fuller** (1710–1790), known as 'the Verginia calculator' of West Africa, **Richard Whately** (1787–1863) of England, **Zerah Colburn** (1804–1840) a math prodigy of U.S.A., **George Parker Bidder** (1806–1878), an engineer of England, **Truman Henry Safford** (1836–1901) of U.S.A., **Jacques Inaudi** (1867–1950) of Italy, and **Shakuntala Devi** (1929–2013) of India (Devi seems to be faster than the fastest computer on earth).

The case of Shakuntala Devi is different from that of Dase. On one day in 1970's, she gave a talk and demonstrated her calculating prowess in the Lecture Hall No. 303 of Power Engineering Building of Indian Institute of Science (IISc), Bangalore. Earlier she had shown her capability of calculating the n-th root of a large positive integer and also of multiplying two large numbers in several universities in Europe. All these were to her "a child's play".

However, a graduate student of the department of Computer science and Automation of IISc wrote a huge arithmetic expression involving terms such as $105^{2.5}/17^{0.5}$ and running from top left corner to the bottom right corner of the fairly long black board. The student had already computed the value of the expression using the then main frame computer IBM 360/44 in IISc. Devi read the expression sequentially (like any other common human) while the student was writing and then within a couple of seconds she told the answer to the audience. The student responded saying that the answer was wrong. She was startled for a moment and almost immediately came out with the answer which was amazingly correct!

Nobody (perhaps including she) knows how she could do the calculation so fast. Probably other states (other than her conscious states) of her mind are involved in the computation. So far as Dase is concerned, he would possibly be using more of his conscious state (than Devi's) in addition to other states of mind. This is one of the author's (Sen's) first-hand experience. She claimed that she was faster than the fastest computer in the world! This has not been disputed anywhere by anybody although her speed and the concerned computer's speed have not been measured and compared by us. She is called a *human computer*. Definitely she has a good knowledge of arithmetic computation.

Which is faster – living computer or non-living computer? Around 1980 one of the authors (Sen) had a chance meeting with Professor Satish Dhawan inside the State Bank of India, IISc campus, Bangalore. He was then the Director of IISc (and also the Chairman of Indian Space Research Organization (ISRO)) and Sen was a faculty member of the then Computer Centre (later called the Supercomputer Education and Research Centre) of IISc. Sen was then designing and developing computer programs for research problems of various engineering departments as well as of the then department of Applied Mathematics on a main frame computer in IISc and also that in Tata Institute of Fundamental Research (TIFR), Mumbai. Many Ph.D.

research students considered him a good programmer and used to come to him to take his help in computer programming to numerically solve their problems.

Professor Dhawan asked him: Tell me, Sen, which is faster – a supercomputer or a human being in computation? He responded readily saying that definitely 'a supercomputer' and felt happy believing that he could answer correctly the too easy a question. By the time one multiplies manually a 4-digit by another 4-digit number, the result of which may not be always correct, the then computer will compute around one hundred thousand multiplications each of 16-digit by 16-digit without any mistake.

He did not realize then that he had given an answer rather too rashly. Professor Dhawan explained: No, Sen, a human being is much faster than a supercomputer. Take, for instance, the photograph of your mother (He has seen numerous people around and their photos/images are stored in his mind). How much time do you take to respond that it is your mother's photo? A fraction of a second. But when you use Cray supercomputer (having a data base consisting of, say, 500 million photos), it would take several seconds to come out with the correct answer.

Sen's implicit ego that he is good in programming is shattered. He started thinking deeply for several days to find out the reason why he committed such a blunder. It appeared to him that there are various (numerous, rather infinite) states of mind – broadly super-conscious, conscious, sub-conscious, and unconscious states (each one of these major four states has infinity of states). In conscious state we are slow (think mostly sequentially with full consciousness) and more prone to error while in sub-conscious state, we are faster (think considerably parallely with some consciousness) and less prone to error and in unconscious state, we are fastest (think mostly parallely without being much conscious of our action) and least prone to error. In super-conscious state, we are superfast and almost/entirely error-free. At the highest state of mind, we will be entirely error-free.

As a driver of a car, when we apply break, we change the gear without being conscious of it. If we attempt to do these consciously then we will be slow and also more likely to commit error (such as pressing the accelerator rather than the brake in a hurry). When one sees his mother's photo, many parallel actions take place very fast, which he is not aware about nor is he knowledgeable about how the whole process of recognition proceeds.

Most of the revelations occur in a super-conscious state of the mind. A scientist is an intense/deep thinker and when he focuses his mind to a specific problem/issue, he often gets the answer/solution. In fact all the knowledge exists in one's mind. All he does is **knowledge mining and consequent inference drawing**.

At some state of the mind, the speed of a living computer is theoretically infinite. There is yet no way to formally compute this speed since we do not know how the biological neural communications at any state of mind take place.

In some state of the mind, the communications are relatively slow, even much slower than a laptop while in some other state it is much faster, even faster than the fastest available computer (over 10^{18} flops) on earth today (2018) or in future.

Which has more storage – living computer or digital computer? A living computer such as a living human being has an infinite storage although the biological neuron count is finite and is around 10^{11} according to neuro-scientists/physicists. We do not know how the information is stored in a neuron nor do we know the storage capacity of a neuron, whether it is truly infinite or not.

An article appeared in the magazine "Scientific American" (April 1, 2015) wherein it is told that in one experiment, it has been proved that the memory is not stored in the neuron. They experimented with a slug ant its nervous system. And the result literally stunned them. They proved that neurons do not store memories. Equivalently, they showed that memories do not live in neuron's synapses (Jacobson 2015). The search for physical storage for memories by the materials scientists is still on unlike that in Vedanta.

Nevertheless an ocean of information resides in one's mind and it is observed/experienced in deep meditation to be limitless. His mind is the store house of the complete ocean of knowledge. His job is to retrieve a specific information/knowledge.

If the physical existence of mind (consisting of infinity of knowledge) is our brain which is physically finite, then it implies that infinity of knowledge exists in a finite space.

In terms of weight, the average adult human brain weighs 1.3–1.4 kg or around 3 lb. In terms of length, the average brain is around **15 cm** long. For comparison, a newborn human baby's brain weighs approximately 350–400 grams or three-quarters of a pound.

Does this imply that, unlike the digital computer storage which can store only a finite amount of information/knowledge (and never infinite, increasing exponentially though), the infinity of knowledge exists in a small finite brain of a living being?

Does a biological neuron store an unknown number of bits of information? Is this unknown number finite or infinite? Is it that the lowest limit of information that the neuron can hold is 1 bit (since the smallest unit of information viz. 1 bit is indivisible)?'

These are the questions posed to the materials/natural scientists who are yet to bring consciousness into their fold.

In the classical mathematics, we define a 'point' as something that exists having no length, no width, and no height. This is a contradictory definition (since some physical space is required for something to exist). However, there is no other way to define a point to convey to the reader its meaning and usage.

Thus, in a real line there exists infinity of points irrespective of whether its length is 500 miles or 10^{-500} cm. Is our knowledge analogous to points requiring no known physical space in a biological neuron?

How fast (quickly) we can access this depends on our state of mind. This state varies from one moment to another. There exists no knowledge outside one's (anybody's) own mind as stated earlier. For a digital computer, the storage is finite and will continue to remain finite eternally although it increases exponentially with time with an upper limit set by laws of physics.

Even if we assume that 1 biological neuron stores 1 bit (the absolute minimum) of information, then a common human being can store at least 10^{11} bits of information.

This (minimal) storage itself is very large and finite. The British library, on the other hand, holds 10^{15} bits of information.

Ramanujan versus Devi and Dase. There are infinity of states of mind of a living computer e.g. a living human being. We may divide these states into 4 categories – unconscious (consciousness is not exactly zero but low), subconscious, conscious, and super-conscious as stated earlier. Each of these categories has (evidently) infinity of possible states as earlier stated.

Most of the common men display or implicitly use certain states in each of the four levels. When a person uses states beyond these certain states, then he/she turns out to us as one distinctly different from a common human being in the context of extraordinary mental ability. These mental ability obviously makes use of the knowledge existing in some of the states of unconscious, conscious, subconscious, and possibly super-conscious states.

Ramanujan (1887–1920) differs from Devi both (i) in the speed of arithmetic computation and (ii) in the retrieval of background knowledge regarding a subject. Devi depicted extraordinary speed of arithmetic computation involving many numbers including fractional ones and their powers/fractional powers and excelled Ramanujan (and perhaps other few human computers).

Ramanujan, on the other hand, has definitely excelled Devi in (specific areas of) numerical/mathematical knowledge and the interpretation (including spiritual ones) that is truly amazing.

The notebooks/scribbling (Sen and Agarwal 2016; Andrews 2012; Peterson 2006; Kanigel 1992) of Ramanujan traced and being researched has been revealing astonishing mental ability of Ramanujan, which is very much well above others concerned. People are still trying to understand Ramanujan and his very intimacy with numbers (as if these are living beings and carry important messages) and getting deeper insight into the mind of this genius.

Dase is significantly different from both Devi and Ramanujan. He uses states of his mind, which are not identical to those of Devi and Ramanujan, but these states are definitely extraordinary. Further, numbers may not be his personal friends as these are so with Ramanujan.

The relationship among numbers and the highly interesting aspects of many of them appear to be an integral part of Ramanujan's mind, but these do not appear to be so with both Devi and Dase.

Other aspects of Ramanujan's life that seem to differ from other human beings with extraordinary computing and other capabilities may be depicted through the well-known and well-researched activities on Ramanujan (available in literature).

Ramanujan: Thought-provoking Questions on 1729, 0, ∞, computation of Pi and Theta. Divinity was manifested through Ramanujan in terms of numbers. We represent some of the numbers and their interesting and deeply significant spiritual interpretation by Ramanujan. The non-conventional way Ramanujan has viewed/experienced the numbers is simply astonishing. One needs to ponder over deeply on whatever has been revealed and uttered by him.

To him, each number is a living being and his personal friend carrying important distinct messages. This is unlike to us, the common humans, who do not feel/view the numbers as beings with interesting characters and significant message carriers and as those which evoke a personal relationship with our emotional attachments.

1. Hardy–Ramanujan Number: $1729 = x^3 + y^3 = u^3 + v^3$. $x = ?$ $y = ?$ $u = ?$ $v = ?$ where x, y, u, and v are all 1- or 2-digit distinct integer. (*Answer without seeing internet or book.*) Use Matlab/any means to find/compute next pair of any k-digit integers (x, y), $k \geq 3$

2. $0 =$ God (Absolute Reality), $\infty =$ His (God's) manifestation. $0 \times \infty =$ not one Number but all Numbers, each of which corresponded to individual acts of creation. Imagine (seriously) everything including your own body is vanishing/has vanished. Will it be dreadful to you? [Thou shalt not divide by zero. Natural Mathematics/Computation has no division by zero.]

3. Compute π, correct at least up to 7 significant digits, using the identity due to Ramanujan

$$4 \Big/ \left[(2 \times \sqrt{2}/9801) \sum_{k=0}^{\infty} 4(k!)(1103 + 26390k)/k!^4 396^{4k} \right]$$

taking, for ∞, (i) 5, (ii) 6, and (iii) 7. Hence compute the relative error in π.

The Matlab commands (to be written in 1 line in the command window) to compute π are

>> **format long g; t1 = (2 * sqrt(2)/9801); t2 = 0; for k = 0 : 6,**
t2 = t2 + (4 * factorial(k) * (1103 + 26390 * k)/(factorial(k)^4 * 396^(4 * k)));
end; pi1 = 4/(t1 * t2)
pi1 = 3.14159272682899 (correct up to 6 significant digits)

4. Ramanujan's general theta function f (a, b) is defined by

$$f(a, b) = \sum_{k=-\infty}^{\infty} a^{k(k+1)/2} b^{k(k-1)/2}$$

where |ab| < 1. Compute, using Matlab/any means. (i) f(0.5, 0.4), (ii) f(0.9, 1), and f(0.01, 0.02) correct up to 5 significant digits. Take, for ∞, minimum number of terms so that you get the required accuracy. Compute the relative error in computing f(a, b).

Squares summation, ∞/non-∞, natural/artificial consciousness/intelligence, living/non-living computers: Questions

1. The sum of n + 1 (n > 0) consecutive squares starting with x = n(2n + 1) is equal to n consecutive squares starting with y = x+n + 1. For example, if n = 1, then x = 3, y = 5 and hence $3^2 + 4^2 = 5^2$. Take (i) n = 2, (ii) n = 3, (iii) n = 4, and (iv) n = 5 and construct the identities.

2. (i) What is the difference between infinity and non-infinity? (∞)
   (ii) What is the difference between Natural Consciousness and Artificial Consciousness? (Natural consciousness is possessed by a living being while artificial consciousness is a simulation of natural consciousness and is depicted in a finite/limited way by a computer)

(iii)   What is the difference between Natural Intelligence and Artificial Intelligence? (Natural intelligence is infinite/unlimited while artificial intelligence is not)

(iv)   Which is faster – Living Computer (e.g. Living Human) or Non-living Computer (e.g. Hyper (exaflops $= 10^{18}$ flops)-computer)? (Living computer)

 (v)   Is there any point in the infinite space/cosmos (universe), where consciousness is exactly zero? (Sen and Agarwal 2016; Dey 1997) (No)

(vi)   What is consciousness? How do you visualize/distinguish consciousness of a living being and that of artifacts? Is the consciousness in a dead (decaying) human body exactly zero? (Sen and Agarwal 2016; Dey 1997) (*Part 1*. There is no unique definition of consciousness. However, it may be viewed as the state of awareness/being aware of an external object or something within (oneself). *Part 2*. Consciousness of a living being is natural and much more vibrant than that of an artefact. *Part 3*. No, but very small as the atoms in the dead body at room temperature are having activities within – electrons moving round the nucleus consisting neutrons/protons)

(vii)   How many states of consciousness do we (living humans) have? Is the number finite (say, 5) or infinite? Why?

Ramanujan versus human computers: Question How does Ramanujan differ from Human Computers such as Johann Martin Zacharias Dase (who multiplied … two 40–digit numbers in 40 min; and two 100–digit numbers in 8 h 45 min. He then calculated a 7–digit logarithm table of the first 1,005,000 numbers during his off—time from 1844 to 1847, …), Jedediah Buxton, Thomas Fuller, Richard Whately, Zerah Colburn, George Parker Bidder, Truman Henry Safford, Jacques Inaudi, and Shakuntala Devi as mentioned earlier?

Limitations of comprehension of a common human A common human being understands 7 ± 2 things at a time. If one reads out a telephone number 23372198, it will not be difficult for him to memorize and reproduce later. If, on the other hand, he reads out a 100-digit number to him, it will not be possible for him to remember and then reproduce later. For an uncommon/extraordinary human, remembering 100 digit (or bigger) number is possible.

One of the authors (Sen) personally had come across a young high school Telugu boy, Kanala Sriharsha Chakravarthy by name, in December, 1994 in the 39th Congress of Indian Society of Theoretical and Applied Mechanics held at Andhra University (Visakhapatnam, India). A number more than 100 digits was read out to him and then readily he reproduced the number absolutely correctly. He told the audience that he could reproduce the number even after one week. He was capable of performing arithmetic computations too with a speed much beyond the capability of a common human.

We have come across extraordinary humans with different kinds of capabilities — not all are necessarily human computers or humans with uncommon memories.

There are personalities such as Albert Einstein and Srinivas Ramanujan who are capable of seeing the physical/mathematical world far beyond that of common humans. Each of these humans has excelled in one (and sometimes more) of the specific mathematical and computational sciences and earned for him a status as an extraordinary human or a genius.

Many of them (not all) are able to bring down their findings/visions to a level which a sincere common man/seeker (of truth) can comprehend/understand. In other animal worlds e.g. in the world of whales/dogs, such capabilities do exist, but we are not enough knowledgeable about their uncommon animal behaviour.

Infinity versus non-infinity "Fast computation" invariably conjures up in a scientist (specifically a computer/computational scientist) a feeling of quick computation in contrast to slow computation. In this context, a digital computer is improving in its computational speed exponentially over certain periods of time, it remains ever finite though.

On the other hand, a living computer at certain states of mind depicts infinite computational speed, not scientifically explainable though. Shakuntala Devi may be considered as an example in this aspect.

Thus the difference between speed of a living computer and a non-living computer is infinite (i.e. infinity minus any finite quantity is always infinity). The difference between natural consciousness and natural intelligence (e.g. of a living human) and artificial consciousness and artificial intelligence (e.g. depicted by a computer) is infinite.

The way Devi does the computation mentally seems to be different from that of Dase. **None of them has come out with the formal procedure or, equivalently, the algorithm that they execute in their mind. In all probability, their computational activities are in different mental states – states which they remain oblivious of. They do these usually too fast (much faster than a hyper-computer) and almost always very impressively correctly.**

Consciousness of living beings versus that of artifacts There is no precise single definition of consciousness. Like the definitions of mathematics, consciousness also is defined/viewed differently by different individuals based on their individual perception (of consciousness).

Consciousness is defined as the state of awareness/being aware of an external object or something within (oneself). It is viewed as sentience, awareness, subjectivity, or to feel wakefulness having a sense of selfhood and the executive control system of the mind.

Many scientists believe that experience is the essence of consciousness and it can only be completely known subjectively from the inside. If consciousness is subjective and not visible from the outside, why do the vast majority of people believe that other people are conscious, but stone and wood are not? This is known as the problem of other minds.

Consciousness has been considered out of bound of materials science such as physics and chemistry and has been avoided as an important research topic by scientists. This is because of a common feeling that a phenomenon defined in a subjective term cannot be meaningfully studied employing an objective experimental procedure.

A psychological study which differentiated between (i) fast, parallel, and extensive unconscious processes and (ii) slow, serial, and limited conscious processes was made by George Mandler in 1975.

Since 1980s, psychologists and neuroscientists developed an area of research called *consciousness studies* resulting many experimental and methodological work published in journals such as *Consciousness and Cognition* and *Journal of Consciousness Studies*.

Besides, regular conferences have been organized by societies such as the Association for the Scientific Study of Consciousness.

Consciousness of an artifact/a computer is just that consciousness that *Paramatma* (universal soul i.e. the soul of the universe, that exists everywhere; there exists no point in infinite space where this consciousness is absent) is having. On the other hand, *Jeevatma* (the individual soul) is having a consciousness much more vibrant than that of an artifact or a computer.

However, the way we attribute consciousness, it is the consciousness that is exclusively possessed by a living being (e.g. a living human). And the consciousness depicted by an artifact is just a simulation of the natural consciousness; hence we can call it artificial consciousness which is no real consciousness of a living being.

Both artifacts and living beings have one thing in common i.e. enormous amount of activities (perpetual/absolute inactivity is absent in both) which are ever present to a varying degree.

Vedic mathematicians (Pandit 1993) are common living humans and do computations using vedic rules usually consciously. They do not, however, have the ability of performing computations which are done by Shakuntala Devi. Their speed of computation cannot match that of Shakuntala Devi. Further, a vedic mathematician cannot compute a truly large arithmetic problem that can be readily solved by Devi.

A difference is that a vedic mathematician consciously knows what he is doing step by step while Devi seems not to know the steps that she performs sequentially/parallely.

Since consciousness is out of bound of materials science such as physics, it has limitations/constraints in terms of explaining the cause of nonstop activity that has been going on eternally throughout the infinite space/endless universe.

Infinity of pockets of mind and extraordinary activity of a pocket in some living being When we ponder over the behaviour of a living being, we will be able to observe that several actions are performed by the being without being conscious or intensely conscious about most or possibly all of them.

In a common living being, we will notice no remarkable extraordinary activity in any pocket of his mind. A pocket may be considered as the virtual location of the mind, in which a mental activity is performed. It is possible that 2 or more activities in 2 or more pockets can be simultaneously going on.

However, in some individuals, a pocket could be extraordinarily active without their being aware of the underlying natural algorithm that is being performed.

Thus Devi and Dase could not tell how they had performed their computational algorithm so fast and almost always so correctly. Some of their distinct pockets of mind are extraordinarily active both in terms of speed as well as in terms of accuracy.

On the other hand Ramanujan had above-normal (but not very extraordinary computational skill like Devi) computational ability and also extraordinary mathematical ability which Devi and Dase did not possess.

Consider Ramanujan's lost notebook. It is the manuscript in which Ramanujan recorded the mathematical discoveries of the last year (1919–1920) of his life. Its whereabouts were unknown to all but a few mathematicians until it was rediscovered by George Andrews in 1976, in a box of effects of G. N. Watson stored at the Wren Library at Trinity College, Cambridge.

The "notebook" is not a book, but consists of loose and unordered sheets of paper — "more than one hundred pages ... in Ramanujan's distinctive handwriting. The sheets contained over six hundred mathematical formulas listed consecutively without formal mathematical proofs." Ramanujan himself could visualize the validity of these formulas intensely.

Such a validity arising from the concerned state of Ramanujan's mind seems much more powerful than the formal mathematical proofs which are not necessarily always free from fuzziness. Not any formula of Ramanujan has been proved to be wrong so far by a mathematician.

George Andrews and Bruce C. Berndt (Andrews and Berndt 2013) have published several books in which they give mathematical proofs for Ramanujan's formulas included in the notebook. Berndt says of the notebooks' discovery: "The discovery of this 'Lost Notebook' caused roughly as much stir in the mathematical world as the discovery of Beethoven's tenth symphony would cause in the musical world (Peterson 2006)."

3 Reading Scientists

3.1 Making a Language Psychologically More Appealing and Easy for Commoners

Ishwar Chandra Vidyasagar (ICV, 1820–1891), born **Ishwar Chandra Bandyopadhyay** was an Indian Bengali polymath and a key figure of the Bengal Renaissance. He was a philosopher, academic educator, writer, translator, printer, publisher, entrepreneur, reformer, and philanthropist. His efforts to simplify and modernize Bengali prose were significant. He also rationalized and simplified the Bengali alphabet and type, which had remained unchanged since Charles Wilkins and Panchanan Karmakar had cut the first (wooden) Bengali type in 1780.

He also forced British to pass widow remarriage act — a courageous step taken by ICV against death threat by several orthodox Hindu so-called Brahmins/priests. The then (19th century and before) social environment was highly contaminated by the misinterpretation of the Hindu scriptures to satisfy the utterly selfish motive of the priests. The consequent customs prescribed by the priesthood resulted in the utter misery of the widows, particularly young ones.

Narendranath Dutta (1863–1902), fondly called 'Naren' – pre-monastic name of Swami Vivekananda (SV) (Dutta 1938), the most famous out of the 16 direct disciples of Sri Ramakrishna Paramahamsa (SR)) in his childhood days studied **Barnaparichay** (Bengali First book both part I and part II written by ICV with a few wise modifications of alphabets and with thoughtful rhymes) as every Bengali kid starts his academic career with studying this first book — the book which is the gateway of learning Bengali language — without exception.

Although several first books did appear, none became popular like Barnaparichay that has tremendous psychological effect on a child's mind.

The learning which was quick, meaningful, and enjoyable remains embedded in the mind for years to come. Even today after over 162 years Barnaparichay is being taught to children and possibly will continue to be taught for decades or even a century or two to come.

One day Naren (SV) sat in the big room of Balaram Bose's house in Kolkata and was reading the preface of the Bengali First book of ICV with undivided attention. His face looked serious. Baburam Maharaj (Swami Premananda (1861–1918) – one of the 16 direct disciples of SR) asked "What? Are you reading once again the First book?" Naren, with eyes wide open looked at Baburam Maharaj and said with a grave tone "Previously I had studied the First book, now I am reading Vidyasagar." Baburam Maharaj stood there startled for a while and then left the place.

Practically the root of all Indian languages is the Sanskrit language, specifically the Vedic Sanskrit language using which Vedas, Vedanta/Upanishads, and Epics such as the Ramayana and the Mahabharata – specifically Srimad Bhagavad Gita (an important part of Mahabharata) – are set in a narrative framework of a dialogue between Pandava prince Arjuna and his guide and charioteer Lord Krishna.

Facing the duty as a warrior to fight the Dharma Yudhha (i.e. righteous war) between Pandavas and Kauravas, Arjuna is counselled by Lord Krishna to fulfil his Kshatriya (warrior) duty as a warrior and establish Dharma.

The *Bhagavad Gita* presents a synthesis of the concept of Dharma, theistic bhakti, the yogic ideals of moksha (salvation/liberation) through jnana yoga (path of knowledge/self-realization), bhakti yoga (Path of Devotion, the feeling of oneness with the Spirit), karma yoga (the path of selfless action and selfless service), and Raja Yoga (science of mind control or, equivalently, Yoga Sutras of Patanjali[ ]), and Samkhya philosophy — one of the most prominent and the oldest of Indian philosophies. The eminent, great sage Kapila was the founder of the Samkhya School).

These constitute the treasure house of ultimate wisdom/knowledge of ancient rishis (spiritual and materials scientists) — the one became grammatically and phonetically so rigid that it became out of bound of the common people and remained the language, both spoken as well as written, for communication only among the truly learned ones (practically in whole of India that has 179 languages being used at her different regions).

An important extraordinary fact of Sanskrit language is conjoining 2 or more words into a single long word. To sieve out the meaning needs prior knowledge of concerned words which have been conjoined. If one does not have this knowledge and is not able to separate the words out of a conjoined one, it might not be possible to get at the true meaning or to interpret the concerned sentence(s).

We do like to have as highly physically compact/concise a sentence as possible, that embodies extraordinary large amount of information. The physically concise verses of the Vedic Sanskrit literature are indeed packed with immense knowledge/wisdom.

The Vedic Sanskrit (including simpler version of Sanskrit mainly meant for not so highly learned people in Sanskrit) has fewer punctuation marks (used to create sense, clarity and stress in sentences) than those of most of the languages in the world. It uses one vertical line (equivalent to period)\, two vertical lines, comma, and semi-colon. There is practically no other punctuation marks such as a question mark (?) and an exclamation mark (!). When we write, for example, 'What is your name?', the question mark (?) is superfluous, isn't it?

In this regard, the 700 verses of Shrimad Bhagavad Gita, for example, written in a booklet is packed with so much of wisdom that needs many times more the physical space than that of the verses for the interpretation/exposition, Thus the Vedic Sanskrit may be considered a highly dense mine of eternal knowledge/wisdom! What an achievement of the Vedic rishis!

This interpretation benefits (in terms of assimilating the import of the ultimate wisdom of ancient rishis) much wider population not well-versed in Sanskrit/Vedic Sanskrit literature.

Anyway for common people this language (Sanskrit), though possibly more scientific and more amenable to machine processing, became gradually the one obsolete for daily usage as a spoken and written language.

It may be seen that the living being e.g. the living human being does usually neither behave/work like a rigid machine nor is he comfortable with any language which is ideally suited for machine processing. We have seen in all live vibrant natural languages, ambiguity, non-rigidity, and receptivity in terms of accepting widely used words in other languages.

As a matter of fact, it is the convenience of the population at large that matters rather than that of the machine (computer).

It may be remarked that computer (programming) languages are completely formal and ideally suited for a machine, but cannot become a language for the masses. It is repulsive for day-to-day use among living humans beings.

Although one may say that a computer programming language consists essentially of imperative sentences (instructions/commands) and hence not suited for daily use. Strictly speaking there is no restriction that a formal language cannot accept assertive, interrogative, and exclamatory sentences. In fact, assertive sentences are often/always used as inputs (input data).

A formal language is interpretation-independent i.e. there is only unique meaning/interpretation of a sentence always unlike that in a natural language. For a diplomat, such a formal language is not at all welcome. He needs to manoeuvre as and when needed through 2 or more possible meanings/interpretations of a sentence. A natural language allows user flexibility rather than rigidity.

The sphere of Sanskrit shrunk and became a language for the highly learned men of Vedic culture/literature, although Sanskrit literature such as Vedas, Vedanta, the Ramayana, and the Mahabharata happen to be the treasure house of unimaginable wisdom and ultimate knowledge contributed at different times by the ancient *rishis* over millennia.

However, there has been a significant effort from the part of the Government of India to actively promote the learning of Sanskrit among people/students so that the eternal wisdom/knowledge in its literature remains widely alive and vibrant for the humanity in its original form.

ICV critically followed the human psychology, specifically the child psychology and made requisite changes/reduced the number of letters of the Bengali alphabet and also introduced the most appropriate and easy-to-remember and enjoyable rhymes that is ideally suited for the beginners — the kids.

Consequently the Bengali language became much easier and very much within the reach of common people. Huge amount of Bengali literature has so far been written and published, many of them have been relatively easily translated in various Indian and non-Indian languages including the English language — considered widely the international language for scientific and literary communications.

The contributions made by ICV can be gauged only if one could read how such a critical mind as that of ICV thoroughly understood the complex human nature/psychology (Dutta 1938; Hadamard 1945) and brought the best of the language for the common masses.

This enabled millions of common as well as highly learned Bengali people (and non-Bengali people through easy translation) to get the taste of numerous memorable writings such as those of Nobel laureate (in literature) Rabindra Nath Tagore (1861–1941), Sarat Chandra Chattapadhyay (1876–1938), and many others following them.

SV was deeply studying this man — his extraordinary mind! What is it that transpired this man to come out with an outstanding language solution with minimal modification ideally suited for the whole of the population, the best one that could not have been made significantly better by anyone so far? How his mind has worked to achieve such a feat!

A clear understanding about the way one's mind had worked to achieve a perpetual/long lasting contribution — scientific or otherwise — for the humanity definitely would influence other minds and encourage emulating one's virtues.

3.2 A Highly Active Mind in a Sickly Body

We have often heard people saying body and mind are closely related and a strong body has a strong vibrant/active mind. We have, however, no dearth of instances where a sickly body is not only capable of intense thinking but also is capable of outstanding innovation.

Zeros of a function and theory of determinants are so widely known that we tend to forget the discoverer viz. **Alexandre–Théophile Vandermonde** (1735–1796) over 2 and a ½ centuries ago, who was born and died in Paris. He was a sickly child, and so his physician father directed him to a career in music, but he later developed an interest in mathematics.

Though sickly, he has a highly energetic and dynamic mind capable of deep thinking. His father's prescription for a career in music finds an outlet in mathematics. This does not violate the discipline of music. Highly vibrant mind of Vandermonde found much deeper form of 'music' in mathematics.

His complete mathematical work consists of only 4 papers published in 1771–1772. These papers include fundamental contributions to the roots of equations, the theory of determinants, and the **knight's tour problem**.

In chess, the knight is the only piece that does not move in a straight line, and the problem is to determine whether a knight can visit each square of a chessboard by a sequence of knight's moves, landing on each square exactly once.

Vandermonde's interest in mathematics lasted for only 2 years. Afterward, he published papers on harmony, experiments with cold, and the manufacture of steel.

He also became interested in politics, joining the cause of the French revolution and holding several different positions in government.

Although Vandermonde is best known for his determinant, it does not appear in any of his four papers. It is believed that someone mistakenly attributed this determinant to him.

3.3   4-in-1 with Critical Analyzing Power as the Common Factor

Marquis de Condorcet (1743–1794), an 18th century French mathematician, enlightenment philosopher, political scientist, and revolutionist, was born at Ribemont, Aisne, France. He lost his father at a very young age, and was raised by his devout mother.

He studied at the Jesuit College in Reims and at the College of Navarre in Paris, where he was known for his diverse talents. In 1759 at the age of 16, his analytical abilities were noticed by d'Alembert (**Jean-Baptiste le Rond d'Alembert** (1717–1783)) and Clairaut (**Alexis Claude Clairaut** (1713–1765)).

Soon d'Alembert became his mentor, and at the age of 22 Condorcet wrote a treatise on the integral calculus that was widely appreciated. Condorcet worked with famous scientists such as Euler, and one of the Founding Fathers of the United States, Benjamin Franklin (1706–1790), who said: "What science can there be more noble, more excellent, more useful… than mathematics".

His most important contribution was in the development of theory of probability and the philosophy of mathematics. In 1785 he wrote an essay on the **Application of Analysis to the Probability of Majority Decisions**, which included

(i) Condorcet's Jury Theorem, that is, if each member of a voting group is more likely than not to make a correct decision, the probability that the highest vote of the group is the correct decision increases as the number of members of the group increases, and

(ii) Condorcet's Paradox, according to which it is possible for an elector to express a preference for A over B, B over C, and C over A, showing that majority preferences could become intransitive with more than two options.

In 1786, at the age of 43, Condorcet married Sophie de Grouchy (1764–1822), an intelligent and highly educated girl more than twenty years younger than him.

Condorcet strongly desired a rationalist reconstruction of the society, championed many liberal causes, and consequently took an active part in the French Revolution that swept France in 1789 like a tempest. He greeted the advent of democracy (rule by the people).

Fearing for his life, he went into hiding. After a few months, in 1794, he was arrested near Paris and sent to prison, where after 3 days he died mysteriously; either he was murdered, committed suicide using a poison, or had a natural death at Bourg–la–Reine, France.

3.4 Newton of France Who Missed Acknowledging Predecessors' Contributions

Pierre Simon de Laplace (1749–1827) was a French mathematician and theoretical astronomer who was so famous in his own time that he was known as the Newton of France.

Laplace was born at Beaumont–en–Auge (a commune in the Calvados department in the Normandy region in north-western France). His father was a small cottager, or perhaps a farm–laborer. Laplace intended to become a theologian, but his interest in mathematics was piqued by his instructors Christopher Gadbled (1734–1782) and Pierre le Canu at the provincial school in Caen.

In 1768, when Laplace was leaving for Paris, Canu wrote him a letter of recommendation to give to d'Alembert; however, d'Alembert sent Laplace away with a problem and told him to come back in a week. Legend has it that Laplace solved it overnight. This impressed d'Alembert, who managed to get Laplace a mathematics teaching position at the Military School only a few days later.

His main interests throughout his life were celestial mechanics, the theory of probability, and personal advancement. At the age of 24 he was already deeply engaged in the detailed application of Newton's law of gravitation to the solar system as a whole, in which the planets and their satellites are not governed by the Sun alone but interact with one another in a bewildering variety of ways.

Even Newton had been of the opinion that divine intervention would occasionally be needed to prevent this complex mechanism from degenerating into chaos. Laplace decided to seek reassurance elsewhere, and succeeded in proving that the ideal solar system of mathematics is a stable dynamic system that will endure unchange for a long time.

This achievement was only part of a long series of triumphs recorded in his monumental treatise **M´ecanique C´eleste** (published in 5 volumes from 1799 to 1825), which summed up the work on gravitation done by several generations of illustrious mathematicians.

Unfortunately for his reputation, he omitted all references to the discoveries of his predecessors and contemporaries, and left it to be inferred that the ideas were entirely his own. The first four volumes were translated into English during 1814–1817 and published in 1829 at his own expense by the American Nathaniel Bowditch (1779–1838), who doubled its length by adding extensive commentary.

Bowditch complained about Laplace's use of "It is easy to see…", for it invariably required several hours of hard work to see what Laplace claimed to be easy. (Bowditch also wrote a classical work on navigation, **New American Practical Navigator**, in 1802. These writings earned him membership of the Royal Society of London, and professorship in mathematics at Harvard, which he declined.)

The principal legacy of the **Mécanique Céleste** to later generations lay in Laplace's development of potential theory, with its far–reaching implications in a dozen different branches of physical science, ranging from gravitation and fluid mechanics to electromagnetism and atomic physics.

Even though he lifted the idea of the potential from Lagrange without acknowledgment, he exploited it so extensively that ever since his time the fundamental differential equation of potential theory has been known as Laplace's equation. However, this equation first appeared in 1752 in a paper by Euler on hydrodynamics.

His other masterpieces included the treatise **Théorie Analytique des Probabilités**, published in 1812, in which he incorporated his own discoveries in probability from the preceding 40 years. Again he failed to acknowledge the many ideas of others' that he mixed in with his own; regardless of this, his book is generally agreed to be the greatest contribution to this part of mathematics by any one man.

In the introduction he says: "At bottom, the theory of probability is only common sense reduced to calculation". This may be so, but the following 700 pages of intricate analysis, in which he freely used Laplace transforms, generating functions, and many other highly nontrivial tools, has been said by some to surpass in complexity even the **Mécanique Céleste**.

After the French Revolution, Laplace's political talents and greed for position came to full flower. His countrymen speak ironically of his "suppleness" and "versatility" as a politician.

What this really means is that each time there was a change of regime (and there were many), Laplace smoothly adapted himself by changing his principles back and forth between fervent republicanism and fawning royalism and each time he emerged with a better job and grander titles.

He has been aptly compared to the apocryphal Vicar of Bray in English literature, who was twice a Catholic and twice a Protestant. The Vicar is said to have replied to the charge of being a turncoat, "Not so, neither, for if I changed my religion, I am sure I kept true to my principle, which is to live and die the Vicar of Bray".

To balance his faults, Laplace was always generous in giving assistance and encouragement to younger scientists. From time to time he helped forward the careers of men like the chemist Gay–Lussac, the traveler and naturalist Alexander von Humboldt (1769–1859), the physicist Poisson, and (appropriately) the young Cauchy, who was destined to become one of the chief architects of 19th century mathematics.

Laplace's last words "what we know is so minute in comparison to what we don't know" reminds us of Plato's remarks on human knowledge: "In the final analysis, the theory of probability is only common sense expressed in numbers", and "All the effects of nature are only mathematical consequences of a small number of immutable laws".

Laplace once remarked that India gave us the ingenious method of expressing all numbers by means of ten symbols, each symbol receiving a value of position as well as an absolute value, a profound and important idea that appears so simple to us now that we ignore its true merit.

But its very simplicity and the great ease with which it lends itself to all computation puts our arithmetic in the first rank of useful inventions, and we shall appreciate the grandeur of this achievement when we remember that it escaped the genius of Archimedes and Apollonius, two of the greatest men of antiquity. Laplace died in Paris exactly a century after the death of Newton, in the same month and year.

3.5 A Shortest Living Teenage Giant and a Victim of Cruel Environment

Évariste Galois (1811–1832) was a French mathematician whose discoveries are considered to be some of the most original mathematical ideas of the 19th century. He was born in Bourg-la-Reine to a happy and respected family. His mother was well-educated, clever, and intellectually sophisticated, and father was a charming, fun-loving man who delighted in making up clever verses about his friends.

Galois' life was one of disillusionment and disappointment. He was tutored at home by his mother, and at the age of 12 Galois was admitted to Louis-le-Grand, a boarding school in Paris. As a boy, he had boundless energy and enthusiasm.

His love for learning led him to read books of mathematics much like other boys would read mysteries. Because of his superior intelligence, school was extremely boring and unchallenging for Galois, and he rebelled against his teacher's harsh, domineering treatment, which he found un-motivating.

Galois loved to do mathematics in his head and would not bother to write out the proofs systematically on paper. This proved to be highly detrimental; he failed the entrance test for the Ecole Polytechnique as a result.

Galois was extremely disappointed by the unjust test results because he did not find the questions difficult. When Galois was 17, the great teacher Professor Louis—Paul–Emile Richard (1795–1849) came to his school. Professor Richard immediately recognized Galois' genius and encouraged him to compile his discoveries and send them to the French Academy, which was France's finest group of scholars.

Galois promptly took up the challenge and sent his papers to Augustin-Louis Cauchy, a professor at Ecole Polytechnique. But, most unfortunately for Galois, Cauchy never read his work and lost his paper as well. Later, Galois decided to take the Ecole Polytechnique entrance test again. Only 2 tries were allowed, and this would be his 2nd and final chance. However, the examiners had already heard of his intelligence and had their minds set against him even before the test.

They taunted him during the oral test so much that Galois lost his temper and hurled an eraser at them. This ended all his hopes for admission into the best school of mathematics in Europe.

At the age of 19, he produced some very important papers on algebra, in particular an important function of algebra to solve equations. Galois discovered the types of equations that can be solved and those that cannot be.

In a moment of optimism, he submitted his work to the Academy of Science in competition for the Grand Prize in mathematics. The Secretary received his work but died before reading it. To make matters worse, when officials went to retrieve the papers, they had mysteriously disappeared.

Disappointments and setbacks finally took their toll on Galois and he became a very bitter man. He developed a distrust for teachers and institutions and got involved in politics instead. He joined the Republicans, which was a forbidden radical group.

Galois was arrested a couple of times but was acquitted of every charge because they were difficult to prove. He was finally jailed for 6 months on the trumped up, trivial charge of wearing the uniform of the nonexistent National Guard.

During his sojourn in jail, Galois renewed his interest in mathematics. Following his release, he sorted out his papers but was unable to complete them before he was forced into a duel over a girl that he was barely acquainted with and hardly cared for. The duel proved to be his end.

The death of Galois was indeed a terrible waste. If he had known that his work would be of such great significance to mathematics, life would have been meaningful for him. As it was, he died a very disappointed man.

It was not until 1846 that his research was published. Mathematicians began to appreciate the importance of Galois' work, which centered on solving equations using groups. Galois found a way to derive a group that corresponds to each equation. This group, called "the group of the equation" contains the important properties of equations. The so-called Galois groups form an important part of modern abstract algebra.

3.6 A Longest Surviving Mathematical Genius for All Seasons

Jacques Salomon Hadamard (1865–1963), an outstanding French mathematician and an FRS, was born to Am´ed´ee Hadamard (a Jewish teacher) and Claire Marie Jeanne Picard (a pianist) in Versailles, France. Jacques studied in Paris at the Lyc´ee Charlemagne, where his father was a teacher.

During his early years, Jacques excelled in all subjects except mathematics. His mathematical knowledge improved remarkably after his admission to the Lyc´ee Louis–le–Grand in 1876. He was awarded prizes in algebra and mechanics in 1883.

In 1884, Jacques secured the first rank in the entrance examinations for the Ecole Polytechnique and the Ecole Normale Superieure. In 1888, at the age of 23, Jacques graduated from the Ecole Normale. He completed his Ph.D. in 1892 and received the Grand Prix of Mathematical Sciences award for his innovative essay on the Riemann zeta function.

Jacques was married in 1892 at the age of 27 to Louise–Anna Tr´enel, who was a Jewish girl. He fathered five children – three sons and two daughters.

In 1893 he was appointed as a lecturer at the University of Bordeaux, where he proved his famous inequality regarding determinants. This resulted in his discovery of Hadamard matrices (for equality). In 1896 he proved the prime number theorem using complex function theory, and he received the prestigious Bordin Prize of the French Academy of Sciences for his original contributions regarding geodesics in the differential geometry of surfaces and dynamic systems.

He became Professor of Astronomy and Rational Mechanics in Bordeaux in the same year. He moved from Bordeaux to Paris in 1897 after resigning his chair at Bordeaux.

In Paris, at the request of Darboux, he published the first volume of a book on 2 dimensional geometry, **Elementary Geometry Lessons** in 1898, which was followed by a second volume on 3 dimensional geometry in 1901. These volumes had a major impact on mathematics education in France. He was awarded the Prix Poncelet in 1898 for his contributions during the preceding 10 years.

Jacques was not limited to his teaching and research activities; he was also involved in politics during his time in Bordeaux. Alfred Dreyfus, his wife's relative and a Jewish army official in the War Ministry, came from Alsace. In 1894, Alfred was accused of selling military secrets to the Germans and was sentenced to life imprisonment.

Initially many people, including Jacques, felt that Alfred was guilty. After moving to Paris in 1897, Jacques discovered that evidence against Alfred had been forged. Jacques could not keep silent, and he became actively involved in rectifying the injustice done to Alfred.

Jacques was elected President of the French Mathematical Society in 1906 and was appointed as the head of mechanics at the Coll´ege de France in 1909.

In 1910 he published **Lectures on the Calculus of Variations**, which was a significant contribution to the development of functional analysis. He was appointed as professor of analysis at the Ecole Polytechnique, where he succeeded Jordan in 1912.

In the same year he was elected to the Academy of Sciences to succeed Poincar´e. The events of World War I brought tragedy to Jacques' life; his two eldest sons, Pierre and Etienne, were killed in action in 1916. Jacques' extraordinary spiritual strength lessened the pain of these tragedies and allowed him to dive deeper into mathematics.

In 1920 he was appointed to the Paul Emile Appell (1855–1930) chair of analysis at the Ecole Centrale while retaining his positions in the Ecole Polytechnique and the Coll´ege de France.

During 1920–1933 he visited the USA, Spain, Czechoslovakia, Italy, Switzerland, Brazil, Argentina, and Egypt. He continued to produce high quality books and research articles. Perhaps the most famous among them was a text that he published in 1922, **Lectures on Cauchy's Problem in Linear Partial Differential Equations**, which was based on a lecture course that he gave at Yale University.

Besides this, he published several papers on new topics such as the theory of probability, specifically Markov chains, and education. At the start of World War II, France was quickly occupied by Nazi Germany in 1940. Jacques escaped to the USA with his family, where he was appointed as a visiting faculty member at Columbia University.

In 1944 he received the devastating news that his third and youngest son, Mathieu, had been killed in the war on July 26, 1943. Jacques left the United States soon afterward, spending about a year in England before returning to Paris after the end of the war in 1945.

After the devastating World War II, he took an active part in the international peace movement. He was allowed to enter the USA due to the extraordinary support of mathematicians there, and he attended the International Congress held in Cambridge, Massachusetts in 1950.

He was eventually made honorary president of the Congress. In 1962, when he was 96, yet another tragedy struck him; his grandson Etienne was killed in a mountaineering accident. This time his spirit could not withstand the tragedy, and afterward he remained in his house to await his death.

The legend known to the world as "Jacques Hadamard" left his mortal body in Paris on October 17, 1963, at the age of about 98, leaving behind the invaluable treasure of his discoveries that enriched mathematics and the mathematical sciences enormously.

Specifically, he is remembered for the Hadamard product, his prime number theorem, the Hadamard matrix, the Cartan–Hadamard theorem, the Cauchy–Hadamard theorem, his infinite product expansion for the Riemann zeta function, Hadamard's inequality, the Hadamard manifold, the Hadamard transform, the Ostrowski–Hadamard (Alexander Markowich Ostrowski, 1893–1986) gap theorem, and Hadamard space.

The famous quotation, "the shortest path between two truths in the real domain passes through the complex domain", is due to him.

3.7 The Scientist of the 20th Century, Who Did Not Want to Live Longer Artificially

Albert Einstein (1879–1955) was a superb German–American theoretical physicist, philosopher, and passionate humanitarian.

He was born in Ulm (Germany) on March 14, 1879. His father was Hermann Einstein (1847–1902), a salesman and electrical engineer. His mother was Pauline Einstein (1858–1920).

Einstein attended a Catholic elementary school during 1884–1889. He ranked 1st in the school although he had childhood speech difficulties. Einstein's father once showed him a pocket compass; Albert realized that there must be something causing the needle to move, despite the apparent empty space. As he grew, he constructed models and mechanical devices and began to show a talent for mathematics.

In 1889, Max Talmud (Talmey), a poor Polish medical student who took meals with the Einsteins on Thursdays for 6 years, introduced the 10-year old Einstein to key texts on science, mathematics, and philosophy. During this time, Talmud wholeheartedly guided Einstein through many secular educational interests.

In 1894, Einstein's family moved to Milan and then to Pavia, Italy, while Albert stayed in Munich, Germany to complete his studies at the Luitpold Gymnasium. His father wanted him to study electrical engineering, but Einstein clashed with authorities and resented the school's regimen and teaching method. In 1895 he withdrew to join his family in Pavia, convincing the school to let him go by using a doctor's note. He later wrote that the spirit of learning and creative thought were lost in strict rote learning.

During this time, Einstein wrote his first scientific work, "The Investigation of the State of Ether in Magnetic Fields".

Einstein applied to the Swiss Federal Institute of Technology (Eidgen̈ossische Polytechnische Schule, also called Polytechnic), Zürich, Switzerland for admission. He appeared for an entrance examination that he could not pass, although he secured excellent scores in physics and mathematics.

The Einsteins sent Albert to Aarau (capital of the northern Swiss canton) to complete secondary schooling. Einstein studied Maxwell's electromagnetic theory there. In 1896, at the age of 17, he graduated and then enrolled in the 4-year teaching diploma program in mathematics and physics at the Polytechnic in Zürich.

At the Polytechnic he met Mileva Maric (1875–1948), his future wife and the only woman among six students in the mathematics and physics group, who enrolled in the same year. Einstein and Maric's friendship grew deeper. They read books together on physics, in which Einstein became increasingly engrossed. Einstein received his teaching diploma from Zürich Polytechnic in 1900, but Maric failed the examination for the "Theory of Functions" course in mathematics.

Einstein published a paper in 1901 on the capillary forces of a straw, entitled **Capillarity**, in the top journal **Annalen der Physik**, 4, pp. 513–523. He worked with Alfred Kleiner (1849–1916), the Professor of Experimental Physics and his Ph.D. advisor, and completed his dissertation entitled **A New Determination of**

Molecular dimensions on April 30, 1905, for which he received his Ph.D. from the University of Zürich.

In his earlier research articles, Einstein attempted to establish the physical reality of atoms and molecules. In 1905, at the age of 26, he published four ground–breaking research articles on Brownian motion (Robert Brown, 1773–1858), the special theory of relativity, photo-electricity, and the equivalence of matter and energy, which later brought him to the attention of the world scientific community.

During the early 20th century, each zigzag motion of a particle suspended in a liquid was believed to have been caused by collision with a single molecule of the liquid. Einstein showed in his article on Brownian motion how this effect could be caused by an uneven bombardment of many tiny molecules. He could also estimate the size of the molecules, which was later verified to be correct by the French physicist Jean Perrin (1870–1942).

Einstein's explanation led to the accurate estimation of the Avogadro number, a constant denoting the number of atoms per mole (an SI unit which measures the number of particles in a specific substance. One mole is equal to $6.02214179 \times 10^{23}$ atoms).

In 1907, Einstein established that mass is related to energy by the simple equation $E = mc^2$.

In 1908 he became a lecturer at the University of Bern, and in 1909 he relinquished the post of lecturer and took the position of physics docent (faculty with mid–level seniority) at the University of Zürich.

In 1911 he became a full professor at Karl–Ferdinand University in Prague, Czech Republic, and during the same year he extended his special theory of relativity and predicted that light from another star would be bent by the gravitational pull of the Sun, which was confirmed through observations made by a British expedition headed by Sir Arthur Stanley Eddington (1882–1944) during the solar eclipse of May 29, 1919.

He refined these ideas in his general theory of relativity in 1915. According to this theory, masses distort the structure of space–time, which explained an anomaly in the orbit of the planet Mercury that could not be accounted for by Newtonian mechanics.

When Erwin Schrodinger (1887-1961) introduced the **probabilistic concepts into Quantum Theory**, Einstein was dismayed and made the famous remark **"God does not play dice with the Universe"**. This famous statement was a rejection of the probabilistic way that quantum mechanics works.

Although the laws of quantum mechanics work incredibly well and one can't successfully explain quantum effects with traditional classical physics, what Einstein meant/said was exact and absolute truth.

Probabilistic approach is a means to get an approximate solution where a deterministic approach is practically out of bound for a better solution because of limited resources available to us and also our limitation to capture exactly all the concerned laws of physics.

Laws of Nature – most of which are unknown/little known to us – are always exactly followed. We have never experienced that a law of nature has been violated.

A tsunami, an earthquake, or a hurricane/cyclone occurs following all laws of nature. If a white flower is seen on a flower plant that produces red flowers traditionally, then it cannot be construed that a law of nature has been violated. The whole environment is too complex to be comprehended by us.

Newton's "System of the World" was based on Euclidean geometry. He viewed space and time as absolute and infinite. Space, to him, was a 3- dimensional entity; time, a separate 1-dimensional entity. This way of looking at the world seemed to be the only possible way. Men were so certain of these assumptions that they never questioned them or regarded them as assumptions. They were truths, incontestable, absolute, and self-evident.

Einstein's system differs from Newton's in 3 major respects: Space is non–Euclidean, space is not absolute, and space is not a separate entity from time. The geometry of space is not Euclidean, according to Einstein, nor is it even a uniform non–Euclidean system.

The presence of matter alters the properties of space in such a way that no one geometry can describe it. That is, the geometry on Earth is not the same as that, for instance, on the Sun. If a man on Earth were to measure a certain distance or period of time on Mars, he would obtain different measurements from those made by a man on Mars that measured that same distance or time.

This discrepancy would occur not because of errors or optical illusions, but because each man's measurements are based on his own local length and local time. Newton's 3 dimensions have been replaced in Einstein's system by 4, the 4th being time.

One of the most important aspects of this new view of the universe is that it dispenses with the approximation of gravity that Newton (Hall 1980) introduced to explain the course of planets, stars, and falling objects. The new theory asserts that all objects move in straight lines, but non–Euclidean straight lines. Thus when planets move in elliptical orbits around the Sun they are simply moving in the "straight lines" of their particular geometry.

In 1914, Einstein was appointed director of the Kaiser Wilhelm Institute for Physics and professor at the Humboldt University of Berlin. He continued at the Institute until 1932.

He became a member of the Prussian Academy of Sciences, and was the President of the German Physical Society during 1916–1918. Einstein was awarded the Nobel Prize in Physics in 1921 for his discovery in the field of photo-electricity.

Based on the theory that light energy is capable of behaving as particles, called photons, he observed that light particles striking the surface of certain metals produces the photoelectric effect: an electron is emitted by the metal that carries energy equal to the energy of the photon that set the electron free. He was one of the first physicists to prove the wave/particle nature of light.

In 1922, Einstein traveled extensively for 6 months in Asia on a lecture tour. He visited Singapore, Ceylon (Sri Lanka), Japan, and Palestine, and was given an exceptionally warm welcome everywhere. In 1925, he was awarded the Copley Medal by the Royal Society of London.

In 1933, Einstein was compelled to emigrate from Germany to the United States of America due to the rise of Nazism under the new chancellor, Hitler, who imposed unacceptable restrictions on Jews (14 Nobel Laureates and 26 of the 60 professors of physics were forced to flee Germany and the neighboring countries).

Einstein did not return to Germany, but accepted a position at the Princeton Institute of Advanced Study where he spent the rest of his life.

A few months before the start of World War II in Europe, Einstein, Edward Teller (1908–2003), Eugene Wigner (1902–1995), the Hungarian physicist Leo Szilard (1898–1964), and others felt that it was their responsibility to alert Americans to the possibility that German scientists might build the atom bomb first and that Hitler would be willing to use such a weapon.

Initially, the President of the United States, Franklin Delano Roosevelt (1882–1945), did not assign this problem much importance, but he changed his mind rather quickly. The United States entered the arms race, and the secret Manhattan Project that drew immense material, financial, and scientific resources was born.

The outcome of this project was that the United States became the only country to successfully develop the atom bomb during World War II.

Einstein lived in the USA for only about 15 years after he became a US citizen in 1940. On April 17, 1955, he experienced internal bleeding due to an abdominal aortic aneurysm, which had been reinforced surgically by Dr. Rudolph Nissen (1896–1981) in 1948.

At the time, he was preparing a speech for a television appearance commemorating Israel's 7th anniversary. He took the draft of his speech along with him to the hospital, but he did not live long enough to complete it.

He refused further surgery, saying ". · · · It is tasteless to prolong life artificially. I have done my share, it is time to go. I will do it elegantly". Early in the morning on April 18, at the age of 76, he breathed his last at Princeton Hospital.

The following 3 quotations are due to him: "So far as the theories of mathematics are about reality, they are not certain; so far as they are certain, they are not about reality"; "Do not worry about your difficulties in mathematics. I assure you that mine are greater", and "Teaching should be such that what is offered is perceived as a valuable gift and not as a hard duty."

The following anecdotes about Einstein, who was named the person of the century by Time magazine in 1999, have been recounted for decades and are both interesting and revealing:

One day during a lecture tour, Einstein's driver, who often sat at the back of the lecture hall during his talks, remarked that he could possibly give the talk himself, having heard it so many times. At the next stop on the tour, Einstein and the driver switched places, with Einstein sitting at the back in his driver's uniform.

Having delivered a flawless talk, the driver was asked a difficult question by a member of the audience. "Well, the answer to the question is quite simple", he casually replied. "I bet my driver, sitting up at the back there, could answer it!"

Einstein's wife often suggested that he dress more professionally when he headed to work. "Why should I?" he would invariably argue. "Everyone knows me there."

When the time came for Einstein to attend his first major conference, she begged him to dress up a bit. "Why should I?" said Einstein. "No one knows me there!"

Einstein was often asked to explain the general theory of relativity. "Put your hand on a hot stove for a minute, and it seems like an hour", he once declared. "Sit with a pretty girl for an hour, and it seems like a minute. That's relativity!"

During an address at the Sorbonne in Paris, Einstein said, "If my theory of relativity is proven successful, Germany will claim me as a German and France will declare that I am a citizen of the world. Should my theory prove untrue, France will say that I am a German and Germany will declare that I am a Jew".

When Einstein was working at Princeton, one day he was going back home and he forgot his address. The driver of the cab did not recognize him. Einstein asked the driver if he knew Einstein's home.

The driver said "Who does not know Einstein's address? Everyone in Princeton knows. Do you want to meet him?" Einstein replied "I am Einstein. I forgot my home address, can you take me there?" The driver took him to his home and did not even collect the fare.

Einstein was once traveling from Princeton on a train when the conductor came down the aisle, punching the tickets of every passenger. When he came to Einstein, Einstein reached in his vest pocket. He couldn't find his ticket, so he reached in his trouser pockets. It wasn't there, so he looked in his briefcase but couldn't find it. Then he looked in the seat beside him. He still couldn't find it.

The conductor said, "Dr. Einstein, I know who you are. We all know who you are. I'm sure you bought a ticket. Don't worry about it". Einstein nodded appreciatively. The conductor continued down the aisle punching tickets.

As he was ready to move to the next car, he turned around and saw the great physicist down on his hands and knees looking under his seat for his ticket. The conductor rushed back and said, "Dr. Einstein, don't worry, I know who you are. No problem. You don't need a ticket. I'm sure you bought one".

Einstein looked at him and said, "Young man, I too, know who I am. What I don't know is where I'm going".

4 Conclusions

We have presented briefly scientific contributions of only a few scientists, and their interaction/reaction with their respective diverse environments – both conducive and significantly hostile. Although there are many other legendary scientists with their gigantic contributions, the few instances mentioned here are good enough to get a feel of the trend of the mindset of the scientific community, in general.

When we critically analyse their visible action vis-à-vis their interaction with the circumstances/surroundings, we are able to observe their nature/character. Their single-mindedness to uplift the humanity/civilization is selfless. Their action is inspiring and their virtues are worthy of emulating.

One thing common in them is intense thinking/concentration/meditation for the problem which they are seeking a solution/answer for. In the process many other questions and sub-questions crop up, which they successively sort out. Finally they get the desired answer.

The whole process needs them to dive deeper and deeper into the ocean of knowledge which they already have, in their mind, and fetch the intended information for the mankind. This process could/would include physical as well as numerical/semi-numerical/non-numerical experimentation within the then available means.

Even in the face of a devastation – personal/otherwise – or of impending death, their mind is so tuned/focussed that it simply does not deter them stopping their quest. Condorcet, Galois, and Hadamard are glaring instances.

Mind remains active (alive with one or more thoughts) all the 24 h on all the days of one's life. While performing activities such as eating, marketing, brushing teeth, and going to bed, mind is active and in a wakeful state. Mind remaining inactive (silent or no-thought state) implies all bodily functions such as blood circulation, breathing, and food digestion halt as there is nobody to perform these tasks.

In a wakeful (conscious) state, the information processing rate (in a common man in a normal environment) is around 11 bits/s (low) while in a sleep/dream state this rate is around 66 bits/s.

A brilliant idea/answer/solution could/would crop up at any time of the day and night when the mind is focussed on a specific question/problem. During any time such as those for eating, travelling (by car or train or aeroplane), walking/jogging, and sleeping, such a revelation viz. the solution/answer could pop up. **Thus research is not limited to a fixed time period in a day. It is truly a 24-h activity**.

While walking through a university/college campus, we see things such as birds, trees, and other humans with the related behaviour/action, we learn, from these, consciously/subconsciously/unconsciously. All these are useful and could add to our innovation/discovery effort for our problem.

Although scientists may not be free from common human weaknesses, their extraordinary contributions can never be undermined. They are our pioneers. We do have a debt of gratitude to these revered immortal souls who continue to inspire generation after generation.

Beyond studying and learning their scientific contributions, getting to know their nature and visible action definitely has extraordinary positive psychological effect in budding scientists.

Acknowledgements We thank Springer for allowing us to borrow some parts of our previously published work. We have also requested Elsevier for the same and expect they will have no objection. Thanks are also due to the reviewers for their constructive suggestions.

Specifically we are indebted to Cristina Flaut, Editor in Chief, Analele Univ. "Ovidius" din Constanta, Math. Series, Faculty of Mathematics and Computer Science, Ovidius University, Bd. Mamaia 124, 900527, Constanta, Romania for her kind invitation to contribute a book-chapter and for her patient guidance whenever we faced a problem.

We have been immensely benefitted from several websites such as en.wikipedia.org and from comments by Susanto Sen, Inventor and Senior Content Editor, TiVo, Bengaluru (Bangalore).

References

Agarwal, R.P., Sen, S.K.: Creators of Mathematical and Computational Sciences. Springer, New York (2014)

Andrews, G.E.: (2012) "The Discovery of Ramanujan's Lost Noteboo" (PDF). The Legacy of Srinivasa Ramanujan: Proceedings of an International Conference in Celebration of the 125th Anniversary of Ramanuja's Birth: University of Delhi: 17–22. Retrieved 29 June 2017

Andrews, G., Berndt, B.C.: Ramanujan's Lost Notebook. vol. 4, Springer (2013)

Dedron, P., Itard, J.: Mathematics and Mathematicians, vols. 2. The Open University Press, Milton Keynes, England (1973)

Dey, S.K.: Analysis of consciousness in Vedanta philosophy. Informatica **21**(3), 405–419 (1997)

Dutta, M.N.: (younger brother of Swami Vivekananda), Vivekananda Swamijir Jiboner Ghatanaboli (Bengali), Part 1 (Page 81), Mohendra Publishing Committee, 3rd edn., Kolkata (1965). (Bengali year 1372); (The English translation of the title is "Events in the life of Swami Vivekananda" first edition was published in 1938 i.e. Bengali year 1345)

Hadamard, J.: An Essay on the Psychology of Invention in the Mathematical Field. Princeton University Press, Princeton (1945)

Hall, A.R.: Philosophers at War: The Quarrel between Newton and Leibniz. Cambridge University Press, Cambridge (1980)

Jacobson, R.: Memories may not live in Neuron's synapses. Scientific American Mind, 1 April 2015

Kanigel, R.: The Man Who Knew Infinity: A Life of the Genius Ramanujan. Washington Square Press, New York (1992)

Lakshmikantham, V., Sen, S.K.: Computational Error and Complexity in Science and Engineering. Elsevier, Amsterdam (2005)

Mathews, J.H., Fink, K.D.: Numerical Methods Using Matlab, 4th edn. PHI Learning Pvt. Ltd., Delhi (2014)

Pandit, M.D: Mathematics as Known to the Vedic Samhitas (Sri Satguru Publications, New Delhi, 298, 1993) 299

Sen, S.K., Agarwal, R.P.: Zero: A Landmark Discovery, the Dreadful Void, and the Ultimate Mind. Academic, New York (2016)

Sen, S.K., Agarwal, R.P.: Pi, e, phi with Matlab: Random and Rational Sequences with Scope in Supercomputing Era. Cambridge Scientific Publishers, U.K. (2011)

Part II
Theories in Social Systems

A General Framework for Individual and Social Choices

Bice Cavallo, Livia D'Apuzzo, Antonio Di Nola, Massimo Squillante
and Gaetano Vitale

Abstract Suitable algebraic structures for individual and social choices are proposed. Some relevant properties are illustrated.

Keywords Algebraic structures · Multi-criteria methods · Preferences modeling
Social choices

1 Introduction

Several modelizations have been proposed for analyzing and interpreting decision procedures for both individual and collective cases (Sixto 1994; Lu et al. 2016; Maturo and Sciarra 2017). Multi-criteria decision methods (MCDMs) represent a research field that has been widely investigated during the last years, by starting from the pioneristic work of Saaty (1977). Many developments have been carried out, for

B. Cavallo (✉) · L. D'Apuzzo
University of Naples, Federico, Italy
e-mail: bice.cavallo@unina.it

L. D'Apuzzo
e-mail: liviadap@unina.it

A. Di Nola · G. Vitale
University of Salerno, Fisciano, Italy
e-mail: adinola@unisa.it

G. Vitale
e-mail: gvitale@unisa.it

M. Squillante
University of Sannio, Benevento, Italy
e-mail: squillan@unisannio.it

© Springer Nature Switzerland AG 2019
C. Flaut et al. (eds.), *Models and Theories in Social Systems*, Studies in Systems,
Decision and Control 179, https://doi.org/10.1007/978-3-030-00084-4_2

both theory and applications (Ishizaka and Nemery 2013; Greco et al. 2016;Greco et al. 2001). An original point of view has been introduced by Cavallo and D'Apuzzo (2009) considering as suitable model for MCDM methods particular ordered and algebraic structures: the Abelian linearly ordered (Alo)-groups. The investigation has been carried out by obtaining further theoretical advancements (Cavallo and D'Apuzzo 2012a) (Cavallo et al. 2012), Cavallo and D'Apuzzo (2015), (Cavallo and D'Apuzzo 2016), (Cavallo et al. 2017), (Cavallo and Brunelli 2018) and related applications (Cavallo et al. 2015) also combining other methodologies as Bayesan networks (Cavallo et al. 2014).

In this chapter, we further focus on algebraic tools to model choices in comparing alternatives according to different criteria. This can be very useful when our choices concern with, e.g. price, utility, life goals, personal values, etc. In these situations we have to balance different criteria. As usual, we need a formalization which gives us tools to well represent data concerning the choice and to solve the above problems. Of course as much sensitive is the representation as correct is the solution. So it is crucial to have many ways to represent the choice process using mathematical suitable properties and even general ones. Then we summarize the main contributes obtained by means of Alo-groups. Among other possible mathematical tools suitable to represent a choice process under many criteria, Riesz spaces represent a quite natural and powerful once. Due at the double nature of both weighted and ordered spaces, Riesz spaces seem to be another suitable way to interpret multi-criteria methods; actually, they may supply a framework to relate, in real problems the computation of weights and the construction of orders. Just we remark that Riesz spaces are already studied and widely applied in economics, and our suggestion to use them in multi-criteria decision making will profit of a large literature concerning Riesz spaces. Here we give the first insight to obtain new results using a new perspective. We exhibit some significant examples to show how properties of Riesz spaces can be used to express preferences. Riesz spaces allow us to combine the advantages of many approaches. We also provide a characterization of collective choice rules which satisfies some classical criteria in social choice theory and an abstract approach to social welfare functions. For further algebraic approaches in social sciences, the reader can see e.g. (Hošková-Mayerová and Maturo 2017, 2018).

The remainder of the chapter is organized as follows: Sect. 2 provides main results obtained by Cavallo and D'Apuzzo (2009), Cavallo and D'Apuzzo (2012a) Cavallo et al. (2012), Cavallo and D'Apuzzo (2015), Cavallo and D'Apuzzo (2016), Cavallo et al. (2017), Cavallo and Brunelli (2018) for dealing with cardinal pairwise comparisons by means of Alo-groups; Sect. 3 provides a first approach proposed by Di Nola et al. (2018) for dealing with social preferences through Riesz spaces; Sect. 4 provides conclusions and future work.

2 Abelian Linearly Ordered Groups for Cardinal Pairwise Comparisons

2.1 Abelian Linearly Ordered Group

As the following definition shows, an Abelian linearly ordered group is a commutative group equipped with an ordering relation:

Definition 1 Let G be a non-empty set, $\odot : G \times G \to G$ a binary operation on G, $\leq$ a weak order on G. Then, $\mathscr{G} = (G, \odot, \leq)$ is an *Abelian linearly ordered group* (*Alo-group* for short), if $(G, \odot)$ is an Abelian group and

$$a \leq b \Rightarrow a \odot c \leq b \odot c. \tag{1}$$

Let us denote with $\div$ the inverse operation of $\odot$ and e the identity element.

An *isomorphism* between two Alo-groups $\mathscr{G} = (G, \odot, \leq)$ and $\mathscr{G}' = (G', \circ, \leq)$ is a bijection $f : G \to G'$ that is both a lattice isomorphism and a group isomorphism, that is:

$$a < b \Leftrightarrow f(a) < f(b) \quad and \quad f(a \odot b) = f(a) \circ f(b). \tag{2}$$

By definition, an Alo-group $\mathscr{G}$ is a *lattice ordered group* (Birkhoff 1984), that is there exists $\max\{a, b\}$, for each $a, b \in G$; thus, Cavallo and D'Apuzzo (2009) provide the notion of $\mathscr{G}$-distance between two elements as follows:

$$d_{\mathscr{G}} : (a, b) \in G \times G \to d_{\mathscr{G}}(a, b) = \max\{a \div b, b \div a\} \in G. \tag{3}$$

Let $n \in \mathbb{N}_0$ and $a \in G$; then, the (n)-*natural-power* (Cavallo and D'Apuzzo 2009) $a^{(n)}$ of $a \in G$ is:

$$a^{(n)} = \begin{cases} e, & if \quad n = 0 \\ \odot_{i=1}^{n} a, & if \quad n \geq 1. \end{cases} \tag{4}$$

Definition 2 (*Cavallo et al. 2012*) Let $z \in \mathbb{Z}$. The (z)-*integer-power* $a^{(z)}$ of $a \in G$ is defined as follows:

$$a^{(z)} = \begin{cases} a^{(n)}, & if \quad z = n \in \mathbb{N}_0 \\ (a^{(n)})^{(-1)} & if \quad z = -n, \quad n \in \mathbb{N}. \end{cases}$$

An Alo-group $\mathscr{G} = (G, \odot, \leq)$ is called *continuous* if the operation $\odot$ is continuous (Cavallo and D'Apuzzo 2009), and *real* if G is a subset of the real line $\mathbb{R}$ and $\leq$ is the weak order on G inherited from the usual order on $\mathbb{R}$. From now on, we will assume that $\mathscr{G} = (G, \odot, \leq)$ is a real continuous Alo-group, with G an open interval. Under these assumptions, $\mathscr{G}$ is a divisible Alo-group and, as a consequence, the equation $x^{(n)} = a$ has a unique solution (Cavallo and D'Apuzzo 2009); thus, it is reasonable to consider the following notions of (n)-*root*, $\mathscr{G}$-*mean*, (q)-*rational-power* and (λ)-*real-power*.

Definition 3 (*Cavallo and D'Apuzzo* 2009) For each $n \in \mathbb{N}$ and $a \in G$, the (n)-*root* of a, denoted by $a^{(\frac{1}{n})}$, is the unique solution of the equation $x^{(n)} = a$, that is:

$$\left(a^{(\frac{1}{n})}\right)^{(n)} = a.$$

Definition 4 (*Cavallo and D'Apuzzo* 2009) The $\mathscr{G}$-*mean* $m_{\mathscr{G}}(a_1, a_2, ..., a_n)$ of the elements $a_1, a_2, ..., a_n$ of G is

$$m_{\mathscr{G}}(a_1, a_2, ..., a_n) = \begin{cases} a_1 & \text{for n = 1}, \\ \left(\bigodot_{i=1}^{n} a_i\right)^{(1/n)} & \text{for } n \geq 2. \end{cases}$$

Definition 5 (*Cavallo et al.* 2012) For each $q = \frac{m}{n}$, with $m \in \mathbb{Z}$ and $n \in \mathbb{N}$, and for each $a \in G$, the (q)-*rational-power* $a^{(q)}$ is defined as follows:

$$a^{(q)} = (a^{(m)})^{(\frac{1}{n})}.$$

Definition 6 (*Cavallo and D'Apuzzo* 2012b) For each $a \in G$ and $\lambda \in \mathbb{R}$, the (λ)-*real-power* of a is:

$$a^{(\lambda)} = \begin{cases} sup_q I_{a,\lambda} \; if a \geq e, \\ inf_q I_{a,\lambda} \; if a \leq e. \end{cases},$$

with:

$$I_{a,\lambda} = \{a^{(q)} : q \in Q \; and \; q < \lambda\}, \quad S_{a,\lambda} = \{a^{(q)} : q \in Q \; and \; q > \lambda\}. \quad (5)$$

Examples of real continuous Alo-groups are the following ones:

Multiplicative Alo-group. $\mathbf{R}^+ = (\mathbb{R}^+, \cdot, \leq)$, where $\cdot$ is the usual multiplication on $\mathbb{R}$, $e = 1$ and $\mathbf{R}^+$-mean operator is the geometric mean,

$$m_{\mathbf{R}^+}(a_1, ..., a_n) = \sqrt[n]{\prod_{i=1}^{n} a_i}.$$

$\mathbf{R}^+$-distance between a and b is $d_{\mathbf{R}^+}(a, b) = \max\{\frac{a}{b}, \frac{b}{a}\}$, and $a^{(\lambda)} = a^\lambda$.

Additive Alo-group. $\mathbf{R} = (\mathbb{R}, +, \leq)$, where $+$ is the usual addition on $\mathbb{R}$, $e = 0$ and $\mathbf{R}$-mean operator is the arithmetic mean,

$$m_{\mathbf{R}}(a_1, ..., a_n) = \frac{\sum_{i=1}^{n} a_i}{n}.$$

$\mathbf{R}$-distance between a and b is $d_{\mathbf{R}}(a, b) = \max\{a - b, b - a\} = |a - b|$, and $a^{(\lambda)} = \lambda a$.

Fuzzy Alo-group. $\mathbf{I} = (]0, 1[, \otimes, \leq)$, where $\otimes :]0, 1[^2 \to]0, 1[$ is the operation defined by

$$a \otimes b = \frac{ab}{ab + (1-a)(1-b)},$$

$e = 0.5$ and $\mathbf{I}$-mean operator is given by the following mean:

$$m_{\mathbf{I}}(a_1, ..., a_n) = \frac{\sqrt[n]{\prod_{i=1}^n a_i}}{\sqrt[n]{\prod_{i=1}^n a_i} + \sqrt[n]{\prod_{i=1}^n (1-a_i)}}.$$

$\mathbf{I}$-distance between a and b is the following one:

$$d_{\mathbf{I}}(a, b) = max \left\{ \frac{a(1-b)}{a(1-b) + (1-a)b}, \frac{b(1-a)}{b(1-a) + (1-b)a} \right\},$$

and

$$a^{(\lambda)} = \frac{a^\lambda}{a^\lambda + (1-a)^\lambda}.$$

Two real continuous Alo-groups $\mathscr{G} = (G, \odot, \leq)$ and $\mathscr{G}' = (G', \circ, \leq)$, with G and G' open intervals, are isomorphic (Cavallo and D'Apuzzo 2009). For example, the function

$$h : x \in]0, +\infty[\mapsto \frac{x}{x+1} \in]0, 1[\tag{6}$$

is an isomorphism between multiplicative and fuzzy Alo-groups, and

$$g : x \in \mathbb{R}^+ \to \log x \in \mathbb{R}. \tag{7}$$

is an isomorphism between multiplicative and additive Alo-groups.

Remark 1 Under our assumptions on $(G, \odot, \leq)$, we stress that $(G, \odot, \leq)$ is an Alo-group if and only if $(G, \odot, e, \bullet, \leq)$ is a Riesz space, with:

$$\lambda \bullet a = a^{(\lambda)}.$$

For definition (Aliprantis and Burkinshaw 2003), a structure $(G, \odot, e, \bullet, \leq)$ is a Riesz space (or a vector lattice) if:

1. $(G, \odot, e, \bullet)$ is a vector space over the field $\mathbb{R}$;
2. $(G, \leq)$ is a lattice;
3. $\forall a, b, c \in G$ if $a \leq b$ then $a \odot c \leq b \odot c$;
4. $\forall \lambda \in \mathbb{R}^+$ if $a \leq b$ then $\lambda \bullet a \leq \lambda \bullet b$.

Riesz spaces will be further analysed in Sect. 3.

2.2 Cardinal Pairwise Comparison Matrices

Let $X = \{x_1, x_2, ..., x_n\}$ be a set of decision elements, such as criteria or alternatives. Popular MCDM method for obtaining cardinal rankings on X is the *Pairwise Comparison Matrix* (*PCM*); PCMs have been a long standing technique for comparing alternatives and their role has been pivotal in the development of modern decision making methods. Entry a_{ij} of the following PCM

$$
A = \begin{array}{c} \\ x_1 \\ x_2 \\ \vdots \\ x_n \end{array}
\begin{array}{cccc}
x_1 & x_2 & \cdots & x_n
\end{array}
\begin{pmatrix}
a_{11} & a_{12} & \cdots & a_{1n} \\
a_{21} & a_{22} & \cdots & a_{2n} \\
\vdots & \vdots & \ddots & \vdots \\
a_{n1} & a_{n2} & \cdots & a_{nn}
\end{pmatrix},
\tag{8}
$$

quantifies the preference intensity of x_i over x_j (i.e. a cardinal pairwise preference). In cardinal PCMs, with respect to the values that a_{ij} can assume and their interpretation, it is fundamental to be aware that various proposals have been presented, studied, and applied in the literature to solve real-world problems. The foremost type of PCM, at least with respect to the number of real-world applications is probably the *multiplicative* PCM, used among others by Saaty (1977) in the theory of Analytic Hierarchy Process (AHP) (Russo and Camanho 2015; Delli Rocili and Maturo 2017). In this sense, pairwise comparisons are expressed as positive real numbers, $a_{ij} \in]0, +\infty[$, and they represent ratios of preferences. We shall note that AHP is not the only method using this scheme for pairwise comparisons. For instance, proponents of Multi Attribute Value Theory (MAVT) such as Keeney and Raiffa (1976) and Belton and Stewart (2002) advocate the use of multiplicative pairwise comparisons to estimate the ratios between weights of criteria when the value function is additive. If $a_{ij} \in \mathbb{R} =]-\infty, +\infty[$ represents a preference difference then A is called *additive* PCM (Barzilai 1998); additive PCMs are used in the Simple Multi-Attribute Rating Technique (SMART) for group decision making by Barzilai and Lootsma (1997). If $a_{ij} \in]0, 1[$ (see Chiclana et al. 2009, Zhang 2016, Tanino 1984 and it reflects a preference degree then A is called *fuzzy* PCM.

In order to generalize and unify the several approaches to cardinal PCMs, Cavallo and D'Apuzzo (2009) propose PCMs, whose entries belong to an Abelian linearly ordered group; this approach has been further discussed by Cavallo (2014) and several scholars are following it (Hou 2016; Xia and Chen 2015).

Currently, PCMs defined over Abelian linearly ordered groups are also used for dealing with uncertainty in decision maker's preferences (see e.g. Ramík 2015 and Cavallo and Brunelli 2018 and for group decision making Cavallo et al. 2017).

2.3 Main Results About PCMs Over Alo-Groups

$A = (a_{ij})$ is a $\mathscr{G}$-reciprocal PCM (8) of order n (Cavallo and D'Apuzzo 2009) if:

$$a_{ij} \in G \quad \text{and} \quad a_{ji} = a_{ij}^{(-1)} \quad \forall\, i, j \in \{1, \ldots, n\}. \tag{9}$$

Let us set (Cavallo and D'Apuzzo 2015):

$$x_i \succ_A x_j \Leftrightarrow a_{ij} > e, \qquad x_i \sim_A x_j \Leftrightarrow a_{ij} = e, \tag{10}$$

where $x_i \succ_A x_j$ and $x_i \sim_A x_j$ stand for "x_i is strictly preferred to x_j"and "x_i is indifferent to x_j", respectively, and

$$x_i \succsim_A x_j \Leftrightarrow (x_i \succ_A x_j \text{ or } x_i \sim_A x_j) \Leftrightarrow a_{ij} \geq e, \tag{11}$$

that stands for "x_i is weakly preferred to x_j".

In order to provide a cardinal ranking on X, for each $\mathscr{G}$-reciprocal $A = (a_{ij})$, Cavallo and D'Apuzzo (2009), Cavallo and D'Apuzzo (2012a) propose the $\mathscr{G}$-mean vector

$$\underline{w}_{m_{\mathscr{G}}}(A) = (m_{\mathscr{G}}(\underline{a}_1), m_{\mathscr{G}}(\underline{a}_2), \cdots, m_{\mathscr{G}}(\underline{a}_n)), \tag{12}$$

with $m_{\mathscr{G}}(\underline{a}_i) = m_{\mathscr{G}}(a_{i1}, a_{i2}, \ldots, a_{in})$ as weighting vector for the elements of X, providing several reasons of this choice.

2.3.1 Coherence Levels

A minimal logical requirement that decision maker's preferences should satisfy is the $\mathscr{G}$-transitivity:

Definition 7 (Cavallo and D'Apuzzo 2015) $A = (a_{ij})$ is $\mathscr{G}$-transitive if it verifies the condition

$$(a_{ij} \geq e, \text{ and } a_{jk} \geq e) \Rightarrow a_{ik} \geq e. \tag{13}$$

$\mathscr{G}$-transitivity in (13) generalizes fuzzy weak transitivity ($a_{ij} \geq 0.5$, $a_{jk} \geq 0.5 \Rightarrow a_{ik} \geq 0.5$) introduced by Tanino (1988) and it generalizes and extends multiplicative transitivity ($a_{ij} > 1$, $a_{jk} > 1 \Rightarrow a_{ik} > 1$) given by Basile and D'Apuzzo (2002).

$\mathscr{G}$-transitivity ensures a rearrangement $(i_1, i_2, \cdots, i_n)$ of $\{1, 2, \cdots, n\}$ such that:

$$x_{i_1} \succsim_A x_{i_2} \succsim_A \ldots \succsim_A x_{i_n}. \tag{14}$$

The ordinal ranking in (14) is called the *actual ranking* on X provided by A.

Proposition 1 (Cavallo and D'Apuzzo 2015) $A = (a_{ij})$ *is $\mathscr{G}$-transitive if and only if there exists a vector $\underline{w} = (w_1, w_2, \ldots, w_n) \in G^n$ satisfying the equivalences:*

$$w_i > w_j \Leftrightarrow a_{ij} > e \quad and \quad w_i = w_j \Leftrightarrow a_{ij} = e. \tag{15}$$

Definition 8 (*Cavallo and D'Apuzzo* 2016) Let $A = (a_{ij})$ be a $\mathscr{G}$-transitive PCM. A vector $\underline{w} = (w_1, w_2, \ldots, w_n) \in G^n$ is an *ordinal evaluation vector* for the actual ranking in (14) (or *coherent priority vector*) if it satisfies equivalences in (15).

The full coherence of the decision maker when stating his/her preferences is instead represented by $\mathscr{G}$-consistency:

Definition 9 (*Cavallo and D'Apuzzo* 2009) $A = (a_{ij})$ is a $\mathscr{G}$-consistent PCM, if verifies the following condition:

$$a_{ik} = a_{ij} \odot a_{jk} \quad \forall i, j, k \in \{1, \ldots, n\}. \tag{16}$$

$\mathscr{G}$-consistency in (16) generalizes multiplicative consistency $a_{ik} = a_{ij}a_{jk}$ (Saaty 1980), additive consistency $a_{ik} = a_{ij} + a_{jk}$ (Barzilai 1998) and fuzzy consistency $a_{ik} = \frac{a_{ij}a_{jk}}{a_{ij}a_{jk}+(1-a_{ij})(1-a_{jk})}$ (Tanino 1984).

A condition stronger than $\mathscr{G}$-transitivity but weaker than $\mathscr{G}$-consistency is the weak $\mathscr{G}$-consistency:

Definition 10 (*Cavallo and D'Apuzzo* 2016) $A = (a_{ij})$ is *weakly $\mathscr{G}$-consistent if*

$$(a_{ij} \geq e \quad and \quad a_{jk} \geq e) \Rightarrow a_{ik} \begin{cases} = \max\{a_{ij}, a_{jk}\} & if \ a_{ij} = e \ or \ a_{jk} = e, \\ > \max\{a_{ij}, a_{jk}\} & otherwise. \end{cases} \tag{17}$$

Weak $\mathscr{G}$-consistency in (17) generalizes and extends the multiplicative weak consistency $(a_{ij} > 1 \ a_{jk} > 1 \Rightarrow a_{ik} > \max\{a_{ij}, a_{jk}\})$ provided by Basile and D'Apuzzo (2002) and it is similar to fuzzy *restricted max-max transitivity* $(a_{ij} \geq 0.5 \ a_{jk} \geq 0.5 \Rightarrow a_{ik} \geq \max\{a_{ij}, a_{jk}\})$ provided by Tanino (1988).

Proposition 2 (Cavallo and D'Apuzzo 2016) *If $A = (a_{ij})$ is a weakly $\mathscr{G}$-consistent PCM, then $\mathscr{G}$-mean vector $\underline{w}_{m_{\mathscr{G}}}(A) = (m_{\mathscr{G}}(\underline{a}_1), \ldots, m_{\mathscr{G}}(\underline{a}_n))$ in (12) is a coherent priority vector (see Definition 8).*

2.3.2 Measuring $\mathscr{G}$-Inconsistency

In order to measure how much a PCM is far from $\mathscr{G}$-consistency, Cavallo and D'Apuzzo (2009) propose a consistency index, which is expressed as a mean of $\mathscr{G}$-distances from $\mathscr{G}$-consistency; i.e. $\mathscr{G}$-distances between a_{ik} and $a_{ij} \odot a_{jk}$.

Definition 11 Let $A = (a_{ij})$ be a $\mathscr{G}$-reciprocal PCM of order $n \geq 3$. Then, its $\mathscr{G}$-consistency index is:

$$I_{\mathscr{G}}(A) = \left(\bigodot_{(i,j,k) \in T} d_{\mathscr{G}}(a_{ik}, a_{ij} \odot a_{jk}) \right)^{\left(\frac{1}{|T|}\right)},$$

with $T = \{(i, j, k) : i < j < k\}$ and $|T| = \frac{n(n-2)(n-1)}{6}$ its cardinality.

Proposition 3 *The following statements hold:*

1. $I_{\mathscr{G}}(A) \geq e$;
2. $I_{\mathscr{G}}(A) = e \Leftrightarrow A$ is $\mathscr{G}$-consistent.

Proposition 4 (Cavallo and D'Apuzzo 2009) *Let* $\mathscr{G} = (G, \odot, \leq)$ *and* $\mathscr{G}' = (G', \circ, \leq)$ *be two real continuous Alo-groups, with* G *and* G' *two open intervals, and* $f : G \to G'$ *an isomorphism. Then, for each PCM* $A = (a_{ij})$ *over* $\mathscr{G} = (G, \odot, \leq)$, *the following equality holds:*

$$f(I_{\mathscr{G}}(A)) = I_{G'}(f(A)),$$

where $f(A) = (f(a_{ij}))$.

Example 1 Let us consider the reciprocal multiplicative PCM (Cavallo and D'Apuzzo 2009):

$$A = \begin{pmatrix} 1 & \frac{1}{7} & \frac{1}{7} & \frac{1}{5} \\ 7 & 1 & \frac{1}{2} & \frac{1}{3} \\ 7 & 2 & 1 & \frac{1}{9} \\ 5 & 3 & 9 & 1 \end{pmatrix};$$

thus, its consistency index is the following one:

$$I_{\mathscr{R}^+}(A) = \sqrt[4]{d_{\mathscr{R}^+}(a_{13}, a_{12} \cdot a_{23}) \cdot d_{\mathscr{R}^+}(a_{14}, a_{12} \cdot a_{24}) \cdot d_{\mathscr{R}^+}(a_{14}, a_{13} \cdot a_{34}) \cdot d_{\mathscr{R}^+}(a_{24}, a_{23} \cdot a_{34})} =$$
$$= \sqrt[4]{2 \cdot 4.2 \cdot 12.6 \cdot 6} = 5.02.$$

Example 2 Let us consider the following reciprocal additive PCM:

$$B = \begin{pmatrix} 0 & 2 & 3 & 7 \\ -2 & 0 & 3 & 5 \\ -3 & -3 & 0 & 2 \\ -7 & -5 & -2 & 0 \end{pmatrix};$$

thus, its consistency index is the following one:

$$I_{\mathscr{R}}(B) = \frac{d_{\mathscr{R}}(a_{13}, a_{12} + a_{23}) + d_{\mathscr{R}}(a_{14}, a_{12} + a_{24}) + d_{\mathscr{R}}(a_{14}, a_{13} + a_{34}) + d_{\mathscr{R}}(a_{24}, a_{23} + a_{34})}{4} =$$
$$= \frac{2 + 0 + 2 + 0}{4} = 1.$$

Interestingly, although they are expressed on two different scales, values of inconsistency indices from different representations of preferences are comparable. For instance, by applying the isomorphism in (7), we have that the consistency index of the multiplicative PCM $g^{-1}(B)$ is $I_{R^+}(g^{-1}(B)) = 2.72$ that is smaller than $I_{R^+}(A)$ in Example 1; thus, B is more consistent than A.

Example 3 Let us consider the following reciprocal fuzzy PCM:

$$C = \begin{pmatrix} 0.5 & 0.67 & 0.86 & 0.67 \\ 0.33 & 0.5 & 0.75 & 0.33 \\ 0.14 & 0.25 & 0.5 & 0.14 \\ 0.33 & 0.67 & 0.86 & 0.5 \end{pmatrix}.$$

Its consistency index can be computed by applying isomorphism in (6); thus, we have:

$$I_{\mathscr{G}}(C) = h(I_{\mathscr{R}^+}(h^{-1}(C))) = 0.59.$$

We stress that $I_{\mathscr{R}^+}(h^{-1}(C)) = 1.41$ is smaller than both $I_{\mathscr{R}^+}(A)$ in Example 1 and $I_{R^+}(g^{-1}(B))$ in Example 2; thus C is more consistent than both A and B.

$\mathscr{G}$-consistency index satisfies properties proposed by Brunelli (2017) and Brunelli and Fedrizzi (2015) (see Cavallo and D'Apuzzo (2012b)), a method for improving $\mathscr{G}$-consistency of a PCM, based on $I_{\mathscr{G}}$, is proposed by Cavallo (2018), and a relation between random consistency index and coherence level of multiplicative PCMs is established by Cavallo (2017).

2.3.3 Dealing with Uncertainty

In order to deal with uncertainty of the decision maker's preferences, Cavallo and Brunelli (2018) propose a general unified approach to Interval Pairwise Comparison Matrices, based on Abelian linearly ordered groups. In this framework, the authors generalize some consistency conditions provided for multiplicative and/or fuzzy interval pairwise comparison matrices and provide inclusion relations between them. Then, they provide a concept of distance between intervals that, together with a notion of mean defined over Abelian linearly ordered groups, allows to provide a consistency index and an indeterminacy index. In this way, by means of suitable isomorphisms between Abelian linearly ordered groups, the authors are able to compare the inconsistency and the indeterminacy of different kinds of Interval Pairwise Comparison Matrices, e.g. multiplicative, additive, and fuzzy, on a unique Cartesian coordinate system.

2.3.4 Group Decision Making

In order to deal with group decision making, Cavallo et al. (2017) propose an aggregation of PCMs and they analyse how the aggregated PCM preserves some coherence levels, such as $\mathscr{G}$-transitivity, weak $\mathscr{G}$-consistency and $\mathscr{G}$-consistency. Then, they reformulate Arrow's conditions (Arrow 1978), (Arrow and Raynaud 1986), (Arrow et al. 2011) in terms of PCMs, and they provide two preference aggregation procedures for representing group preferences that give a social PCM and a social cardinal ranking, respectively. Finally, they analyse how these preference aggregation procedures satisfy reformulated Arrow's conditions.

3 Riesz Spaces for Social Preferences

Our choices are strictly related to our ability to compare alternatives according to different criteria, e.g. price, utility, feelings, life goals, social conventions, personal values, etc. This means that in each situation we have different *best alternatives* with respect to many criteria; usually, the context gives us the *most suitable* criteria, but no one says that there is a unique criterion. Even when we want to make a decision according to the opinions of the experts in a field we may not have a unique advice. To sum up, we have to be able to define our *balance* between different criteria and opinions, to give to each comparison a weight which describes the importance, credibility or goodness and then to include all these information in a mixed criteria. As usual, we need a formalization which gives us tools to solve these problems; properties of this formalization are well summarized by Saaty (1990), according to whom

> [it] must include enough relevant detail to: represent the problem as thoroughly as possible, but not so thoroughly as to lose sensitivity to change in the elements; consider the environment surrounding the problem; identify the issues or attributes that contribute to the solution; identify the participants associated with the problem.

Riesz spaces, with their double nature of both weighted and ordered spaces, seem to be the natural framework to deal with multi-criteria methods; in fact, in real problems we want to obtain an order starting from weights and to compute weights having an order.

We remark that Riesz spaces are already studied and widely applied in economics, mainly supported by works of Aliprantis (see Abramovich et al. (1995); Aliprantis and Brown (1983); Aliprantis and Burkinshaw (2003)). The proofs of results presented in this section are shown by Di Nola et al. (2018).

3.1 Riesz Spaces

Definition 12 A structure $\mathcal{R} = (R, +, \cdot, \bar{0}, \preceq)$ is a Riesz space (or a vector lattice) if and only if:

- $\mathcal{R} = (R, +, \cdot, \bar{0})$ is a vector space over the field $\mathbb{R}$;
- $(R, \preceq)$ is a lattice;
- $\forall a, b, c \in R$ if $a \preceq b$ then $a + c \preceq b + c$;
- $\forall \lambda \in \mathbb{R}^+$ if $a \preceq b$ then $\lambda \cdot a \preceq \lambda \cdot b$.

A Riesz space $(R, +, \cdot, \bar{0}, \preceq)$ is said to be *archimedean* iff for every $x, y \in R$ with $n \cdot x \preceq y$ for every $n \in \mathbb{N}$ we have $x \preceq \bar{0}$. A Riesz space $(R, +, \cdot, \bar{0}, \preceq)$ is said to be linearly ordered iff $(R, \preceq)$ is totally ordered. We will denote by R^+ the subset of positive elements of R Riesz space (the *positive cone*), i.e. $R^+ = \{a \in R \mid \bar{0} \preceq a\}$. We say that u is a *strong unit* of R iff for every $a \in R$ there is a positive integer n with $|a| \leq n \cdot u$, where $|a| = (a) \vee (-a)$.

Examples:

1. An example of non-linearly ordered Riesz space is the vector space $\mathbb{R}^n$ equipped with the order $\preceq$ such that $(a_1, \ldots, a_n) \preceq (b_1, \ldots, b_n)$ if and only if $a_i \leq b_i$ for all $i = 1, \ldots, n$; it is also possible to consider $(1, \ldots, 1)$ as strong unit.
2. A non-archimedean example is $\mathbb{R} \times_{LEX} \mathbb{R}$ with the lexicographical order, i.e. $(a_1, a_2) \preceq (b_1, b_2)$ if and only if $a_1 < b_1$ or $(a_1 = b_1$ and $a_2 \leq b_2)$; in this case $(1, 0)$ is a strong unit.
3. $(\mathbb{R}, +, \cdot, 0, \leq)$, which is the only (up to isomorphism) archimedean linearly ordered Riesz space, as showed by Labuschagne and Van Alten (2007); obviously 1 can be seen as the standard strong unit.
4. $(\mathbb{R}^C, +, \cdot, \mathbf{0}, \preceq)$ the space of (not necessarily continuous) functions from C compact subset of $\mathbb{R}$, e.g. the closed interval $[0, 1]$, to $\mathbb{R}$, such that for every $f, g \in \mathbb{R}^C$ and $\alpha \in \mathbb{R}$ we have $(f + g)(x) = f(x) + g(x)$, $(\alpha \cdot f)(x) = \alpha f(x)$, $f \preceq g \Leftrightarrow f(x) \leq g(x) \; \forall x \in C$ and $\mathbf{0}$ is the zero-constant function; if we consider continuous functions the one-constant function $\mathbf{1}$ is a strong unit.
5. $(M_n(R), +, \cdot, 0_{n \times n}, \preceq)$ the space of $n \times n$ matrices over R Riesz space with component-wise operations and order as in example (1).

Definition 13 A cone in R^n is a subset K of R^n which is invariant under multiplication by positive scalars. A polyhedral cone is convex if it is obtained by finite intersections of half-spaces.

Cones play a crucial role in Riesz spaces theory, as showed by Aliprantis and Tourky (2007) with also some applications (e.g. to linear programming Aliprantis and Tourky 2007([Corollary 3.43]). Another remarkable example of this fruitful tool is the well-known Baker–Beynon duality (Beynon 1975), which shows that the category of finitely presented Riesz spaces is dually equivalent to the category of (polyhedral) cones in some Euclidean space. Analogously to Euclidean spaces, in R^n (with R generic Riesz space) we can consider *orthants*, i.e. a subset of R^n defined by constraining each Cartesian coordinate to be $x_i \preceq \bar{0}$ or $x_i \succeq \bar{0}$. Here we introduce the definition of *TP-cones*, which will be useful in the sequel.

Definition 14 Let us consider L cone. We say that L is a *TP-cone* if it is the empty-set, or an orthant or an intersection of them.

Why should we use an element of a Riesz space to express the intensity of a preference? Riesz spaces provide a general framework to present at-once all approaches and to describe properties in the context of PCMs. Preferences via Riesz spaces are *universal*, in the sense that *(I)* they can express a ratio or a difference or a fuzzy relation, *(II)* the obtained results are true in every formalization and *(III)* Riesz spaces are a common language which can be used as a bridge between different points of view.

What does it mean non-linear intensity? In multi-criteria methods decision makers deals with many (maybe conflicting) objectives and intensity of preferences is expressed by a (real) number in each criteria. In AHP we have different PCMs, which describe different criteria; if we consider $\mathbb{R}^n$ [see example (2) above] we are just writing all these matrices as a unique matrix with vectors as elements. Actually, we can consider each component of a vector as the standard way to represent the intensity preference and the vector itself as the natural representation of multidimensional (i.e. multi-criteria) comparison. This construction has its highest expression in the subfield of MCDM called Multi-Attribute Decision Making, which has several models and applications in military system efficiency, facility location, investment decision making and many others (e.g. see Belton 1986, Torrance et al. 1996, Xu 2015, Zanakis et al. 1998)

Does it make sense to consider non-archimedean Riesz space in this context? Let us consider the following example. A worker with economic problems has to buy a car. We can consider the following hierarchy:

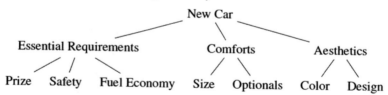

It is clear that Essential Requirements (ER), Comforts (C) and Aesthetics (A) cannot be just weighted and combined as usual. In fact, we may have the following two cases:

- we put probability different to zero on (C) and (A) and in the process can happen that the selected car is not the most economically convenient or even too expensive for him (remember that the worker has a low budget and he has to buy a car), and this is an undesired result.
- conversely, to skip the case above, we can just consider (ER) as unique criterion and neglect (C) and (A). Also in this case we have a non-realistic model, indeed our hierarchy does not take into account that if two cars have the same rank in (ER)

then the worker will choose the car with more optionals or with a comfortable size for his purposes.

In a such situation it seems to be natural to consider a lexicographic order [see example (2) above] such as $(\mathbb{R} \times_{LEX} \mathbb{R}) \times_{LEX} \mathbb{R}$, where each component of a vector $(x, y, z) \in (\mathbb{R} \times_{LEX} \mathbb{R}) \times_{LEX} \mathbb{R}$ is a preference intensity in (RE), (C) and (A) respectively (we may shortly indicate the hierarchy with $(RE) \times_{LEX} (C) \times_{LEX} (A)$). We remark that *lexicographic preferences* cannot be represented by any continuous utility function (see Debreu 1954).

Which kinds of intensity can we express with functions? This approach is one of the most popular and widely studied one, under the definition of *utility functions*. These functions provide a cardinal presentation of preferences, which allows to work with choices using a plethora of different tools, related to the model (e.g. see Harsanyi 1953, Houthakker 1950, Levy and Markowitz 1979). We want to stress that in example (4) we consider functions from a compact to $\mathbb{R}$, without giving a meaning of the domain, which can be seen as a time interval, i.e. in this framework it is also possible to deal with Discounted Utility Model and intertemporal choices (e.g. see Frederick et al. 2002). Manipulation of a particular class of these functions (i.e. piecewise-linear functions defined over $[0, 1]^n$) in the context of Riesz MV-algebras is presented by Di Nola et al. (2016). Furthermore, it is possible to consider more complex examples, for instance we can consider the space $\mathbb{R}^F$ of functionals, where F is a general archimedean Riesz space with strong unit (e.g. see Cerreia-Vioglio et al. 2015).

3.2   On Collective Choice Rules for PCMs and Arrow's Axioms

In this section we want to formalize and characterize Collective Choice Rules f in the context of *generalized PCMs*, i.e. PCMs with elements in a Riesz space, which satisfy classical conditions in social choice theory.

Let R be a Riesz space. Let us consider m experts/decision makers and n alternatives. A collective choice rule f is a function

$$f : GM_n^m \rightarrow GM_n$$

such that

$$f(X^{(1)}, \ldots, X^{(m)}) = X$$

where X is a *social* matrix, GM_n is the set of all matrices (PCMs) over R with n alternatives such that for every $i \in \{1, \ldots, n\}$ $x_{ii} = \bar{0}$. f can be seen also as follows:

$$f = (\tilde{f}_{ij})_{1 \leq i, j \leq n},$$

where

$$\tilde{f}_{ij} : GM_n^m \to R.$$

Note that GM_n is a subspace of $M_n(R)$ (see example (5)), i.e. it is a Riesz space. Let us introduce properties related with axioms of democratic legitimacy and informational efficiency required in Arrow's theorem.

$\forall i, j \ (\exists f_{ij} : R^m \to R : \tilde{f}_{ij}(X^{(1)}, \dots, X^{(m)}) = f_{ij}(x_{ij}^{(1)}, \dots, x_{ij}^{(m)}))$ (Property I^*)

$\forall i, j \ (f_{ij}((R^m)^+) \subseteq R^+)$ (Property P^*)

$\nexists i \in \{1, \dots, m\} : \forall X^{(j)}, \text{ with } j \neq i \ (f(X^{(1)}, \dots, X^{(i)}, \dots, X^{(m)}) = X)$ (Property D^*)

Theorem 1 *Let R be a Riesz space and let f be a function $f : (R^{n^2})^m \to R^{n^2}$. f is a collective choice rule satisfying Axioms of Arrow's theorem if and only if f has properties I^*, P^* and D^*.*

3.3 On Social Welfare Function Features

Social welfare functions (SWFs) are all the collective choice rules which provide a total preorder on the set of alternatives. We can decompose a SWF g as follows:

$$g = \omega \circ f,$$

where f is a collective choice rule having properties I^*, P^* and D^*, and ω is a function such that

$$\omega : GM_n \to \mathbf{TP},$$

where **TP** is the set of total preorders on the set of alternatives. Let us consider a social matrix $X = f(X^{(1)}, \dots, X^{(m)})$. We want to characterize property of ω such that g is a social welfare function.

It is trivial to check that if X is *transitive*, then it is possible to directly compute an order which expresses the preferences over alternatives. In fact, let X be a GM_n, it has two properties:

$(\rho) \ x_{ii} = \bar{0},$ (Reflexivity)

$(\gamma) \ \forall i, j \in \{1, \dots, n\} \ x_{ij} \in R.$ (Completeness)

If we have also that

$$(\tau) \; (\bar{0} \preceq x_{ij} \text{ and } \bar{0} \preceq x_{jk}) \Rightarrow \bar{0} \preceq x_{ik} \qquad \text{(Transitivity)}$$

We say that an order $\lesssim_X$ is *compatible with* X if and only if we have that:

$$\bar{0} \preceq x_{ij} \quad \Leftrightarrow \quad j \lesssim_X i.$$

An analogous definition is proposed by Trockel (1998) in the context of utility functions.

Proposition 5 *Let X be a transitive GM_n (TGM_n) then there exists a unique total preorder $\lesssim_X$ compatible with X. Or equivalently, the correspondence*

$$\theta : TGM_n \rightarrow \mathbf{TP}$$

which associates to each $X \in TGM_n$ a preorder $\lesssim_X$ compatible with X itself is a surjective function. Moreover $\lesssim_X \equiv \lesssim_{\alpha \cdot X}$ for every $\alpha \in \mathbb{R}^+$, and $\lesssim_X \equiv \gtrsim_{\alpha \cdot X}$ for every $\alpha \in \mathbb{R}^-$.

Let $\mathscr{C}(R) = \{A \subseteq R \mid A \text{ is a cone}\}$ be the set of all closed cones of R Riesz space. By Proposition 5 we can consider the function Φ

$$\Phi : \mathbf{TP} \rightarrow \mathscr{C}(TGM_n)$$

such that

$$\Phi(\lesssim) = \{X \in TGM_n \mid \lesssim \text{ is compatible with } X\}$$

Proposition 6 *The function Φ is injective.*

We can define an order relation $\ll$ over $\mathbf{TP}$ as follows:

$$\lesssim_1 \ll \lesssim_2 \quad \Leftrightarrow \quad i \lesssim_2 j \; \rightarrow \; i \lesssim_1 j .$$

It is also possible to denote with $\lesssim = \lesssim_1 \vee \lesssim_2$ as the total preorder such that

$$i \lesssim j \quad \Leftrightarrow \quad i \lesssim_1 j \text{ and } i \lesssim_2 j.$$

Remark 2 By easy considerations, we have that $\Phi(\lesssim_1) \cap \Phi(\lesssim_2) = \Phi(\lesssim_1 \vee \lesssim_2)$. Moreover, note that $\mathbf{TP}$ is closed with respect to $\vee$, i.e. ($\mathbf{TP}$, $\vee$) is a *join-semilattice*.

Examples

Let us consider n alternatives. The spaces of total preorder with $n = 2$ and $n = 3$ have the following configurations:

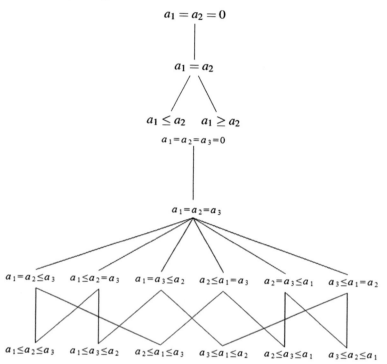

Note that in each space we have exactly one atom which expresses indifference. We call *basic total preorder* an element which is minimal in $(\mathbf{TP}, \ll)$.

Remark 3 In order to deal with aggregation of many TGM_n we added a root ($\top$), which can be interpreted as impossibility to make a social decision (related to *Condorcet's paradox* and *Arrow's impossibility theorem* in the context of PCMs). We put

$$\Phi(\top) = \emptyset.$$

Proposition 7 *Every $\lesssim$ total preorder different from $\top$ can be written as $\bigvee_i \lesssim_i$, where $\lesssim_i$ are basic total preorders.*

Proposition 8 *Let $\lesssim$ be a basic total preorder over n elements. We have that $\Phi(\lesssim)$ is an orthant in TGM_n.*

Analogously to θ we can define Θ in this way:

$$\Theta : \mathscr{C}(TGM_n) \ \rightarrow \ \mathbf{TP}$$

where $\Theta(\emptyset) = \top$ and

$$\Theta(K) = \Phi^{-1}\left(\bigcap_{\substack{C \in \Phi(\mathbf{TP}) \\ C \cap K \neq \emptyset}} C\right).$$

By Remark 2 we have that the function is well-defined.

Definition 15 Let $(A, \leq_A)$ and $(B, \leq_B)$ be two partially ordered sets. An antitone Galois correspondence consists of two monotone functions: $F : A \to B$ and $G : B \to A$, such that for all a in A and b in B, we have $F(a) \leq_B b \quad \Leftrightarrow \quad a \geq_A G(b)$.

Now we can state the following result.

Theorem 2 *The couple* (Θ, Φ) *is an antitone Galois correspondence between* $(\mathscr{C}(TGM_n), \subseteq)$ *and* $(\mathbf{TP}, \ll)$.

We denote by K_n the subset of $\mathscr{C}(TGM_n)$ of all the cones L such that $L \in \Phi(\mathbf{TP})$.

Proposition 9 *Let L be a cone of TGM_n. We have that*

$$L \in \Phi(\mathbf{TP}) \quad \Leftrightarrow \quad L \text{ is a } TP - cone.$$

Let us define the categories $\mathbb{TP}_n$ (of total preorders) and $\mathbb{K}_n$ (of TP-cones in TGM_n). In $\mathbb{TP}_n$ the objects are total preorder on n elements and arrows are defined by order $\ll$, i.e.

$$\precsim_1 \to \precsim_2 \quad \Leftrightarrow \quad \precsim_1 \ll \precsim_2 .$$

In a similar way we define $\mathbb{K}_n$ whose objects are TP-cones in the space TGM_n and arrows are defined by inclusion.

Theorem 3 *Categories of preorders and of TP-cones are dually isomorphic.*

4 Conclusions and Future Work

In this chapter, we considered some models for both individual and social choices. The structures that we propose allow us to unify some previous models analyzed in the literature and also to overcome or interpret some critical issues or paradoxes. In the future work we will carry out the research by classifying decision procedures by means of logical and algebraic structures and applying these tools to further well known themes of decision theory.

References

Abramovich, Y., Aliprantis, C., Zame, W.: A representation theorem for riesz spaces and its applications to economics. Econ. Theory **5**(3), 527–535 (1995)

Aliprantis, C.D., Brown, D.J.: Equilibria in markets with a riesz space of commodities. J. Math. Econ. **11**(2), 189–207 (1983)

Aliprantis C.D., Burkinshaw, O.: Locally Solid Riesz Spaces With Applications to Economics, vol. 105. American Mathematical Society (2003)

Aliprantis, C.D., Tourky, R.: Cones and Duality, vol. 84. American Mathematical Society (2007)

Arrow, K.: Social Choice and Individual Values. Yale University Press, New Haven (1978)

Arrow, K.J., Raynaud, H.: Social Choice and Multicriterion Decision-Making. MIT Press, Cambridge (1986)

Arrow, K.J., Sen, A., Suzumura, K.E: Handbook of Social Choice and Welfare vol. 2, 1st edn. North Holland (2011)

Barzilai, J.: Consistency measures for pairwise comparison matrices. J. Multi-Criteria Decis. Anal. **7**(3), 123–132 (1998)

Barzilai, J., Lootsma, F.A.: Power relations and group aggregation in the multiplicative AHP and SMART. J. Multi-Criteria Decis. Anal. **6**(3), 155–165 (1997)

Basile, L., D'Apuzzo, L.: Weak consistency and quasi-linear means imply the actual ranking. Int. J. Uncertain. Fuzziness Knowl. Based Syst. **10**(3), 227–239 (2002)

Belton, V.: A comparison of the analytic hierarchy process and a simple multi-attribute value function. Eur. J. Oper. Res. **26**(1), 7–21 (1986)

Belton, V., Stewart, T.: Multiple Criteria Decision Analysis: An Integrated Approach. Springer Science & Business Media, Berlin (2002)

Beynon, W.M.: Duality theorems for finitely generated vector lattices. In: Proceedings of the London Mathematical Society, vol. s3-31, pp. 114–128 (1975)

Birkhoff, G.: Lattice Theory, vol. 25. American Mathematical Society (1984)

Brunelli, M.: Studying a set of properties of inconsistency indices for pairwise comparisons. Ann. Oper. Res. **248**(1), 143–161 (2017)

Brunelli, M., Fedrizzi, M.: Axiomatic properties of inconsistency indices for pairwise comparisons. J. Oper. Res. Soc. **66**(1), 1–15 (2015)

Cavallo, B.: A further discussion of a semiring-based study of judgment matrices: properties and models. Inf. Sci. **181**, 2166–2176 (2011); Inf. Sci. **287**, 61–67 (2014)

Cavallo, B.: Computing random consistency indices and assessing priority vectors reliability. Inf. Sci. **420**, 532–542 (2017)

Cavallo, B.: $\mathscr{G}$-distance and $\mathscr{G}$-decomposition for improving $\mathscr{G}$-consistency of a pairwise comparison matrix. Fuzzy Optim. Decis. Mak. (2018). https://doi.org/10.1007/s10700-018-9286-3

Cavallo, B., Brunelli, M.: A general unified framework for interval pairwise comparison matrices. Int. J. Approx. Reason. **93**, 178–198 (2018)

Cavallo, B., D'Apuzzo, L.: A general unified framework for pairwise comparison matrices in multicriterial methods. Int. J. Intell. Syst. **24**(4), 377–398 (2009)

Cavallo, B., D'Apuzzo, L.: Deriving weights from a pairwise comparison matrix over an alo-group. Soft Comput. **16**(2), 353–366 (2012a)

Cavallo, B., D'Apuzzo, L.: Investigating properties of the $\odot$-consistency index. In: Advances in Computational Intelligence, pp. 315–327 Springer, Berlin (2012b)

Cavallo, B., D'Apuzzo, L.: Reciprocal transitive matrices over abelian linearly ordered groups: characterizations and application to multi-criteria decision problems. Fuzzy Sets Syst. **266**, 33–46 (2015)

Cavallo, B., D'Apuzzo, L.: Ensuring reliability of the weighting vector: weak consistent pairwise comparison matrices. Fuzzy Sets Syst. **296**, 21–34 (2016)

Cavallo, B., D'Apuzzo, L., Squillante, M.: About a consistency index for pairwise comparison matrices over a divisible alo-group. Int. J. Intell. Syst. **27**(2), 153–175 (2012)

Cavallo, B., Canfora, G., D'Apuzzo, L., Squillante, M.: Reasoning under uncertainty and multi-criteria decision making in data privacy. Qual. Quant. **48**(4), 1957–1972 (2014)

Cavallo, B., D'Apuzzo, L., Squillante, M.: A multi-criteria decision making method for sustainable development of naples port city-area. Qual. Quant. **49**(4), 1647–1659 (2015)

Cavallo, B., D'Apuzzo, L., Vitale, G.: Reformulating arrow's conditions in terms of cardinal pairwise comparison matrices defined over a general framework. Group Decis. Negot. **27**, 107–127 (2018)

Cerreia-Vioglio, S., Maccheroni, F., Marinacci, M., Montrucchio, L.: Choquet integration on riesz spaces and dual comonotonicity. Trans. Am. Math. Soc. **367**(12), 8521–8542 (2015)

Chiclana, F., Herrera-Viedma, E., Alonso, S., Herrera, F.: Cardinal consistency of reciprocal preference relations: a characterization of multiplicative transitivity. IEEE Trans. Fuzzy Syst. **17**(1), 14–23 (2009)

D'Apuzzo, L., Marcarelli, G., Squillante, M.: Analysis of Qualitative and Quantitative Rankings in Multicriteria Decision Making, pp. 157–170. Springer, Milano (2009)

Debreu, G.: Representation of a preference ordering by a numerical function. Decis. Process. **3**, 159–165 (1954)

Delli Rocili. L., Maturo, A.: Social problems and decision making for teaching approaches and relationship management in an elementary school. In: Hošková-Mayerová, Š., Maturo, F., Kacprzyk, J. (eds.) Mathematical-Statistical Models andQualitative Theories for Economic and Social Sciences, Studies in Systems, Decision and Control, vol 104, pp. 81–93. Springer International Publishing, (2017)

Di Nola, A., Lenzi, G., Vitale, G.: Riesz–mcnaughton functions and riesz mv-algebras of nonlinear functions. Fuzzy Sets Syst. (2016)

Di Nola, A., Squillante, M., Vitale, G.: Social preferences through riesz spaces: a first approach. In: Soft Computing Applications for Group Decision-making and Consensus Modeling, pp 113–127. Springer, Berlin (2018)

Frederick, S., Loewenstein, G.: O'donoghue, T.: Time discounting and time preference: a critical review. J. Econ. Lit. **40**(2), 351–401 (2002)

Greco, S., Matarazzo, B., Slowinski, R.: Rough sets theory for multicriteria decision analysis. Eur. J. Oper. Res. **129**(1), 1–47 (2001)

Greco, S., Ehrgott, M., Figueira, J.: Multiple Criteria Decision Analysis: State of the Art Surveys. Springer, Berlin (2016)

Harsanyi, J.C.: Cardinal utility in welfare economics and in the theory of risk-taking. J. Polit. Econ. **61**(5), 434–435 (1953)

Hou, F.: A multiplicative alo-group based hierarchical decision model and application. Commun. Stat. Simul. Comput. **45**(8), (2016)

Houthakker, H.S.: Revealed preference and the utility function. Economica **17**(66), 159–174 (1950)

Hošková-Mayerová, Š., Maturo, A.: Fuzzy sets and algebraic hyperoperations to model interpersonal relations. In: Maturo, A., Hošková-Mayerová, Š., Soitu, D.T., Kacprzyk, J. (eds) Recent Trends in Social Systems: Quantitative Theories andQuantitative Models, Studies in Systems, Decision and Control, vol. 66, pp. 211–221 (2017)

Hošková-Mayerová, Š., Maturo, A.: Decision-making process using hyperstructures and fuzzy structures in social sciences. In: Collan, M., Kacprzyk, J. (eds.) Soft Computing Applications for Group Decision-making and Consensus Modeling, Studies in Fuzziness and Soft Computing, vol. 357, pp. 103–111. Springer International Publishing, (2018)

Ishizaka, A., Nemery, P.: Multi-criteria Decision Analysis: Methods and Software. Wiley, New York (2013)

Keeney, R.L., Raiffa, H.: Decisions with Multiple Objectives: Preferences and Value Tradeoffs. Wiley, New York (1976)

Labuschagne, C., Van Alten, C.: On the variety of riesz spaces. Indag. Math. **18**(1), 61–68 (2007)

Levy, H., Markowitz, H.M.: Approximating expected utility by a function of mean and variance. Am. Econ. Rev. **69**(3), 308–317 (1979)

Lu, J., Han, J., Hu, Y., Zhang, G.: Multilevel decision-making: a survey. Inf. Sci. **346–347**, 463–487 (2016)

Maturo, A., Sciarra, A.: Reasoning and decision making in practicing counseling. In: Hošková-Mayerová Š, Maturo, F., Kacprzyk, J. (eds.) Mathematical-Statistical Models and Qualitative Theories for Economic and Social Sciences, Studies in Systems, Decision and Control, vol. 104, Springer International Publishing, pp. 265–281 (2017)

Ramík, J.: Isomorphisms between fuzzy pairwise comparison matrices. Fuzzy Optim. Decis. Making 14(2), 199–209 (2015)

Russo, R.F.S.M., Camanho, R.: Criteria in ahp: a systematic review of literature. Procedia Comput. Sci. 55, 1123–1132 (2015)

Saaty, T.L.: A scaling method for priorities in hierarchical structures. J. Math. Psychol. 15, 234–281 (1977)

Saaty, T.L.: The Analytic Hierarchy Process. McGraw-Hill, New York (1980)

Saaty, T.L.: How to make a decision: the analytic hierarchy process. Eur. J. Oper. Res. 48(1), 9–26 (1990)

Sixto, R.: Decision theory and decision analysis: trends and challenges. Springer, Netherlands (1994)

Tanino, T.: Fuzzy preference orderings in group decision making. Fuzzy Sets Syst. 12(2), 117–131 (1984)

Tanino, T.: Fuzzy preference relations in group decision making. In: Non-Conventional Preference Relations in Decision-Making, pp. 54–71. Springer, Heidelberg (1988)

Torrance, G.W., Feeny, D.H., Furlong, W.J., Barr, R.D., Zhang, Y., Wang, Q.: Multiattribute utility function for a comprehensive health status classification system: Health utilities index mark 2. Med. Care 34(7), 702–722 (1996)

Trockel, W.: Group actions on spaces of preferences and economic applications. Util. Funct. Ordered Spaces 5, 159–175 (1998)

Ventre, A., Maturo, A., Hoskova-Mayerova, S., Kacprzyk, J.: Multicriteria and multiagent decision making with applications to economic and social sciences. In: Studies in Fuzziness and Soft Computing, vol. 315. Springer-Verlag, Berlin, Heidelberg (2013)

Xia, M., Chen, J.: Consistency and consensus improving methods for pairwise comparison matrices based on abelian linearly ordered group. Fuzzy Sets Syst. 266, 1–32 (2015)

Xu, Z.: Uncertain Multi-attribute Decision Making: Methods and Applications. Springer, Berlin (2015)

Zanakis, S.H., Solomon, A., Wishart, N., Dublish, S.: Multi-attribute decision making: a simulation comparison of select methods. Eur. J. Oper. Res. 107(3), 507–529 (1998)

Zhang, H.: Group decision making based on multiplicative consistent reciprocal preference relations. Fuzzy Sets Syst. 282, 31–46 (2016)

Nationality Board Diversity in Organizations: A Brief Review and Future Research Directions

Fabrizio Maturo, Stefania Migliori and Francesco Paolone

Abstract In recent decades, many studies have focused on the issue of diversity in organizations due to the increasing interest on its possible impact on firms' performance and other organizational aspects. Because of the growing globalized markets, greater mobility, and demographic developments which are bringing more people to work with others who differ in cultural, demographic, informational, and personality attributes, scholars have concentrated on various type of team variety within organizations, e.g. gender, nationality, educational, age, ethnicity, and race. Despite the different goals of these researches, most insiders agree that diversity plays an essential role for the efficacy of management and decisions, long term prosperity, and ability to develop relationships with the external environment. Nevertheless, the findings of many studies are conflicting and suggest a negative role of diversity. For these reasons, this paper, focusing on one aspect of organizational variety, i.e. nationality diversity, aims to propose a brief review of the main existing studies on this topic. The final purpose of this contribution is to discuss and understand what are the results and limits of these studies.

Keywords Nationality board diversity · Board diversity · Firm diversity
Firm heterogeneity · Corporate diversity

The authors are entered in alphabetical order by their last names. The individual contributions of the authors are the following. F. Maturo: Introduction, Conclusions, Table 6. S. Migliori: Sects. 2.1 and 2.3. F. Paolone: Abstract, Sect. 2.2.

F. Maturo (✉) · S. Migliori
Department of Management and Business Administration,
"G. D' Annunzio" University of Chieti-Pescara, V.le Pindaro 42, Pescara, Italy
e-mail: f.maturo@unich.it

S. Migliori
e-mail: stefania.migliori@unich.it

F. Paolone
Department of Business Administration, "Parthenope" University of Naples,
Naples, Italy
e-mail: francesco.paolone@uniparthenope.it

1 Introduction

In today's global economy, diversity is relevant for the purpose of avoiding the homogenization phenomenon within organizations as well as favouring the integration. The existence of diversity in an organization also represents the source of a constant process of innovation that would maintain incentives to investment. Thus, diversity has becoming a distinguishing feature of contemporary organizations due to two main determinants: external factors, such as the process of globalization of markets, advances in technology and communications, greater mobility (Vallejo Andrada et al. 2017; Urban and Šarka Hošková-Mayerová 2017) and the enactment of rules designed to safeguard the principle of equality in hiring practices (Biemann and Kearney 2010); and internal factors relating to the company's vision that considers diversity as an avenue through which to bring different skills, knowledge, experience and perspective within a company, which can enhance more creativity and innovative solutions to problem and decision and consequently the firm's competitive advantage (van Knippenberg and Schippers 2007; Harrison and Klein 2007).

Also at institutional level, the adoption of diverse teams in organizations is widely promoted: *managing diversity and promoting inclusion has become part of the business world's strategic agenda as a response to a more diversified society where knowledge and innovation are essential for obtaining competitive advantages in a globalised economy* (European Commission 2015). For these reasons, to date, there have been many researches and theoretical discussions on this topic and the concept of diversity. However there is still a general consensus about the difficulty of synthesizing and comparing the results. This is due to several factors.

The first reason is certainly that the precise meaning of "diversity" is not clear because many scholars have casually used this term as synonymous of heterogeneity, variety, dissimilarity, and dispersion (Harrison and Klein 2007). The second cause is that several different theoretical perspectives have been used to guide diversity research; often, these perspectives suggest contradictory effects and are supported by empirical studies. The third problem is that the diversity literature itself is very diverse because of the use of plurality of demographic variables (such as gender, race, ethnicity, age, tenure, education, and marital status) and non-demographic variables (values, attitudes, conscientiousness, individual performance, and pay). The fourth reason is that many indices have been proposed but no universally accepted indicators for measuring and monitoring diversity has yet been established (Ricotta et al. 2003).

Hence, a key role is played by the methodological approach adopted to operationalize the concept of diversity. The importance of this aspect is highlighted by the study of Harrison and Klein (2007), which draw attention to the need for *matching a specific operationalization of diversity to a specific conceptualization of diversity*. Specifically, the adoption of a correct measurement depends on the concept of diversity that researches have in mind to observe. In this perspective, the authors have highlighted that diversity is a theoretical construct, broken down into three distinct aspects: separation (differences in position or opinion among members), variety (differences in information, knowledge or experience among members) and disparity (differences in concentration of valued social assets or resources).

Due to the multidimensional aspect of diversity, this research focuses on one type of organizational variety, i.e. nationality board diversity (NBD). Our purpose is to provide a brief review of the main existing studies on this topic. The final aim of this contribution is to discuss the main results and limits of previous studies, and why the literature is composed by conflicting theories and findings even if we only consider NBD.

In the literature, NBD is considered one of the most important diversity characteristic of organizations (Ararat et al. 2015). Many scholars state that NBD enhances board's independence and effectiveness. For example, Ruigrok et al. (2007) consider that the board effectiveness may increase as a consequence of the presence of foreigners on board (Ruigrok et al. 2007; Shehata 2013). Brickley and Zimmerman (2010) argue that, despite their monitoring deficiencies, foreign independent directors may enhance the advisory capability of boards. In effect, many other scholars confirm that board diversity may increase board independence (e.g. Ayuso and Argandoña 2009) because foreigners on board support corporate social responsibility reporting. Moreover, according to the studies of (Ujunwa 2012), the presence of foreign board members, in line with the resource dependency theory (RDT), may bring a large stock of qualified candidates for the board with higher experiences than domestic members. In addition, foreigners members can also improve minority investors' confidence regarding the professional management of organizations (Oxelheim and Randøy 2003).

On the contrary, following the Hahn and Lasfer (2016) perspective, the presence of foreign directors may impair the internal governance due to a lower number of board meetings, thus bringing to a weak monitoring by the board. Therefore, even if foreign directors have special international expertise, the cost and benefit tradeoff of appointing foreign directors compared to local directors decreases the firm's governance effectiveness (Hahn and Lasfer 2016). Moreover, Barako and Brown (2008), focusing on the Kenyan banking sector, have claimed that the presence of different nationalities on boards may push up to cross-cultural communication problems.

This article is structured as follows. The introduction presents the main issues which have caused, for many decades, inconsistent results of the literature regarding organizational diversity in general and NBD. The second section proposes a brief review of the studies regarding NBD. Specifically, we divide this section to give special attention to three different aspects: theories, methods, and findings. In the last section, we present our discussion and conclusions.

2 A Brief Review on Theories, Methodology, and Findings Regarding Nationality Board Diversity in Organizations

This section proposes a brief review on theories, methodology, and findings regarding NBD in organizations. At this purpose, the "SCOPUS" database has been used. The research of the papers, which are presented in this study, is dated 2017/11/15. Our query on the"SCOPUS" database was based on the keyword *board diversity*; all the papers responding to this request has been exported and consulted to understand if they concerned NBD. The paper regarding other types of diversity have been excluded. The full list of the collected studies is composed by 23 papers and presented in Table 6. We stress that the following investigation is limited on the consulted literature, and it does not pretend to be neither exhaustive nor general.

2.1 Theoretical Frameworks

The analysis of the consulted literature shows that NBD has been studied through the use of different theoretical perspectives, very often in combination with each other (thus, more than one theory for paper). This highlights the multidisciplinary approach that characterizes these studies. Table 1 lists the theoretical perspectives used in the selected studies. It refers only to the studies that explicitly indicate the theories adopted. We note that about 52% of the papers does not explicitly indicate the theoretical perspective used whereas the others highlight the adoption of a variety of theoretical approaches; however, the principal agent theory (PAT), resource dependency theory (RDT), stewardship theory, and theories of social psychology (e.g. similarity attraction theory (SAT)) appear to be the most adopted.

Table 1 remarks that the most used theoretical perspective is the PAT, and this is certainly linked to the fact that the concept of national board (NBD) is related to the board which is a key variable of the corporate governance systems of the companies. The PAT (Berle and Means 1932; Jensen and Meckling 1976), focused on the conflict of interests between the (principal) property and the management (agent), attributes to the board the control function on management in order to avoid opportunistic behaviours that may result in lower performance. In particular, the introduction in the board composition of independent directors brings a greater degree of independence and objectivity to the board in the process of monitoring the results produced by management, as well as greater tension towards the achievement of performance

Table 1 Theoretical perspectives explicitly adopted by research on NBD

Theoretical perspective	Number of studies
PAT	7
RDT	6
Stewardship theory	2
Theories of social psychology (social identity (SCT), similarity-attraction (SAT), homosocial reproduction)	2
Stakeholder theory	1
Resource-based view theory	1
Theories of organizational behavior	1

goals. This theory is considered an appropriate approach for analysing NBD and its effect on firm performance (Eulerich et al. 2014). In the selected sample, the PAT has been used in seven studies, and always in combination with other theories (mainly RDT). Among these studies, the majority of them (86%) has referred to NBD as an independent variable (IV) and analyzes its effects on various types of dependent variables (DV).

Randøy et al. (2006) and Ararat et al. (2015), respectively, have analysed the effects of the NBD on stock market performance and other accounting performance (e.g. return on equity). Instead, two other studies have focused on the effects of the NBD on corporate disclosure. In particular, the study by Hoang et al. (2016) has shown a positive effect of NBD on the quality and quantity of the corporate social disclosure of Vietnamese listed firms, while Shehata (2013) has provided a theoretical analysis on the (positive) relationship between diversity and corporate information disclosure. Finally, the study by Ben-Amar et al. (2011) have dealt with the effect of NBD on the success of the strategic merger and acquisition decisions (M&A) on a sample of 289 M&A decisions undertaken by Canadian firms. In addition, Santen and Donker (2009) have studied NBD in the perspective of the financial distress of the firm highlighting the existence of a positive relationship on a sample of 58 listed companies on the Amsterdam Stock Exchange. Only one research (Matari et al. 2014), based on PAT and RDT, have considered NBD as a variable that moderates the effect of the board of directors (the board size and board meeting) on firm performance.

The RDT is the second most embraced theory into the selected studies. This theory suggests that the determining factor of firms' survival capability is their ability to acquire and maintain resources (Pfeffer and Langton 1993). Consequently, the company's ability to survive in the long term is determined by its ability to create relationships and links with the external environment in order to acquire the resources considered critical for its survival. Therefore, the relationships that a company establishes, in the environment of which it is an integral part, should be aimed at creating stable flows of resources and reducing the degree of uncertainty. The diversity in board composition regarding age, nationality, culture, gender, education etc, is a way able to bring to the firm diverging resources that are advantage for the firm.

The literature highlights that foreign members with their different backgrounds can add to the board diverse expertise which domestic members do not possess (Chen et al. 2008). Foreign board members can assure foreign minority investors that the company is managed also in their interests (Oxelheim and Randøy 2003). Moreover, diversity may increase the effectiveness of the decision process in the board room trough the discussion or the higher information processing (Hillman et al. 2000). This theory is used (together with the PAT) in various studies (e.g. Matari et al. 2014; Randøy et al. 2006; Hoang et al. 2016; Santen and Donker 2009) and it is followed in combination with other theories in only two studies. In particular, the study by Eulerich et al. (2014) which has also taken up the stewardship theory and that of Arnegger et al. (2013) that has embraced the RDT theory in joint with the germane theories of organizational behaviour (information processing theory, human cognition theory, similarity attraction paradigm, social categorization theory).

Both PAT and RDT have been adopted with the stewardship theory (Donaldson and Davis 1991, 1997) in two studies. The latter, in contrast to the PAT, is based on the assumption that the management board's members not act opportunistically because their interests correspond with those of the shareholder meeting. Specifically, it has been used in only two studies: that of Ben-Amar et al. (2011) in order to explain why in some cases "it is not obvious that independence (without knowledge or incentives) leads to better director performance than knowledge and strong incentives (without independence)" (Ben-Amar et al. 2011) and the one of Eulerich et al. (2014) that highlights the negative effect of NBD on firm performance.

Finally, other theories that appear in two studies are the theories of social psychology. Particularly, the article of Kaczmarek et al. (2012), using three concepts derived from the social psychology theories (similarity-attraction, homosocial reproduction, and social identity) have used NBD as a dependent variable. This research has underlined that increasing the presence of non-British members on the nomination committees is likely to have a positive impact on the level of NBD. Moreover, the concept of similarity-attraction (SAT) has been selected by Arnegger et al. (2013) to hypothesize that the relation between firm size and international background diversity in the board is not linear.

2.2 Indices

Regarding the methodological approaches that have been adopted by researchers for assessing NBD, we also discover several different choices. Table 2 displays a synthetic list of the various indices that have been used for NBD in the 23 selected organizational studies.

Table 2 Different indices that have been used in assessing NBD in organizational studies

Index	Number of studies
Proportions	10
Blau-Simpson index	4
Absolute abundance	3
Composite indicator	2
Dummy variable	1
Richness index	1
Diversity profiles	1
No indices, only theory	1
Total	23

The great part of the literature (10 + 3 + 1 = 14 papers on NBD) has focused on very simple indices such as proportions (ratio of the total number of foreign members to the total number), absolute abundances, and dummy variables. The richness index has been used by only one research (Eulerich et al. 2014) whereas no one has selected the Shannon–Wiener index (Shannon 1948) despite it is one of the most important indicators of variety. The richness index is very simpe because it is the sum of the modalities (total number of nationalities) but it does not take into account even-ness (the distribution of the different species). Instead, the Shannon-Wiener index (Shannon 1948) is given by:

$$\triangle_{Sh} = -\sum_{i=1}^{k} f_i \log(f_i) \tag{1}$$

The Shannon–Wiener index is also known as Teachman's index (Teachman 1980). It can range from zero to $log(k)$, thus its maximum is a function of the number of nationalities and its interpretation is similar to the Blau-Simpson index. The Shannon-Wiener index is affected by both the number of nationalities and their evenness; however, it is particularly sensitive to the presence of rare nationalities in an orga-nization. Despite this considerations, it is very strange that no one has decided to refer to this known index of diversity. Indeed, in many contexts (e.g. Ecology, Statis-tics, Medicine, Social Science, also Management in general), it is one of the most commonly used.

Four studies have chosen the Blau-Simpson index. In addition, to solve the prob-lem of the multidimensionality of the diversity concept, recent studies have tried to adopt composite indices encompassing also age, gender, education, and indepen-dence (Ararat et al. 2015); however, even if they have used a standardized Blau's values for each attribute to compose their NBD index, the limits to give the same weight to each attribute and blend quantitative and qualitative data still remain (age is classified into five categories with a consequent inevitable loss of information on the variability of the data).

The Blau–Simpson (1977) index is given by:

$$\triangle_K = 1 - \sum_{i=1}^{k} f_i^2 \tag{2}$$

where f_i is the relative frequency of the i-th nationality and k is the total number of nationalities. The Blau-Simpson index is also known as Herfindahl (1950) index and Hirschman (1964) index, but it was originally proposed by Simpson (1949) as a measure of species diversity in an ecosystem. It ranges from zero to $(k-1)/k$. Both evenness and richness contribute to a higher Blau's index and it is a good indicator of the dominance of one or few nationalities on the others; however, it is not a good predictor of richness because it is particularly sensitive to changes in the relative abundances of the most dominant nationalities.

Recently, Maturo et al. (2015, 2017) have proposed the use of diversity profiles in the context of NBD. Diversity profiles have been proposed in different contexts first by Hill (1973) and then by Patil and Taillie (1979). They are a coherent system for diversity estimates, which provide numbers that reflect both evenness and richness, and include variants of the richness, Shannon-Wiener and Blau-Simpson indices. They consist of a sequence of measurements considering different aspects of NBD to be encompassed in a single diversity spectrum. The diversity profile (Patil and Taillie 1979; Di Battista et al. 2016a) is given by:

$$\triangle_\beta = \sum_{i=1}^{k} \frac{(1 - f_i^\beta)}{\beta} f_i \quad \beta \geq -1 \tag{3}$$

where f_i is the relative abundance of the nationality i, β denotes the relative importance of richness and evenness, and the restriction that $\beta \geq -1$ assures that $\triangle_\beta$ has certain desirable properties (e.g. Patil and Taillie 1979; Di Battista et al. 2014, 2016b; Maturo and Di Battista 2018). The plot of $\triangle_\beta$ versus β provides the diversity profile which is a curve in the domain of β. The most used indices of diversity are special cases of Eq. 3: for $\beta = -1$, we get the richness index; for $\lim_{\beta \to 0}$, we obtain the Shannon-Wiener diversity index (Eq. 1); and for $\beta = 1$, we get the Blau-Simpson index (Eq. 2). Therefore, diversity profiles are functions dependent on a parameter that reflects the sensitivities to rare and abundant nationalities, and provide a continuum of possible diversity measures (e.g. Ricotta et al. 2003; Maturo et al. 2016; Di Battista et al. 2017) giving a an interesting graphical representation of NBD.

2.3 Results

The analysis of the results of the selected studies shows that most of them have analyzed NBD as an independent variable that can influence different aspects of companies (performance, decision-making processes, disclosure, ..., etc.) (see Table 3).

Table 3 highlights that most of the selected studies have focused on the analysis of the effects of NBD on the companies' performance. Specifically, there is an increasing number of empirical research that detects a positive influence of foreign board members, e.g. the studies of Estélyi and Nisar (2016), Ujunwa (2012), Rose (2007), Bronzetti et al. (2010), Ararat et al. (2015). Contrary, a negative link between NBD and performance has been found by García-Meca et al. (2015) for banks, and Eulerich et al. (2014) for firms in the German two-tier system. Only two studies do not found a significant relationship between NBD and performance. Specifically, the study provided by Darmadi (2011) for the Indonesian context and the report of Randøy et al. (2006) for the 500 largest Danish, Norwegian and Swedish companies. In only one study (Matari et al. 2014), NBD has been analyzed as a variable that

Table 3 Results of the studies on NBD

Effect of NBD on	Number of studies
Performance	10
CSR disclosure	2
Board meeting frequency	1
Corporate expropriation	1
Financial distress	1
Corporate disclosure	1
Strategic merger and acquisition performance	1
Total	17

Table 4 NBD as a moderator or dependent variable

Role of NBD in the analysis	Number of studies
Moderating effect of NBD on firm performance	1
Variables that can influence NBD	
Foreign shareholders	1
Governance regime of the country	2
Company characteristic	1
Presence of non-nationals on the nomination committees	1
Firm size	1
Total	7

moderates the relationship between the executive committee and the performance (see Table 4). Our research also shows that only 26% of the selected studies has analyzed NBD as a dependent variable. In particular, the antecedents of NBD are: the presence of foreign shareholders, the governance regime of the country, the firm size, the company characteristic and the presence of non-nationals on the nomination committees (see Table 4).

The selected studies investigated NBD using data of a large number of very different contexts. Table 5 shows the list of the different countries that these analyzes have treated. The contexts that have been most analyzed are: North Countries, Germany, Netherlands, and UK. Other contexts have been analyzed in a single study, while few studies have analysed and compared more contexts at the same time. This, in some cases, makes some results not fully generalizable or leads to conflicting results.

Table 5 Contexts that have been analyzed in the selected studies

Contexts	Number of studies
Australia	1
Belgium	1
Brazil	1
Canada	2
China	1
France	2
Germany	5
Hong Kong	1
India	1
Indonesia	1
Ireland	1
Israel	1
Italy	2
Japan	1
Kenya	1
Luxembourg	1
Malaysia	2
Mexico	1
Monaco	1
Netherlands	4
Nigeria	1
Northern Countries	5
Poland	1
Qatar	1
Russia	1
Singapore	1
South Africa	1
Spain	2
Switzerland	1
Taiwan	1
Turkey	1
UK	4
USA	2
Vietnam	1
Zimbabwe	1

3 Discussion and Conclusions

Voluntary initiatives and legal requirements for promoting "diversity and inclusion" in today's organizations have stimulated the debate of insiders and institutions on this topic. To date, in organizational literature, many theoretical discussions and empirical studies on the concept of diversity and its effects have been proposed. However, there is still a general consensus about the difficulty of comparing findings.

The first cause of confusion in organizational research is that the term "diversity" is often used as a synonym of "variability" but "diversity" should be considered only as a synonym of "variety" (Patil and Taillie 1982; Pielou 1975; Ricotta et al. 2003; Mayo et al. (2016). "Variability" generally refers to any type of data (numerical and categorical data) whereas the concept of "variety" is focuses on differences in kind or category. To avoid confusion, the term "diversity" should not be used for indicating the variability of numerical variables (e.g. it does not make sense saying that a 40-year-old person is "diverse" from a 41-year-old); the terms "dispersion", "disparity", and "inequality" are more appropriate for numerical data. The second problem is that several different theoretical perspectives have been used into diversity research, and often these perspectives suggest contradictory effects. The third problem is that many different variables have been considered for dealing with diversity, e.g. gender, race, ethnicity, age, tenure, education, and marital status, values, attitudes, conscientiousness, individual performance, and pay.

Therefore, it is clear that diversity is a multidimensional concept. For this reason, this study limits its proposes on a brief review of the main existing studies on national board diversity. Despite focusing only on NBD, the literature is composed by conflicting results. Considering that we focus only on NDB that is a clear kind on diversity and is a categoric variable, we should state that the cause of conflicting results are; the use of different theories and the implementation of different methodologies in measuring diversity.

With regard to former issue, many scholars suggest a positive value of NBD because they argue that when the distribution of variety is maximum everybody is different and everybody has a unique viewpoint to offer. However, other theories suggest a negative impact of NDB because each member is different from everyone else, and members can not discuss and optimally coordinate with others. Hence, it seems that maximum or minimum variety are not convenient; thus, we can deduce that a moderate variety may be an optimal strategy for each organization. However, the problem remains unresolved from a theoretical perspective because every theory has its reasons and they all are shareable and reasonable.

Concerning the methodological drawback, there is a great problem in the NBD literature, i.e. many indices have been used but no universally accepted measure exists. Researchers adopt several indices, but practically these latter depict different aspects of diversity. Proportion and total abundance are simple to use but they neglect the whole distribution of the data. On the other hand, the traditional diversity indices, e.g. richness, Shannon and Simpson indices, are also limited; indeed, some of them measure only richness or evenness whereas others measure both aspects but with different shades. For example, the richness index is sensitive only on the total number of nationalities. Consequently, different indices can lead to different finding and various studies can not absolutely be compared. In our opinion, this is the greatest problem of the studies on NBD, and also on diversity research in general.

Paying our attention on the presented findings, this review allows us to make some observations (see Table 6). First, we observe the lack of studies that analyze the antecedents of NBD. In order to better understand the relevance of the presence of foreign members in corporate governance system and how to stimulate NBD, it would be desirable to develop more studies focused on which factors can push or inhibit the NBD. Secondly, most research has analyzed the effect of NBD on firm performance. In order to better understand the role of governance mechanisms, it would be necessary to develop more studies using NBD as a moderation or mediation variable in the relationships between governance mechanisms and performance. This would allow us to better understand the complex relationship between NBD, governance, and performance. Finally, only few contexts have been widely analyzed. There are many other countries that have been analyzed only by one research, and therefore further analysis are needed to better interpret the presented results.

Table 6 Review of studies on NBD in organizations. PAT: principal agent theory. RDT: resource dependence theory. SAT: similarity-attraction theory. IPT: information processing theory. HCT: human cognition theory. SCT: social categorization theory

Paper—Authors—Year—Journal	Theory	Methods	Findings
Diverse boards: why do firms get foreign nationals on their boards?—K. Sághy Estélyia, T. M. Nisar—2016—Journal of corporate finance	Not explicit	Prop.	NBD is positively associated with shareholder heterogeneity, international market operations, and performance
Do institutional or foreign shareholders influence national board diversity? assessing board diversity through functional data analysis—F. Maturo, S. Migliori, and F. Paolone—2017—Studies in systems, decision and control	Not explicit	Diversity profiles	Institutional shareholders do not influence NBD, while foreign shareholders strongly affect it
Governance regimes and nationality diversity in corporate boards: a comparative study of Germany, the Netherlands and the UK—K. van Veen and J. Elbertsen—2008—Corporate governance	Not explicit	Prop.	Diversity of a corporate board is strongly directly dependent on the governance regime of the country of origin of the company
How international are executive boards of European MNCs? Nationality diversity in 15 European countries—K. van Veen, and I. Marsman—2008—European management journal	Not explicit	Prop.	Differences in NBD result from unique differences in governance regimes within the individual countries, and to a lesser extend the result of company characteristic

(continued)

Table 6 (continued)

Paper—Authors—Year—Journal	Theory	Methods	Findings
Impact of foreign directors on board meeting frequency—P. D. Hahn, and M. Lasfer—2016—International review of financial analysis	Not explicit	Prop.	A trade-off between increased NBD coupled with reduced monitoring through fewer meetings, weakens the internal governance mechanism, reduces the advisory role benefits of foreign nonexecutive directors and significantly exacerbate agency conflicts
Antecedents of board composition: the role of nomination committees—S. Kaczmarek, S. Kimino, and A. Pye—2012—Corporate governance: an international review	SAT, homosocial reproduction, SCT	Compos. measure of faultlines	The presence on the NC of females or non-British nationals has a positive impact on the level of board gender and NBD
Board characteristics and the financial performance of Nigerian quoted firms—A. Ujunwa—2012—Corporate governance: the international journal of business in society	Not explicit	Prop.	NBD impacts positively on firm performance
Board diversity and corporate expropriation—A. Husni Hamzah, and A. H. Zulkafli—2014—International conference on accounting studies	Not explicit	Absolute number	There is a negative relationship between foreign directors and corporate expropriation
Board diversity and corporate social disclosure: evidence from Vietnam—T. C. Hoang, I. Abeysekera, and S. Ma—2016—J Bus Ethics	RDT, PAT	Blau index	Positive effect of NBD on corporate social disclosure
Board diversity and firm performance: the Indonesian evidence—S. Darmadi—2011—Corporate ownership and control	Not explicit	Dummy variable	NBD has no influence on firm performance
Board diversity and its effects on bank performance: an international analysis—E. García-Meca and I.-M. García-Sánchez and J. Martínez-Ferrero—2015—Journal of banking & finance		Prop.	NBD inhibits bank performance
Board diversity in the perspective of financial distress: empirical evidence from The Netherlands—B. Santen, and H. Donker—2009—Corporate board: role, duties & composition	RDT, PAT	Absolute number	Positive relationship between the presence of foreign non-executive directors and financial distress
Comprehensive board diversity and quality of corporate social responsibility disclosure: evidence from an emerging market—N. Katmon, Z. Zuriyati Mohamad, N. Mat Norwani, and O. Al Farooque—2016—J Bus Ethics	Resource-based view theory	Blau index	The quality of CSR disclosure is significantly negatively associated with NBD

(continued)

Table 6 (continued)

Paper—Authors—Year—Journal	Theory	Methods	Findings
Corporate board diversity in Malaysia: a longitudinal analysis of gender and nationality diversity—D. Zainal, N. Zulkifli, and Z. Saleh—2013—International journal of academic research in accounting, finance and management sciences	Not explicit	Absolute number	It only examines the trend of NBD
Does board diversity really matter? gender does not, but citizenship does—C. Rose, P. Munch-Madsen, and M. Funch—2013—Int. journal of business science and applied management	Not explicit	Prop.	Impact of board citizenship on performance: board members with a background from common law have a positive influence on corporate performance (ROA, ROE and ROCE)
Firm size and board diversity—M. Arnegger, C.Hofmann, K. Pull, and K. Vetter—2014.—J. Manag. Gov.	RDT, IPT, HCT, SAT, SCT	Blau	International background diversity increases with increasing firm size, but international background diversity does so at decreasing rates
Governance, board diversity and firm value: the case of the Italian publicly listed firms—G. Bronzetti, R. Mazzotta, and G. Sicoli—2010—Corporate ownership & Control	Not explicit	Compos. index	Board diversity positively affect performance
How board diversity affects firm performance in emerging markets: evidence on channels in controlled firms—M. Ararat, M. Aksu, and A. Tansel Cetin—2015—Corporate governance: an international review	PAT, theories of social psychology	Blau	Positive and non-linear relationship between demographic diversity and performance, mediated by the board's monitoring efforts
How could board diversity influence Corporate disclosure?—N. F. Shehata—2013—Corporate board: role, duties & composition	PAT, stakeholder	No measure, Only theory	Positive influence of NBD on corporate disclosure
The impact of management board diversity on corporate performance—an empirical analysis for the German two-tier system—M. Eulerich, P. Velte, and C. van Uum—2014—Problems and perspectives in management	PAT, stewardship, RDT	Richness	Negative effects of NBD on corporate performance
The moderating effect of board diversity on the relationship between executive committee characteristics and firm performance in Oman: empirical study—E. M. Al-Matari, A. K. Al-Swidi, and F. H. Bt Fadzil—2014—Asian Social Science	PAT, RDT	Prop.	NBD moderates the effect of the board of directors (the board size, the board independence and the board meeting) on firm performance (ROA)

(continued)

Table 6 (continued)

Paper—Authors—Year—Journal	Theory	Methods	Findings
A nordic perspective on corporate board diversity—T. Randøy, S. Thomsen, and L. Oxelheim—2006—Report	PAT, RDT	Prop.	No significant diversity effect of gender, age, and NBD on stock market performance or on ROA
What makes better boards? a closer look at diversity and ownership—W. Ben-Amar, C. Francoeur, T. Hafsi, and R. Labelle—2013—British journal of management	PAT, stewardship theory	Prop.	Demographic diversity (also NBD) is found to have a clear and non-linear effect on strategic merger and acquisition performance

References

Ararat, M., Aksu, M., Tansel Cetin, A.: How board diversity affects firm performance in emerging markets: evidence on channels in controlled firms. Corp. Gov. (Oxford) **23**(2), 83–103 (2015). www.scopus.com

Arnegger, M., Hofmann, C., Pull, K., Vetter, K.: Firm size and board diversity. J. Manage. Gov. **18**(4), 1109–1135 (May 2013). https://doi.org/10.1007/s10997-013-9273-6

Ayuso, S., Argandoña, A.: Responsible corporate governance: towards a stakeholder board of directors? Corp. Owner. Control **6**(4) (2009). https://doi.org/10.22495/cocv6i4p1

Barako, G., Brown, A.: Corporate social reporting and board representation: evidence from the Kenyan banking sector. J. Manage. Gov. **12**(4), 309–324 (May 2008). https://doi.org/10.1007/s10997-008-9053-x

Ben-Amar, W., Francoeur, C., Hafsi, T., Labelle, R.: What makes better boards? a closer look at diversity and ownership. Brit. J. Manage. **24**(1), 85–101 (Nov 2011). https://doi.org/10.1111/j.1467-8551.2011.00789.x

Berle, A., Means, G.: The modern corporation and private property. Macmillan, N.Y. (1932)

Biemann, T., Kearney, E.: Size does matter: how varying group sizes in a sample affect the most common measures of group diversity. Org. Res. Methods **13**(3), 582–599 (Jul 2010). https://doi.org/10.1177/1094428109338875

Blau, P.: Inequality and heterogeneity. Free Press, New York (1977)

Brickley, J.A., Zimmerman, J.L.: Corporate governance myths: Comments on Armstrong, Guay, and Weber. SSRN Electron. J. (2010). https://doi.org/10.2139/ssrn.1681030

Bronzetti, G., Mazzotta, R., Sicoli, G.: Governance, board diversity and firm value: the case of the Italian publicly listed firms. Corp. Owner. Control **8**(1) (2010). https://doi.org/10.22495/cocv8i1c6p5

Chen, C.-W., Lin, J.B., Yi, B.: CEO duality and firm performance—an endogenous issue. Corp. Owner. Control (2008). https://doi.org/10.22495/cocv6i1p6

Darmadi, S.: Board diversity and firm performance: the Indonesian evidence. Corp. Owner. Control **8**(2) (2011). https://doi.org/10.22495/cocv8i2c4p4

Davis, J., Donaldson, F.S.L.: Toward a stewardship theory of management. Acad. Manage. Rev. **22**(1), 20–47 (1997)

Di Battista, T., Fortuna, F., Maturo, F.: Parametric functional analysis of variance for fish biodiversity. In: International Conference on Marine and Freshwater Environments, iMFE 2014 (2014). www.scopus.com

Di Battista, T., Fortuna, F., Maturo, F.: Environmental monitoring through functional biodiversity tools. Ecol. Ind. **60**, 237–247 (2016a)

Di Battista, T., Fortuna, F., Maturo, F.: Parametric functional analysis of variance for fish biodiversity assessment. J. Environ. Inf. **28**(2), 101–109 (2016b). https://doi.org/10.3808/jei.201600348

Di Battista, T., Fortuna, F., Maturo, F.: BioFTF: an R package for biodiversity assessment with the functional data analysis approach. Ecol. Ind. **73**, 726–732 (Feb 2017). https://doi.org/10.1016/j.ecolind.2016.10.032

Donaldson, L., Davis, J.: Stewardship theory or agency theory: CEO governance and shareholder returns. Aust. J. Manage. **16**(1), 49–64 (1991). https://doi.org/10.1177/031289629101600103

Estélyi, K.S., Nisar, T.M.: Diverse boards: why do firms get foreign nationals on their boards? J. Corp. Fin. **39**, 174–192 (Aug 2016). https://doi.org/10.1016/j.jcorpfin.2016.02.006

Eulerich, M., Velte, P., Uum van, C.: The impact of management board diversity on corporate performance—an empirical analysis for the German two-tier system. Prob. Persp. Manage. **12**(1), 25–39 (2014). ISSN 1727-7051

European Commission. Diversity within small and medium-sized enterprises—best practices and approaches for moving ahead. Brussels (2015)

García-Meca, E., García-Sánchez, I.-M., Martínez-Ferrero, J.: Board diversity and its effects on bank performance: an international analysis. J. Bank. Fin. **53**, 202–214 (Apr 2015). https://doi.org/10.1016/j.jbankfin.2014.12.002

Hahn, P., Lasfer, M.: Impact of foreign directors on board meeting frequency. Int. Rev. Fin. Anal. **46**, 295–308 (Jul 2016). https://doi.org/10.1016/j.irfa.2015.11.004

Harrison, D.A., Klein, K.J.: What's the difference? diversity constructs as separation, variety, or disparity in organizations. Acad. Manage. Rev. **32**(4), 1199–1228 (2007). www.scopus.com

Herfindahl, O.: Concentration in the Steel Industry. Columbia University (1950). https://books.google.it/books?id=vsZ4YgEACAAJ

Hill, M.: Diversity and evenness: a unifying notation and its consequences. Ecology **54**, 427–432 (1973)

Hillman, A.J., Cannella, A.A., Paetzold, R.L.: The resource dependence role of corporate directors: strategic adaptation of board composition in response to environmental change. J. Manage. Stud. **37**(2), 235–256 (Mar 2000). https://doi.org/10.1111/1467-6486.00179

Hirschman, A.O.: The paternity of an index. Am. Econ. Rev. **54**(5), 761–762 (1964). ISSN 00028282. https://doi.org/10.2307/1818582

Hoang, T.C., Abeysekera, I., Ma, S.: The effect of board diversity on earnings quality: an empirical study of listed firms in Vietnam. Aust. Acc. Rev. **27**(2), 146–163 (Dec 2016). https://doi.org/10.1111/auar.12128

Jensen, M.C., Meckling, W.H.: Theory of the firm: managerial behavior, agency costs and ownership structure. J. Fin. Econ. **3**(4), 305–360 (Oct 1976). ISSN 0304405X. https://doi.org/10.1016/0304-405x(76)90026-x

Kaczmarek, S., Kimino, S., Pye, A.: Antecedents of board composition: the role of nomination committees. Corp. Gov. Int. Rev. **20**(5), 474–489 (May 2012). https://doi.org/10.1111/j.1467-8683.2012.00913.x

Matari, E.M.A., Swidi, A.K.A., Fadzil, F.H.B.: The moderating effect of board diversity on the relationship between executive committee characteristics and firm performance in Oman: empirical study. Asian Soc. Sci. **10**(12) (May 2014). https://doi.org/10.5539/ass.v10n12p6

Maturo, F., Di Battista, T.: A functional approach to Hill's numbers for assessing changes in species variety of ecological communities over time. Ecol. Ind. **84**, 70 – 81 (2018). ISSN 1470-160X. https://doi.org/10.1016/j.ecolind.2017.08.016, http://www.sciencedirect.com/science/article/pii/S1470160X17304934

Maturo, F., Migliori, S., Consorti, A.: Corporate board diversity. In: Haller, A., Galea, M. (eds.), Proceedings of the International Conference "Humanities and Social Sciences Today." Economics, pp. 113–124. Pro Universitaria, Bucharest (2015)

Maturo, F., Di Battista, T., Fortuna, F.: Biodiversity assessment using functional tool. R package BioFTF (2016)

Maturo, F., Migliori, S., Paolone, F.: Do institutional or foreign shareholders influence national board diversity? assessing board diversity through functional data analysis, pp. 199–217. Springer International Publishing, Cham (2017). https://doi.org/10.1007/978-3-319-54819-7_14

Mayo, M., van Knippenberg, D., Guillén, L., Firfiray, S.: Team diversity and categorization salience. Org. Res. Methods 19(3), 433–474 (Jul 2016). https://doi.org/10.1177/1094428116639130

Oxelheim, L., Randøy, T.: The impact of foreign board membership on firm value. J. Bank. Fin. 27(12), 2369–2392 (2003). https://doi.org/10.1016/s0378-4266(02)00395-3

Patil, G., Taillie, C.: An overview of diversity. In: Grassle, J., Patil, G., Smith, W., Taillie, C. (eds.) Ecological Diversity in Theory and Practice, pp. 23–48. International Co-operative Publishing House, Fairland, MD (1979)

Patil, G., Taillie, C.: Diversity as a concept and its measurement. J. Am. Stat. Ass. 77, 548–567 (1982)

Pfeffer, J., Langton, N.: The effect of wage dispersion on satisfaction, productivity, and working collaboratively: evidence from college and university faculty. Adm. Sci. Quart. 38(3), 382 (Sept 1993). https://doi.org/10.2307/2393373

Pielou, E.: Ecological Diversity. Wiley, New York (1975)

Randøy, T., Thomsen, S., Oxelheim, L.: A nordic perspective on corporate board diversity. Age 390(0.5428) (2006)

Ricotta, C., Corona, P., Marchetti, M., Chirici, G., Innamorati, S.: LaDy: software for assessing local landscape diversity profiles of raster land cover maps using geographic windows. Environ. Model. Softw. 18, 373–378 (2003)

Rose, C.: Does female board representation influence firm performance? the Danish evidence. Corp. Gov. Int. Rev. 15(2), 404–413 (Mar 2007). https://doi.org/10.1111/j.1467-8683.2007.00570.x

Ruigrok, W., Peck, S., Tacheva, S.: Nationality and gender diversity on Swiss corporate boards. Corp. Gov. Int. Rev. 15(4), 546–557 (Jul 2007). https://doi.org/10.1111/j.1467-8683.2007.00587.x

Santen, B., Donker, H.: Board diversity in the perspective of financial distress: empirical evidence from the Netherlands. Corp. Board Role Duties Comp. 5(2) (2009). https://doi.org/10.22495/cbv5i2art3

Shannon, C.: A mathematical theory of communication. Bell Syst. Tech. J. 27, 379–423 (1948)

Shehata, N.F.: How could board diversity influence corporate disclosure? Corp. Board Role Duties Comp. 9(3) (2013). https://doi.org/10.22495/cbv9i3art4

Simpson, E.: Measurement of diversity. Nat. 163, 688 (1949)

Teachman, J.D.: Analysis of population diversity. Sociol. Methods Res. 8(3), 341–362 (Feb 1980). https://doi.org/10.1177/004912418000800305

Ujunwa, A.: Board characteristics and the financial performance of Nigerian quoted firms. Corp. Gov. Int. J. Bus. Soc. 12(5), 656–674 (Oct 2012). https://doi.org/10.1108/14720701211275587

Urban, R., Hošková-Mayerová, Š.: Threat life cycle and its dynamics. Deturope 9(2), 93–109 (2017). ISSN 1821-2506

Vallejo Andrada, A., Hošková-Mayerová, Š., Krahulec, J., Sarasola Sanchez-Serrano, J.L.: Risks associated with reality: how society views the current wave of migration: one common problem—two different solutions. Springer International Publishing, pp. 283–305 (2017). https://doi.org/10.1007/978-3-319-54819-7_19

van Knippenberg, D., Schippers, M.C.: Work group diversity. Ann. Rev. Psychol. 58(1), 515–541 (Jan 2007). https://doi.org/10.1146/annurev.psych.58.110405.085546

Mathematical Modeling of Some Physical Phenomena Through Dynamical Systems

Olivia Ana Florea

Abstract The differential equations and system of differential equations represent the kernel of the mathematical modeling, offering tools to predict the natural phenomena from science, technics, medicine, biology, etc. In this chapter we will analyze the phase portraits of different dynamical systems linear and non-linear, the lagrangian formalism of a problem encountered in aerodynamics and averaging method for nonlinear differential equation.

1 Introduction

The main aim of the present chapter is to highlight the study of the dynamical systems stability that model different phenomena from engineering, chemistry, physics, biology, medicine (Urban and Hoskova-Mayerova 2017). The most important property of alive systems is the rhythm, being encountered at all unicellular and multicellular organisms. The first oscillation chemical reaction was developed by Belousov (1959). The molecules with complex structure, e.g enzymes, have the rhythm property, also. The first kinetic model of oscillating reactions was studied by Lotka (1920) and Volterra (1926) who proposed similar equations in ecology to study the oscillations that appear in a population of prey-predator type. The theory of nonlinear systems has a particular interest in the study of the phase trajectories of limit cycle type, that describes a periodical isolated behavior of a physical system.

All mechanism that evolve in time can be represented through a dynamical system. Elementary examples can be found in mechanics, computer science and medicine. The most important thing is the evolution of the system, that is represented by the functions that describe the state of the system as a function of time and satisfy the

O. A. Florea (✉)
Faculty of Mathematics and Computer Science, Transilvania University of Braşov,
Bd. Iuliu Maniu, no 50, Braşov, Romania
e-mail: olivia.florea@unitbv.ro

© Springer Nature Switzerland AG 2019
C. Flaut et al. (eds.), *Models and Theories in Social Systems*, Studies in Systems, Decision and Control 179, https://doi.org/10.1007/978-3-030-00084-4_4

equation of motion of the system, Lungu and Chisalita (2007), Bârzu et al. (2003), Broer and Takens (2009). The dynamical systems are encountered also in chemistry. In the paper (Florea and Purcaru 2012) is studied a chemical phenomenon, an example of an autocatalytic reaction. Using the stability in first approximation and the theory of bifurcations is studied the stability of the autocatalytic reaction. The fractals can be also interpreted as dynamical systems. Their geometry can be seen as a language that describes models and analyzes complex forms from nature. The basics of fractal geometry are algorithms that can be visualized as structures and different forms using the computer, Carstea et al. (2014), Enache-David and Sangeorzan (2016), Sangeorzan and David (2000).

The present chapter is structured as follows: in the second section are presented the used theorems in the stability study in first approximation enunciating the linearizing principles Lyapunov–Perron and Hartman–Grobman. In the third section is presented the classification of phase portraits in case of linear bi-dimensional systems, this section is ending with a model of the alcohol influence in the human blood and graphical representation of its trajectory. The behavior of the nonlinear dynamical systems with practical exemplifications from dynamics of the material point and oscillations is realized in the forth section of the present chapter. The study of Lagrangian formalism is realized in the Sect. 5 of this chapter through a practical problem from aerodynamics. The chapter is ending with averaging methods for the differential non-linear equations and the stability study of their solutions in first approximation.

2   Analysis Methods for Continuous Dynamical Systems

To characterize a physical system there are necessary some concepts that model the associated dynamical system namely the stability, attractivity and bifurcation. The stability is a property of a dynamical system that characterize the phase portrait in a vicinity of an invariant set A. The most studied type of stability is the Lyapunov stability.

Let be the continuous dynamical system associated to the differential equation of first order:

$$\dot{x} = f(x), x \in \mathbb{R}^n, f = (f_1, \ldots, f_n) \tag{1}$$

and $\phi_t : \mathbb{R}^n \to \mathbb{R}^n, \phi_t(x_0) = x(t, x_0)$ is the solution of the Cauchy problem attached to the Eq. (1) with $x(0) = x_0$.

Definition 4.1 Let x_0 be an equilibrium point for the Eq. (1), such as $f(x_0) = 0$. We state that the equilibrium point x_0 is Lyapunov stable if $(\forall)\varepsilon > 0, (\exists)\delta_\varepsilon > 0$ such as for any x that verifies $|x - x_0| < \delta_\varepsilon$ we have that $|\phi_t(x) - \phi_t(x_0)| < \varepsilon, (\forall)t \geq 0$.

Definition 4.2 We state that the equilibrium point x_0 is attractor if there exists a vicinity U of x_0 and a time $\tau > 0$ having the property that for $(\forall)x \in U, (\forall)t \geq$

τ, $\phi_t(x) \in U$ and $\lim\limits_{t\to\infty} |\phi_t(x) - \phi_t(x_0)| = 0$. An equilibrium point x_0 that is stable and atractor is called asymptotically stable.

The theorems in first approximations known as linearizing principles are very important in the theory of dynamical systems because they reduce the local study of a nonlinear dynamical system around of an equilibrium point to the study of the stability of the linear dynamical system. They have a very important purpose in the bifurcation and stability theory. The stability of an equilibrium point of a linear dynamical system is characterized by the following theorems.

Theorem 4.1 *Let be the linear dynamical system given in the matriceal form:*

$$\dot{X} = A \cdot X \tag{2}$$

1. *If all eigenvalues λ_k of the matrix A have the real part negative, $Re(\lambda_k) < 0$, $k = \overline{1,n}$ then the constants $C, \alpha > 0$ exist such that for any $x_0 \in \mathbb{R}^n$ the next relations take place:*

$$|x(t, x_0)| \le C|x_0|e^{-\alpha t}, (\forall)t \ge 0 \text{ and } \lim\limits_{t\to\infty} x(t, x_0) = 0 \tag{3}$$

   This represents that the origin is an asymptotic stable point for the linear system.
2. *If all eigenvalues λ_k of the matrix A have the real part not positive $Re(\lambda_k) \le 0$, $k = \overline{1,n}$ and those with the real part null $Re(\lambda_k) = 0$ are simple, then the positive constant $C > 0$ exists such that for any $x_0 \in \mathbb{R}^n$ the next relation be fulfilled:*

$$|x(t, x_0)| \le C|x_0|, (\forall)t \ge 0 \tag{4}$$

   In this situation the origin is an equilibrium simple stable point for the given system.
3. *If there exists an eigenvalue λ_j of the matrix A with the real part positive $Re(\lambda_k) > 0$ then $(\forall)\varepsilon > 0$, $(\exists)x_0$ with $|x_0| < \varepsilon$ such that $\lim\limits_{t\to\infty} |x(t, x_0)| = \infty$, therefore the origin is an unstable equilibrium point for the linear system. Likewise if $Re(\lambda_k) \le 0$, $k = \overline{1,n}$, but there is a multiple eigenvalue λ_j with $Re(\lambda_j) = 0$, then the origin is an unstable equilibrium point.*

Theorem 4.2 *Let $\dot{x} = Ax + B(t)x$ be the n- dimensional linear differential equation, with $B \in C^0$ a continuous function for $t \ge 0$. Then:*

1. *If all eigenvalues λ_k of the matrix A verify the relation $Re(\lambda_k) < 0, k = \overline{1,n}$ and the supremum norm of B function has the limit: $\lim\limits_{t\to\infty} |B(t)| = 0$, then all solutions for the Cauchy problems with initial condition verify that $\lim\limits_{t\to\infty} |x(t, x_0)| = 0$, and $x = 0$ is asymptotically stable.*
2. *If exists an eigenvalue λ_j of the matrix A with $Re(\lambda_j) > 0$ and $\lim\limits_{t\to\infty} |B(t)| = 0$, then $(\forall)\varepsilon < 0$, $(\exists)x_0$ with $|x_0| < \varepsilon$ such that $\lim\limits_{t\to\infty} |x(t, x_0|) = \infty$, therefore $x = 0$ is unstable.*

Next we will enunciate the linearisation principle the Lyapunov–Perron that relies on the above two theorems, (Kuznetsov 2008).

Theorem 4.3 *Let be the nonlinear dynamical system associated to the differential equation* (1), *where the function* f *is at least of class* C^1, *and* x_0 *is an equilibrium point of the system* $f(x_0) = 0$. $A = f'(x_0)$ *is the Jacobi's matrix associated of* f *in* x_0 *that defines the linear system.*

1. *If all eigenvalues* λ_k *of the matrix* A *with* $Re(\lambda_k) < 0$ *then* x_0 *is a stable asymptotically equilibrium point.*
2. *If at least one eigenvalue* λ_k *of the matrix* A *has* $Re(\lambda_k) > 0$, *then* x_0 *is an unstable equilibrium point.*

Definition 4.3 An equilibrium point x_0 for the system (1) is named hyperbolic equilibrium point if the Jacobi's matrix $f'(x_0)$ has all eigenvalues with the real part different by zero, $Re(\lambda_k) \neq 0$.

We can draw the conclusion that the principle of Lyapunov–Perorn characterize the stability of a hyperbolic equilibrium point. The theorem Hartman–Grobman is also a linearizing principle that realize the connection between the behavior of phase portrait around of a hyperbolic equilibrium point of a nonlinear dynamical system and that of a linear system. Next we will enunciate the theorem of Hartman–Grobman (Quandt 1986; Coayla-Teran et al. 2007).

Theorem 4.4 *Let be* x_0 *a hyperbolic equilibrium point for the nonlinear dynamical system. Then there exist two vicinities* $U \in V(x_0)$ *in the phase space corresponding to the nonlinear vector field* f *and* $V \in V(0)$ *in the phase space corresponding to the linearized vector field around* x_0, *respectively and there is the homeomorphism* $h : V \to U$ *that keeps the sense of the orbits. Locally, we state that the phase portraits of the two systems are equivalent topologically.*

3   Phase Portraits of the Bi-Dimensional Linear Dynamical Systems

We consider the bi-dimensional linear dynamical system determined by the differential equations:

$$\begin{cases} \dot{x} = ax + by \\ \dot{y} = cx + dy \end{cases} \text{ with } (x, y) \in \mathbb{R}^2, \text{ and } a, b, c, d \in \mathbb{R} \tag{5}$$

The system can be written in the equivalent form (2), where $A = \begin{pmatrix} a & b \\ c & d \end{pmatrix}$. The eigenvalues of this matrix are the roots of characteristic equation associate to the system (5): $\det(A - \lambda I_2) = 0$ that is equivalent with $\lambda^2 - \nabla \lambda + \Delta = 0$, where $\nabla = a + d = \text{Tr}A$, $\Delta = \det A$. Based on the sign and the nature of the eigenvalues, the bi-dimensional linear dynamical system could have only one equilibrium point $(0, 0)$ or an infinity of equilibrium points. The properties of stability and attractivity, also the type of the equilibrium point depend on the eigenvalues $\lambda_{1,2}$ of the matrix A. We will present the most important case encountered in the study of equilibrium points of bi-dimensional linear dynamical systems:

1. $\lambda_{1,2} \in \mathbb{R}, \lambda_{1,2} > 0$. In this situation we have two particular cases:

   (a) if $\lambda_1 \neq \lambda_2$ then the critical point is a non degenerated repulsive node and it is unstable;
   (b) if $\lambda_1 = \lambda_2$ then the critical point is a degenerated repulsive node and it is unstable or it could be a source;

2. $\lambda_{1,2} \in \mathbb{R}, \lambda_{1,2} < 0$. In this situation we have two particular cases:

   (a) if $\lambda_1 \neq \lambda_2$ then the critical point is a non degenerated attractive node and it is stable;
   (b) if $\lambda_1 = \lambda_2$ then the critical point is a degenerated attractive node and it is stable or it could be a pit;

3. $\lambda_{1,2} \in \mathbb{R}, \lambda_1 < 0, \lambda_2 > 0$ then the critical point is saddle and it is an unstable point.
4. $\lambda_{1,2} \in \mathbb{C} \backslash \mathbb{R}$ with $\text{Re}(\lambda_{1,2}) > 0$. In this case the trajectories are spirals that move away from origin. The critical point is a repulsive focus.
5. $\lambda_{1,2} \in \mathbb{C} \backslash \mathbb{R}$ with $\text{Re}(\lambda_{1,2}) < 0$. In this case the trajectories are spirals that move toward origin. The critical point is a repulsive focus.
6. $\lambda_{1,2} \in \mathbb{C} \backslash \mathbb{R}$ with $\text{Re}(\lambda_{1,2}) = 0$. In this case the trajectories are ellipses and the critical point is a center. The critical point is simple stable because the trajectories remain in its vicinity, but they do not tend to it when $t \rightarrow \infty$.
7. if $\lambda_1 = 0, \lambda_2 \neq 0$ then the critical points are on an infinite line. If $\lambda_2 > 0$ then the points are unstable, if $\lambda_2 < 0$ then the points are stable.
8. if $\lambda_{1,2} = 0$ then we have the following situations:

   (a) if the matrix A is not identically null, then the equilibrium points are on an infinite line without direction;
   (b) if the matrix A is identically null then the entire phase plane is formed by the equilibrium points.

There are many bi-dimensional models that are encountered in our life. Next we will present a particular case of the influence of the alcohol in the human blood. In this model we consider that x represents the concentration of alcohol in stomach and y represents the concentration of alcohol in blood. Hence we have the bi-dimensional linear dynamical system:

$$\begin{cases} \dot{x} = ax - by \\ \dot{y} = bx + ay \end{cases}, \text{ where } a, b > 0 \tag{6}$$

Fig. 1 The trajectory of the alcohol in human blood

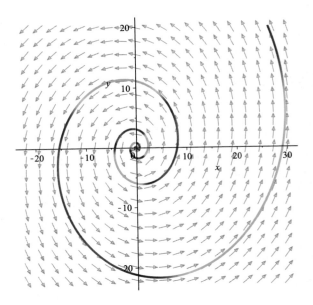

The eigenvalues of the matrix A associated with the dynamical systems are: $\lambda_{1,2} = a \pm bi$. We observe that the roots of characteristic equations are complex with $\mathrm{Re}(\lambda_{1,2}) > 0$. Based on the classification above we can mention that the critical point $(0, 0)$ is an unstable focus (Fig. 1).

The solution of the dynamical system with the initial conditions: $x(0) = x_0$, $y(0) = y_0$ will be:

$$\begin{cases} x(t) = e^{at}(x_0 \cos bt - y_0 \sin bt) \\ y(t) = e^{at}(x_0 \sin bt + y_0 \cos bt) \end{cases}$$

Next we want to deduce the equation of the trajectory. For this we will consider the general form of a complex number: $z = x + iy$, where $z = z(t)$. Using its derivative in report with time we have: $\dot{z} = \dot{x} + i\dot{y}$, but taking into consideration the equations from the dynamical system, we will obtain the expression:

$$\dot{z} = az + ibz$$

The exponential form of a complex number is: $z = r \cdot e^{i\theta}$, where $r = r(t)$, $\theta = \theta(t)$. Using its derivative in report with time we have:

$$\dot{z} = \dot{r}e^{i\theta} + ir\dot{\theta}e^{i\theta}$$

Equaling the last two expressions of $\dot{z}$ we obtain the next differential equations of first degree:

$$\dot{r} = ar; \qquad \dot{\theta} = b$$

Taking into account the initial conditions: $r(0) = r_0$, $\theta(0) = \theta_0$ then the solutions are:

$$r = r_0 e^{at}, \qquad \theta = bt + \theta_0$$

From these two solutions, we can express the equation of the trajectory:

$$r = r_0 e^{a \cdot \frac{\theta - \theta_0}{b}}.$$

The solutions of the dynamical system in polar coordinates will be:

$$\begin{cases} x = r \cos\theta = r_0 e^{at} \cos(bt + \theta_0) \\ y = r \sin\theta = r_0 e^{at} \sin(bt + \theta_0) \end{cases}.$$

4 Phase Portraits of the Bi-Dimensional Non-linear Dynamical Systems

In this section we will consider a bi-dimensional non-linear dynamical system:

$$\begin{cases} \dot{x} = f(x, y) \\ \dot{y} = g(x, y) \end{cases} \tag{7}$$

The equilibrium point (x_0, y_0) is determined by solving the system of equations: $f(x, y) = 0$; $g(x, y) = 0$. Using the translation $X = x - x_0$, $Y = y - y_0$ the study of stability of the equilibrium point is reduced to the stability of $(0, 0)$. We will consider the linearized system for that we will compute the Jacobi's matrix in the equilibrium point (x_0, y_0). This strategy is based on the theorem of Hartman–Grobman such that the phase portrait for the linear case is given by the approximation of the equilibrium point for the non-linear case.

In the case of non-linear dynamical system it is possible to appear limit cycles. The problem of establishing the number of the limit cycles in a dynamical system represents a very important step in the study of stability. This problem, known as the 16th problem of Hilbert it was solved only partially for simple dynamical systems it represents an open problem (Wilso 1978; Li 2003).

Theorem 4.5 *If the functions from (7) are polynomials then there is a finite number of limit cycles. Inside of a domain bounded by a limit cycle there is at least an equilibrium point.*

A very important criterion to determine the existence of limit cycles is given by Bendixson: the sufficient condition that in a region from the phase plane do not exist periodical solutions for the dynamical system is that limit cycles do not exist. (Li Y. et al. 1993)

Theorem 4.6 *In the case of autonomous dynamical system* (7) *with* $f, g : D \subset \mathbb{R}^2 \to \mathbb{R}$, *where* D *is a simple domain, and the functions* f, g *are differentiable on* D, $f, g \in C^1(D)$, *if the divergence of the vector fields of the system:* $div \vec{v} = \frac{\partial f}{\partial x} + \frac{\partial g}{\partial y} \neq 0$, *keeps a constant sign on* D, *then there is none closed phase trajectory completely contains in* D.

4.1 Dynamical Systems from Mechanics Modeled Through Differential Equation

The fundamental equation of dynamics is:

$$m\vec{a} = \vec{F} \Leftrightarrow m\ddot{\vec{r}} = \vec{F}(t, \vec{r}, \dot{\vec{r}}) \tag{8}$$

In this section we will study the motion in the phase plane $(x, \dot{x})$. Therefore, the differential equation of second degree $m\ddot{x} = F(x)$ is equivalent with the system:

$$\begin{cases} \dot{x} = v \\ \dot{v} = \frac{1}{m}F(x) \end{cases} \tag{9}$$

The equilibrium point is $A(a, 0)$ where $F(a) = 0$. Making the translation: $x = X + a, v = V$ we will study the stability in origin of the solution of the differential equation system:

$$\begin{cases} \dot{X} = V = f(X, V) \\ \dot{V} = \frac{1}{m}F(X + a) = g(X, V) \end{cases} \tag{10}$$

The characteristic polynomial associated to the Jacobi matrix of the system computed in origin will lead the the equation: $\lambda^2 - \frac{1}{m}F'(a) = 0$. If $F'(a) > 0$ then the eigenvalues are reals: $\lambda_1 < 0, \lambda_2 > 0$ hence, the equilibrium point will be unstable.

Next we will consider the case of the harmonic damping oscillator, with the dynamic equation:

$$m\ddot{x} = F_e + F_r + F_p$$

where $F_e = -kx$, k is the spring coefficient and $F_r = -c\dot{x}$, c is the resistant coefficient of the force F_r and F_p is the periodic force that maintains the motion $F_p = A \cos \omega_0 t$. The differential equation of motion is:

$$\ddot{x} + 2\alpha^2 \dot{x} + \omega^2 x = B \cos \omega_0 t \tag{11}$$

where $\frac{c}{m} = 2\alpha^2$; $\frac{k}{m} = \omega^2$; $\frac{A}{m} = B$. The characteristic equation is $\lambda^2 + 2\alpha^2 \lambda + \omega^2 = 0$ with the discriminant $\Delta = 4(\alpha^4 - \omega^2)$. We will study all the possible situations:

1. $\Delta > 0$. In this situation the eigenvalues are real, because $S = -2\alpha^2 < 0$ and $P = \omega^2 > 0$ the eigenvalues are negative and the solution for homogeneous equation is a-periodic because for $t \to \infty$, $x_0(t) = C_1 e^{\lambda_1 t} + C_2 e^{\lambda_2 t} \to 0$.

2. $\Delta = 0$. In this case the eigenvalues are real, equal and negative: $\lambda_{1,2} = -\alpha^2$. Also, the motion is a-periodic because for $t \to \infty$, the homogeneous solution $x_o(t) = e^{-\alpha^2 t}(C_1 + C_2 t) \to 0$.

3. $\Delta < 0$. We note by $\alpha^4 - \omega^2 = -b^2$ and the eigenvalues are complex with the real part positive. In this case the motion is harmonic because the homogeneous solution has the general form: $x_o(t) = A_0 e^{\alpha^2 t} \sin(bt + \beta)$

The particular solution for (11) should be searched with the following form: $x_p = a \cos \omega_0 t + b \sin \omega_0 t$ where $a = \frac{B(\omega^2 - \omega_0^2)}{(\omega^2 - \omega_0^2)^2 + 4\alpha^4 \omega_0^2}$, $b = \frac{2B\alpha^2 \omega_0}{(\omega^2 - \omega_0^2)^2 + 4\alpha^4 \omega_0^2}$.

5 Lagrangian Formalism of the Dynamical Problems

The Lagrangian formalism is used for the study of the stability of the dynamical systems of the inverse problems. Considering the dynamical system, (Obadeanu and Neamtu 1999):

$$\dot{x} = f(t, x), \qquad x \in \mathbb{R}^n, t \in \mathbb{R} \tag{12}$$

The main problem consists in finding of some conditions such that the above system admits the variational principle for which there exist the functions: $A_i(t, x)$, $B(t, x)$. Therefore the Lagrangian formalism is:

$$L(t, x) = \sum_{i=1}^{n} A_i(t, x)\dot{x}_i + B(t, x) \tag{13}$$

The Euler–Lagrange equations associated to the Lagrangian are:

$$\frac{d}{dt}\left(\frac{\partial L}{\partial \dot{x}_i}\right) - \frac{\partial L}{\partial x_i} = 0, i = \overline{1, n} \tag{14}$$

The above equation could be written in the equivalent form:

$$\sum_{j=1}^{n}\left(\frac{\partial A_i}{\partial x_j} - \frac{\partial A_j}{\partial x_i}\right)\dot{x}_j + \frac{\partial A_i}{\partial t} - \frac{\partial B}{\partial x_i} = 0, i = \overline{1, n} \tag{15}$$

Noting by:

$$C_{ij} = \frac{\partial A_i}{\partial x_j} - \frac{\partial A_j}{\partial x_i}, \qquad D_i = \frac{\partial A_i}{\partial t} - \frac{\partial B}{\partial x_i} \tag{16}$$

we will obtain the next system:

$$\sum_{j=1}^{n} C_{ij}\dot{x}_j + D_i = 0, i = \overline{1,n} \tag{17}$$

We consider 1-differential form $\phi = \sum_i A_i dx_i + Bdt$ with the exterior derivative:

$$\Omega = d\phi = \frac{1}{2}\sum_{i,j}\left(\frac{\partial A_i}{\partial x_j} - \frac{\partial A_j}{\partial x_i}\right)dx_i \wedge dx_j + \sum_i\left(\frac{\partial A_i}{\partial t} - \frac{\partial B}{\partial x_i}\right)dx_i \wedge dt =$$

$$= \frac{1}{2}\sum_{i,j} C_{ij}dx_i \wedge dx_j + \sum_i D_i dx_i \wedge dt$$

Using Poincare lemma it is sufficient that $d\Omega = 0$ and we obtain the auto-adjunct conditions:

$$\begin{cases} C_{ij} + C_{ji} = 0 \\ \frac{\partial C_{ij}}{\partial x_k} + \frac{\partial C_{jk}}{\partial x_i} + \frac{\partial C_{ki}}{\partial x_j} = 0 \\ \frac{\partial C_{ij}}{\partial t} = \frac{\partial D_i}{\partial x_j} - \frac{\partial D_j}{\partial x_i} \end{cases} \tag{18}$$

If the system (17) verifies the conditions (18) then it comes from a Lagrangian. For the system (14) we search an integrant factor anti-symmetric with the general form $C_{ij} + C_{ji} = 0$. In this situation we have:

$$\sum_i C_{ij}(\dot{x}_j - f_j) = 0 \tag{19}$$

Let us consider the motion of a body in a vertical plane under the action of the gravitational force in a resistant environment. The resistance force that acts on the body has the expression: $\vec{R} = -mg\phi(v)\frac{\vec{v}}{v}$. In case of the motion in the air, the function $\phi(v)$ is obtained experimentally through aerodynamics characteristics. The equation of motion is:

$$m\vec{a} = \vec{G} + \vec{R}. \tag{20}$$

We consider that the resistance force is directly proportional with v^2, and its projection on the axis are: $R_x = k_1 v^2$, $R_y = k_2 v^2$, where k_1, k_2 are the coefficients of resistance and friction with the air. The equation on motion projected on the axis will have the form:

$$\begin{cases} m\ddot{x} = -mg\sin\theta - k_1 v^2 \\ m\ddot{y} = -mg\cos\theta + k_2 v^2 \end{cases} \tag{21}$$

where $\dot{x} = v\cos\theta$; $\dot{y} = v\sin\theta$.

Using the derivatives in report with time we have: $\ddot{x} = \dot{v}\cos\theta - v\dot{\theta}\sin\theta$, $\ddot{y} = \dot{v}\sin\theta + v\dot{\theta}\cos\theta$. Noting by: $V = \frac{v}{g}$, $a = k_1\frac{g}{m}$, $b = k_2\frac{g}{m}$ the system (21) will have the following form:

$$\begin{cases} \dot{V} = -\sin\theta - aV^2 \\ \dot{\theta} = \frac{bV^2 - \cos\theta}{V} \end{cases} \tag{22}$$

In the case when $a = 0, b = 1$. Lupu et al. (2001) the system (22) can be written in the following form:

$$\begin{cases} \frac{dV}{dt} = -\sin\theta \\ \frac{d\theta}{dt} = \frac{V^2 - \cos\theta}{V} \end{cases} \tag{23}$$

Multiplying the above system with an antisymmetric factor $C = C(t, V, \theta) : C = \begin{pmatrix} 0 & C \\ -C & 0 \end{pmatrix}$ we will obtain:

$$\begin{aligned} C\left(\dot{\theta} - \frac{V^2 - \cos\theta}{V}\right) &= 0 \\ -C(\dot{V} + \sin\theta) &= 0 \end{aligned} \tag{24}$$

The auto-adjonction conditions are:

$$\frac{\partial C}{\partial t} - \sin\theta \frac{\partial C}{\partial V} + \frac{V^2 - \cos\theta}{V}\frac{\partial C}{\partial\theta} + c\frac{\sin\theta}{V} = 0 \tag{25}$$

In the case when $C = C(V)$ we have:

$$-\sin\theta \frac{\partial C}{\partial V} + C\frac{\sin\theta}{V} = 0$$

Solving the symmetric system associated for the partial quasi-linear equation we obtain the prime integral: $C = kV$. Based on (15) and (16) and solve the following system:

$$\frac{\partial A_1}{\partial\theta} - \frac{\partial A_2}{\partial V} = V; \frac{\partial A_1}{\partial t} - \frac{\partial B}{\partial V} = -V^2 + \cos\theta; \frac{\partial A_2}{\partial t} - \frac{\partial B}{\partial\theta} = -V\sin\theta$$

we obtain:

$$A_1 = \theta(V + 1); A_2 = V; B = \frac{V^3}{3} - V\cos\theta$$

and the associated Lagrangian for the system (23) will be:

$$L = \theta(V + 1)\dot{V} + V\dot{\theta} + \frac{V^3}{3} - V \cos\theta$$

In the case when the resistance coefficient $a \neq 0$ we will study the stability in first approximation. The equilibrium point is determined solving the system:

$$\sin\theta + aV^2 = 0 \tag{26}$$

$$\frac{V^2 - \cos\theta}{V} = 0 \tag{27}$$

From the second equation we have: $V^2 = \cos\theta$ and replacing into the first equation we obtain $\theta = -\text{arctg}a$, therefore $V^2 = \cos(\text{arctg}a)$ that is equivalent with $V^4 = \cos^2(\text{arctg}a) = \frac{1}{1+a^2}$. The equilibrium point is: $M\left(-\text{arctg}a, \sqrt[4]{\frac{1}{1+a^2}}\right)$. The study of stability will be realized in origin, and for this it is necessary to make the translation: $T = \theta - \theta_0 = \theta + \text{arctg}a$, $U = V - V_0 = V - \sqrt[4]{\frac{1}{1+a^2}}$. The new form of the system will be:

$$\begin{cases} \dot{U} = -\sin(T - \text{arctg}a) - a\left(U + \sqrt[4]{\frac{1}{1+a^2}}\right) = f(T, U) \\ \dot{T} = \frac{1}{\left(U + \sqrt[4]{\frac{1}{1+a^2}}\right)}\left[\left(U + \sqrt[4]{\frac{1}{1+a^2}}\right)^2 - \cos(T - \text{arctg}a)\right] = g(T, U) \end{cases} \tag{28}$$

Using the Jacobi's matrix, the linear form of the system is:

$$\begin{cases} \dot{U} = \frac{\partial f}{\partial U}U + \frac{\partial f}{\partial T}T = -\frac{2a}{\sqrt[4]{1+a^2}}U - \frac{1}{\sqrt{1+a^2}}T \\ \dot{T} = \frac{\partial g}{\partial U}U + \frac{\partial g}{\partial T}T = 2U - \frac{a}{\sqrt[4]{1+a^2}}T \end{cases} \tag{29}$$

The characteristic equation of the differential system is: $\lambda^2 + \frac{3a}{\sqrt[4]{1+a^2}}\lambda + 2\sqrt{1+a^2} = 0$ with the discriminant: $\Delta = \frac{a^2-8}{\sqrt{1+a^2}}$. The study of stability in the phase plane will give us the following situations:

1. if $\Delta > 0$ then $a \in (2\sqrt{2}, \infty)$. The eigenvalues are real with $S < 0$, $P > 0$ that means the eigenvalues are negative and the equilibrium point is an attractive node.
2. if $\Delta = 0$ then $a = 2\sqrt{2}$. The eigenvalues are real, equal and negative and the equilibrium point is degenerative attractive.
3. if $\Delta < 0$ the $a \in (0, 2\sqrt{2})$. The eigenvalues are complex with real part negative. In this case the equilibrium point is an attractive focus.

6 The Averaging Method for Nonlinear Differential Equation

Let us consider the nonlinear differential equation:

$$\ddot{x} + \varepsilon f(\dot{x}, x) + x = 0 \tag{30}$$

The nonlinear differential equation of second degree could be written under the form of a differential system of first degree:

$$\begin{cases} \dot{x} = y \Rightarrow \ddot{x} + \varepsilon f(y, x) + x = 0 \\ \dot{y} = -x - \varepsilon f(x, y) \end{cases} \tag{31}$$

We want to find the solution having the form:

$$\begin{cases} x = r \cos(t + \phi) \\ y = -r \sin(y + \phi) \end{cases}$$

where r and ϕ are functions that depend on time: $r = r(t)$, $\phi = \phi(t)$. The derivatives in report with time are:

$$\begin{cases} \dot{x} = \dot{r} \cos(t + \phi) - r(1 + \dot{\phi}) \sin(t + \phi) \\ \dot{y} = -\dot{r} \sin(t + \phi) - r(1 + \dot{\phi}) \cos(t + \phi) \end{cases}$$

Replacing the derivatives into the system (31) and after some elementary computations we obtain:

$$\dot{\phi} = -\frac{\varepsilon}{r} F(t, r, \phi) \cos(t + \phi) \tag{32}$$

$$\dot{r} = -\varepsilon F(t, r, \phi) \sin(t + \phi)$$

where $F(t, r, \phi) = f(r \cos(t + \phi), -r \sin(t + \phi))$. For the trigonometric functions it is used the averaging method for $t \in [0, 2\pi]$ for all functions that have the form $R(\sin^n(t + \phi), \cos^m(t + \phi))$ in order to obtain the linear differential equations. A particular expression for $f(x, \dot{x})$ is $x^2 - \dot{x}$ with $F(t, r, \phi) = r^2 \cos^2(t + \phi) + r \sin(t + \phi)$ and the Eq. (32) become:

$$\dot{\phi} = -\varepsilon \left[r \cos^3(t + \phi) + \sin(t + \phi) \cos(t + \phi) \right] \tag{33}$$

$$\dot{r} = -\varepsilon \left[r^2 \cos^2(t + \phi) \sin(t + \phi) + r \sin^2(t + \phi) \right]$$

The averaging method gives us the following results:

$$\frac{1}{2\pi}\int_0^{2\pi}\sin^2(t+\phi)dt=\frac{1}{2};\quad \frac{1}{2\pi}\int_0^{2\pi}\cos^2(t+\phi)\sin(t+\phi)dt=0$$

$$\frac{1}{2\pi}\int_0^{2\pi}\cos(t+\phi)\sin(t+\phi)dt=0;\quad \frac{1}{2\pi}\int_0^{2\pi}\cos^3(t+\phi)dt=0$$

Therefore our differential equation of first degree will be: $\dot{\phi}=0, \dot{r}=-\varepsilon\frac{r}{2}$. Taking into account the initial conditions: $\phi(0)=\phi_0, r(0)=r_0$ the solution is: $\phi=\phi_0$, $r=r_0e^{-\varepsilon\frac{t}{2}}$. The parametric equations of trajectory is:

$$x=r_0e^{-\varepsilon\frac{t}{2}}\cos(t+\phi_0);\quad y=-r_0e^{-\varepsilon\frac{t}{2}}\sin(t+\phi_0)$$

For the study of stability it is necessary to find the critical points of the system:

$$\begin{cases}\dot{x}=y\\ \dot{y}=-x+\varepsilon(-y+x^2)\end{cases} \tag{34}$$

These points are obtained from the equations: $y=0$ and $-x+\varepsilon(-y+x^2)=0$ and they are: $O(0,0)$ and $A\left(\frac{1}{\varepsilon},0\right)$. The Jacobi's matrix is: $J(x,y)=\begin{pmatrix}0 & 1\\ -1+2\varepsilon x & -\varepsilon\end{pmatrix}$.

- the study of stability for $O(0,0)$. The characteristic equation is: $\lambda^2+\varepsilon\lambda+1=0$ with the discriminant $\Delta=\varepsilon^2-4$. If $\varepsilon\in(0,2)$ then the eigenvalues are complex with real part negative and therefore the critical point is a stable focus, if $\varepsilon\in(-2,0)$ then the eigenvalues are complex with the real part positive and the critical point is an unstable focus. If $\varepsilon\in(-\infty,-2)$ then the eigenvalues are real and positive and the critical point is an unstable node; for $\varepsilon\in(2,\infty)$ the critical point is a stable node. (Fig. 2)
- the study of stability for $A\left(\frac{1}{\varepsilon},0\right)$. The characteristic equation is $\lambda^2+\varepsilon\lambda-1=0$. The eigenvalues are positive $(\forall)\varepsilon\in\mathbb{R}$, and the critical point is a unstable saddle. (Fig. 3)

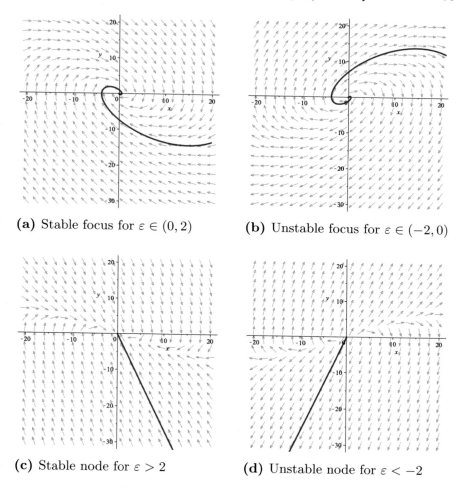

(a) Stable focus for $\varepsilon \in (0, 2)$

(b) Unstable focus for $\varepsilon \in (-2, 0)$

(c) Stable node for $\varepsilon > 2$

(d) Unstable node for $\varepsilon < -2$

Fig. 2 Stability for the critical point $O(0, 0)$

Fig. 3 Stability of critical point $A\left(\frac{1}{\varepsilon}, 0\right)$

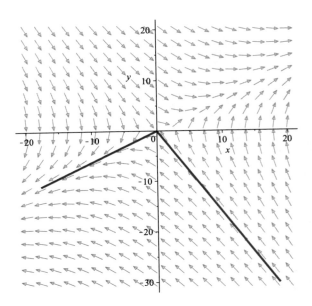

7 Conclusions

The characterization of a material system is realized based on some concepts that model the associate dynamical system such as the stability, the attraction, the bifurcation. The stability is a property of the dynamical system that characterizes the phase portrait in the vicinity of an invariant set. This concept is associated of an invariant set from phase space and it appears in the case of small perturbations of some characteristics of the dynamical system. The aim of this chapter is to highlight the utility of the stability of dynamical systems that model different phenomena from technical engineering, chemistry, physics, biology, medicine. In this chapter were approached some practical dynamical systems encountered in the real life. The study of stability of equilibrium points was realized using the methods in the first approximation. The phase portraits for the linear and nonlinear dynamical systems were represented using Maple software. Using the variational principle the equations Euler–Lagrange associated to the formalism Lagrangian are obtained. This analysis is used to study the dynamical systems for the inverse problems. The study of stability for differential nonlinear equations is realized using the averaging method.

References

Belousov, B.P.: Periodicheski deistvuyushchaya reaktsia i ee mekhanism [Periodically acting reaction and its mechanism]. In: Sbornik referatov po radiotsionnoi meditsine [Collection of Abstracts on Radiation Medicine], 1958, pp. 14–147. Moscow, Medgiz (1959)

Bârzu, A., Bourceanu, G., Onel, L.: Nonlinear dynamics. MatrixRom, Bucharest (2003)

Broer, H., Takens, F.: Dynamical Systems and Chaos. Springer, Berlin (2009)

Carstea, C., Enache-David, N., Sangeorzan, L.: Fractal model for simulation and inflation control. In: Bulletin of the Transilvania University of Brasov. Series III: Mathematics, Informatics, Physics, vol. 7(56), issue 2, pp. 161–168 (2014)

Coayla-Teran, E.A., Mohammed, S.-E.A., Ruffino, P.R.C.: Theorems along hyperbolic stationary trajectories. South. Ill. Univ. Carbondale OpenSIUC **2**, 1–18 (2007)

Enache-David, N., Sangeorzan, L.: An application on web path personalization. In: Proceedings of the 27th International Business Information Management Association Conference - Innovation Management and Education Excellence Vision 2020: From Regional Development Sustainability to Global Economic Growth, Milan, Italy, pp. 2843–2848 (2016)

Florea, O., Purcaru, M.: Mathematical modeling and the stability study of some chemical phenomena. Proc. AFASES **1**, 525–529 (2012)

Kuznetsov, N.V.: Stability and Oscillations of Dynamical Systems: Theory and Applications. Jyvaskyla Studies of Computing (2008)

Li, J.: Hilbert's 16th problem and bifurcations of planar polynomial vector fields. Int. J. Bifurc. Chaos **13**(1), 47–106 (2003)

Lotka, A.J.: Undamped oscillations derived from the law of mass action. J. Am. Chem. Soc. **42**, 1595–1599 (1920)

Lungu, N., Chisalita, A.: Dynamical Systems And Chaos. MatrixRom, Bucharest (2007)

Lupu, M., Postelnicu, A., Deaconu, A.: A Study on the flutter stability of dynamic aeroelastic systems. In: Proceedings CDM , pp. 37–46. Brasov (2001)

Obadeanu, V., Neamtu, M.: Systemes dynamiques differentielles a controle optimal formulation lagrangienne (II). Novi Sad J. Math. **29**(3), 211–220 (1999)

Quandt, J.: On the Hartman–Grobman theorem for maps. J. Differ. Equ. **64**(2), 154–164 (1986)

Sangeorzan, L., David, N.: Some methods of generating fractals and encoding images. In: Proceedings of the 2nd International Conference on Symmetry and Antisymmetry in Mathematics, Formal Languages and Computer Science, Satellite Conference of 3ECM, Brasov, Romania, Transilvania University Publishing House, pp. 337–342 (2000)

Urban, R., Hoskova-Mayerova, S.: Threat life cycle and its dynamics. Deturope **9**(2), 93–109 (2017)

Volterra, V.: Fluctuations in the abundance of a species considered mathematically. Nature **118**, 558–560 (1926)

Wilson, G.: Hilbert's sixteeen problem. Topology **17**(1), 55–73 (1978)

Methods for Improving the Quality of Image Reconstruction in Computerized Tomography

Doina Carp, Constantin Popa and Cristina Şerban

Abstract Appeared in early 50s in medical applications, but essentially developed since early 70s Computerized Tomography (CT) has became nowadays one of the most powerful investigation tool in science and technology. In this chapter we present several classes of methods which can rise the efficiency and/or robustness of classical projection-based algorithms (as Kaczmarz and Cimmino type algorithms) for algebraic reconstruction of images in Computerized Tomography.

1 Introduction

In this chapter we will describe projection-based methods for image reconstruction in Computerized Tomography (CT). In medical applications CT has been the basis for interventional work like CT guided biopsy and minimally invasive therapy, tumors detecting in the head and brain, and blood clots and blood vessel defects. CT images are also used as basis for planning radiotherapy cancer treatment. Because CT imaging provides both good soft tissue resolution (contrast) as well as high spatial resolution it is successfully used in orthopedic medicine and imaging of bony structures including vertebral discs, imaging of complex joints like the shoulder or hip as a functional unit and fractures, especially those affecting the spine. But, CT has also extremely important applications in industrial real world problems. Typical areas of use for CT in industry are in the detection of flaws such as voids and cracks, and particle analysis in materials. In metrology, CT allows measurements of the external as well as the internal geometry of complex parts. So far, CT metrology is the only technology able to measure as well the inner as the outer geometry of

D. Carp · C. Popa (✉) · C. Şerban
Faculty of Mathematics and Computer Science, Ovidius University,
Mamaia Blvd. 124, 900527 Constanta, Romania
e-mail: cpopa1956@gmail.com

D. Carp
e-mail: doina.carp@gmail.com

C. Şerban
e-mail: cgherghina@gmail.com

© Springer Nature Switzerland AG 2019
C. Flaut et al. (eds.), *Models and Theories in Social Systems*, Studies in Systems,
Decision and Control 179, https://doi.org/10.1007/978-3-030-00084-4_5

95

a component without need to cut it through and destroy it. As such, it is the only technology for industrial quality control of workpieces having non-accessible internal features (e.g. components produced by additive manufacturing) or multi-material components (e.g. two-component injection molded plastic parts or plastic parts with metallic inserts), as well as coordinate metrology.

The classical X-ray imaging techniques are based on the absorption of X rays as they pass through the different parts of a patient's body. Depending on the amount absorbed in a particular tissue such as muscle or lung, a different amount of X rays will pass through and exit the body. During conventional X-ray imaging, the exiting X rays interact with a detection device (X-ray film or other image receptor) and provide a 2-dimensional projection image of the tissues within the patient's body—an X-ray produced "photograph" called a "radiography". Although also based on the variable absorption of X rays by different tissues, computed tomography (CT) imaging, also known as "CAT scanning" (Computerized Axial Tomography), provides a different form of imaging known as cross-sectional imaging. A CT imaging system produces cross-sectional images or "slices" of anatomy, like the slices in a loaf of bread. These cross-sectional images are used for a variety of diagnostic and therapeutic purposes. The most recent CT systems are capable of "spiral" (also called "helical") scanning as well as scanning in the formerly more conventional "axial" mode. In addition, many CT systems are capable of imaging multiple slices simultaneously. Such advances allow relatively larger volumes of anatomy to be imaged in relatively less time.

A CT scanner looks like a big, square doughnut (Fig. 1, left). The patient aperture (opening) is 60–70 cm (24"–28") in diameter. Inside the covers of the CT scanner is a rotating frame which has an x-ray tube mounted on one side and the banana shaped detector mounted on the opposite side (Fig. 1, right). A fan beam of X-ray is created as the rotating frame spins the X-ray tube and detector around the patient

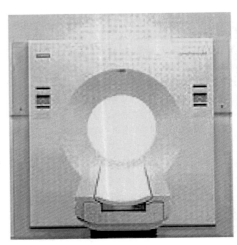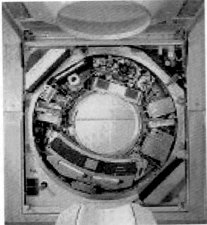

Fig. 1 Left: CT scanner—frontal view; right: CT scanner—X ray tube

Fig. 2 CT environment

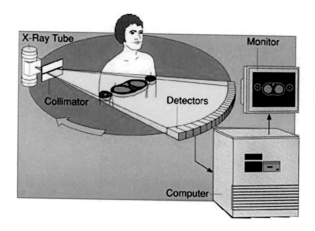

(Fig. 2). Each time the X-ray tube and detector make a 360° rotation, an image or "slice" has been acquired. This "slice" is collimated (focused) to a thickness between 1 and 10 mm using lead shutters in front of the X-ray tube and X-ray detector. As the X-ray tube and detector make this 360° rotation, the detector takes numerous snapshots (called profiles) of the attenuated X-ray beam. Typically, in one 360° lap, about 1,000 profiles are sampled. Each profile is subdivided spatially (divided into partitions) by the detectors and fed into about 700 individual channels. Each profile is then backwards reconstructed (or "back projected") by a dedicated computer into a two-dimensional image of the "slice" that was scanned.

Note: The above information have been taken from http://www.imaginis.com/ct-scan/how-does-ct-work.

In this chapter we will be concerned with what happens inside the "dedicated computer" (Fig. 2), which transforms the data taken by the detectors computer into a two-dimensional image. From the view point of its mathematical model (see e.g. Herman 1980) the CT image reconstruction problem can be formulated as follows: let $D \subset \mathbb{R}^2$ be a bounded domain and $f : D \longrightarrow \mathbb{R}^2$ a function such that we know the value of the line integral $\int_{AB \cap D} f(s) ds$ for any line AB in $\mathbb{R}^2$; then, can we find the values $f(x, y)$, for any point $(x, y) \in D$? The problem was solved as a mathematical problem by Johann Radon in 1917. He provided a formula which expresses $f(x, y)$, called the "inverse Radon transform". Based on this formula has been developed the filtered back projection method for reconstructing the CT images; this method is implemented in almost all the CT devices. We will not be concerned with this method, but with an alternative called Algebraic Reconstruction Technique (ART, see e.g. Herman 1980). In the ART framework the CT image reconstruction problem is discretized and transformed into a system of linear equations

$$Ax = b, \tag{1}$$

Fig. 3 Construction of the
scanning matrix

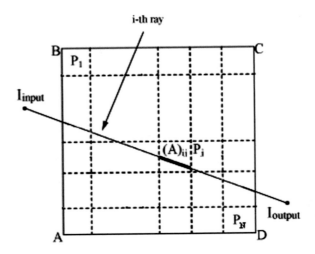

where A is an $m \times n$ (real) matrix and $b \in \mathbb{R}^m$. Because m is the number of scanning rays and n is the number of pixels (image resolution), A is usually a (very) large matrix. Its coefficients A_{ij} are computed as the length of the segment representing the intersection of the i-th scanning X-ray with the j-th pixel of the image. For a square $N = q \times q$ image (see Fig. 3) any X-ray will intersect at most $2q - 1$ pixels, which tells us that A is also (very) sparse. Moreover, if A_i, A^j are the i-th row, and j-th column of A, respectively we will suppose on the whole chapter that

$$A_i \neq 0, \forall i = 1, \ldots, m; \; A^j \neq 0, \forall j = 1, \ldots, n, \tag{2}$$

without restricting the generality of the system (1). Indeed if it exists a zero row $A_{i_0} = 0 \Leftrightarrow A_{i_0 j} = 0, \forall j = 1, \ldots, n$, it would mean that the X-ray i_0 does not intersect any pixel P_j, therefore it can be eliminated from the scanning process, i.e. from the matrix A. If it would exist a zero column $A^{j_0} = 0 \Leftrightarrow A_{i j_0} = 0, \forall i = 1, \ldots, m$ it would mean that the pixel j_0 is not intersected by any scanning ray, thus we can fix it to any value/color because it will not contribute in the reconstructed image. From a computational/algorithmic view point in such a situation we can skip working with the column A^{j_0} in the corresponding numerical algorithm. Unfortunately, due to errors induced by continuous mathematical modelling assumptions, discretization and measurements errors, the discrete ART mathematical model is no longer a consistent system of linear equations as (1), and it must be reformulated as a linear least squares problem (see for details e.g. Censor and Stavros 1997): find $x \in \mathbb{R}^n$ such that

$$\| Ax - b \| = \min\{\| Az - b \|, \; z \in \mathbb{R}^n\}, \tag{3}$$

where $\| \cdot \|$ is the Euclidean norm. Accordingly to all the above aspects, our chapter is organised as follows. Sections 1, 2 and 3 present the Kaczmarz and Cimmino projection methods for consistent systems of linear equations. In Sect. 4 we describe the extensions of algorithms from previous sections to inconsistent linear least squares problems. In Sect. 5 we present Kaczmarz and Cimmino type algorithms with oblique and generalized oblique projections, whereas in Sect. 6 we describe various constraining strategies for improving the reconstruction together with some considerations about possible "hybrid" type projection algorithms: Kaczmarz with Conjugate Gradient, Kaczmarz with various Genetic type algorithms, Kaczmarz as a direct solver. The chapter ends with Sect. 7 in which we present numerical experiments with some of the methods described. The experiments show that in some cases the quality of the reconstructed image can be much improved.

We will end this introductory section with the principal definitions and notations that will be used in the rest of the chapter. The euclidean scalar product and norm on some vector space $\mathbb{R}^q$ will be denoted by $\langle \cdot, \cdot \rangle$, $\| \cdot \|$. By $x \perp y$ we will denote the orthogonality between the vectors x and y, whereas $S^\perp$ will denote the orthogonal complement of a vector subspace $S \subset \mathbb{R}^q$. If $C \subset \mathbb{R}^q$ is a nonempty closed convex set, by $P_C(x)$ we will denote the orthogonal projection of $x \in \mathbb{R}^q$ on C (with respect to the euclidean scalar product). If A is an $m \times n$ (real) matrix we will denote by A^T, A^+, $\mathcal{N}(A)$, $\mathcal{R}(A)$, $rank(A)$, $\| \cdot \|_2$ its transpose, Moore–Penrose pseudoinverse, null space, range, rank and spectral norm, respectively. If A_i is the i-th row of A we define the sets H_i, S_i and the projections P_{H_i}, P_{S_i} by

$$H_i = \{x \in \mathbb{R}^n, \langle x, A_i \rangle = b_i\}, \quad S_i = \{x \in \mathbb{R}^n, \langle x, A_i \rangle = 0\} \tag{4}$$

$$P_{H_i}(x) = x - \frac{\langle x, A_i \rangle - b_i}{\| A_i \|^2} A_i, \quad P_{S_i}(x) = x - \frac{\langle x, A_i \rangle}{\| A_i \|^2} A_i. \tag{5}$$

By $D = diag(\alpha_1, \alpha_2, \ldots, \alpha_n)$ we will denote the diagonal matrix with $\alpha_1, \ldots, \alpha_n$ on the diagonal, whereas $B = col(B^1, \ldots, B^q)$ will be the matrix with the columns $B^1, \ldots, B^q$.

2 Kaczmarz's Projections Algorithm

In Kaczmarz (1937) (see also its english translation Kaczmarz 1993) S. Kaczmarz proposed his famous projection based iterative algorithm.

Algorithm Kaczmarz

Initialization: $x^0 \in \mathbb{R}^n$
Iterative step: select an index $i_k \in \{1, \ldots, m\}$ and perform the projection based iterative method. In a modern description it looks as follows.

$$x^{k+1} = P_{H_{i_k}}(b; x^k), \quad k \geq 0. \tag{6}$$

The selection procedure used by Kaczmarz was the **cyclic** one, i.e.

$$i_k = k(\text{mod } m) + 1, \quad k \geq 0. \tag{7}$$

For developments and extensions of this version of Kaczmarz algorithm see e.g. Herman (1980), Censor and Stavros (1997) and references therein. In the present section we will be interested in another version of Kaczmarz's algorithm, analysed in Tanabe (1971) by K. Tanabe.

Algorithm Kaczmarz-Tanabe (KT)

Initialization: $x^0 \in \mathbb{R}^n$
Iterative step:
$$x^{k+1} = P_{H_1} \circ \cdots \circ P_{H_m}(b; x^k), \quad k \geq 0. \tag{8}$$

In the paper Tanabe (1971), the author did a complete analysis of the convergence properties of the algorithm KT. In what follows we shall briefly present some of them which are necessary in the developments from the other sections of the chapter. Let $Q_i, i = 1, \ldots, m - 1, Q$ be the $n \times n$ matrices and R the $n \times m$ one, defined by

$$Q_0 = I, \quad Q_i = P_1 P_2 \ldots P_i, \quad Q = P_1 \ldots P_m, \tag{9}$$

$$R = \text{col}\left(\frac{1}{\| A_1 \|^2} Q_0 A_1, \ldots, \frac{1}{\| A_m \|^2} Q_{m-1} A_m\right), \tag{10}$$

where I is the unit matrix. Then

$$x^{k+1} = Qx^k + Rb, \, (\forall)x \in \mathbb{R}^n; \, Q + RA = I. \tag{11}$$

Theorem 1 *For any matrix A and any vector $b \in \mathcal{R}(A)$, the sequence $(x^k)_{k \geq 0}$ generated by the KT algorithm (8) converges and*

$$\lim_{k \to \infty} x^k = P_{\mathcal{N}(A)}(x^0) + x_{LS} \in S(A; b). \tag{12}$$

A very important particular case of the above result is the following. Let us consider the system (which is always consistent!)

$$A^T y = 0, \tag{13}$$

and define the projections $\varphi_j : \mathbb{R}^m \longrightarrow \mathbb{R}^m$ by

$$\varphi_j(y) = y - \frac{\langle y, A^j \rangle}{\| A^j \|^2} A^j, \, j = 1, \ldots, n. \tag{14}$$

Proposition 1 *If* $(y^k)_{k \geq 0}$ *is the sequence generated by*

$$y^{k+1} = \varphi_1 \circ \cdots \circ \varphi_n(y^k), k \geq 0, \tag{15}$$

with $y^0 \in \mathbb{R}^m$ *arbitrary, then*

$$\lim_{k \to \infty} y^k = P_{\mathcal{N}(A^T)}(y^0). \tag{16}$$

Moreover, for $y^0 = b$ *in (15)*

$$\lim_{k \to \infty} (b - y^k) = b - P_{\mathcal{N}(A^T)}(b) = P_{\mathcal{R}(A)}(b). \tag{17}$$

Remark 1 If $b \notin \mathcal{R}(A)$ the sequence $(x^k)_{k \geq 0}$ generated by the KT algorithm (8) still converges and

$$\lim_{k \to \infty} x^k = P_{\mathcal{N}(A)}(x^0) + Gb, \tag{18}$$

where G is an $n \times m$ matrix with the properties

$$AGA = A, \quad GAG = G, \quad (GA)^T = GA. \tag{19}$$

But, the limit in (18) is at a certain distance from the set $LSS(A; b)$ (see Popa and Zdunek (2004)).

In Natterer (1986) was proposed a version of KT algorithm with a relaxation parameter. In this respect, for $\omega \neq 0$ arbitrary fixed we define the application

$$P_{H_i}^{\omega}(b; x) = x - \omega P_{H_i}(x), i = 1, \ldots, m. \tag{20}$$

Then we can write Kaczmarz's iteration with relaxation parameter ω (RK, for short) for the problem (3) as follows.

Algorithm RK

Initialization: $\omega > 0, x^0 \in \mathbb{R}^n$;
Iterative step:
$$x^{k+1} = (P_{H_1}^{\omega} \circ \cdots \circ P_{H_m}^{\omega})(b; x)(b; x^k), \ k \geq 0. \tag{21}$$

We present below the convergence result corresponding to the algorithm RK.

Theorem 2 *If the problem (3) is consistent,* $\omega \in (0, 2)$ *and* $x^0 \in \mathbb{R}^n$ *then, the sequence* $(x^k)_{k \geq 0}$ *generated with the algorithm RK (21) converges and*

$$\lim_{k \to \infty} x^k = P_{\mathcal{N}(A)}(x^0) + x_{LS}. \tag{22}$$

Moreover, if $x^0 \in \mathcal{R}(A^T)$, *then the limit in (22) is exactly* x_{LS}.

3 Cimmino's Reflections Algorithm

As mentioned at the beginning of the chapter, Cimmino's reflections algorithm was proposed by its author in Cimmino (1938). There are the following two main differences with respect to Kaczmarz's one:

1. Instead of Kaczmarz's orthogonal projections P_{H_i}, Cimmino used orthogonal reflections

$$R_{H_i}(x) = x - 2\frac{\langle x, A_i \rangle - b_i}{\| A_i \|^2} A_i \tag{23}$$

2. The reflections $R_{H_i}(x)$ are computed simultaneously, not successively as in (8) and a strict convex combination of them provides the next approximation (see below (25)–(26) and Fig. 4).

In this section, the system (1) will be rectangular, but consistent. Moreover, we will suppose that

$$rank(A) \geq 2, \tag{24}$$

which is not restrictive (the case $rank(A) = 1$ can be solved by hand).

Algorithm Cimmino

Initialization: $x^0 \in \mathbb{R}^n$
Iterative step:

$$x^k = x^{k-1} - \frac{2}{\omega} \sum_{i=1}^{m} \omega_i \frac{\langle x^{k-1}, A_i \rangle - b_i}{\| A_i \|^2} A_i, \quad k \geq 0 \tag{25}$$

Fig. 4 The Cimmino's reflections iteration

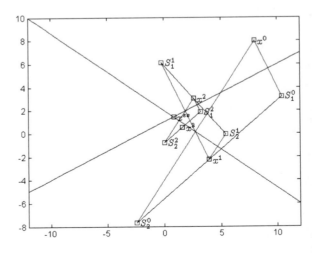

where

$$\omega_i > 0, \ \forall i = 1, \ldots, m, \ \ \omega = \sum_{i=1}^{m} \omega_i. \tag{26}$$

Theorem 3 *In the above hypothesis on the system (1) and its matrix A, for any $x^0 \in \mathbb{R}^n$ and the sequence $(x^k)_{k \geq 0}$ generated by the algorithm (25)–(26) converges to one of its solutions.*

Remark 2 In the second part of his paper Cimmino (1938), the author observed that the sequence $(x^\nu)_{\nu \geq 0}$ from (25) converges also in the general (possibly inconsistent) case of the (3), but to a solution $\tilde{x} \in \mathbb{R}^n$ of its "weighted" formulation

$$\| \tilde{A}\tilde{x} - \tilde{b} \| = \min\{\| \tilde{A}z - \tilde{b} \|, z \in \mathbb{R}^n\}, \tag{27}$$

with

$$\tilde{A} = \tilde{D}A, \ \tilde{b} = \tilde{D}b, \ \tilde{D} = \text{diag}\left(\frac{\sqrt{\omega_1}}{\|A_1\|}, \ldots, \frac{\sqrt{\omega_m}}{\|A_m\|}\right). \tag{28}$$

Using this important result by Cimmino, we can define a particular version of (25) which generates a sequence $(x^\nu)_{\nu \geq 0}$ convergent to a least squares solution of (3) in the inconsistent case by just defining the weights ω_h in (25) as

$$\omega_i = \| A_i \|^2, \ i = 1, \ldots, m. \tag{29}$$

With this choice the matrix $\tilde{D}$ from (28) becomes the identity, thus (27) will be exactly (3). Moreover, with this particular choice for ω_h Cimmino becomes the first projection-based method which approximates the least squares solutions of (3).

4 Extensions to Inconsistent Linear Least Squares Problems

In this section we will consider the inconsistent case of (3), i.e. $P_{\mathcal{N}}(A^T) \neq 0$. Then, accordingly to Proposition 1, the sequence $(y^k)_{k \geq 0}, y^0 = b$ generated in Eq. 15 converges to $P_{\mathcal{N}(A^T)}(b)$ (see (16)) therefore $b^k = b - y^k$ will converge to $P_{\mathcal{R}(A)}(b)$. Thus, "at limit" the system $Ax = b^k$ becomes $Ax = P_{\mathcal{R}(A)}(b)$, and in particular consistent; hence we can solve it by the classical KT algorithm. This theoretical procedure allows us to propose te following Extended KT algorithm for the numerical solution of the (inconsistent) least squares problem problem (3).

Algorithm (EKT)

Initialization: $y^0 = b, x^0 \in \mathbb{R}^n$
Iterative step:

$$y^{k+1} = \varphi_1 \circ \cdots \circ \varphi_n(y^k), \tag{30}$$

$$b^{k+1} = b - y^{k+1}, \tag{31}$$

$$x^{k+1} = \Phi_1 \circ \cdots \circ \Phi_m(b^{k+1}; x^k), \quad k \geq 0. \tag{32}$$

Theorem 4 *(i) For any matrix A, any vector $b \in \mathbb{R}^m$ and any initial approximation x^0 the sequence generated with the algorithm* **(KE)** *is convergent and*

$$\lim_{k \to \infty} x^k = P_{N(A)}(x^0) + Gb_A.$$

(ii) We have the equality

$$LSS(A; b) = \{P_{N(A)}(x^0) + Gb_A, x^0 \in \mathbb{R}^n\}.$$

(iii) The minimal norm solution x_{LS} is obtained as a limit-point of (30)–(32) if and only if $x^0 \in R(A^T)$. In this case we have

$$x_{LS} = Gb_A.$$

In the extended case we can also consider the introduction of relaxation parameters. In this respect, for $\alpha \in (0, 2)$ we first define the applications $\varphi_j(\alpha, y) = (1 - \alpha)y + \alpha\varphi_j(y)$ and then consider the version of EKT algorithm with Relaxation Parameters.

Algorithm (EKTRP)

Initialization: $x^0 \in \mathbb{R}^n$, $y^0 = b$
Iterative step:

$$y^{k+1} = \varphi_1 \circ \cdots \circ \varphi_n(\alpha; y^k); \tag{33}$$

$$b^{k+1} = b - y^{k+1} \tag{34}$$

$$x^{k+1} = P_{H_1}^\omega \circ \cdots \circ P_{H_m}^\omega(b^{k+1}; x^k), \quad k \geq 0. \tag{35}$$

The following convergence result of the type in Theorem 4 is valid in this case.

Theorem 5 *(i) For any matrix A, any vector $b \in \mathbb{R}^m$, any $\alpha, \omega \in (0, 2)$, and any initial approximation x^0, the sequence $(x^k)_{k \geq 0}$ generated with the algorithm* **(EKTRP)** *is convergent and*

$$\lim_{k \to \infty} x^k = P_{N(A)}(x^0) + Gb_A \in LSS(A; b).$$

(ii) If $x^0 \in \mathcal{R}(A^T)$ then the above limit equals x_{LS}.

Using the same procedure as in EKTRP, we can design an extension of Cimmino's algorithm (25)–(26) as presented below.

Algorithm Extended Cimmino (EC)

Initialization: $\alpha_i > 0, \omega_i > 0, \alpha = \sum_{j=1}^{n} \alpha_j, \omega = \sum_{i=1}^{m} \omega_i, y^0 = b, x^0 \in \mathbb{R}^n$
Iterative step:

$$y^{k+1} = y^k - \frac{2}{\alpha} \sum_{j=1}^{m} \omega_j \frac{\langle y^k, A^j \rangle}{\| A^j \|^2} A^j, \tag{36}$$

$$b^{k+1} = b - y^{k+1} \tag{37}$$

$$x^{k+1} = x^k - \frac{2}{\omega} \sum_{i=1}^{m} \omega_i \frac{\langle x^k, A_i \rangle - b_i^k}{\| A_i \|^2} A_i, \quad k \geq 0. \tag{38}$$

The convergence result for the algorithm EC, proved in Nicola et al. (2012) is presented below.

Theorem 6 *For any matrix A satisfying (24), any $b \in \mathbb{R}^m$, any $\alpha_j > 0, \omega_i > 0$ and any x^0, the sequence $(x^k)_{k \geq 0}$ generated with the algorithm EC converges and*

$$\lim_{k \to \infty} x^k = P_{N(A)}(x^0) + Gb_A \in LSS(A; b).$$

5 Oblique and Generalized Oblique Projections

In the present section of the chapter we will present some versions of the algorithms in Sects. 2 and 3, in which the classical orthogonal projection onto the system hyperplanes are replaced with oblique ones, defined by SPD and SPSD matrices. The main reasons in using these oblique projections would be the following:

- to get an (asymptotic) acceleration of convergence
- to get an enough good approximation of a solution in only few iterations.

5.1 Kaczmarz with Oblique Projections

Let D be an $n \times n$ SPD matrix. We will first define the energy inner product and norm, induced by D

$$\langle x, y \rangle_D = \langle Dx, y \rangle, \| x \|_D^2 = \langle x, x \rangle_D. \tag{39}$$

This allows us to construct the oblique projection onto the hyperplane H_i by

$$P_{H_i}^D(x) = x - \frac{\langle x, A_i \rangle}{\langle D^{-1} A_i, A_i \rangle} D^{-1} A_i, \tag{40}$$

and the Kaczmarz with Oblique projections algorithm (KO) .

Algorithm KO

Initialization: $x^0 \in \mathbb{R}^n$
Iterative step:

$$x^{k+1} = P_{H_1}^D \circ \cdots \circ P_{H_m}^D (x^k), \quad k \geq 0. \tag{41}$$

We will first connect this algorithm with a reformulation of the problem (3) as

$$\| \bar{A}\bar{x} - b \| = \min\{\| \bar{A}z - b \|, z \in \mathbb{R}^n\}, \text{ where } \bar{A} = AD^{-\frac{1}{2}}. \tag{42}$$

Then, the two problems are connected through the result from below.

Proposition 2 *(i) The rows of the matrix $\bar{A}$ from (42) are given by*

$$\bar{A}_i = D^{-\frac{1}{2}} A_i, \ i = 1, \ldots, m. \tag{43}$$

(ii) $LSS(A; b) = D^{-\frac{1}{2}} \left(LSS(\bar{A}; b) \right)$
(iii) If $n \leq m$ and $\text{rank}(A) = n$ then

$$x_{LS} = D^{-\frac{1}{2}} \bar{x}_{LS}. \tag{44}$$

Remark 3 The same result remains true also in the consistent case of the problem (3) (and so will be (42)).

Moreover, the convergence result for the algorithm KO in the consistent case of the problem (3) is presented in the result from below.

Theorem 7 *For any $x^0 \in \mathbb{R}^n$ a given initial approximation, the sequence $(x^k)_{k \geq 0}$ generated by the algorithm KO (41) converges and $\lim_{k \to \infty} x^k = x^*(x^0) \in S(A, b)$.*

5.2 Diagonal Weighting Algorithm

The considerations from this section follow the original paper Censor et al. (2001). We start with the particular case $\omega_h = 1, \forall h = 1, \ldots, m$ for which Cimmino's algorithm (25)–(26) becomes

$$x^{k+1} = x^k - \frac{2}{m} \sum_{i=1}^{m} \frac{\langle x^k, A_i \rangle - b_i}{\|A_i\|^2} A_i. \tag{45}$$

If we write this component wise we get

$$x_j^{k+1} = x_j^k - \frac{2}{m} \sum_{i=1}^{m} \frac{\langle x^k, A_i \rangle - b_i}{\|A_i\|^2} A_{ij}, \quad j = 1, \ldots, n. \tag{46}$$

From the above relations we get an information about the practically observed "slowness" of Cimmino's algorithm. Indeed, for a fixed $j \in \{1, \ldots, n\}$, when A is sparse only a relatively"small" number of the elements $A_{1j}, A_{2j}, \ldots, A_{mj}$ are nonzero, but in (46) the sum of their contributions is divided by the relatively"large" m. This observation, made by the authors in Censor et al. (2001), led them to consider a modification of the algorithm, in which the factor $\frac{1}{m}$ in (45) is replaced by a factor that depends only on the sparsity structure of A. For doing this, they first proposed a more general form of Cimmino's algorithm (25) which we shall describe in what follows. In this respect, they considered a set of symmetric and positive semidefinite $n \times n$ matrices G_i, $i = 1, \ldots, m$, such that their Moore-Penrose pseudoinverses G_i^+ satisfy

$$\langle G_i^+ A_i, A_i \rangle \neq 0, \quad \forall i = 1, \ldots, m. \tag{47}$$

With these matrices G_i we define the generalized oblique projections $P_{H_i}^{G_i}$ (see (40)).

$$P_{H_i}^{G_i} x = x + \frac{b_i - \langle x, A_i \rangle}{\langle G_i^+ A_i, A_i \rangle} G_i^+ A_i. \tag{48}$$

The algorithm proposed in Censor et al. (2001) is the following.

Algorithm Diagonal Weighting (DW)

Initialization: $x^0 \in \mathbb{R}^n$
Iterative step:

$$x^{k+1} = x^k + \sum_{i=1}^m G_i \left(P_{H_i}^{G_i} x^k - x^k \right), \quad k \geq 0. \tag{49}$$

Remark 4 A particular case in which G_i^+ can be easily computed is

$$G_i = \mathrm{diag}(g_{i1}, g_{i2}, \ldots, g_{in}), \quad g_{ij} \geq 0, \tag{50}$$

for which

$$G_i^+ = \mathrm{diag}(\gamma_{i1}, \gamma_{i2}, \ldots, \gamma_{in}), \quad \gamma_{ij} = \begin{cases} 1/g_{ij}, & g_{ij} \neq 0 \\ 0, & g_{ij} = 0 \end{cases}. \tag{51}$$

In this case the algorithm (49) has the form (component wise)

$$x_j^{k+1} = x_j^k - \sum_{i, A_{ij} \neq 0} \frac{\langle x^k, A_i \rangle - b_i}{\sum_{k, A_{ik} \neq 0} \frac{A_{ik}^2}{g_{ik}}} A_{ij}, \quad j = 1, \ldots, n. \tag{52}$$

Remark 5 If we define the weights g_{ij} in the above particular case (50) by

$$g_{ij} = \frac{2\omega_i}{\omega}, \forall j = 1, \ldots, n, \ \forall i = 1, \ldots, m \tag{53}$$

then (47) is satisfied and in (48) we get $P_{H_i}^{G_i} x = x - \frac{\langle x, A_i \rangle - b_i}{\langle A_i, A_i \rangle} A_i$, thus (49) becomes exactly (25), i.e. the algorithm DW is a generalization of Cimmino's one.

For connecting the family $\{(G_i)_{i=1,...,m}\}$ to the sparsity structure of A we consider the following additional properties for the family of matrices $\mathcal{G} = \{G_i, i = 1, \ldots, m\}$. We will say that $\mathcal{G}$ is **Sparsity Pattern Oriented (SPO)** if

$$g_{ij} \geq 0, \quad g_{ij} = 0 \text{ if and only if } A_{ij} = 0, \text{ and } \sum_{i=1}^{m} G_i = I. \tag{54}$$

Example 1 (Censor et al. 2001) For an arbitrary fixed $j \in \{1, 2, \ldots, n\}$

$$g_{ij} = \begin{cases} 1/s_j, & A_{ij} \neq 0 \\ 0, & A_{ij} = 0 \end{cases}, \quad i = 1, \ldots, m, \tag{55}$$

where s_j is the number of nonzero elements in the set $\{A_{1j}, A_{2j}, \ldots, A_{mj}\}$ (i.e. in the j-th column of A).

Example 2 For an arbitrary fixed $j \in \{1, 2, \ldots, n\}$

$$g_{ij} = \frac{|A_{ij}|}{\sum_{k=1}^{m} |A_{kj}|}, \quad i = 1, \ldots, m. \tag{56}$$

Example 3 For an arbitrary fixed $j \in \{1, 2, \ldots, n\}$

$$g_{ij} = \frac{A_{ij}^2}{\| A^j \|^2}, \quad i = 1, \ldots, m. \tag{57}$$

In order to analyse the convergence of the algorithm DW we consider its *proximity function* $F : \mathbb{R}^n \longrightarrow \mathbb{R}$ defined by

$$F(x) = \sum_{i=1}^{m} \| P_{H_i}^{G_i} x - x \|_{G_i}^2 \tag{58}$$

and the set of its minimizers, Φ

$$\Phi = \{x \in \mathbb{R}^n, F(x) \leq F(z), \forall z \in \mathbb{R}^n\}. \tag{59}$$

By direct computation and using the properties from below of the SPO families

$$0 \leq g_{ij} \leq 1, \quad G_i G_i^+ A_i = A_i, \text{ and } \| x \|^2 = \sum_{i=1}^{m} \| x \|_{G_i}^2,$$

we can show that

$$F(x) = \sum_{i=1}^{m} \frac{1}{\langle G_i^+ A_i, A_i \rangle} (b_i - \langle x, A_i \rangle)^2 = \| Ax - b \|_M^2, \tag{60}$$

where M is the $m \times m$ SPD matrix

$$M = \text{diag}(\frac{1}{\langle G_1^+ A_1, A_1 \rangle}, \ldots, \frac{1}{\langle G_m^+ A_m, A_m \rangle}). \tag{61}$$

This tells us that we always have minimizers for F, i.e. $\Phi \neq \emptyset$. Moreover, if the initial problem (3) is consistent, so will be the weighted one $\| Ax - b \|_M^2 = \min!$ (see (39) and Proposition 2), and they are equivalent, i.e. in this case $\Phi = S(A; b)$.

Theorem 8 *In the above hypothesis for the SPO family $(G_i)_{i=1,\ldots,m}$, for any initial approximation $x^0 \in \mathbb{R}^n$ the sequence generated with the algorithm (49) converges to an element of Φ, which depends on the initial approximation x^0. If the problem (3) is consistent then this limit belongs to the set of its solutions.*

According to the results in Sect. 4 we can extend the algorithm DW (49) for inconsistent problems, following the procedure in (30)–(32). First of all the column version of DW algorithm (ColDW), for the system $A^T y = 0$ is the following.

Algorithm ColDW

Initialization: $y^0 = b$;
Iterative step:

$$y^{k+1} = y^k + \sum_{j=1}^{n} \Gamma_j \left(P_{V_j}^{\Gamma_j} y^k - y^k \right). \tag{62}$$

$$P_{V_j}^{\Gamma_j} y = y - \frac{\langle y, A^j \rangle}{\langle \Gamma_j^+ A^j, A^j \rangle} \Gamma_j^+ A^j, \quad j = 1, \ldots, n, \ y \in \mathbb{R}^m, \ k \geq 0 \tag{63}$$

where $(\Gamma_j)_j$ is a family of $m \times m$ diagonal matrices of the form

$$\Gamma_j = \text{diag}(\gamma_{j1}, \ldots, \gamma_{jm}). \tag{64}$$

We will suppose that $(\Gamma_j)_j$ is also Sparsity Pattern Oriented, but with respect to the matrix A^T, i.e. (by analogy with (54))

$$\gamma_{ji} \geq 0, \quad \gamma_{ji} = 0 \Leftrightarrow A_{ij} = 0, \quad \sum_{j=1}^{n} \Gamma_j = I. \tag{65}$$

The families similar with those from Examples 1–3 are presented below.

Example 4 For an arbitrary fixed $i \in \{1, 2, \ldots, m\}$

$$\gamma_{ji} = \begin{cases} 1/\sigma_i, & A_{ij} \neq 0 \\ 0, & A_{ij} = 0 \end{cases}, \quad j = 1, \ldots, n, \tag{66}$$

where σ_i is the number of nonzero elements in the set $\{A_{i1}, A_{i2}, \ldots, A_{in}\}$ (i.e. in the i-th row of A).

Example 5 For an arbitrary fixed $i \in \{1, 2, \ldots, m\}$

$$\gamma_{ji} = \frac{|A_{ij}|}{\sum_{k=1}^{n} |A_{ik}|}, \quad j = 1, \ldots, n. \tag{67}$$

Example 6 For an arbitrary fixed $i \in \{1, 2, \ldots, m\}$

$$\gamma_{ji} = \frac{A_{ij}^2}{\| A_i \|^2}, \quad j = 1, \ldots, n. \tag{68}$$

Then, the Extended DW (EDW) algorithm is the following (see also (48)–(49) and (62)–(63)).

Algorithm EDW

Initialization: $x^0 \in \mathbb{R}^n$, $y^0 = b$;
Iterative step:

$$y^{k+1} = y^k + \sum_{j=1}^{n} \Gamma_j \left(\frac{\langle y, A^j \rangle}{\langle \Gamma_j^+ A^j, A^j \rangle} \Gamma_j^+ A^j \right), \tag{69}$$

$$b^{k+1} = b - y^{k+1}, \tag{70}$$

$$x^{k+1} = x^{k+1} = x^k + \sum_{i=1}^{m} G_i \left(\frac{b_i^{k+1} - \langle x, A_i \rangle}{\langle G_i^+ A_i, A_i \rangle} G_i^+ A_i \right), \quad k \geq 0. \tag{71}$$

A convergence result as in Theorem 5 holds for the algorithm EDW (69)–(71).

6 Constraining Strategies

A very important class of problems which suggested the use of constraining strategies was the algebraic reconstruction of images in computerized tomography. This was imposed by the fact that the components x_i of the exact image $x \in \mathbb{R}^n$ lie in a given interval $[a_i, b_i]$, $i = 1, \ldots, n$, see e.g Herman (1980). But, when applying a reconstruction algorithm (let's call it **ALG**), even if the initial approximation x^0 satisfies $x_i^0 \in [a_i, b_i]$, $i = 1, \ldots, n$, the other approximations x^k, $k \geq 1$ can run out of those intervals. This is because (e.g. in the consistent case), with $x^0 \in \mathcal{R}(A^T)$

most of the algorithms **ALG** converge to x_{LS}, and this solution can be out of the "box" $B = [a_1, b_1] \times \cdots \times [a_n, b_n]$, as shown in the example below.

Example 7 For the consistent system ($\alpha \in \mathbb{R}$)

$$\begin{bmatrix} 1 & \alpha \\ 2 & 2\alpha \end{bmatrix} \begin{bmatrix} x_1 \\ x_2 \end{bmatrix} = \begin{bmatrix} 1 + \alpha \\ 2 + 2\alpha \end{bmatrix},$$

for $\alpha = \sqrt{2}$, the minimal norm solution is $x_{LS} = (\frac{1+\sqrt{2}}{3}, \frac{2+\sqrt{2}}{3})^T$ and does not belong to the box $[0, 1] \times [0, 1]$.

Then, for an enough large $k \geq 0$, the approximations x^k will come closer to x_{LS} and runs out of the corresponding box, giving a non appropriate reconstruction of the exact image.

6.1 A Single Constraining Function

In order to overcome this situation and to keep x^k in the constraints set, we must introduce a constraining strategy in **ALG**, and call it **Constrained ALG (CALG)**. This was firstly done in the paper Koltracht and Lancaster (1990) where the authors considered a constraining function $C : \mathbb{R}^n \longrightarrow \mathbb{R}^n$ with closed image $\text{Im}(C) \subset \mathbb{R}^n$ and the properties

$$\| Cx - Cy \| \leq \| x - y \|, \tag{72}$$

$$\text{if} \quad \| Cx - Cy \| = \| x - y \| \quad \text{then} \quad Cx - Cy = x - y, \tag{73}$$

$$\text{if} \quad y \in \text{Im}(C) \quad \text{then} \quad y = Cy. \tag{74}$$

Example 8 If $[a_i, b_i] \subset \mathbb{R}$ are some given intervals, $i = 1, \ldots, n$ we define C_{box} as the orthogonal projection operator onto the closed convex set $[a, b] = [a_1, b_1] \times \cdots \times [a_n, b_n] \subset \mathbb{R}^n$,

$$(C_{box}x)_i = \begin{cases} x_i, & \text{if } x_i \in [a_i, b_i] \\ a_i, & \text{if } x_i < a_i \\ b_i, & \text{if } x_i > b_i. \end{cases} \tag{75}$$

Then, the constrained algorithm is defined as follows.

Algorithm CALG

Initialization: $x^0 \in \text{Im}(C)$;
Iterative step:

$$x^{k+1} = C[ALG(b; x^k)], \quad k \geq 0. \tag{76}$$

We may provide the following general convergence result.

Theorem 9 *If the problem (3) is consistent and the constraining function satisfy all the assumptions (72)–(74), the sequence $(x^k)_{k \geq 0}$ generated by the algorithm* **CALG**, *converges to a solution $x^* \in S(A; b) \cap Im(C)$.*

Remark 6 If **ALG** is Kaczmarz, Cimmino, DW, the corresponding convergence result from the above theorem was proved in Koltracht and Lancaster (1990), Nicola et al. (2012) and Popa (2010), respectively.

Let us now consider the inconsistent case of (3), and the corresponding version of **ALG** (called **EALG**), according to the procedure (30)–(32).

Algorithm (EALG)

Initialization: $y^0 = b, x^0 \in \mathbb{R}^n$
Iterative step:

$$y^{k+1} = ALG_{col}(y^k), \tag{77}$$
$$b^{k+1} = b - y^{k+1}, \tag{78}$$
$$x^{k+1} = ALG(b^{k+1}; x^k), \quad k \geq 0, \tag{79}$$

where ALG_{col} is the column version of ALG, applied to the system $A^T y = 0$, with $y^0 = b$. The corresponding Extended version of CALG, called **CEALG** is the following.

Algorithm (CEALG)

Initialization: $y^0 = b, x^0 \in \mathbb{R}^n$
Iterative step:

$$y^{k+1} = ALG_{col}(y^k), \tag{80}$$
$$b^{k+1} = b - y^{k+1}, \tag{81}$$
$$x^{k+1} = C\left[ALG(b^{k+1}; x^k)\right], \quad k \geq 0. \tag{82}$$

Theorem 10 *If the constraining function satisfy (only!) the assumptions (72) and (74), the sequence $(x^k)_{k \geq 0}$ generated by the algorithm* **ECALG**, *converges to a solution $x^* \in LSS(A; b) \cap Im(C)$.*

Remark 7 If **EALG** is Extended Kaczmarz, Extended Cimmino, Extended DW, the corresponding convergence result from the above theorem was proved in Popa (2008), Nicola et al. (2012) and Popa (2010), respectively.

6.2 A Family of Constraining Functions

The above constraining strategies in which only one constraining function is used, become inefficient for other classes of problems, in which e.g. we are not interested in well reconstructing the whole image, but only to identify and correctly count some particles in a given domain (the $\| x \|_0$ least squares formulations in particle velocimetry; (see e.g. Butz et al. 2016; Petra et al. 2009). In this situation we have to use a family of constraining functions, which are iteration dependent (a Family Constrained ALG).

Algorithm FCALG

Initialization: $x^0 \in \text{Im}(C_0)$;
Iterative step:

$$x^{k+1} = C_{k+1}[ALG(b; x^k)], \quad k \geq 0 \tag{83}$$

where $(C_k)_{k \geq 0}$ is a family of constraining functions of the type C from (72)–(74). In the consistent case of (3), if this family satisfies certain assumptions and the set $V = \cap_{k \geq 0} Im(C_k)$ is nonempty, the sequence $(x^k)_{k \geq 0}$ generated by **FCALG** converges toward a solution $x^* \in S(A; b) \cap V$ (see for details Pantelimon and Popa 2013; Censor et al. 2014).

6.3 Hybrid Projection Algorithms

Depending on the nature of the problem, various combinations of classical projection methods with different algorithms are possible. In this respect we can mention a mixed Kaczmarz—Conjugate Gradient method, which gave very good results in reconstruction of real medical CT images—see for details Popa (2010). Moreover, by introducing supplementary directions for projection, Kaczmarz can be transformed in a "direct solver", which gives the solution in only one iteration—see details and efficient applications to rigid body dynamics problems in Köstler et al. (2012). And, last but not least, the possible combination of Kaczmarz method with genetic type algorithms, with good results in image reconstruction from limited data—see details in Bautu et al. (2006).

7 Numerical Experiments

7.1 Extended and Constrained Kaczmarz-Tanabe Algorithm

Experiment 1 We constructed two model (exact) images: $image_1$ with resolution 10×10, $image_2$ with resolution 20×20 (see Figs. 5 and 6).

The scanning matrices were constructed as in geotomography (see Fig. 7), $A_1 : 100 \times 100$, $A_2 : 400 \times 400$, and the corresponding consistent right hand sides b^1, b^2, i.e. $b^i = A_i \, image_i$, $i = 1, 2$. Then we applied algorithms K and CK (with a box constraining function, as in (75), with $a_i = 0$, $b_i = 1$, $i = 1, \ldots, m$) from

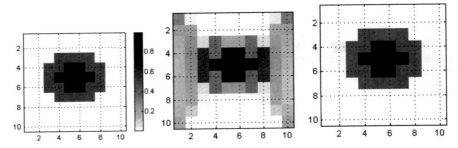

Fig. 5 Left: original; middle: algorithm K; right: algorithm CK

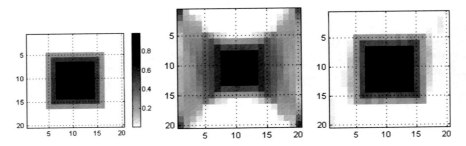

Fig. 6 Left: original; middle: algorithm K; right: algorithm CK

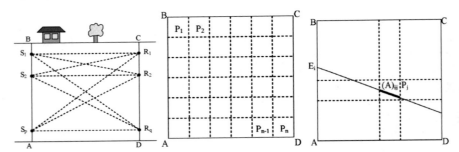

Fig. 7 Scanning procedure in electromagnetic geotomography

Sect. 6.1, with $x^0 = 0$ and 90 iterations. The reconstructions are presented in Figs. 5 and 6. We observe the much better accuracy for the constrained algorithms.

Experiment 2 We perturbed the right hand side $b = b^1$ or $b = b^2$ from the previous experiment with a 5% randomly generated noise, as presented in the MATLAB code from below ($N = 100$ or $N = 400$ is the resolution of the image)

- $rand('state', 0); pert = rand(N, 1); np = norm(pert);$
- $pert = (1/np) * pert; b = b + 0.05 * pert.$

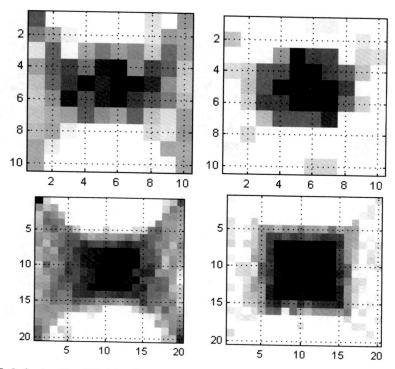

Fig. 8 Left: algorithm KE; right: algorithm CKE

Then we applied algorithms KE and CKE, with $x^0 = 0$ and 90 iterations. The reconstructed images are presented in Fig. 8. We can observe the improvement due to the constraining step.

Experiment 3 The advantage of the KE algorithm against the classical K for "much more" inconsistent problems will become clear from our third set of experiments. In this case the data (provided by Prof. Dr. Joachim Hornegger and Dr. Marcus Prümmer from University of Erlangen-Nürnberg, Germany) refer to a 2D-head phantom of resolution 64×64, which represents a "slice" from a 3D reconstruction. This 2D problem is defined by a matrix A of dimensions $m \times n$ ($m = 149.740, n = 4096 = 64 \times 64$) and a right hand side $b \in \mathbb{R}^n$. The system $Ax = b$ is "highly" inconsistent ($\| Ax_{LS} - b \| = \| P_{N(A^T)}(b) \| \approx 629.2$, by comparing it with the norm of the right hand side $\| b \| = 858.17$), thus we reformulated it as a least squares problem of the form (3). We applied to this problem the CK, KE and CKE algorithms, starting with $x^0 = 0$, terminating after 20 iterations. The reconstructions are presented in Fig. 9. We can see that, due to the "high" level of inconsistency, the CK algorithm cannot produce a reliable result, whereas KE and CKE algorithms give us the satisfactory middle and right images.

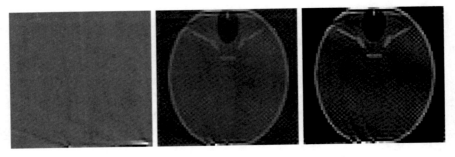

Fig. 9 Left: algorithm CK; middle: algorithm KE; right: algorithm CKE

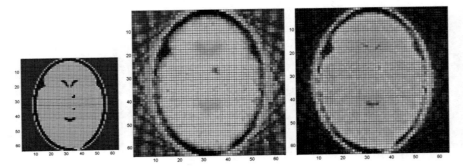

Fig. 10 Consistent case, $x^0 = 0$, 500 iterations; left: exact, middle: C, right: CC

7.2 Extended and Constrained Cimmino Algorithm

Experiment 1 In this case the data consist on a head phantom image (provided by Prof. Dr. Tommy Elfving from University of Linköpping, Sweden): 63×63 pixels resolution with a scanning matrix with 1376 rays (i.e. the problem matrix A is 1376×3969). A consistent and an inconsistent right hand side b was used in our reconstruction experiments. We applied algorithms C and CC, with $x^0 = 0$ and 500 iterations in the consistent case. The results are given in Fig. 10; a small improvement can be observed for the CC algorithm.

Experiment 2 With the same data, but the inconsistent right hand side b, we applied the algorithms C and EC, $x^0 = 0$ and 2000 iterations. We observe in Fig. 11 a better reconstruction with the EC method.

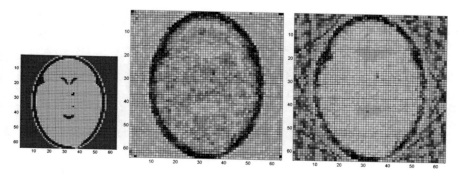

Fig. 11 Inconsistent case, $x^0 = 0$, 2000 iterations; left: exact, middle: C, right: EC

7.3 Cimmino Versus Diagonal Weighting

Experiment 1 In this case the data are those from Sect. 7.2. We applied algorithms C, DW, CC and CDW, with $x^0 = 0$ and 60 iterations in the consistent case. The results are given in Figs. 12 and 13; a good improvement can be observed for the DW and CDW algorithms.

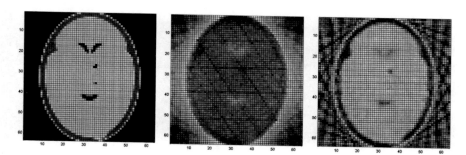

Fig. 12 No constraints; original image, C, DW

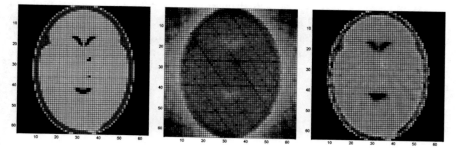

Fig. 13 With constraints; original image, CC, CDW

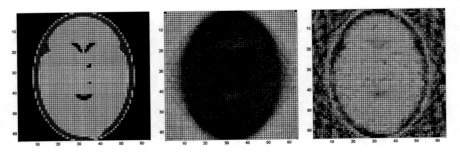

Fig. 14 No constraints; original image, EC, EDW

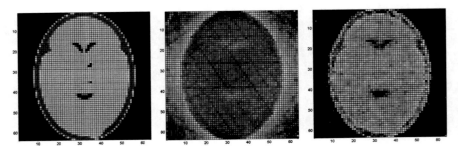

Fig. 15 With constraints; original image, CEC, CEDW

Experiment 2 With the same data, but the inconsistent right hand side b, we applied the algorithms C, CEC, EDW and CEDW with $x^0 = 0$ and 60 iterations. We observe in Figs. 14 and 15 a better reconstruction with the EDW and CEDW methods.

7.4 Family Constrained Algorithms

For this section the data come from a particular image reconstruction inverse problem discussed in Petra et al. (2009) and Pantelimon and Popa (2013). This problem arises from 3D Tomographic Particle Image Velocimetry (TomoPIV), which is an optical method for measuring velocities of fluids. Details about the construction of the constraining family from (83) can be found in the paper Pantelimon and Popa (2013). We presented here only the results from Fig. 16. We observe the much better reconstruction when using a family of constraints.

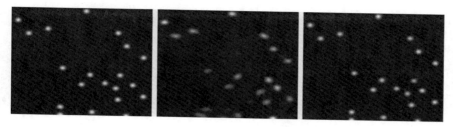

Fig. 16 Left: original image; middle: single constraint reconstruction; right: family of constraints reconstruction

References

Bautu, A., Bautu, E., Popa, C.: Evolutionary algorithms in image reconstruction from limited data. In: Proceedings of the 4th Workshop on Mathematical Modelling on Environmental and Life Sciences Problems, 7–10 Sept 2005. Constanta, Romania, Publishing House of the Romanian Academy, Bucharest, Romania, pp. 15–26 (2006)

Butz F., Fügenschuh, Wood J.N., Breuer M.: Particle-image velocimetry and the assignement problem. In: Fink, A., Fgenschuh, A., Geiger, M. (eds.), Operations Research Proceedings (GOR (Gesellschaft fr Operations Research e.V.)). Springer, Cham, pp. 243–249 (2016)

Censor, Y., Stavros, A.Z.: Parallel optimization: theory, algorithms and applications. In: Numerical Mathematics and Scientific Computation Series. Oxford Univ. Press, New York (1997)

Censor, Y., Gordon, D., Gordon, R.: Component averaging: an efficient iterative parallel algorithm for large and sparse unstructured problems. Parallel Comput. **27**, 777–808 (2001)

Censor, Y., Pantelimon, I., Popa, C.: Family constraining of iterative algorithms. Numer. Alg. **66**(2), 323–338 (2014)

Cimmino, G.: Calcolo approssiomatto per le soluzioni dei sistemi di equazioni lineari. Ric. Sci. progr. tecn. econom. naz. **1**, 326–333 (1938)

Herman, G.T.: Image reconstruction from projections. In: The Fundamentals of Computerized Tomography. Academic Press, New York (1980)

Kaczmarz, S.: Angenaherte Auflosung von Systemen linearer Gleichungen. Bull. Acad. Polonaise Sci. et Lettres **A** , 355–357 (1937)

Kaczmarz, S.: Approximate solution of systems of linear equations. Int. J. Control. **57**(6), 1269–1271 (1993)

Koltracht, I., Lancaster, P.: Constraining strategies for linear iterative processes. IMA J. Numer. Anal. **10**, 555–567 (1990)

Köstler, H., Popa, C., Preclik, T., Rüde, U.: On Kaczmarz's projection iteration as a direct solver for linear least squares problems. Linear Alg. Appl. **436**(2), 389–404 (2012)

Natterer, F.: The Mathematics of Computerized Tomography. Wiley, New York (1986)

Nicola, A., Petra, S., Popa, C., Schnörr, C.: A general extending and constraining procedure for linear iterative methods. Int. J. Comput. Math. **89**(2), 231–253 (2012)

Pantelimon, I., Popa, C.: Constraining by a family of strictly nonexpansive idempotent functions with applications in image reconstruction. BIT Numer. Math. **53**, 527–544 (2013)

Petra, S., Popa, C., Schnörr, C.: Accelerating Constrained SIRT with Applications in Tomographic Particle Image Reconstruction (2009). http://www.ub.uni-heidelberg.de/archiv/9477

Popa, C., Zdunek, R.: Kaczmarz extended algorithm for tomographic image reconstruction from limited-data. Math. Comput. Simul. **65**, 579–598 (2004)

Popa, C.: Constrained Kaczmarz extended algorithm for image reconstruction. Linear Alg. Appl. **429**(8–9), 2247–2267 (2008)

Popa, C.: A hybrid Kaczmarz—conjugate gradient algorithm for image reconstruction. Math. Comput. Simul. **80**(12), 2272–2285 (2010)

Popa, C.: Extended and constrained diagonal weighting algorithm with application to inverse problems in image reconstruction. Inv. Probl. **26**(6), 17p (2010)

Tanabe, K.: Projection method for solving a singular system of linear equations and its applications. Numer. Math. **17**, 203–214 (1971)

Questionnaires, Bar and Hyperstructures

Pipina Nikolaidou

Abstract An application of hyperstructure theory on social sciences is presented. The bar is a tool proposed in questionnaires instead of Likert Scale. This new tool gives the researchers the opportunity to correct any kind of tendency maybe appeared in the results, helping them to obtain more accurate results, or come to a conclusion in cases the results from the Likert Scale give them no information. Moreover, the filling questionnaire procedure is being accomplished using the bar, instead of Likert scale, on computers where the results are saved automatically so they are ready for research.

Keywords Bar · Questionnaires

AMS S. Classification 20N20 · 16Y99

1 Introduction

The class of hyperstructures called H_v-structures introduced in 1990 (Vougiouklis 1991) and they satisfy the weak axioms where the non-empty intersection replaces the equality. Some basic definitions are the following:

In a set H equipped with a hyperoperation (abbreviation *hyperoperation = hope*)

$$\cdot : H \times H \to P(H) - \{\varnothing\},$$

we abbreviate by

WASS the *weak associativity*: $(xy)z \cap x(yz) \neq \varnothing, \forall x, y, z \in H$ and by
COW the *weak commutativity*: $xy \cap yx \neq \varnothing, \forall x, y \in H$.

The hyperstructure $(H, \cdot)$ is called H_v-*semigroup* if it is *WASS* and it is called H_v-*group* if it is reproductive H_v-semigroup, i.e.,

P. Nikolaidou (✉)
School of Education, Democritus University of Thrace, 68 100 Alexandroupolis, Greece
e-mail: pnikolai@eled.duth.gr

© Springer Nature Switzerland AG 2019
C. Flaut et al. (eds.), *Models and Theories in Social Systems*, Studies in Systems, Decision and Control 179, https://doi.org/10.1007/978-3-030-00084-4_6

$$xH = Hx = H, \forall x \in H.$$

Motivation. In the classical theory the quotient of a group with respect to an invariant subgroup is a group. F. Marty from 1934, states that, the quotient of a group with respect to any subgroup is a hypergroup. Finally, the quotient of a group with respect to any partition (or equivalently to any equivalence relation) is an H_v-group. This is the motivation to introduce the H_v-structures (Vougiouklis 1991, 1994).

For more definitions and applications on H_v-structures one can see Corsini and Leoreanu (2003), Davvaz (2003, 2013), Davvaz and Leoreanu (2007), Hoskova (2005), Nikolaidou and Vougiouklis (2012), Kaplani and Vougiouklis (2017).

An important application which can be used in social sciences is the combination of hyperstructure theory with fuzzy theory. In 2008, Vougiouklis and Vougiouklis proposed the replacement of the Likert scale, usually used in questionnaires with the bar (Kambaki-Vougioukli and Vougiouklis 2008):

Definition 1 In every question substitute the Likert scale with 'the bar' whose poles are defined with '0' on the left end, and '1' on the right end:

$$0 \underline{\hspace{5cm}} 1$$

The subjects/participants are asked instead of deciding and checking a specific grade on the scale, to cut the bar at any point s/he feels expresses her/his answer to the specific question.

The final suggested length of the bar according to the Golden Ratio is 6.2 cm (Kambaki-Vougioukli et al. 2017; Vougiouklis 2011). The bar proposal is closely related with the fuzzy theory, as the Likert Scale represents a discrete situation, but with the bar we move from a discrete situation to a continuous one (Kambaki-Vougioukli et al. 2011).

There are certain shortcomings usually identified in Likert scales, that are based in the fact that in Likert scales we deal with a discrete situation. On the other hand, the use of the bar is based in fuzzy logic. It is an application of hyperstructure theory and fuzzy theory in social sciences. There are identified certain advantages concerning the use of the bar compared to that of a scale, during both stages of filling, as well as, processing a questionnaire. The basic advantage, though, is that the use of the bar instead of the Likert scale gives to researchers the opportunity to elaborate the questionnaires in many different ways. The way of the elaboration depends each time on the filled questionnaires but of course also on the problem. The use of the bar can open new perspectives in questionnaire elaboration and generally help the researchers for more accurate results (Vougiouklis and Kambaki-Vougioukli 2011; Vougiouklis and Vougiouklis 2015; Nikolaidou 2017; Markos 2017).

2 A New Filling Questionnaire Procedure

There are several advantages of the bar, the only disadvantage is the data collection. In order to overcome the serious bar shortcoming, the filling questionnaire procedure has been improved using technology; in this way the procedure is much faster for the participants and the results are saved automatically, for the researchers. The questionnaires can be answered on a software developed for this purpose (Nikolaidou and Vougiouklis 2012).

Using this software the results can automatically be transferred for research elaboration. The implemented program has been developed to overcome the problems raised during the data collection, inputting of data from questionnaires to processing. It eliminates the time of data collection, transferring data directly for any kind of elaboration.

More specifically, the answers are saved in multiple ways in order to give the researchers the opportunity to choose among numerous ways, which is one of the advantages of the bar. The results are being saved on a simple database indicating the exact point (in mm) each participant has 'cut' the bar, but also the segment chosen if the bar was divided in 3, 4,..., 10 equal segments. This way is used in order to make the first elaboration, as with Likert Scale. Using this software, though, gives the researcher the opportunity choose another subdivision, i.e in 5, 6,..., 9 equal-area segments (Vougiouklis and Kambaki-Vougioukli 2013) depending on the experiment or on the tendency of the answers:

1. the Gauss distribution
2. the increasing low parabola: $x = y^2$
3. the increasing upper parabola: $1 - y = (1 - x)^2$
4. the decreasing low parabola: $y = (1 - x)^2$
5. the decreasing upper parabola: $1 - y = x^2$

The use of the bar combined with this software eliminates the time of filling the questionnaires, making the procedure easier for the participants (Nikolaidou and Vougiouklis 2014).

3 Evaluation

The following survey has been established in the department of Elementary Education of Democritus University of Thrace, in the frame of course evaluation, and especially of Geometry of first semester. The sample was 159 students, who were asked to answer questions related to the course, to the teacher, to the teaching of the course and to questions about themselves.

The questionnaire used was the same used for the course evaluation. The only difference was that the six-grade Likert Scale was substituted by the bar, which was firstly divided into six equal-segments. In the next step, the use of histograms arise if the answers follow any kind of distribution or they present any kind of tendency. In these cases, the bar was redivided into equal-area segments according to the distribution or the tendency, for more accurate results.

The filling questionnaire procedure has been accomplished using computers, and especially the described software.

3.1 Course

The first question category is about the Course and consists of 9 question. The histograms obtained when the bar was divided into 6 equal-segments are the following:

It is obvious, that there is an upward trend in most questions, but in question 6 the answers follow the increasing-low parabola ($x = y^2$). In order to obtain more accurate results, we move on a second subdvision of the bar and as it is already mentioned, the bar is now divided into six equal-area segments according to the increasing-low parabola:

As we can see, the second histogram is very different from the first one. That means that despite the fact that according to the first subdivision most answers were in the last two scales, the second histogram proves the opposite (Figs. 1, 2, 3, 4, 5, 6, 7 and 8).

3.2 Teaching

The second question category is about the Teaching and the subjects answered 5 questions. The histograms obtained once again following the 6-equal-segments subdivision are the following:

In the last question the answers follow the increasing upper parabola ($1 - y = (1 - x)^2$) and the histogram obtained dividing the bar into equal-area segments is the following:

From the new histogram there is no more information obtained. Anyone can realize that the first presented upward trend is confirmed with the second subdivision.

3.3 Teacher

The question category concering the Teacher is composed of 3 questions:

In the first question the bar can be once again divided according to the increasing upper parabola:

The results from the first subdivision are the same with the second.

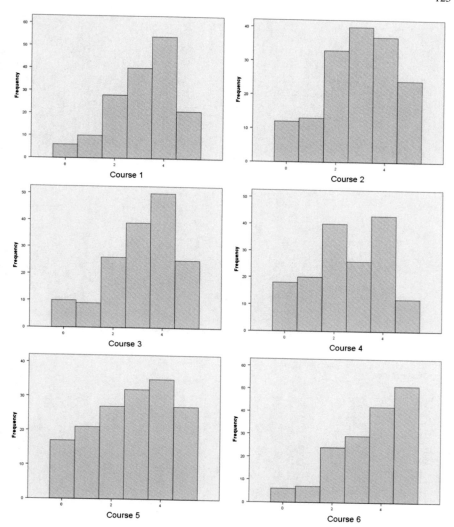

Fig. 1 Equal segments

3.4 Students

In the final question category about the students, we have 4 questions:

In this category it is obvious that all questions present a kind of tendency, so we can use an equal-area subdivision according to Gauss or parabola. The histograms make this decision easy; so for the first and the last question, the answers follow the

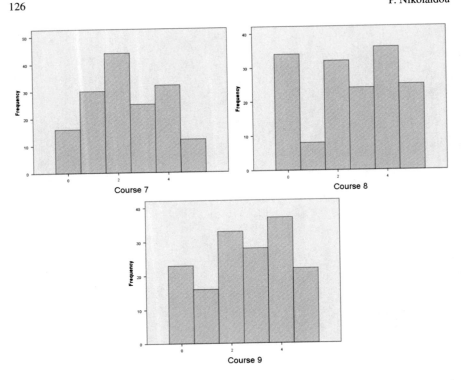

Fig. 1 (continued)

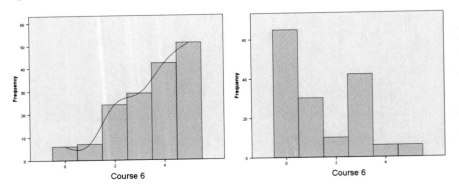

Fig. 2 Equal segments and equal-area segments

Gauss distribution, but in the second and the third question the bar will be divided following the parabola. The new histograms are the following:

As it is obvious, in the first question although in the first subdivision the answers were concentrated in the middle of the bar, the second subdivision prove an upward trend of the answers.

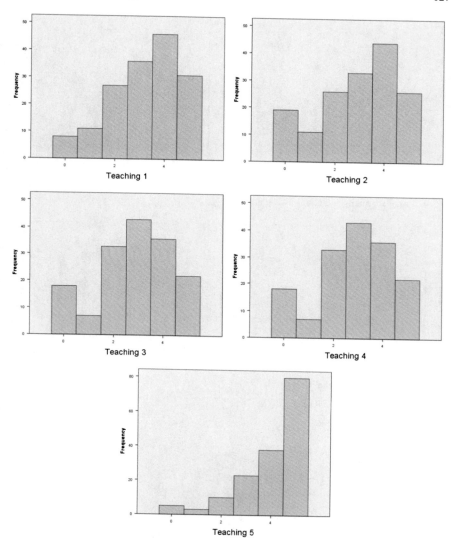

Fig. 3 Equal segmentsa

In question 2 the results are just confirmed, as the two histograms are almost the same.

From question's 3 second subdivision, the new histogram obtained arises an upward trend while in the first histogram the answers were following the increasing low parabola.

Finally, in question 4, although in the first histogram most of the answers seem to be in the center of the bar, in the new subdivision according to Gauss distribution most of the answers are in the first and the last scale.

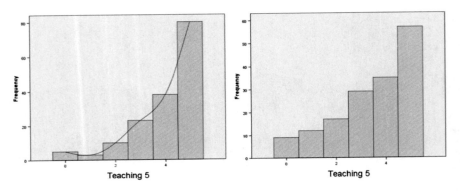

Fig. 4 Equal segments and equal-area segments

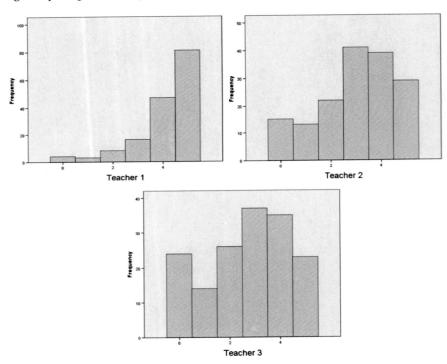

Fig. 5 Equal segments

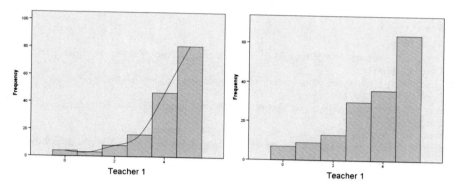

Fig. 6 Equal segments and equal-area segments

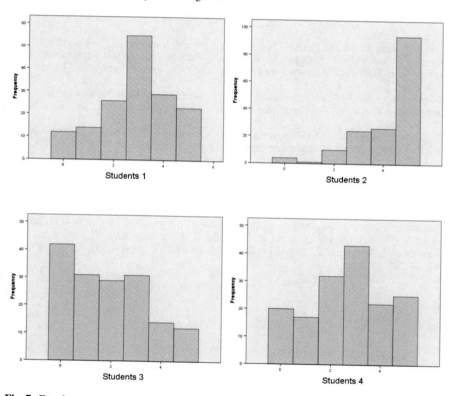

Fig. 7 Equal segments

References

Corsini, P., Leoreanu, V.: Applications of Hyperstructure Theory. Kluwer Academic Publisher, Dordrecht (2003)

Davvaz, B.: A brief survey of the theory of H_v structures. In: 8th AHA Congress, pp. 39–70. Spanidis Press (2003)

Davvaz, B., Leoreanu, V.: Hyperring Theory and Applications. International Academic Press, Cambridge (2007)

Davvaz, B.: Polygroup Theory and Related Systems. World Scientific, Singapore (2013)

Hoskova, S.: H_v structure is fifteen. In: Proceedings of 4th International Mathematical Workshop FAST VUT Brno, pp. 55–57. Czech Republic (2005). http://math.fce.vutbr.cz/pribyl/workshop2005/prispevky/Hoskova.pdf

Kambaki-Vougioukli, P., Karakos, A., Lygeros, N., Vougiouklis, T.: Fuzzy instead of discrete. Ann. Fuzzy Math. Inf. (AFMI) 2(1), 81–89 (2011)

Kambaki-Vougiouklis, P., Nikolaidou, P., Vougiouklis, T.: Questionnaires in Linguistics Using the Bar and the H_v-Structures. Studies in Systems, Decision and Control, vol. 66, pp. 257–266. Springer, Berlin (2017)

Kambaki-Vougioukli, P., Vougiouklis, T.: Bar instead of scale. Ratio Sociol. 3, 49–56 (2008)

Kaplani, T., Vougiouklis, T.: Finite H_v-fields with strong-inverses. Ratio Math. 33, 115–126 (2017)

Markos, A.: A fuzzy coding approach to data processing using the bar. Ratio Math. 33, 127–138 (2017)

Nikolaidou, P.: Multiple ways of processing in questionnaires. Ratio Math. 33, 139–150 (2017)

Nikolaidou, P., Vougiouklis, T.: H_v-structures and the bar in questionnaires. Ital. J. Pure Appl. Math. 29, 341–350 (2012)

Nikolaidou, P., Vougiouklis, T.: Hyperstructures in questionnaires. In: Proceedings of 12th AHA, Algebraic Hyperstructures and its Applications. Xanthi, Greece (2014)

Vougiouklis, T.: The fundamental relation in hyperrings. The general hyperfield. In: 4th AHA Congress, pp. 203–211. World Scientific, Singapore (1991)

Vougiouklis, T.: Hyperstructures and their Representations. Monographs in Mathematics. Hadronic Press, Florida (1994)

Vougiouklis, T.: Bar and theta hyperoperations. Ratio Math. 21, 27–42 (2011)

Vougiouklis, T., Kambaki-Vougioukli, P.: Use Bar China-USA Bus. Rev. 10(6), 484–489 (2011)

Vougiouklis, T., Kambakis-Vougiouklis, P.: Bar Quest. Chin. Bus. Rev. 12(10) (2013)

Vougiouklis, T., Vougiouklis, P.: Questionnaires with the bar in social sciences. Sci. Philos. 3(2), 47–58 (2015)

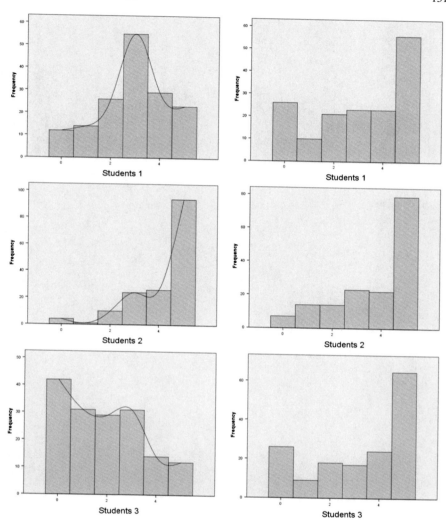

Fig. 8 Equal segments and equal-area segments

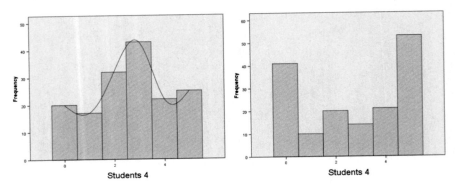

Fig. 8 (continued)

Micropolar Thermoelasticity with Voids Using Fractional Order Strain

Lavinia Codarcea-Munteanu and Marin Marin

Abstract The chapter is dealing with the study of the thermoelasticity of the micropolar materials with voids that uses the fractional order strain, in order to determine some equations of this linear thermoelasticity theory, as well as of a reciprocity relation for the mentioned bodies. Finding the form of the constitutive equations and using them for analyzing the reciprocity, toghether with obtaining the equation of thermal conductivity under the terms of our theory is the main purpose, realizing a parallel between classical theory and this specific case, leading to a better understanding of the behaviour of these materials.

Keywords Micropolar materials with voids · Fractional derivative
Thermoelasticity · Reciprocity

1 Introduction

The theory of materials with voids are present in various areas of our life, applications of porous materials being present in fields such as petroleum industry, materials science, drugs and medical devices industry or biology. Applications of this theory are also found in geology, more precisely in the study of rocks and soil, as well as, in the manufacturing of porous materials, such as, ceramics, granular materials superimposed on solid materials, or mineral wool.

Porous media can not be ignored, because we find them everywhere in our environment, starting from building materials, such as, wood, sand, cement, bricks or concrete, continuing with biological tissues, such as, bones, skin, lung or hair and reaching artificial materials, for example, metallic foam or pressed powders.

L. Codarcea-Munteanu (✉) · M. Marin
Department of Mathematics and Computer Science, Transilvania University of Braşov, Bd.
Eroilor 29, 500036 Braşov, Romania
e-mail: codarcealavinia@unitbv.ro

M. Marin
e-mail: m.marin@unitbv.ro

© Springer Nature Switzerland AG 2019
C. Flaut et al. (eds.), *Models and Theories in Social Systems*, Studies in Systems,
Decision and Control 179, https://doi.org/10.1007/978-3-030-00084-4_7

The great diversity of porous media has allowed the study of applications in multiple directions, from biomedicine, where dental and medical prostheses use these materials, to engineering, where soil natural deposits modeling or solar and geothermal energy are just a few examples of porous media theory application in ours everyday life.

The starting point of this theory lies with the authors M.A. Goodman ans S.C. Cowin, who, in their important scientific work (Goodman and Cowin 1972) have introduced an additional degree of kinematic freedom in order to develop the theory of continuous granular flowing materials.

The theory of elastic materials with voids was developed in the linear case by S.C. Cowin and J.W. Nunziato in Cowin and Nunziato (1983), where the authors demonstrated the weak solution uniqueness, as well as the weak stability of the solutions, theory which is in fact the linear version of the nonlinear theory of the porous solids behaviour studied by them in Nunziato and Cowin (1979).

Another version of the linear theory of elastic materials with voids, called the theory of dilatation in elasticity, was proposed by Markov in (1981), where the author presents the mechanical behaviour of an elastic solid containing a large number of micropores and the material specific constants are calculated including this theory.

A study of the thermoelastic materials, taking into account the interaction between the thermal and mechanical fields, is highlighted by Ieşan in (1980), where theories of deformable solids with particles having more than three degrees of freedom are presented, the author expanding later this theory on the thermoelasticity of materials with voids, in Ieşan (1986).

In the last mentioned paper, the author approaches basic problems of thermoelasticity for three models of continuous mediums: materials with voids, micropolar bodies and non-simple materials. Microcontinuum field theories in Eringen (1999) represent an extension of classical field theories, material bodies being considered as collections of large numbers of deformable particles.

Outstanding results have been achieved in the materials with voids area, with many studies and works devoted to this theme. For example, Marin in (1994, 1998, 2016), an the one hand develops previous results, and on the other hand extends the study on specific materials. In the works (Marin 1997, 2010a, b; Marin et al. 2017), Marin and the coauthors deepen and develop new perspectives on addressing extremely different problems of elasticity and thermoelasticity of these materials, creating in some cases, models that lead to a better understanding of material processes, because, precisely this is a great part of the beauty of mathematics, that is to create patterns of description of some material structures or natural processes that surround us, being connected to the life and environment that define us, going beyond the limits of pure theory.

Generalized from the micropolar theory introduced by Cosserat, the micropolar with voids theory has been the subject of multiple studies and papers, some examples being Iovane and Passarella (2002), Ciarletta and Scalia (2004), Lianngenga and Lalawmpuia (2015), Lianngenga (2016).

Regarding the theory of thermoelasticity of continuous bodies with voids, M. Aouadi has devoted many of his studies to the development of this theory (Aouadi

2010), including the effects of diffusion as a natural consequence of the necessity imposed by the evolution of hight-class technology, where the effects of temperature and diffusion can not be ignored.

In the above mentioned paper, the author has established a reciprocal relation involving two diffusion processes, describing that, at low and high temperatures, the heat and mass transfer processes play an important role in satellite issues, space shuttles re-entry and landing on water or soil, applications being also found in the oil field, where the process of thermodiffusion is of interest for more efficient extraction of oil at a higher concentration.

A history of fractional calculus is presented in Malinowska et al. (2015) from its beginnings, three hundred years ago, triggered by a question from Leibnitz to l'Hôpital, until today, describing the multiple different ways of introducing fractional differentiation, those introduced by Riemann–Liouville, Caputo or Miller–Ross being just a few examples.

Many approaches have been made recently with results including different fractional derivative-dependent problems, a perspective that unifies these modalities is where fractional operators depend on general kernels, several authors developing this subject, see Agrawal (2010).

Even though the fractional derivative theory has been approached for three centuries as a pure theory, the last decades have shown that this theory has been applied successfully in practice, being widely used in different fields such as classical or quantum physics, thermodynamics, chemistry, engineering and control theory, see Băleanu et al. (2012).

Applying the theory of fractional derivatives and integrals, starting with Abel and continuing with the development known in the nineteenth century, opened the way for creating models for describing physical and natural processes, see Sherief et al. (2010).

2 Theory

We consider an open region $\mathscr{S}$ of the three-dimensional Euclidean space, which corresponds in its reference configuration to a micropolar body with voids.

It is supposed to be a regular region, having the boundary $\partial\mathscr{S}$ closed and bounded, and for the closure of which we use the notation $\overline{\mathscr{S}}$. During this paper, we will report the continuum evolution to a fixed system of rectangular Cartesian coordinates and we will use the tensor Cartesian notation. At the some time, we will note the points in $\mathscr{S}$ with x_i ($i = 1, 2, 3$) and the time with $t \in [0, \infty)$. Both arguments, those representing spatial coordinates and time, will be lacking when confusion can not be made.

We will also adopt the well-known convention, according to which, an indice that repeats itself in a monom leads to the summation of Einstein and an indice preceded by a comma represents the partial differentiation corresponding to that Cartesian coordinate. In addition, we will take into account that the point above the function

f (scalar, vectorial or tensor) will be used to refer to the material derivate, the Latin and Greek indices take the value 1, 2, 3, respectively 1, 2, In the following, we will consider the thermoelasticity theory for the micropolar bodies with voids, the equation and the basic conditions that govern this theory being described below.

A micropolar thermoelastic body with voids has the behaviour characterized by the following independent variables, see Marin (2016):

- $u_i(x, t)$, $\varphi_i(x, t)$ are the components of displacement and microrotation vector field from the reference configuration
- θ is the variation of the body temperature from the absolute temperature T_0, that it has in reference configuration i.e.,

$$\theta(x, t) = T(x, t) - T_0 \,,$$

- σ is the variation of the volume fraction from the reference configuration volume fraction ν_0 i.e.,

$$\sigma(x, t) = \nu(x, t) - \nu_0 \,,$$

- ε_{ij}, γ_{ij} as tensors, and Φ_i as vector, are the kinematic characteristics of the strain, having the geometric equations as follows:

$$\begin{aligned}
\varepsilon_{ij} &= u_{j,i} + \varepsilon_{jik}\varphi_k \\
\gamma_{ij} &= \varphi_{j,i} \\
\Phi_i &= \sigma_{,i} \,,
\end{aligned} \tag{1}$$

where ε_{jik} is Ricci's tensor, called also, the alternating symbol.

As a reference is the fact that the next balances of momentum refer to linear and angular momentum respectively.

The equations that govern the thermoelasticity theory of micropolar bodies with voids are derived, following the J.W. Nunziato and S.C. Cowin method in Nunziato and Cowin (1979), as below:

- the equations of motion (the balances of momentum):

$$\begin{aligned}
t_{ji,j} + \rho_0 F_i &= \rho_0 \ddot{u}_i \,, \text{ in } \mathscr{S} \times (0, \infty) \\
m_{ji,j} + \varepsilon_{ijk}t_{jk} + \rho_0 L_i &= I_{ij}\ddot{\varphi}_j \,, \text{ in } \mathscr{S} \times (0, \infty);
\end{aligned} \tag{2}$$

- the balance of equilibrated forces:

$$h_{i,i} + \xi + \rho_0 H = \rho_0 k\ddot{\sigma}, \text{ in } \mathscr{S} \times (0, \infty) \tag{3}$$

$$\rho_0 T\dot{\eta} = q_{i,i} + \rho_0 S \,. \tag{4}$$

The notations that have been used in the previous equations are presented below:

- t_{ji}, m_{ji} are the components of the stress tensor, respectively, of the couple stress tensor,
- h_i are the components of the equilibrated stress vector,
- q_i are the components of the heat flux vector,
- ρ_0 is the mass density in the reference configuration,
- F_i, L_i are the components of the body force vector, respectively, of the couple body force,
- S is the heat supply per unit mass,
- k is the coefficient of equilibrated inertia,
- I_{ij} are the coefficients of microinertia,
- η is the specific entropy,
- ξ is the intrinsic equilibrated body force ,
- H is the extrinsic equilibrated body force.

Together with the system consisting of the Eqs. (2), (3) and (4), the initial conditions are considered as follows:

$$
\begin{aligned}
u_i(x,0) &= u_i^0(x), \ \dot{u}_i(x,0) = u_i^1(x), \ x \in \overline{\mathscr{S}} \\
\varphi_i(x,0) &= \varphi_i^0(x), \ \dot{\varphi}_i(x,0) = \varphi_i^1(x), \ x \in \overline{\mathscr{S}} \\
\sigma(x,0) &= \sigma^0(x), \ \dot{\sigma}(x,0) = \sigma^1(x), \ x \in \overline{\mathscr{S}} \\
\theta(x,0) &= \theta^0(x), \ \eta(x,0) = \eta^0(x), \ x \in \overline{\mathscr{S}}
\end{aligned}
\tag{5}
$$

besides the boundary conditions, which are given in the form:

$$
\begin{aligned}
u_i(x,t) &= \tilde{u}_i \ \text{on} \ \partial\mathscr{S}_1 \times [0,\infty) \\
\varphi_i(x,t) &= \tilde{\varphi}_i \ \text{on} \ \partial\mathscr{S}_2 \times [0,\infty) \\
\sigma(x,t) &= \tilde{\sigma} \ \text{on} \ \partial\mathscr{S}_3 \times [0,\infty) \\
\theta(x,t) &= \tilde{\theta} \ \text{on} \ \partial\mathscr{S}_4 \times [0,\infty) \\
t_i(x,s) &:= t_{ji}(x,s)n_j(x) = \tilde{t}_i \ \text{on} \ \partial\mathscr{S}_1^c \times [0,\infty) \\
m_i(x,s) &:= m_{ji}(x,s)n_j(x) = \tilde{m}_i \ \text{on} \ \partial\mathscr{S}_2^c \times [0,\infty) \\
h(x,s) &:= h_i(x,s)n_i(x) = \tilde{h} \ \text{on} \ \partial\mathscr{S}_3^c \times [0,\infty) \\
q(x,s) &:= q_i(x,s)n_i(x) = \tilde{q} \ \text{on} \ \partial\mathscr{S}_4^c \times [0,\infty)
\end{aligned}
\tag{6}
$$

where $\partial\mathscr{S}_1$, $\partial\mathscr{S}_2$, $\partial\mathscr{S}_3$, $\partial\mathscr{S}_4$ and their respective complements $\partial\mathscr{S}_1^c$, $\partial\mathscr{S}_2^c$, $\partial\mathscr{S}_3^c$ and $\partial\mathscr{S}_4^c$ are subsets of the surface $\partial\mathscr{S}$ such that

$$
\begin{aligned}
\partial\mathscr{S}_1 \cup \partial\mathscr{S}_1^c &= \partial\mathscr{S}_2 \cup \partial\mathscr{S}_2^c = \partial\mathscr{S}_3 \cup \partial\mathscr{S}_3^c = \partial\mathscr{S}_4 \cup \partial\mathscr{S}_4^c = \partial\mathscr{S}, \\
\partial\mathscr{S}_1 \cap \partial\mathscr{S}_1^c &= \partial\mathscr{S}_2 \cap \partial\mathscr{S}_2^c = \partial\mathscr{S}_3 \cap \partial\mathscr{S}_3^c = \partial\mathscr{S}_4 \cap \partial\mathscr{S}_4^c = \varnothing,
\end{aligned}
$$

n_i are the components of the unit outward normal vector to the surface $\partial\mathscr{S}$ and $u_i^0, u_i^1, \varphi_i^0, \varphi_i^1, \sigma^0, \sigma^1, \theta^0, \eta^0, \tilde{u}_i, \tilde{\varphi}_i, \tilde{\sigma}, \tilde{\theta}, \tilde{t}_i, \tilde{m}_i, \tilde{h}, \tilde{q}$ are prescribed continuous functions in their domains.

The energy balance of micropolar thermoelastic materials with voids, following Aouadi (2010), Marin (1998), can be postulated in the form:

$$\int_{\mathscr{S}} (\rho_0 \dot{u}_i \ddot{u}_i + I_{ij}\dot{\varphi}_i\ddot{\varphi}_i + \rho_0 k\dot{\sigma}\ddot{\sigma} + \rho_0 \dot{e})dV =$$
$$\int_{\mathscr{S}} \rho_0 (F_i\dot{u}_i + L_i\dot{\varphi}_i + H\dot{\sigma} + S)dV + \int_{\partial\mathscr{S}} (t_i\dot{u}_i + m_i\dot{\varphi}_i + h\dot{\sigma} + q)dA, \tag{7}$$

where e is the internal energy per unit mass. In the following, will be used the Caputo definition, introduced in Caputo (1967), and used in Podlubny (1998), El-Karamany and Ezzat (2011), Sheoran and Kundu (2016), Youssef (2016), for the fractional derivative of order $\alpha \in (0, 1]$ with respect to time t, given by:

$$D_t^\alpha f(x,t) = \frac{\partial^\alpha f(x,t)}{\partial t^\alpha} = \frac{1}{\Gamma(1-\alpha)} \int_0^t (t-s)^{-\alpha} \frac{\partial f(x,t)}{\partial s} ds \tag{8}$$

where $\Gamma(\cdot)$ is the Gamma function.

The function $f(x,t)$ is supposed to be absolutely continuous with respect to time t, so:

$$\lim_{\alpha \to 1} \frac{\partial^\alpha f(x,t)}{\partial t^\alpha} = \frac{\partial f(x,t)}{\partial t} \tag{9}$$

At the base of the Cattaneo model, see Cattaneo (1948), Hetnarski (1996), Hetnarski and Ignaczak (1999), Postvenko (2009) is the notion of the relaxation time, this model being originally developed by Cattaneo for gaseous media. According to this model, the heat flux is given by:

$$q_i + \tau_0 \dot{q}_i = -K_{ij}T_{,j} , \quad i, j = 1, 2, 3 \tag{10}$$

where τ_0 represents the relaxation time and K_{ij} represents the tensor of thermal conductivity. The Helmholtz free energy is

$$\Phi = e - T\eta , \tag{11}$$

where e is the internal energy and η is the entropy function.

3   Result

In the sequel, using the method presented in Youssef (2016) and taking into consideration that the state of thermoelastic material is adopted as below:

$$\Phi = \Phi(\widetilde{\varepsilon}_{ij}, \gamma_{ij}, T, T_{,i}, \sigma, \sigma_{,i}),$$
$$e = e(\widetilde{\varepsilon}_{ij}, \gamma_{ij}, T, T_{,i}, \sigma, \sigma_{,i})$$
$$\eta = \eta(\widetilde{\varepsilon}_{ij}, \gamma_{ij}, T, T_{,i}, \sigma, \sigma_{,i})$$
$$q = q(\widetilde{\varepsilon}_{ij}, \gamma_{ij}, T, T_{,i}, \sigma, \sigma_{,i})$$

(12)

where

$$\widetilde{\varepsilon}_{ij} = (1 + \tau^\alpha D_t^\alpha)\varepsilon_{ij}$$

(13)

τ being the constant parameter of mechanical relaxation time, the immediate goal is to determine the derivative formula of the free energy with respect to time. For this, following Chirilă (2017), see also Codarcea-Munteanu and Marin (2017), the balance of momentum $(2)_1$, the balance angular momentum $(2)_2$ and the balance of equilibrated forces (3) equations are multiplied by $\dot{u}_i$, $\dot{\varphi}_i$ respectively by $\dot{\sigma}$. Then, the relations obtained previously are integrated over $\mathscr{S}$, and after that, they are introduced into the principle of conservation of energy (7), getting:

$$\int_{\mathscr{S}} (t_{ji,j}\dot{u}_i + m_{ji,j}\dot{\varphi}_i + \varepsilon_{ijk}t_{jk}\dot{\varphi}_i + h_{i,i}\dot{\sigma} + \xi\dot{\sigma})dV + \int_{\mathscr{S}} \rho_0 \dot{e}dV =$$
$$= \int_{\mathscr{S}} \rho_0 SdV + \int_{\partial\mathscr{S}} (t_{ji}\dot{u}_i n_j + m_{ji}\dot{\varphi}_i n_i + h_i\dot{\sigma}n_i + q_i n_i)dA.$$

(14)

Using the theorem of divergence, the above relation (14) becomes:

$$\int_{\mathscr{S}} t_{ij}(\dot{u}_{j,i} + \varepsilon_{jik}\dot{\varphi}_k) + m_{ij}\dot{\varphi}_{j,i} + h_i\dot{\sigma}_{,i} + q_{i,i} + \rho_0 S - \xi\dot{\sigma} - \rho_0\dot{e})dV = 0 \,. \quad (15)$$

Due to the continuity of the integrated functions and to the fact that the integration domain is arbitrary, we have the pontwise form from (15) relation as:

$$\rho_0\dot{e} = t_{ij}(\dot{u}_{j,i} + \varepsilon_{jik}\dot{\varphi}_k) + m_{ij}\dot{\varphi}_{j,i} + h_i\dot{\sigma}_{,i} + q_{i,i} + \rho_0 S - \xi\dot{\sigma} \,. \quad (16)$$

But, from relation (11), we get that

$$\dot{\Phi} = \dot{e} - \dot{T}\eta - T\dot{\eta} \,, \quad (17)$$

then, using the relations $(1)_1$, (17) and taking into account the fractional order strain, we obtain:

$$\rho_0\dot{\Phi} = t_{ij}\dot{\widetilde{\varepsilon}}_{ij} + m_{ij}\dot{\gamma}_{ij} + h_i\dot{\Phi}_i + q_{i,i} + \rho_0 S - \xi\dot{\sigma} - \rho_0\dot{T}\eta - \rho_0 T\dot{\eta} \,, \quad (18)$$

which is exactly the wanted formula for the free energy derivative using respect to time using fractional order strain.

From the relation (12) and the chain rule, we have:

$$\rho_0 \dot{\Phi} = \rho_0 \frac{\partial \Phi}{\partial \widetilde{\varepsilon}_{ij}} \dot{\widetilde{\varepsilon}}_{ij} + \rho_0 \frac{\partial \Phi}{\partial \gamma_{ij}} \dot{\gamma}_{ij} + \rho_0 \frac{\partial \Phi}{\partial T} \dot{T} + \rho_0 \frac{\partial \Phi}{\partial T_{,i}} \dot{T}_{,i} + \rho_0 \frac{\partial \Phi}{\partial \sigma} \dot{\sigma} + \rho_0 \frac{\partial \Phi}{\partial \Phi_i} \dot{\Phi}_i .$$
(19)

Comparing (18) and (19) relations, we deduce:

$$t_{ij} = \rho_0 \frac{\partial \Phi}{\partial \widetilde{\varepsilon}_{ij}} , \quad m_{ij} = \rho_0 \frac{\partial \Phi}{\partial \gamma_{ij}} , \quad h_i = \rho_0 \frac{\partial \Phi}{\partial \Phi_i}$$

$$\eta = -\frac{\partial \Phi}{\partial T} , \quad \xi = -\rho_0 \frac{\partial \Phi}{\partial \sigma} , \quad \frac{\partial \Phi}{\partial T_{,i}} = 0$$
(20)

and

$$q_{i,i} + \rho_0 S = \rho_0 T \dot{\eta}$$
(21)

which is exactly the equation of energy in this case, specified by us.

The form of the free energy, following Marin (2016), considering the situation in which the initial body is stress free, with a null intrinsic equilibrated body force and a null flux rate, according to the linear theory, is:

$$\rho_0 \Phi (\widetilde{\varepsilon}_{ij}, \gamma_{ij}, \theta, \sigma, \Phi_i) = \frac{1}{2} A_{ijmn} \widetilde{\varepsilon}_{ij} \widetilde{\varepsilon}_{mn} + B_{ijmn} \widetilde{\varepsilon}_{ij} \gamma_{mn} + \frac{1}{2} C_{ijmn} \gamma_{ij} \gamma_{mn} +$$

$$+ B_{ij} \widetilde{\varepsilon}_{ij} \sigma + C_{ij} \gamma_{ij} \sigma + D_{ijk} \widetilde{\varepsilon}_{ij} \Phi_k + E_{ijk} \gamma_{ij} \Phi_k + \frac{1}{2} A_{ij} \Phi_i \Phi_j + d_i \Phi_i \sigma + \frac{1}{2} g \sigma^2 -$$

$$- \alpha_{ij} \widetilde{\varepsilon}_{ij} \theta - \beta_{ij} \gamma_{ij} \theta - \gamma_i \Phi_i \theta - m \sigma \theta - \frac{1}{2} a \theta^2 - \frac{1}{2} \omega \dot{\sigma}^2 .$$
(22)

In the previous relation, $f = -\omega \dot{\sigma}$ represents the dissipation that relates to the inelastic behaviour of the voids. The coefficient ω, as well as the coefficient of equilibrated inertia k, must be nonnegative in order to satisfy a dissipation inequality which is deduced from the Second Law of Thermodynamics, see Cowin and Nunziato (1983), Marin (2016).

In terms of coefficients, they are prescribed, signifying the material characteristics functions, which fulfills the following symmetry relations:

$$A_{ijmn} = A_{mnij}, \quad C_{ijmn} = C_{mnij}, \quad A_{ij} = A_{ji}, \quad I_{ij} = I_{ji}$$
(23)

Using the relations (21) and (22), the following constitutive equations are obtained:

$$t_{ij} = A_{ijmn} \widetilde{\varepsilon}_{mn} + B_{ijmn} \gamma_{mn} + B_{ij} \sigma + D_{ijk} \Phi_k - \alpha_{ij} \theta =$$
$$= A_{ijmn} (1 + \tau^\alpha D_t^\alpha) \varepsilon_{mn} + B_{ijmn} \gamma_{mn} + B_{ij} \sigma + D_{ijk} \Phi_k - \alpha_{ij} \theta ,$$

$$m_{ij} = B_{mnij} \widetilde{\varepsilon}_{mn} + C_{ijmn} \gamma_{mn} + C_{ij} \sigma + E_{ijk} \Phi_k - \beta_{ij} \theta =$$
$$= B_{mnij} (1 + \tau^\alpha D_t^\alpha) \varepsilon_{mn} + C_{ijmn} \gamma_{mn} + C_{ij} \sigma + E_{ijk} \Phi_k - \beta_{ij} \theta ,$$

$$h_i = D_{mni}\widetilde{\varepsilon}_{mn} + E_{mni}\gamma_{mn} + A_{ij}\Phi_j + d_i\sigma - \gamma_i\theta =$$
$$= D_{mni}(1 + \tau^\alpha D_t^\alpha)\varepsilon_{mn} + E_{mni}\gamma_{mn} + A_{ij}\Phi_j + d_i\sigma - \gamma_i\theta \,,$$
$$\rho_0\eta = \alpha_{ij}\widetilde{\varepsilon}_{ij} + \beta_{ij}\gamma_{ij} + \gamma_i\Phi_i + m\sigma + a\theta =$$
$$= \alpha_{ij}(1 + \tau^\alpha D_t^\alpha)\varepsilon_{ij} + \beta_{ij}\gamma_{ij} + \gamma_i\Phi_i + m\sigma + a\theta \,, \tag{24}$$
$$\xi = -B_{ij}\widetilde{\varepsilon}_{ij} - C_{ij}\gamma_{ij} - d_i\Phi_i - g\sigma + m\theta =$$
$$= -B_{ij}(1 + \tau^\alpha D_t^\alpha)\varepsilon_{ij} - C_{ij}\gamma_{ij} - d_i\Phi_i - g\sigma + m\theta \,.$$

So, we have deduced the following theorem:

Theorem 1 *The theory of generalized thermoelasticity with fractional order strain for micropolar bodies with voids has the constitutive equations given by (23) relations.*

For the purpose of determining the non-Fourier heat equation under the terms of our theory, we use the relation (20)$_4$ and according to it, the (21) relation can be written in the new developed form, see Youssef (2016):

$$q_{i,i} + \rho_0 S = -\rho_0 T \left(\frac{\partial^2 \Phi}{\partial T \partial \widetilde{\varepsilon}_{ij}}\widetilde{\varepsilon}_{ij} + \frac{\partial^2 \Phi}{\partial T \partial \gamma_{ij}}\dot{\gamma}_{ij} + \frac{\partial^2 \Phi}{\partial T^2}\dot{T} + \frac{\partial^2 \Phi}{\partial T \partial \sigma}\dot{\sigma} + \frac{\partial^2 \Phi}{\partial T \partial \Phi_i}\dot{\Phi}_i \right)$$
$$\tag{25}$$

that is equivalent to:

$$q_{i,i} + \rho_0 S = -T \left[\frac{\partial}{\partial T}\left(\rho_0 \frac{\partial \Phi}{\partial \widetilde{\varepsilon}_{ij}} \right)\widetilde{\varepsilon}_{ij} + \frac{\partial}{\partial T}\left(\rho_0 \frac{\partial \Phi}{\partial \gamma_{ij}} \right)\dot{\gamma}_{ij} - \rho_0 \frac{\partial}{\partial T}\left(-\frac{\partial \Phi}{\partial T} \right)\dot{T} - \right.$$
$$\left. - \frac{\partial}{\partial T}\left(-\rho_0 \frac{\partial \Phi}{\partial \sigma} \right)\dot{\sigma} + \frac{\partial}{\partial T}\left(\rho_0 \frac{\partial \Phi}{\partial \Phi_i} \right)\dot{\Phi}_i \right] .$$
$$\tag{26}$$

Using the (20) relations in the previous relation, it becomes:

$$q_{i,i} = -\rho_0 S - T \left(\frac{\partial t_{ij}}{\partial T}\widetilde{\varepsilon}_{ij} + \frac{\partial m_{ij}}{\partial T}\dot{\gamma}_{ij} - \rho_0 \frac{\partial \eta}{\partial T}\dot{T} - \frac{\partial \xi}{\partial T}\dot{\sigma} + \frac{\partial h_i}{\partial T}\dot{\Phi}_i \right) . \tag{27}$$

Taking into account the constitutive equations (24), the relation (27) leads to:

$$q_{i,i} = -\rho_0 S + \alpha_{ij}T\widetilde{\varepsilon}_{ij} + \beta_{ij}T\dot{\gamma}_{ij} + aT\dot{T} + mT\dot{\sigma} + \gamma_i T\dot{\Phi}_i \tag{28}$$

According to linearity, it is considered $T \approx T_0$, so from (28) relation, we get:

$$q_{i,i} = -\rho_0 S + \alpha_{ij}T_0(1 + \tau^\alpha D_t^\alpha)\dot{\varepsilon}_{ij} + \beta_{ij}T_0\dot{\gamma}_{ij} + aT_0\dot{T} + mT_0\dot{\sigma} + \gamma_i T_0\dot{\Phi}_i \,. \tag{29}$$

The previous relation, toghether with the relation obtained from the Cattaneo heat flux, that is:

$$(-K_{ij}T_{,j})_{,i} = q_{i,i} + \tau_0\dot{q}_{i,i} \,, \qquad i, j = 1, 2, 3 \tag{30}$$

leads us to

$$(-K_{ij}T_{,j})_{,i} = -\rho_0 S + \alpha_{ij}T_0(1 + \tau^\alpha D_t^\alpha)\dot{\varepsilon}_{ij} + \beta_{ij}T_0\dot{\gamma}_{ij} + aT_0\dot{T} + mT_0\dot{\sigma} + \gamma_i T_0\dot{\Phi}_i -$$
$$- \tau_0[\rho_0\dot{S} - \alpha_{ij}T_0(1 + \tau^\alpha D_t^\alpha)\ddot{\varepsilon}_{ij} - \beta_{ij}T_0\ddot{\gamma}_{ij} - aT_0\ddot{T} - mT_0\ddot{\sigma} - \gamma_i T_0\ddot{\Phi}_i] \tag{31}$$

The above relation is equivalent to:

$$(K_{ij}T_{,j})_{,i} = \rho_0\left(1 + \tau_0\frac{\partial}{\partial t}\right)S + \left(\frac{\partial}{\partial t} + \tau_0\frac{\partial^2}{\partial t^2}\right)[-\alpha_{ij}T_0(1 + \tau^\alpha D_t^\alpha)\varepsilon_{ij} - \tag{32}$$
$$-\beta_{ij}T_0\gamma_{ij} - aT_0T - mT_0T - \gamma_i T_0\Phi_i]$$

a relation that represents, according to the proposed aim, precisely, the non-Fourier heat equation under the terms of our theory.

Reciprocity

With the purpose of facilitate writing, taking the model from Ieşan (1980), we rewrite the equations of motion in the form:

$$t_{ji,j} + F_i = \rho_0\ddot{u}_i , \quad \text{in } \mathscr{S} \times (0, \infty)$$
$$m_{ji,j} + \varepsilon_{ijk}t_{jk} + L_i = I_{ij}\ddot{\varphi}_j , \quad \text{in } \mathscr{S} \times (0, \infty) , \tag{33}$$

the balance of equilibrated forces as:

$$h_{i,i} + \xi + H = \rho_0 k\ddot{\sigma} , \tag{34}$$

and, noting $Q = \rho_0 S$, the equation of energy as follows:

$$\rho_0 T_0\dot{\eta} = q_{i,i} + Q . \tag{35}$$

The approach of reciprocity will be made from the perspective of including the initial conditions by the basic equations.

At is it known, the convolution product of the functions u and v, defined on $\mathscr{S} \times [0, \infty)$, that are continuous in relation to time, is given by:

$$(u * v)(x, t) = \int_0^t u(x, t - s)v(x, s)ds, \tag{36}$$

where $(x, t) \in \mathscr{S} \times [0, \infty)$.

It is immediate to demonstrate the theorem below:

Theorem 2 *The functions u_i, φ_j, σ and η verify the Eqs. (33), (34), (35) and the initial conditions (5), if and only if the following relations are valid:*

$$g * t_{ji,j} + \mathscr{F}_i = \rho_0 u_i$$
$$g * (m_{ji,j} + \varepsilon_{ijk} t_{jk} + \mathscr{L}_i) = I_{ij} \varphi_j$$
$$g * (h_{i,i} + \xi) + \mathscr{H} = \rho_0 k \sigma \qquad in \; \mathscr{S} \times [0, \infty) \qquad (37)$$
$$\rho_0 \eta = \frac{1}{T_0} l * q_{i,i} + \mathscr{W},$$

where

$$\begin{aligned} g(t) &= (l * l)(t) \\ l(t) &= 1, \end{aligned} \qquad t \in [0, \infty) \qquad (38)$$

$$\begin{aligned} \mathscr{F}_i &= g * F_i + \rho_0 (t u_i^1 + u_i^0) \\ \mathscr{L}_i &= g * L_i + I_{ij} (t \varphi_j^1 + \varphi_j^0) \\ \mathscr{H} &= g * H + \rho_0 k (t \sigma^1 + \sigma^0) \\ \mathscr{W} &= \frac{1}{T_0} l * Q + \rho_0 \eta^0 \, . \end{aligned} \qquad (39)$$

Considering two external data systems $A^{(\delta)}$ that act successively on the thermoelastic micropolar material with voids, defined by

$$A^{(\delta)} = \Big\{ F_i^{(\delta)}, L_i^{(\delta)}, H^{(\delta)}, Q^{(\delta)}, \tilde{u}_i^{(\delta)}, \tilde{\varphi}_i^{(\delta)}, \tilde{\sigma}^{(\delta)}, \tilde{t}_i^{(\delta)}, \tilde{m}_i^{(\delta)}, \tilde{h}^{(\delta)}, \tilde{\theta}^{(\delta)}, \tilde{q}^{(\delta)},$$
$$u_i^{0(\delta)}, u_i^{1(\delta)}, \varphi_i^{0(\delta)}, \varphi_i^{1(\delta)}, \sigma^{0(\delta)}, \sigma^{1(\delta)}, \theta^{0(\delta)}, \eta^{0(\delta)} \Big\}, \quad \delta = 1, 2, \qquad (40)$$

we will note by

$$a^{(\delta)} = (u_i^{(\delta)}, \varphi_i^{(\delta)}, \sigma^{(\delta)}, \theta^{(\delta)}), \qquad \delta = 1, 2, \qquad (41)$$

a solution of the mixed problem, related to $A^{(\delta)}$. We will also have:

$$\begin{aligned} t_i^{(\delta)} &= t_{ji}^{(\delta)} n_j, \quad m_i^{(\delta)} = m_{ji}^{(\delta)} n_j \\ h^{(\delta)} &= h_i^{(\delta)} n_i, \quad q^{(\delta)} = q_i^{(\delta)} n_i \\ \mathscr{W}^{(\delta)} &= \frac{1}{T_0} l * Q^{(\delta)} + \rho_0 \eta^{0(\delta)} \end{aligned} \qquad (42)$$

Lemma 1 *Assuming that the symmetry relations (23) are fulfilled and considering the functions $\tilde{E}_{\delta\mu}$ defined by*

$$\tilde{E}_{\delta\mu} = t_{ij}^{(\delta)} * \tilde{\varepsilon}_{ij}^{(\mu)} + m_{ij}^{(\delta)} * \gamma_{ij}^{(\mu)} + h_i^{(\delta)} * \Phi_i^{(\mu)} - [\rho_0 \eta^{(\delta)}] * \theta^{(\mu)} - \xi^{(\delta)} * \sigma^{(\mu)}, \qquad (43)$$

the following symmetry relation is deduced:

$$\tilde{E}_{\delta\mu} = \tilde{E}_{\mu\delta} \, for \; \delta = 1, 2, \; \mu = 1, 2 \qquad (44)$$

Proof Using the first symmetry relation $(23)_1$ and the first constitutive equation $(24)_1$, we have:

$$
\begin{aligned}
t_{ij}^{(\delta)} * \widetilde{\varepsilon}_{ij}^{(\mu)} - [B_{ijmn}\gamma_{mn}^{(\delta)} + B_{ij}\sigma^{(\delta)} + D_{ijk}\Phi_k^{(\delta)} - \alpha_{ij}\theta^{(\delta)}] * \widetilde{\varepsilon}_{ij}^{(\mu)} = \\
= t_{ij}^{(\mu)} * \widetilde{\varepsilon}_{ij}^{(\delta)} - [B_{ijmn}\gamma_{mn}^{(\mu)} + B_{ij}\sigma^{(\mu)} + D_{ijk}\Phi_k^{(\mu)} - \alpha_{ij}\theta^{(\mu)}] * \widetilde{\varepsilon}_{ij}^{(\delta)} .
\end{aligned} \tag{45}
$$

From the second symmetry relation $(23)_2$ and the second constitutive equation $(24)_2$, we obtain:

$$
\begin{aligned}
m_{ij}^{(\delta)} * \gamma_{ij}^{(\mu)} - [B_{mnij}\widetilde{\varepsilon}_{mn}^{(\delta)} + C_{ij}\sigma^{(\delta)} + E_{ijk}\Phi_k^{(\delta)} - B_{ij}\theta^{(\delta)}] * \gamma_{ij}^{(\mu)} = \\
= m_{ij}^{(\mu)} * \gamma^{(\delta)} - [B_{mnij}\widetilde{\varepsilon}_{mn}^{(\mu)} + C_{ij}\sigma^{(\mu)} + E_{ijk}\Phi_k^{(\mu)} - B_{ij}\theta^{(\mu)}] * \gamma_{ij}^{(\delta)} .
\end{aligned} \tag{46}
$$

Taking into account the third symmetry relation $(23)_3$ and third constitutive equation $(24)_3$, we obtain:

$$
\begin{aligned}
h_i^{(\delta)} * \Phi_i^{(\mu)} - [D_{mni}\widetilde{\varepsilon}_{mn}^{(\delta)} + E_{mni}\gamma_{mn}^{(\delta)} + d_i\sigma^{(\delta)} - \gamma_i\theta^{(\delta)}] * \Phi_i^{(\mu)} = \\
= h_i^{(\mu)} * \Phi_i^{(\delta)} - [D_{mni}\widetilde{\varepsilon}_{mn}^{(\mu)} + E_{mni}\gamma_{mn}^{(\mu)} + d_i\sigma^{(\mu)} - \gamma_i\theta^{(\mu)}] * \Phi_i^{(\delta)} .
\end{aligned} \tag{47}
$$

The fourth constitutive equation $(24)_4$ leads us to the next relation:

$$
\begin{aligned}
[\rho_0\eta^{(\delta)}] * \theta^{(\mu)} - [\alpha_{ij}\widetilde{\varepsilon}_{ij}^{(\delta)} + \beta_{ij}\gamma_{ij}^{(\delta)} + \gamma_i\Phi_i^{(\delta)} + m\sigma^{(\delta)}] * \theta^{(\mu)} = \\
= [\rho_0\eta^{(\mu)}] * \theta^{(\delta)} - [\alpha_{ij}\widetilde{\varepsilon}_{ij}^{(\mu)} + \beta_{ij}\gamma_{ij}^{(\mu)} + \gamma_i\Phi_i^{(\mu)} + m\sigma^{(\mu)}] * \theta^{(\delta)} ,
\end{aligned} \tag{48}
$$

and from the last constitutive equation $(24)_5$, we get the relation below:

$$
\begin{aligned}
\xi^{(\delta)} * \sigma^{(\mu)} + [B_{ij}\widetilde{\varepsilon}_{ij}^{(\delta)} + C_{ij}\gamma_{ij}^{(\delta)} + d_i\Phi_i^{(\delta)} - m\theta^{(\delta)}] * \sigma^{(\mu)} = \\
= \xi^{(\mu)} * \sigma^{(\delta)} + [B_{ij}\widetilde{\varepsilon}_{ij}^{(\mu)} + C_{ij}\gamma_{ij}^{(\mu)} + d_i\Phi_i^{(\mu)} - m\theta^{(\mu)}] * \sigma^{(\delta)} .
\end{aligned} \tag{49}
$$

Summing member by member the relations (45), (46), (47) and substrating the relations (48), (49), we will obtain precisely the desired symmetry relation

$$
\widetilde{E}_{\delta\mu} = \widetilde{E}_{\mu\delta} , \qquad \delta = 1, 2, \ \mu = 1, 2.
$$

∎

We keep in mind the previous proven property, and, at the same time, we make an incursion into the classical thermoelasticity of the micropolar materials with voids, without using fractional order strain.

To demonstrate the Betti-type reciprocal relation in this case, we also rely on a symmetry relation given by:

Lemma 2 *The functions* $E_{\delta\mu}$, *defined by*

$$E_{\delta\mu} = t_{ij}^{(\delta)} * \varepsilon_{ij}^{(\mu)} + m_{ij}^{(\delta)} * \gamma_{ij}^{(\mu)} + h_i^{(\delta)} * \Phi_i^{(\mu)} - [\rho_0 \eta^{(\delta)}] * \theta^{(\mu)} - \xi^{(\delta)} * \sigma^{(\mu)} , \quad (50)$$

under the conditions in which the symmetry relation (23) are valid, verify the symmetry relations:

$$E_{\delta\mu} = E_{\mu\delta} , \quad \delta = 1, 2 , \quad \mu = 1, 2. \tag{51}$$

As a remark, its demonstration is obtained using the classical corresponding constitutive equations. Comparing the two symmetry relations (44) and (51), verified by the functions $\widetilde{E}_{\delta\mu}$ and $E_{\delta\mu}$, it is noticed that, in fact, the fractional derivative does not influence the classical theory of micropolar bodies with voids thermoelasticity, observing that the classical symmetry relation (51) is verified by $\widetilde{\varepsilon}_{ij}$.

Starting from this conclusion, we can express under another form $E_{\delta\mu}$, with the help of geometric equations and (37)$_4$ relation, as:

$$E_{\delta\mu} = \left[t_{ij}^{(\delta)} * u_j^{(\mu)} + m_{ij}^{(\delta)} * \varphi_j^{(\mu)} + h_i^{(\delta)} * \sigma^{(\mu)} - \left(\frac{1}{T_0} l * q_i^{(\delta)} \right) * \theta^{(\mu)} \right]_{,i} - t_{ij,i}^{(\delta)} * u_j^{(\mu)} -$$
$$- \left[m_{ij,i}^{(\delta)} + \varepsilon_{jik} t_{ik}^{(\delta)} \right] * \varphi_j^{(\mu)} - \left[h_{i,i}^{(\delta)} + \xi^{(\delta)} \right] * \sigma^{(\mu)} + \left(\frac{1}{T_0} l * q_i^{(\delta)} \right) * \theta_{,i}^{(\mu)} - \mathscr{W}^{(\delta)} * \theta^{(\mu)}$$

$$(52)$$

Taking into account the above relation, as well as the fact that in linear theory $q_i = K_{ij}\theta_{,j}$, and considering also the properties of the convolution product, the divergence theorem and the relations (37), (42), we deduce:

$$\int_{\mathscr{S}} (g * E_{\delta\mu}) dV = \int_{\partial\mathscr{S}} g * \left[t_i^{(\delta)} * u_i^{(\mu)} + m_i^{(\delta)} * \varphi_i^{(\mu)} + h^{(\delta)} * \sigma^{(\mu)} - \frac{1}{T_0} l * q^{(\delta)} * \right.$$
$$\left. * \theta^{(\mu)} \right] dA + \int_{\mathscr{S}} \left[\mathscr{F}_i^{(\delta)} * u_i^{(\mu)} + \mathscr{L}_i^{(\delta)} * \varphi_i^{(\mu)} + \mathscr{H}^{(\delta)} * \sigma^{(\mu)} - g * \mathscr{W}^{(\delta)} * \theta^{(\mu)} - \right.$$
$$\left. - \rho_0 u_i^{(\delta)} * u_i^{(\mu)} - I_{ij} \varphi_j^{(\delta)} * \varphi_i^{(\mu)} - \rho_0 k \sigma^{(\delta)} * \sigma^{(\mu)} + \frac{1}{T_0} g * l * \left(K_{ij} \theta_{,j}^{(\delta)} * \theta_{,i}^{(\mu)} \right) \right] dV,$$

$$(53)$$

where

$$\mathscr{F}_i^{(\delta)} = g * F_i^{(\delta)} + \rho_0 \left[t u_i^{1(\delta)} + u_i^{0(\delta)} \right]$$
$$\mathscr{L}_i^{(\delta)} = g * L_i^{(\delta)} + I_{ij} \left[t \varphi_j^{1(\delta)} + \varphi_j^{0(\delta)} \right]$$
$$\mathscr{H}^{(\delta)} = g * H^{(\delta)} + \rho_0 k \left[t^{1(\delta)} + \sigma^{0(\delta)} \right] \tag{54}$$
$$\mathscr{W}^{(\delta)} = \frac{1}{T_0} l * Q^{(\delta)} + \rho_0 \eta^{0(\delta)}$$

All of this, naturally leads us to the conclusion that the reciprocity theorem in thermoelasticity micropolar materials with voids using fractional order strain, coincides in fact with the reciprocity theorem of the classical theory, which means that we have the following theorem valid in the theory presented during this work.

Theorem 3 (Reciprocal Theorem) *Suppose that the symmetry relations (23) are holding, $a^{(\delta)}$ is the corresponding solution of the mixed problem to the external system $A^{(\delta)}$, $\delta = 1, 2$, and the coefficients of microinertia I_{ij} and thermal conductivity K_{ij} tensors are symmetrical, we have the following relation of reciprocity:*

$$
\int_{\mathscr{S}} \left[\mathscr{F}_i^{(1)} * u_i^{(2)} + \mathscr{L}_i^{(1)} * \varphi_i^{(2)} + \mathscr{H}^{(1)} * \sigma^{(2)} - g * \mathscr{W}^{(1)} * \theta^{(2)} \right] dV +
$$
$$
+ \int_{\partial \mathscr{S}} g * \left[t_i^{(1)} * u_i^{(2)} + m_i^{(1)} * \varphi_i^{(2)} + h^{(1)} * \sigma^{(2)} - \frac{1}{T_0} l * g^{(1)} * \theta^{(2)} \right] dA =
$$
$$
= \int_{\mathscr{S}} \left[\mathscr{F}_i^{(2)} * u_i^{(1)} + \mathscr{L}_i^{(2)} * \varphi_i^{(1)} + \mathscr{H}^{(2)} * \sigma^{(1)} - g * \mathscr{W}^{(2)} * \theta^{(1)} \right] dV +
$$
$$
+ \int_{\partial \mathscr{S}} g * \left[t_i^{(2)} * u_i^{(1)} + m_i^{(2)} * \varphi_i^{(1)} + h^{(2)} * \sigma^{(1)} - \frac{1}{T_0} l * q^{(2)} * \theta^{(1)} \right] dA
$$

(55)

4 Discussion and Conclusion

Developing, throughout this chapter, different theorems of thermoelasticity of micropolar materials with voids, including fractional order strain, we find that there are segments in which classical theory overlaps with the theory we have studied in present paper, for example, the reciprocity relation being valid in both situations.

It is a starting point for developing other studies on the thermoelasticity of micropolar bodies with voids, which leads to a better connection between pure theory and its application in the world around us.

References

Agrawal, O.P.: Generalized variational problems and Euler - Lagrange equations. Comput. Math. Appl. **59**(5), 1852–1864 (2010). https://doi.org/10.1016/j.camwa.2009.08.029

Aouadi, M.: A theory of thermoelastic materials with voids. Z. Angew. Math. Phys. **61**, 357–379 (2010). https://doi.org/10.1007/s00033-009-0016-0

Băleanu, D., Diethelm, K., Scalas, E., Trujillo, J.J.: Fractional Calculus: Models and Numerical Methods. Series on Complexity, Nonlinearity and Chaos, vol. 5, 2nd edn. World Scientific, Singapore (2012). https://doi.org/10.1142/10044

Caputo, M.: Linear models of dissipation whose Q is almost frequency independent-II. Geophys. J. R. Astron. Soc. **13**(5), 529–539 (1967). https://doi.org/10.1111/j.1365-246X.1967.tb02303.x

Cattaneo, C.: Sulla conduzione del calore. Atti. Sem. Mat. Fis. Univ. Modena **3**, 83–101 (1948)

Chirilă, A.: Generalized micropolar thermoelasticity with fractional order strain. Bull. Transilv. Univ. Braşov, Ser. III: Math. Inf. Phys. **10**(59)(1), 83–90 (2017)

Ciarletta, M., Scalia, A.: Some results in linear theory of thermomicrostretch elastic solids. Meccanica **39**, 191–206 (2004)

Codarcea-Munteanu, L., Marin, M: Thermoelasticity with fractional order strain for dipolar materials with voids. Bull. Transilv. Univ. Braşov, Ser. III: Math. Inf. Phys. (2017) (accepted)

Cowin, S.C., Nunziato, J.W.: Linear elastic materials with voids. J. Elast. **13**, 125–147 (1983)

El-Karamany, A.S., Ezzat, M.A.: On fractional thermoelasticity. Math. Mech. Solids **16**(3), 334–346 (2011). https://doi.org/10.1177/1081286510397228

Eringen, A.C.: Microcontinuum Field Theories I. Foundations and Solids. Springer, New York (1999)

Goodman, M.A., Cowin, S.C.: A continuum theory for granular materials. Arch. Ration. Mech. Anal. **44**, 249–266 (1972)

Hetnarski, R.B.: Thermal Stresses IV. Elsevier, Amsterdam (1996)

Hetnarski, R.B., Ignaczak, J.: Generalized thermoelasticity. J. Therm. Stress. **22**, 451–476 (1999)

Ieşan, D.: Generalized mechanics of solids. Univ. Al. I. Cuza, Centrul de multiplicare, Iaşi (1980)

Ieşan, D.: A theory of thermoelastic materials with voids. Acta Mech. **60**(1–2), 67–89 (1986)

Iovane, G., Passarella, F.: Some theorems in thermoelasticity for micropolar porous media. Rev. Roum. Sci. Tech. Mech. Appl. **46**(1–6), 9–18 (2002)

Lianngenga, R.: Theory of micropolar thermoelastic materials with voids. IJPAMS **9**(1), 1–8 (2016)

Lianngenga, R., Lalawmpuia: Micropolar elasticity containing voids. IJISET **2**(12), 838–844 (2015)

Malinowska, A.B., Odzijewicz, T., Torres, D.F.M.: Advanced Methods in the Fractional Calculus of Variations. Springer Briefs in Applied Sciences and Technology. Springer, Cham (2015). https://doi.org/10.1007/978-3-319-14756-7

Marin, M.: The lagrange identity method in thermoelasticity of bodies with microstructure. Int. J. Eng. Sci. **32**(8), 1229–1240 (1994)

Marin, M.: On weak solutions in elasticity of dipolar bodies with voids. J. Comput. Appl. Math. **82**(1–2), 291–297 (1997). https://doi.org/10.1016/s0377-0427(97)00047-2

Marin, M.: A temporally evolutionary equation in elasticity of micropolar bodies with voids. UPB Sci. Bull. Ser. A Appl. Math. Phys. **60**(3–4), 67–78 (1998)

Marin, M.: Harmonic vibrations in thermoelasticity of microstretch materials. J. Vib. Acoust. **132**(4), 044501,6 (2010a)

Marin, M.: Some estimates on vibrations in thermoelasticity of dipolar bodies. J. Vib. Control **16**(1), 33–47 (2010b)

Marin, M.: An approach of a heat-flux dependent theory for micropolar porous media. Meccanica **51**, 1127–1133 (2016). https://doi.org/10.1007/s11012-015-0265-2

Marin, M., Codarcea, L., Chirilă, A.: Qualitative results on mixed problem of micropolar bodies with microtemperatures. Appl. Appl. Math. (2017) (accepted)

Markov, K.Z.: On the dilatation theory of elasticity. ZAMM Z. Angew. Math. Mech. **61**(8), 349–358 (1981)

Nunziato, J.W., Cowin, S.C.: A nonlinear theory of elastic materials with voids. Arch. Ration. Mech. Anal. **72**, 175–201 (1979)

Podlubny, I.: Fractional Differential Equation: An Introduction to Fractional Derivatives, Fractional Differential Equations, to Methods of Their Solution and some of Their Applications. Academic Press, New York (1998)

Postvenko, Y.Z.: Thermoelasticity that uses fractional heat conduction equation. J. Math. Sci. **162**(2), 296–305 (2009)

Sheoran, S.S., Kundu, P.: Fractional order generalized thermoelasticity theories: a review. Int. J. Adv. Appl. Math. Mech. **3**(4), 76–81 (2016)

Sherief, H.H., El-Sayed, A.M.A., Abd El-Latief, A.M.: Fractional order theory of thermoelasticity. Int. J. Solids Struct. **47**, 269–275 (2010)

Youssef, H.M.: Theory of generalized thermoelasticity with fractional order strain. J. Vib. Control **22**(18), 3840–3857 (2016)

Diffusion in Microstretch Thermoelasticity with Microtemperatures and Microconcentrations

Adina Chirilă and Marin Marin

Abstract This chapter is dealing with the linear theory of microstretch thermoelasticity for materials whose particles have microelements that are equipped with microtemperatures and microconcentrations. The focus is on isotropic and homogeneous bodies, for which we derive the field equations and the constitutive equations. Then we introduce some dimensionless quantities and establish the continuous dependence of solutions upon initial data and body loads by means of the Gronwall inequality. This extension of mechanics of generalized continua that includes both thermal and diffusion effects aims at providing a rigorous mathematical model with various possible applications in materials science, engineering and even biology.

Keywords Microstretch thermoelasticity · Microtemperatures · Microconcentrations · Mechanics of generalized continua

1 Introduction

According to Aouadi et al. (2017), it is important both from a theoretical and a practical point of view to study models of continuum mechanics describing interaction of several physical fields due to a large number of applications in materials science, chemical industry, aviation and biology. Mechanics of generalized continua was introduced since classical continuum mechanics fails to accurately describe the behaviour of materials with microstructure. This subject is indeed of interest nowadays because there are interactions between mechanical processes at different spatial scales in virtually all physical natural or man-made materials and systems.

A. Chirilă · M. Marin (✉)
Department of Mathematics and Computer Science, Transilvania University of Braşov, Bd. Eroilor 29, 500036 Braşov, Romania
e-mail: m.marin@unitbv.ro

A. Chirilă
e-mail: adina.chirila@unitbv.ro

© Springer Nature Switzerland AG 2019
C. Flaut et al. (eds.), *Models and Theories in Social Systems*, Studies in Systems, Decision and Control 179, https://doi.org/10.1007/978-3-030-00084-4_8

149

For example, the concept of generalized continua seems to be very promising in finding appropriate modelling approaches for the rapidly advancing nanotechnologies. The importance of this subject is also emphasized by the need to introduce multiscale numerical techniques in order to couple different spatial scales in one numerical scheme. In fact, it seems to contribute to the better understanding of the world around us by providing improved mathematical models compared to classical continuum mechanics, thus influencing our lives.

A prominent example of generalized continua is the Cosserat model and it emerged from the seminal work (Cosserat et al. 1909), which made a historic contribution to materials science. According to Maugin et al. (2010), the revolutionary contribution of this book is that material points of an elastic solid are considered equipped with directors, which give rise to the concept of couple stress and a new conservation law for the moment of momentum. But the Cosserats did not give constitutive equations. According to Altenbach et al. (2011), these contributions underlined the idea that in a continuum translations and rotations should be defined independently. According to Maugin et al. (2010), it is a challenge to experimentally verify the theoretical methods of Cosserat continuum modelling since it is difficult to produce materials with noticeable rotational effects and well controlled microstructure in order to independently determine the Cosserat parameters. Further models of generalized continua were introduced in Mindlin (1963), Green (1965) and Green et al. (1965).

Other models of mechanics of generalized continua are studied in Ciarletta et al. (1993). Most recently, dipolar and micropolar or Cosserat thermoelasticity are studied, for instance, in Marin et al. (2017) and Chirilă et al. (2017). Generalizations of thermoelasticity with fractional order strain may be found in Chirilă (2017) and Chirilă et al. (2018). Other generalizations are approached in Marin and Baleanu (2016) and Marin (2016).

This chapter discusses microstretch thermoelasticity, which is a generalization of the Cosserat model and was introduced in Eringen (1990). The concept of microtemperatures was introduced in Wozniak (1967). In Grot (1969), the points of a generic microelement were assumed to have different temperatures. Microstretch thermoelasticity with microtemperatures was studied in Ieşan (2007), Aouadi (2008), Scalia et al. (2011), Bitsadze (2016). Following Aouadi et al. (2017), this chapter considers both the diffusion effect and the thermal effect in the microelements of a microstretch elastic solid.

According to Aouadi et al. (2017), the processes of heat and mass diffusion play an important role in many engineering applications, such as satellite problems, returning space vehicles and aircraft landing on water or land. In particular, the process of diffusion raised the interest of oil companies, which aim at improving the conditions of oil extractions, and of the manufacturers of integrated circuits, integrated resistors or semiconductor substrates.

2 Theory

We consider a body that at some moment in time occupies a bounded region Ω of the three-dimensional Euclidean space with the piecewise smooth boundary $\partial \Omega$. We refer the motion of the body to a fixed system of rectangular cartesian axes $Ox_i, i = 1, 2, 3$.

The usual summation and differentiation conventions are used. Latin subscripts range over the integers 1, 2, 3 and the Einstein summation convention is employed over repeated indices. The material time derivative is denoted by a superposed dot and we use a comma followed by a subscript in order to represent partial differentiation with respect to the corresponding spatial coordinate.

Following Ieşan (2007), we present the linear theory of microstretch thermoelasticity. We consider u_i to be a displacement vector field over Ω. Therefore, we may write the balance of the linear momentum in the form

$$t_{ji,j} + \rho f_i = \rho \ddot{u}_i \tag{1}$$

where t_{ij} is the stress tensor, ρ is the reference mass density and f_i is the body force. The balance of the first stress moments reduces to

$$h_{k,k} + g + \rho l = J \ddot{\varphi} \tag{2}$$

and

$$m_{ji,j} + \varepsilon_{irs} t_{rs} + \rho g_i = I_{ij} \ddot{\varphi}_j \tag{3}$$

Here φ is the microdilatation function, φ_i is the microrotation vector, h_j is the microstretch vector, m_{ij} is the couple stress tensor, g is the internal body force, l is the external microstretch body load and g_i is the body couple density. Moreover, ε_{ijk} is the alternating symbol and δ_{ij} will be Kronecker's delta. According to Aouadi et al. (2017), we have

$$\dot{C} = \eta_{i,i} \tag{4}$$

where η_i is the flux vector of mass diffusion and C is the concentration. The strain tensors are

$$e_{ij} = u_{j,i} + \varepsilon_{jik} \varphi_k \qquad \kappa_{ij} = \varphi_{j,i} \qquad \zeta_i = \varphi_{,i} \tag{5}$$

Let e be the internal energy density per unit mass and let ε_i denote the first moment of energy vector. The balance of energy and the balance of the first moment of energy can be expressed as

$$\rho \dot{e} = t_{ij} \dot{e}_{ij} + m_{ij} \dot{\kappa}_{ij} + h_j \dot{\zeta}_j - g \dot{\varphi} + q_{j,j} + \rho s \tag{6}$$

and

$$\rho \dot{\varepsilon}_i = q_{ji,j} + q_i - Q_i + \rho G_i \tag{7}$$

Here q_i is the heat flux vector, s is the heat supply per unit mass, q_{ij} is the first heat flux moment tensor, Q_i is the microheat flux average and G_i is the first heat supply moment vector. According to Aouadi et al. (2017), we introduce the first moment of mass diffusion by

$$\rho \dot{\omega}_i = \eta_{ji,j} + \eta_i - \tilde{\sigma}_i \tag{8}$$

where η_{ij} is the first mass diffusion flux moment tensor and $\tilde{\sigma}_i$ is the micromass diffusion flux average.

As in Aouadi et al. (2017), we have

$$\rho \dot{S} - \left(\frac{q_k}{T} + \frac{1}{T}T_m q_{km}\right)_{,k} + \left(\frac{P\eta_k}{T} + \frac{P}{T}T_m \eta_{km}\right)_{,k} - \rho\left(\frac{s}{T} + \frac{1}{T}T_k G_k\right) \geq 0 \tag{9}$$

and

$$\dot{C} = (\eta_k + C_m \eta_{km})_{,k} \tag{10}$$

Here T is the absolute temperature, S is the microentropy and P is the particle chemical potential. We call the functions T_i and C_i microtemperatures and microconcentrations, respectively. We multiply (9) by T, then we replace ρs from (6) and ρG_i from (7) in the resulting inequality and add up with $\eta_{ji,j}C_i + (\eta_i - \tilde{\sigma}_i)C_i - \rho\dot{\omega}_i C_i = 0$, which results from (8). Next we introduce the function Ψ by

$$\Psi = e + T_i \varepsilon_i + \omega_i C_i - TS \tag{11}$$

and substitute it in the resulting relation, along with $\eta_{k,k} = \dot{C} - C_{m,k}\eta_{km} - C_m \eta_{km,k}$, which results from (10). We consider that $\theta = T - T_0$, where T_0 is the absolute temperature in the reference configuration. In the context of the linear theory, inequality (9) becomes

$$\rho\left(-\dot{\Psi} + \dot{T}_i \varepsilon_i + \omega_i \dot{C}_i - \dot{\theta}S\right) + \eta_{ji,j}C_i + (\eta_i - \tilde{\sigma}_i)C_i + q_k \frac{\theta_{,k}}{T_0} - T_{m,k}q_{km} +$$
$$+ P_{,k}\eta_k + P\dot{C} + t_{ij}\dot{e}_{ij} + m_{ij}\dot{\kappa}_{ij} + h_j\dot{\zeta}_j - g\dot{\varphi} + T_i(q_i - Q_i) \geq 0 \tag{12}$$

As in Aouadi et al. (2017), the constitutive equations are

$$\Psi = \hat{\Psi}(A) \quad t_{ij} = \hat{t}_{ij}(A) \quad m_{ij} = \hat{m}_{ij}(A) \quad h_i = \hat{h}_i(A) \quad g = \hat{g}(A)$$
$$\varepsilon_i = \hat{\varepsilon}_i(A) \quad S = \hat{S}(A) \quad q_i = \hat{q}_i(A) \quad q_{ij} = \hat{q}_{ij}(A) \quad Q_i = \hat{Q}_i(A) \tag{13}$$
$$\omega_i = \hat{\omega}_i(A) \quad P = \hat{P}(A) \quad \eta_i = \hat{\eta}_i(A) \quad \eta_{ij} = \hat{\eta}_{ij}(A) \quad \tilde{\sigma}_i = \hat{\sigma}_i(A)$$

where

$$A = \left(e_{ij}, \kappa_{ij}, \zeta_i, \varphi, \theta, \theta_{,i}, T_i, T_{i,j}, C, C_{,i}, C_i, C_{i,j}\right) \tag{14}$$

We introduce the notation $\rho\Psi = \tilde{\sigma}$. It follows from (12) and (13) that

$$\tilde{\sigma} = \hat{\sigma}(e_{ij}, \kappa_{ij}, \zeta_i, \varphi, \theta, T_i, C, C_i) \tag{15}$$

and

$$t_{ij} = \frac{\partial\tilde{\sigma}}{\partial e_{ij}} \quad m_{ij} = \frac{\partial\tilde{\sigma}}{\partial\kappa_{ij}} \quad h_i = \frac{\partial\tilde{\sigma}}{\partial\zeta_i} \quad g = -\frac{\partial\tilde{\sigma}}{\partial\varphi}$$

$$\rho S = -\frac{\partial\tilde{\sigma}}{\partial\theta} \quad \rho\varepsilon_i = \frac{\partial\tilde{\sigma}}{\partial T_i} \quad P = \frac{\partial\tilde{\sigma}}{\partial C} \quad \rho\omega_i = \frac{\partial\tilde{\sigma}}{\partial C_i} \tag{16}$$

and

$$\eta_{ji,j}C_i + (\eta_i - \tilde{\sigma}_i)C_i + q_k\frac{\theta_{,k}}{T_0} - T_{m,k}q_{km} + P_{,k}\eta_k + T_i(q_i - Q_i) \geq 0 \tag{17}$$

We combine the approaches from Aouadi et al. (2017) and Ieşan (2007), so we consider $\tilde{\sigma}$ in the linear theory for the anisotropic case as

$$\begin{aligned}
2\tilde{\sigma} &= A_{ijrs}e_{ij}e_{rs} + 2B_{ijrs}e_{ij}\kappa_{rs} + C_{ijrs}\kappa_{ij}\kappa_{rs} + 2D_{ij}e_{ij}\varphi + \\
&+ 2E_{ij}\kappa_{ij}\varphi + 2F_{ijk}e_{ij}\zeta_k - 2a_{ij}e_{ij}\theta - 2L_{ijk}e_{ij}T_k + \\
&+ 2G_{ijk}\kappa_{ij}\zeta_k + A_{ij}\zeta_i\zeta_j - 2b_{ij}\kappa_{ij}\theta + 2M_{ijk}\kappa_{ij}T_k + 2B_i\zeta_i\varphi - \\
&- 2d_i\zeta_i\theta - 2N_{ij}\zeta_iT_j + \xi\varphi^2 - 2F\varphi\theta + 2R_i\varphi T_i - a\theta^2 - \\
&- 2b_i\theta T_i - B_{ij}T_iT_j - 2\varpi\theta C + \varrho C^2 - 2R_{ij}T_iC_j - C_{ij}C_iC_j + \\
&+ 2d_{ij}e_{ij}C + 2f_{ij}\kappa_{ij}C + 2\tilde{f}_i\zeta_iC + 2\tilde{g}_1\varphi C
\end{aligned} \tag{18}$$

where the constitutive coefficients satisfy the following symmetries

$$\begin{aligned}
A_{ijrs} &= A_{rsij} \quad C_{ijrs} = C_{rsij} \quad A_{ij} = A_{ji} \quad B_{ij} = B_{ji} \\
C_{ij} &= C_{ji} \quad R_{ij} = R_{ji}
\end{aligned} \tag{19}$$

The constants ϖ and ϱ are the measures of the thermodiffusion and diffusive effects, respectively. B_{ij} and C_{ij} are tensors of microthermal and microdiffusion conductivity, respectively. The tensor R_{ij} is a measure of microthermodiffusion.

3 Result

In the sequel, we follow the strategy from Aouadi et al. (2017) and introduce diffusion and microconcentrations in the mathematical model proposed in Ieşan (2007) for microstretch thermoelasticity with microtemperatures. For the linear theory of diffusion in microstretch thermoelasticity with microtemperatures and microconcentrations we derive the constitutive equations below by (16) and (18) in the anisotropic case

$$t_{ij} = A_{ijrs}e_{rs} + B_{ijrs}\kappa_{rs} + D_{ij}\varphi + F_{ijk}\zeta_k + L_{ijk}T_k - a_{ij}\theta + d_{ij}C$$

$$m_{ij} = B_{rsij}e_{rs} + C_{ijrs}\kappa_{rs} + E_{ij}\varphi + G_{ijk}\zeta_k + M_{ijk}T_k - b_{ij}\theta + f_{ij}C$$

$$h_i = F_{rsi}e_{rs} + G_{rsi}\kappa_{rs} + A_{ij}\zeta_j + B_i\varphi - N_{ij}T_j - d_i\theta + \tilde{f}_iC$$

$$g = -D_{ij}e_{ij} - E_{ij}\kappa_{ij} - B_i\zeta_i - \xi\varphi - R_iT_i + F\theta - \tilde{g}_1C \qquad (20)$$

$$\rho S = a_{ij}e_{ij} + b_{ij}\kappa_{ij} + d_i\zeta_i + F\varphi + b_iT_i + a\theta + \varpi C$$

$$\rho\varepsilon_i = L_{rsi}e_{rs} + M_{rsi}\kappa_{rs} - N_{ji}\zeta_j + R_i\varphi - B_{ij}T_j - b_i\theta - R_{ij}C_j$$

$$P = d_{ij}e_{ij} + f_{ij}\kappa_{ij} + \tilde{f}_i\zeta_i + \tilde{g}_1\varphi - \varpi\theta + \varrho C$$

$$\rho\omega_i = -C_{ij}C_j - R_{ji}T_j$$

The same as in Aouadi et al. (2017), we have from (17)

$$q_i = k_{ij}\theta_{,j} + K_{ij}T_j \qquad \eta_i = h_{ij}P_{,j} + H_{ij}C_j$$

$$q_{ij} = -P_{ijkl}T_{l,k} \qquad \eta_{ij} = -F_{ijkl}C_{l,k}$$

$$Q_i = (k_{ij} - \tilde{k}_{ij})\theta_{,j} + (K_{ij} - \tilde{K}_{ij})T_j \qquad \tilde{\sigma}_i = (h_{ij} - \tilde{h}_{ij})P_{,j} + (H_{ij} - \tilde{H}_{ij})C_j \qquad (21)$$

where the following inequality

$$\frac{1}{T_0}k_{ij}\theta_{,i}\theta_{,j} + \left(\frac{1}{T_0}K_{ij} + \tilde{k}_{ij}\right)\theta_{,i}T_j + P_{ijkl}T_{l,k}T_{j,i} + \tilde{K}_{ij}T_iT_j + \qquad (22)$$

$$+ \tilde{H}_{ij}C_iC_j + (\tilde{h}_{ij} + H_{ij})P_{,i}C_j + h_{ij}P_{,i}P_{,j} - F_{jikl}C_{l,kj}C_i \geq 0$$

and symmetry conditions

$$P_{ijkl} = P_{klij} \qquad F_{ijkl} = F_{klij} \qquad K_{ij} = K_{ji} \qquad H_{ij} = H_{ji} \qquad k_{ij} = k_{ji} \qquad (23)$$

$$h_{ij} = h_{ji} \qquad \tilde{K}_{ij} = \tilde{K}_{ji} \qquad \tilde{H}_{ij} = \tilde{H}_{ji} \qquad \tilde{k}_{ij} = \tilde{k}_{ji} \qquad \tilde{h}_{ij} = \tilde{h}_{ji}$$

hold true. As in Aouadi et al. (2017) and Ieşan (2007), we consider that

$$\rho T_0\dot{S} = q_{i,i} + \rho s \qquad (24)$$

The equation of heat conduction follows by substituting (20)$_5$ and (21)$_1$ in (24)

$$T_0a_{ij}\dot{e}_{ij} + T_0b_{ij}\dot{\kappa}_{ij} + T_0d_i\dot{\zeta}_i + T_0F\dot{\varphi} + T_0b_i\dot{T}_i + T_0a\dot{\theta} + T_0\varpi\dot{C} - \rho s = k_{ij}\theta_{,ji} + K_{ij}T_{j,i} \qquad (25)$$

By (21)$_1$ and (4), we obtain

$$\dot{C} = h_{ij}P_{,ji} + H_{ij}C_{j,i} \qquad (26)$$

Then, by substituting into the equation above relation (20)$_7$, we get

$$\dot{C} = h_{ij}\left(d_{ij}e_{ij} + f_{ij}\kappa_{ij} + \tilde{f}_i\zeta_i + \tilde{g}_1\varphi - \varpi\theta + \varrho C\right)_{,ji} + H_{ij}C_{j,i} \qquad (27)$$

Hence, the basic equations of the theory of diffusion in microstretch thermoelasticity with microtemperatures and microconcentrations are the equations of motion (1), (2), (3), the energy equations (25) and (7), the diffusion equations (26) and (8), the constitutive equations (20), (21) and the geometrical equations (5). Moreover, we need to consider initial conditions and boundary conditions. The initial conditions are

$$u_i(x, 0) = u_i^0(x) \quad \dot{u}_i(x, 0) = u_i^1(x) \quad \varphi_i(x, 0) = \varphi_i^0(x) \quad \dot{\varphi}_i(x, 0) = \varphi_i^1(x)$$

$$\varphi(x, 0) = \varphi^0(x) \quad \dot{\varphi}(x, 0) = \varphi^1(x) \quad \theta(x, 0) = \theta^0(x) \quad T_i(x, 0) = T_i^0(x)$$

$$P(x, 0) = P^0(x) \quad C_i(x, 0) = C_i^0(x) \quad x \in \bar{\Omega}$$

$$(28)$$

where $u_i^0, u_i^1, \varphi_i^0, \varphi_i^1, \varphi^0, \varphi^1, \theta^0, T_i^0, P^0$ and C_i^0 are given. Let $S_r, r = 1, \ldots, 14$ be subsets of $\partial\Omega$ such that $\bar{S}_1 \cup S_2 = \bar{S}_3 \cup S_4 = \bar{S}_5 \cup S_6 = \bar{S}_7 \cup S_8 = \bar{S}_9 \cup S_{10} = \bar{S}_{11} \cup S_{12} = \bar{S}_{13} \cup S_{14} = \partial\Omega$ and $S_1 \cap S_2 = S_3 \cap S_4 = S_5 \cap S_6 = S_7 \cap S_8 = S_9 \cap S_{10} = S_{11} \cap S_{12} = S_{13} \cap S_{14} = \varnothing$. The boundary conditions are

$$u_i = \bar{u}_i \text{ on } \bar{S}_1 \times I \quad \varphi_i = \bar{\varphi}_i \text{ on } \bar{S}_3 \times I \quad \varphi = \bar{\varphi} \text{ on } \bar{S}_5 \times I$$

$$\theta = \bar{\theta} \text{ on } \bar{S}_7 \times I \quad T_i = \bar{T}_i \text{ on } \bar{S}_9 \times I \quad P = \bar{P} \text{ on } \bar{S}_{11} \times I$$

$$C_i = \bar{C}_i \text{ on } \bar{S}_{13} \times I \quad t_{ji}n_j = \bar{t}_i \text{ on } S_2 \times I \quad m_{ji}n_j = \bar{m}_i \text{ on } S_4 \times I \quad (29)$$

$$h_k n_k = \bar{h} \text{ on } S_6 \times I \quad q_j n_j = \bar{q} \text{ on } S_8 \times I \quad q_{ki}n_k = \bar{q}_i \text{ on } S_{10} \times I$$

$$\eta_i n_i = \bar{\eta} \text{ on } S_{12} \times I \quad \eta_{ki}n_k = \bar{\eta}_i \text{ on } S_{14} \times I$$

where $\bar{u}_i, \bar{\varphi}_i, \bar{\varphi}, \bar{\theta}, \bar{T}_i, \bar{P}, \bar{C}_i, \bar{t}_i, \bar{m}_i, \bar{h}, \bar{q}, \bar{q}_i, \bar{\eta}, \bar{\eta}_i$ are prescribed functions and $I = (0, \infty)$.

In the following, we want to derive the constitutive equations in the case of isotropic and homogeneous bodies. We follow Kearsley et al. (1975) and consider

$$A_{ijkm} = \lambda\delta_{ij}\delta_{km} + (\mu + \kappa)\delta_{ik}\delta_{jm} + \mu\delta_{im}\delta_{jk}$$

$$B_{ijkm} = \tilde{B}_1\delta_{ij}\delta_{km} + \tilde{B}_2\delta_{ik}\delta_{jm} + \tilde{B}_3\delta_{im}\delta_{jk}$$

$$C_{ijkm} = \tilde{C}_1\delta_{ij}\delta_{km} + \tilde{C}_2\delta_{ik}\delta_{jm} + \tilde{C}_3\delta_{im}\delta_{jk}$$

$$D_{ij} = D_1\delta_{ij} \quad E_{ij} = E\delta_{ij} \quad F_{ijk} = F_1\varepsilon_{ijk} \quad a_{ij} = A_4\delta_{ij} \quad L_{ijk} = L_1\varepsilon_{ijk}$$

$$G_{ijk} = \tilde{G}_1\varepsilon_{ijk} \quad A_{ij} = A_5\delta_{ij} \quad b_{ij} = \tilde{B}_4\delta_{ij} \quad M_{ijk} = M_1\varepsilon_{ijk} \quad N_{ij} = N_1\delta_{ij}$$

$$B_{ij} = \tilde{B}_5\delta_{ij} \quad R_{ij} = R\delta_{ij} \quad C_{ij} = \tilde{C}_4\delta_{ij} \quad d_{ij} = D_2\delta_{ij} \quad f_{ij} = F_2\delta_{ij}$$

$$(30)$$

According to Eringen (1999), $\tilde{B}_1, \tilde{B}_2, \tilde{B}_3, E, F_1, \tilde{B}_4$ are equal to zero. According to Ieşan (2007), L_1, R_i, b_i are zero. Therefore, the isotropic form of the free energy becomes

$$2\tilde{\sigma} = \lambda e_{ii}e_{kk} + (\mu + \kappa)e_{ij}e_{ij} + \mu e_{ij}e_{ji} + \tilde{C}_1\kappa_{ii}\kappa_{kk} + \tilde{C}_2\kappa_{ij}\kappa_{ij} +$$
$$+ \tilde{C}_3\kappa_{ij}\kappa_{ji} + 2D_1e_{ii}\varphi - 2A_4e_{ii}\theta + 2L_1\varepsilon_{ijk}e_{ij}T_k + 2\tilde{G}_1\varepsilon_{ijk}\kappa_{ij}\zeta_k +$$
$$+ A_5\zeta_i\zeta_i + 2M_1\varepsilon_{ijk}\kappa_{ij}T_k + 2B_i\zeta_i\varphi - 2d_i\zeta_i\theta - 2N_1\zeta_iT_i + \xi\varphi^2 - \tag{31}$$
$$- 2F\varphi\theta + 2R_i\varphi T_i - a\theta^2 - 2b_i\theta T_i - \tilde{B}_5T_iT_i - 2\varpi\theta C + \varrho C^2 -$$
$$- 2RT_iC_i - \tilde{C}_4C_i^2 + 2D_2e_{ii}C + 2F_2\kappa_{ii}C + 2\tilde{f}_i\zeta_iC + 2\tilde{g}_1\varphi C$$

Furthermore, the isotropic form of the constitutive equations becomes

$$t_{pq} = \lambda\delta_{pq}e_{kk} + (\mu + \kappa)e_{pq} + \mu e_{qp} + D_1\delta_{pq}\varphi - A_4\delta_{pq}\theta + D_2\delta_{pq}C$$
$$m_{pq} = \tilde{C}_1\delta_{pq}\kappa_{kk} + \tilde{C}_2\kappa_{pq} + \tilde{C}_3\kappa_{qp} + \tilde{G}_1\varepsilon_{pqk}\zeta_k + M_1\varepsilon_{pqk}T_k + F_2\delta_{pq}C$$
$$h_p = \tilde{G}_1\varepsilon_{ijp}\kappa_{ij} + A_5\zeta_p - N_1T_p + \tilde{f}_pC$$
$$g = -D_1e_{ii} - \xi\varphi + F\theta - \tilde{g}_1C \tag{32}$$
$$\rho S = A_4e_{ii} + F\varphi + a\theta + \varpi C$$
$$\rho\varepsilon_p = M_1\varepsilon_{ijp}\kappa_{ij} - N_1\zeta_p - \tilde{B}_5T_p - RC_p$$
$$P = -\varpi\theta + \varrho C + D_2e_{ii} + F_2\kappa_{ii} + \tilde{f}_i\zeta_i + \tilde{g}_1\varphi$$
$$\rho\omega_p = -RT_p - \tilde{C}_4C_p$$

The same as in Aouadi et al. (2017), we have

$$q_i = k\theta_{,i} + k_1T_i \qquad \eta_i = hP_{,i} + h_1C_i$$
$$Q_i = (k - k_3)\theta_{,i} + (k_1 - k_2)T_i \qquad \tilde{\sigma}_i = (h - h_3)P_{,i} + (h_1 - h_2)C_i \tag{33}$$
$$q_{ij} = -k_4T_{k,k}\delta_{ij} - k_5T_{i,j} - k_6T_{j,i} \qquad \eta_{ij} = -h_4C_{k,k}\delta_{ij} - h_5C_{i,j} - h_6C_{j,i}$$

Below we consider that the Cauchy–Schwarz rule holds true for the considered functions. Hence, the field equations of the theory of homogeneous and isotropic bodies are

$$(\mu + \kappa)\Delta u_i + (\lambda + \mu)u_{j,ji} + \kappa\varepsilon_{ijk}\varphi_{k,j} + D_1\varphi_{,i} - A_4\theta_{,i} + D_2C_{,i} + \rho f_i = \rho\ddot{u}_i$$
$$(A_5\Delta - \xi)\varphi - D_1u_{i,i} - N_1T_{k,k} + F\theta + \tilde{f}_kC_{,k} - \tilde{g}_1C + \rho l = J\ddot{\varphi}$$
$$\tilde{C}_2\Delta\varphi_i + (\tilde{C}_1 + \tilde{C}_3)\varphi_{j,ji} + \kappa\varepsilon_{ijk}u_{k,j} - 2\kappa\varphi_i + M_1\varepsilon_{jik}T_{k,j} + F_2C_{,i} + \rho g_i = I_1\ddot{\varphi}_i$$
$$k\Delta\theta + k_1T_{i,i} - A_4T_0\dot{u}_{i,i} - FT_0\dot{\varphi} - aT_0\dot{\theta} - \varpi T_0\dot{C} = -\rho s$$
$$k_6\Delta T_k + (k_4 + k_5)T_{j,jk} + M_1\varepsilon_{ijk}\dot{\varphi}_{j,i} - N_1\dot{\varphi}_{,k} - \tilde{B}_5\dot{T}_k -$$
$$- R\dot{C}_k - k_3\theta_{,k} - k_2T_k = \rho G_k$$
$$R\dot{T}_i + \tilde{C}_4\dot{C}_i = h_6\Delta C_i + (h_4 + h_5)C_{j,ji} - h_3P_{,i} - h_2C_i$$
$$\dot{C} = h\Delta P + h_1C_{i,i}$$

$$\tag{34}$$

We introduce a new formulation in which the chemical potential becomes the new state variable instead of the concentration. Therefore, we express C as a function of P

$$C = \frac{1}{\varrho}P + \frac{\varpi}{\varrho}\theta - \frac{D_2}{\varrho}u_{k,k} - \frac{F_2}{\varrho}\varphi_{k,k} - \frac{\tilde{f}_k}{\varrho}\varphi_{,k} - \frac{\tilde{g}_1}{\varrho}\varphi \tag{35}$$

Hence, the field equations in terms of the displacement, the microdilatation, the microrotation, the temperature, the chemical potential, the microtemperatures and the microconcentrations may be expressed in the following form

$$(\mu + \kappa)\Delta u_i + \left(\lambda + \mu - \frac{D_2^2}{\varrho}\right)u_{j,ji} + \kappa\varepsilon_{ijk}\varphi_{k,j} - \frac{D_2 F_2}{\varrho}\varphi_{k,ki} + \left(D_1 - \frac{\tilde{g}_1 D_2}{\varrho}\right)\varphi_{,i} -$$
$$- \frac{D_2 \tilde{f}_k}{\varrho}\varphi_{,ki} + \left(\frac{D_2\varpi}{\varrho} - A_4\right)\theta_{,i} + \frac{D_2}{\varrho}P_{,i} + \rho f_i = \rho\ddot{u}_i \tag{36}$$

$$(A_5\Delta - \xi)\varphi + \left(\frac{\tilde{g}_1 D_2}{\varrho} - D_1\right)u_{i,i} - N_1 T_{i,i} + \left(F - \frac{\tilde{g}_1\varpi}{\varrho}\right)\theta + \frac{\tilde{f}_i}{\varrho}P_{,i} + \frac{\tilde{f}_i\varpi}{\varrho}\theta_{,i} -$$
$$- \frac{\tilde{f}_i D_2}{\varrho}u_{k,ki} - \frac{\tilde{f}_i F_2}{\varrho}\varphi_{k,ki} - \frac{\tilde{f}_i \tilde{f}_k}{\varrho}\varphi_{,ki} - \frac{\tilde{g}_1}{\varrho}P + \frac{\tilde{g}_1 F_2}{\varrho}\varphi_{k,k} + \frac{\tilde{g}_1^2}{\varrho}\varphi + \rho l = J\ddot{\varphi} \tag{37}$$

$$\tilde{C}_2\Delta\varphi_i + \left(\tilde{C}_1 + \tilde{C}_3 - \frac{F_2^2}{\varrho}\right)\varphi_{j,ji} + \kappa\varepsilon_{ijk}u_{k,j} - \frac{F_2 D_2}{\varrho}u_{k,ki} - 2\kappa\varphi_i -$$
$$- \frac{F_2 \tilde{f}_k}{\varrho}\varphi_{,ki} - \frac{F_2 \tilde{g}_1}{\varrho}\varphi_{,i} + M_1\varepsilon_{jik}T_{k,j} + \frac{F_2}{\varrho}P_{,i} + \frac{F_2\varpi}{\varrho}\theta_{,i} + \rho g_i = I_1\ddot{\varphi}_i \tag{38}$$

$$k\Delta\theta + k_1 T_{i,i} + \left(\frac{\varpi D_2}{\varrho} - A_4\right)T_0\dot{u}_{i,i} + \left(\frac{\varpi\tilde{g}_1}{\varrho} - F\right)T_0\dot{\varphi} -$$
$$- \left(a + \frac{\varpi^2}{\varrho}\right)T_0\dot{\theta} - \frac{\varpi T_0}{\varrho}\dot{P} + \frac{\varpi T_0 F_2}{\varrho}\dot{\varphi}_{k,k} + \frac{\varpi T_0 \tilde{f}_k}{\varrho}\dot{\varphi}_{,k} = -\rho s \tag{39}$$

$$\frac{1}{\varrho}\dot{P} + \frac{\varpi}{\varrho}\dot{\theta} - \frac{D_2}{\varrho}\dot{u}_{k,k} - \frac{F_2}{\varrho}\dot{\varphi}_{k,k} - \frac{\tilde{f}_k}{\varrho}\dot{\varphi}_{,k} - \frac{\tilde{g}_1}{\varrho}\dot{\varphi} = h\Delta P + h_1 C_{i,i} \tag{40}$$

along with $(34)_5$ and $(34)_6$, which remain unchanged.

Following Ieşan (2007), we use the dimensionless quantities

$$x_i' = \frac{1}{l_0}x_i \quad t' = \frac{c_1}{l_0}t \quad u_i' = \frac{1}{l_0}u_i \quad \varphi_i' = \varphi_i \quad \varphi' = \varphi$$
$$\theta' = \frac{1}{T_0}\theta \quad T_i' = T_i l_0 \quad P' = \frac{1}{P_0}P \quad C_i' = C_i l_0 \tag{41}$$

where l_0 is a standard length. By suppressing primes, the field equations in non-dimensional form become

$$\alpha_1 \Delta u_i + \tilde{\alpha}_1 u_{j,ji} + \gamma_1 \varepsilon_{ijk}\varphi_{k,j} - \gamma_3 \varphi_{k,ki} + \gamma_2 \varphi_{,i} - \tilde{\bar{\gamma}}_k \varphi_{,ik} + \tilde{k}_1 \theta_{,i} + \gamma_5 P_{,i} + \tilde{F}_i = \ddot{u}_i \tag{42}$$

$$\alpha_4 \Delta \varphi - \gamma_2 u_{i,i} - \kappa_3 T_{,i} + \kappa_4 \theta + p_i P_{,i} + \tau_i \theta_{,i} - \tilde{\bar{\gamma}}_i u_{j,ji} - \\ - \sigma_i \varphi_{k,ki} - \sigma_{ik}\varphi_{,ki} - pP + \sigma \varphi_{k,k} + \alpha_5 \varphi + H = J_0 \ddot{\varphi} \tag{43}$$

$$\alpha_2 \Delta \varphi_i + \alpha_3 \varphi_{j,ji} + \gamma_1 \varepsilon_{ijk} u_{k,j} - \gamma_3 u_{j,ji} - 2\gamma_1 \varphi_i - \sigma_k \varphi_{,ki} - \\ - \sigma \varphi_{,i} + \kappa_2 \varepsilon_{jik} T_{k,j} + \tilde{p} P_{,i} + \tau \theta_{,i} + \Phi_i = I_0 \ddot{\varphi}_i \tag{44}$$

$$K \Delta \theta + \kappa_5 T_{,i} + \tilde{\kappa}_1 \dot{u}_{i,i} - \kappa_4 \dot{\varphi} - \xi_1 \dot{\theta} - \gamma_6 \dot{P} + \tau \dot{\varphi}_{k,k} + \tau_k \dot{\varphi}_{,k} = -Q_0 \tag{45}$$

$$\kappa_6 \Delta T_k + \kappa_7 T_{i,ik} + \kappa_2 \varepsilon_{ijk}\dot{\varphi}_{j,i} - \kappa_3 \dot{\varphi}_{,k} - \xi_3 \dot{T}_k - \tilde{c} \dot{C}_k - \xi_4 \theta_{,k} - \xi_2 T_k = \tilde{R}_k \tag{46}$$

$$\tilde{c} \dot{T}_i + H_0 \dot{C}_i = H_6 \Delta C_i + H_4 C_{j,ji} - H_3 P_{,i} - H_2 C_i \tag{47}$$

$$q_1 \dot{P} + \gamma_6 \dot{\theta} - \gamma_5 \dot{u}_{i,i} - \tilde{p} \dot{\varphi}_{k,k} - p_k \dot{\varphi}_{,k} - p \dot{\varphi} = q_7 \Delta P + q_8 C_{i,i} \tag{48}$$

where the coefficients have the following expressions

$$\alpha_1 = \frac{\mu + \kappa}{\rho c_1^2} \qquad \tilde{\alpha}_1 = \frac{1}{\rho c_1^2}\left(\lambda + \mu - \frac{D_2^2}{\varrho}\right) \qquad \gamma_1 = \frac{\kappa}{\rho c_1^2} \qquad \gamma_3 = \frac{D_2 F_2}{\varrho \rho c_1^2 l_0}$$

$$\gamma_2 = \frac{1}{\rho c_1^2}\left(D_1 - \frac{\tilde{g}_1 D_2}{\varrho}\right) \qquad \tilde{\bar{\gamma}}_k = \frac{D_2 \tilde{f}_k}{\varrho \rho c_1^2 l_0} \qquad \tilde{\kappa}_1 = \frac{T_0}{\rho c_1^2}\left(\frac{D_2 \varpi}{\varrho} - A_4\right)$$

$$\gamma_5 = \frac{D_2 P_0}{\varrho \rho c_1^2} \qquad \tilde{F}_i = \frac{l_0 f_i}{c_1^2} \qquad \alpha_4 = \frac{A_5}{l_0^2 c_1^2 \rho} \qquad \kappa_3 = \frac{N_1}{l_0^2 c_1^2 \rho} \qquad p_i = \frac{\tilde{f}_i P_0}{\varrho l_0 c_1^2 \rho}$$

$$\kappa_4 = \frac{T_0}{c_1^2 \rho}\left(F - \frac{\tilde{g}_1 \varpi}{\varrho}\right) \qquad \tau_i = \frac{\tilde{f}_i \varpi T_0}{\varrho l_0 c_1^2 \rho} \qquad \sigma_i = \frac{\tilde{f}_i F_2}{\varrho l_0^2 c_1^2 \rho} \qquad \sigma_{ik} = \frac{\tilde{f}_i \tilde{f}_k}{\varrho l_0^2 c_1^2 \rho}$$

$$\sigma = \frac{\tilde{g}_1 F_2}{\varrho l_0 c_1^2 \rho} \qquad \alpha_5 = \frac{1}{c_1^2 \rho}\left(\frac{\tilde{g}_1^2}{\varrho} - \xi\right) \qquad H = \frac{l}{c_1^2} \qquad J_0 = \frac{J}{l_0^2 \rho}$$

$$p = \frac{\tilde{g}_1 P_0}{\varrho c_1^2 \rho} \qquad \alpha_2 = \frac{\tilde{C}_2}{l_0^2 c_1^2 \rho} \qquad \alpha_3 = \frac{1}{l_0^2 c_1^2 \rho}\left(\tilde{C}_1 + \tilde{C}_3 - \frac{F_2^2}{\varrho}\right) \qquad \Phi_i = \frac{g_i}{c_1^2} \tag{49}$$

$$\kappa_2 = \frac{M_1}{l_0^2 c_1^2 \rho} \qquad \tilde{p} = \frac{F_2 P_0}{\varrho l_0 c_1^2 \rho} \qquad \tau = \frac{F_2 \varpi T_0}{\varrho l_0 c_1^2 \rho} \qquad I_0 = \frac{I_1}{\rho l_0^2} \qquad K = \frac{k T_0}{l_0 \rho c_1^3}$$

$$\kappa_5 = \frac{k_1}{l_0 \rho c_1^3} \qquad \xi_1 = \frac{T_0^2}{\rho c_1^2}\left(a + \frac{\varpi^2}{\varrho}\right) \qquad \gamma_6 = \frac{\varpi T_0 P_0}{\varrho \rho c_1^2} \qquad Q_0 = \frac{l_0 s}{c_1^3}$$

$$\kappa_6 = \frac{k_6}{\rho l_0^3 c_1^3} \quad \kappa_7 = \frac{k_4 + k_5}{\rho l_0^3 c_1^3} \quad \xi_3 = \frac{\tilde{B}_5}{\rho l_0^2 c_1^2} \quad \tilde{c} = \frac{R}{\rho l_0^2 c_1^2} \quad \tilde{R}_k = \frac{G_k}{c_1^3}$$

$$\xi_4 = \frac{k_3 T_0}{\rho l_0 c_1^3} \quad \xi_2 = \frac{k_2}{\rho l_0 c_1^3} \quad H_0 = \frac{\tilde{C}_4}{l_0^2 \rho c_1^2} \quad H_6 = \frac{h_6}{l_0^3 \rho c_1^3} \quad H_2 = \frac{h_2}{l_0 \rho c_1^3}$$

$$H_4 = \frac{h_4 + h_5}{l_0^3 \rho c_1^3} \quad H_3 = \frac{h_3 P_0}{l_0 \rho c_1^3} \quad q_1 = \frac{P_0^2}{\varrho \rho c_1^2} \quad q_7 = \frac{h P_0^2}{l_0 \rho c_1^3} \quad q_8 = \frac{h_1 P_0}{l_0 \rho c_1^3}$$

To the equations above we add the initial conditions (28) and the boundary conditions

$$u_i = \bar{u}_i \quad \varphi_i = \bar{\varphi}_i \quad \varphi = \bar{\varphi} \quad \theta = \bar{\theta} \quad T_i = \bar{T}_i \quad P = \bar{P} \quad C_i = \bar{C}_i \qquad (50)$$

on $\partial\Omega \times [0, t_1]$, where $\bar{u}_i, \bar{\varphi}_i, \bar{\varphi}, \bar{\theta}, \bar{T}_i, \bar{P}, \bar{C}_i$ are prescribed functions and t_1 is a given positive constant. We consider two solutions $\{u_i^{(\alpha)}, \varphi_i^{(\alpha)}, \varphi^{(\alpha)}, \theta^{(\alpha)}, T_i^{(\alpha)}, P^{(\alpha)}, C_i^{(\alpha)}\}$ of the equations above corresponding to the external data system $\mathcal{I}^{(\alpha)} = \{\tilde{F}_i^{(\alpha)}, \Phi_i^{(\alpha)}, H^{(\alpha)}, Q_0^{(\alpha)}, \tilde{R}_i^{(\alpha)}, \bar{u}_i, \bar{\varphi}_i, \bar{\varphi}, \bar{\theta}, \bar{T}_i, \bar{P}, \bar{C}_i, u_i^{0(\alpha)}, u_i^{1(\alpha)}, \varphi_i^{0(\alpha)}, \varphi_i^{1(\alpha)}, \varphi^{0(\alpha)}, \varphi^{1(\alpha)}, \theta^{0(\alpha)}, T_i^{0(\alpha)}, P^{0(\alpha)}, C_i^{0(\alpha)}\}$, where $\alpha = 1, 2$. We introduce the functions $u_i = u_i^{(1)} - u_i^{(2)}$, $\varphi_i = \varphi_i^{(1)} - \varphi_i^{(2)}$, $\varphi = \varphi^{(1)} - \varphi^{(2)}$, $\theta = \theta^{(1)} - \theta^{(2)}$, $T_i = T_i^{(1)} - T_i^{(2)}$, $P = P^{(1)} - P^{(2)}$, $C_i = C_i^{(1)} - C_i^{(2)}$. Hence $\{u_i, \varphi_i, \varphi, \theta, T_i, P, C_i\}$ is a solution of the problem corresponding to the external data system $\mathcal{I} = \{\tilde{F}_i, \Phi_i, H, Q_0, \tilde{R}_i, 0, 0, 0, 0, 0, 0, 0, u_i^0, u_i^1, \varphi_i^0, \varphi_i^1, \varphi^0, \varphi^1, \theta^0, T_i^0, P^0, C_i^0\}$ where $\tilde{F}_i = \tilde{F}_i^{(1)} - \tilde{F}_i^{(2)}$, $\Phi_i = \Phi_i^{(1)} - \Phi_i^{(2)}$, $H = H^{(1)} - H^{(2)}$, $Q_0 = Q_0^{(1)} - Q_0^{(2)}$, $\tilde{R}_i = \tilde{R}_i^{(1)} - \tilde{R}_i^{(2)}$, $u_i^0 = u_i^{0(1)} - u_i^{0(2)}$, $u_i^1 = u_i^{1(1)} - u_i^{1(2)}$, $\varphi_i^0 = \varphi_i^{0(1)} - \varphi_i^{0(2)}$, $\varphi_i^1 = \varphi_i^{1(1)} - \varphi_i^{1(2)}$, $\varphi^0 = \varphi^{0(1)} - \varphi^{0(2)}$, $\varphi^1 = \varphi^{1(1)} - \varphi^{1(2)}$, $\theta^0 = \theta^{0(1)} - \theta^{0(2)}$, $T_i^0 = T_i^{0(1)} - T_i^{0(2)}$, $P^0 = P^{0(1)} - P^{0(2)}$, $C_i^0 = C_i^{0(1)} - C_i^{0(2)}$. This problem will be denoted by $\mathcal{P}$.

In the following we would like to establish a continuous dependence result for the problem $\mathcal{P}$. We may write the equations above in the form described in the following lemma.

Lemma 3.1 *The equations of diffusion in microstretch thermoelasticity with microtemperatures and microconcentrations in the isotropic case are*

$$\pi_{ji,j} + \tilde{F}_i = \ddot{u}_i$$
$$s_{j,j} + f + H = J_0 \ddot{\varphi}$$
$$\mu_{ji,j} + \varepsilon_{ijk}\pi_{jk} + \Phi_i = I_0 \ddot{\varphi}_i$$
$$\dot{\Pi} = \Sigma_{j,j} + Q_0 \qquad (51)$$
$$\dot{\chi}_k = \lambda_{rk,r} + \Sigma_k - \Gamma_k + \tilde{R}_k$$
$$\dot{\Omega}_i = \tilde{\eta}_{ji,j} + \tilde{\eta}_i - \tilde{\Sigma}_i$$
$$\dot{Z} = \tilde{\eta}_{li,i} \tilde{Q}$$

where

$$\pi_{ji} = (\tilde{\alpha}_1 - \alpha_1 + \gamma_1)e_{rr}\delta_{ij} + \alpha_1 e_{ji} + (\alpha_1 - \gamma_1)e_{ij} - \gamma_3\kappa_{rr}\delta_{ij} +$$
$$+ \gamma_2\varphi\delta_{ij} - \tilde{\tilde{\gamma}}_r\zeta_r\delta_{ij} + \tilde{\kappa}_1\theta\delta_{ij} + \gamma_5 P\delta_{ij}$$
$$f = -\gamma_2 e_{rr} + \kappa_4\theta - pP + \sigma\kappa_{rr} + \alpha_5\varphi + \nu_i\zeta_i$$
$$s_i = \alpha_4\zeta_i - \kappa_3 T_i + p_i P + \tau_i\theta - \tilde{\tilde{\gamma}}_i e_{rr} - \sigma_i\kappa_{rr} - \sigma_{ik}\zeta_k + \varsigma\varepsilon_{rsi}\kappa_{rs} - \nu_i\varphi$$
$$\mu_{ji} = \tilde{\tau}\kappa_{rr}\delta_{ij} + (\alpha_3 - \tilde{\tau})\kappa_{ij} + \alpha_2\kappa_{ji} + \varsigma\varepsilon_{jik}\zeta_k + \kappa_2\varepsilon_{jik}T_k + \tilde{p}\delta_{ij}P +$$
$$+ \tau\delta_{ij}\theta - \gamma_3\delta_{ij}e_{rr} - \sigma_k\delta_{ij}\zeta_k - \sigma\delta_{ij}\varphi$$
$$\Sigma_i = K\theta_{,i} + \kappa_5 T_i$$
$$\Pi = -\tilde{\kappa}_1 e_{rr} + \kappa_4\varphi + \xi_1\theta + \gamma_6 P - \tau\kappa_{rr} - \tau_k\zeta_k \tag{52}$$
$$\chi_k = \kappa_2\varepsilon_{rsk}\kappa_{rs} - \kappa_3\zeta_k - \xi_3 T_k - \tilde{c}C_k$$
$$\Gamma_k = (K - \xi_4)\theta_{,k} + (\kappa_5 - \xi_2)T_k$$
$$\lambda_{jk} = -\kappa_6 T_{k,j} - (\kappa_7 - \kappa_8)T_{j,k} - \kappa_8 T_{r,r}\delta_{jk}$$
$$\Omega_i = -\tilde{c}T_i - H_0 C_i$$
$$\tilde{\eta}_i = H_8 P_{,i} + H_9 C_i$$
$$\tilde{\Sigma}_i = (H_8 - H_3)P_{,i} + (H_9 - H_2)C_i$$
$$\tilde{\eta}_{ji} = -H_6 C_{i,j} - H_{10}C_{j,i} - (H_4 - H_{10})C_{k,k}\delta_{ij}$$
$$Z = q_1 P + \gamma_6\theta - \gamma_5 e_{rr} - \tilde{p}\kappa_{rr} - p_k\zeta_k - p\varphi$$

We define the following functions, which will prove useful in the sequel

$$2U = (\tilde{\alpha}_1 - \alpha_1 + \gamma_1)e_{rr}e_{ss} + \alpha_1 e_{ij}e_{ij} + (\alpha_1 - \gamma_1)e_{ji}e_{ij} - 2\gamma_3\kappa_{rr}e_{ss} +$$
$$+ 2\gamma_2\varphi e_{rr} - 2\tilde{\tilde{\gamma}}_i\zeta_i e_{rr} + \tau\kappa_{rr}\kappa_{ss} + (\alpha_3 - \tilde{\tau})\kappa_{ji}\kappa_{ij} + \alpha_2\kappa_{ij}\kappa_{ij} -$$
$$- 2\sigma_i\zeta_i\kappa_{rr} - \sigma_{ik}\zeta_k\zeta_i - 2\sigma\varphi\kappa_{rr} - \alpha_5\varphi^2 - 2\nu_i\varphi\zeta_i + \tag{53}$$
$$+ 2\varsigma\zeta_k\kappa_{ij}\varepsilon_{ijk} + \alpha_4\zeta_i^2 + 2\tilde{c}C_i T_i + 2P\theta\gamma_6$$

and

$$\mathcal{D} = K\theta_{,i}\theta_{,i} + (\kappa_5 + \xi_4)\theta_{,i}T_i + \xi_2 T_i T_i + \kappa_8 T_{r,r}T_{s,s} + (\kappa_7 - \kappa_8)T_{j,i}T_{i,j} +$$
$$+ \kappa_6 T_{i,j}T_{i,j} + (\tilde{Q}H_9 + H_3)P_{,i}C_i + H_2 C_i C_i + \tilde{Q}H_8 P_{,i}P_{,i} + \tag{54}$$
$$+ H_6 C_{i,j}C_{i,j} + H_{10}C_{j,i}C_{i,j} + (H_4 - H_{10})C_{r,r}C_{s,s}$$

Then we may introduce the function ψ on $[0, t_1]$ as

$$\psi = \frac{1}{2}\int_\Omega \left(\dot{u}_i\dot{u}_i + I_0\dot{\varphi}_i\dot{\varphi}_i + J_0\dot{\varphi}^2 + 2U + \xi_1\theta^2 + \xi_3 T_i T_i + \right.$$
$$\left. + H_0 C_i^2 + q_1 P^2 + 2\int_0^t \mathcal{D}dt \right)dV \tag{55}$$

Moreover, we need to assume that U and $\mathcal{D}$ are positive definite quadratic forms. So we define the positive constants $\vartheta_1, \vartheta_2, \Lambda_1, \Lambda_2$ such that

$$\vartheta_1(\theta_{,i}\theta_{,i} + T_iT_i + T_{i,j}T_{i,j} + P_{,i}P_{,i} + C_iC_i + C_{i,j}C_{i,j}) \leq \mathcal{D} \leq$$
$$\leq \vartheta_2(\theta_{,i}\theta_{,i} + T_iT_i + T_{i,j}T_{i,j} + P_{,i}P_{,i} + C_iC_i + C_{i,j}C_{i,j}) \quad (56)$$

$$\Lambda_1(e_{ij}e_{ij} + \kappa_{ij}\kappa_{ij} + \zeta_i\zeta_i + \varphi^2) \leq U \leq \Lambda_2(e_{ij}e_{ij} + \kappa_{ij}\kappa_{ij} + \zeta_i\zeta_i + \varphi^2)$$

for all variables $e_{ij}, \kappa_{ij}, \zeta_i, \varphi, \theta_{,i}, T_i, T_{i,j}, P_{,i}, C_i, C_{i,j}$ and any $t \in [0, t_1]$.

Theorem 3.1 *Let $\{u_i, \varphi_i, \varphi, \theta, T_i, P, C_i\}$ be a solution of the problem $\mathcal{P}$. Then*

$$\dot{\psi} = \int_\Omega \left(\tilde{F}_i\dot{u}_i + \Phi_i\dot{\varphi}_i + H\dot{\varphi} + Q_0\theta - \tilde{R}_iT_i \right) dV \quad (57)$$

Proof We have from (52) and (53)

$$\pi_{ji}\dot{e}_{ji} + \mu_{ij}\dot{\kappa}_{ij} + s_i\dot{\zeta}_i - f\dot{\varphi} + \dot{\Pi}\theta - \dot{\chi}_iT_i - \dot{\Omega}_iC_i + \dot{Z}P =$$
$$= \frac{1}{2}\frac{\partial}{\partial t}\left(2U + \xi_1\theta^2 + \xi_3T_iT_i + H_0C_i^2 + q_1P^2\right) \quad (58)$$

Then, by multiplying the equations in (51) by $\dot{u}_i, \dot{\varphi}, \dot{\varphi}_i, \theta, T_k, C_i$ and P, respectively and by adding them up, we obtain

$$\pi_{ji}\dot{e}_{ji} + \mu_{ij}\dot{\kappa}_{ij} + s_i\dot{\zeta}_i - f\dot{\varphi} + \dot{\Pi}\theta - \dot{\chi}_iT_i - \dot{\Omega}_iC_i + \dot{Z}P =$$
$$= \left(\pi_{ji}\dot{u}_i + \mu_{ji}\dot{\varphi}_i + s_j\dot{\varphi} + \Sigma_j\theta - \lambda_{ji}T_i - \tilde{\eta}_{ji}C_i + \tilde{Q}\tilde{\eta}_jP\right)_{,j} + \quad (59)$$
$$+ \tilde{F}_i\dot{u}_i + \Phi_i\dot{\varphi}_i + H\dot{\varphi} + Q_0\theta - \tilde{R}_iT_i - \ddot{u}_i\dot{u}_i - I_0\ddot{\varphi}_i\dot{\varphi}_i - J_0\ddot{\varphi}\dot{\varphi} - \mathcal{D}$$

Hence

$$\frac{1}{2}\frac{\partial}{\partial t}\left(2U + \xi_1\theta^2 + \xi_3T_iT_i + H_0C_i^2 + q_1P^2 + \dot{u}_i\dot{u}_i + I_0\dot{\varphi}_i\dot{\varphi}_i + J_0\dot{\varphi}^2 + 2\int_0^t \mathcal{D}dt\right) =$$
$$= \left(\pi_{ji}\dot{u}_i + \mu_{ji}\dot{\varphi}_i + s_j\dot{\varphi} + \Sigma_j\theta - \lambda_{ji}T_i - \tilde{\eta}_{ji}C_i + \tilde{Q}\tilde{\eta}_jP\right)_{,j} +$$
$$+ \tilde{F}_i\dot{u}_i + \Phi_i\dot{\varphi}_i + H\dot{\varphi} + Q_0\theta - \tilde{R}_iT_i$$

$$(60)$$

If we integrate the relation above over Ω and use the divergence theorem and the boundary conditions, then we obtain the desired equality. ∎

For the final theorem we need to introduce the following functions on the interval $[0, t_1]$

$$\tilde{P} = \left[\int_\Omega \left(\tilde{F}_i\tilde{F}_i + \Phi_i\Phi_i + H^2 + Q_0^2 + \tilde{R}_i\tilde{R}_i\right) dV\right]^{1/2} \quad (61)$$

and

$$\Upsilon = \left\{ \int_\Omega \left[\dot{u}_i \dot{u}_i + \dot{\varphi}_i \dot{\varphi}_i + \dot{\varphi}^2 + e_{ij} e_{ij} + \kappa_{ij} \kappa_{ij} + \zeta_i \zeta_i + \varphi^2 + C_i^2 + P^2 + \theta^2 + T_i T_i + \right. \right.$$
$$\left. \left. + \int_0^t \left(\theta_{,i} \theta_{,i} + T_i T_i + T_{i,j} T_{i,j} + P_{,i} P_{,i} + C_i C_i + C_{i,j} C_{i,j} \right) dt \right] dV \right\}^{1/2}$$

(62)

Theorem 3.2 *If the constants ρ, I_0, J_0, ξ_1, ξ_3, H_0, q_1 are strictly positive and U and D are positive definite, then there exist the positive constants ρ_1 and ρ_2 such that*

$$\Upsilon(t) \le \rho_1 \Upsilon(0) + \rho_2 \int_0^t \tilde{P}(s) ds \tag{63}$$

for $t \in [0, t_1)$.

Proof The inequality of Schwartz yields by (57) and (61)

$$\dot{\psi} \le \tilde{P} \left[\int_\Omega \left(\dot{u}_i \dot{u}_i + \dot{\varphi}_i \dot{\varphi}_i + \dot{\varphi}^2 + \theta^2 + T_i T_i \right) dV \right]^{1/2} \tag{64}$$

Hence, the inequality above leads to

$$\dot{\psi} \le \tilde{P} \Upsilon \tag{65}$$

so that

$$\psi(t) \le \psi(0) + \int_0^t \tilde{P}(s) \Upsilon(s) ds, \quad t \in [0, t_1] \tag{66}$$

From (55), (56) and (62) we infer that

$$\psi(t) \ge \varsigma_1 \Upsilon^2(t) \qquad \psi(0) \le \varsigma_2 \Upsilon^2(0) \qquad t \in [0, t_1] \tag{67}$$

where

$$\varsigma_1 = \frac{1}{2} min(1, I_0, J_0, 2\Lambda_1, \xi_1, \xi_3, 2\vartheta_1, H_0, q_1)$$
$$\varsigma_2 = \frac{1}{2} max(1, I_0, J_0, 2\Lambda_2, \xi_1, \xi_3, 2\vartheta_2, H_0, q_1) \tag{68}$$

By (66) and (67) we find that

$$\Upsilon^2(t) \le \rho_1^2 \Upsilon^2(0) + 2\rho_2 \int_0^t \tilde{P}(s) \Upsilon(s) ds \qquad t \in [0, t_1] \tag{69}$$

where

$$\rho_1 = \left(\frac{\varsigma_2}{\varsigma_1}\right)^{1/2} \qquad \rho_2 = \frac{1}{2\varsigma_1} \qquad (70)$$

The Gronwall inequality yields the final result. ∎

4 Discussion and Conclusion

The novelty of this chapter consists in the introduction of the concepts of diffusion and microconcentrations in the mathematical model of microstretch thermoelasticity with microtemperatures. This is the first time one deals with this subject and this work may be continued by studying thoroughly all the characteristic properties of the model. As a starting point, we established a continuous dependence result upon initial data and body loads for the solutions of the problem of diffusion in microstretch thermoelasticity with microtemperatures and microconcentrations.

This topic in mechanics of generalized continua is important since classical elasticity proves to be inappropriate in modelling the behaviour of materials possessing internal structure, such as porous solids with spherical inclusions and pressurized voids. In fact, as mentioned in Eringen (1999), animal bone carrying bone marrow and many other biological tissues subject to micropressures arising from spherical microexpansions or contractions may be modelled by microstretch elasticity.

References

Altenbach, H., Maugin, G.A., Erofeev, V. (eds.): Mechanics of Generalized Continua. Advanced Structured Materials. Springer, Berlin (2011)

Aouadi, M.: Some theorems in the isotropic theory of microstretch thermoelasticity with microtemperatures. J. Therm. Stress. **31**(7), 649–662 (2008)

Aouadi, M., Ciarletta, M., Tibullo, V.: A thermoelastic diffusion theory with microtemperatures and microconcentrations. J. Therm. Stress. **40**(4), 486–501 (2017)

Bitsadze, L.: The dirichlet BVP of the theory of thermoelasticity with microtemperatures for microstretch sphere. J. Therm. Stress. **39**(9), 1074–1083 (2016)

Chirilă, A.: Generalized micropolar thermoelasticity with fractional order strain. Bull. Transilv. Univ. Braşov Ser. III: Math. Inf. Phys. **10**(59)(1), 83–90 (2017)

Chirilă, A., Agarwal, R.P., Marin, M.: Proving uniqueness for the solution of the problem of homogeneous and anisotropic micropolar thermoelasticity. Bound. Value Probl. **3**, 1–14 (2017)

Chirilă, A., Marin, M.: The theory of generalized thermoelasticity with fractional order strain for dipolar materials with double porosity. J. Mater. Sci. **53**(5), 3470–3482 (2018)

Ciarletta, M., Ieşan, D.: Non-classical Elastic Solids. Pitman Research Notes In Mathematics Series. Longman Scientific and Technical, London (1993)

Cosserat, E., Cosserat, F.: Sur la théorie des corps deformables. Dunod, Paris (1909)

Eringen, A.C.: Microcontinuum Field Theories: I. Foundations and Solids. Springer, New-York (1999)

Eringen, A.C.: Theory of thermo-microstretch elastic solids. Int. J. Eng. Sci. **28**, 1291–1301 (1990)

Green, A.E.: Micro-materials and multipolar continuum mechanics. Int. J. Eng. Sci. **3**, 533–537 (1965)

Green, A.E., Rivlin, R.S.: Multipolar continuum mechanics: functional theory I. Proc. R. Soc. Lond. A **284**, 303–324 (1965)

Grot, R.: Thermodynamics of a continuum with microstructure. Int. J. Eng. Sci. **7**, 801–814 (1969)

Ieşan, D.: Thermoelasticity of bodies with microstructure and microtemperatures. Int. J. Solids Struct. **44**(25–26), 8648–8662 (2007)

Kearsley, E.A., Fong, J.T.: Linearly independent sets of isotropic cartesian tensors of ranks up to eight. J. Res. Natl. Stand. Sec. B **79B**(1–2), 49–58 (1975)

Marin, M.: An approach of a heat-flux dependent theory for micropolar porous media. Meccanica **51**(5), 1127–1133 (2016)

Marin, M., Ellahi, R., Chirilă, A.: On solutions of Saint-Venant's problem for elastic dipolar bodies with voids. Carpathian J. Math. **33**(2), 199–212 (2017)

Marin, M., Baleanu, D.: On vibrations in thermoelasticity without energy dissipation for micropolar bodies. Bound. Value Probl. **111**, 1–19 (2016)

Maugin, G.A., Metrikine, A.V. (eds.): Mechanics of Generalized Continua: One Hundred Years After the Cosserats. Advances in Mechanics and Mathematics. Springer, New-York (2010)

Mindlin, R.D.: Microstructure in Linear Elasticity. Columbia University, New York (1963)

Scalia, A., Svanadze, M.: Uniqueness theorems in the equilibrium theory of thermoelasticity with microtemperatures for microstretch solids. J. Mech. Mater. Struct. **6**(9–10), 1295–1311 (2011)

Wozniak, C.: Thermoelasticity of the bodies with microstructure. Arch. Mech. Stos. **19**, 355 (1967)

Axial-Symmetric Potential Flows

S. A. Plaksa

Dedicated to memory of my colleague and friend Igor Mel'nichenko on the occasion of his 80th birthday.

Abstract We consider axial-symmetric stationary flows of the ideal incompressible fluid as an important case of potential solenoid vector fields. We establish relations between axial-symmetric potential solenoid fields and principal extensions of complex analytic functions into a special topological vector space containing an infinite-dimensional commutative Banach algebra. In such a way we substantiate a method for explicit constructing axial-symmetric potentials and Stokes flow functions by means of components of the mentioned principal extensions and establish integral expressions for axial-symmetric potentials and Stokes flow functions in an arbitrary simply connected domain symmetric with respect to an axis. The obtained integral expression of Stokes flow function is applied for solving boundary problem about a streamline of the ideal incompressible fluid along an axial-symmetric body. We obtain criteria of solvability of the problem by means distributions of sources and dipoles on the axis of symmetry and construct unknown solutions using multipoles together with dipoles distributed on the axis.

Keywords Laplace Equation · Axial-symmetric potential · Stokes flow function Streamline · Monogenic function · Analytic function

S. A. Plaksa (✉)
Department of Complex Analysis and Potential Theory, Institute of Mathematics of the National Academy of Science of Ukraine, 3, Tereshchenkivs'ka st., Kyiv 01004, Ukraine
e-mail: plaksa62@gmail.com; plaksa@imath.kiev.ua

© Springer Nature Switzerland AG 2019
C. Flaut et al. (eds.), *Models and Theories in Social Systems*, Studies in Systems, Decision and Control 179, https://doi.org/10.1007/978-3-030-00084-4_9

1 Potential Solenoid Fields and Flows

Consider a spatial stationary vector field defined by means the vector-function $\mathbf{V} \equiv \mathbf{V}(x, y, z)$ of the Cartesian coordinates x, y, z. The vector $\mathbf{V}$ is defined by means three real scalar functions $v_1 := v_1(x, y, z), v_2 := v_2(x, y, z), v_3 := v_3(x, y, z)$ which give its coordinates in the point (x, y, z), videlicet: $\mathbf{V} = (v_1, v_2, v_3)$.

Defining a *potential solenoid* field in a simply connected domain Q of the three-dimensional real space $\mathbb{R}^3$, the vector-function $\mathbf{V}$ satisfies the system of equations

$$\operatorname{div} \mathbf{V} = 0, \qquad \operatorname{rot} \mathbf{V} = 0, \tag{1}$$

where the divergence and the rotor are defined by the following equalities, respectively:

$$\operatorname{div} \mathbf{V} := \frac{\partial v_1}{\partial x} + \frac{\partial v_2}{\partial y} + \frac{\partial v_3}{\partial z},$$

$$\operatorname{rot} \mathbf{V} := \left(\frac{\partial v_3}{\partial y} - \frac{\partial v_2}{\partial z}, \ \frac{\partial v_1}{\partial z} - \frac{\partial v_3}{\partial x}, \ \frac{\partial v_2}{\partial x} - \frac{\partial v_1}{\partial y} \right).$$

In particular, the velocity field of stationary flow of the ideal incompressible fluid satisfies Eq. (1) and is an important case of potential solenoid vector field.

For a potential solenoid field there exists a scalar potential function $u(x, y, z)$ such that

$$\mathbf{V} = \operatorname{grad} u := \left(\frac{\partial u}{\partial x}, \frac{\partial u}{\partial y}, \frac{\partial u}{\partial z} \right),$$

and u satisfies the three-dimensional Laplace equation

$$\Delta_3 u(x, y, z) := \left(\frac{\partial^2}{\partial x^2} + \frac{\partial^2}{\partial y^2} + \frac{\partial^2}{\partial z^2} \right) u(x, y, z) = 0. \tag{2}$$

2 Plane Potential Solenoid Fields and a Complex Analytic Method of Their Description

In the case where the potential function u does not depend on the coordinate z, the field is called *plane* stationary potential solenoid field. In this case the potential function $u(x, y)$ satisfies the two-dimensional Laplace equation

$$\Delta_2 u(x, y) := \left(\frac{\partial^2}{\partial x^2} + \frac{\partial^2}{\partial y^2} \right) u(x, y) = 0.$$

An important achievement of mathematics is the description of plane potential fields by means of analytic functions of complex variable.

A potential $u(x, y)$ and a flow function $v(x, y)$ of plane stationary potential solenoid field satisfy the Cauchy–Riemann conditions

$$\frac{\partial u(x, y)}{\partial x} = \frac{\partial v(x, y)}{\partial y}, \qquad \frac{\partial u(x, y)}{\partial y} = -\frac{\partial v(x, y)}{\partial x}, \qquad (3)$$

and they form the complex potential $F(x + iy) = u(x, y) + iv(x, y)$ being an analytic function of complex variable $x + iy$. In turn, every analytic function $F(x + iy)$ satisfies the two-dimensional Laplace equation

$$\Delta_2 F(x + iy) \equiv F''(x + iy)\,(1^2 + i^2) = 0$$

due to the equality $1^2 + i^2 = 0$ for the unit 1 and the imaginary unit i of the algebra of complex numbers.

Many applied problems for plane potential flows are naturally formulated in terms of flow function, and it promotes their effective solving as well as the very well advanced methods of analytic functions in the complex plane (see, e.g., Lavrentyev and Shabat 1987).

3 Axial-Symmetric Potential Solenoid Fields and Flows

In the case where a spatial potential field is symmetric with respect to the axis Ox, a potential function $u(x, y, z)$ satisfying Eq. (2) is also symmetric with respect to the axis Ox, i.e. $u(x, y, z) = \varphi(x, r) = \varphi(x, -r)$, where $r := \sqrt{y^2 + z^2}$, and φ is known as the *axial-symmetric potential*. Then in a meridian plane xOr there exists a function $\psi(x, r)$ known as the *Stokes flow function* such that the functions φ and ψ satisfy the following system of equations degenerating on the axis Ox:

$$r\,\frac{\partial \varphi(x, r)}{\partial x} = \frac{\partial \psi(x, r)}{\partial r}, \qquad r\,\frac{\partial \varphi(x, r)}{\partial r} = -\frac{\partial \psi(x, r)}{\partial x}. \qquad (4)$$

Under the condition that there exist continuous second-order partial derivatives of the functions $\varphi(x, r)$ and $\psi(x, r)$, the system (4) implies the equation

$$r\Delta\varphi(x, r) + \frac{\partial \varphi(x, r)}{\partial r} = 0 \qquad (5)$$

for the axial-symmetric potential and the equation

$$r\Delta\psi(x, r) - \frac{\partial \psi(x, r)}{\partial r} = 0 \qquad (6)$$

for the Stokes flow function, where $\Delta := \dfrac{\partial^2}{\partial x^2} + \dfrac{\partial^2}{\partial r^2}.$

Equations (5), (6) are particular cases of the equation

$$r \,\Delta U(x, r) + m \,\frac{\partial U(x, r)}{\partial r} = 0 \qquad (7)$$

for generalized axial-symmetric potential $U(x, r)$, where $m = \text{const} \neq 0$.

An axial-symmetric flow is one of the most widespread kinds of spatial flows. For instance, such flows are axial-symmetric flows along fuselage of aeroplanes, missiles and dirigible balloons, cumulative charges, the movement of fluids and gases in channels with round profiles etc. (cf., e.g., Lavrentyev 1957; Lavrentyev and Shabat 1977, 1987; Batchelor 1970; Loitsyanskii 1987).

In view of degeneration of Eq. (4) on the axis Ox, the theory is developed considerably worse for solutions of system (4) than for solutions of system (3), i.e. complex analytic functions (see Lavrentyev and Shabat 1977, p. 18).

4 Hypercomplex Methods of Research of Spatial Potentials

Analytic function methods in the complex plane for plane potential fields inspire searching analogous effective methods for solving spatial and multidimensional problems of mathematical physics. Many such methods are based on mappings of hypercomplex algebras.

Hamilton (1866), Moisil and Theodoresco (1931), Fueter (1935), Sudbery (1979), Gürlebeck and Sprössig (1997), Kravchenko and Shapiro (1996), Leutwiler (1992), Ryan (1998), Colombo et al. (2011) and many other developed methods which are based on mappings of noncommutative algebras.

P. Ketchum (1928, 1929), Ringleb (1933), Sobrero (1934), Lorch (1943), Wagner (1948), Ward (1953), Riley (1953), Blum (1955), Roşculeţ (1955), Kunz (1971), Edenhofer (1976), Snyder (1982), I. Mel'nichenko (1975, 1986, 2003), Kovalev and Mel'nichenko (1981), Mel'nichenko and Plaksa (2008) and many other developed methods which are based on mappings of commutative algebras.

Last decades, the hypercomplex analysis in both commutative and nonconnutative algebras is very intensively developing. Its applications are developed for constructing solutions of equations of mathematical physics (especially, the multidimensional Laplace equation (see, e.g., Plaksa and Shpakovskii 2011; Plaksa 2012; Plaksa and Shpakivskyi 2012, 2014; 2017; Plaksa and Pukhtaievych 2013, 2014; Shpakivskyi 2016), the Helmholtz equation (see, e.g., Kravchenko and Shapiro 1996), the Klein–Gordon equation (see, e.g., Kravchenko 2009), the Navier–Stokes equation (see, e.g., Cerejeiras and Kähler 2000; Binlin Zhang et al. 2014; Gürlebeck et al. 2016; Grigor'ev 2017), the biharmonic equation (see, e.g., Gryshchuk and Plaksa 2009, 2013, 2016, 2017), the equations for axial-symmetric potential and for generalized axial-symmetric potential, the equation for Stokes flow function and other elliptic equations degenerating on an axis (see, e.g., Mel'nichenko and Plaksa 1996, 1997, 2004, 2008; Plaksa 2009, 2012, 2013). In fact, studying analytic functions of a complex variable, hyperholomorphic and monogenic functions defined in commutative and noncommutative algebras discovers a way to develop effective analytic methods for solving various problems of mathematical physics.

In particular, we proved in the papers Mel'nichenko and Plaksa (1997, 2008) that solutions of the system (4) in a domain convex in the direction of the axis Or are constructed by means components of principal extensions of analytic functions of a complex variable into a corresponding domain of a special two-dimensional vector manifold in an infinite-dimensional commutative Banach algebra.

5 An Infinite-Dimensional Commutative Banach Algebra $\mathbb{H}_{\mathbb{C}}$ and a Topological Vector Space $\widetilde{\mathbb{H}}_{\mathbb{C}}$ Containing the Algebra $\mathbb{H}_{\mathbb{C}}$

Let $\mathbb{H}_{\mathbb{C}} := \{a = \sum_{k=1}^{\infty} a_k e_k \ : \ a_k \in \mathbb{C}, \ \sum_{k=1}^{\infty} |a_k| < \infty\}$ be a commutative associative Banach algebra over the field of complex numbers $\mathbb{C}$ with the norm $\|a\|_{\mathbb{H}_{\mathbb{C}}} := \sum_{k=1}^{\infty} |a_k|$ and the following multiplication table for elements of the basis $\{e_k\}_{k=1}^{\infty}$:

$$e_n e_1 = e_n, \quad e_m e_n = \frac{1}{2}\left(e_{m+n-1} + (-1)^{n-1} e_{m-n+1}\right) \quad \forall \ m \geq n \geq 1$$

(cf., e.g., Mel'nichenko and Plaksa 1997, 2008). The multiplication table was offered by Mel'nichenko 1984.

The algebra $\mathbb{H}_{\mathbb{C}}$ is isomorphic to the algebra $\mathbf{F}_{\cos}$ of absolutely convergent trigonometric Fourier series

$$c(\tau) = \sum_{k=1}^{\infty} c_k \, i^{k-1} \cos(k-1)\tau$$

with complex coefficients c_k and the norm $\|c\|_{\mathbf{F}_{\cos}} := \sum_{k=1}^{\infty} |c_k|$. In this case, we have the isomorphism $e_{2k-1} \longleftrightarrow i^{k-1} \cos(k-1)\tau$ between basic elements.

Consider the Cartesian plane $\mu := \{\zeta = xe_1 + re_2 \ : \ x, r \in \mathbb{R}\}$ which is a linear span of the elements e_1, e_2 over the field of real numbers $\mathbb{R}$.

For a domain $D \subset \mathbb{R}^2$ we use consentaneous denotations for congruent domains of the plane μ and the complex plane $\mathbb{C}$, namely: $D_\zeta := \{\zeta = xe_1 + re_2 \ : \ (x, r) \in D\} \subset \mu$ and $D_z := \{z = x + ir \ : \ (x, r) \in D\} \subset \mathbb{C}$.

We proved in the papers Mel'nichenko and Plaksa (1997, 2008) that in the case where the domain D is convex in the direction of the axis Or, solutions of the system

(4) can be constructed by means components of principal extensions of complex functions analytic in D_z into the domain D_ζ.

To generalize such a relation between solutions of the system (4) and hyper-complex functions for domains of more general form, let us insert the algebra $\mathbb{H}_\mathbb{C}$ in the topological vector space $\widetilde{\mathbb{H}}_\mathbb{C} := \{g = \sum_{k=1}^{\infty} c_k e_k : c_k \in \mathbb{C}\}$ with the topology of coordinate-wise convergence.

Note that $\widetilde{\mathbb{H}}_\mathbb{C}$ is not an algebra because the product of elements $g_1, g_2 \in \widetilde{\mathbb{H}}_\mathbb{C}$ is defined not always. But for each $g = \sum_{k=1}^{\infty} c_k e_k \in \widetilde{\mathbb{H}}_\mathbb{C}$ and $\tilde\zeta = (x + iy)e_1 + re_2$, where $x, y, r \in \mathbb{R}$, one can define the product

$$g\tilde\zeta \equiv \tilde\zeta g := \left(c_1(x + iy) - \frac{c_2}{2}r\right)e_1 + \left(c_2(x + iy) + \left(c_1 - \frac{c_3}{2}\right)r\right)e_2 +$$

$$+ \sum_{k=3}^{\infty} \left(c_k(x + iy) + \frac{1}{2}\left(c_{k-1} - c_{k+1}\right)r\right)e_k.$$

6 Monogenic Functions Taking Values in the Space $\widetilde{\mathbb{H}}_\mathbb{C}$

Below, we shall consider functions given in domains of the plane μ and taking values in the space $\widetilde{\mathbb{H}}_\mathbb{C}$.

We say that a continuous function $\Phi : Q_\zeta \to \widetilde{\mathbb{H}}_\mathbb{C}$ is *monogenic* in a domain $Q_\zeta \subset \mu$ if Φ is differentiable in the sense of Gateaux in every point of Q_ζ, i.e. if for every $\zeta \in Q_\zeta$ there exists an element $\Phi'(\zeta) \in \widetilde{\mathbb{H}}_\mathbb{C}$ such that

$$\lim_{\varepsilon \to 0+0} (\Phi(\zeta + \varepsilon h) - \Phi(\zeta))\varepsilon^{-1} = h\,\Phi'(\zeta) \quad \forall h \in \mu. \tag{8}$$

Let us note that we use the notion of monogenic function in the sense of exis-tence of derived numbers for this function in the domain Q_ζ (cf. Goursat 1910; Trokhimchuk 1964).

Consider the decomposition

$$\Phi(\zeta) = \sum_{k=1}^{\infty} U_k(x, r)\, e_k, \qquad \zeta = xe_1 + re_2, \tag{9}$$

of a function $\Phi : Q_\zeta \to \widetilde{\mathbb{H}}_\mathbb{C}$ with respect to the basis $\{e_k\}_{k=1}^{\infty}$.

Below, we suppose that the functions $U_k : Q \to \mathbb{C}$ are $\mathbb{R}$-differentiable in the domain Q, i.e.

$$U_k(x + \Delta x, r + \Delta r) - U_k(x, r) = \frac{\partial U_k(x, r)}{\partial x} \Delta x + \frac{\partial U_k(x, r)}{\partial r} \Delta r +$$

$$+ o\left(\sqrt{(\Delta x)^2 + (\Delta r)^2}\right), \qquad (\Delta x)^2 + (\Delta r)^2 \to 0,$$

for all $(x, r) \in Q$. Evidently, such an assumption implies the fact that the function (9) is continuous in $Q_{\tilde{\zeta}}$.

In the following theorem we establish the necessary and sufficient conditions for a function $\Phi : \Omega_\zeta \to \widetilde{\mathbb{H}}_\mathbb{C}$ be monogenic in a domain $\Omega_\zeta \subset \mu$.

Theorem 1 *Let in the decomposition (9) of a function $\Phi : Q_\zeta \to \widetilde{\mathbb{H}}_\mathbb{C}$ the functions $U_k : Q \to \mathbb{C}$ be $\mathbb{R}$-differentiable in Q. In order the function Φ be monogenic in the domain Q_ζ, it is necessary and sufficient that the following Cauchy–Riemann condition be satisfied in Q_ζ:*

$$\frac{\partial \Phi}{\partial r} = \frac{\partial \Phi}{\partial x} e_2. \tag{10}$$

Proof Necessity. If the function $\Phi : Q_\zeta \to \widetilde{\mathbb{H}}_\mathbb{C}$ is monogenic in the domain Q_ζ, then for $h = e_1$ the equality (8) turns into the equality

$$\Phi'(\zeta) = \frac{\partial \Phi(\zeta)}{\partial x}.$$

Now, setting this expression and $h = e_2$ in the equality (8), we obtain the condition (10).

Sufficiency. Let us write the conditions (10) in expanded form:

$$\frac{\partial U_1(x, r)}{\partial r} = -\frac{1}{2} \frac{\partial U_2(x, r)}{\partial x},$$
$$\frac{\partial U_2(x, r)}{\partial r} = \frac{\partial U_1(x, r)}{\partial x} - \frac{1}{2} \frac{\partial U_3(x, r)}{\partial x}, \tag{11}$$
$$\frac{\partial U_k(x, r)}{\partial r} = \frac{1}{2} \frac{\partial U_{k-1}(x, r)}{\partial x} - \frac{1}{2} \frac{\partial U_{k+1}(x, r)}{\partial x}, \quad k = 3, 4, \dots,$$

Let $\varepsilon > 0$ and $h := h_1 e_1 + h_2 e_2$, where $h_1, h_2 \in \mathbb{R}$. Taking into account the equalities (11), we have

$$\left(\Phi(\zeta + \varepsilon h) - \Phi(\zeta)\right) \varepsilon^{-1} - h \, \Phi'(\zeta) =$$

$$= \left(\sum_{k=1}^{\infty} \left(U_k(x + \varepsilon h_1, r + \varepsilon h_2) - U_k(x, r)\right) e_k -\right.$$

$$\left. - \varepsilon (h_1 e_1 + h_2 e_2) \sum_{k=1}^{\infty} \frac{\partial U_k(x, r)}{\partial x} e_k \right) \varepsilon^{-1} =$$

$$= \sum_{k=1}^{\infty} \varepsilon^{-1} \left(U_k(x + \varepsilon h_1, r + \varepsilon h_2) - U_k(x, r) - \right.$$

$$\left. - \frac{\partial U_k(x, r)}{\partial x} \varepsilon h_1 - \frac{\partial U_k(x, r)}{\partial r} \varepsilon h_2 \right) e_k . \qquad (12)$$

Inasmuch as the functions U_k are $\mathbb{R}$-differentiable in Q, the last series converges coordinate-wise to zero, i.e. the function Φ is monogenic in Q_ς. Theorem is proved.

7 Principal Extensions of Complex Analytic Functions and Its Relations to Axial-Symmetric Potential Fields

Let us construct for every complex analytic function a special monogenic function taking values in the space $\widetilde{\mathbb{H}}_{\mathbb{C}}$. Such a monogenic function is a generalization of the principal extension of complex analytic function into a commutative Banach algebra. We establish below relations between generalized principal extensions of complex analytic functions and axial-symmetric potential solenoid fields. In such a way we substantiate a method for explicit constructing axial-symmetric potentials and Stokes flow functions in an arbitrary simply connected domain symmetric with respect to the axis Ox by means of components of the mentioned principal extensions.

Let the domain $D \subset \mathbb{R}^2$ be symmetric with respect to the axis Ox and the domain D_z be simply connected. Let the boundary ∂D_z of domain D_z cross the real axis at the points b_1 and b_2. We assume that $b_1 < b_2$.

For every $z \in D_z \setminus \mathbb{R}$ let us fix an arbitrary Jordan rectifiable curve $\Gamma_{z\bar{z}}$ in the domain D_z that is symmetric with respect to the real axis $\mathbb{R}$ and connects the points z and $\bar{z}$. In addition, we shall agree that in the case where the domain D_z is unbounded, the curve $\Gamma_{z\bar{z}}$ crosses the real axis $\mathbb{R}$ on the interval $(-\infty, b_1)$.

For $z \in D_z \setminus \mathbb{R}$ let $\sqrt{(t-z)(t-\bar{z})}$ be that continuous branch of the analytic function $G(t) = \sqrt{(t-z)(t-\bar{z})}$ outside of the cut along $\Gamma_{z\bar{z}}$ for which $G(b_2) > 0$.

For every $z \in D_z$ with $\operatorname{Im} z = 0$, we define by continuity $\sqrt{(t-z)(t-\bar{z})} := t - z$ for $z < b_2$, and $\sqrt{(t-z)(t-\bar{z})} := -(t-z)$ for $z > b_2$.

For every function $F : D_z \to \mathbb{C}$ analytic in a domain D_z consider the function

$$\frac{1}{2\pi i} \int_{\gamma} (t e_1 - \varsigma)^{-1} F(t) \, dt = U_1(x, r) e_1 + 2 \sum_{k=2}^{\infty} U_k(x, r) e_{2k-1} \qquad (13)$$

given in D_ς and taking values in $\widetilde{\mathbb{H}}_{\mathbb{C}}$, where

$$U_k(x, r) := \frac{1}{2\pi i} \int_{\gamma} \frac{F(t)}{\sqrt{(t-z)(t-\bar{z})}} \left(\frac{\sqrt{(t-z)(t-\bar{z})} - (t-x)}{r} \right)^{k-1} dt, \qquad (14)$$

$\zeta = xe_1 + re_3$ and $z = x + ir$ for $(x, r) \in D$, and γ is an arbitrary closed Jordan rectifiable curve in D_z which embraces $\Gamma_{z\bar{z}}$.

Let us note that if to take a domain $D' \subset D$ which is symmetric with respect to the axis Ox and convex in the direction of the axis Or, and to fix the segment connecting the points z and $\bar{z}$ as the curve $\Gamma_{z\bar{z}}$ for every $z \in D'_z \setminus \mathbb{R}$, then the function (13) turns into the principal extension of the analytic function F into the domain D'_ζ (see Hille and Phillips 1957, p. 165).

Therefore, we shall consider the function (13) as a *principal extension* of analytic function $F : D_z \to \mathbb{C}$ into the domain D_ζ.

The following theorem generalizes Theorem 2.6 in Mel'nichenko and Plaksa (2008) (cf. also Theorem 18 in Mel'nichenko and Plaksa 1997), which describes relations between principal extensions of analytic functions into the plane μ and solutions of the system (4) in domains convex in the direction of the axis Or.

Theorem 2 *Let the domain $D \subset \mathbb{R}^2$ be symmetric with respect to the axis Ox and the domain D_z be simply connected. If $F : D_z \to \mathbb{C}$ is an analytic function in a domain D_z, then the first and the second components of the function (13) generate the solutions φ and ψ of system (4) in D by the formulas*

$$\varphi(x, r) = U_1(x, r), \quad \psi(x, r) = r\, U_2(x, r). \tag{15}$$

Moreover, the functions φ and ψ defined by the formulas (15) are solutions in D of Eqs. (5) and (6), respectively.

Proof In view of the equality (14) and Cauchy theorem, the functions (15) have the form

$$\varphi(x, r) = \frac{1}{2\pi i} \int_\gamma \frac{F(t)}{\sqrt{(t - z)(t - \bar{z})}}\, dt, \tag{16}$$

$$\psi(x, r) = -\frac{1}{2\pi i} \int_\gamma \frac{F(t)\,(t - x)}{\sqrt{(t - z)(t - \bar{z})}}\, dt. \tag{17}$$

Now, substituting the partial derivatives of functions (16) and (17) into the equations of system (4) and Eqs. (5) and (6), one can see that the mentioned equations become identities in the domain D. Theorem is proved.

Thus, the formulas (15) enable to construct axial-symmetric potentials and Stokes flow functions by means of components of principal extensions of complex analytic functions into the plane μ.

In particular, elementary functions of the variable $\zeta = xe_1 + re_2$ are principal extensions of corresponding elementary functions of a complex variable. Let us write expansions with respect to the basis $\{e_k\}_{k=1}^\infty$ of some elementary functions of the variable $\zeta = xe_1 + re_2$. Note that in view of isomorphism between the algebras $\mathbb{H}_\mathbb{C}$ and $\mathbf{F}_{\cos}$, the construction of expansions of this sort is reduced to the determination of relevant Fourier coefficients.

The expansion of a power function has the form (see Mel'nichenko and Plaksa 1997; 2008)

$$
\zeta^n = (x^2 + r^2)^{n/2} \left(P_n(\cos \vartheta)\, e_1 + 2 \sum_{k=1}^{n} \frac{(\operatorname{sgn} r)^k\, n!}{(n+k)!}\, P_n^k(\cos \vartheta)\, e_{k+1} \right),
$$

where n is a positive integer, $\quad \cos \vartheta := x(x^2 + r^2)^{-1/2}$,

$$
\operatorname{sgn} r := \begin{cases} 1 & \text{for } r \geq 0, \\ -1 & \text{for } r < 0, \end{cases}
$$

and Legendre polynomials P_n and associated Legendre polynomials P_n^m are defined be the equalities

$$
P_n(t) := \frac{1}{2^n\, n!} \frac{d^n}{dt^n} (t^2 - 1)^n, \qquad P_n^m(t) := (1 - t^2)^{m/2} \frac{d^m}{dt^m} P_n(t) .
$$

For the functions e^ζ, $\sin \zeta$ and $\cos \zeta$ we have

$$
e^\zeta = e^x \left(J_0(r)\, e_1 + 2 \sum_{k=1}^{\infty} J_k(r)\, e_{k+1} \right),
$$

$$
\sin \zeta = \sin x \left(J_0(ir)\, e_1 + 2 \sum_{k=1}^{\infty} J_{2k}(ir)\, e_{2k+1} \right) - 2i\, \cos x \sum_{k=1}^{\infty} J_{2k-1}(ir)\, e_{2k},
$$

$$
\cos \zeta = \cos x \left(J_0(ir)\, e_1 + 2 \sum_{k=1}^{\infty} J_{2k}(ir)\, e_{2k+1} \right) + 2i\, \sin x \sum_{k=1}^{\infty} J_{2k-1}(ir)\, e_{2k},
$$

where Bessel functions J_m are defined by the equality

$$
J_m(t) := \frac{(-1)^m}{\pi} \int_0^\pi e^{it \cos \tau}\, \cos m\tau\, d\tau .
$$

In the following theorem we describe relations between components U_k of hyper-complex monogenic function (13) and solutions of elliptic equations degenerating on the axis Ox.

Theorem 3 *Let the domain $D \subset \mathbb{R}^2$ be symmetric with respect to the axis Ox and the domain D_z be simply connected. If $F : D_z \to \mathbb{C}$ is an analytic function in a domain D_z, then the components U_k of principal extension (13) of function F into the domain D_ζ satisfy the equations*

$$r^2 \Delta U_k(x,r) + r \frac{\partial U_k(x,r)}{\partial r} - (k-1)^2 U_k(x,r) = 0, \quad k = 1,2,\ldots,$$

in the domain D. In addition, the function

$$\psi_k(x,r) := r^{k-1} U_k(x,r)$$

is a solution in D of the equation

$$r \Delta \psi_k(x,r) - (2k-3) \frac{\partial \psi_k(x,r)}{\partial r} = 0, \quad k = 1,2,\ldots.$$

Theorem 2 follows from the equalities (13), (14) and Theorem 3.1 in Plaksa (2009).

8 Integral Expressions for Axial-Symmetric Potentials and Stokes Flow Functions in Boundary Value Problems

Boundary value problems for solutions of elliptic equations have numerous applications in mathematical physics. For the two- and three-dimensional Laplace equations, various methods for the efficient solving of boundary value problems are developed. However, the direct application of these methods to solving of boundary value problems for axial-symmetric potentials and Stokes flow functions is a quite complicated problem due to a degeneration of the Eqs. (5), (6) on the axis Ox.

In the paper Keldysh (1951), some correct statements of boundary value problems for an elliptic equation with a degeneration on a straight line are described. They have shown certain differences of these problems from boundary value problems for elliptic equations without degeneration.

Therefore, for solving of boundary value problems in a meridian plane of an axial-symmetric potential field, it is necessary to develop special methods that take into account the nature and specific features of axial-symmetric problems.

One of ways for researching axial-symmetric problems is based on representation of its solutions in the form of potentials of a simple or double layer. With this purpose, fundamental solutions of the appropriate equations with partial derivatives are used (cf., e.g., Weinstein 1948, 1953, 1962; Mikhailov and Rajabov 1972; Rajabov 1974). For instance, in such a way in the paper Rajabov (1974), the main boundary value problems for solutions of equation (7) in a domain with the Lyapunov boundary was reduced to the Fredholm integral equations.

Many methods of research of elliptic equations are based on integral expressions of solutions via analytic functions of a complex variable (cf. Whittaker and Watson 1927; Bateman 1944; Henrici 1953, 1957; Huber 1954; Mackie 1955; Erdelyi 1956; Gilbert 1969; Krivenkov 1957; 1960; Rajabov 1965, 1968; Polozhii 1973; Polozhii and Ulitko 1965; Kapshivyi 1972; Aleksandrov and Soloviev 1979; Mel'nichenko 1984).

8.1 Integral Expressions for Axial-Symmetric Potentials and Stokes Flow Functions

The formulas (16) and (17) generate axial-symmetric potentials and Stokes flow functions in an arbitrary simply connected domain symmetric with respect to the axis Ox.

Below, we formulate four statements on the representability of axial-symmetric potentials and Stokes flow functions by the formulas (16) and (17), respectively. The cases of a bounded domain D and an unbounded domain D are considered separately.

In the case of a bounded domain D, the following two statements are true.

Theorem 4 (Plaksa 2001, 2003; Mel'nichenko and Plaksa 2008) *Suppose that a function $\varphi(x, r)$ is even with respect to the variable r and satisfies Eq. (5) in a bounded domain D symmetric with respect to the axis Ox. In this case, there exists the unique function F analytic in the domain D_z and satisfying the condition*

$$F(\bar{z}) = \overline{F(z)} \quad \forall z \in D_z \tag{18}$$

and such that the equality (16) is fulfilled for all $(x, r) \in D$.

Theorem 5 (Plaksa 2003; Mel'nichenko and Plaksa 2008) *Suppose that the function $\psi(x, y)$ is even with respect to the variable r and satisfies Eq. (6) in a bounded domain D symmetric with respect to the axis Ox and the additional assumption*

$$\psi(x, 0) \equiv 0 \quad \forall (x, 0) \in D. \tag{19}$$

In this case, there exists a function F_0 analytic in the domain D_z such that the equality (17) is fulfilled with $F = F_0$ for all $(x, r) \in D$. Moreover, any analytic function F which satisfies the condition (18) and the equality (17) for all $(x, r) \in D$ is expressed in the form $F(z) = F_0(z) + C$, where C is a real constant.

The requirement (19) is natural. For example, for the model of steady flow of an ideal incompressible fluid without sources and vortexes it means that the axis Ox is a line of flow.

In the case of an unbounded domain D with the bounded Jordan boundary, the following two statements are true.

Theorem 6 (Plaksa 2002; Mel'nichenko and Plaksa 2008) *Suppose that a function $\varphi(x, r)$ satisfies Eq. (5) in an unbounded domain D with the bounded Jordan boundary symmetric with respect to the axis Ox. Suppose also that the function $\varphi(x, r)$ is even with respect to the variable r and is vanishing at infinity. In this case, there exists the unique analytic in D_z function F vanishing at infinity and satisfying the condition (18) and such that the equality (16) is fulfilled for all $(x, r) \in D$.*

Theorem 7 (Plaksa 2003, Mel'nichenko and Plaksa 2008) *Suppose that a function* $\psi(x, r)$ *satisfies Eq. (6) and the condition (19) in an unbounded domain D with the bounded Jordan boundary symmetric with respect to the axis Ox. Suppose also that the function* $\varphi(x, r)$ *is even with respect to the variable r and is vanishing at infinity. In this case, there exists the unique analytic in* D_z *function F vanishing at infinity and satisfying the condition (18) and such that the equality (17) is fulfilled for all* $(x, r) \in D$. *Moreover, the function F has a zero at least of the second order at infinity.*

It follows from Theorem 2 that the functions (16), (17) satisfies the system (4) in the domain D for every function F analytic in D_z. But these functions takes real values if and only if the condition (18) is satisfied.

Thus, all axial-symmetric potentials and Stokes flow functions, i.e. solutions of the system (4) in D with a physical interpretation, are represented by the integral expressions (16), (17) which can be used for solving boundary value problems for axial-symmetric potential solenoid fields.

Let us note that if the boundary ∂D_z is a Jordan rectifiable curve and the function F belongs to the Smirnov class E_1 (see, e.g., Privalov 1950, p. 205) in the domain D_z, then the formulas (16) and (17) can be transformed to the form

$$\varphi(x, r) = \frac{1}{2\pi i} \int_{\partial D_z} \frac{F(t)}{\sqrt{(t - z)(t - \bar{z})}} \, dt \,, \tag{20}$$

$$\psi(x, r) = -\frac{1}{2\pi i} \int_{\partial D_z} \frac{F(t)(t - x)}{\sqrt{(t - z)(t - \bar{z})}} \, dt \tag{21}$$

for all $(x, r) \in D$, where $F(t)$ are the angular boundary values of the function F which, as it is known (see, e.g., Privalov 1950, p. 205), exist at almost all points $t \in \partial D_z$. In the case where D is an unbounded domain, we admit additionally that the function F is vanishing at infinity for obtaining the formula (20) or the function F has a zero at least of the second order at infinity for obtaining the formula (21).

In the papers Plaksa (2001), Mel'nichenko and Plaksa (2008) we established sufficient conditions for continuous continuations of the functions (20), (21) on the boundary ∂D of a domain D and obtained estimations for modules of continuity of boundary values of the mentioned functions.

Using the integral expressions (20), (21) of axial-symmetric potentials and Stockes flow functions, we developed a functional analytic method for effective solving boundary problems in a meridian plane of spatial axial-symmetric potential field (see Plaksa 2001, 2002, 2003; Mel'nichenko and Plaksa 2008).

8.2 Integral Equation for an Outer Dirichlet Boundary Value Problem for the Stokes Flow Function

Below, let D be an unbounded domain in the meridian plane xOr and the boundary ∂D is a closed Jordan rectifiable curve symmetric with respect to the axis Ox. The closure of domain D is denoted by $\overline{D}$.

Let us consider the following *outer Dirichlet boundary value problem for the Stokes flow function*: to find a continuous in $\overline{D}$ function $\psi(x, r)$ which is a solution of Eq. (6) in D when its boundary values $\psi_{\partial D}(x, r)$ are given on the boundary ∂D, i.e. $\psi(x, r) = \psi_{\partial D}(x, r)$ for all $(x, r) \in \partial D$. It is also required that the function ψ is vanishing at infinity and satisfies the identity (19).

Note that, vanishing at infinity and satisfying the identity (19), the Stokes flow function $\psi(x, r)$ satisfies the maximum principle in the domain D. It follows from the maximum principle that a solution of the mentioned Dirichlet problem is unique.

Let us remind that the boundary ∂D_z of domain D_z cross the real axis at the points b_1, b_2 and $b_1 < b_2$.

For every $z \in \partial D_z \setminus \mathbb{R}$, by $\Gamma_{z\bar{z}}$ we denote that Jordan subarc of the boundary ∂D_z with the end points z and $\bar{z}$ which contains the point b_1. For $z \in \partial D_z \setminus \mathbb{R}$ let $\sqrt{(t - z)(t - \bar{z})}$ be that continuous branch of the analytic function $G(t) = \sqrt{(t - z)(t - \bar{z})}$ outside of the cut along $\Gamma_{z\bar{z}}$ for which $G(b_2) > 0$.

The direction of the circuit of ∂D_z with the domain D_z to the left is taken to be the positive direction.

If the function F has the properties stipulated in Theorem 7, then we shall call F the *creative function* for the function $\psi(x, r)$.

It is established in the papers Plaksa (2003), Mel'nichenko and Plaksa (2008) that the solution of the mentioned Dirichlet problem is expressed in the form (21), where the creative for ψ function F is a solution of the integral equation

$$-\frac{1}{2\pi i} \int\limits_{\partial D_z} \frac{F(t)(t - x)}{\sqrt{(t - z)(t - \bar{z})}} \, dt = \psi_{\partial D}(x, r), \quad \forall (x, r) \in \partial D. \tag{22}$$

Here values of the function $\sqrt{(t - z)(t - \bar{z})}$ for $t \in \Gamma_{z\bar{z}}$ are taken on the right side of the cut $\Gamma_{z\bar{z}}$.

In the papers Plaksa (2003), Mel'nichenko and Plaksa (2008) we developed a method for a transition of Eq. (22) to the Cauchy singular integral equation on the real axis.

In a case important for applications where ∂D_z is a smooth curve satisfying certain additional requirements, then the mentioned singular integral equation is reduced to the Fredholm integral equation of the second kind. Moreover, it is established in Plaksa (2003), Mel'nichenko and Plaksa (2008) that in this case there exists the unique function F which satisfies Eq. (22) and is creative for the solution $\psi(x, r)$ of the mentioned Dirichlet problem.

Let us note that we obtained the Fredholm integral equation for the Dirichlet boundary value problem for the Stokes flow function in the case where the boundary ∂D_z belongs to a class being wider than the class of Lyapunov curves.

Let us note else that in the case where ∂D_z is a circle, the solution F of Eq. (22) is obtained explicitly (see Plaksa 2003; Mel'nichenko and Plaksa 2008).

9 Boundary Value Problem About a Steady Streamline of the Ideal Incompressible Fluid Along an Axial-Symmetric Body

Consider an outer boundary problem having important applications in the hydrodynamics of potential flows.

Let us consider the following *problem about a steady streamline* of the ideal incompressible fluid along an axial-symmetric body: to find a solution $\psi_1(x, r)$ of Eq. (6) in D that satisfies the condition

$$\psi_1(x, r) = 0 \quad \forall \ (x, r) \in \partial D \cup \{(x, r) \in D \ : \ r = 0\} \tag{23}$$

and have the following asymptotic

$$\psi_1(x, r) = \frac{1}{2} v_\infty r^2 + o(1), \quad x^2 + r^2 \to \infty, \ v_\infty > 0. \tag{24}$$

For the model of steady flow of the ideal incompressible fluid the condition (23) means that the boundary ∂D and the axis Ox are lines of flow. In the asymptotic (24) v_∞ is a velocity of unbounded flow at infinity.

We note that explicit solutions of such a problem are known in certain particular cases of steady streamline along an axial-symmetric body (see Lavrentyev and Shabat 1977; Loitsyanskii 1987; Batchelor 1970; Weinstein 1953; Mel'nichenko and Pik 1973a, b, 1975).

Inasmuch as the Stokes flow function

$$\psi(x, r) = \psi_1(x, r) - \frac{1}{2} v_\infty r^2$$

is vanishing at infinity, we can apply the integral expression (21) for solving the boundary value problem about a steady streamline of the ideal incompressible fluid along an axial-symmetric body.

To solve this problem in such a way, we obtain the integral equation

$$\frac{1}{\pi i} \int\limits_{\partial D_z} \frac{F(t)(t - x)}{\sqrt{(t - z)(t - \bar{z})}} \, dt = v_\infty r^2, \quad (x, r) \in \partial D : r \neq 0, \tag{25}$$

where it is necessary to find the function F creative for $\psi(x, r)$. Evidently, Eq. (25) is a particular case of Eq. (22).

In addition, using the integral expression (21) for the Stokes flow function, in the papers Mel'nichenko and Plaksa (2003, 2008) we obtained some results having a natural physical interpretation. Namely, for a boundary problem about a streamline of the ideal incompressible fluid along an axial-symmetric body, we obtained criteria of solvability by means distributions of sources and dipoles on the axis of symmetry and constructed unknown solutions using multipoles together with dipoles distributed on the axis.

9.1 Expressions of Solutions via Distributions of Sources on the Axis

Consider a source located on the axis Ox at the point $(x_0, 0)$ with the intensity q. Such a source is simulated by means of analytic function $F(t) = q/(t - x_0)$ for which the following flow function corresponds by the formula (17):

$$\psi(x, r) = -q \, \frac{x - x_0}{\sqrt{(x - x_0)^2 + r^2}} \, .$$

As a result of an interaction between an flow of the ideal incompressible fluid oncoming with the velocity $v_\infty > 0$ and a source with the intensity $q > 0$ located at the point $(x_1, 0)$ and a source with the intensity $-q$ (i.e. a sink) located at the point $(x_2, 0)$ in the case $x_1 < x_2$ one can obtain the solution

$$\psi_1(x, r) = \frac{1}{2} v_\infty r^2 - q \, \frac{x - x_1}{\sqrt{(x - x_1)^2 + r^2}} + q \, \frac{x - x_2}{\sqrt{(x - x_1 2)^2 + r^2}} \quad (26)$$

of problem about a steady streamline of the ideal incompressible fluid along an axial-symmetric oval body, for which the boundary points satisfy the equality $\psi_1(x, r) = 0$. Lines of flow are given by the equations $\psi_1(x, r) = $ const (see, e.g., Lavrentyev and Shabat 1977, p. 201).

In the case where $x_1 > x_2$, the function (26) is no solution of problem about a steady streamline because, evidently, there are exist no points satisfying the equality $\psi_1(x, r) = 0$ for any $x \in (x_2, x_1)$.

If sources with intensities q_k are located on the axis Ox in points $(x_k, 0)$, respectively, then in order the function

$$\psi_1(x, r) = \frac{1}{2} v_\infty r^2 - \sum_{k=1}^{n} q_k \, \frac{x - x_k}{\sqrt{(x - x_k)^2 + r^2}}$$

be a solution of the problem about a steady streamline, it is necessary (but it is not sufficient, generally speaking) that the total intensity of sources be

$$\sum_{k=1}^{n} q_k = 0 .$$ (27)

Now, suppose that the function $F(z)$ is analytic in $\mathbb{C} \setminus [a_1, a_2]$, where $[a_1, a_2]$ is a segment of the real axis. Let $F^+(t)$ or $F^-(t)$ denote its angular boundary values on (a_1, a_2) when $z \to t$ from a half-plane upper or lower with respect to the real axis, respectively. Denote by $L_p[a_1, a_2]$ the set of functions summable on $[a_1, a_2]$ to the p th power.

Using the Cauchy theorem, it is easy to prove the following theorem having a natural physical interpretation.

Theorem 8 *Suppose that the solution F of Eq. (25) has the form*

$$F(z) = \sum_{k=1}^{n} \frac{q_k}{z - x_{1,k}} + F_1(z) ,$$ (28)

where all $x_{1,k}$ belong to a segment $[a_1, a_2]$, the equality (27) is fulfilled and the function F_1 can be continued to an analytic function outside of the segment $[a_1, a_2]$ and its boundary values $F_1^+(t)$, $F_1^-(t)$ belong to $L_p[a_1, a_2]$, $p > 1$. Then the solution of the problem about steady streamline is given by the formula

$$\psi_1(x, r) = \frac{v_\infty r^2}{2} - \sum_{k=1}^{n} q_k \frac{x - x_{1,k}}{\sqrt{(x - x_{1,k})^2 + r^2}} +$$

$$+ \int_{a_1}^{a_2} \frac{q(t)(t - x)}{\sqrt{(t - x)^2 + r^2}} \, dt \quad \forall \, (x, r) \in D ,$$ (29)

where

$$q(t) := -\frac{1}{2\pi i} \, (F_1^+(t) - F_1^-(t)) \equiv -\frac{1}{\pi} \, \text{Im} \, F_1^+(t)$$

is the distribution density of intensity of sources on $[a_1, a_2]$ which correspond to the function F_1. Moreover, the total intensity of such sources is

$$\int_{a_1}^{a_2} q(t) \, dt = 0.$$ (30)

Theorem 8 generalizes the corresponding theorem in Mel'nichenko and Plaksa (2003, 2008), where the case $q_k \equiv 0$ was considered. Let us note that the formula

(29) with $q_k \equiv 0$ is well-known classical result (see Lavrentyev and Shabat 1977, p. 201). At the same time, in this case, Theorem 8 enables to find the distribution density of sources intensity via boundary values of the creative function F_1 on the set of sources distribution.

It is easy to prove the following theorem converse to Theorem 8.

Theorem 9 *Suppose that the solution of the problem about steady streamline is given by the formula (29), where $q(t) \in L_p[a_1, a_2]$, $p > 1$, and the equalities (27), (30) are fulfilled. Then the solution F of Eq. (25) has the form (28), where the function F_1 can be continued to the function analytic outside of the segment $[a_1, a_2]$ and its boundary values $F_1^+(t)$, $F_1^-(t)$ belong to $L_p[a_1, a_2]$. Moreover, in this case*

$$F_1(z) = -\int_{a_1}^{a_2} \frac{q(t)}{t - z} \, dt \quad \forall \, z \in \mathbb{C} \setminus [a_1, a_2]. \tag{31}$$

Theorem 9 generalizes the corresponding theorem in Mel'nichenko and Plaksa (2003, 2008), where the case $q_k \equiv 0$ was considered.

It is possible to rewrite the formula (29) in a more short form if to introduce a Lebesgue–Stieltjes measure generated by the following function of bounded variation:

$$\mu(t) := \int_{a_1}^{t} q(\tau) \, d\tau + \sum_{k=1}^{n} q_k \, \theta(t - x_{1,k}), \tag{32}$$

where

$$\theta(\tau) := \begin{cases} 1 \text{ for } \tau \geq 0, \\ 0 \text{ for } \tau < 0 \end{cases}$$

is the Heaviside function. Then the formula (29) can be rewritten as

$$\psi_1(x, r) = \frac{v_\infty r^2}{2} + \int_{a_1}^{a_2} \frac{(t - x)}{\sqrt{(t - x)^2 + r^2}} \, d\mu(t) \quad \forall \, (x, r) \in D.$$

We can use as well the following formal equality:

$$d\mu(t) = \left(q(t) + \sum_{k=1}^{n} q_k \, \delta(t - x_{1,k}) \right) dt,$$

where δ is the Dirac delta function.

Thus, for all $(x, r) \in D$, the formula (29) can be also rewritten as

$$\psi_1(x, r) = \frac{v_\infty r^2}{2} + \int_{a_1}^{a_2} \left(q(t) + \sum_{k=1}^{n} q_k \delta(t - x_{1,k}) \right) \frac{(t - x)}{\sqrt{(t - x)^2 + r^2}} \, dt \ .$$

Let's agree that integrals on unlimited intervals of the real axis are understood in the sense of the principal value.

The following theorem was essentially proved in Mel'nichenko and Plaksa (2003, 2008).

Theorem 10 *The distribution density $q(t)$ of intensity of sources, which correspond to the function F_1, is expressed via values of the function (31) on the set $(-\infty, b_1) \cup (b_2, \infty)$ in the form of the repeated integral*

$$q(t) = \frac{b_2 - b_1}{2\pi^2 \sqrt{(b_2 - t)(t - b_1)}} \int_{-\infty}^{\infty} A(t, \xi) \int_{-\infty}^{\infty} B(\xi, \tau) \, d\tau \, d\xi \quad \forall t \in [b_1, b_2]$$

(33)

in the case when it exists. Here

$$A(t, \xi) := \mathrm{ch}\,(\pi\xi) \exp\left(i\xi \ln \frac{t - b_1}{b_2 - t} \right),$$

$$B(\xi, \tau) := -\frac{F(b_1 + (b_2 - b_1)(\mathrm{cth}\,\frac{\tau}{2} + 1)/2)}{\exp(\tau) - 1} \exp(-i\tau\xi).$$

Proof Consider the integral equation (31), in which $a_1 = b_1, a_2 = b_2$ and $z \in (-\infty, b_1) \cup (b_2, \infty)$. Using the change of variables $t = b_1 + (b_2 - b_1)\frac{\tau}{\tau+1}$ and $z = b_1 + (b_2 - b_1)\frac{\xi}{\xi-1}$, we transform it to the integral equation

$$\int_{0}^{\infty} \frac{q_*(\tau)}{\tau + \xi} \, d\tau = F_*(\xi), \qquad \xi > 0,$$

(34)

in which

$$F_*(\xi) := -F\left(b_1 + (b_2 - b_1)\frac{\xi}{\xi - 1} \right) / (\xi - 1)$$

and

$$q_*(\tau) := q\left(b_1 + (b_2 - b_1)\frac{\tau}{\tau + 1} \right) / (\tau + 1).$$

Equation (34) is solvable explicitly (see, e.g., Zabreiko and al 1968, p. 30). As a result of the inversion of integral operator in Eq. (34), we obtain the expression (33) for the distribution density $q(t)$ of intensity of sources on the segment $[b_1, b_2]$. The theorem is proved.

9.2 Expressions of Solutions via Distributions of Dipoles on the Axis

Consider a dipole located on the axis Ox at the point $(x_0, 0)$. Let the moment p of dipole be directed along the axis Ox. We shall also call p by the intensity of dipole. Such a dipole is simulated by means of analytic function $F(t) = p/(t - x_0)^2$ for which the following flow function corresponds by the formula (17):

$$\psi(x, r) = p \frac{r^2}{((x - x_0)^2 + r^2)^{3/2}} \cdot$$

It is well known that as a result of an interaction between an flow of the ideal incompressible fluid oncoming with the velocity $v_\infty > 0$ and a dipole with the intensity $-p$ located at the point $(0, 0)$, one can obtain the picture of steady streamline along a ball with the radius $R = \sqrt[3]{\frac{2p}{v_\infty}}$ and the center in the origin (see, e.g., Lavrentyev and Shabat 1977, p. 200) Lines of flow are given by the equations

$$\psi_1(x, r) = \frac{v_\infty r^2}{2} - p \frac{r^2}{(x^2 + r^2)^{3/2}} = \text{const.}$$

Let us note that inasmuch as the solution F of Eq. (25) has a zero at least of the second order at infinity, then F has a primitive function $\mathcal{F}$ in D which is vanishing at infinity.

Now, using the Cauchy theorem, it is easy to prove the following theorem having also a natural physical interpretation.

Theorem 11 *Suppose that the solution F of Eq. (25) has the form*

$$F(z) = \sum_{k=1}^{m} \frac{p_k}{(z - x_{2,k})^2} + F_2(z), \tag{35}$$

where all $x_{2,k}$ belong to a segment $[a_1, a_2]$ and a primitive function $\mathcal{F}_2$ for the function F_2 can be continued to an analytic function outside of the segment $[a_1, a_2]$ and its boundary values $\mathcal{F}_2^+(t)$, $\mathcal{F}_2^-(t)$ belong to $L_p[a_1, a_2]$, $p > 1$. Then the solution of the problem about steady streamline is given by the formula

$$\psi_1(x, r) = \frac{v_\infty r^2}{2} + \sum_{k=1}^{m} p_k \frac{r^2}{((x - x_{2,k})^2 + r^2)^{3/2}} -$$

$$- r^2 \int_{a_1}^{a_2} \frac{p(t)}{((t - x)^2 + r^2)^{3/2}} dt \quad \forall \, (x, r) \in D, \tag{36}$$

where

$$p(t) := -\frac{1}{2\pi i} \left(\mathcal{F}_2^+(t) - \mathcal{F}_2^-(t)\right) \equiv -\frac{1}{\pi} \operatorname{Im} \mathcal{F}_2^+(t)$$

is the distribution density of intensity of dipoles on $[a_1, a_2]$ which correspond to the function $\mathcal{F}_2$.

It is also easy to prove the following theorem converse to Theorem 11.

Theorem 12 *Suppose that the solution of the problem about steady streamline is given by the formula (36), where $p(t) \in L_p[a_1, a_2]$, $p > 1$. Then the solution F of Eq. (25) has the form (35), where the function F_2 has a primitive function $\mathcal{F}_2$ that can be continued to an function analytic outside of the segment $[a_1, a_2]$ and its boundary values $\mathcal{F}_2^+(t)$, $\mathcal{F}_2^-(t)$ belong to $L_p[a_1, a_2]$. Moreover, in this case*

$$\mathcal{F}_2(z) = -\int_{a_1}^{a_2} \frac{p(t)}{t - z}\, dt \quad \forall\, z \in \mathbb{C} \setminus [a_1, a_2]. \tag{37}$$

The following theorem is proved similarly to Theorem 10.

Theorem 13 *The distribution density $p(t)$ of intensity of dipoles, which correspond to the function $\mathcal{F}_2$, is expressed via values of the function (37) on the set $(-\infty, b_1) \cup (b_2, \infty)$ in the form of the repeated integral*

$$p(t) = \frac{b_2 - b_1}{2\pi^2 \sqrt{(b_2 - t)(t - b_1)}} \int_{-\infty}^{\infty} A(t, \xi) \int_{-\infty}^{\infty} C(\xi, \tau)\, d\tau\, d\xi \quad \forall\, t \in [b_1, b_2]$$

in the case when it exists. Here the function $A(t, \xi)$ is defined in Theorem 10 and

$$C(\xi, \tau) := -\frac{\mathcal{F}_2(b_1 + (b_2 - b_1)(\operatorname{cth} \frac{\tau}{2} + 1)/2)}{\exp(\tau) - 1}\, \exp(-i\tau\xi).$$

Theorem 11–13 generalize the corresponding theorem in Mel'nichenko and Plaksa (2003, 2008), where the case $p_k \equiv 0$ was considered.

Let us note that for a source with the intensity $q > 0$ located at the point $(x_1, 0)$ and a source with the intensity $-q$ located at the point $(x_2, 0)$ in the case $x_1 < x_2$, the following equality holds:

$$-q\, \frac{x - x_1}{\sqrt{(x - x_1)^2 + r^2}} + q\, \frac{x - x_2}{\sqrt{(x - x_2)^2 + r^2}} =$$

$$= -r^2 \int_{x_1}^{x_2} \frac{q}{((t - x)^2 + r^2)^{3/2}}\, dt \quad \forall\, (x, r) \in D.$$

Therefore, such a pair of sources can be replaced by dipoles located on the segment $[x_1, x_2]$ with the distribution density of their intensity $p(t) = q$ for all $t \in [x_1, x_2]$.

Taking into account this note, it is easy to conclude that every solution of the problem about steady streamline of the form (29) is expressed also by the formula (36), where $p_k \equiv 0$, $p(t) = \mu(t)$ and the function $\mu(t)$ is defined by the equality (32).

But among domains D for which the solution of the problem about steady streamline is expressed by the formula (36), there are domains for which the function ψ_1 can not be expressed as (29). For example, the last statement is true in the case where there exists $p_k \neq 0$ in the equality (36). Thus, the class of domains for which the solution of the problem about steady streamline is given by the formula (36) is wider than the class of domains for which the solution of the mentioned problem is given by the formula (29).

If to introduce the function

$$\widetilde{p}(t) := p(t) - \sum_{k=1}^{m} p_k \, \delta(t - x_{2,k})$$

and a Lebesgue–Stieltjes measure generated by the following function of bounded variation:

$$\nu(t) := \int_{a_1}^{t} p(\tau) \, d\tau - \sum_{k=1}^{m} p_k \, \theta(t - x_{2,k}),$$

then the formula (36) can be rewritten as

$$\psi_1(x, r) = \frac{v_\infty r^2}{2} - r^2 \int_{a_1}^{a_2} \frac{d\nu(t)}{((t - x)^2 + r^2)^{3/2}} =$$

$$= \frac{v_\infty r^2}{2} - r^2 \int_{a_1}^{a_2} \frac{\widetilde{p}(t)}{((t - x)^2 + r^2)^{3/2}} \, dt \quad \forall \, (x, r) \in D . \tag{38}$$

Let us note that if the function (38) is the solution of the problem about steady streamline and, in addition, $\widetilde{p}(t) \geq 0$ for all $t \in [a_1, a_2]$, then the axial-symmetric body $\mathbb{R}^2 \setminus D$ is convex in the direction of the axis Or. It follows evidently from a monotonicity with respect to r^2 of the integral in the formula (38).

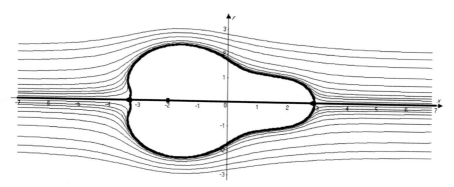

Fig. 1 The streamline along a "pear"

9.3 Using Multipoles for a Construction of Solutions of the Problem About Steady Streamline

At the same time, there are also domains for which it is necessary to use multipoles together with dipoles to obtain the solution of the problem about steady streamline.

For example, the streamline along a "pear" is represented on Fig. 1. In this case, lines of flow are given by the equations

$$\psi_1(x, r) = r^2 \left(1, 7 - \frac{44(x + 2)}{((x + 2)^2 + r^2)^{5/2}} - \frac{20}{((x + 2, 245)^2 + r^2)^{3/2}} - \right.$$

$$\left. - \frac{1}{((x - 1)^2 + r^2)^{3/2}} - \frac{1}{((x - 2)^2 + r^2)^{3/2}} \right) = \text{const.}$$

The solution ψ_1 is obtained by means of three dipoles located in the points $(-2, 245; 0), (1; 0), (2; 0)$ and a quadrupole located in the point $(-2; 0)$. To construct this solution, we use the formula (17) in which the flow function for a quadrupole

$$\psi(x, r) = -\frac{44(x + 2)r^2}{((x + 2)^2 + r^2)^{5/2}}$$

corresponds to the creative function $F(t) = -\frac{88}{3(t+2)^3}$. Let us note that in this case a "pear" $\mathbb{R}^2 \setminus D$ is not convex in the direction of the axis Or.

The streamline along a "matreshka" is represented on Fig. 2. In this case, it turns out already that the body $\mathbb{R}^2 \setminus D$ is convex in the direction of the axis Or. Lines of flow are given by the equations

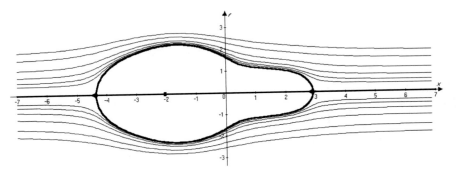

Fig. 2 The streamline along a "matreshka"

$$\psi_1(x, r) = r^2 \left(1,7 - \frac{44(x+2)}{((x+2)^2 + r^2)^{5/2}} - \frac{20}{((x+2,5)^2 + r^2)^{3/2}} - \right.$$

$$\left. - \frac{1}{((x-1)^2 + r^2)^{3/2}} - \frac{1}{((x-2)^2 + r^2)^{3/2}} \right) = \text{const.}$$

The change of the streamline picture is obtained due to a displacement of a dipole from the point $(-2, 245; 0)$ into the point $(-2, 5; 0)$.

An essential specificity for applications is a fact that an use of multipoles can give no solution of the problem about steady streamline. A combination of dipoles and multipoles gives a streamline picture only if certain relations between their intensities are fulfilled.

9.4 Interaction Between a Flow and a Pair "Dipole and Quadrupole"

Let us consider an interaction between a flow of the ideal incompressible fluid oncoming with the velocity $v_\infty > 0$ and a dipole and a quadrupole, which are located on the axis Ox.

Let a quadrupole of intensity m be located at the point $(0; 0)$ and a dipole of intensity p be located at the point $(x_0; 0)$, $x_0 \neq 0$.

We consider two cases: $x_0 > 0$ (see Fig. 3) and $x_0 < 0$ (see. Fig. 4).

(1) In the case where $x_0 > 0$ we use the function

$$\psi(x, r) = -m\frac{3xr^2}{(x^2 + r^2)^{5/2}} - p\frac{r^2}{((x - x_0)^2 + r^2)^{3/2}} \tag{39}$$

that corresponds to the creative function $F(t) = -\frac{2m}{t^3} - \frac{p}{(t-x_0)^2}$ in accordance with the formula (17).

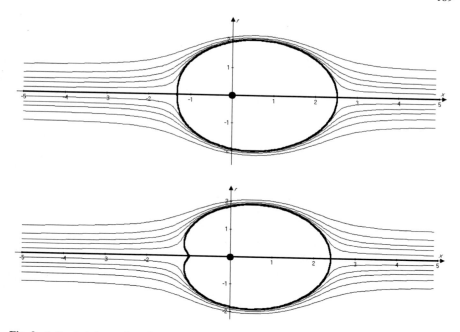

Fig. 3 A dipole is located on the right of a quadrupole

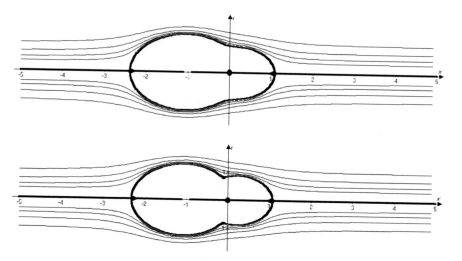

Fig. 4 A dipole is located on the left of a quadrupole

If the intensity p of the dipole is small in comparison with the intensity m of the quadrupole (the quantitative relation is formulated below), then the singularity $(0, 0)$ of the function $\psi(x, r)$ is located on the boundary that has the equation

$$\frac{v_\infty}{2} + \frac{\psi(x, r)}{r^2} = 0$$

which, taking into account the equality (39), we rewrite as

$$\frac{v_\infty}{2} - \frac{3xm}{(x^2 + r^2)^{5/2}} - \frac{p}{((x - x_0)^2 + r^2)^{3/2}} = 0. \tag{40}$$

Therefore, the function

$$\psi_1(x, r) = \frac{v_\infty r^2}{2} - m\frac{3xr^2}{(x^2 + r^2)^{5/2}} - p\frac{r^2}{((x - x_0)^2 + r^2)^{3/2}} \tag{41}$$

is no solution of the problem about steady streamline.

An increase of the dipole intensity results finally in formation of a closed contour Γ for which coordinates of points satisfy the Eq. (40) and, moreover, the point $(0, 0)$ is located inside of the domain bounded by Γ. Then the equality (40) is fulfilled at a point $(x, 0)$ with $x < 0$, i.e. we have the following equality:

$$\frac{v_\infty}{2} + \frac{3m}{x^4} - \frac{p}{(x_0 - x)^3} = 0,$$

from which we find

$$p = \frac{v_\infty}{2}(x_0 - x)^3 \left(1 + \frac{6m}{v_\infty x^4}\right). \tag{42}$$

The equality (42) is fulfilled at a point $(x, 0)$ with $x < 0$ if and only if

$$p \geq \frac{v_\infty}{2} \min_{x < 0}\left((x_0 - x)^3 \left(1 + \frac{6m}{v_\infty x^4}\right)\right) =: c_1,$$

and in this case the function (41) gives the solution of the problem about steady streamline for unbounded domain with the boundary Γ.

Now, we can assert:

(a) if $p \geq c_1$, then the function (41) is the solution of the problem about steady streamline for a certain domain D;

(b) if $p < c_1$, then there is no domain D, for which the function (41) would be the solution of the problem about steady streamline.

One can see the examples of streamline picture in the cases $p > c_1$ (the upper picture on Fig. 3) and $p = c_1$ (the lower picture on Fig. 3).

(2) In the case where $x_0 < 0$ we conclude that if the point $(0, 0)$ is located inside of the domain bounded by the contour Γ, then for each $x \in [x_0, 0]$ there exists a point (x, r) with $r > 0$ that the equality (40) is fulfilled, from which we find

$$p = \frac{v_\infty}{2}((x - x_0)^2 + r^2)^{3/2} \left(1 - \frac{6mx}{v_\infty(x^2 + r^2)^{5/2}}\right). \tag{43}$$

For each fixed $x \in [x_0, 0]$, the equality (43) is fulfilled at a point (x, r) with $r > 0$ if and only if

$$p \geq \frac{v_\infty}{2} \min_{r \geq 0} \left((x - x_0)^2 + r^2)^{3/2} \left(1 - \frac{6mx}{v_\infty(x^2 + r^2)^{5/2}}\right)\right).$$

Finally, for each $x \in [x_0, 0]$ there exists a point (x, r) with $r > 0$ in which the equality (43) is fulfilled if and only if

$$p \geq \frac{v_\infty}{2} \max_{x \in [x_0, 0]} \min_{r \geq 0} \left(((x - x_0)^2 + r^2)^{3/2} \left(1 - \frac{6mx}{v_\infty(x^2 + r^2)^{5/2}}\right)\right) =: c_2.$$

Thus, the following statements are true:

(a) if $p \geq c_2$, then the function (41) is the solution of the problem about steady streamline for a certain domain D;

(b) if $p < c_2$, then there is no domain D, for which the function (41) would be the solution of the problem about steady streamline.

One can see the examples of streamline picture in the cases $p > c_2$ (the upper picture on Fig. 4) and $p = c_2$ (the lower picture on Fig. 4).

Let us note that the boundary ∂D is piecewise-smooth if $p = c_1$ in the case where $x_0 > 0$ (see the lower picture on Fig. 3) or $p = c_2$ in the case where $x_0 < 0$ (see the lower picture on Fig. 4).

Acknowledgements This research is partially supported by Grant of Ministry of Education and Science of Ukraine (Project No. 0116U001528).

References

Aleksandrov, A.Ya., Soloviev, Yu.P.: Three-Dimensional Problems of the Theory of Elasticity. Moscow, Nauka (in Russian) (1979)

Batchelor, G.K.: An Introduction to Fluid Dynamics. Cambridge (1970)

Bateman, H.: Partial Differential Equations of Mathematical Physics. Dover, New York (1944)

Blum, E.K.: A theory of analytic functions in Banach algebras. Trans. Am. Math. Soc. **78**, 343–370 (1955)

Cerejeiras, P., Kähler, U.: Elliptic boundary value problems of fluid dynamics over unbounded domains. Math. Methods Appl. Sci. **23**, 81–101 (2000)

Colombo, F., Sabadini, I., Struppa, D.C.: Progress in mathematics. In: Noncommutative Functional Calculus: Theory and Applications of Slice Hyperholomorphic Functions, vol. 289 (2011)

Edenhofer, J.: A solution of the biharmonic Dirichlet problem by means of hypercomplex analytic functions. In: Meister, V.E., Wendland, W.L., Weck, N. (eds.) Functional Theoretic Methods for Partial Differential Equations (Proceedings of International Symposium Held at Darmstand,

Germany, April 12–15, 1976). Lecture Notes in Mathematics, vol. 561, pp. 192–202. Springer, Berlin (1976)

Erdelyi, A.: Singularities of generalized axially symmetric potentials. Commun. Pure Appl. Math. 9(3), 403–414 (1956)

Fueter, R.: Die Funktionentheorie der Differentialgleichungen $\Delta u = 0$ und $\Delta \Delta u = 0$ mit vier reellen Variablen. Comment. Math. Helv. 7, 307–330 (1935)

Goursat, E.: Cours d'analyse mathematique, vol. 2. Gauthier-Villars, Paris (1910)

Gilbert, R.P.: Function Theoretic Methods in Partial Differential Equations. Academic, New York, London (1969)

Grigorev, Y.: Quaternionic functions and their applications in a viscous fluid flow. Complex Anal. Oper. Theory 12, 491–508 (2017). https://doi.org/10.1007/s11785-017-0715-z

Gryshchuk, S.V., Plaksa, S.A.: Monogenic functions in a biharmonic algebra. Ukr. Math. J. 61(12), 1865–1876 (2009)

Gryshchuk, S.V., Plaksa, S.A.: Basic properties of monogenic functions in a biharmonic plane. In: Complex Analysis and Dynamical Systems V, Contemporary Mathematics, vol. 591, pp. 127–134. Providence, R.I. (2013)

Gryshchuk, S.V., Plaksa, S.A.: Schwartz-type integrals in a biharmonic plane. Int. J. Pure Appl. Math. 83(1), 193–211 (2013)

Gryshchuk, S.V., Plaksa, S.A.: Monogenic functions in the biharmonic boundary value problem. Math. Methods Appl. Sci. 39(11), 2939–2952 (2016)

Gryshchuk, S.V., Plaksa, S.A.: A Schwartz-type boundary value problem in a biharmonic plane. Lobachevskii J. Math. 38(3), 435–442 (2017)

Gryshchuk, S.A.: Reduction of a Schwartz-type boundary value problem for biharmonic monogenic functions to Fredholm integral equations. Open Math. 15(1), 374–381 (2017)

Gürlebeck, K., Habetha, K., Sprößig, W.: Application of Holomorphic Functions in Two and Higher Dimensions. Birkhäuser, Boston (2016)

Gürlebeck, K., Sprössig, W.: Quaternionic and Clifford Calculus for Physicists and Engineers. Wiley, New York (1997)

Hamilton, W.: Elements of Quaternions. University of Dublin press, Dublin (1866)

Henrici, P.: Zur Funktionentheory der Wellengleichung. Comment. Math. Helv. 27(3–4), 235–293 (1953)

Henrici, P.: On the domain of regularity of generalized axially symmetric potentials. Proc. Am. Math. Soc. 8(1), 29–31 (1957)

Hille, E., Phillips, R.S.: Functional Analysis and Semi-Groups. American Mathematical Society, Providence (1957)

Huber, A.: On the uniqueness of generalized axially symmetric potentials. Ann. Math. 60(2), 351–358 (1954)

Kapshivyi, A.A.: On a fundamental integral representation of x-analytic functions and its application to solution of some integral equations. In: Mathematical Physics, Kiev, vol. 12, pp. 38–46 (in Russian) (1972)

Keldysh, M.V.: On some cases of degeneration of an equation of elliptic type on the boundary of a domain. Dokl. Akad. Nauk SSSR 77(2), 181–183 (1951). in Russian

Ketchum, P.W.: Analytic functions of hypercomplex variables. Trans. Am. Math. Soc. 30(4), 641–667 (1928)

Ketchum, P.W.: A complete solution of Laplace's equation by an infinite hypervariable. Am. J. Math. 51, 179–188 (1929)

Kovalev, V.F., Mel'nichenko, I.P.: Biharmonic functions on biharmonic plane. Dop. AN Ukr. Ser. A 8, 25–27 (1981). in Russian

Kravchenko, V.V.: Applied Pseudoanalytic Function Theory. Birkhäuser, Boston (2009)

Kravchenko, V.V., Shapiro, M.V. In: Integral Representations for Spatial Models of Mathematical Physics. Pitman Research Notes in Mathematics. Addison Wesley Longman Inc, Menlo Park (1996)

Krivenkov, Y.P.: On one representation of solutions of the EulerPoisson-Darboux equation. Dokl. Akad. Nauk SSSR **116**(3), 351–354 (1957)

Krivenkov, Y.P.: Representation of solutions of the Euler–Poisson–Darboux equation via analytic functions. Dokl. Akad. Nauk SSSR **116**(4), 545–548 (1957)

Krivenkov, Y.P.: Problem D for the Euler–Poisson–Darboux equation. Investig. Mech. Appl. Math. (5), 134–145 (1960)

Kunz, K.S.: Application of an algebraic technique to the solution of Laplace's equation in three dimensions. SIAM J. Appl. Math. **21**(3), 425–441 (1971)

Lavrentyev, M.A.: Cumulative charge and the principles of its operation. Uspekhi matematicheskikh nauk **12**(4), 41–56 (1957)

Lavrentyev, M.A., Shabat, B.V.: Problems of Hydrodynamics and Theirs Mathematical Models. Nauka, Moscow (in Russian) (1977)

Lavrentyev, M.A., Shabat, B.V.: Methods of the Theory of Functions of a Complex Variable. Nauka, Moscow (in Russian) (1987)

Leutwiler, H.: Modified quaternionic analysis in $\mathbb{R}^3$. Complex Var. Theory Appl. **20**, 19–51 (1992)

Loitsyanskii, L.G.: Mechanics of Liquids and Gases. Nauka, Moscow (in Russian) (1987)

Lorch, E.R.: The theory of analytic function in normed abelin vector rings. Trans. Am. Math. Soc. **54**, 414–425 (1943)

Mackie, A.G.: Contour integral solutions of a class of differential equations. J. Ration. Mech. Anal. **4**(5), 733–750 (1955)

Mel'nichenko, I.P.: The representation of harmonic mappings by monogenic functions. Ukr. Math. J. **27**(5), 499–505 (1975)

Mel'nichenko, I.P.: On a Method of Description of Potential Fields with Axial Symmetry, Contemporary Questions of Real and Complex Analysis, pp. 98–102. Institute of Mathematics of Ukrainian Academy of Sciences, Kiev (1984)

Mel'nichenko, I.P.: Biharmonic bases in algebras of the second rank. Ukr. Math. J. **38**(2), 224–226 (1986)

Mel'nichenko, I.P.: Algebras of functionally invariant solutions of the three-dimensional Laplace equation. Ukr. Math. J. **55**(9), 1551–1557 (2003)

Mel'nichenko, I.P., Pik, E.M.: On a method for obtaining axial-symmetric flows. Dop. AN Ukr. Ser. A **2**, 152–155 (1973a)

Mel'nichenko, I.P., Pik, E.M.: Quaternion equations and hypercomplex potentials in the mechanics of a continuous medium. Sov. Appl. Mech. **9**(4), 383–387 (1973b)

Mel'nichenko, I.P., Pik, E.M.: Quaternion potential of the ideal noncomprssible fluid. Prikl. Mech. **11**(1), 125–128 (1975)

Mel'nichenko, I.P., Plaksa, S.A.: Potential fields with axial symmetry and algebras of monogenic functions of vector variable, I. Ukr. Math. J. **48**(11), 1717–1730 (1996)

Mel'nichenko, I.P., Plaksa, S.A.: Potential fields with axial symmetry and algebras of monogenic functions of vector variable, II. Ukr. Math. J. **48**(12), 1916–1926 (1996)

Mel'nichenko, I.P., Plaksa, S.A.: Potential fields with axial symmetry and algebras of monogenic functions of vector variable, III. Ukr. Math. J. **49**(2), 253–268 (1997)

Mel'nichenko, I.P., Plaksa, S.A.: Outer Boundary Problems for the Stokes Flow Function and Steady Streamline Along Axial-symmetric Bodies, Complex Analysis and Potential Theory, pp. 82–91. Institute of Mathematics of the National Academy of Sciences of Ukraine, Kiev (2003)

Mel'nichenko, I.P., Plaksa, S.A.: Commutative algebra of hypercomplex analytic functions and solutions of elliptic equations degenerating on an axis. Zb. Pr. Inst. Mat. NAN Ukr. **1**(3), 144–150 (2004)

Mel'nichenko, I.P., Plaksa, S.A.: Commutative Algebras and Spatial Potential Fields. Institute of Mathematics NAS of Ukraine, Kiev (2008)

Mikhailov, L.G., Rajabov, N.: An analog of the Poisson formula for certain second-order equations with singular line. Dokl. Akad. Nauk Tadzh. SSR **15**(11), 6–9 (1972)

Moisil, G.C., Theodoresco, N.: Functions holomorphes dans l'espace. Mathematica (Cluj) **5**, 142–159 (1931)

Plaksa, S.A.: On integral representations of an axisymmetric potential and the Stokes flow function in domains of the meridian plane, I. Ukr. Math. J. **53**(5), 726–743 (2001)

Plaksa, S.A.: On integral representations of an axisymmetric potential and the Stokes flow function in domains of the meridian plane, II. Ukr. Math. J. **53**(6), 938–950 (2001)

Plaksa, S.A.: Dirichlet problem for an axisymmetric potential in a simply connected domain of the meridian plane. Ukr. Math. J. **53**(12), 1976–1997 (2001)

Plaksa, S.A.: On an outer Dirichlet problem solving for the axial-symmetric potential. Ukr. Math. J. **54**(12), 1982–1991 (2002)

Plaksa, S.A.: Dirichlet problem for the Stokes flow function in a simply connected domain of the meridian plane. Ukr. Math. J. **55**(2), 241–281 (2003)

Plaksa, S.A.: Singular and Fredholm integral equations for Dirichlet boundary problems for axial-symmetric potential fields. In: Factotization, Singular Operators and Related Problems: Proceedings of the Conference in Honour of Professor Georgii Litvinchuk, Funchal, Jan 28–Feb 1 (2002), pp. 219–235. Kluwer Academic publishers, Dordrecht (2003)

Plaksa, S.: Commutative algebras of hypercomplex monogenic functions and solutions of elliptic type equations degenerating on an axis. In: More Progress in Analysis: Proceedings of 5th International ISAAC Congress, Catania, July 25–30 (2005), pp. 977–986. World Scientific, Singapore (2009)

Plaksa, S.A.: Commutative algebras associated with classic equations of mathematical physics. In: Advances in Applied Analysis. Trends in Mathematics, pp. 177–223. Springer, Basel (2012)

Plaksa, S.A.: Integral theorems for monogenic functions in an infinite-dimensional space with a commutative multiplication. Zb. Pr. Inst. Mat. NAN Ukr. **10**(4–5), 306–319 (2013)

Plaksa, S.A., Pukhtaievych, R.P.: Constructive description of monogenic functions in a three-dimensional harmonic algebra with one-dimensional radical. Ukr. Math. J. **65**(5), 740–751 (2013)

Plaksa, S.A., Pukhtaievych, R.P.: Monogenic functions in a finite-dimensional semi-simple commutative algebra. An. Şt. Univ. Ovidius Constanţa **22**(1), 221–235 (2014)

Plaksa, S.A., Shpakovskii, V.S.: Constructive description of monogenic functions in a harmonic algebra of the third rank. Ukr. Math. J. **62**(8), 1251–1266 (2011)

Plaksa, S.A., Shpakivskyi, V.S.: A description of spatial potential fields by means of monogenic functions in infinite-dimensional spaces with a commutative multiplication. Bull. Soc. Sci. Lett. Łódź **62**(2), 55–65 (2012)

Plaksa, S.A., Shpakivskyi, V.S.: Cauchy theorem for a surface integral in commutative algebras. Complex Var. Elliptic Equ. **59**(1), 110–119 (2014)

Plaksa, S.A., Shpakivskyi, V.S.: Monogenic functions in a finite-dimensional algebra with unit and radical of maximal dimensionality. J. Alger. Math. Soc. **1**, 1–13 (2014)

Plaksa, S.A., Shpakivskyi, V.S.: An extension of monogenic functions and spatial potentials. Lobschevskii J. Math. **38**(2), 330–337 (2017)

Polozhii, G.N.: Theory and Application of p-Analytic and (p, q)-Analytic Functions. Kiev, Naukova Dumka (1973)

Polozhii, G.N., Ulitko, A.F.: On formulas for an inversion of the main integral representation of p-analiytic function with the characteristic $p = x^k$. Prikl. Mekhanika **1**(1), 39–51 (1965)

Privalov, I.I.: Boundary Properties of Analytic Functions. Gostekhizdat, Moscow (1950)

Rajabov, N.R.: Some boundary-value problems for an equation of the axisymmetric field theory. In: Investigations on Boundary-Value Problems in the Theory of Functions and Differential Equations, pp. 79–128. Academy of Sciences of Tadzhik SSR, Dushanbe (1965)

Rajabov, N.R.: Integral representations and their inversion for a generalized Cauchy–Riemann system with singular line. Dokl. Akad. Nauk Tadzh. SSR **11**(4), 14–18 (1968)

Rajabov, N.R.: Construction of potentials and investigation of inner and outer boundary problems of Dirichlet and Neumann types for the Euler– Poisson– Darboux equations on the plane. Dokl. Akad. Nauk Tadzh. SSR **17**(8), 7–11 (1974)

Riley, J.D.: Contributions to the theory of functions of a bicomplex variable. Tohoku Math. J. **5**(2), 132–165 (1953)

Ringleb, F.: Beiträge zur funktionentheorie in hyperkomplexen systemen, I. Rend. Circ. Mat. Palermo **57**(1), 311–340 (1933)

Roşculeţ, M.N.: Algebre infinite associate la ecuaţii cu derivate parţiale, omogene, cu coeficienţi constanţi de ordin oarecare. Studii şi Cercetări Matematice **6**(3–4), 567–643 (1955)

Ryan, J.: Dirac operators, conformal transformations and aspects of classical harmonic analysis. J. Lie Theory **8**, 67–82 (1998)

Shpakivskyi, V.S.: Constructive description of monogenic functions in a finite-dimensional commutative associative algebra. Adv. Pure Appl. Math. **7**(1), 63–76 (2016)

Shpakivskyi, V.S.: Curvilinear integral theorems for monogenic functions in commutative associative algebras. Adv. Appl. Clifford Algebras **26**, 417–434 (2016)

Snyder, H.H.: An introduction to theories of regular functions on linear associative algebras. In: Rihard, N.D. (ed.) Commutative Algebra. Analytic methods /Lecture Notes in Pure and Applied Mathematics, vol. 68, pp. 75–94. Marcel Dekker inc., New York, Basel, (1982)

Sobrero, L.: Nuovo metodo per lo studio dei problemi di elasticità, con applicazione al problema della piastra forata. Ricerche di Ingegneria **13**(2), 255–264 (1934)

Sudbery, A.: Quaternionic analysis. Math. Proc. Camb. Phil. Soc. **85**, 199–225 (1979)

Trokhimchuk, Ju.Ju: Continuous Mappings and Conditions of Monogeneity. Israel Program for Scientific Translations, Jerusalem; Daniel Davey & Co., Inc, New York (1964)

Wagner, R.D.: The generalized Laplace equations in a function theory for commutative algebras. Duke Math. J. **15**, 455–461 (1948)

Ward, J.A.: From generalized Cauchy–Riemann equations to linear algebras. Proc. Am. Math. Soc. **4**, 456–461 (1953)

Weinstein, A.: Discontinuous integrals and generalized potential theory. Trans. Am. Math. Soc. **63**(2), 342–354 (1948)

Weinstein, A.: Generalized axially symmetric potential theory. Bull. Am. Math. Soc. **59**(1), 20–38 (1953)

Weinstein, A.: Singular partial differential equations and their applications. In: Proceedings of the Symposium University of Maryland (1961); Fluid dynamic and applied mathematics, pp. 29–49 (1962)

Whittaker, E.T., Watson, G.N.: A Course of Modern Analysis, vol. 2. Cambridge University Press, Cambridge (1927)

Zabreiko, P.P. et al.: Integral Equations. Nauka, Moscow (1968)

Zhang, B., Fu, Y., Rădulescu, V.D.: The stationary Navier-Stokes equations in variable exponent spaces of Clifford-valued functions. Adv. Appl. Clifford Algebras **24**, 231–252 (2014)

Monte Carlo Method Application and Generation of Random Numbers by Usage of Numerical Methods

Dušan Knežo and Alena Vagaská

Abstract Social scientists increasingly use statistical simulation techniques to help them understand the social processes they care about and the statistical methods used to study them. There are two types of computer simulation techniques, which are quickly becoming essential tools for empirical social scientists: Monte Carlo simulation and resampling methods. This chapter is dealing with Monte Carlo method that is very often used for simulating systems with many coupled degrees of freedom, for simulation of experiments in many areas of research, for investigation of processes with a random character and is most useful when it is difficult to use other approaches. The chapter presents some methods of generating random numbers by usage of standard numerical methods for various probability distributions types as well as application possibility of Monte Carlo method for a CA-simulation of the random processes.

Keywords Monte carlo method · Random numbers · Numerical methods

1 Introduction

The Monte Carlo Method (MCM), also known as Monte Carlo Simulation (MCS) is one of the stochastic methods. It is very often used in the cases, when using of analytical methods is too complicated or difficult. Monte Carlo simulation is a computerized mathematical technique and, for this reason, its usage has gone hand-in-hand with the increasing effectiveness of computers. Since computer technology has been developing at high speed in recent decades, the importance of the Monte Carlo method is growing. With increasing memory capacity and growing computer

D. Knežo · A. Vagaská (✉)
Faculty of Manufacturing Technologies, Department of Natural Sciences and Humanities, Bayerova 1, 080 01 Prešov, Slovak Republic
e-mail: alena.vagaska@tuke.sk

D. Knežo
e-mail: dusan.knezo@tuke.sk

© Springer Nature Switzerland AG 2019
C. Flaut et al. (eds.), *Models and Theories in Social Systems*, Studies in Systems, Decision and Control 179, https://doi.org/10.1007/978-3-030-00084-4_10

performance, we can now make a huge amount of simulations in a relatively short time, so simulation of processes and systems by usage of Monte Carlo method has proved to be effective in many areas of real life. Professionals in such widely disparate fields as finance, project management, energy, manufacturing, engineering, research and development, insurance, traffic flow, environment and of course social science use the technique of Monte Carlo Simulation (Dagpunar 2007; Fishman 1996; McLeish 2005; Glasserman 2003; Woch et al. 2017).

We are permanently confronted with uncertainty, ambiguity, and variability. In addition, even though we have unprecedented access to information, we cannot accurately predict the future. For every decision, what we make, we have to consider some risks, so it is necessary to perform risk analysis. Monte Carlo simulation lets us see all the possible outcomes of our decisions and assess the impact of risk, allowing for better decision making under uncertainty. Monte Carlo simulations is a computer-intensive technique that allows people to account for risk in quantitative analysis and decision-making. Monte Carlo simulation provides the decision-maker with a range of possible outcomes and the probabilities they will occur for any choice of action. It shows the extreme possibilities—the outcomes of going for broke and for the most conservative decision—along with all possible consequences for middle-of-the-road decisions (Čičmanec and Bořil 2017; Jirgl et al. 2016; Kubíček et al. 2014).

Monte Carlo simulation performs risk analysis by building models of possible results by substituting a range of values—a probability distribution—for any factor that has inherent uncertainty. It then calculates results over and over, each time using a different set of random values from the probability functions. Depending upon the number of uncertainties and the ranges specified for them, a Monte Carlo simulation could involve thousands or tens of thousands of recalculations before it is complete. Monte Carlo simulation produces distributions of possible outcome values. By using probability distributions, variables can have different probabilities of different outcomes occurring. Probability distributions are a much more realistic way of describing uncertainty in variables of a risk analysis.

MCS is playing an increasing role in commercial applications, including marketing and Customer Relationship Management (CRM), risk analysis in global finance. It provides an efficient way to simulate processes involving chance and uncertainty and can be applied in areas as diverse as market sizing, customer lifetime value measurement and customer service management.

Scientists working on the atom bomb first used this technique; it was named for Monte Carlo, the Monaco resort town renowned for its casinos. Since its introduction in World War II, Monte Carlo simulation has been used to model a variety of social, physical, chemical and other conceptual systems.

The Monte Carlo methods are very useful for simulating in many areas of research, for example Monte Carlo simulations are widely used in engineering for sensitivity analysis, in computational physics, physical chemistry, computational biology, finance and business, artificial intelligence and of course in applied mathematics and statistics (Glasserman 2003; Knežo 2012a, b, 2014; Sobol 1971; Hubacek 2017; Svatošová 2017).

This method is based on generating of random numbers or pseudorandom numbers reflected on probability distribution (Gentle 2003). Uses of Monte Carlo methods require large amounts of random numbers and it was that their use led to development of pseudorandom number generators. Specific algorithms and functions are used for this purpose as a part of common or specific software.

Let us assume that we have some generator of random or pseudorandom numbers following the standard uniform distribution $U(0, 1)$. The *Free Pascal* programming language provides us the function *Random*, which is able to generate pseudorandom numbers from the interval $[0, 1)$. This function generates pseudorandom numbers based on the Mersenne Twister algorithm. The *Free Pascal* was used for performed methods presented in this chapter.

2 Methods of Random Numbers Generating

Assuming we have a generator of pseudorandom number from the standard uniform distribution $U(0, 1)$, we can use some various methods for generating of random or pseudorandom numbers satisfying continuous distribution D.

The Inverse method is used very often. Let us assume that we want to generate random numbers following distribution D with the probability density function f and distribution function F. As we have a random number r_U from the standard uniform distribution $U(0, 1)$, so for the number r_U we can write

$$r_D = F^{-1}(r_U),\tag{1}$$

where F^{-1} is an inverse function to the function F.

The typical example for this case is generation of random numbers from exponential distribution with the probability density function

$$f(x) = \lambda e^{-\lambda x},\ x \geq 0,\tag{2}$$

and distribution function

$$F(x) = 1 - e^{-\lambda x},\ x \geq 0.\tag{3}$$

Because

$$F^{-1}(x) = -\frac{1}{\lambda}\ln x,\tag{4}$$

so the number r_E

$$r_E = -\frac{1}{\lambda}\ln r_U,\tag{5}$$

is a random number from the exponential distribution.

The Acceptance-Rejection Method is additional very often used method, which is based on the ability to generate random numbers from the distribution P with the probability density function

$$h(x) = \frac{1}{c}g(x), x \in M,$$ (6)

where

$$g(x) \geq f(x), x \in M,$$ (7)

and the number c satisfies

$$c = \int_M g(x)dx \geq \int_M f(x)dx = 1.$$ (8)

Generation of random number r_D runs over three steps. In the first step the random number r_D from the standard uniform distribution $U(0, 1)$ is generated. In the second step, we generate the random number r_P from the distribution P. In the last step we make decision about acceptance-rejection of r_D. If

$$r(u) \leq g(r_P),$$ (9)

than we set $r_D = r_P$, otherwise we repeat this algorithms.

The Box–Muller transformation is also suitable method for generating random numbers r_N from the standardized normal distribution $N(0, 1)$. This method provides us to generate couple random numbers r_{1N} and r_{2N} based on the known random numbers r_{1U} and r_{2U} by usage of the next formulas

$$r_{1N} = \sqrt{-2 \ln r_{1U}} \cdot \cos(2\pi \cdot r_{2U}),$$ (10)

$$r_{2N} = \sqrt{-2 \ln r_{1U}} \cdot \sin(2\pi \cdot r_{2U}).$$ (11)

3   Generating of Random Numbers from Standardized Normal Distribution by Inverse Transformation Method

In order to generate random numbers from the standard normal distribution, the usage of Inverse Transformation Method is useful. To solve the Eq. (1) we have used standard numerical methods.

The distribution function Φ of the standard normal distribution $N(0, 1)$ may also be expressed as follows

$$\Phi(x) = \begin{cases} 0.5 - \frac{1}{\sqrt{2\pi}} \int\limits_{0}^{|x|} e^{-\frac{t^2}{2}} dt, & x < 0, \\[3mm] 0.5 + \frac{1}{\sqrt{2\pi}} \int\limits_{0}^{x} e^{-\frac{t^2}{2}} dt, & x \geq 0, \end{cases} \tag{12}$$

so it is enough to calculate integral (13) during the process of numerical expression of value of distribution function

$$\int\limits_{0}^{x} \varphi(t)dt = \frac{1}{\sqrt{2\pi}} \int\limits_{0}^{x} e^{-\frac{t^2}{2}} dt, \ x \geq 0, \tag{13}$$

It is possible to prove that for all x is satisfied

$$\left| \varphi^4(x) \right| \leq \varphi^4(0) = \frac{3}{\sqrt{2\pi}}, \tag{14}$$

so for numerical computation of integral (13) by usage of Simpson's rule with accuracy ε, the step h is chosen according to the next condition

$$h \leq \sqrt[4]{\frac{60\varepsilon \cdot \sqrt{2\pi}}{x}}. \tag{15}$$

Based on the known value of r_U we can calculate the corresponding value r_N as the solution of the equation

$$\Phi(r_N) = r_U, \tag{16}$$

where the Eq. (16) can be calculated by the bisection method.

In the case that we want to generate a random number from the distribution $N(\mu, \sigma^2)$, it is possible to use the next formula (17) after generating a random number from the distribution $N(0, 1)$

$$r_{N(\mu,\sigma^2)} = \mu + \sigma \cdot r_N. \tag{17}$$

The calculations were carried out using the Free Pascal program, while to calculate integral and to solve the Eq. (16) the accuracy of 10^{-11} was used.

In the Fig. 1, it can be seen a comparison of 10^6 the generated random numbers and the probability density function of the standard normal distribution. Coincidence was also confirmed by statistical tests.

Fig. 1 Comparison of
random number histogram
with probability density
$N(0,1)$ *Source* own

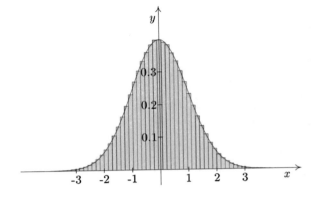

4 Generating of Random Numbers from Log-Normal Distribution by Usage of Inverse Transformation Method

The probability density function of log-normal distribution $LN(\mu, \sigma^2)$ is given by the formula

$$f(x) = \frac{1}{x \cdot \sqrt{2\pi}} \cdot e^{-\frac{(\ln x - \mu)^2}{2\sigma^2}}. \tag{18}$$

As in the case of normal distribution, based on the known value of r_U we can calculate the corresponding value r_{LN} as the solution of the equation

$$F(r_{LN}) = r_U, \tag{19}$$

where Richardson extrapolation was used to calculate the value of the distribution function F, i.e. for numerical calculation of definite integral.

The calculations were carried out using the Free Pascal program, while to calculate integral and to solve the Eq. (19) the accuracy of 10^{-8} was used. A comparison of 10^5 the generated random numbers and the probability density function of log-normal distribution $LN(0, 1)$ can be seen in the Fig. 2. Coincidence was also confirmed by statistical tests.

5 Generating of Random Numbers from Other Distribution by Usage of Inverse Transrormation Method

The procedure, which was used for generating of random numbers from the standard normal distribution and log-normal distribution, can be used analogously also for Student t-distribution $T(n)$, χ^2 - distribution $\chi^2(n)$ and Fisher Snedecor F - distribution $F(m, n)$.

Fig. 2 Comparison of random number histogram with probability density $LN(0,1)$ *Source* own

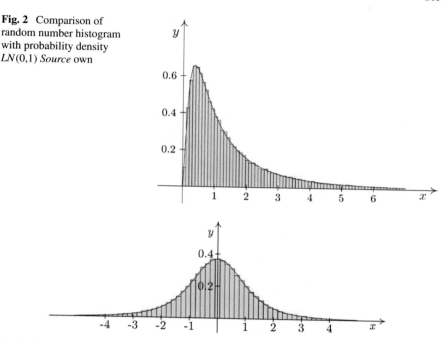

Fig. 3 Comparison of random number histogram with probability density $T(3)$ *Source* own

The Fig. 3 shows a comparison of 10^5 generated random numbers and the probability density function of Student t-distribution $T(3)$.

6 Application of Monte Carlo Method to Estimate Costs Concerning Prosthetic and Orthotic Aids

The cost value of sold prosthetic and orthotic devices depends on many random factors, so it is useful to apply Monte Carlo method when estimate this value. Based on the data obtained from real life, namely data expressing the quantity of prosthetic and orthotic aids sold during one year 2011 in chosen region in Slovakia, the simulation of costs w performed.

From the obtained data, we have taken into account only data relating to the four basic categories of aids:

- orthotic shoes and orthotic inserts
- soft orthoses for lower and upper limbs
- splints for lower and upper limbs
- torso orthoses.

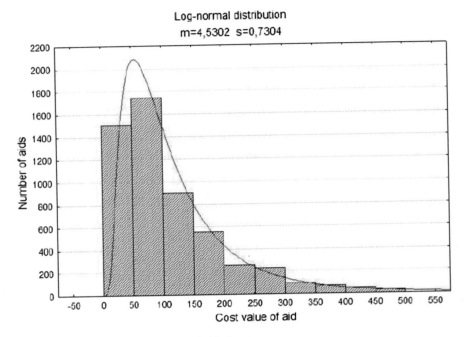

Fig. 4 Histogram of cost value of issued aids *Source* own

In these categories, a total number of 5620 devices were issued in the selected region, with a total value of 666 059 €.

Assuming that the value of devices issued in the given year is from the Poisson probability distribution with parameter $\lambda = 56$ (the number of devices expressed in hundreds), so

$$P(x = k) = \frac{56^k \cdot e^{-56}}{k!}, \tag{20}$$

we have generated 100 random values $N_1, N_2, \ldots, N_{100}$ obeying Poisson distribution given by formula (5). These values represent the number of instruments issued for a variety of 100 years. Random values were generated in two steps:

- generation of random number p from interval [0, 1],
- determination of non-negative integer x meeting the condition $x = F^{-1}(p)$, where F^{-1} is the inverse function of the Poisson distribution function with the parameter $\lambda = 56$

In the next step, we estimated the type of distribution of our data obtained from practice (including estimation of parameters). By testing, we have concluded that empirical data are corresponding to logarithmic-normal distribution with parameters $\mu = 4.5302$ and $\sigma = 0.7304$, (see Fig. 4) and with probability density function

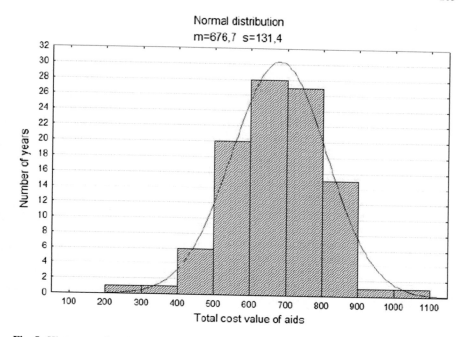

Fig. 5 Histogram of total cost value of aids issued during one year *Source* own

$$f(x) = \frac{1}{0.7304\, x\, \sqrt{2\pi}} e^{-\frac{1}{2} \left(\frac{\ln x - 4.5302}{0.7304} \right)^2}, x > 0 \qquad (21)$$

For each value from values $N_1, N_2, \ldots, N_{100}$, we generated N_i random values obeying logarithmic-normal distribution with a density given by formula (21), what represent $P_1, P_2, \ldots, P_{N_i}$ values of the individual devices issued in individual years. We were following the same steps as generating data about amount of issued devices.

In the next step, sums $S_1, S_2, \ldots, S_{100}$ of values of issued devices for individual years (expressed in thousands of euros) and we estimated the type of distribution from which these sums of data are. By testing, we were able to state and conclude, that sums of values obey normal distribution with parameters $\mu = 676.7$ and $\sigma = 131.4$.

Based on this, it is possible to determine the interval estimates and of course, other estimates of the value of issued aids in the given region. This can provide a basis for the planning of production and storage capacities, as well as for the estimation of the financial costs associated with securing the devices to the extent necessary (Urban and Hošková-Mayerová 2017). For more information and graphical representation of the total cost value of aids issued in a given year, the reader can find in Knežo 2012b (Fig. 5).

7 Conclusion

The method, presented in this chapter, is efficient application of numerical methods to generate random or pseudorandom numbers needed to simulate experiments using the Monte Carlo method. The advantage of the method is in particular the fact that its use can be done by generating random numbers from various distributions in automated calculations. So this means that the random number generating procedure can be a direct part of the application that simulates the experiment. Therefore, it is not necessary to call external program resources or databases. Its use is particularly useful where other more efficient methods are not available. Its usage is particularly useful in these case where another more effective methods are not available. The usage of this method requires not only sufficient knowledge of users about accuracy of individual numerical methods but also about accuracy with which the software is able to calculate, because in some cases, especially when demanding high precision, we work with numbers which at the borders of computer options.

Monte Carlo simulation provides a way for analysing complicated systems in which chance plays a key role. Therefore, for social scientists this method provides a strong tool when they observe and simulate social processes. Monte Carlo simulation provides a way for analysing marketing processes, there it has a range of important applications, including the integration of customer behaviour models, the modelling of customer service processes and the evaluation and selection of analytical software tools.

Acknowledgements The work presented in this chapter has been supported by the project KEGA 026TUKE-4/2016.

References

Čičmanec, L., Bořil, J.: Assessment of tactical mission simulation exercise. In: Transport Means 2017. Kaunas University of Technology, Kaunas, pp. 729–734 (2017). ISSN 1822-296X

Dagpunar, J.S.: Simulation and Monte Carlo: with applications in finance and MCMC. Wiley, ACCEM (2008), New York, (2007), 348 p. ISBN 978-0-470-85494-5

Fishman, G.S.: Monte Carlo Concepts, Algorithms and Applications. Springer, Berlin (1996)

Gentle, J.E.: Random Number Generation and Monte Carlo Methods. Springer, Berlin (2003)

Glasserman, P.: Monte Carlo Methods in Financial Engineering. Springer, Berlin (2003)

Hubacek, M., Vrab, V.: Cost assessment of training using constructive simulation. eBook: Mathematical-Statistical Models and Qualitative Theories for Economic and Social Sciences, vol. 104. Springer International Publishing AG, New York (2017). ISSN 2198-4182

Jirgl, M., Bořil, J., Jalovecký, R.: Assessing quality of pilot training with use of mathematical analyses. In: Proceedings of the 2016 17th International Conference on Mechatronics - Mechatronika (ME), Czech Technical University in Prague, Prague, pp. 91–96 (2016). ISBN 978-80-01-05882-4

Knežo, D.: About the method of Monte Carlo and its applications. Transf. Innov. **24**, 178–181 (2012a)

Knežo, D.: Estimation of costs concerning prothetic and orthotic aids using the Monte Carlo method. In: AEI'2012: International Conference on Applied Electrical Engineering and Informatics (2012b)

Knežo, D.: Inverse transformation method for normal distribution and the standard numerical methods. Int. J. Interdiscip. Theory Pract. **5**(2014), 6–10 (2014)

Kubíček, P., Šašinka, Č., Stachoň, Z.: Selected cognitive issues of positional uncertainty in geographical data. Geografie - Sborník České geografické společnosti, Česká geografická společnost **119**(1), 67–90 (2014). ISSN 1212-0014

McLeish, D.L.: Monte Carlo Simulation and Finance. Wiley, New York (2005)

Sobol, I.M.: Die Monte-Carlo-Methode. VEB Deutscher Verlag der Wissenschaften, Berlin (1971)

Svatoňová, H., Šikl, R.: Cognitive aspects of interpretation of image data. eBook: Mathematical-Statistical Models and Qualitative Theories for Economic and Social Sciences, vol. 104. Springer International Publishing AG, New York (2017) ISSN 2198-4182

Urban, R., Hošková-Mayerová, Š.: Threat life cycle and its dynamics. Deturope **9**(2), 93–109 (2017)

Woch, M., Zieja, M.,Tomaszewska, J.: Analysis of the time between failures of aircrafts. In: 2nd International Conference on System Reliability and Science (ICSRS 2017), pp. 112–118 (2017)

Part III
Models in Social Systems

Rolling Circles of Motions: Yesterday and Today

Murat Tosun and Soley Ersoy

Abstract In this chapter, we give a short historical survey of basic events which had happened during the development of models depend on a rolling circles around circles. The first seeds of these models can be seen at the couple of Tusi which was derived for stating his astronomical theory in the 13th century. Tusi's model generates just a straight line. Nowadays, it is well known the curves traced out by a point on a circle rolling on the inside and outside of another circle are hypocycloids and epicycloids, respectively. These curves are used for practical engineering problems such as the slider-crank mechanism or design of the rotary engine. One of the degenerate case of hypocycloidal motion is elliptic motion and recently it is called Cardan motion. In this chapter, we deal with elliptic and cycloidal (epicycloid or hypocycloid) motion by use of the complex forms of Bottema's instantaneous invariants characterizing the infinitesimal properties of motion.

1 Introduction

Nasir Al-Din Tusi (1201–1274), the famous astronomer, devised a model of a combination of uniform circular motions to generate the linear motion of the planets by taking a circle rolling inside a twice larger circle. Then any point belonging to the circumference of the smaller circle traces out a straight line segment along a diameter of the larger circle, see Fig. 1. Now this construction is known as Tusi couple. Through the instrumentality of this construction Tusi refined the model for planetary motions from the time of Ptolemy (Hetherington 2006; Sayılı 1998; Ragep 1993).

Nicolaus Copernicus (1473–1543) used this model for the same purpose that to explain the motion of celestial objects (Roberts 1957; Kennedy and Roberts 1959; Hartner 1973). Tusi Couple was reappeared in Copernicus's De revolutionibus

M. Tosun · S. Ersoy (✉)
Faculty of Arts and Sciences, Department of Mathematics, Sakarya University, Sakarya 54187, Turkey
e-mail: sersoy@sakarya.edu.tr

M. Tosun
e-mail: tosun@sakarya.edu.tr

© Springer Nature Switzerland AG 2019
C. Flaut et al. (eds.), *Models and Theories in Social Systems*, Studies in Systems, Decision and Control 179, https://doi.org/10.1007/978-3-030-00084-4_11

Fig. 1 Tusi couple from Al-Tadhkirah fi'ilm al-hay'ah—A memoir on the science of astronomy

Fig. 2 The Tusi couple in the Copernicus's De revolutionibus

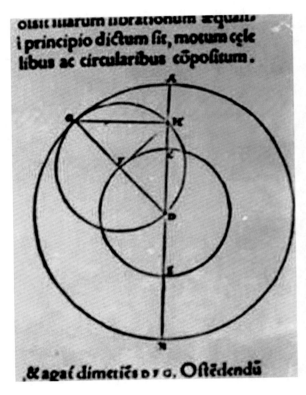

orbium coelestium (On the Revolutions of the Celestial Spheres) with almost exactly the same Latinized alphabet markings on his diagrams as of Tusi's *Al-Tadhkirah fi'ilm al-hay'ah*, see Fig. 2.

Today it is known that the linear motion is a particular case of the family of hypocycloid curves (Hilbert and Cohn-Vossen 1952).

Fig. 3 La Hire's
Hypocycloidal Train
Mechanism in the Le
Magasin Pittoresque, Vol.
VII, 1840, 117

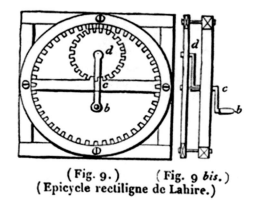

(Fig. 9.)　　(Fig. 9 *bis.*)
(Epicycle rectiligne de Lahire.)

Also, Cardan motion is a special hypocycloid motion in which the moving centrode has half the diameter of the fixed centrode. In this instance, especially in kinematics, these two circular centrodes (or planetary gears pair of 2:1 ratio) are called Cardan circles or circles of Geronimo Cardano, (1501–1576). This classical motion is characterized by instantaneous invariants of Bottema and has attracted much attention of in the realm of kinematics (Rauh et al. 1938; Bottema 1949; Freudenstein 1960; Bottema 1979; Sekulić 1998). It is one of the most simple mechanism for generating ellipses and the isosceles slider crank, trammel of Archimedes, swinging scotch yoke mechanisms and their inversions (Kiper et al. 2007; Lee and Hervé 2012; Liu and McCarthy 2017).

Moreover, the same construction is ascribed to another distinguished mathematician Philippe De La Hire (1640–1719). He studied on conic sections and epicycloids in the same manner as Desargues and Pascal. His famous lemma expresses that the trajectory of any point belonging to the circumference of the smaller circle is a straight line segment along a diameter of the larger circle. Any point on the moving body of the rolling circle, which is not on the circumference, moves on an ellipse (La Hire 1706; Boyer 1947; Stachel 1999). A demonstration can be seen in the Fig. 3.

In this chapter, we reassume the elliptic and cycloidal (epicycloid or hypocycloid) motion in complex numbers form since the complex formulation of position is more convenient than vector formulation in the kinematics of planar motion. It provides an analytically efficient technique method of calculating velocities and accelerations by expressing rotations by complex exponential function. Moreover, the instantaneous invariants characterizing the infinitesimal properties of the elliptic and cycloidal motions are obtained analogous with the method of (Bottema 1979) but in complex number formulation.

2 Preliminaries

Let $\varphi(t)$ be time dependent angle of rotation of a complex plane m in continuous motion relative to a fixed complex plane f. The transformation between the coincident points of these planes is expressed by

$$Z = ze^{i\varphi} + q \tag{1}$$

where $z = x + iy$, $Z = X(\varphi) + iY(\varphi)$ and $q = a(\varphi) + ib(\varphi)$. The general form for motion of the origin in m becomes $\frac{d^j q}{d\varphi^j} = q_j$ with $j = 1, 2, 3, \ldots$ specifying a given motion up to n-th order. The subscript j denotes the j-th derivative of a function throughout this chapter.

The complex plane m is chosen to rotate with a constant angular velocity relative to fixed complex plane f, that is, $\frac{d\varphi}{dt} = 1$ and $\frac{d^j \varphi}{dt^j} = 0$ for $j = 2, 3, \ldots$ to study the kinematic geometry of the motion of complex plane m relative to f. The concept of instantaneous invariants was formally introduced by Bottema (1961) and was called B-invariants (Bottema-invariants) by Veldkamp (1963). The complex forms of Bottema's instantaneous invariants characterizing the infinitesimal properties of motion up the n-th order are given in Eren and Ersoy (2017) as follows;

$$q_0 = q_1 = Re(q_2) = 0 \quad \text{and} \quad Im(q_2) > 0. \tag{2}$$

Thus, the Eq. (1) and its derivatives are determined as

$$Z_0 = z, \quad Z_1 = iz, \quad Z_2 = -z + \left(\frac{q_2 - \bar{q}_2}{2}\right), \ldots, \quad Z_n = (i)^n z + q_n, \quad (n > 2). \tag{3}$$

3 Rolling Circles in Complex Plane

The theory in this section is not new, but the proofs given by use of the complex numbers to make them a more accessible tool for problem solving in the field of kinematics and shorter than those in print.

3.1 Elliptic Motion in Complex Plane

For the purpose of studying elliptic motion with complex numbers, let $A_1 = (-R, 0)$ and $A_2 = (R, 0)$ be two points of f remaining on the imaginary and real axes of

Fig. 4 The moving and fixed complex planes

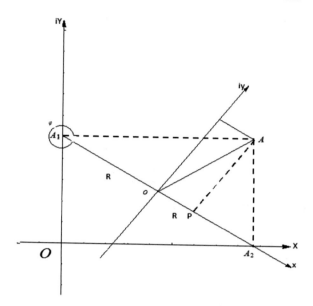

m, respectively, that is, $|A_1A_2| = 2R$. Consider a point A (x, y) rotating with angle -2φ while the complex plane m rotates relative to f with φ, see Fig. 4.

Thus, the trajectory of A is written in the form

$$Z = ze^{i\varphi} + Re^{-i\varphi}. \tag{4}$$

From the differentiation of the Eq. (4) we get the equations of the moving and fixed centrodes as $z_P = Re^{-2i\varphi}$ and $Z_P = 2Re^{-i\varphi}$, respectively, Fig. 5.

Thereby if we eliminate φ from the Eq. (4) we get an general ellipse equation as follows;

$$\left(|z|^2 + R^2\right)|Z|^2 - R\bar{z}Z^2 - Rz\bar{Z}^2 - \left(|z|^2 - R^2\right)^2 = 0 \tag{5}$$

with respect to position of z, Fig. 6.

For the degenerate case of $z = Re^{i\theta}$, the Eq. (5) becomes

$$2|Z|^2 - \left(Ze^{-i\theta/2}\right)^2 - \left(\bar{Z}e^{i\theta/2}\right)^2 = 0,$$

that is,

$$i^2\left(e^{i\theta/2} - e^{-i\theta/2}\right)^2 X^2 + 2iXY\left(e^{i\theta} - e^{-i\theta}\right) + \left(e^{i\theta/2} + e^{-i\theta/2}\right)Y^2 = 0.$$

This means that the points on the rolling circle move on the straight line

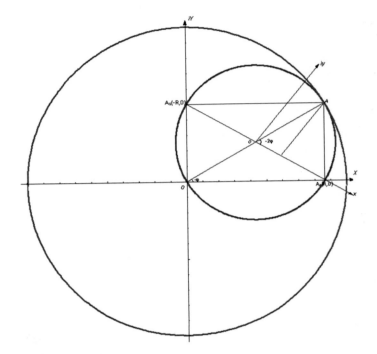

Fig. 5 The moving and fixed centrodes

Fig. 6 The trajectory of the point A

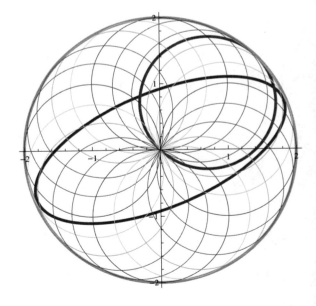

Fig. 7 The trajectory of the
point A on the moving
centrode

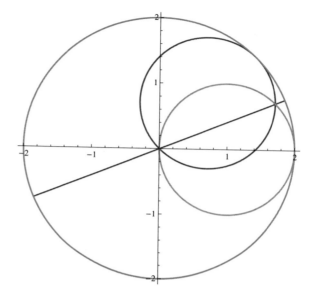

$$Y = -i\,\frac{e^{i\theta/2} - e^{-i\theta/2}}{e^{i\theta/2} + e^{-i\theta/2}}\,X = \tan\frac{\theta}{2}\,X$$

passing through the center of the fixed circle (Fig. 7).

Considering $z = o$ means that it is on the center of the moving centrode then the Eq. (5) becomes $|Z| = R$, then the points on the center of rolling circle move on a circle (Fig. 8).

As a consequence, the center of the moving centrode traces a circle, any point on the moving centrode traces a straight line and any other point describes an ellipse (Bottema 1949).

If it is desired to characterize the elliptic motion by the complex form of instantaneous invariants, then the Eq. (4) can be transformed by the equations

$$Z' = -e^{i\pi/2}\,(Z - 2R) = -i\,(Z - 2R)$$

and

$$z' = -e^{i\pi/2}\,(z - R) = -i\,(z - R)\,.$$

Hence, if these two equations are substituted into the Eq. (4), then the elliptic motion is represented by

$$Z' = z'e^{i\varphi} - iR\left(e^{i\varphi/2} - e^{-i\varphi/2}\right)^2. \tag{6}$$

If we compare this last equality with the Eq. (1) of a motion in any case, we get the complex forms of the Bottema's instantaneous invariants of elliptic motion as follows;

Fig. 8 The trajectory of the
point A coincident with the
center of the moving
centrode

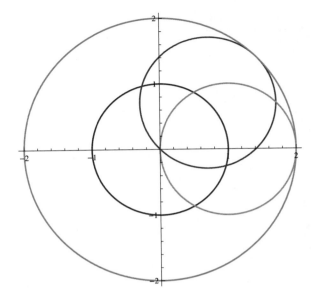

$$q_0 = -iR\left(e^{i\varphi/2} - e^{-i\varphi/2}\right)^2, \quad q_1 = R\left(e^{i\varphi} - e^{-i\varphi}\right), \quad q_2 = iR\left(e^{i\varphi} + e^{-i\varphi}\right), \ldots$$

and goes on.

The infinitesimal properties of elliptic motion up the k-th order are characterized by

$$q_k = (i)^{k-1}R\left(e^{i\varphi} + (-1)^k e^{-i\varphi}\right), \quad k \in \mathbb{N}.$$

At the reference position while $\varphi = 0$, the instantaneous invariants are determined as

$$q_0 = q_1 = Re(q_2) = 0, \quad Im(q_2) = 2R, \ldots$$

$Im(q_2)$ was expressed as the diameter inflection circle (Eren and Ersoy 2017). The inflection circle is the locus of the points with zero curvature.

In conclusion, the complex forms of the Bottema's instantaneous invariants of elliptic motion at the reference position are obtained as

$$q_{2k+1} = Re(q_{2k}) = 0, \quad Im(q_{2k}) = (-1)^{k-1}2R, \quad k \in \mathbb{N}.$$

On the other hand, the inverse of the elliptic motion given by the Eq. (4) is Cardioid motion with the equation $z = Ze^{-i\varphi} + Re^{-2i\varphi}$ such that the point $Z = X + iY$ is constant, $z = x(\varphi) + iy(\varphi)$ is sliding with respect to φ, see Fig. 9. The trajectory of the point z will be Limacon in general (Bottema 1979).

Fig. 9 The inverse of the
elliptic motion

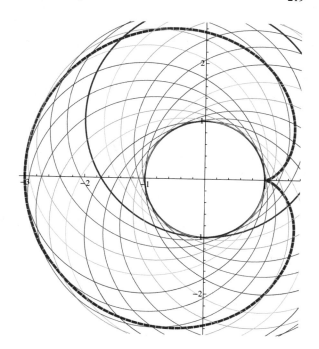

3.2 Cycloidal Motion in Complex Plane

Let the fixed and moving centrodes be $|Z|^2 = R^2$ and $|z|^2 = r^2$, respectively, in order to study the motion of rolling circles in general. If these circles are positioned as tangent to each other at the point given by $Z = iR$ and $z = -ir$, then the equation of the motion becomes

$$Z = ze^{i\varphi} + i(R + r)e^{\left(\frac{ir}{R+r}\right)\varphi}. \tag{7}$$

Here the trajectory of a point is hypocycloid while $R > 0$ and $r > 0$ or epicycloid while $R > 0$ and $r < 0$, see the examples in Fig. 10.

For the purposes of constituting the complex instantaneous invariants of cycloidal motion, let $Z' = Z - iR$ and $z' = z + ir$. If we substitute these equalities in the Eq. (7), we get

$$Z' = z'e^{i\varphi} - i\left(R + re^{i\varphi} - (R+r)e^{\left(\frac{ir}{R+r}\right)\varphi}\right). \tag{8}$$

The comparison of the Eqs. (1) and (8) states that

$$q_0 = -i\left(R + re^{i\varphi} - (R+r)e^{\left(\frac{ir}{R+r}\right)\varphi}\right).$$

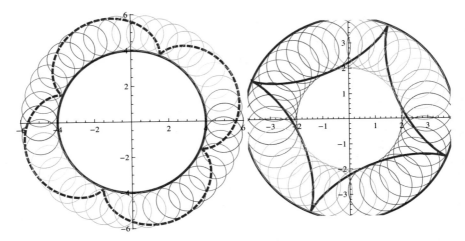

Fig. 10 Epicycloidal motion for $R = 4$ and $r = 1$ and hypocycloid motion $R = 4$ and $r = -1$

The successive differentiations of this last equation are

$$q_1 = r \left(e^{i\varphi} - e^{\left(\frac{ir}{R+r}\right)\varphi} \right),$$
$$q_2 = ir \left(e^{i\varphi} - \frac{r}{R+r} e^{\left(\frac{ir}{R+r}\right)\varphi} \right),$$
$$q_3 = -r \left(e^{i\varphi} - \left(\frac{r}{R+r}\right)^2 e^{\left(\frac{ir}{R+r}\right)\varphi} \right),$$
$$\vdots$$

Then infinitesimal properties of cycloidal motion up the k-th order are characterized by

$$q_k = (i)^{k-1} r \left(e^{i\varphi} - \left(\frac{r}{R+r}\right)^{k-1} e^{\left(\frac{ir}{R+r}\right)\varphi} \right), \quad k \in \mathbb{N}.$$

As a consequence the complex instantaneous invariants of cycloidal motion are

$$q_0 = q_1 = Re(q_2) = 0, \quad Im(q_2) = \frac{rR}{R+r}, \quad q_k = (i)^{k-1} r \left(1 - \left(\frac{r}{R+r}\right)^{k-1} \right), \quad k \in \mathbb{N}$$

where $\varphi = 0$ at the reference position.

If this motion is specialized to aforementioned elliptic motion then $R = -2r$, the complex instantaneous invariants of elliptical motion are

$$q_0 = q_1 = Re(q_2) = 0, \quad Im(q_2) = 2r, \quad q_{2k+1} = Re(q_{2k}) = 0, \quad Im(q_{2k}) = (-1)^k 2r, \quad k \in \mathbb{N}$$

where $\varphi = 0$ at the reference position.

4 Conclusions

The historical continuum from Tusi to La Hire shows us that the systems of astronomy are stimulated a thorough study of examples of hypocycloids. At the present time, we know that hypocycloids and epicycloids are curves traced out by a point on a circle rolling on the inside or outside of another circle. A change in relative sizes and speeds of the circles will produce multi-cusped curves. The most common types of roulettes used in engineering applications are hypocycloids and epicycloids and several examples of mechanisms depend on cycloidal motions.

References

Bottema, O.: On Cardan positions for the plane motion of a rigid body, Proceedings of Koninklijke Nederlandsche Akademie van Wetenschappen, (Royal Netherland Academy of Sciences), Series A, **52**(6), 643–651 (1949)

Bottema, O.: On instantaneous invariants. In: Proceedings of the International Conference for Teachers of Mechanisms, pp. 159–164. New Haven, (CT), Yale University (1961)

Bottema, O., Roth, B.: Theoretical Kinematics. North-Holland, Amsterdam (1979)

Boyer, C.B.: Note on epicycles & ellipse from Copernicus to Lahire. Isis **38**(1/2), 54–56 (1947)

Eren, K., Ersoy, S.: Revisiting Burmester theory with complex forms of Bottema's instantaneous invariants. Complex Var. Elliptic Equ. **62**(4), 431–437 (2017)

Freudenstein, F.: The Cardan position of a plane. In: Transactions of the Sixth Conference on Mechanisms, pp. 129–133. Purdue University, West Lafayette, IN, 10–11 Oct 1960

Glaeser, G., Stachel, H.: Open Geometry: OpenGL + Advanced geometry. Springer, New York (1999)

Hartner, W.: Copernicus, the man, the work, and its history. Proc. Am. Philos. Soc. **117**, 416–417 (1973)

Hetherington, N.: Planetary Motions: a historical perspective. Westport. Greenwood Press, CT and London (2006)

Hilbert, D., Cohn-Vossen, S.: Geometry and the Imagination. Chelsea Publishing Company, New York (1952)

Kennedy, E.S., Roberts, V.: The planetary theory of Ibn ash-Shâtir. Isis **50**, 227–235 (1959)

Kiper, G., Söylemez, E., Kişisel, A.Ö.: Polyhedral linkages synthesized using Cardan motion along radial axes. In: 12th IFToMM World Congress. Besançon, France (2007)

La H., De, P.: Traité des roulettes, Académie des Sciences, Mémoires, (edition of 1707, 350–352.), 340–349 (1706)

Lee, C.C., Hervé, J.M.: On the helical Cardan motion and related paradoxical chains. Mech. Mach. Theory **52**, 94–105 (2012)

Liu, Y., McCarthy, J.M.: Design of mechanisms to draw trigonometric plane Curves. J. Mech. Robot. **9**(2), 024503-1–7 (2017)

Roberts, V.: The solar and lunar theory of Ibn ash-Shâtir. Isis **48**, 428–432 (1957)

Ragep, F.J.: Nasir al-Din al-Tusi's Memoir on Astronomy (al Tadhkira fi'ilm al-hay'a), Sources in the History of Mathematics and Physical Sciences, 12, vol. 1, Springer (1993)

Rauh, K., Marks, H., Bündgens, M., Otto, K.: Kardanbewegung und Koppelbewegung (Cardan motion and coupler motion), Schriftenreihe Praktische Getriebetechnik, Heft 2. VDI-Verlag, Berlin (1938)

Sayılı, A.: The Observatory in Islam and its place in the general history of the observatory, Institute for the History of Arabic-Islamic Science at the Johann Wolfgang Goethe University, Frankfurt am Main (1998)

Sekulić, A.: Method of synthesis of Cardanic motion. Univ. Niš, Facta Univ. Ser. Mech. Eng. **1**(5), 565–572 (1998)

Veldkamp, G.R.: Curvature theory in plane kinematics [Doctoral dissertation]. T.H. Delft, Groningen (1963)

Some Remarks on Social Life in Romanian Towns and Cities in the 1930s, Based on Statistical Data

Daniel Flaut and Enache Tuşa

Abstract In this short chapter, we present some aspects regarding social life in Romanian towns and cities, based on statistical data, in the fourth decade of the 20th century, a period marked by economic crises, social problems, and the imminent outbreak of World War II.

Keywords Urban population · Standard of living · Dwellings · Health care
Birth and death rates

1 Introduction

According to data published by the National Institute of Statistics, on the 1st of January 2017 Romania had a resident population of 19,638,309 people. Its 320 towns and cities had a resident population of 10,528,473 people, with 47.62% men and 52.38% women. Urban dwellers represented 53.61% of the resident population. As far as distribution by age is concerned, it can be noted that the young population (0–14 years old) was more numerous in urban (1,556,205 people, 14.78%) than in rural areas (1,498,323 people). In towns and cities, the percentage of residents between the ages of 15 and 64 amounted to 69.41%, a higher figure than in the countryside (63.47%). On the other hand, the percentage of elderly residents (65 or above) was lower in urban (15.81%) than in rural areas (20%) (http://statistici.insse.ro/shop/index.jsp?page=tempo3&lang=ro&ind=POP105A).

This is some of the statistical data reflecting the present situation of urban Romanian residents. It is interesting to observe the state of affairs in Romanian towns and cities many decades ago, in the 1930s.

D. Flaut (✉) · E. Tuşa
Faculty of History and Political Sciences, Ovidius University, Constanţa, Romania
e-mail: daniel_flaut@yahoo.com

E. Tuşa
e-mail: tusaenache@yahoo.com

© Springer Nature Switzerland AG 2019
C. Flaut et al. (eds.), *Models and Theories in Social Systems*, Studies in Systems,
Decision and Control 179, https://doi.org/10.1007/978-3-030-00084-4_12

For Romania, the 1930s represented a decade of considerable turmoil, with alternating periods of crisis and development. First there was the economic crisis of 1929–1933, characterised in Romania by lower prices for agricultural and industrial products, lower consumption, salary cuts, reduced incomes for agricultural producers and urban dwellers alike, a lower standard of living, bankruptcies, unemployment, all of these resulting in riots, strikes, and workers' protests. This was followed by economic recovery and a certain leap forward between 1933 and 1937. Towards the end of 1937 and the beginning of 1938 there was a new crisis, followed by the first intimations of the Second World War (Axenciuc 1999, pp. 237–238).

As far as these years are concerned, Romania can be regarded as a representative example of societal adaptation to such challenges. In this chapter we aim to present some statistical data regarding the urban population of Romania and various aspects of daily life in urban areas: dwellings, health care, birth and death rates, wages and prices, the cost of living. In order to write this chapter we used statistical data from the Statistical Yearbook of Romania 1939 and 1940 (Anuarul Statistic al României 1939 şi 1940), as well as a series of studies and papers, particularly those authored in the 1930s and 1940s by Sabin Manuila, Dumitru C. Georgescu, I. Measnicov and G. Banu.

2 Population

In the 1930s Romania continued to have a low level of urbanization. In 1930 only 20.1% of the country's 18,052,896 inhabitants lived in towns and cities (Manuila and Georgescu 1938, p. 17). By 1939 the population of Romania had reached 19,933,802 inhabitants, but the percentage of urban dwellers had dropped to 18.2%, due to population growth in rural areas (Anuarul statistic al României 1940, pp. 150–154).

In 1930 Romania had 172 towns and cities, with a total population of 3,632,178 inhabitants and an average number of inhabitants of 21,117. Towns and cities in Muntenia had the highest average population (39,040 inhabitants), followed by Crişana and Maramureş (34,389 inhabitants), whereas towns and cities in Oltenia and Dobruja had the lowest average population (12,387 and 10,743 inhabitants respectively). The high average number associated with town and cities in Muntenia was due to the population of Bucharest, the capital city. In Crişana and Maramureş the high average number was due to the fact that in the respective provinces market towns were regarded as rural communes. In Oltenia and particularly in Dobruja, small market towns, with a low population and more or less rural features, were administered, for political reasons, as urban communes (Manuila and Georgescu 1938, p. 13). Dobruja had 18 towns and cities, with a total population of 196,478 inhabitants (Anuarul statistic al României 1940, p. 48; Limona 2009, p. 40).

At province level, Muntenia (27.1%), Bukovina (26.6%), Moldavia (24.3%) and Dobruja (23.8%) accounted for the highest percentages of urban dwellers. The lowest percentages corresponded to Bessarabia (12.9%) and Oltenia (13%). 20 of the urban settlements in these areas were municipalities, totalling 1,970,877 inhabitants, or

54.3% of the country's urban population. The remaining 1,661,301 inhabitants lived in the other 152 towns and cities, 51 of which, together with the municipalities, also served as county seats (Manuila and Georgescu 1938, pp. 17–18). Among the 71 county seats, Bucharest occupied first place in terms of population figures (with 639,040 inhabitants), followed by Kishinev with 114,896, Chernivtsi with 112,427, Iaşi with 102,872, Galaţi with 100,611 and Cluj with 100,844 inhabitants, whereas the bottom of the list featured Zalău, with 8,340 inhabitants, Făgăraş, with 7,841, Diciosânmartin, with 6.567, and Miercurea Ciuc, with 4,807 inhabitants (Anuarul statistic al României 1940, 48–49).

In 1930 the 172 towns and cities contained 21% of the total number of households and 45.3% of the total number of business firms in Romania (Manuila and Georgescu 1938, p. 18).

In 1930, 58.6% of urban dwellers were Romanians, a lower percentage than the one in rural areas, due to the fact that minority populations often prevailed in the urban settlements of Transylvania, Banat and Bessarabia. As far as religion was concerned, according to the 1930 census, Romanian urban dwellers comprised 60.9% Eastern Orthodox, 4.6% Greek Catholics, 10.4% Roman Catholics, 4.9% Reformed, 2.6% Evangelicals, 0.7% Unitarians, 0.3% Armenian Gregorians, 0.3% Lipovans, 0.1% Baptists, 14.3% Mosaic, and 1% Mohammedans (Manuila and Georgescu 1938, pp. 68–69; Banu 1944, pp. 336–337; Georgescu 1937, p. 70).

There were fewer men than women in towns and cities, as was also the case in the countryside. However, the percentage of men was higher in urban (49.6%) than in rural areas (49%), mainly due to the continuous migration of the male population from villages to urban spaces and also, to a large extent, to the presence in towns and cities of (predominantly male) school-goers and military personnel (Manuila and Georgescu 1938, pp. 23–24; Georgescu 1937, p. 69).

As regards the birthplace of urban dwellers, it can be noted that barely more than 50% were autochthonous. Unable to practise agriculture in a different location, peasants had to leave their home village, change their occupation and move to urban areas, where they could earn more. However, sources of income were not available everywhere. In smaller towns, where there was limited industrial development, there were more natives than in rural communes. On the other hand, in big cities the percentage of natives was low. In 1930 the city of Constanţa had the highest immigrant population (73.6%) and Kishinev had the highest percentage of natives (61.9%). At province level, the highest percentage of natives in 1930 was recorded in urban spaces in Bessarabia (64.4% men, 71.1% women) and the lowest in urban spaces in Banat (36.3% men, 38.4% women) (Measnicov 1942, pp. 401–403).

As far as distribution by age is concerned, it can be noted that in 1930 the urban population between 0 and 19 only amounted to 39.07%, being less numerous than in rural areas (48.2%). A significant percentage of urban dwellers consisted of people between the ages of 15–19 (12.39%) and between the ages of 20–24 respectively (12.54%), due to the presence in towns and cities of school-goers and military personnel. In towns and cities the most numerous age group was the one between 15 and 30 (41.75%) (Manuila and Georgescu 1938, pp. 26–29; Banu 1944, p. 332).

Table 1 The percentage of married citizens per 1000 inhabitants and the percentage of divorcees per 100 marriages in Romanian urban areas between 1930 and 1939

	1930	1931	1932	1933	1934	1935	1936	1937	1938	1939
Percentage of married citizens per 1000 inhabitants	16.2	15.5	16.7	17.1	18.9	19.4	20	20.5	19.8	21.9
Percentage of divorcees per 100 marriages	9.6	10	9.6	9.6	9.9	10.7	11.1	11.8	12.7	10.1

Source Anuarul statistic al României (1940, p. 143)

Table 2 The number of marriages and divorces in Romanian urban areas between 1930 and 1939

	1930	1931	1932	1933	1934	1935	1936	1937	1938	1939
Marriages	28,278	27,243	29,356	30,100	33,375	34,460	35,723	36,988	35,790	39,734
Divorces	2,723	2,821	2,820	2,895	3,294	3,676	3,962	4,352	4,532	4,010

Source Anuarul statistic al României (1940, p. 143)

Data collected in the 1930 census regarding the civil status of Romanian residents shows that at the time 38.5% of urban dwellers were single, 50.6% married, 9.5% widowed, 1.1% divorced and 0.3% had not declared their status. The percentage of single people in towns and cities was 10% higher than in the countryside, whereas the percentage of married people was 11.5% lower. The percentage of divorcees was twice as high in urban areas than in the countryside (Banu 1944,p. 302; Georgescu 1937, pp. 69–70).

In the 1930s the highest percentage of married people per 1000 urban dwellers (21.9) was recorded in 1939 and the lowest (15.5) in 1931. The lowest percentages of divorcees per 100 marriages in urban areas were recorded in 1930, 1932 and 1933 (9.6) and the highest in 1938 (12.7) (Table 1).

331,047 marriages and 35,085 divorces were recorded in urban areas between 1930 and 1939. The lowest number of marriages (27,243) took place in 1931 and the highest (39,734) in 1939. In the 1930s the highest number of divorces (4,532) took place in 1938 and the lowest (2,723) in 1930 (Table 2).

In 1930 the percentage of marriages in Romanian urban areas reached 8 per 1000 inhabitants. In the fourth decade of the 20th century, the percentage of marriages followed a downward curve, whereas that of divorces was on the increase (Banu 1944, pp. 302–303; Georgescu 1941, p. 45).

3   The Standard of Living

The standard of living of urban dwellers in the 1930s is quite difficult to assess. Opinions differ, with some writers emphasising how good conditions were whereas others saw shortages and poverty everywhere (Iacob 2016, p. 80). In the following section we are going to focus on several aspects of daily life in urban areas: housing, health care, birth and death rates, wages and prices, the cost of living.

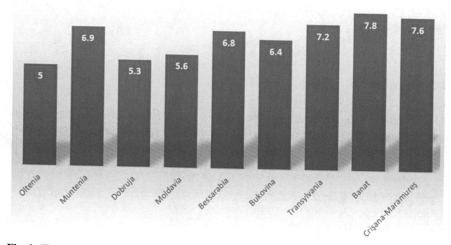

Fig. 1 The number of urban inhabitants per building at province level in 1930. *Source* Manuila and Georgescu (1938, p. 14)

3.1 Housing

In 1930, the 3,632,178 urban dwellers in Romania occupied 560,558 buildings. Population density per building was higher in urban (6.5) than in rural (4.5) areas. The most densely populated houses were to be found in Banat, with 7.8 inhabitants per building, Crişana and Maramureş with 7.6, and Transylvania with 7.2. Urban areas in Oltenia had the lowest number of inhabitants per building (5.0). Dobruja was characterised by the same average number of inhabitants per building in both urban and rural areas (5.3), which highlights the semi-rural nature of towns in this province, with the exception of the city of Constanţa (Fig. 1). Bucharest, the country's largest city, also had the highest average number of inhabitants per building (9.3) (Manuila and Georgescu 1938, pp. 14–17).

There were no significant changes in terms of urban living conditions in the 1930s. A process of polarization took place. Luxury houses and apartment blocks were built in city centres. In some cities there was also an increase in the number of "cheap" housing for employees, with access to modern facilities (running water, sewerage, electricity, and in some cases even central heating). At the lower end of the scale were the improvised dwellings erected in the poor neighbourhoods on the outskirts of big cities. Inhabited by people who had moved into urban areas seeking a source of livelihood, such homes had one or two rooms, a wooden skeleton and a cardboard roof. These neighbourhoods lacked even elementary sanitation and living conditions. The roads were not paved, there was no sewerage and no electricity. In terms of construction materials, urban dwellings fell into the following categories: approximately 50% were made of brick and stone, 20% were made of wood (especially in hillside and mountain towns), and 30% were made of scrap lumber, twigs and clay (Axenciuc 1999, p. 381).

3.2 Health Care, Birth and Death Rates

Although only 20% of the country's population lived in towns or cities, the majority of health care facilities were concentrated in urban areas. In the interwar period, 234 surgeries were set up in towns and cities, with an average of 12,500 inhabitants for each physician (Axenciuc 1999, p. 382). In 1934, the relevant ministry's network comprised 2,400 physicians, but only 984 of them resided in rural communes. At the beginning of 1937, Romania had 7,162 physicians, including 990 in Bucharest alone (Şandru 1980, pp. 266–267). In 1938, for example, physicians from municipal surgeries consulted 430,840 patients; 385,666 called at the surgery and 45,174 received house calls (Anuarul statistic al României 1940, p. 206).

Approximately 600,000 births per year were recorded in the 1930s, only 75,000–80,000 of which took place in urban areas. The highest number of live births in urban areas was recorded in 1930 (82,013), and the lowest in 1933 (73,240) (Table 3).

Birth rates in Romanian towns and cities followed a downward curve, decreasing from 23.6‰ in 1930 to 20.9‰ in 1939 (Fig. 2). This decrease was caused exclusively by socio-economic factors which resulted in a higher percentage of single people, delayed marriages, a limited number of births to improve the family's standard of living, etc. (Banu 1944, p. 310).

Fertility was much lower in urban than in rural areas. In 1932, for instance, the number of live births per 1000 women aged 14 to 44 was 79.5, that is half of what it was in the countryside (173.1) (Banu 1944, p. 310).

Although the death rate followed a slightly downward curve in Romania, its values went up in urban areas (Fig. 2). Urban mortality ranged from 18‰ between 1930 and 1934 to 19.6‰ between 1935 and 1939 (Banu 1944, p. 313). In the 1930s the highest number of deaths in urban areas took place in 1938 (71,997) and the lowest (60,588) in 1933 (Table 3).

Natural increase rates in Romanian towns and cities dropped from 5.9‰ in 1930 to 1.9‰ in 1939 (Fig. 2). The lowest rate of natural increase was recorded in 1935 (Table 3, Fig. 2).

Table 3 Live births, deaths and natural increase in Romanian urban areas between 1930 and 1939

	1930	1931	1932	1933	1934	1935	1936	1937	1938	1939
Live births	82,013	77,989	78,541	73,240	74,017	73,602	76,320	77,311	77,456	75,691
Deaths	61,606	65,881	66,538	60,588	63,453	68,471	69,267	69,325	71,997	68,791
Natural increase	20,407	12,108	12,003	12,652	10,564	5,131	7,053	7,986	5,457	6,900

Source Anuarul statistic al României (1940, p. 143)

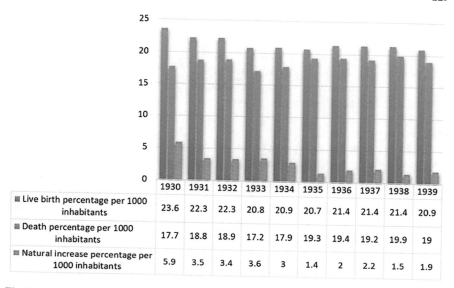

	1930	1931	1932	1933	1934	1935	1936	1937	1938	1939
▪ Live birth percentage per 1000 inhabitants	23.6	22.3	22.3	20.8	20.9	20.7	21.4	21.4	21.4	20.9
▪ Death percentage per 1000 inhabitants	17.7	18.8	18.9	17.2	17.9	19.3	19.4	19.2	19.9	19
▪ Natural increase percentage per 1000 inhabitants	5.9	3.5	3.4	3.6	3	1.4	2	2.2	1.5	1.9

Fig. 2 Live births, deaths and natural increase in Romanian urban areas between 1930 and 1939 expressed as percentages per 1000 inhabitants. *Source* Anuarul statistic al României (1940, p. 143)

3.3 Wages and Prices

Because they were subject to a bout of inflation between 1935 and 1938, the nominal Romanian wages of the 1930s cannot be compared or discussed in terms of main tendencies (Axenciuc 1999, p. 301). Between 1930 and 1938, wages were higher in the public than in the private sector (Iacob and Iacob 1995, p. 188).

Salary thresholds (expressed in lei) for jobs provided in the general state budget between 1934 and 1935 had the following values: health care worker (third class), 3,015–2,582; archbishop, 27,050; lawyer (third class), 8,300–5,650; librarian (second class), 8,600; baker, 2,650–750; cook, 3,600–500; army captain, 9,350–8,500; police quaestor, 10,600–10,300; shoemaker, 1,000; colonel, 18,050–16,700; associate professor (with 6 pay scale levels), 19,150–17,750; consul general (first class), 17,400–11,900; chief accountant (third class), 10,440–7,700; customs officer, 5,650–5,200; electrician (second class), 3,300; public guardian, 2,300; division general, 25,450; signalman (second class), 4,800–2,250; engineer, 19,500–3,300; institute teacher (with 3 pay scale levels), 10,500–5,550; tenured schoolteacher (with 3 pay scale levels), 4,700–3,700; judge, 19,150–9,250; lieutenant, 7,800–7,200; marshal, 35,100; train mechanic (second class), 5,535–4,288; physician, 11,900–8,500; minister, 30.400; metropolitan bishop, 28,350–23,500; notary public, 4,000–1,900; patriarch, 31,550; station sergeant, 3,400; county prefect, 17,250–16,750; prefect of police, 19,900–10,950, university professor (with 6 pay scale levels), 29,550–25,350; chauffeur, 5,350–1,900; usher, 3,240–2,400; mason, 2,650 (Iacob 2006, pp. 87–88).

It can be noted, for example, that as far as salaries were concerned, higher education senior teaching staff belonged to the upper class. The salary of a university professor was almost as high as that of a minister, and the salary of an associate professor was higher than that of a county prefect.

In 1937, the (starting-maximum) monthly salaries of university teaching staff were as follows: tenured professor, 13,000–29,500 lei; associate professor, 8,650–19,150 lei; lecturer, 7,800–17,500 lei; assistant professor, 6,150–12,500 lei (Scurtu et al. 2001, p. 167).

In 1938 gross nominal wages in the lowest paid industries were as follows: lumber - 1,654 lei; food – 1,715 lei; textile – 2,023 lei; glass – 2,032 lei, etc. In the course of that year, out of a total of 1,068,000 industry and trade employees, 12.8% had a monthly income of up to 600 lei (20 lei/day); 23% between 601 and 1,125 lei (20–37.50 lei/day); 29.6% between 1,126 and 1,975 lei (37.50–65.80 lei/day); 10.6% between 1,976–2,475 lei (65.80–82.50 lei/day and 23.2% over 2,476 lei (82.50 lei/day). It can be noted that almost two thirds of them had salaries of up to 1,975 de lei (Axenciuc 1999, pp. 301; 443).

In those days (as is still the case nowadays), Romanians spent most of their income on food and other living expenses. Here are some of the average retail prices in lei of several food products, items of clothing and footwear, fuel and rents in 1934, 1937 and 1938 (Table 4).

If we take a look at the situation in 1934, we can note, for instance, that an electrician, who had an average salary of 3,300 lei, would have been able to buy approximately 237 kg of beef, or 72 kg of charcuterie, or 32 kg of coffee, etc.

As far as the year 1938 is concerned, if we were to calculate product consumption and costs (for an average family of 4.3 members), we would note that most employees lived more or less on the poverty line. For 338,000 employees (35.8%) with incomes ranging from 20 to 37.5 lei/day, there was hardly enough money for daily food (Axenciuc 1999, p. 302). However, 23.2% of the employees had higher incomes. Out of a salary of 2,476 lei, for instance, it was possible to buy 110 kg of beef, or approximately 75 kg of pork, or approximately 45 kg of charcuterie, etc.

A side-by-side analysis of population incomes and rental costs reveals that the salary of a junior clerk or workman was insufficient for them to afford adequate accommodation.

3.4   The Cost of Living

The cost of living varied from one Romanian city to another, as some cities were more expensive than others to live in. In 1940, the Central Institute of Statistics published a series of statistical data regarding the cost of living from 1933 to 1934 and from 1937 to 1939 in 71 Romanian county seats. Out of the total number of 71, we have selected 10 to exemplify the situation: the top 5 (entries 1–5) and the last 5 (entries 67–71). The monthly cost of living in these towns and cities can be observed in Table 5.

Table 4 The country average retail prices in lei of several food products, items of clothing and footwear, fuel and rents in 1934, 1937 and 1938

	1934	1937	1938
Foods of animal origin			
Beef/1 kg	13.90	21.30	22.50
Pork/1 kg	22.20	27.90	29.20
Charcuterie/1 kg	45.65	51.75	55.15
Milk/1 litre	4.75	5.05	5.40
Eggs (each)	1.19	1.56	1.73
Traditional sheep milk cheese/1 kg	43.25	48.95	54.20
Chickens, hens (each)	31.65	44.65	49.80
Foods of vegetable origin			
White bread/1 kg	7.15	8.50	8.70
Wholegrain bread/1 kg	5.85	7.10	7.20
000 wheat flour/1 kg	9.15	11.15	11.40
Polenta/1 kg	3.30	4.30	4.65
Caster sugar/1 kg	24.15	29.20	31.05
Coffee/1 kg	102.80	146.15	157.00
Rice/1 kg	18.25	27.75	29.00
Sunflower oil/1 litre	26.75	34.95	34.40
Table wine/1 litre	–	16.30	17.65
Baking chocolate/1 kg	–	114.85	118.65
Fruit and vegetables			
Potatoes/1 kg	2.35	2.75	3.25
Dried beans/1 kg	4.50	8.90	11.00
Apples/1 kg	–	10.75	11.50
Clothing and footwear			
Suit fabric/1 m	428.00	477.30	534.00
Men's shoes/1 pair	417.00	553.25	605.00
Women's shoes/1 pair	395.00	507.25	559.00
Fuel			
Oak firewood/100 kg	60.40	63.70	74.05
Beech firewood/100 kg	53.10	63.10	74.40
Kerosene/1 litre	4.35	4.95	3.90
Rents			
Three-room apartment/month	1,360	1,550	1,595

Source Anuarul statistic al României (1940, p. 632)

Table 5 The monthly cost of living in some county seats between 1933 and 1934, and between 1937 and 1939

Ent. no.	1933		1934		1937		1938		1939	
	Town/City	Costs per month (in lei)	Town/City	Costs per month (in lei)	Town/City	Costs per month (in lei)	Town/City	Costs per month (in lei)	Town/City	Costs per month (in lei)
1.	Bucharest	11,099	Bucharest	10,596	Bucharest	12,421	Bucharest	13,016	Bucharest	14,194
2.	Cluj	9,403	Cluj	8,908	Cluj	11,096	Braşov	11,722	Braşov	12,430
3.	Timişoara	9,263	Braşov	8,846	Braşov	10,814	Cluj	11,706	Constanţa	12,302
4.	Braşov	9,165	Timişoara	8,813	Timişoara	10,316	Timişoara	11,097	Cluj	11,983
5.	Sibiu	9,143	Arad	8,674	Constanţa	10,187	Constanţa	10,734	Timişoara	11,551
67.	Tulcea	5,868	Târgu Jiu	5,719	Huşi	7,056	Tulcea	7,594	Vaslui	8,043
68.	Huşi	5,770	Huşi	5,620	Tulcea	7,021	Huşi	7,429	Fălticeni	8,000
69.	Dorohoi	5,698	Tulcea	5,560	Târgu Jiu	7,001	Târgu Jiu	7,351	Botoşani	7,835
70.	Râmnicu Sărat	5,697	Râmnicu Sărat	5,507	Fălticeni	6,935	Fălticeni	7,326	Dorohoi	7,754
71.	Fălticeni	5,640	Dorohoi	5,460	Dorohoi	6,635	Dorohoi	7,185	Huşi	7,714

Source Anuarul statistic al României (1940, p. 650)

According to the data featured in Table 5, it can be noted that in the periods under discussion, 1933–1934 and 1937–1939, Bucharest had the highest cost of living of all Romanian cities. The top of the list also includes highly urbanized cities such as Cluj, Braşov, Timişoara, Constanţa, Sibiu, Arad (Georgescu-Roegen 1937, p. 276).

As far as the cost of living was concerned, the capital was followed by Transylvanian municipalities (Cluj in 1933, 1934 and 1937, Braşov in 1938 and 1939). The high cost of living in Transylvanian cities was due to the high rental costs in that area. The explanation lies in the fact that these cities did not benefit from construction projects in which new buildings were erected, as had been the case in the Romanian Old Kingdom in the wake of the First World War (Georgescu 1934).

Among the 71 county seats, the lowest cost of living (entry 71) was to be found in Moldavian towns (Fălticeni in 1933, Dorohoi in 1934, 1937 and 1938, Huşi in 1939). In 1939, all 5 cheapest county seats were situated in Moldavia.

In comparison with Bucharest, the 1933–1934 and 1937–1939 cost of living levels in some county seats were as follows (Table 6).

According to the data featured in Table 6, it can be noted that in the periods under discussion, 1933–1934 and 1937–1939, the annual cost of living in some cheaper county seats was about 40–50% lower than the annual cost of living in the capital.

In 1939, the annual cost of living in lei in some county seats could be broken down into the following categories (Table 7).

According to the data featured in Table 7, it can be noted that rental costs were extremely high in cities with a high cost of living. In Constanţa, for instance, these expenses amounted to 42,000 lei/year, approximately 45% of the total yearly expenses in Huşi. In expensive cities, the percentage of total annual expenses that went towards rent was very high (25.8% in Bucharest, 26.8% in Braşov, 28.5% in Constanţa, 25% in Cluj, 23.4% in Timişoara). It can also be noted that in some cheaper county seats rental costs were lower. The cost of a year's rent in Vaslui, Dorohoi, Botoşani or Huşi was the equivalent of four months' rent in Cluj, for example. In these cheaper county seats, the percentage of annual expenses that went towards rent was low (approximately 12–13%).

According to the data featured in the same table, it can also be noted that transport expenses were high in expensive cities. In Timişoara, for instance, these expenses amounted to 7,920 lei/year, the equivalent of 8 months' rent in towns such as Vaslui, Botoşani, Dorohoi or Huşi. The percentages of total annual expenses that went towards transport were as follows: 9.6% in Bucharest, 4.4% in Braşov, 4.5% in Constanţa, 4.6% in Cluj and 5.7% in Timişoara. It can also be noted that in cheaper county seats transport expenses were very low. For instance, the annual cost of transport in Vaslui, Fălticeni, Botoşani, Dorohoi and Huşi together amounted to the same sum as 8 months' transport expenses in cities such as Braşov, Constanţa or Cluj. The percentage of annual expenses that went towards transport was as low as 0.7–1.3% in the above mentioned cheaper county seats.

Cost of living levels in some county seats in comparison with the same expense categories in Bucharest in 1939 were as follows (Table 8):

According to the data featured in Table 8, it can be noted that in expensive cities people spent almost as much money on food, clothing and footwear, textiles, light-

Table 6 Cost of living levels in some county seats in comparison with Bucharest in 1933–1934 and 1937–1939

Ent. no.	1933		1934		1937		1938		1939	
	Town/City	Living cost level	Town/City	Living cost level	Town/City	Living cost level	Town/City	Living cost level	Town/City	Living cost level
1.	Bucharest	100	Bucharest	100	Bucharest	100	Bucharest	100	Bucharest	100
2.	Cluj	84.7	Cluj	84.1	Cluj	89.3	Braşov	90.1	Braşov	87.6
3.	Timişoara	83.5	Braşov	83.5	Braşov	87.1	Cluj	89.9	Constanţa	86.7
4.	Braşov	82.6	Timişoara	83.2	Timişoara	83.1	Timişoara	85.3	Cluj	84.4
5.	Sibiu	82.4	Arad	81.9	Constanţa	82,0	Constanţa	82.5	Timişoara	81.4
67.	Tulcea	52.9	Târgu Jiu	54.0	Huşi	56,8	Tulcea	58.3	Vaslui	56.7
68.	Huşi	52.0	Huşi	53.0	Tulcea	56.5	Huşi	57.1	Fălticeni	56.4
69.	Dorohoi	51.3	Tulcea	52.5	Târgu Jiu	56.4	Târgu Jiu	56.5	Botoşani	55.2
70.	Râmnicu Sărat	51.3	Râmnicu Sărat	52.0	Fălticeni	55.8	Fălticeni	56.3	Dorohoi	54.6
71.	Fălticeni	50.8	Dorohoi	51.5	Dorohoi	53.4	Dorohoi	55.2	Huşi	54.3

Source Anuarul statistic al României (1940, pp. 650–651)

Table 7 The annual cost of living in lei in some county seats broken down into expense categories in 1939

Entry no.	Town/City	Total	Foods of animal origin	Foods of vegetable origin	Fruit	Clothing and footwear	Textiles	Lighting, water and fuel	Miscellaneous	Transport	Entertainment	Rent
1.	Bucharest	170,333	35,618	21,155	5,205	16,196	7,800	13,566	4,617	16,416	5,760	44,000
2.	Brașov	149,154	33,947	21,062	4,953	15,173	6,900	12,688	4,471	6,600	3,360	40,000
3.	Constanța	147,626	32,912	19,022	4,370	13,819	6,300	15,037	4,590	6,600	2,976	42,000
4.	Cluj	143,799	33,217	20,425	4,027	15,248	6,500	12,934	4,144	6,600	4,704	36,000
5.	Timișoara	138,623	31,846	19,924	3,090	14,890	6,400	14,541	4,252	7,920	3,360	32,400
67.	Vaslui	96,510	26,885	16,984	3,019	13,508	5,100	12,079	3,911	720	2,304	12,000
68.	Fălticeni	96,002	25,543	18,844	3,684	13,767	4,500	8,949	3,835	720	2,160	14,000
69.	Botoșani	94,023	26,129	16,896	2,921	13,979	4,500	10,654	3,872	960	2,112	12,000
70.	Dorohoi	93,050	25,971	17,948	3,364	13,180	4,600	9,783	3,564	720	1,920	12,000
71	Huși	92,568	27,009	17,354	2,772	13,212	4,200	8,694	3,777	1,200	2,400	12,000

Source Anuarul statistic al României (1940, pp. 652–653)

Table 8 Cost of living levels in some county seats in comparison with the same expense categories in Bucharest in 1939

Entry no.	Town/City	Total	Foods of animal origin	Foods of vegetable origin	Fruit	Clothing and footwear	Textiles	Lighting, water and fuel	Miscellaneous	Transport	Entertainment	Rent
1.	Bucharest	100.0	100.0	100.0	100.0	100.0	100.0	100.0	100.0	100.0	100.0	100.0
2.	Braşov	87.6	95.3	99.6	95.2	93.7	88.5	93.5	96.8	40.2	58.3	90.9
3.	Constanţa	86.7	92.4	89.9	84.0	85.3	80.8	110.8	99.4	40.2	51.7	95.5
4.	Cluj	84.4	93.3	96.5	77.4	94.1	83.3	95.3	89.8	40.2	81.7	81.8
5.	Timişoara	81.4	89.4	94.2	59.4	91.9	82.1	107.2	92.1	48.2	58.3	73.6
67.	Vaslui	56.7	75.5	80.3	58.0	83.4	65.4	89.0	84.7	4.4	40.0	27.3
68.	Fălticeni	56.4	71.7	89.1	70.8	85.0	57.7	66.0	83.1	4.4	37.5	31.8
69.	Botoşani	55.2	73.4	79.9	56.1	86.3	57.7	78.5	83.9	5.8	36.7	27.3
70.	Dorohoi	54.6	72.9	84.8	64.6	81.4	59.0	72.1	77.2	4.4	33.3	27.3
71	Huşi	54.3	75.8	82.0	52.3	81.6	53.8	64.1	81.8	7.3	41.7	27.3

Source Anuarul statistic al României (1940, pp. 656–657)

ing, water, fuel and rent as the inhabitants of Bucharest. In some cities (Constanţa and Timişoara, for instance), lighting, water and fuel expenses were even higher than in the capital. On the other hand, transport expenses in expensive cities were approximately 50–60% lower than in Bucharest. If we draw a comparison between Braşov and Bucharest for example, we can note that the difference between these two cities was primarily due to the fact that transport expenses were 59.8% lower and entertainment expenses were 41.7% lower in the Transylvanian municipality. As regards the other categories, Braşov expenses were up to 10% lower than the ones in the capital.

According to the data featured in the same table, it can also be noted that in some cheaper county seats (Vaslui, Fălticeni, Botoşani, Dorohoi and Huşi) people spent up to 28.30% less on foods of animal origin, 20.1% less on foods of vegetable origin, 95.6% less on transport, 66.7% less on entertainment and 72.7% less on rent than the inhabitants of Bucharest.

4 Conclusions

In this short chapter, we have presented some statistical data regarding the urban population of Romania and the standard of living in the fourth decade of the 20th century, a period marked by economic crises, social problems, and the imminent outbreak of World War II. We can conclude that Romanian society continued to move forward, despite all hardships. Even if Romania continued to have a low level of urbanization, there has been a lot of progress in urban modernization, trying to bring the Romanian urban areas closer to the Western urban areas (Axenciuc 1999, p. 383).

References

Anuarul Statistic al României 1939 şi 1940, 762 pp. M.O. Imprimeria Naţională, Bucureşti (1940)

Axenciuc, V.: Introducere în istoria economică a României. Epoca modernă şi contemporană, partea I., 448 pp. Editura Fundaţiei "România de mâine", Bucureşti (1999)

Banu, G.: Tratat de asistenţă medicală, volumul I. Medicina socială ca ştiinţă. Eugenia. Demografia, 430 pp. Casa Şcoalelor, Bucureşti (1944)

Georgescu, D.C.: Populaţia satelor româneşti. Sociologie românească, Anul II, nr. 2-3, februarie-martie, 68–79 (1937)

Georgescu, D.C.: Despre descreşterea poporului român. In Recensământul general al României-din 1941. Lămurirea opiniei publice. Proclamaţii şi apeluri. Studii, articole, reportaje, umor, insigne, medalii, afişe, pp. 43–48. Monitorul Oficial şi Imprimeriile Statului. Imprimeria Centrală, Bucureşti (1941)

Georgescu, N. St.: Introducere. In: Statistica preţurilor şi a costului vieţii pentru 1933. Monitorul Oficial şi Imprimeriile Statului. Imprimeria Naţională, Bucureşti (1934)

Georgescu-Roegen, N.: Costul vieţii în oraşele româneşti. Sociologie românească, Anul II, nr.7-8, februarie-martie, 75–77 (1937)

Iacob, Gh.: România în secolul XX Politică şi societate, Partea I. (2016). http://history.uaic.ro/wp-content/uploads/2012/12/2015sem2-Partea-I-Romania-in-secolul-XX.pdf. Accessed 14 Sep 2017

Iacob, Gh., Iacob, L.: Modernizare-europenism: Ritmul şi strategia modernizării, 300 pp. Editura Universităţii "Al. I. Cuza", Iaşi (1995)

Limona, R.: Populaţia Dobrogei în perioada interbelică, 240 pp. Semănătorul- Editura-online, Bucharest (2009)

Manuila, S., Georgescu, D.C.: Populaţia României, 311 pp. Editura Institutului Central de Statistică, Bucureşti (1938)

Measnicov, I.: Migraţiunile interioare în România. Sociologie românească, Anul IV, nr. 7-12, iulie-decembrie, 392–411 (1942)

Populaţia rezidentă la 1 ianuarie 2017 pe grupe de vârstă şi vârste, sexe şi medii de rezidenţă, macroregiuni, regiuni de dezvoltare şi judeţe. http://statistici.insse.ro/shop/index.jsp?page=tempo3&lang=ro&ind=POP105A. Accessed 18 Sep 2017

Scurtu, I. (coord.), Stănescu-Stanciu, T., Scurtu, G.M.: România între anii 1918–1940. Documente şi materiale, 168 pp. Editura Universităţii Bucureşti, Bucureşti (2001)

Şandru, D.: Politica sanitară a României între cele două războaie mondiale. Vrancea-Studii şi comunicări III, 259–276 (1980)

Developments in Decision-Making Process Within the European Union System

Dan Vătăman

Abstract The decisions taken at European Union level influence the lives of more than 500 million inhabitants of the Member States. Therefore, as European citizens we all should be interested in the decision-making procedures especially that, as shown in the European Treaties, the citizens have the right to participate directly in the democratic life of the Union, including by a request to the European Commission to make a legislative proposal. Consequently, this chapter is dealing with developments in decision-making process within European Union, the aim being to raise public awareness on the way in which the European legislation is adopted and also to clarify some principles, theories and technical issues regarding decision-making process within the European Union system.

Keywords Citizens' initiative · Decision-making process · Derogatory clauses
EU's institutional system · Legislative procedures
Normative and jurisprudential developments

1 Preliminary Considerations

The European Union (EU) is founded on common values to all Member States, especially respect for the rule of law which is a prerequisite for the protection of all other fundamental values. Taking into account the European Treaties provisions according to which all decisions "shall be taken as openly and as closely as possible to the citizen" (Article 10 (3) TEU) the objective of this chapter is to increase the public's level of understanding about the way in which the European legislation is adopted, especially that the Treaties and the legislative acts adopted on their basis have primacy over the law of Member States. For this purpose, in a first stage are highlighted the historical evolution of decision-making process with reference to the provisions

D. Vătăman (✉)
Doctoral School of Humanities, Ovidius University of Constanța, Constanța, Romania
e-mail: vataman.dan@gmail.com
URL: https://www.researchgate.net/profile/Vataman_Dan

© Springer Nature Switzerland AG 2019
C. Flaut et al. (eds.), *Models and Theories in Social Systems*, Studies in Systems,
Decision and Control 179, https://doi.org/10.1007/978-3-030-00084-4_13

239

of the European Treaties adopted over time and, at the same time, are emphasized the novelties introduced by Lisbon Treaty about the legislative process, especially the "citizens' initiative" and the increased role for National Parliaments. In the next step are analysed the institutional interaction within the decision-making process at European Union level, the analysis being concentrated on applicable principles and competencies of each institution which is involved in the decision-making process. For a better understanding of this process there are also analysed the ordinary legislative procedure and the special legislative procedures as well as the derogatory clauses from the legislative procedures laid down in the European Treaties.

2 Historical Evolution of Decision-Making Process Within European Union

2.1 Milestones in the Evolution of the Decision-Making Process

Along with the evolution in time of the European Communities the decision-making process has also evolved and the procedures have become increasingly complex.

Thus, if in early years of the European Communities the Council was the main decision-making institution (as shown in Article 28 of ECSC Treaty, Articles 148 and 149 of EEC Treaty, Articles 118 and 119 of EAEC Treaty), with the entry into force of the Single European Act (1987) it was established the cooperation procedure through which it has been achieved a real concertation between the Council, the Commission and the European Parliament (Article 6 of Single European Act).

A significant evolution in this process has occurred following the entry into force of the Maastricht Treaty (1993) when the legislative powers of the European Parliament were increased by introducing the co-decision procedure and extending the cooperation procedure (Articles 189a, 189b and 189c of EC Treaty). After the introduction of the co-decision procedure in the Maastricht Treaty an interinstitutional agreement was negotiated in 1993 which spelled out in greater detail the functioning of the Conciliation Committee, but did not cover other aspects of codecision (OJ C 329, 6.12.1993, p. 141).

The Amsterdam Treaty (1999) made a substantial progress by simplifying the co-decision procedure making it faster, more efficient and more transparent and also by extending it to new policy areas (Vătăman 2011, p. 57). In this respect, Declaration no. 34 (annexed to the Final Act of the Intergovernmental Conference which drafted the Amsterdam Treaty) called on the European Parliament, the Council and the Commission "to make every effort to ensure that the co-decision procedure operates as expeditiously as possible" (OJ C 340, 10.11.1997, p. 137). As a result, in May 1999, the three institutions adopted a Joint Declaration on Practical Arrangements for the Co-decision Procedure which was much more explicit about the role of each of them at each phase of the procedure (OJ C 148, 28.5.1999, p. 1).

After the entry into force of the Nice Treaty (2003) the European Parliament's legislative power were considerably increased by expanding the co-decision procedure to a larger number of matters and also by the assent required to establish enhanced co-operation in an area covered by the co-decision process (Article 1 point 9 of Nice Treaty).

The greatest milestone in the evolution of the decision-making process was the Lisbon Treaty (entered into force in 2009) which represents a major step towards full equality between European Parliament and Council. Continuing on from previous Treaties, the Lisbon Treaty not only confirmed that the co-decision procedure is the most legitimate from a democratic point of view but also renamed it the "ordinary legislative procedure" and establishing it as a common law procedure (Article 2 point 239 of Lisbon Treaty). At the same time, the Lisbon Treaty has introduced a series of other novelties about the legislative process, especially the "citizens' initiative" (Article 1 point 12 of Lisbon Treaty) and, very important, set out for the first time strengthened skills for national Parliaments enhancing their ability to express views on draft legislative acts of the Union (Protocol on the role of National Parliaments in the European Union annexed to Lisbon Treaty).

2.2 Citizens' Initiative - A New Right for European Citizens

Through the reform made by Lisbon Treaty has been reinforced European citizenship and enhanced further the democratic functioning of the Union. Thus, according to the consolidated version of the Treaty on European Union (TEU) every citizen shall have the right to participate in the democratic life of the Union and decisions shall be taken as openly and as closely as possible to the citizen. To this end, the Treaty allows European citizens to express their concerns by way of a citizens' initiative which affords them the possibility of directly approaching the Commission with a request inviting it to submit a proposal for a legal act of the Union for the purpose of implementing the Treaties similar to the right conferred on the European Parliament and on the Council under the Treaty on the Functioning of the European Union (TFEU) provisions (Articles 225 and 241 of TFEU).

The rules and procedures governing the citizens' initiative are set out in Regulation (EU) no. 211/2011 of the European Parliament and of the Council of 16 February 2011 (OJ L 65, 11.3.2011, pp. 1–22) in accordance with Article 11 of TEU and Article 24 of TFEU. The legal framework for the European Citizens' Initiative has been completed by Commission Implementing Regulation (EU) No 1179/2011 of 17 November 2011 laying down technical specifications for online collection systems pursuant to Regulation (EU) No 211/2011 (OJ L 301, 18.11.2011, pp. 3–9).

Almost six years after its implementation, the European Citizens' Initiative (ECI) has not reached its full potential to foster citizen participation in the democratic life of the European Union, which is why the European Commission announced its intention to revise the Regulation on the citizens' initiative. For this purpose, the European Commission made a legislative proposal which aims to improve the functioning of the ECI by addressing the shortcomings identified in its implementation, with as main policy objectives to making the ECI more accessible, less burdensome and easier to use for organisers and supporters and, at the same time, to achieving the full potential of the ECI as a tool to foster debate and participation at European level, including of young people, and bring the European Union closer to its citizens. According to the European Commission, to achieve those objectives, the procedures and conditions required for the ECI should be clear, simple, user-friendly and proportionate to the nature of this instrument and also should strike a judicious balance between rights and obligations (COM/2017/0482 final).

Despite the challenges in ECI implementation it can be said that this new tool is an important step towards strengthening European citizen's voice on issues that affect the quality of daily life. Thereby, the efforts made for ECI improvement will lead to an enhanced citizens' involvement and participation in actions and policies of the Union and, through this, increasing the democratic legitimacy and credibility at a time when the European Union continues to face a series of crises, the most important being the crisis of confidence.

2.3 Enhanced Role for the National Parliaments in EU Decision-Making Process

In the early years of the European Communities, Members of the European Parliament (MEPs) were appointed by each of the Member States' National Parliaments, reason for which there was a close link between the European Parliamentary Forum and the National Parliaments (NPs). This connection was broken in 1979 when it was made transition from appointed Assembly to elected Parliament through introduction of elections by direct universal suffrage.

This situation continued until the Treaty of Maastricht was adopted. As is apparent from two declarations annexed to the Treaty, the signatory states considered it necessary to encourage greater involvement of National Parliaments in the activities of the European Union and, to this end, the exchange of information between National Parliaments and the European Parliament should be stepped up. It was also considered that it would be good the European Parliament and the National Parliaments to meet as necessary as a Conference of the Parliaments (OJ C 191, 29.7.1992, pp. 100–101).

The protocol on the role of the National Parliaments annexed to the Treaty of Amsterdam encouraged greater involvement of national parliaments in the activities of the European Union and enhanced their ability to express their views on matters which may be of particular interest to them (OJ C 340, 10.11.1997, p. 113).

At long last, the Lisbon Treaty clarified the role of national parliaments in the new institutional structure of the European Union and at the same time conferred them new powers and a series of strengthened competences in decision-making process. Thus, according to the Treaty on European Union (consolidated versions), National

Parliaments contribute actively to the good functioning of the Union: through being informed by the institutions of the Union and having draft legislative acts of the Union forwarded to them; by seeing to it that the principle of subsidiarity is respected; by taking part, within the framework of the area of freedom, security and justice, in the evaluation mechanisms for the implementation of the Union policies in that area; by taking part in the revision procedures of the Treaties; by being notified of applications for accession to the Union; by taking part in the inter-parliamentary cooperation between national Parliaments and with the European Parliament (Article 12 TEU). The procedures for involving National Parliaments in this process are detailed in Protocol no. 1 on the role of National Parliaments in the European Union (OJ C 202, 7.06.2016, pp. 203–205) and Protocol no. 2 on the application of the principles of subsidiarity and proportionality (OJ C 202, 7.06.2016, pp. 206–209).

The importance of close cooperation and continuous exchanges of information between the National Parliaments and the European Parliament was also highlighted in the Treaty on Stability, Coordination and Governance in the Economic and Monetary Union (2012) which states that "the European Parliament and the national Parliaments of the Contracting Parties will together determine the organisation and promotion of a conference of representatives of the relevant committees of the national Parliaments and representatives of the relevant committees of the European Parliament in order to discuss budgetary policies and other issues covered by this Treaty" (Article 13 TSCG).

3 Institutional Interaction Within the Decision-Making Process at European Union Level

3.1 Applicable Principles

Under the provisions of the Treaties on which the European Union is founded can be configured a set of principles that underpin the functioning of the institutions, among which: representation of interests, conferral of competences, institutional balance and cooperation between institutions.

Representation of interests in the legislative process is a principle not expressly mentioned in the Treaties but it results from the analysis of the attributions of each institution. Therefore, the European Parliament is made up of Members directly elected in the Member States of the European Union (Article 14 TEU) and consequently represent the interests of the citizens which gives legitimacy and constitutes the democratic basis of the whole Union. In the case of the Council, this institution brings together representatives of Member State at ministerial level, who have the authority to commit their governments to the actions agreed on in the meetings and therefore represent the interests of the Member States (Article 16 TEU). The European Commission's mission is to promote the general interest of the European Union by proposing and enforcing legislation as well as by implementing policies and the budget (Article 17 TEU).

The limits of institutions competences are governed by the principle of conferral. Thus, according to TEU provisions, each institution shall act within the limits of the powers conferred on it in the Treaties, and in conformity with the procedures, conditions and objectives set out in them (Article 13 (2) TEU).

The principle of institutional balance implies that each institution of European Union has to act in accordance with the competences conferred on it by the Treaties, without hindering in any way the exercise of the competences attributed to the other institutions. The concept of institutional balance as the legal principle was established by the Court of Justice in the cases C-9/56 and C-10/56, Meroni v. High Authority (European Court reports, English special edition 1957–1958, p. 133).

The principle of cooperation is explicitly addressed in Article 4 of TEU which establishes that the Union and the Member States shall, in full mutual respect, assist each other in carrying out tasks which flow from the Treaties. Regarding coopera-tion between institutions, the Treaty states that the institutions shall practice *mutual sincere cooperation* (Article 13 (2) TEU). Also, TFEU states that the European Par-liament, the Council and the Commission shall consult each other and by common agreement make arrangements for their cooperation and, to that end, they may, in compliance with the Treaties, conclude interinstitutional agreements which may be of a binding nature (Article 295 TFEU).

3.2   Institutions Involved in the Decision-Making Process

The European Commission, the European Parliament and the Council of the Euro-pean Union are the three central legislative institutions of the European Union and are often referred to as the "institutional triangle". In this process the European Com-mission represents the interests of the European Union as a whole, the European Parliament represents the interests of European Citizens and the Council represents the member state governments.

These three institutions are the most important but are not the only institutions that have a role in the decision-making process. Thus, we should also consider: the European Economic and Social Committee, which is an advisory body representing workers' and employers' organisations and other interest groups; the Committee of the Regions, which is an advisory body representing Europe's regional and local authorities (Article 13 (4) TEU and Article 300 (1) TFEU).

As a rule the European Commission has a quasi-monopoly on the legislative initiative, the Treaty on European Union stipulating that the Union legislative acts may only be adopted on the basis of a Commission proposal, except where the Treaties provide otherwise (Article 17 (2) TEU). The Commission is therefore responsible for preparing almost all proposed legislative acts, in particular those under the codecision procedure. Nevertheless, in the specific cases provided for by the Treaties, legislative acts may be adopted on the initiative of a group of Member States or of the European

Parliament, on a recommendation from the European Central Bank or at the request of the Court of Justice or the European Investment Bank (Article 289 (4) TFEU).

According to the TEU provisions the European Parliament shall, jointly with the Council, exercise the legislative function (Articles 14 (1) and 16 (1) TEU). The procedure followed for a legislative proposal depends on the type and subject of the proposal. The vast majority of EU legislative acts are jointly adopted by the European Parliament and Council, while in specific cases a single institution can adopt alone.

The European Parliament, the Council and the Commission shall consult each other and by common agreement make arrangements for their cooperation. To that end, they may, in compliance with the Treaties, conclude interinstitutional agreements which may be of a binding nature (Article 295 TFEU).

The three institutions involved in the decision-making process are assisted by an Economic and Social Committee and a Committee of the Regions, two bodies which acting in an advisory capacity. According to TFEU provisions, the Economic and Social Committee shall be consulted by the European Parliament, by the Council or by the Commission where the Treaties so provide, as well as in all cases in which these institutions consider it appropriate. Also, this body may issue an opinion on its own initiative in cases in which it considers such action appropriate (Article 304 TFEU). On its turn, the Committee of the Regions shall be consulted by the European Parliament, by the Council or by the Commission where the Treaties so provide and in all other cases (in particular those which concern cross-border cooperation) in which one of these institutions considers it appropriate (Article 307 TFEU).

4 Legal Acts of the European Union and the Procedures for Adopting Them

4.1 Main Characteristics of the European Union Legal Acts

Depending on their nature, the legal acts may be legislative or non-legislative and can have a binding force or not. In this respect, Articles 289, 290 and 291 of TFEU clarify the differences between legislative acts, delegated acts and implementing acts. Thus, the legislative acts are adopted by ordinary legislative procedure or special legislative procedures. A legislative act adopted by the European Parliament and the Council may delegate to the Commission the power to adopt non-legislative acts of general application to supplement or amend certain non-essential elements of the legislative act. Implementing acts are generally adopted by the Commission, which is competent to do so in cases where uniform conditions for implementing legally binding Union acts are needed. Implementing acts are a matter for the Council only in duly justified specific cases and in the cases.

Article 288 of TFEU establishes that to exercise the European Union's competences, the institutions shall adopt regulations, directives, decisions, recommendations and opinions. In accordance with the TFEU provisions, the regulations have general application, are binding in their entirety and directly applicable in all Member States. In their turn, the directives are binding, as to the result to be achieved, upon each Member State to which it is addressed, but leave to the national authorities the choice of form and methods. Thus, for a directive to take effect at national level, Member States must adopt a law to transpose it (According to Article 4 (3) of TEU, the Member States shall take any appropriate measure, general or particular, to ensure fulfilment of the obligations arising out of the Treaties or resulting from the acts of the institutions of the Union). As respects the decisions, these are binding in their entirety which is why they cannot be applied incompletely, selectively or partially. A decision which specifies those to whom it is addressed shall be binding only on them. Recommendations and opinions shall have no binding force on those to whom they are addressed, but may provide guidance as to the interpretation and content of Union law.

In the majority of cases, the Treaties detail the type of act which should be used. In these cases, Article 296 (1) of TFEU states that where the Treaties do not specify the type of act to be adopted, the institutions shall select it on a case-by-case basis, in compliance with the applicable procedures and with the principle of proportionality.

Under the Article 297 of TFEU, regulations and directives which are addressed to all Member States, as well as decisions which do not specify to whom they are addressed, shall be published in the Official Journal of the European Union. They shall enter into force on the date specified in them or, in the absence thereof, on the twentieth day following that of their publication. Other directives, and decisions which specify to whom they are addressed, shall be notified to those to whom they are addressed and shall take effect upon such notification.

4.2 Legislative Procedures Laid Down in the European Treaties

In light of the TFEU provisions can be identified an ordinary legislative procedure (formerly the co-decision procedure) and certain special legislative procedures (which replaced the former consultative, cooperation and assent procedures).

4.2.1 Ordinary Legislative Procedure

Under the reform made by the Lisbon Treaty the co-decision procedure has become the "ordinary legislative procedure" and also its application area has been extended to a significant number of Union activity areas, such as: services of general economic interest (Article 316 TFEU); citizens' initiative (Article 316 TFEU); legislation concerning the common agricultural policy (Article 43 TFEU); exclusion in a Member State of certain activities from the application of provisions on the right of establishment (Article 51 TFEU); extending provisions on freedom to provide services to service providers who are nationals of a third State and who are established within the Union (Article 56 TFEU); adoption of other measures on the movement of capital to and from third countries (Article 64 TFEU); administrative measures relating to capital movements in connection with preventing and combating crime and terrorism (Article 75 TFEU); visas, border checks, free movement of nationals of non-member countries, management of external frontiers, absence of controls at internal frontiers (Article 77 TFEU); asylum, temporary protection or subsidiary protection for nationals of third countries (Article 78 TFEU); judicial cooperation in criminal matters – procedures, cooperation, training, settlement of conflicts, minimum rules for recognition of judgments (Article 82 TFEU) and others. A full list of the 85 legal bases subject to the ordinary legislative procedure can be found in Report of European Parliament on the Treaty of Lisbon 2007/2286 (INI).

The main characteristic of the ordinary legislative procedure is the joint adoption by the European Parliament and the Council of a regulation, directive or decision on a proposal from the Commission. According to the Article 294 TFEU, the ordinary legislative procedure consists of three stages (readings), with the possibility for the European Parliament and the Council to conclude at any reading, if they reach an overall agreement in the form of a joint text. The Joint Declaration of three institutions adopted in 2007 has revised the practical arrangements on the operation of the co-decision procedure, explicitly recognising the importance of the "trilogue system" throughout the procedure. Thus, such trilogues increase significantly the possibilities for agreement at first and second reading stages, as well as contributing to the preparation of the work of the Conciliation Committee - European Parliament, Council and Commission, Joint Declaration on Practical Arrangements for the Codecision Procedure (OJ C 145, 30.6. 2007, pp. 5–9).

At the stage of *first reading*, the European Parliament and the Council examine in parallel the Commission's proposal.

The European Parliament acts first and delivers a position with regard to the Commission's proposal without amendments, amending it, including as a result of a first reading agreement, or rejecting it. This position, prepared by a rapporteur, is discussed and amended within the relevant parliamentary committee, then debated in plenary session, where it is adopted by a simple majority (Vătăman 2015, p. 175).

In its turn, the Council adopts its position on the basis of the Commission's proposal, amended where necessary, in the light of the European Parliament's first reading. According to the TFEU, if the Council approves the European Parliament's position, the act concerned shall be adopted in the wording which corresponds to the position of the European Parliament and the legislative procedure is thus closed at this stage. In the case in which the Council does not approve the European Parliament's position, it shall adopt its position at first reading and communicate it to the European Parliament in order to proceed to the second reading (Article 294 (4) and (5) TFEU).

In the *second reading* stage the European Parliament can approve, reject or amend the Council's position at first reading, generally within three-month time limit. According to Article 294 (14) of TFEU, the periods of three months and six weeks referred to in the first and second reading stages of ordinary legislative procedure shall be extended by a maximum of one month and two weeks respectively at the initiative of the European Parliament or the Council. Thus, if the European Parliament approves the Council's common position or does not take a decision within the stipulated deadline, the act concerned shall be deemed to have been adopted in the wording which corresponds to the position of the Council. If the European Parliament rejects the Council's common position by an absolute majority of its members, the proposed act shall be deemed not to have been adopted (Article 294 (7)(a) and (b) of TFEU). Also, the European Parliament may propose amendments to the common position by an absolute majority of its members and the text thus amended is forwarded to the Council and the Commission. In this case TFEU specifically requires the Commission to deliver an opinion on the European Parliament's amendments. The Commission's opinion is very important because the Council will have to act unanimously on the amendments on which the Commission has delivered a negative opinion (Article 294 (9) of TFEU).

Following receipt of the European Parliament's second reading position, the Council has a period of three months (or four, if an extension has been requested) to conclude its second reading. In this stage the Council may approve the Parliament's amendments by a qualified majority (or by unanimity where appropriate), case in which act in question shall be deemed to have been adopted. If the Council does not accept all of Parliament's amendments, the President of the Council, in agreement with the President of the European Parliament, shall within six weeks (or eight weeks, if an extension has been agreed) convene a meeting of the Conciliation Committee (Article 294 (8)(b) of TFEU).

The Conciliation Committee involves members of the Council or their representatives and an equal number of members representing the European Parliament, which have the task of reaching agreement on a joint text, by a qualified majority of the members of the Council or their representatives and by a majority of the members representing the European Parliament within six weeks of its being convened, on the basis of the positions of the European Parliament and the Council at second reading. The Commission shall take part in the Conciliation Committee's proceedings and shall take all necessary initiatives with a view to reconciling the positions of the European Parliament and the Council, in order to facilitate the adoption of the legislative act. If, within six weeks (or eight weeks if it was decided an extension) the Conciliation Committee does not approve the joint text, the proposed act shall be deemed not to have been adopted (Article 294 (10), (11) and (12) of TFEU).

Third reading stage is required when the Conciliation Committee approves a joint text. In this situation the European Parliament, acting by a majority of the votes cast, and the Council, acting by a qualified majority, shall each have a period of six weeks from that approval in which to adopt the act in question in accordance with the joint text. If they fail to do so, the proposed act shall be deemed not to have been adopted and the procedure is ended (Article 294 (13) of TFEU).

4.2.2 Special Legislative Procedures

The special legislative procedures derogate from the ordinary legislative procedure and therefore constitutes some exceptions. As opposed to ordinary legislative procedure, special legislative procedures are not described precisely by the European Treaties, which just provides their application in specific cases (Article 289 (2) of TFEU).

In the case of special legislative procedures the decision belongs to the Council, the European Parliament's role being limited to give its consent to a legislative proposal or be consulted on it. Thus, two types of procedures can be identified, namely: *consent* and *consultation*.

Under the consent procedure, the Council must act unanimously on a proposal from the Commission and after obtaining the consent of the European Parliament (Article 352 (1) of TFEU). In this case the Parliament has the power to accept or reject a proposed act by an absolute majority vote, but cannot amend it. Where the European Parliament does not give its consent, the proposal cannot be adopted.

As a legislative procedure, the consent it is used:

- when the Council take an appropriate action to combat discrimination based on sex, racial or ethnic origin, religion or belief, disability, age or sexual orientation (Article 19 (1) of TFEU);
- when the subsidiary general legal basis is applied in line with Article 352 of TFEU.

As a non-legislative procedure, the consent it is usually applied:

- when the Council make a determination that there is a clear risk of a serious breach by a Member State of the values on which the Union is founded or in the case when the European Council determine the existence of a serious and persistent breach by a Member State of the Union's values (Article 7(1) and (2) of TEU);
- for the adoption of the decisions by the European Council on the proposals for the amendment of the Treaties (Article 48(7) of TEU);
- when the Council shall act on the request of an European State to become a member of the Union (Article 49 of TEU) as well as when is concluded an agreement with a Member State which decides to withdraw from the Union (Article 50(2) of TEU);
- when the Council adopt a regulation laying down the multiannual financial framework which ensuring that Union expenditure develops in an orderly manner and within the limits of its own resources (Article 312(2) of TFEU).

Under the consultation procedure the Council adopts a legislative proposal after the European Parliament has submitted its opinion on it. In this procedure the Parliament may approve, reject or propose amendments to a legislative proposal. The Council is not legally obliged to take into account the Parliament's opinion, but according to the case-law of the Court of Justice, it must not take a decision without having received it. Thus, adoption of a legislative act by violating the obligation of consultation represents an infringement of an essential procedural requirement, situation in which can be brought an action before the Court of Justice of the European

Union for annulment of the act (Article 263 TFEU). Example in this regard are judgments of the Court of Justice of 29 October 1980 in Case138/79, Roquette Frères v. Council (European Court reports 1980, p. 3333) and Case139/79, Maizena GmbH v. Council (European Court reports 1980, p. 3393), cases in which the Court declared a Council regulation invalid because the Council was in breach of its obligation to consult Parliament, arguing that "the consultation is the means which allows the Parliament to play an actual part in the legislative process, such power represents an essential factor in the institutional balance intended by the treaty".

Pursuant to the provisions of the European Treaties, the consultation procedure is now applicable in a limited number of legislative areas, such as:

- adoption of directives establishing the coordination and cooperation measures necessary to facilitate protection of European citizens in the territory of a third country in which the Member State of which he is a national is not represented, by the diplomatic or consular authorities of any Member State, on the same conditions as the nationals of that State (Article 23 of TFEU);
- application of the TFEU provisions on aids granted by Member States (Article 109 of TFEU);
- adoption of provisions for the harmonisation of legislation concerning turnover taxes, excise duties and other forms of indirect taxation to the extent that such harmonisation is necessary to ensure the establishment and the functioning of the internal market and to avoid distortion of competition (Article 113 of TFEU);
- establishing an Employment Committee with advisory status to promote coordination between Member States on employment and labour market policies (Article 150 of TFEU).
- implementing measures for the Union's own resources system (Article 311 of TFEU).

Also, another situation when is required consultation refers to the obligation of the High Representative of the Union for Foreign Affairs and Security Policy to regularly consult the European Parliament on the main aspects and the basic choices of the common foreign and security policy and the common security and defence policy and inform it of how those policies evolve. Moreover, the Treaty provides that he shall ensure that the views of the European Parliament are duly taken into consideration (Article 36 of TEU).

4.3 Derogatory Clauses in Legislative Procedures

Despite the fact that through reform made by Lisbon Treaty has been expanded the scope of the ordinary legislative procedure in a number of new areas and, therefore, qualified majority replaces unanimity, some areas are still considered "sensitive", reason for that Member States are little inclined to relinquish part of their power of opposition in certain policy areas.

As shown above in the section about special legislative procedures, certain areas remain subject to unanimity in whole or in part, as they are particularly important: adoption of the Union's multiannual financial framework; harmonisation of national legislation on indirect taxation; harmonisation of national legislation in the field of social security and social protection; certain provisions in the field of justice and home affairs; common foreign and security policy. Given the difficulty of reaching agreements in these areas, by the Lisbon Treaty were introduced several types of clauses aimed to simplify the EU's decision-making process by to applying the ordinary legislative procedure to areas for which the Treaties had laid down a special legislative procedure. Thus, the derogatory clauses are classified into three categories, namely: *passerelle clauses*, *brake clauses* and *accelerator clauses*.

4.3.1 Passerelle Clauses

Passerelle clauses allow derogation from the legislative procedures laid down in the Treaties, making it possible to apply the ordinary legislative procedure to areas for which was necessary a special legislative procedure, thus the vote with qualified majority replaces unanimity. However, activation of a passerelle clause depends on the consensus of Member States, in all cases being necessary a unanimous decision by the European Council or the Council of the European Union.

There is a general passerelle clause applicable to all policies and actions of the European Union and, also, a number of other specific passerelle clauses to certain European policies.

The general passerelle clause is applicable when the European Treaties provides for the Council to act by unanimity in a given area or case, situation in which the European Council may adopt a decision authorising the Council to act by a qualified majority in that area or in that case. Also, where the European Treaties provides for legislative acts to be adopted by the Council in accordance with a special legislative procedure, the European Council may adopt a decision allowing for the adoption of such acts in accordance with the ordinary legislative procedure. In both cases, the European Council shall act by unanimity after obtaining the consent of the European Parliament, which shall be given by a majority of its component members. It should be noted that any initiative taken by the European Council shall be notified to the national Parliaments which have the opportunity to make known their opposition, thus hindering the clause application (Article 48(7) of TEU).

The specific passerelle clauses have a series of particularities and their implementation may vary from case to case. Thus, according to the European Treaties provisions, can be identified six specific passerelle clauses which are applicable to:

- Common Foreign and Security Policy, except decisions having military or defence implications (Article 31 of TEU);
- multiannual financial framework of the Union (Article 312(2) of TFEU);
- achieve the objectives of the Union policy on the environment (Article 192(2) of TFEU);

- social policy of the Union (Article 153 of TFEU);
- adopt measures concerning family law with cross-border implications (Article 81(3) of TFEU);
- where a provision of the Treaties which may be applied in the context of enhanced cooperation, in this case all members of the Council may participate in its deliberations, but only members of the Council representing the Member States participating in enhanced cooperation shall take part in the vote (Article 333 of TFEU).

4.3.2 Brake Clauses

So-called brake clauses have been created in order to enable a Member State to temper the ordinary legislative procedure by a braking mechanism, if it considers that the fundamental principles of its social security system or its criminal justice system are threatened by the draft legislation in the course of being adopted.

Thus, where a member of the Council declares that a draft legislative act would affect important aspects of its social security system, including its scope, cost or financial structure, or would affect the financial balance of that system, it may request that the matter be referred to the European Council. In that case, the ordinary legislative procedure shall be suspended. After debates, the European Council shall refer the draft back to the Council (in that case which shall terminate the suspension of the ordinary legislative procedure) or take no action or request the Commission to submit a new proposal, situation in which it is considered that the draft legislative act was not adopted (Article 48 of TFEU).

In the cases of adoption the necessary measures to facilitate mutual recognition of judgments and judicial decisions and police and judicial cooperation in criminal matters having a cross-border dimension (measures adopted in accordance with the ordinary legislative procedure), if a member of the Council considers that a draft legislative act would affect fundamental aspects of its criminal justice system, it may request that the draft be referred to the European Council. After the necessary debates, and in case of a consensus, the European Council shall, within four months of this suspension, refer the draft back to the Council, which shall terminate the suspension of the ordinary legislative procedure. In case of disagreement, and if at least nine Member States wish to establish enhanced cooperation on the basis of the draft legislative act concerned, they may turn to the accelerator clause (Article 82(3) and Article 83(3) of TFEU).

4.3.3 Accelerator Clauses

Accelerator clauses facilitates collaboration among certain Member States as they allow derogation from the engagement procedure for enhanced cooperation in some areas provided by the Treaties, particularly those concerning criminal offences, cooperation in criminal matters and police cooperation.

In cases where a Member State has used the brake clause to oppose the adoption of a legislative act relating judicial cooperation in criminal matters or establishment of common rules for criminal offences and sanctions in the areas of particularly serious crime with a cross-border dimension, the other States who wish may turn to the accelerator clause and thus continue and conclude the legislative procedure between them, under the framework of enhanced cooperation. According to the TFEU provisions, where at least nine Member States wish to establish enhanced cooperation on the basis of the draft directive concerned, they shall notify the European Parliament, the Council and the Commission accordingly. In such a case, the authorisation to proceed with enhanced cooperation referred to in Article 20(2) of TEU and Article 329(1) of TFEU shall be deemed to be granted and the provisions on enhanced cooperation shall apply (Article 82(3) and Article 83(3) of TFEU).

An accelerator clause can be applied in the case in which the Council decides by means of regulation to establish a European Public Prosecutor's Office. Given that such a decision must be adopted unanimously by the Council, in the absence of unanimity in the Council, a group of at least nine Member States may request that the draft regulation be referred to the European Council. In that case, the procedure in the Council shall be suspended. After debates, and in case of a consensus, the European Council shall, within four months of this suspension, refer the draft back to the Council for adoption. Within the same timeframe, in case of disagreement, and if at least nine Member States wish to establish enhanced cooperation on the basis of the draft regulation concerned, they shall notify the European Parliament, the Council and the Commission accordingly. In such a case, the authorisation to proceed with enhanced cooperation shall be deemed to be granted and the Treaties provisions on enhanced cooperation shall apply (Article 86(1) of TFEU).

Also, an accelerator clause can be applied in the police cooperation, field in which are involving all the Member States' competent authorities, including police, customs and other specialised law enforcement services in relation to the prevention, detection and investigation of criminal offences. The Council must act unanimously and, if it is otherwise, a group of at least nine Member States may request that the draft regulation be referred to the European Council. In this case also, the procedure in the Council shall be suspended. After debates, and in case of a consensus, the European Council shall, within four months of this suspension, refer the draft back to the Council for adoption. Within the same timeframe, in case of disagreement, and if at least nine Member States wish to establish enhanced cooperation on the basis of the draft regulation concerned, they shall notify the European Parliament, the Council and the Commission accordingly. In such a case, the authorisation to proceed with enhanced cooperation shall be deemed to be granted and the Treaties provisions on enhanced cooperation shall apply (Article 87(3) TFEU).

5   Conclusions

Along the evolution of the European Communities to the European Union, the decision-making process at European level has become more complex and transparent, decisions being taken today as openly and as closely as possible to the citizens.

The Lisbon Treaty has introduced a series of novelties about the legislative process, especially the "citizens' initiative" and, very important, set out for the first time strengthened skills for national Parliaments enhancing their ability to express views on draft legislative acts of the Union. More than that, the Lisbon Treaty not only confirmed that the co-decision procedure is the most legitimate from a democratic point of view but the same time renamed it the "ordinary legislative procedure" and establishing it as a general rule for passing legislation at European Union level.

Even if through this reform has been expanded the scope of the ordinary legislative procedure in a number of new areas and, therefore, qualified majority replaces unanimity, some areas are still considered "sensitive" by Member States, reason for that sought to preserve their right of veto in certain policy areas. Consequently, in order to apply the ordinary legislative procedure to areas for which the Treaties had laid down a special legislative procedure, were introduced some derogatory clauses which creates the premises for a flexible decision-making process and thus contributes to strengthening the Union's capacity to act. Even so, the decision-making process is quite difficult because their implementation is not possible without consent of all Member States, especially that activation of a derogatory clause is always dependent on a unanimous decision of the Council or the European Council.

Taking into account the complexities and dynamics of the EU decision-making process and moreover the fact that these decisions influence the lives of more than 500 million European citizens, any opportunity should be used to raise public awareness on the way in which the European legislation is adopted. It is more than obvious that through enhancing social awareness with regard to EU decision-making process not only is added a new dimension to European democracy but also is intensified the public debate on European policies, which would lead to greater confidence and optimism about the future of European Union.

References

Declaration on respect for time limits under the co-decision procedure, OJ C 340, 10.11.1997
European Parliament, Council and Commission, Joint Declaration on Practical Arrangements for the New Codecision Procedure (Art. 251 of the Treaty Establishing the European Community), OJ C 148, 28.5.1999
Declaration no. 13 on the role of National Parliaments in the European Union and Declaration no. 14 on the Conference of the Parliaments, OJ C 191, 29.7.1992
European Parliament, Council and Commission, Joint Declaration on Practical Arrangements for the Codecision Procedure (Art. 251 of the EC Treaty), OJ C 145, 30.6.2007

Interinstitutional agreement on the phase preceding the adoption of a common position by the Council and on arrangements for proceedings in the Conciliation Committee provided for in Article 189b, OJ C 329, 6.12.1993

Judgment of the Court of 29 October 1980 (Case 138/79), SA Roquette Frères v. Council of the European Communities, European Court reports 1980. http://eur-lex.europa.eu/

Judgment of the Court of 29 October 1980 (Case 139/79), Maizena GmbH v. Council of the European Communities, European Court reports 1980. http://eur-lex.europa.eu/

Proposal for a Regulation of the European Parliament and of the Council on the European citizens' initiative, COM/2017/0482 final. http://ec.europa.eu/citizens-initiative/public/regulation-review?lg=en

Protocol no. 13 on the role of National Parliaments in the European Union, OJ C 340, 10.11.1997

Protocol on the role of National Parliaments in the European Union annexed to Lisbon Treaty, OJ C 306, 17.12.2007

Single European Act, OJ L 169, 29.6.1987

Treaty establishing the European Coal and Steel Community (ECSC). http://eur-lex.europa.eu

Treaty establishing the European Economic Community (EEC). http://eur-lex.europa.eu

Treaty establishing the European Atomic Energy Community (EAEC). http://eur-lex.europa.eu

Treaty of Maastricht, OJ C 224, 31.8.1992

Treaty of Amsterdam, OJ C 340, 10.11.1997

Treaty of Nice, OJ C 80, 10.3.2001

Treaty of Lisbon, OJ C 306, 17.12.2007

Treaty on European Union (TEU), OJ C 202, 7.6.2016

Treaty on the Functioning of the European Union (TFEU), OJ C 202, 7.6.2016

Treaty on Stability, Coordination and Governance in the Economic and Monetary Union (TSCG). http://eur-lex.europa.eu

Vătăman, D.: History of the European Union. Pro Universitaria Publishing House, Bucharest (2011)

Vătăman, D.: European Union: Specialized Practical Guide. Pro Universitaria Publishing House, Bucharest (2015)

Assessment of the Efficiency of Respiratory Protection Devices Against Lead Oxide Nanoparticles

Eva Kellnerová, Kristýna Binková and Šárka Hošková-Mayerová

Abstract Chapter evaluates the current state of health and safety related problems of people exposed to environmental burden due to occupational requirements. In such conditions, particulate nanoscale aerosols play an important role, therefore the survey is focused on personal protective equipment used against inhalation of pollutants from the air. Respiratory protective filters work against inhalation of pollutants by exposed persons. Filter efficiency is determined according to standardized methods given by the standardized Czech technical norms. However, such rehearsals are not specifically focused on an ultrafine aerosol with the content of nanoparticles in the range of 7.6–299.6 nm. The chapter evaluates permeability of one of the most often used protective filter OF-90 against ultrafine aerosol lead oxide with predetermined characteristics.

Keywords Nanoparticles · Ultrafine aerosol · Filter efficiency

1 Introduction

Clean air and good working conditions are very important for human health. Pollution of ambient air can affect not only human health but also entire ecosystems and materials that surround us. The level of pollution is affected by contamination by substances discharged from different sources due to human activities (e.g. transport, incineration, industrial production, etc.). There are many types of pollutants with

E. Kellnerová (✉)
Department of Emergency Management, University of Defence, Brno, Czech Republic
e-mail: eva.kellnerova@unob.cz

K. Binková
Department of Economics, University of Defence, Brno, Czech Republic
e-mail: kristyna.binkova@unob.cz

Š. Hošková-Mayerová
Department of Mathematics and Physics, University of Defence, Brno, Czech Republic
e-mail: sarka.mayerova@unob.cz

variable toxic effects. At present, the amount of ultrafine particles released from human activity (including nanoparticles) in the air is increasing. An effort must be given to protect surrounding environment. Nevertheless, there may be situations, whether operational, chemical, transport or even of war, when it is not possible. Given the increasing production of commercially available nanotechnology-based products, exposure to nanomaterials for employees, consumers and the environment seems inevitable. Free nanoparticles enter the environment by the gradual decomposition of nanomaterials (degradation due to UV radiation, metabolism of living organisms, imperfect decontamination, etc.). Exposure to nanoparticles is a key factor in assessing the risk of nanomaterials from the point of view of their potential toxicological and ecotoxicological effects. Military technology produces a number of pollutants, especially during the usage of military technology (such as tanks, combat vehicles, armoured personnel carriers, etc.) and the introduction of new weapon systems and technologies. (Otřísal and Florus 2014; Otřísal et al. 2018).

Soldiers commonly encounter a high mental and physical burden during their professional life. This requires high psychological and physical endurance, but this may not be enough when combining several difficulties at the same time. The occupational workload of military personnel often involves exhausting working conditions (handling with dangerous chemical substances, working in a polluted environment, lack of opportunities for rest, isolation, lack of sleep, etc.). During these situations, soldiers are constrained to move in the atmosphere with increased concentrations of harmful substances. In case of fire interventions, other hazardous combustion products may be added to the fire (such as diesel fuel, exhaust fumes, formaldehyde, acrolein, benzene, as well as flame retardants in subsequent fire extinguishing) (Amster et al. 2013; Otřísal et al. 2017). Soldiers are also ranked among professions endangered by higher heat loads. The heat emission load is determined by measuring the WBGT index (Wet-Bulb-Globe-Temperature) according to the European standards DIN 33403 and ISO 7243. This standard also sets the maximum allowable time of exposure to heat (Xiang et al. 2014).

Increased occupational burden itself poses a risk to the physical health and mental condition of army members involved in field interventions (Amster et al. 2013). Soldiers carrying out physically demanding work and weaponry exercises are at increased risk of injury, hospitalization, and disability compared to other army disciplines, mainly due to their frequent deployment in combat and interventions (Anderson et al. 2015). The presence of people in an environment with an increased amount of particles (especially with the size of $1-100$ nm) have also been identified as a significant risk factor, which have negative influence on their health status (Korzeniewski et al. 2015; Nindl et al. 2013; Amster et al. 2013; Martinello et al. 2014; Mosteanu et al. 2017).

1.1   Nanoparticles in the Air

Nanoparticles in the air are particulate matter (PM), which may cause toxic effects to human health in some circumstances. Nanomaterial is defined as a natural, random or manufactured material containing particles in unbound state or as an aggregate or agglomerate in which 50% or more particles with one or more external dimensions are in the range of 1–100 nm (EC 2011). Trace elements, especially those bound to ultra-fine particulate material (V, Cr, Mn, Fe, Pb, Ni, Cu, Zn, other heavy metals, etc.), can play an important role in the harmful effects and toxicity of PM. Typical are combined loads of several metallic elements at the same time (e.g., PbO, CuO and ZnO) of less than 100 nm. The chemical composition and quantitative relationships between such nanoparticles vary and depend on the specific technology used during their formation and the subsequent interaction among them (Minigalieva et al. 2017).

New scientific findings show that the formation and distribution behaviour of polycyclic aromatic hydrocarbons (PAHs) and polychlorinated dibenzodioxins and dibenzofurans (PCDD/F) are significantly influenced by the presence of heavy metal oxides in fires. Individual heavy metals exhibit different distribution behaviour under the same experimental conditions, mainly influenced by the organic component of the combustion mixture (Wobst et al. 2003).

1.2   Behaviour of Nanoparticles in Human Body

Recent studies have shown that particle size is a key factor affecting not only the concentration distribution of individual elements but also the bioavailability of these trace elements. Therefore, particle size is a critical parameter for assessing the health effects of particles in the air, that depends on individual perceptions of a person to a large extent (Niu et al. 2010; Kendall and Holgate 2012).

The size of nanoparticles allows them to be more easily absorbed by the human body (skin, lungs and digestive system) than the same substance in macromolecule. Regarding the volume, ultrafine nanoparticles have a very high particle concentration. However, their behaviour inside the body has not been fully explored. The main focus of current research is on the accumulation of undegradable or slowly degradable nanoparticles in organs (epithelium, endothelium, neurons, etc.). The behaviour of nanoparticles is a function of their size, shape and surface reactivity with surrounding tissues. In an in vivo health risk study, nanoparticles produced from smoke induced increased inhibition and apoptosis of endothelial cells and partly also production of certain phospholipids (Pedata et al. 2013).

Nanoparticles can be a trigger for inflammation-induced stress reactions and decreasing of pathogen resistance. The unique properties of nanoparticles (extremely low in size, high reactivity, specific shape, surface, etc.) may cause the risk of creation free radicals in the body when inhaled (Niu et al. 2010). Extremely small particles also affect regulatory mechanisms of enzymes and other proteins. The chemical composition of such substances contributes to their toxic action (Pedata et al. 2013). Non-specific symptoms such as flu, cold, fatigue, dizziness, weakness, disorientation, confusion, migraine, or cognitive impairment are also among the potential health effects (Savabieasfahani et al. 2015).

From the medical point of view, the negative influence of nanoparticles on the pulmonary and cardiovascular system is important. However, this process is still the subject of intensive research. The predominant hypothesis states that inhaled particles trigger an inflammatory response in the alveoli with a consequent systemic inflammatory effect resulting in damage to the cardiovascular system. Affected persons that move closely to nanoparticle sources are therefore exposed to the mentioned health risks to a higher extent than common population (Pedata et al. 2013; Amster et al. 2013).

1.3 Impact of the Health Status of Soldiers on Their Professional Abilities

Professionals with a higher risk of inhalation of harmful nanoparticles have their career specialization mostly focused on the maintenance of weapons, combat vehicles, cannons, various special weapon systems and military equipment. Health symptoms occurring among people present in a polluted environment can be chronic and life-long. Combined multiple physical and mental disabilities can lead to psychological problems or even to disability in performing their profession. In contrast, health problems caused by inhalation of ultrafine particles are often not treated as a disease. This problem has the greatest impact on soldiers during their duties and plays an important role in the concentration and accuracy when performing the profession. Such health problems do not affect military personnel only during their active careers, but they also limit the choice of employment in civilian sector after termination of active military service. All soldiers will find themselves in a dynamic and unstable civilian sector that is different from the strict and hierarchical military milieu. The chances of former soldiers to succeed in a corresponding job position at the start of their second career may be declined by their impaired health. In case a soldier leaves an army with health problems, the situation has far more serious consequences.

1.4 Methodology for Allocation of Personal Protective Equipment

The procedure for allocation of personal protective equipment (PPE) is governed by certified procedures and guidelines (Technological Agency of the Czech Republic 2016). The methodology is designed to protect respiratory organs from the effects of nanoparticles in the work environment, where nanoparticles can be assumed to be

			Risks						
			Chemical						
			Aerosols			Liquids			
		Marking	Dust, Fibers	Smoke, fog	Solids	Immersion	Splash	Gases, vapours	Nanoparticles
			12	13	14	15	16	17	22
Head	Skull	A							
	Hearing	B							
	Vision	C							
	Respiratory organs	D							
	Face	E							
	Whole head	F							
Upper limbs	Hands	G							
	Arms (parts)	H							
Lower limbs	Foot	I							
	Legs (parts)	J							
Other	Skin	K							
	Body/belly	L							
	Parenteral routes	M							
	Whole body	N							

Fig. 1 Table for evaluation of risks for allocation and usage of PPE. *Source* Technological Agency of the Czech Republic (2010)

purposefully or unintentionally produced (Schulte et al. 2010). In justified cases, for reasons of occupational health and safety (OHS), the employer is obliged to provide the employees with appropriate PPE.

The risk assessment is based on the risk assessment procedure for the selection and use of the PPE or respiratory protective equipment (RPE), respectively. Figure 1 shows the parts of the body in the lines and the possible exposure risks in the columns.

However, the risk distribution provided by this regulation is very general. In many cases, employers do not have a further risk assessment procedure, which is necessary for the proper and targeted selection of the PPE. It should also be noted that the problem is not only the inaccurate differentiation of risks, but also the absence of other risks – such as the risks associated with the occurrence of nanoparticles. Therefore, the Czech Occupational Safety Research Institute recommends supplementation of this assessment by the following steps (Technological Agency of the Czech Republic 2016):

- step 1 - defining the work system, processing the list of workplaces and activities and identifying the hazards,
- step 2 - setting the exposure time,

Fig. 2 Evaluation of risks
related to nanoparticles.
Source Technological
Agency of the Czech
Republic (2010)

Exposure time category	Hazard group				
	I.	II.	III.	IV.	V.
1	1	1	1	2	2
2	1	1	2	2	3
3	1	2	2	3	3
4	2	2	3	3	3

- step 3 - hazard identification - the hazard group is selected depending on the chemical composition of nanoparticles and their size,
- step 4 - assessment of the risks associated with nanoparticle activity depending on the classification according to the exposure time and hazard group of the given type of nanoparticles, see Fig. 2:
  - value 1 - no significant impact on health is implied by nanoparticles, the risk may be perceived as acceptable, use of RPE is appropriate,
  - value 2 - it is impossible to exclude negative impacts of nanoparticles on health of an employee, the risk is conditionally acceptable, it is recommended to use RPE,
  - value 3 - impacts of nanoparticles on health are assumed as negative, the risk is unacceptable, the reliability of the used RPE must always be tested for filter efficiency,
- step 5 - risk elimination/limitation and allocation of RPE - RPE available on the Czech and European markets can be considered as suitable for use for protection against nanoparticles. In the case of extreme exposure to nanoparticles, it is advisable to prioritize employee protection by technical means, using collective protection. If the identified risk belongs to the hazard group of 3 or 2 and when it is associated with the action of nanoparticles, it is appropriate to consider also other possibilities, such as appropriate eye protection etc.

The methodology highlights the need of accurately identified and described nature of all pollutants occurring in the workplace and the need to determine their maximum permitted concentrations and exposition limits. In the same way, it is necessary to verify the ability of the PPE to protect employees from the risks associated with the occurrence of ultra-fine particles. The decision process can not only be based on the product specification parameters given by the manufacturer, but it is necessary to verify the protective properties of the PPE at a particular workplace. In case of nanoparticle control, the situation is complicated by the fact that due to unique properties and behaviour of nanoparticles, the results acquired by the different laboratories should not be compared. Defining stringent laboratory conditions for nanoparticle capture tests could produce comparable results, but these would be torn apart from real-world conditions at the workplace (Technology Agency of the Czech Republic 2016).

1.4.1   Methodology for Allocation of RPE in the Army

In the Czech military environment, the methodology for the testing of the content of pollutants in the operation and shooting of military vehicles was introduced (such as military combat vehicles, armoured combat vehicles, armoured personnel carriers, tanks, tank cannons, ejectors, tank machine guns, etc.). This methodology does not indicate any specific means intended to provide protection against inhalation of pollutants, but evaluates the effect of the pollutants in the affected area in a complex way (effect on the whole body). A prerequisite for successful testing is compliance with the rules for the operation of vehicles and firearms. Requirements for measuring procedures in a given location are laid down by the Governmental Act 361/2007 that defines the occupational conditions for the protection of worker's health (Czech Republic 2007). The overall risk assessment is carried out in a comprehensive way and it is taken into account especially (Vávra and Braun 2007):

- the measured concentrations of chemical pollutants in the working atmosphere compared to the maximum admissible concentrations for the given chemicals,
- to what extent was the maximum permissible concentration in the working atmosphere exceeded and what is the frequency of such results,
- the duration of the harmful effects to the worker,
- the overall hygiene situation at the workplace,
- the nature of toxic effects of chemicals (acute, chronic, cumulative, etc.).

    In addition, it is necessary to take into account the energy output of individual workers (minute ventilation) and the supportive action of some physical and chemical factors for accelerated penetration of chemicals into the organism, respectively their excretion or detoxification, duration of action, rate of release of the substance into the environment, etc. Evaluation of the results of these tests is carried out according to the requirements laid down by Governmental Act 361/2007. (Vávra and Braun 2007; Czech Republic 2007).

1.4.2   Means of Individual Protection of Persons

Means of individual protection of persons are means protecting the persons against the harmful substances against which they are intended. They are part of the personal equipment of the individual. The protection is provided for respiratory organs and other organs on the sensitive parts of the head, depending on the extent of the head covering (e.g. skin of the face, eyes, mouth, etc.). The use of filter respirators differs according to the type of pollutant, its concentration, the ability of the filter to capture the pollutant, the oxygen content in the atmosphere (at least 17 vol. %), the type of activity performed, etc. (Martínek and Linhart 2006).

    In the army environment, respiratory masks with suitable filtering media are commonly used as protection of respiratory organs against inhalation of contaminated air, classified as individual protection devices. Filter respirators consist of a cheekpiece that is connected to one or several filters. Fittings of filter respirators include quarter masks, half masks, face masks and cheekpieces from the filter material. These devices ensure the protection of their users in distinct degrees. In practical use, the face mask and full-head mask are referred to as the protective mask with marking of its type (Martínek and Linhart 2006).

Protective Masks

Protective mask with effective filter is a device designed to protect the respiratory tract, face, eyes and covered areas of the skin against chemical agents, radioactive substances, combat biological agents and other toxic substances. Filters attached to masks may be structurally divisible (Hylák and Pivovarník 2016):

- cheekpiece filter - a type of non-encapsulated filter inserted in the cheekpiece usually from the inside and in the course of wearing the face mask it can not be replaced without removing the mask,
- housing filter - a filter with functional parts stored in a plastic or metal housing, connected to the protective mask with a thread and can be replaced during the wearing process.

Filters for Protective Masks

Protective mask filters should not be used in environments with a high content of pollutants (above 0.5 vol. %) and in an environment where the oxygen content in the air drops below 17 vol. %. In these cases, insulating devices (air or oxygen) are used. The protective filter is a lightweight but sufficiently mechanically durable and totally gas-tight metal case with a high-quality anticorrosive protection, in which a one-piece or two-part cartridge is placed. One of the charge layers (located directly above the filter inlet) is filter, the other is sorptive. The filter layer cleans the filtered air from the dust and aerosols and is made from a special filter paper folded into organ shapes with the largest possible area. The sorptive layers catch the pollutants in the form of gases and vapours from the filtered air. They are formed by a sorbent that is fastened between two perforated sheet meshes (Hylák and Pivovarník 2016). In accordance with the technical standards of the Czech Republic, the particle filters are separate filters or are usually part of combined filters that are capable of entrapping dispersed solid and liquid particles or gases and vapours from the passing air.

An experiment assessing the efficacy and permeability of ultra-fine aerosols of lead oxide by a military filter was conducted to verify whether RPE introduced in the ACR are capable of removing hazardous ultra-fine particles in the range of 7.6–299.6 nm on an acceptable level. Conclusions of such research will provide information on whether are exposed persons moving in environments with high concentrations of ultrafine aerosol of pollutants endangered by the risk of adverse health effects when using the tested OF-90 respiratory filter as a protection against inhalation of pollutants.

2 Experimental Part

The particle trap can be flat or deep. Surface filtration mechanically captures each particle larger than the space between the fibers of the filter. On the contrary, in the case of depth filtration, the particle size is smaller than the space between the fibers and therefore several filtering mechanisms are applied (Van der Waals force, electrostatic force and force due to surface tension, etc.) (Hrůza 2005).

$$K_P = \frac{G_1}{G_2} \cdot 100$$

To calculate the permeability coefficient (K_p) [%], the calculation given in Czech Defence Standard 841503 (CDS 841503), where G_1 is the amount of particles behind filter, G_2 is the total amount of particles before entering the filter, and G_1/G_2 is the penetration (P). The maximum allowed paraffin oil aerosol permeability coefficient for protective filters designed for all-arm protection masks is allowed according to the CDS 841503 up to 0.001% (Czech Defence Standard 2007):

The total efficiency of capturing ultrafine aerosols (E) [%] is a combination of all capture principles and their share on overall effectiveness varies with the size and shape of each particle (Hrůza 2005):

$$E = \left(1 - \frac{G_1}{G_2}\right) \cdot 100$$

3 Materials and Methods

In cooperation with the Institute of Analytical Chemistry of the Czech Academy of Sciences (IACH CAS), equipment for filtration efficiency testing was designed to verify the permeability and quality of selected respiratory protective filter.

3.1 Methodology of the Testing

Lead oxide (PbO) nanoparticles were generated continuously in situ in a tube reactor using an evaporation-oxidation-condensation technique. A ceramic crucible containing a small amount of lead wire was placed inside a ceramic working tube of a vertically oriented furnace (Karbolit TZF 15/50/610). The molten portion of lead was evaporated at the center of the furnace at 830 °C. Formed metal vapours were transferred from the furnace with an inert stream of nitrogen and diluted with air flow, after which the oxygen then oxidized lead to lead oxide. Both flows were set at 3 l.min^{-1} with mass flow controllers (MFC). The lead oxide nanoparticles were diluted in the second step with air flow (20 l.min^{-1}) and then passed through a fil-

Fig. 3 Sampling device. *Source* own

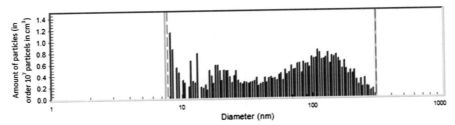

Fig. 4 Particle distribution before entering the filter (total concentration of 6.3×10^7 particles.cm^{-3}). *Source* TSI (2010)

ter firmly located in a specially developed sampling device (see Fig. 3). During the measurement, the aerosol flow rate was set to 95 l.min^{-1}.

3.2 Testing Conditions

With respect to particle concentration, nanoparticle distribution was measured directly using the scanning mobility particle sizer (SMPS, model 3936L72, TSI Inc., Shoreview, USA) continuously in the sizes 7.6–299.6 nm. Nanoparticle measurements were performed using SMPS before and after aerosol passed through the filter (see Figs. 4 and 6). Samples were measured four times at five-minute intervals (total ten to twenty-five minutes).

Filter	Diameter	Height	Weight	Material	Usage against	Kp_{max} [%]
OF-90	110 mm	90 mm	260 g	Special plastic	Solid and liquid substances and aerosols of poisonous substances according to CSN EN 143 class P3, biological aerosols and radioactive dusts	0,001[*]

* Footnote: According to the Czech Defence Standard 841503 (Czech Defence Standard 2007).

Fig. 5 Tested filter. *Source* (SIGMA GROUP a.s.)

3.3 *The Filter Tested with Lead Oxide Nanoparticles*

In the Czech Armed Forces (CAF), a combined anti-particle filter OF-90 belongs among the most frequently used protective mask filters, because it is the part of the standard equipment of each soldier. The description of the tested filter is stated in the Fig. 5.

4 Results and Discussion

SMPS measured the results in a particle size range of 7.6–299.6 nm. The results were evaluated using the Aerosol Instrument Manager software, which calculated the particle size distribution by dividing the particle size range into 64 identical channels. In addition, the total particle count (C_N) was calculated, where n means the total number of channels and (C_i) [N.cm^{-3}] means the number of particles in the ith channel:

$$C_N = \sum_{i=1}^{n} C_i$$

To calculate the efficiency of the filter and the permeability coefficient of the ultra-fine PbO test aerosol, the C_i values were used. The mean particle diameter (MD) [nm] was calculated according to the relationship, where (D_i) is the particle diameter:

$$MD = \frac{1}{C_N} \sum_{i=1}^{n} D_i C_i$$

Based on the assumption about spherical particle shape, the total particle surface (SA) [$\mu m^2.cm^{-3}$] and the total particle volume [$\mu m^2.cm^{-3}$] were then determined from the relationships, where R_i is the radius of the particles measured in ith channel:

$$SA = 4\pi \sum_{i=1}^{n} R_i^2 C_i, \quad V = \frac{4}{3}\pi \sum_{i=1}^{n} R_i^3 C_i$$

Subsequently, the specific particle surface area was calculated (SSA) [μm^{-1}] according to the formula:

$$SSA = \frac{SA}{V}$$

In addition, the mass concentration of particles (C_M) [$\mu g.m^{-3}$] and specific surface energy (SSE) [$J.cm^{-3}$] were estimated on the basis of measured data according to the following formulas:

$$C_M = V\rho, \quad SSE = SSA\sigma$$

where ρ means average particle density and σ means surface tension. For illustrative purposes, $\rho = 1.5$ g.cm^{-3} was chosen, which corresponds to the values found in various urban aerosol studies and $\sigma = 0.0728$ N.m^{-1}, corresponding to the tabulated value of surface tension of water at 20 °C (Zhao et al. 2016).

4.1 Acquired Measured Results

Particle concentrations before and after passage through the combined filter OF-90 were performed at a concentration of 1.2×10^6 particles.cm^{-3} per channel. Four measurements were made at 5 min intervals, and the average particle concentration was determined. Concentration values were then recalculated to particle penetration through the filter, or the permeability coefficient K_p, respectively. Figure 6 illustrates an example of obtained values for total particle number concentrations after 15 mins of ventilation through the filter OF-90.

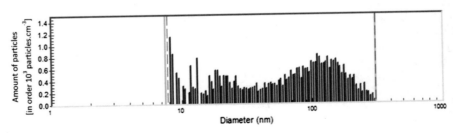

Fig. 6 Particle size distribution behind the tested filter OF-90 after 15 mins of PbO aerosol ventilation. *Source* TSI (2010)

	Number particles size	Diameter particles size	Surface particles size	Volume particles size	Mass particles size
Median (nm)	9.66	10.8	14.7	174.2	174.2
Mean (nm)	11.3	15.1	44.6	153.2	153.2
Geometrical mean (nm)	10.6	12.3	21.3	98.6	98.6
Mode (nm)	7.91	7.91	7.91	250.3	250.3
Geometrical standard deviation	1.36	1.62	2.85	3.2	3.2
Total concentration	9.98×10^5 (particles.cm^3)	11.3 (mm.cm^3)	5.34×10^8 (nm.cm^3)	3.98×10^9 (nm.cm^3)	5.96 (µg/m^3)

Fig. 7 Statistical report of results from measurement before entering the filter. *Source* TSI (2010)

	Number particles size	Diameter particles size	Surface particles size	Volume particles size	Mass particles size
Median (nm)	71.9	131.5	168.0	198.1	198.1
Mean (nm)	84.5	137.3	171.6	195.6	195.6
Geometrical mean (nm)	56.7	115.3	157.1	184.9	184.9
Mode (nm)	8.2	145.9	250.3	250.3	250.3
Geometrical standard deviation	2.67	1.96	1.58	1.43	1.43
Total concentration	569.7 (particles.cm^3)	$4.81.10^{-2}$ (mm.cm^3)	$2.08.10^7$ (nm.cm^3)	$5.94.10^8$ (nm.cm^3)	0.891 (µg/m^3)

Fig. 8 Statistical report of results from measurement behind the filter. *Source* TSI (2010)

	MD [nm]	C_N [N.cm^{-3}]	C_M [µg.m^{-3}]	SA [µm^2.cm^{-3}]	V [µm^3.cm^{-3}]	SSA [µm^{-1}]	SSE [J.cm^{-3}]
Before filter	0.53	$9.98.10^5$	$3.82.10^{11}$	$3.42.10^{10}$	$2.54.10^{11}$	0.13	$9.77.10^{-3}$
Behind filter	$6.19.10^{-4}$	$5.70.10^2$	$5.70.10^{10}$	$1.33.10^9$	$3.80.10^{10}$	$3.50.10^{-2}$	$2.55.10^{-3}$

Footnote: MD - average particle size measured by SMPS, CN - total particle count, CM - mass particle concentration, SA - total particle surface, V - total particle volume, SSA - particle specific surface, SSE - specific surface energy of particles.

Fig. 9 Comparison of nanoparticle values (7.6−299.6 nm) before and behind the OF-90 filter. *Source* own

Figures 7 and 8 illustrate the results obtained by the Aerosol Instrument Manager software (TSI 2010). The geometric mean of particle concentrations was calculated as the nth root of their multiple in the measured particle size distribution range (n).

The obtained data show a significant reduction of the nanoparticle concentration in the measured air by the OF-90 filter after 15 mins ventilation as compared to the original concentration, see Fig. 9.

4.2   Permeability of Nanoparticles Through the Filter OF-90

The average permeability coefficient of all four measurements (in 10, 15, 20 and 25 min) was 0.06%, with a total filtration efficiency of 99.94%. The resulting values exceed the requirement of Czech army standard for the maximum permeability coefficient K_p at the utmost level of 0.001% (Czech Defence Standard 2007).

5   Conclusion

The aim of the contribution was to assess the safety and health of the personnel of the CAF with a focus on the protection against inhalation of pollutants present in the ambient air. The experiment was conducted to evaluate the permeability of the predetermined ultra-fine PbO aerosol through the military filter OF-90, which is one of the most commonly used anti-particle filters by soldiers in practice. The detected outputs were evaluated according to the requirements of the technical standards of the Czech Republic with a focus on the military environment. The research extended knowledge about the solved issue in the field of respiratory protection in the CAF. Acquired results should help improve the working and living conditions of the personnel at the exposed workplaces of the military units.

The filtration efficiency of the OF-90 anti-particulate filter was measured with the use of ultra-fine lead oxide aerosol with a particle size in the range of 7.6–299.6 nm. The testing was carried out in the laboratories of the IACH CAS.

The experiment showed that the required aerosol permeability coefficient limit (0.001%) was exceeded at the original concentration of 1.2×10^6 particles.cm^{-3} (Czech Defence Standard 2007). The reason is probably the modified method used for testing compared to technical standards, where lead oxide aerosol and different range of measurement conditions were used (focused on the area of ultra-fine aerosols). Different test conditions (i.e., precursor combustion temperature, PbO particle distribution profile, etc.), different filter load or capture efficiency of the filter, which may not be specific to the tested ultrafine aerosol may contribute to results that exceeded the limit set by the CDS 841503. However, further studies will be needed to determine the reasons for the high permeability coefficient of the testing aerosol. Concentrations exceeding the approved respiratory filter efficiency limit may pose the risk of acute and chronic health effects, which can be a serious obstacle for professional active career of soldiers or those starting their second career. For further developments of this study, it could be interesting to focus future research on this topic using fuzzy theory (e.g. Maturo 2016; Hošková-Mayerová 2017) and particularly fuzzy regression models (e.g. Maturo and Maturo 2013; Maturo 2016, Maturo and Hošková-Mayerová 2017) for assessing the efficiency of respiratory protection devices against lead oxide nanoparticles.

Acknowledgements The work presented in this paper was supported by a grant project run at the IACH CAS (GAP503/11/2315). The financial support of the University of Defence thanks to the specific research project (SV16-FVL-K106-KELL) is also valued.

References

Amster, E.D., et al.: Occupational exposures and symptoms among firefighters and police during the carmel cohort study. Isr. Med. Assoc. J. **15**, 288–292 (2013)

Anderson, M.K., et al.: Occupation and other risk factors for injury among enlisted U.S. Army Soldiers. Public Health **129**, 531–538 (2015)

CR–Technology Agency of the Czech Republic: Certified methodology for providing personal protective equipment in a nanoparticle environment. 15 (2016)

CR: Government Decree No. 361/2007 Coll., Laying down the conditions for the protection of the health of workers at work (2007)

Czech Defence Standard: Respiratory devices for individual protection - Military protective masks. Office for Defence Standardization Cataloging and State Quality Assurance 20 (2007)

EC: Official Journal of the European Recommendations on the definition of nano-material. http://eur-lex.europa.eu/LexUriServ/LexUriServ.do?uri=OJ:L:2011:275:0038:0040:EN:PDF (2011) Accessed 30 Jan 2018

Hrůza, J.: Improvement of Filtration Features of Materials, 1st edn. Technical university of Liberec, Liberec (2005)

Hylák, Č., Pivovarník, J.: Individual and Collective Civil Protection of the Czech Republic, 1st edn. Ministry of the Interior of the Czech Republic, Prague (2016). ISBN 978-80-87544-18-1

Hošková-Mayerová, S.: An overview of topological and fuzzy topological hypergroupoids. Ratio Math. **33**, 21–38 (2017)

Kendall, M., Holgate, S.: Health impact and toxicological effects of nanomaterials in the lung. Respirology **17**, 743–758 (2012)

Korzeniewski, K., Nitsch-Osuch, A., Konior, M., Lass, A.: Respiratory tract infections in the military environment. Respir. Physiol. Neurobiol. **209**, 76–80 (2015)

Martínek, B., Linhart, P.: Civil Protection, 1st edn. Ministry of Interior – The General Directorate of the Fire Rescue Brigade of the Czech Republic, Prague (2006)

Martinello, K., et al.: Direct identification of hazardous elements in ultra-fine and nanominerals from coal fly ash produced during diesel co-firing. Sci. Total Environ. **470–471**, 444–452 (2014)

Maturo, F.: Dealing with randomness and vagueness in business and management sciences: the fuzzy-probabilistic approach as a tool for the study of statistical relationships between imprecise variables. Ratio Math. **30**(1), 45–58 (2016)

Maturo, F., Hošková-Mayerová, Š.: Fuzzy Regression Models and Alternative Operations for Economic and Social Sciences. Studies in Systems, Decision and Control, pp. 235–247 (2017). https://doi.org/10.1007/978-3-319-40585-8_21

Maturo, F., Fortuna, F.: Bell-shaped fuzzy numbers associated with the normal curve. Topics on Methodological and Applied Statistical Inference, pp. 131–144 (2016). https://doi.org/10.1007/978-3-319-44093-4_13

Maturo, A., Maturo, F.: Research in Social Sciences: Fuzzy Regression and Causal Complexity. Studies in Fuzziness and Soft Computing, pp. 237–249 (2013). https://doi.org/10.1007/978-3-642-35635-3_18

Minigalieva, I.A., et al.: In vivo toxicity of copper oxide, lead oxide and zinc oxide nanoparticles acting in different combinations and its attenuation with a complex of innocuous bio-protectors. Toxicology **380**, 72–93 (2017)

Mosteanu, D., Barsan, G., Otřísal, P., Giurgiu, L., Oancea, R.: Obtaining the Volatile oils from wormwood and tarragon plants by a new microwave hydrodistillation method. Revista de Chimie **68**(11), 2499–2502 (2017). ISSN 0034-7752

Nindl, B.C., et al.: Physiological Employment Standards III: physiological challenges and consequences encountered during international military deployments. Eur. J. Appl. Physiol. **113**, 2655–2672 (2013)

Niu, J., Rasmussen, P.E., Hassan, N.M., Vincent, R.: Concentration distribution and bioaccessibility of trace elements in nano and fine urban airborne particulate matter: influence of particle size. Water Air Soil Pollut. **213**, 211–225 (2010)

Otřísal, P., Florus, S.: Resistance of the isolative protective garment designated for specialists´ protection against selected chlorinated hydrocarbons. Vojenské zdravotnické listy **83**(1), 11–17 (2014). ISSN 0372-7025

Otřísal, P., Florus, S., Barsan, G., Mosteanu, D.: Employment of simulants for testing constructive materials designed for body surface isolative protection in relation to chemical warfare agents. Revista de Chimie **69**(2), 300–304 (2018). ISSN 0034-7752

Otřísal, P., Florus, S., Švorc, Ľ., Barsan, G., Mosteanu, D.A.: New colorimetric assay for determination of selected toxic vapors and liquids permeation through barrier materials using the minitest device. Revistad De Materiale Plastice **54**(4), 748–751 (2017). ISSN 0025-5289

Pedata, P., et al.: Apoptotic and proinflammatory effect of combustion generated organic nanoparticles in endothelial cells. Toxicol. Lett. **219**, 307–314 (2013)

Savabieasfahani, M., et al.: Elevated titanium levels in Iraqi children with neurodevelopmental disorders echo findings in occupation soldiers. Environ. Monit. Assess. **187**(4127), 1–11 (2015)

Schulte, P.A., et al.: Occupational exposure limits for nanomaterials: state of the art. J. Nanopart. Res. **12**, 1971–1987 (2010)

TACZ 2010: Technological Agency of the Czech Republic, https://www.tacr.cz/index.php/en/

TSI: Aerosol Instrument Manager® Software, USA, TSI (2010)

Vávra, S., Braun, P.: Methodology of testing the content of pollutants during operation and shooting of military vehicles of the CAF, 1st edn. Military Technical Institute Division, Vyškov (2007)

Wobst, M., Wichmann, H., Bahadir, M.: Influence of heavy metals on the formation and the distribution behavior of PAH and PCDD/F during simulated fires. Chemosphere **51**, 109–115 (2003)

Xiang, J., BI, P., Pisaniello, D., Hansen, A.: Health impacts of workplace heat exposure. Ind. Health **52**, 91–101 (2014)

Zhao, S., Yu, Y., Yin, D., He, J.: Effective density of submicron aerosol particles in a typical valley city, Western China. Aerosol. Air Qual. Res. **17**, 1–13 (2016)

Community Detection in Social Networks

Fataneh Dabaghi-Zarandi and Marjan Kuchaki Rafsanjani

Abstract Over recent years, the usage of social, biologic, communication and the World Wide Web networks is widely increased. Each of these networks consists of many complex and various data that can be modeled as a graph. This graph is composed of a set of nodes and edges that each node model an entity in these networks and connection between two entities is defined as an edge. In this regard, to have a better understanding of organizations and functions in these networks, graph nodes can be classified in different groups. Each group of nodes is called community that its nodes have more similarity with each other. Therefore, community detection is an important field to understand the topology and functions of networks. In this chapter, we introduce several method in community detection and compare them together.

Keywords Community detection · Social networks · Structural similarity
Attribute-structural similarity · Modularity

1 Introduction

Nowadays, there exist many complex networks such as World Wide Web, Social networks, Transport and so on that are used them a lot. In this regard, these networks produce complex and various data and can be modeled by graph structure. The entities of each network are represented by nodes and communications between two entities is defined an edge. Commonly, these networks consist of a big and complex communication graph resulting in difficult management. Therefore, graph topology is divided into several groups. Each group contains a number of similar nodes and is considered as a community. The nodes in each community have densely connected

F. Dabaghi-Zarandi · M. Kuchaki Rafsanjani (✉)
Department of Computer Science, Faculty of Mathematics and Computer,
Shahid Bahonar University of Kerman, Kerman, Iran
e-mail: kuchaki@uk.ac.ir

F. Dabaghi-Zarandi
e-mail: dabaghi_fataneh@math.uk.ac.ir

© Springer Nature Switzerland AG 2019
C. Flaut et al. (eds.), *Models and Theories in Social Systems*, Studies in Systems,
Decision and Control 179, https://doi.org/10.1007/978-3-030-00084-4_15

with each other and have a sparse connection to the rest of the network. In recent years, community detection has been an attractive topic in data mining, data retrieval and social network analysis (Bullmore and Sporns 2009; Fortunato 2010; Leskovec et al. 2010; Urban and Hošková-Mayerová 2017).

Commonly, a community is considered a group of nodes that their similarities are maximal. Similarity of nodes is computed using three measures *structural similarity, attribute similarity* and *attribute-structural similarity*. Structural similarity is included the topology of the network and attribute similarity is contained internal characteristics of individual nodes without considering network topology. Attribute-structural similarity is combining structural similarity and attribute similarity (Kim et al. 2010; Fortunato 2010). Therefore, attribute-structural similarity is computed based on topology and attributes of the network. This similarity is good measure for community detection in social networks, because most of social networks are included structural similarity and attribute similarity. For example, a social network, interactions or relationships between a group of people are considered as structural similarities and characteristics or roles of a person such as age and occupation are considered as attribute similarities (Boobalan et al. 2016).

Many community detection methods are based on the quality function optimization. Therefore, the most famous of these methods is Newman-Girvan modularity (Newman and Girvan 2004). In the other words, modularity is used in optimization methods to determine the quality of communities in complex networks. In addition, the larger value of modularity displays the best quality in the detected communities (Fortunato 2010; Blondel et al. 2008).

In this chapter, we represent a survey in community detection. In this regard, we express methods in two group structural similarity and attribute-structural similarity. Methods based on structural similarity compute communities using topology network, even if the network has attributes. Methods based on attribute-structural similarity calculate communities using structural similarity and attribute similarity.

The remainder of this chapter is organized as follows. Section 2 provides basic definitions and related fundamentals. Section 3 summarizes some related community detection approaches. Section 4 explains evaluation measures. Section 5 represents dataset that used in community detection. Finally, Sect. 6 concludes the chapter.

2 Background

In this section, we introduce some concepts and phrases that are needed to understand community detection. In the following, these concepts including, complex networks, mathematical modeling of complex networks, similarity, community, community detection, overlapping community, modularity and so on will be discussed in detail.

2.1 Complex Networks

Almost all of the complex systems such as biology, sociology and engineering can be considered as complex networks. Examples of these networks include the Internet, World Wide Web, power networks, social networks, protected networks, collaboration, transportation and communication networks.

2.2 Mathematical Modeling of Complex Networks

Each complex network can be modeled by $G(V, E)$. In this regard, $V(G) = \{v_1, v_2, \ldots, v_n\}$ is the set of nodes and $E(G) = \{(v_i, v_j)|v_i, v_j \in V(G)\}$ is the set of edges. Each edge is between two nodes v_i and v_j. The number of nodes and edges are represent n and m parameters respectively, and calculated by Eqs. (1) and (2).

$$n = |V| \tag{1}$$
$$m = |E| \tag{2}$$

2.3 Structural Similarity

Similarity is a measure that computed the likeness between a pair of nodes. Therefore, nodes that are similar can be putted in a same community (Ester et al. 1996; Huang et al. 2011). The approaches such as common neighbors and k-distance neighborhood can be computed structural similarity (Boobalan et al. 2016; Arab and Afsharchi 2014).

2.4 Attribute Similarity

Attribute similarity is compared based on internal characters of nodes and the graph topology does not play any role in its calculation.

2.5 Community

The community is set of nodes that these nodes have characters and roles similar. Nodes in each community have connection a lot that produce high density, whereas connection these nodes with the rest of network is very low. Therefore, Community structure is very important in complex networks.

2.6 Overlapping Communities

Communities have overlapping if some nodes in the network belong two or several communities. Commonly, overlapping community exist in very social networks. So, discover and recognize these communities is very important field in social network analysis (Leskovec et al. 2009; Li 2012).

2.7 Community Detection

Community detection is an important subject to discover community structure (Easley and Kleinberg 2010). Also, community detection is considered as one of the clustering problems in the graph.

2.8 Inter-community Edges and Intra-community Edges

When communities of a network are detected, network edges are divided into two groups: *inter-community edges* and *intra-community edges*. As shown in Fig. 1, inter-community edges are included the edges between communities and intra-community edges are included the edges inside communities.

Fig. 1 Inter-community edges and intra-community edges

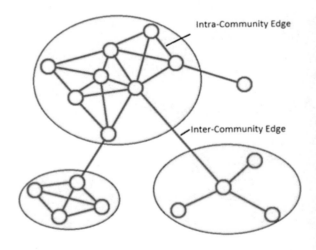

2.9 Modularity

Modularity is a quality function that can compute the quality of communities. In this regard, the famous quality function is Newman-Girvan modularity that is represented by Eq. (3).

$$Q = \sum e_{ii} - a_i^2 \qquad (3)$$

where e_{ii} is a fraction of edges that are in the community i and a_i represents fraction of node degree that are in the community i.

3 Related Works

3.1 Structural Similarity

In this section, we explain several methods in community detection based on structural similarity.

3.1.1 Minimum Spanning Tree and Modularity

In Saoud and Moussaoui (2016), an approach is represented based on Minimum Spanning Tree and Modularity (MST-M). This method gets a network graph and convert it to a tree. Then, some edges are removed from the tree and produced several disconnect components. Each disconnect components is considered as a primary community. In the following, two primary communities that have maximum similarity are chosen and merged. The modularity of existing communities is computed and this modularity along with communities are stored. Therefore, the steps of merging, calculating modularity and storing are repeated to reach the best communities. Finally, communities with maximum modularity are selected as community detection. In the following, these steps are explained in detail.

- *Minimum Spanning Tree*: The goal of this step is converting network graph to a tree. For each edge is considered a weight based on Eq. (4). Therefore, this weight represents the dissimilarity between nodes.

$$w(e_l) = w(v_i, v_j) = \frac{|\Gamma(v_i)| + |\Gamma(v_j)|}{|\Gamma(v_i) \cap \Gamma(v_j)|} \qquad (4)$$

where e_l represents edge between two nodes v_i and v_j, $\Gamma(v_i)$ is a set of nodes that connect to node i and also include node i.
Minimum spanning tree cashes graph $G(V, E)$ along with $w(e_l)$ for each edge

and converts to a tree $T(V', E')$ that sum of its edges weight $(W(T) = \sum_{(v_i, v_j) \in T} w(v_i, vj))$ is minimal. Finally, the number of edges and nodes T is calculated by Eq. (5).

$$V' = V, |V'| = |V| = n, E' \subset E, |E'| = n - 1 \qquad (5)$$

- *Generating Several Disconnect Components*: The set R_e including $\frac{n-1}{2}$ edges that have the least similarity in the graph are selected. Then, these edges are removed from T and produced $\frac{n+1}{2}$ disconnect components. In this regard, each disconnect component is considered as a primary community. Therefore, there exits $k = \frac{n+1}{2}$ primary communities that can be represented as $C_1, C_2, \ldots, C_k$.
- *Merging Communities*: Two communities are selected based on S_{ij} similarity are merged, so that this similarity is maximal. S_{ij} is computed by Eq. (6).

$$S_{ij} = \frac{\text{the number of inter-community edges between } C_i \text{ and } C_j}{\sqrt{d_{C_i} d_{C_j}}} \qquad (6)$$

$$d_{C_i} = \sum_{j=1}^{|C_i|} degree(v_i) \qquad (7)$$

where d_{C_i} is the sum of nodes degree in community C_i. In the following, the modularity of these communities is computed using Eq. (8).

$$Q = \frac{1}{2} \sum_{ij} \left[A_{ij} - \frac{d_i d_j}{2m} \right] \delta(C_i, C_j) \qquad (8)$$

$$\delta(C_i, C_j) = \begin{cases} 1 & \text{if } C_i = C_j \\ 0 & \text{Otherwise} \end{cases} \qquad (9)$$

where, A represents the adjacency matrix, m is the total of network edges, d_i and d_j are the number of nodes degree v_i and v_j respectively, and C_i and C_j display i and j communities.

Merging operation be continued to reach one community. After each merging, the modularity is calculated and stored along with existing communities. Finally, the step is selected as the best community detection that have the highest modularity.

If, there exit single-member communities in detecting communities, are merged by communities that connect the highest edges.

- Review and Evaluation

    - In this approach, no parameters such as the number of communities did not get from the input.

- The communities do not have any overlap.
- The time complexity of MST-M is $O(2m + n(\log n + (\frac{n-1}{2})m + |C_r|))$.

3.1.2 Hybrid Merging of Sub-communities

In Arab and Afsharchi (2014), an approach is represented based on Hybrid Merging of Sub-communities (HMS). In this approach, at the first, a weight is assigned to each edge in the graph based on common neighbors. In this regard, the edges are arranged in descending order according to their weight. Therefore, primary communities consist of one or two nodes are generated. Then, primary communities are merged using hybrid merging to achieve communities close to real communities. In the following, steps of detecting community are explained in details.

- *Computing Nodes Similarity*: Nodes similarity is calculated based on cosine similarity that represented by Eq. 10.

$$S_{cosine} = \frac{|\Gamma_i \cap \Gamma_j|}{\sqrt{|\Gamma_i||\Gamma_j|}} \tag{10}$$

where Γ_i is the set of nodes that connect v_i along with v_i.

- *Weighting The Edges*: Each edge is assigned a weight based on the cosine similarity. Then, edges are put on an array list and arranged in descending order using their weights.
- *Generating Primary Communities*: At the first, all nodes are considered unsigned and edges select one by one from top of array list. Therefore, for each edge, the nodes attached to it are assigned and become members in a new community. Moreover, if at least one of nodes attached to edge is assigned, no community will be formed.

  In the following, if there exit nodes that not belong to any community, then are put in a new community separately. Finally, primary communities are formed with one or two nodes. An example of primary communities is shown in Fig. 2. As can be seen in the figure, all of primary communities are included one or two nodes.
- *Hybrid Merging*: Primary communities are merged to achieve detecting communities close to real communities. In this regard, merging operation uses two merge approaches is called *pairwise merging* and *single neighbor merging*. Pairwise merging, according to Eq. (11), two communities that have ΔQ maximal are selected to merge. This merging method increase modularity, but its speed is very low. Single neighbor merging, communities connect to only one community, are merged with that corresponding community. This merging approach has high speed, but may do not increase the modularity. In the following, communities are merged in R_m times that $\frac{R_m}{F}$ times are used pairwise merging and the rest of it are used single neighbor merging.

$$\Delta Q = \frac{E_{ij}}{m} - 2 \times \frac{2E_i + Ext_i}{2m} \frac{2E_j + Ext_j}{2m} \tag{11}$$

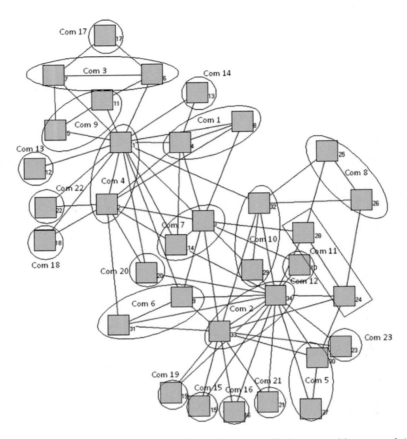

Fig. 2 An example of primary communities in hybrid merging of sub-communities approach (Arab and Afsharchi (2014))

- Review and Evaluation

  – In this approach, two parameters R_m and F are considered as input. R_m is the number of mering and F is denoted fraction of R_m for the number of pairwise merging.
  – The communities do not have any overlap.
  – The time complexity of HMS is $O(n \log n)$.

3.1.3 Local Neighbor Method

In Eustace et al. (2015), a method is represented based on Local Neighbor (LN). In this method, nodes are included in one community that the similarity between them is greater than the average similarity of graph nodes. In the following, steps of creating communities are explained.

- *Similarity Measure*: Similarity measure between two nodes is computed by Eq. (12).

$$\frac{|\Gamma(v_i) \cap \Gamma(v_j)|}{|\Gamma(v_i)|} \tag{12}$$

$\Gamma(v_i)$ is a set that included all nodes adjacent to node v_i along with itself node v_i.
- *Creating Local Community*: At first, the function that called α-close is represented. This function is used to produce local community in the network [Eq. (13)].

$$\frac{|\Gamma(v_i) \cap \Gamma(v_j)|}{\min(|\Gamma(v_i)|, |\Gamma(v_j)|)} \geq \alpha \tag{13}$$

where α is the average of all neighborhood ratio of all nodes in the network. Therefore, v_i and v_j are included same community if only if satisfy in Eq. (13). Finally, a set of local community is obtained as Eq. (14).

$$lc = \{lc(v_1), lc(v_2), \ldots, lc(v_k)\} \tag{14}$$

- *Community*: A community is included several local communities that have more similarity. The similarity between two local communities are computed by Eq. (15).

$$\frac{|\Gamma(lc(v_i)) \cap \Gamma(lc(v_j))|}{|\Gamma(lc(v_i))|} \tag{15}$$

Finally, the functions that called β-close is defined to produce network communities. This function is represented in Eq. (16).

$$\frac{|\Gamma(lc(v_i)) \cap \Gamma(lc(v_j))|}{\min(|\Gamma(lc(v_i))|, |\Gamma(lc(v_j))|)} \geq \beta \tag{16}$$

where β is the minimum ratio of closely related nodes between two local communities. As a result, a community is included several local communities that satisfy in β-close.

- Review and Evaluation

  - In this method, no parameters are not received from the input.
  - The communities do not have any overlap.
  - The time complexity of α-close and β-close are $O(|V|)$ and $O(|E|)$ respectively. Finally, the time complexity of this method is $O(m + n)$. Where, n and m are the total nodes and edges respectively.

3.1.4   Label Propagation Algorithm-Stepping Method

In Li et al. (2017), a method is represented based on Label Propagation Algorithm-Stepping (LPA-S). This method is performed in two consecutive phases. In the first phase, primary communities are formed and in the second phase, primary communities are merged to achieve communities close to real communities of the network. In this regards, primary communities randomly are chosen to merge. Therefore, repeating the steps of LPA-S method are produced several different community structures. Finally, evaluation function used for choosing the best community structure. In the following, these phases are explained.

1. Each node is assigned with a unique label.
2. Similarity between each two nodes is computed by Eq. (17).

$$S_{ij} = \sum_{Z \in \Gamma(i) \cap \Gamma(j)} \frac{1}{k(Z)} \tag{17}$$

where $\Gamma(i) \cap \Gamma(j)$ is a set of common neighbors of nodes i and j and $k(Z)$ is degree of node Z.
3. A random visiting order is produced for all nodes in the network.
4. Update every node's label to the neighbor that has the highest value of similarity S.
5. If each node with the most similar neighbors has the same label, the process stops and nodes with the same label belong to one sub-network. Otherwise, repeat steps 3 and 4.
   After that, primary communities are generated and these communities are merged to achieve communities close real communities.
6. Similarity between each pair of primary communities is calculated by Eq. (18).

$$S_{sub,x,y} = \frac{\sum_{i \in x, j \in y} S_{ij}}{\min(\sum_{i \in x} k(i), \sum_{j \in y} k(j))} \tag{18}$$

where x and y are communities and $k(i)$ is degree of node i.
7. A random visiting order is generated for all sub-communities.
8. Nodes' label of sub-communities are changed based on the most similar neighbor sub-communities. After that, similarity matrix is updated.
9. If two communities remain or no label is updated, the process is stopped and nodes with the same label are belong one community. Otherwise, steps 7–8 are repeated.

In this method, a random visiting order is generated and by repeating steps, LPA-S are achieved several different community structures, so that one of the community structure is the best of them. Therefore, the best community structure is selected using the evaluation function introduced in Eq. (19).

$$DN = \frac{FID}{FIN} \qquad (19)$$

$$FID = \sum_{x=1}^{n^{'}} \frac{k_{inx}}{k_x - k_{inx}} \qquad (20)$$

$$FIN = \sum_{x=1}^{n^{'}} \frac{n_x}{n - n_x} \qquad (21)$$

where x is a label of community, k_{inx} is denoted the total internal degree of nodes in community x, k_x is the total degree of nodes in community x, n_x is the number of nodes in community x, FID and FIN are the fraction of inner degree and the fraction of inner nodes respectively.

Finally, the value of DN is computed for all community structure resulting LPA-S method. Then, a set of community structures is selected as the best communities that have the maximum value of DN.

- Review and Evaluation

  - In this method, no parameters such as the number of communities did not receive from the input.
  - The communities do not have any overlap.
  - The time complexity of LPA-S method is $O(n^2)$. Although, the time complexity of this method is not linear, but is better than all approaches (Newman and Girvan 2004; Girvan and Newman 2002; Lu and Wei 2012) that have time complexity $O(n^3)$.

3.1.5 Compare Methods

In Table 1, all methods that expressed based on structural similarity are compared. In this regard, these methods are compared based on properties including time complexity, overlapping and inputs. As you can see in the table, communities have no overlap in all the methods. The only approach that receives input variable is HMS.

Table 1 Compare the discussed structural similarity methods

Method	Time complexity	Overlapping	Inputs
MST-M (Saoud and Moussaoui 2016)	$O(2m + n(\log n + (\frac{n-1}{2})m + |C_r|))$	No	No
HMS (Arab and Afsharchi 2014)	$O(n \log n)$	No	R_m and F
LN (Eustace et al. 2015)	$O(m + n)$	No	No
LPA-S (Li et al. 2017)	$O(n^2)$	No	No

3.2 Attribute-Structural Similarity

In this section, approaches that are based on attribute and structural measures are explained. These approaches are very efficient for networks that nodes have internal characters such as social networks. In the following, two methods in this field are represented.

3.2.1 KNAS Method

In Boobalan et al. (2016), a method is represented based on K-Neighbourhood Attribute-Structural similarity (KNAS). In this method, at first, a network topology is separated to several communities using structural similarity. Then, each of these communities is divided into several smaller communities based on attribute similarity. Finally, these small communities are merged to obtained communities close to real communities.

Goal of KNAS method is detecting communities based on structural similarity and attribute similarity. In Fig. 3 community detection is shown. In this regard, a network topology is displayed in part (a). Nodes are shown by capital letters (A, B, C and D) and x and y are attributes of each node. In part (b), communities are detected using structural similarity. In part (c), communities are detected using attribute similarity. As you see in figure, nodes with same attribute are in the same community. In part (d), communities are detected using attribute-structural similarity. Therefore, nodes with same structural and attribute similarity are in the same community. In the following, steps of this method are explained.

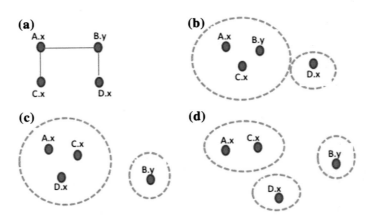

Fig. 3 An example of social network and kinds of community detection (Boobalan et al. 2016)

- *Computing Primary Communities' Center*: By three levels are selected centers of communities.

1. *Computing Neighborhood of Nodes*: Neighborhood of nodes that called N are calculated by Eq. 22.

$$N_k(x) = \{y | y \in D, dist(x, y) \le k\} \tag{22}$$

where k denote Manhattan distance that is taken from the user and $dist(x, y)$ determine distance between two nodes x and y.

2. *Computing Relative Density Function*: For each node x is computed density relative $(rd_k(x))$ using neighborhood of nodes that represented in the following.

$$rd_k(x) = \left(\frac{\sum_{(y \in N_k(x))} dist_k(x, y)}{|N_k(x)|} \right)^{-1} \tag{23}$$

where $|N_k(x)|$ denote the number of neighbor nodes x that their distant are less than k.

3. *Local Outlier Factor (LOF)*: This measure can be calculated by the average local density of node x with node y on k-nearest neighbours of x.

$$LOF_k(x) = \frac{\sum_{y \in N_k(x)} \frac{rd_k(y)}{rd_k(x)}}{|N_k(x)|} \tag{24}$$

Nodes are arranged ascending order based on their *LOF* values. Therefore, each nodes that have the *LOF* value less than or equal one is selected as initial centers. Finally, m communities in the zero level will have as $\{C_1^0, \ldots, C_m^0\}$.

- *Appending Each of Nodes into Communities*: Each node is joined a community that its Manhattan distance to the center of the community is less than the other centers. This matter are explained in Eq. (25).

$$C^* = \{v_i | v_i \in V, \min d(v_i, c_j^n)\} \tag{25}$$
$$C^* \in \{C_1^0, \ldots, C_m^0\} \tag{26}$$

- *Dividing Communities into Smaller Communities Based on the Attribute Similarity*: At first, attribute similarity of each communities is computed. Therefore, if this attribute similarity is less than intended threshold, relevant community is divided into small communities. In this regard, using similarity score function is calculated attribute similarity that displayed in Eq. (27).

$$SC(V_i, V_j) = \sum_{v_i \in V_i, v_j \in V_j} \frac{S_{ij}(v_i, v_j)}{|V_i| \times |V_j|} \tag{27}$$

where S_{ij} denote attribute similarity matrix that calculated by Eq. 28, V_i and V_j are two communities, v_i and v_j are two node that exist in community V_i and V_j respectively.

$$S_{ij} = \frac{1}{L}\sum_{a=0}^{L} S_{ija} \tag{28}$$

where L is set of attribute in the network and S_{ija} is computed in the following.

$$S_{ija} = \begin{cases} 1 & \text{If both of } v_i \text{ and } v_j \text{ have attribute } a \text{ or do not have attribute } a \\ 0 & \text{Otherwise} \end{cases} \tag{29}$$

where a is a th attribute in set L.

- *Merging Communities*: Two communities will merged if Manhattan distant of their centers have the lowest amount. For example, if Manhattan distant C_1 and C_2 is less than Manhattan distant C_1 and C_3, then two communities C_1 and C_2 are merged. Finally, merging operation is continued until the number of communities will reach to m.
- *Updating Community Centers*: The average distance between the nodes of same community is calculated using Eq. (30).

$$Avg(V_i) = \frac{1}{|V_i|}\sum_{j\in V_i} d(V_j, Vi), v_i \in V \tag{30}$$

Therefore, the center of community i in $(n+1)$ th repeat is computed using Eq. (31).

$$C_i^{n+1} = \min_{v_j \in V_i} ||v_j - Avg(V_i)|| \tag{31}$$

- *Repeating Previous Steps*: The steps from "Appending Each of Nodes into Communities" to "Updating Community Centers" are repeated until nodes included each community have much similarity together. Therefore, repeating operation is stopped when objective function that represented in Eq. (32) is converged.

$$O(\{V_i\}_{i=1}^{k} = \sum_{i=1}^{k} SC(V_i, V_i) \tag{32}$$

- Review and Evaluation

  - In this method, centers of communities are selected based on structural similarity density. So, the runtime has decreased relative to the random selection of centers.
  - The communities have overlap.

- Parameter k is considered as input. This parameter is denoted the number of neighborhood steps.
- The time complexity of this method is $O(e(nm + n^2 l + nms + m^2))$.

3.2.2 SA-Cluster

In Zhou et al. (2009), an approach is represented using Structural and Attribute similarities (SA-Cluster) that detected communities. In this approach, at first, each of the edges is assigned a weight where computed based on common attributes of two nodes that connected to it. Therefore, using the density of edges, k nodes are selected as centers of communities and then communities will be created. In the following, steps of SA-Cluster are explained in detail.

1. At first, weight of all edges is considered one.
2. Matrix of random walk distance is created by Eq. (33).

$$R_A^l = \sum_{\gamma=1}^{l} c(1 - c)^\gamma P_A^\gamma \qquad (33)$$

where P_A is a transmission probability matrix for an attribute network (G_a) and c determine by user.

3. k nodes are selected as centers of communities based on highest density values. In this regard, density function is represented by Eq. (34).

$$f_B^D(v_i) = \sum_{v_j \in V} \left(1 - e^{-\frac{d(v_i, v_j)^2}{2\sigma^2}} \right) \qquad (34)$$

where $d(v_i, v_j)$ represent the distant between two nodes v_i and v_j and σ is parameter that determined by user. In the following, nodes are arranged as descending order based on their density value. Finally, k nodes with highest density are selected as primary centers and k communities are formed $(\{C_1^0, \ldots, C_k^0\})$.

4. k communities are created based on k centers. Each node becomes a member of a community that has the least distance to center of the corresponding community.

5. Centers of communities are updated using Eq. (35).

$$c_i^{t+1} = \operatorname{argmin}_{v_j \in V_i} ||R_A^l(v_j) - R_A^l(\overline{v_i})|| \qquad (35)$$

where c_i^{t+1} is denoted center of community i in repeater $t + 1$th, V_i is community ith and $R_A^l(\overline{v_i})$ is the average random walk of nodes in community V_i that is calculated by Eq. (36).

$$R_A^l(\overline{v_i}) = \frac{1}{|V_i|} \sum_{v_k \in V_i} R_A^l(v_k, v_j), \forall v_j \in V \qquad (36)$$

where V is the set of all nodes.

6. All weights $(w_1, \ldots, w_m)$ are updated using Eq. (37).

$$w_i^t = \frac{1}{2}\left(w_i^t + \frac{m\sum_{j=1}^{k}\sum_{v\in V_j} vote_i(c_j, v)}{\sum_{p=1}^{m}\sum_{j=1}^{k}\sum_{v\in V_j} vote_p(c_j, v)}\right) \tag{37}$$

where k and m denote the number of communities and edges respectively, C_j is community j and $vote_p(c_j, v)$ is attribute similarity that computed as follows:

$$vote_i(v_p, v_q) = \begin{cases} 1, & \text{If both of } v_p \text{ and } v_q \text{ have attribute } a \text{ or do not have attribute } a \\ 0, & \text{Otherwise} \end{cases}$$

$$\tag{38}$$

7. Matrix R_A^l is calculated again.
8. Until the objective function presented in Eq. (39) is not convergent, the third to seventh steps are repeated.

$$O(\{V_i\}_{i=1}^{k}) = \sum_{i=1}^{k} d(V_i, V_j) \tag{39}$$

$$d(V_1, V_2) = \sum_{v_i\in V_1, v_j\in V_2} \frac{d(v_i, v_j)}{|V_1| \times |V_2|} \tag{40}$$

$$d(v_i, v_j) = R_A^l(v_i, v_j) \tag{41}$$

9. $V_1, \ldots, V_k$ are considered as outputs.

- Review and Evaluation

  - The communities do not have any overlap.
  - Several parameters are determined by user as input.
  - The time complexity is $O(n^2)$, Where n is the total of nodes.

3.3 Comparing Methods in Attribute-Structural Similarity

In Table 2, methods that are in attribute-structural similarity field are represented and compared based on characters including time complexity, overlapping and inputs.

Table 2 Compare the discussed attribute-structural similarity methods

Method	Time complexity	Overlapping	Inputs
kNAS (Boobalan et al. 2016)	$O(e(nm + n^2 l + nms + m^2))$	Yes	k
SA-Cluster (Zhou et al. 2009)	$O(n^2)$	No	c, σ, l and k

4 Evaluation Measures

In this section, evaluation measures that employed for methods in community detection are explained. In the following, evaluation measures that use for methods based on structural similarity and attribute-structural similarity are represent in Sects. 4.1 and 4.2 respectively.

4.1 Evaluation Measures in Structural Similarity Methods

• *Normalized Mutual Information (NMI)*: For evaluating methods in community detection, there exist real networks that their communities are known. In this regard, NMI is a measure to evaluate real communities with communities that obtained from community detection methods. Therefore, if the obtained communities be exactly with real communities, the value of NMI is equal one. Moreover, if the obtained communities be exactly one community, the value of NMI is equal zero (Danon et al. 2005).

To computing *NMI*, a confusion matrix N is defined. The first row of this matrix is real communities and its first column included communities obtained from the proposed method. Therefore, each N_{ij} is the number of nodes that are in both communities i and j. In the following, *NMI* function is represented in Eq. (42).

$$NMI(A, B) = \frac{-2 \sum_{i=1}^{C_A} \sum_{j=1}^{C_B} N_{ij} \log\left(\frac{N_{ij} n}{N_{i.} N_{.j}}\right)}{\sum_{i=1}^{C_A} N_{i.} \log\left(\frac{N_{i.}}{n}\right) + \sum_{j=1}^{C_B} N_{.j} \log\left(\frac{N_{.j}}{n}\right)} \tag{42}$$

where A and B are set of real communities and communities obtained from the proposed method respectively, C_A and C_B denote the number of real communities and communities obtained from the proposed method respectively, n is the total nodes in the network, $N_{i.}$ is the number of nodes in community i from real communities and $N_{.j}$ is the number of nodes in community j from the proposed method.

• *Modularity*: This measure is an optimization of quality function that evaluates quality of communities. As mentioned, the larger value of modularity inducts the best quality of detected communities. In the following, modularity is represented by Eq. (43) (Newman and Girvan 2004).

$$Q = \sum_i e_{ii} - a_i^2 \tag{43}$$

$$e_{ii} = \frac{E_i}{m} \tag{44}$$

$$a_i^2 = \frac{\sum_{v \in C_i} d_v}{\sum_{v \in G} d_v} \tag{45}$$

4.2 Evaluation Measures in Attribute-Structural Similarity Methods

- *Density*: Density function that represented in Eq. (46) computes the total density of edges into each of the community. In this regard, edges' density in each community is a fraction of intra-Community edges to total edges.

$$D(\{V_i\}_{i=1}^k) = \sum_{i=1}^k \frac{|\{(v_x, v_y)|v_x, v_y \in V_i, (v_x, v_y) \in E\}|}{|E|} \tag{46}$$

where $\{V_i\}_{i=1}^k$ is set of k communities that obtain performing any method of community detection, v_x and v_y are two nodes that belong to the same community V_i and $|E|$ is the total edges.

- *Tanimoto Coefficient (TC)*: TC is used to find the similarity between all the nodes with the centroid within the community where the value ranges from 0 to 1 (Lipkus 1999).

$$TC_{AB} = \frac{c}{a + b - c} \tag{47}$$

where a is the number 1's in attributes of A node, b denotes the number of 1's in attributes of B node and c is number of common 1's in attributes of A and B node.

5 Datasets

In this section, datasets that employed to test the methods in community detection are expressed. There exist real and artificial networks that are explained.

5.1 Real Networks

There are several networks that their communities are known and also there exist real networks with unknown communities.

- *Real networks that their communities are known*: For example Zachary's Karate Club network, Dolphin Social network, US College Football network (Newman 2017).
- *Other Real Networks*: For example Jazz musicians network, E-mail network (Meyerhenke 2017) that these networks have unknown communities.

5.2 Artificial Networks

Artificial networks have to generate and do not exist in the real world. These networks are defined in the following.

- *Girvan-Newman (GN) networks*: In GN networks, each network is composed of 128 nodes and is divided into four communities of 32 nodes. Additionally, the average degree of each network is 16 and almost all nodes have the same degree. Moreover, the number of connections that each node has with other nodes located outside of its community is considered to parameter K_{out}. If K_{out} is lower than eight, each node has more connections with its community nodes than the other nodes in the network. Therefore, K_{out} is changed between one to eight in order to produce various networks.
- *Lancichinetti—Fortunato—Radicchi (LFR) Networks*: The LFR benchmark networks (Lancichinetti et al. 2008) is another standard dataset for testing community detection. They have heterogeneous distributions for node degree and community size, both of which are power law distributions. Several parameters must be set before generating an LFR network: the number of nodes N, average degree $< k >$, maximum degree k_{max}, minimum community size c_{min}, maximum community size c_{max}, exponent γ for the degree distribution, exponent β for the community size distribution, and mixing parameter μ.

6 Conclusion

In this chapter, a survey on community detection in social networks is presented. Basic concepts such as complex networks, structural similarity, attribute similarity, community, overlapping community are defined. These methods are divided into two groups, structural similarity and attribute-structural similarity. Structural similarity is computed according to topology similarity. Attribute-structural similarity is calculated based on topology and characterizes of the networks. Then, several methods in community detection are investigated and compared to each others. In addition, datasets and evaluation measures that used in community detection are defined. In this chapter is tried to define all of the requirements to understand community detection in social networks.

References

Arab, M., Afsharchi, M.: Community detection in social networks using hybrid merging of sub-communities. J. Netw. Comput. Appl. **40**, 73–84 (2014)

Blondel, V.D., Guillaume, J.-L., Lambiotte, R., Lefebvre, E.: Fast unfolding of communities in large networks. J. Stat. Mech. Theory Exp. **2008**(10), 10008 (2008)

Boobalan, M.P., Lopez, D., Gao, X.Z.: Graph clustering using k-neighbourhood attribute structural similarity. Appl. Soft Comput. **47**, 216–223 (2016)

Bullmore, E., Sporns, O.: Complex brain networks: graph theoretical analysis of structural and functional systems. Nat. Rev. Neurosci. **10**(3), 186–198 (2009)

Danon, L., Diaz-Guilera, A., Duch, J., Arenas, A.: Comparing community structure identification. J. Stat. Mech. Theory Exp. **2005**(09), P09008 (2005)

Easley, D., Kleinberg, J.: Networks, Crowds, and Markets: Reasoning About a Highly Connected World. Cambridge University Press (2010)

Ester, M., Kriegel, H.-P., Sander, J., Xu, X.: A density-based algorithm for discovering clusters in large spatial databases with noise. Knowl. Discov. Data Min. (KDD) **96**(34), 226–231 (1996)

Eustace, J., Wang, X., Cui, Y.: Community detection using local neighborhood in complex networks. Phys. A: Stat. Mech. Appl. **436**, 665–677 (2015)

Fortunato, S.: Community detection in graphs. Phys. Rep. **486**(3), 75–174 (2010)

Girvan, M., Newman, M.E.: Community structure in social and biological networks. Proc. Natl. Acad. Sci. **99**(12), 7821–7826 (2002)

Huang, J., Sun, H., Han, J., Feng, B.: Density-based shrinkage for revealing hierarchical and overlapping community structure in networks. Phys. A: Stat. Mech. Appl. **390**(11), 2160–2171 (2011)

Kim, Y., Son, S.-W., Jeong, H.: Finding communities in directed networks. Phys. Rev. E **81**(1), 016103 (2010)

Lancichinetti, A., Fortunato, S., Radicchi, F.: Benchmark graphs for testing community detection algorithms. Phys. Rev. E **78**(4), 046110 (2008)

Leskovec, J., Lang, K.J., Dasgupta, A., Mahoney, M.W.: Community structure in large networks: natural cluster sizes and the absence of large well-defined clusters. Int. Math. **6**(1), 29–123 (2009)

Leskovec, J., Lang, K.J., Mahoney, M.: Empirical comparison of algorithms for network community detection. In: Proceedings of the 19th ACM International Conference on World Wide Web, Raleigh, North Carolina, USA, pp. 631–640 (2010)

Li, C.: Study on overlapping community structure. Procedia Eng. **29**, 4244–4248 (2012)

Li, W., Huang, C., Wang, M., Chen, X.: Stepping community detection algorithm based on label propagation and similarity. Phys. A: Stat. Mech. Appl. **472**, 145–155 (2017)

Lipkus, A.H.: A proof of the triangle inequality for the tanimoto distance. J. Math. Chem. **26**(1), 263–265 (1999)

Lu, H., Wei, H.: Detection of community structure in networks based on community coefficients. Phys. A: Stat. Mech. Appl. **391**(23), 6156–6164 (2012)

Meyerhenke, H.: Dimacs (2017). Accessed 10 June 2017

Newman, M.: Network data (2017). Accessed 10 June 2017

Newman, M.E., Girvan, M.: Finding and evaluating community structure in networks. Phys. Rev. E **69**(2), 026113 (2004)

Saoud, B., Moussaoui, A.: Community detection in networks based on minimum spanning tree and modularity. Phys. A: Stat. Mech. Appl. **460**, 230–234 (2016)

Urban, R., Hošková-Mayerová, S.: Threat life cycle and its dynamics. DETUROPE-Cent. Eur. J. Reg. Dev. Tourism **9**(2), 93–109 (2017)

Zhou, Y., Cheng, H., Yu, J.X.: Graph clustering based on structural/attribute similarities. Proc. VLDB Endow. **2**(1), 718–729 (2009)

The E-learning System for Teaching Bridging Mathematics Course to Applied Degree Studies

Maria Antonietta Lepellere, Irina Cristea and Donatella Gubiani

Abstract Over the last decade it is becoming more common that traditional universities embrace also the e-learning system, because most students are working students (being not able to attend the courses) and most teachers noticed that using only their handouts or classical books is not enough for a good preparation of these students. The interactive materials, offered by the on-line tools, encourage faster and more intuitive learning. Therefore more dissemination articles, describing the experience of those universities/teachers exploring/using the e-learning system, are more common in literature. Also the current chapter aims to follow the same direction. In this work, we present a comparison between two universities, an Italian one (University of Udine) and a Slovenian one (University of Nova Gorica), situated at less than 50 km far away from one another, similar as the educational system, but different as their dimension (in numbers of students and professors/researchers). Our study is focussed on the e-learning system for teaching the math bridge-course to applied degrees studies.

Keywords E-learning · Bridging math course · Moodle platform

M. A. Lepellere
Department of Agricultural, Food, Environmental and Animal Sciences,
University of Udine, Udine, Italy
e-mail: maria.lepellere@uniud.it

I. Cristea (✉) · D. Gubiani
Center for Information Technologies and Applied Mathematics,
University of Nova Gorica, Nova Gorica, Slovenia
e-mail: irina.cristea@ung.si; irinacri@yahoo.co.uk

D. Gubiani
e-mail: donatella.gubiani@ung.si

© Springer Nature Switzerland AG 2019
C. Flaut et al. (eds.), *Models and Theories in Social Systems*, Studies in Systems,
Decision and Control 179, https://doi.org/10.1007/978-3-030-00084-4_16

1 Introduction

The study of students' math difficulties in passing from secondary school to university has been the subject of various researches (e.g. Alcock and Simpson 2002; Gueudet 2008) for its impacts at the individual and social level: in particular, students spend more time completing their scientific study or decide to abandon it. Many universities are moving in an attempt to ease the secondary-tertiary transition by providing: more structured guidance, linking courses, lowering the level of maths taught, sometimes even lowering exam requirements, even though the standard program is maintained. However, these actions have proved to be ineffective. Besides, traditional bridge courses, being limited in time and not customized to meet the individual's difficulties, have often proved to be ineffective. Moreover, there are more and more students who come to higher education institutions with a differentiated background than in the past, with different and vague visions of mathematics, its learning and role in their future careers and in their life (Kajander and Lovric 2005). Students often have a distorted perception of what learning is and what it involves: acquiring knowledge often is seen as a collection of facts that can be absorbed passively. In addition, university math in its offerings and learning opportunities to it hardly fits the diversity of the student body.

Higher education is undergoing great developments made necessary by changing social and economic demands as well as the profound changes in how knowledge is produced and organized. Teachers are asked to offer a wider range of topics and consequently students are increasingly expected to have more autonomy of learning. Regarding this, online teaching resources offer enormous potential but at the same time it is recognized that students require different skills than traditional courses (Romiszowski 2004; Hošková-Mayerová and Rosicka 2015) and greater motivation and commitment in learning.

As mathematics studies demonstrate, mathematical teaching processes are complex and planning a course should not underestimate the three levels that influence students' behavior: the cognitive level, which concerns the learning of specific concepts and methods discipline; the meta-cognitive level, which concerns the control of subjects on their learning processes; and finally the affective or non-cognitive level, which takes into account the beliefs, emotions and attitudes of the learner (Zan 2006). "Technology can play a role in each of these levels, including the non-cognitive one, as it can deeply influence the beliefs, emotions and attitudes of learners, on the other hand, it is itself subject to rooted convictions and can provoke strong emotions" (Albano and Ferrari 2008).

These systems today, in different university contexts, are not an alternative to traditional didactics, but rather an integration to it. The current trend is the Blended Learning: the traditional classroom lesson is enriched by the powerful tools available from the various e-learning platforms.

In this work, we present a comparison between two European universities, focussing on the e-learning course prepared as a support for the traditional bridging mathematics course. These universities are: the University of Udine (UNIUD), an

Italian one, and a Slovenian one, the University of Nova Gorica (UNG), located at less than 50 km far away from one another, adopting a similar traditional educational system, but being different with respect to the number of students and professors/researchers.

The University of Udine was founded in 1978 as part of the reconstruction plan of Friuli after the earthquake in 1976. Its aim was to provide the Friulian community with an independent centre for advanced training in cultural and scientific studies. UNIUD is the youngest university in north-east part of Italy (Friuli Venezia Giulia), with a bidding offer for the academic year 2016–2017 of 36 Bachelor's Degrees (7 inter university), 32 Master's degrees (7 inter university) and 4 single cycle degrees (1 inter university) for a total of 72 degree courses and 15,385 enrolled students—4,598 students being enrolled in the first year at 1st and 31st of December 2016, respectively. Besides 13 inter-university courses are with the University of Trieste, one with the University of Trento and one with the Universities of Padua, Verona and Bolzano. The first-level study programmes (Bachelor) are divided into four main areas: Scientific, Humanistic and Teacher training, Economic and Law, and Medical. The Scientific area contains 4 micro areas: Agricultural Sciences; Biotechnology; Engineering and Architecture; Science (that are managed by the corresponding Departments).

On the other hand, University of Nova Gorica is a different reality: born in 1995 as an international postgraduate School of Environmental Sciences, developed into the Nova Gorica Polytechnic in 1999, it became the University of Nova Gorica in 2006. It is the youngest and smallest university in Slovenia (among other 3 Slovenian universities), aiming to be an international creative environment for researchers, professors and students, promoting interdisciplinary and practically-oriented programmes covering three main areas: Scientific, Humanistic and Artistic. The Scientific area is represented by four faculties: School of Engineering and Management, School of Environmental Sciences, School of Science, and School of Viticulture and Enology. In the academic year 2016/17, it offered 7 first-level study programmes (Bachelor's), 9 second-level study programmes (Master's) and 8 third-level study programmes (Ph.D.), for a total of 488 enrolled students (158 students being enrolled in the first year at the first level), including 194 foreign students.

Both universities have chosen as e-learning environment the Moodle platform (Modular Object Oriented Dynamic Learning Environment, https://moodle.org), implemented in PHP (Hypertext Preprocessor) and JavaScript (object oriented scripting language). The Moodle platform provides users with different applications that, if they are used in an integrated way, they could change the organization of the learning. The e-courses are provided with forums to discuss the topics addressed so as to make participants more and more involved in deepening and studying during the activity. It is an open source software with the GNU General Public License (GPL), so it's free and editable by any programmer. This philosophy has resulted in the formation of an international community of people working on the platform or using it for their activities, ensuring a constant upgrade. It has now become a worldwide standard for teledidactics, as it is shown by the number of countries in which it is used.

Moodle always seeks to integrate tools that showcase the topics dealt with effectively, but at the same time manages to develop the social potential of this type of learning by incorporating tools that make possible the continuous interaction between students and teachers and among students themselves. Moodle allows the teachers to customize their courses, integrate the traditional lessons with on-line materials devoted to the broad development of media. However, the users have richer contents, but at the same time calibrated for their learning, without reaching the paradox in which the increase in theoretical information reduces the ability to solve real problems (Piccinini et al. 2016).

2 Mathematics Bridging Course

Nowadays all technical study programmes contain, already in the first year, various courses of Mathematics (Hoškova 2009; Hoškova and Račkova 2007). It is known that they are the basis for other technical subjects on one hand, and the students arrive from secondary schools with different profiles, on the other hand. Thereby several traditional universities include in their curricula a short, but intensive, course of basic mathematics, called a mathematics bridging course (see for example Sosnovskyet et al. 2013; Bardelle and Di Martino 2012), usually organized during the summer period or in September. It is intended as a refreshment course, where the main chapters in Arithmetics, Algebra, Geometry, and Trigonometry are recalled, in order to allow the new students to rich a certain level of mathematics. Moreover, attending such a course, the new students can explore how much mathematics is required and which are their own lacks, having the possibility to get a better preparation for the regular Math courses (i.e. Linear Algebra, Calculus).

2.1 Why an E-course of Basic Mathematics at a Traditional University?

Both universities involved in our study (UNIUD and UNG) have a traditional mathematics bridging course in their educational curricula, but differently designed, as we will describe here below, and supported by an e-learning course, launched in the academic year 2016/17 at UNIUD, and in 2014/15 at UNG, respectively.

The enrollment to all degree courses in the field of science at UNIUD is preceded by an entrance test that contains the prerequisites in various disciplines, that are considered fundamental to successfully face up the study course, mathematics always being present in all the tests. In all students' guides of the various study courses in the scientific area, the need for the initial preparation of the students is highlighted, in particular in the guide of Engineering programmes, a detailed indication (syllable) about the initial essential knowledge in the fields of Mathematics and Science has been included. Given the importance of this basic mathematical preparation, for all degree courses in the Scientific Area, at the beginning of the academic year, an additional compulsory basic mathematics course (a bridging course) is required for those who did not reach a specific threshold in the math section at the entrance test. At the end of the course there is an examination and, in some cases, its failure represents a block for continuing the studies. For example, for Engineering degree there is an additional training requirement that must be completed by passing the "Basic Mathematics" exam, which is proposed in several sessions during the academic year. There is also an agreement between the Department of Polytechnic Engineering and some secondary schools to provide mathematics courses in schools for university access, at the end the student can access to the same exam. For Architecture degree, passing this exam is essential to supporting the 2nd year examinations. In other cases, students must repeat the admission test until they pass it, otherwise they will not be able to enroll in the second year.

In order to continuously improve the teaching methods and encourage a better and more effective students learning, starting with the academic year 2004/2005 some of the courses in the various degree programmes use the University's e-learning portal to supplement lessons with additional documentations (notes, slides, links to external materials) or activities (exercises, partial audits, videos, etc.) as a support to the traditional teaching. In the Strategic Plan of the University, it is stated that it is necessary to "provide the student with all the educational tools necessary for the progressive formation of self-learning, in particular through e-learning". Since the academic year 2015/2016, the Moodle platform (https://elearning.uniud.it/moodle/) is used for this purpose and, one year later, a bridge course in mathematics (called "Interactive path of mathematical literacy") was settled up on the platform. At UNIUD the idea to use a bridge e-course born because the mathematics courses in the first years at the applied degree programmes were suffering, with a low percentage of students passing the relative exams. The rector of the university asked the Department of Mathematics, Computer Science and Physics to take care of this problem. The Department, thinking that part of the problem was due to a lack of initial mathematics preparation of the students, appointed a committee for creating the online preparatory materials related to the math bridging course.

At UNG, the enrollment is not based on an entrance exam, but in order to register as a student at the first level of some study programmes, the students have to successfully graduate from secondary school by passing the final general examination (that in most of the cases contains also a mathematics exam). In September, a package of 90 h (30 h for each bridging course, i.e. Mathematics, Physics, and Chemistry) is offered in advance to the future enrolled students to UNG, in order to provide them with a certain knowledge in these three main subjects. These refreshment courses are not mandatory, a student can decide to attend them or not. We noticed that the attendance at all bridging courses is very low, because most of the students are outside Nova Gorica, and moreover, in the last two years, outside Slovenia (around one third of the students in the first year comes from abroad). Thus for them the attendance in September means a supplementary cost (translated in one month more in the student house and life cost). Motivated by this fact and the students' needs, since the academic year 2014/2015, in the UNG on-line system (https://moodle.ung.si/)

based on the Moodle platform, an on-line course of basic mathematics, as a support of the face-to-face refreshment mathematics course, was created.

2.2 The Management of the E-course of Basic Mathematics

The students can use the e-course of basic mathematics already in September, before the beginning of the other courses, in parallel with the traditional bridging math course, but also (and this is strongly recommended to them) during the entire academic period, anytime they need basic aspects, that are not discussed in detailed during the standard courses.

As a general rule, the students can access the courses that are in Moodle only if they are enrolled in those courses. Thereby the enrollment is a key activity in the e-learning process, having a strong influence on the use of the e-courses. There are mainly two different ways of enrollment: the students enroll themselves, or the teachers enroll the students. At UNIUD, in the technical Bachelor programmes, the number of the students is high. This is the main reason for using the self-enrollment in the math e-course in the Moodle platform: a student has the possibility to enroll in this course whenever he/she needs it during the university study. Conversely, at UNG, where the number of the students is lower, as soon as the students' office has the list of the future students, the system administrator insert them in the Moodle portal and the responsible teacher of the e-learning math course enrolls them into the e-course, where they remain till their graduation. Similarly, those students that don't attend the bridging course are usually enrolled to this course later on, when the instructors of the math courses advertise the Moodle platform and its content.

An analysis of this aspect in both universities, together with its direct consequences is presented in Sect. 3.

2.3 The Structure and the Content of the E-course of Basic Mathematics

Each e-course in a Moodle platform can be differently settled up and can contain different types of up-loaded materials. Here we concentrate only on the e-course of basic mathematics, proposed at UNIUD and UNG. Again we can notice similarities and differences in organizing this e-course at both universities. The chapters proposed to be revised during the math bridging course (and also in the electronic version) are, more or less, the same, covering the main aspects in:

(a) *Arithmetics*: sets and number sets.
(b) *Algebra:* linear and quadratic equations/disequations; rational disequations; equations/disequations with roots; equations/disequations with logarithms; equations/ disequations with absolute value; polynomials; functions.

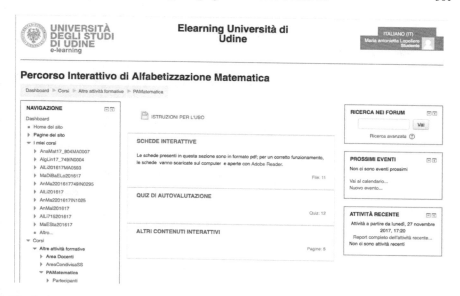

Fig. 1 Content of the math e-course at UNIUD: organized by types of the resources

(c) *Trigonometry:* (only at UNG, while at UNIUD this chapter is in preparation) the main trigonometrical functions and simple equations involving them.
(d) *Analytical Geometry:* (only at UNIUD) the line in the plane; conics.

The structure of the e-course is different, as we can notice in Figs. 1 and 2. At UNIUD, the e-course is structured in three parts: the first one contains "Interactive files", that are pdf files with theoretical aspects and also some preliminary exercises, followed by two parts of exercises grouped in "Self-evaluation quizzes" part, and "Other interactive contents". Differently, at UNG the topic format was adopted, meaning that the content is divided in four chapters, each of them containing both theory and exercises.

Since one main role of e-courses is to make the course more attractive, and, if it is possible, to offer to the users also a visual perspective on the content, different types of interactive materials can be up-loaded/created in the Moodle platform. At UNIUD, the interactive theoretical files are pdf files (see Fig. 3, the left part) containing the description of the basic concepts and short quizzes with questions of true/false or multiple response, reporting the error and the correct answer. The same type of questions are included in the Self-evaluation quiz part, but including the final score calculation and the error verification, meaning that the teacher has a view on students' preparation. The third part of the course, called "Other interactive contents", includes several solved exercises using the on-line tool of GeoGebra (https://www.geogebra.org/materials/).

On the other side, at UNG, the interactive materials are focused on quizzes, links to external Slovenian websites containing interactive materials prepared for students

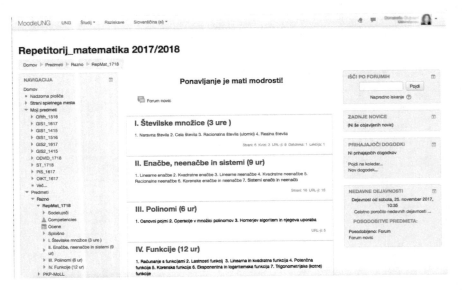

Fig. 2 Content of the math e-course at UNG: organized by topics

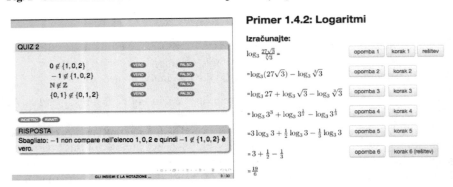

Fig. 3 Different kinds of interactive contents: quizzes in pdf files (UNIUD) and exercises step by step in web pages (UNG)

of secondary schools, exercises with full solutions, presented step by step, with hints on how to continue the exercise, helping the students to understand the correct way of completing it, without showing them (from the beginning) the whole solution (see Fig. 3, the right part). In particular, this second kind of resources have been created including Moodle pages (web pages) that integrate simple JavaScript codes to support dynamic interaction with students that need hints, solutions by steps, and/or the final solution.

3 Descriptive Statistical Analysis: Students' Accesses

A crucial point in the management of an e-learning course is the possibility given to the teachers to have a complete report of the activities carried out by the students enrolled in the course, so as to be able to check their progress at intermediate steps, their assiduity in consulting the available materials, and the level of participation in any collaborative activity. The verification of any anomalies allows the teacher to promptly react, soliciting the students to better participate, providing additional materials or giving them a direct support before the end of the course.

In this preliminary study, our main objective is related to the distribution of the students' access into the math bridging course in Moodle and to its resources. In order to achieve this goal, we used only the data automatically provided by Moodle in its sections "Users" and "Activity report". The data analysed in this work are downloaded at the end of November 2017 for both universities.

The first aspect that we analyze concerns the period of the students' activities in the bridging math e-course in Moodle. More clearly, we calculate the *interval activity* of all students enrolled in this course for both universities. It is determined by its initial point, corresponding to the first access into this e-course, and by its final point, represented by the final access. Graphically, this is represented by a vertical bar, that must be read not as a continuous access in time, but it only indicates the first and the last access of a particular student to the course, without any other information regarding his/her activity in the middle of this period. Therefore, on the x-axis we indicate the access dates, while on the y-axis the enrolled students to this e-course. For privacy reasons, the name of the students are replaced by a numerical value created on the downloaded data, without a particular meaning for our analysis. Moreover, for a better visualization, the students are ordered dependently on their first access.

In the following, we will separately present the descriptive statistical analysis of the above mentioned aspect, starting with the **University of Udine (UNIUD)**. Here the attention has been paid on the students of the Bachelor programmes in Agricultural Sciences (denoted by short as AGR) where, on November, 29th 2017, there were 177 students enrolled in the e-course of basic mathematics, representing 40% of the enrolled students in these agricultural programmes. A lower percentage of enrolled students in this e-course characterizes the study programmes in Science (SCI) (22.5%) and Engineering (ENG) (8.5%), respectively.

Figure 4 represents the *interval activity* for each student involved in the AGR Bachelor study programmes and enrolled in the e-course of basic mathematics, from the 1st October 2016, till the 29th November 2017. One clearly notices in the graphic two intensive periods of accesses, related to the months of September and October, representing the beginning of each academic year, when the regular courses start and the e-course of basic mathematics is advertised by the math teachers. Unfortunately, currently there is no connection between the face-to-face bridging math course and the similar one in the Moodle platform. For this reason, the students start to access the e-course in different temporal periods.

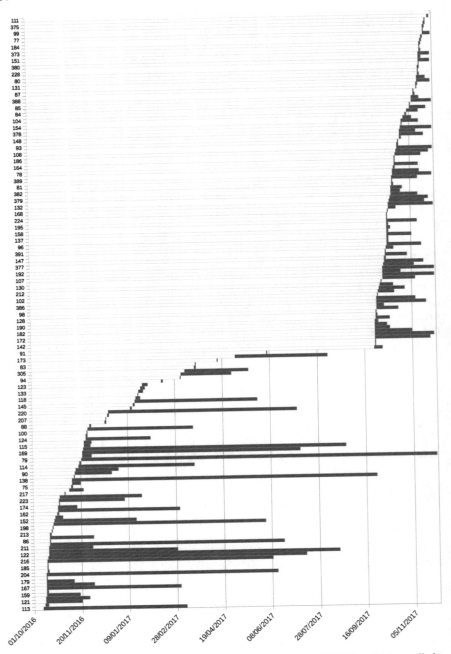

Fig. 4 Intervals activity into the e-course of basic mathematics of the UNIUD students enrolled at the AGR study programmes: from their first registration to their last activity in the e-course

In the academic year 2016/17, there were 655 AGR students enrolled in the first year. Thus, this is the minimum number of hypothetical users of the e-course, since additional students, enrolled in the previous years, could join them. We observed that only 75 students (representing 11.5% of the AGR students enrolled in the first year) made the first login into this e-course before August 2017, besides them 54 students accessing the e-course already in October or November 2016 (this means 8.2% of the total number of AGR enrolled students). The situation in the current academic year (2017/18) is changed. The AGR students enrolled in the first year are 520 and 19.4% of them have logged in the e-course since the end of September to the end of November 2017. It is evident a consistent increasing of the number of the enrolled students in the academic year 2017/18 with respect to the previous one.

Moreover, taking into account all enrolled AGR users of the e-course of basic mathematics, we can observe an average of the interval activity of 35.8 days, while for several students this period is very short: its length is only of 1 or 2 days for 44.3% of the users. This means that the students access the course, without using all its potentialities. This depends on the fact that most of the resources are pdf files, that can be downloaded once and used in the future period, without being necessary to make a login to the Moodle platform. Besides, it could happen that the students will access again the course in future, when they will need other materials, but this will be registered in the next months/years.

Considering now the other study programmes at UNIUD and the number of the students enrolled in the e-course of basic mathematics, we can notice a similar situation at SCI, while this scenario is different at ENG. Indeed, at the Engineering Bachelor's programmes, the total number of the students enrolled in the first year in the two academic years 2016/17 and 2017/18, is 547 and 450, respectively; among them just 24 students (representing 4.4%) have accessed the e-course during the first year and 13 students (representing 2.9%) during the second one. This situation is strongly influenced principally by the fact that the bridging math e-course was meant more for the students in Agricultural Sciences, being requested by their math teachers (since the initial difficulties in mathematics of the students are more pronounced). Secondly, the instructors of this e-course are from the Science area. By consequence, the e-course of basic mathematics was more advertised at these two study programmes, than at the Engineering one.

A similar analysis at the **University of Nova Gorica (UNG)** has been conducted, but in a different scenario, since the e-course of basic mathematics was settled up in the academic year 2014/15. Firstly, we can examine the students' activity in this e-course for their entire Bachelor's degree, having more information with respect to this. Secondly, it must be recalled here, that the students' enrollment into the e-course is made by the teacher, often at the beginning of the face-to-face course of basic mathematics. Another important aspect regards the construction of the e-course, meaning that at UNG each academic year a new e-course of basic mathematics is created, containing only the students enrolled in that specific academic year. In particular, in the following we will describe the activity in this e-course of the UNG students enrolled to the university in the academic year 2014/15.

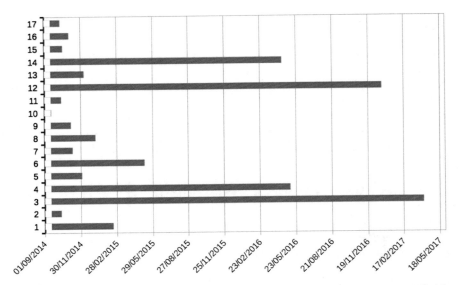

Fig. 5 Intervals activity into the e-course of basic mathematics of the UNG students enrolled in the academic year 2014/2015: from their registration to their last activity in the e-course

As visible in Fig. 5, the number of the enrolled students is lower with respect to UNIUD, accordingly with the dimension of the universities, as it has been explained in Sect. 1. The Moodle system registered a total of 24 enrolled students but, only 17 of them have effectively used the e-course. For this reason in Fig. 5 we reported only their activity. Taking into account all these students and their entire study period (obviously, some of them have not already finished their studies, while some others have abandoned the university), we can observe an average of the interval activity of 227 days, spanning from a minimum of 1 day (only for 1 user) to a maximum of 934 days. A more detailed analysis shows that, most of them have used the e-course in the first 2–3 months of the academic year, during the face-to-face course of basic mathematics and the math course in the first semester. It is also evident that 4 students (17%) have used the course also during the second year, and 2 also in the third and last year (8.5%).

Making a comparison between the above two analyses, nevertheless the number of the students at the University of Udine is higher than those at University of Nova Gorica, having a great impact on the method of the enrollment (self-enrollment at UNIUD and enrollment by the teacher at UNG), the number of the active users of the e-course of basic mathematics is very small. In general, we notice that in both universities more and more students use the materials included in these e-courses during their entire Bachelor study programme: most of them in the first semester of the first year (as a support for the mathematics standard courses), but also during the first year, the second one and, as already visible at UNG, in the third one. It worth to stress also the fact that most of the viewed materials are those presented in the first part of the e-course (both pdf files or quizzes), meaning that the students are probably not enough motivated to use all materials. This will be the subject of a future work.

4 Conclusions and Further Work

In this chapter we have conducted a descriptive statistical analysis regarding the usage of the e-course of basic mathematics, created as a support of the traditional face-to-face bridging math course for applied degree programmes, making a parallel between two different university environments: the University of Udine and University of Nova Gorica. Since both universities aim to enhance the level of teaching, by using the opportunities offered by new information and communication technologies, in particular by combining the traditional face-to-face learning process with the e-learning one, we want to emphasize the important role of the virtual classrooms in this process. By presenting this study, we hope to have identified the difficulties encountered by the users (students and teachers) of the e-course of basic mathematics in both universities and propose ideas for overcoming them, improving in the same time the quality of the learning process.

The first aspect that has arisen is connected with the low percentage of students that properly use this e-course. It is clear that a better advertisement of the e-course at the level of the university is needed, and in the same time, a stronger collaboration between the teachers of the face-to-face bridging math course and the e-course of basic mathematics. Secondly, having in mind that active didactic courses can not be only the reproduction in the electronic format of teachers' slides and exercises, but they must contain interactive materials, teachers will need additional information and training in e-learning pedagogy and didactics, about blended learning and flipped classrooms in particular. They will also need additional information how to efficiently use and reuse open educational resources, with a regular opportunity for peer to peer sharing of good practices and advice (Gubiani et al. 2018).

In addition to this, e-learning requires more attention to the preparation of materials: the course content should be self-explanatory for students; this is very different from normal classroom work in which it is possible to clarify the answers personally by directly answering student questions. This is particularly important in structuring bridging courses both for the particularity of the material that must meet the very diversified needs of the students, and because it must be taken as a reference point if the student deems it useful to review certain topics to better follow the mathematics courses at university. Active Teaching means building courses that are not only delivered when they are fully realized with presentations, exercises, tests, but courses that progressively enrich with new teaching materials, Internet resources, teacher observations, student feedbacks. What we could certainly do is to guide the students who attend the course in the use of its materials. It would be desirable to have tutors to take care of the forum for all students who need support. Another possibility could be to create forums for each single course. As far as Udine concerns, since several agreements with secondary schools in the region already exist, involving the teachers of these schools in the use and improvement of the platform could be also proposed (see Di Luca et al. 2012).

Furthermore we would like, in both universities, to investigate the students' opinion on the offered materials through a questionnaire. In the light of their results, it will be possible to review the existing materials and the content of the course. We realize that students' difficulties go even beyond mathematical content. For example, we aim to find answers to the following questions. Do the students know how to deal with a mathematical problem? How to use notations and mathematical representations? How to use basic mathematical knowledge to solve a problem? This type of study will also be the subject of our future research, together with a project about joining forces for a common redesign of the e-course of basic mathematics, with evident benefits for both universities, in the Moodle platform.

Acknowledgements We are grateful to all instructors who participated for the creation of the e-course of basic mathematics at the University of Udine and allowed us to collect the data necessary for this preliminary analysis. We also thank the referees for their useful comments and suggestions for possible future research developments.

References

Albano, G., Ferrari, P.L.: Integrating technology and research in mathematics education: the case of e-learning. In: Garcia Penalvo, F.J. (ed.) Advances in E-learning: Experiences and Methodologies, pp. 132–148. Information Science Reference (IGI Global), Hershey (PA-USA) (2008)

Alcock, L., Simpson, A.: Definitions: dealing with categories mathematically. Learn. Math. **22**(2), 28–34 (2002)

Bardelle, C., Di Martino, P.: E-learning in secondary-tertiary transition in mathematics: for what purpose? ZDM **44**(6), 787–800 (2012)

Di Luca, M., Vitacolonna, E., Genovese, L., Bolondi, G., Polcini, F.: Piattaforma E-Learning per una didattica per competenze in matematica. In: Atti del convegno AICA Didamatica 2012, Taranto, Italy (2012)

Gubiani, D., Cristea, I., Urbančič, T.: Introducing e-learning to a traditional university: a case-study. In: Qualitative and Quantitative Models in Socio-Economic Systems and Social Work. Studies in Systems, Decision and Control (2018)

Gueudet, G.: Investigating the secondary tertiary transition. Educ. Stud. Math. **67**, 237–254 (2008)

Hošková, S.: Experience with Blended (Distance) Learning Study Materials, International Conference on Distance Learning, Simulation and Communication CATE 2009, Brno, Czech Republic, pp. 70–77 (2009)

Hošková, S., Račková, P.: Experience of mathematics blended-learning at university of defence adult learnings of mathematics. In: Proceedings of Contributions, ALM 14—Adults Learning Mathematics (ALM)—A Research Forum, Ireland, pp. 177–194 (2007). ISBN 0-9552717-3-8 Paperback book, ISBN 0-9552717-5-4, Format CDROM,

Hošková-Mayerová, S., Rosická, Z.: E-Learning pros and cons: active learning culture? Procedia—Soc. Behav. Sci. **191**(2 June 2015), 958–962 (2015). ISSN 1877-0428

Kajander, A., Lovric, M.: Transition from secondary to tertiary mathematics: McMaster University experience. Int. J. Math. Educ. Sci. Technol. **36**(2–3), 149–160 (2005)

Piccinini, L.C., Lepellere, M.A., Chang, T.F.M., Iseppi, L.: Structured knowledge in the frame of Bak-Sneppen models. Ital. J. Pure Appl. Math. **36** (2016). ISSN 2239-0227

Romiszowski, A.J.: How's the e-learning baby? Factors leading to success of failure of an educational technology innovation. Educ. Technol. (Saddle Brook Then Englewood Cliffs NJ) **44**(1), 5–27 (2004)

Sosnovsky, S., Dietrich, M., Andres, E., Goguadze, G., Winterstein, S., Libbrecht, P., et al.: Mathbridge: bridging the gaps in European remedial mathematics with technology-enhanced learning. In: Fischer, P., Hochmuth, R., Frischemeier, D., Wassong, T. (eds.) Using Tools for Learning Mathematics and Statistics, pp. 437–451. Springer, Berlin, Heidelberg (2013)

Zan, R.: Difficoltà in matematica. Osservare, interpretare, intervenire. Springer, Milano (2006)

Applications of Multi-Agent Systems in Social Sciences: Virtual Enterprises as an Example

Anata-Flavia Ionescu and Dorin-Mircea Popovici

Abstract When it comes to modeling social life, few paradigms are more suitable than multi-agent systems (MASs). MASs have helped gain a deeper understanding of countless social phenomena (through agent-based social simulation) that were hard to study using analytical approaches. More recently, MASs have evolved into handy assistants to the human users they can act on behalf of. One example of such use of MASs is in implementing virtual enterprises (VEs), i.e. temporary alliances of geographically distributed organizations created in order to exploit a specific business opportunity. Competition, negotiation and cooperation mechanisms are usually copresent in the creation and operation of VEs, with individual members being simultaneously autonomous self-interested entities and parts of a larger whole, the benefits of which they are also trying to maximize. In turn, VEs have application areas such as commerce, manufacturing, and tourism, offering important contributions to the universal contemporary strife to obtain customer-driven products while optimizing time, costs, and quality. In this chapter, we aim to review the most important applications of MASs in social sciences, with a focus on VEs as a special case.

1 Introduction

Agents and multi-agent systems (MASs) are very seminal paradigms in computer science, as they provide abstractions for the behavior of living beings at the individual and group level, respectively. Real-life behavior has thus been extrapolated to hardware devices and software components that emulate intelligence and social abilities, regularly acting on behalf of their flesh-and-blood owners or users. The decentralized approach underlying MAS development offers several advantages such as (Huhns and Stephens 1999):

A.-F. Ionescu (✉) · D.-M. Popovici
Faculty of Mathematics and Informatics, Ovidius University of Constanta, 124 Mamaia Bd, 900527 Constanta, Romania
e-mail: anata.ionescu@univ-ovidius.ro

D.-M. Popovici
e-mail: dmpopovici@univ-ovidius.ro

© Springer Nature Switzerland AG 2019
C. Flaut et al. (eds.), *Models and Theories in Social Systems*, Studies in Systems, Decision and Control 179, https://doi.org/10.1007/978-3-030-00084-4_17

- Solutions for problems in information environments that are too complex and/or dynamic to centralize;
- Fault tolerance (through redundancy);
- Integrating expert knowledge from multiple sources, which may result in faster and qualitatively superior decisions;
- Reusability;
- Allowing private information of the collaborating entities (i.e., agents) to remain private.

The expansion of MASs is illustrated by the existence of a dedicated programming paradigm—agent oriented programming (AOP; Shoham and Leyton-Brown 2008)—alongside a plethora of dedicated programming languages—e.g., AGENT0 (Shoham and Leyton-Brown 2008), MetateM (Fisher and Hepple 2009), Brahms (Sierhuis et al. 2009), GOAL (Hindriks 2009)—as well as MAS development tools, such as debuggers (see Poutakidis et al. 2009) or integrated development environments (see Pokahr and Braubach 2009). In this chapter, we frame MASs as tools for modeling and simulating social processes. Accordingly, we cover MAS applications such as electronic commerce, social simulation, and business process management.

The chapter is structured as follows: in the first section, we offer a short review of MAS theory and of VEs as an application area; the second section provides an exemplification of VE architectures; we discuss the theory and its applications in a third section, and, in the final section, we abstract a number of relevant conclusions and survey promising avenues for future research.

2 Theory

MASs address problems that warrant distributed solutions and are too complex to be centralized. They have therefore found applicability in numerous areas, among which we mention (Weiss 1999; Wooldridge 2009): e-commerce, telecommunications, traffic control systems, transportation systems, travel planning, web information gathering and filtering, social simulation, simulation of business processes, and virtual enterprises or organizations.

In this section, we provide a short review of agents and MAS theory and their applications in social and organizational sciences.

2.1 Agents and MASs: An Overview

Most definitions of **agents** emphasize two main characteristics (Weiss 1999; Wooldridge 2009):

1. *Autonomy*, i.e., the agent's ability to decide on its own account;
2. Ability to *interact* with other agents.

An intelligent agent is typically endowed with three cardinal features (Wooldridge 2009):

1. *Reactivity*—its ability to adapt to changes in the environment;
2. *Proactivity*—its ability to take the initiative in adapting its behavior to anticipated changes in the environment;
3. *Social ability*—the ability to intelligently interact with other agents.

Agents are autonomous in the sense that, to a certain degree, they regulate their own behavior (e.g., acting according to their own experience, possibly to rules they acquired through learning from their interactions with the environment, etc.), and intelligent in the sense that they reason in order to optimize some performance measures in unpredictable environments where success (i.e., thoroughly achieving the agent's goals) is not guaranteed (see Weiss 1999; Wooldridge 2009).

Agents "perceive" (or "see") their environment using sensors, then make a decision based on some form of reasoning (e.g., deductive, practical, etc., depending upon their architecture—see Wooldridge 2009), and finally act (using their effectors, or actuators) according to the decision they reached.

Agents can have various architectures, depending primarily upon the form of reasoning they must implement. An agent architecture that models practical reasoning—perhaps the most natural and common in human thinking—is the **Belief-Desire-Intention (BDI)** architecture, described in detail in Wooldridge (2009). Practical reasoning is a process consisting of the following two phases:

1. **Deliberation**, i.e. setting the agent's goals. These goals, to which the agent will *commit*, will constitute the agent's *intentions*;
2. **Means-ends reasoning**, whereby the agent decides how to reach the goals established through deliberation—i.e., to materialize its intentions.

Intentions persist until the goals that generated them are either believed by the agent to have been achieved, or dropped by the agent after reconsideration. Once committed to, intentions constrain future deliberation and influence the agent's beliefs.

Let *Per* be the set of all perceptions, *Bel* be the set of all possible beliefs, *Des* be the set of all possible desires, and *Int* be the set of all possible intentions. Then, deliberation generally comprises the following steps:

- The agent updates its beliefs depending upon its current beliefs and perceptions, according to a *belief revision function*:

$$brf : 2^{Bel} \times Per \to 2^{Bel}$$

- The agent updates its desires depending upon its current beliefs and intentions, using a function that generates *options*:

$$options : 2^{Bel} \times 2^{Int} \to 2^{Des}$$

- Finally, the agent applies a *filter function* that maps its current beliefs, desires, and intentions to a new subset of intentions:

$$filter : 2^{Bel} \times 2^{Des} \times 2^{Int} \to 2^{Int}$$

A MAS is a collection of agents that interact. MASs may be regarded as special cases of distributed systems, but deviate from common definitions (or at least from common implementations) of distributed systems in that synchronization and coordination happen at run-time (instead of design-time), and entities maintain self-interested economic interactions (Wooldridge 2009).

The key to functional agent societies—as in the case of human societies—is coordination (Huhns and Stephens 1999). Coordination regards two major categories of agent behavior:

- **Cooperation**, which is achieved through *planning*;
- **Competition**, mainly concerned with *negotiation*.

A framework for coordination in MASs is offered by Computational Organization Theory (COT), a field that studies organizations as computational entities, drawing on social and organizational sciences (Carley and Gasser 1999). COT theorists view social and organizational intelligence of a MAS as distinct from the capabilities of the agents that compose it, transcending the limits of individual agency. The fundamental mechanism underlying coordination is communication between agents (explicit "speech acts" in agent communication languages such as FIPA-ACL, or implicit communication) (Huhns and Stephens 1999; Wooldridge 2009). Information sharing through communication allows MASs to act as corporate entities, exhibiting emergent (spontaneous or imposed) organizational design, which is thus reduced to an information processing problem (Carley and Gasser 1999).

Note from the above review that MAS scholars inevitably refer to social sciences at some point in grounding and defining MAS theory. First, MASs may be regarded as a natural metaphor for artificial intelligent social systems (Wooldridge 2009). Furthermore, MASs and social sciences support each other's advancement, being bound by a two-way connection: while MAS theory capitalizes on concepts, principles, laws, etc. from social sciences in defining agents and agent interactions, which in turn bring about advancements in software development in general (and particularly in the development of massive open distributed systems such as the Internet), social sciences make use of MAS applications for modeling and simulating social phenomena so as to better understand—or complement—human social behavior (Weiss 1999; Wooldridge 2009). The following subsection is dedicated to the latter direction of the connection.

2.2   Applications of MASs in Social Sciences

As discussed earlier in this chapter, agents are built to use intelligence for the optimization of the design objectives they serve. Among the key characteristics that enable intelligent behavior in agents, we mentioned social ability. Yet, one might question the appropriateness of such a paradigm in the first place: *why* would we need a social approach? Second, what makes agents social entities, i.e., *how* is social ability implemented? In the following, we offer answers to these two questions.

2.2.1   Why Social Ability?

So, what are the problems we would choose to solve by reproducing patterns of social interaction and why? It turns out that having non-cooperative entities cooperate yields rapid and valid solutions to a wide range of problems, pertaining to domains such as the application areas mentioned in the previous section. For example, VEs emerged and developed because software agents representing businesses and/or individuals can save time and costs by collaborating (Sandholm 1999). Hence, the very end of many MASs is to provide better solutions to given problems by distributing decision making processes to self-interested agents.

The social principles they build upon allow MASs to have a wide area of applications in social sciences. MAS uses in social and organizational sciences typically serve one of two main purposes:

1. They emulate biological and socio-psychological processes that are hard to explain and predict using analytical approaches, in order to afford the scientific community a deeper understanding of these processes.
2. They extend or enhance human activities—most often work activities, compensating for limitations in terms of resources such as space and time.

Serving the first purpose, *multi-agent social simulation* is a handy tool for deriving properties of social reality that are hard to establish analytically, hard to observe, or hard to manipulate in nature. MASs therefore allow social scientists not only to intimately observe existing scenarios otherwise difficult to analyze, but also to test potential scenarios and thus explore phenomena that might not exist in actuality, but are theoretically possible.

An already classic example of using MASs for social simulation is the EOS project (Doran et al. 1994). The aim of EOS was to explain the rapid growth in social complexity in Upper Paleolithic France. The project led to the discovery of social phenomena such as overcrowding or clobbering. Second, simulation results suggested that coalition formation was facilitated by perceptual abilities and hampered by resource complexity. Furthermore, it advanced the understanding of the mechanisms through which individual beliefs influence society.

Social simulation using MASs has also allowed policy makers to relatively accurately anticipate (predict) social effects of certain policies (e.g., the Freshwater Integrated Resource Management with Agents (FIRMA) Project; see Wooldridge 2009).

The other major purpose of the MAS paradigm in social sciences we mentioned regarded *virtualizations of organizational structure and processes through MASs*. For the rest of this chapter, we will dwell on this second purpose.

2.2.2 Social Ability in Agent Design

Regarding how agents are endowed with social ability, first note that agent-oriented programming (AOP) mimics social interactions: as in a human society, they influence one another's mental states (consisting of beliefs and commitments) through communication acts, and make new commitments based on their current mental states (Shoham and Leyton-Brown 2008).

Then again, how do we make agents cooperate? Note that agents are not designed with a built-in concern for the "greater good" of the artificial societies they are part of—on the contrary, they are conceptualized as non-cooperative, strategic entities (Sandholm 1999). Therefore, doesn't cooperation in agents appear to contradict the very definition of agency? Inspiration for addressing this issue can also be found in the way human societies are governed: negotiation protocols and strategies in MASs can be designed such that agents obtain benefits from collaborations, thus inducing agents to cooperate. As is the case with individual choice in human societies, agents often maximize their own benefits by choosing to adhere to groups or teams (thus forming coalitions). Cooperation is *individually rational* for an agent if the agent is not worse off in a coalition than on his own. This results in agents opportunistically forming or adhering to network structures whenever such alliances are susceptible of generating added value. This behavioral pattern constitutes the object of coalitional (cooperative) game theory (Shoham and Leyton-Brown 2008).

As artificial societies, MASs are designed to serve the global good of the population as a whole, given individual *social choice* (usually expressed through voting). To this end, MAS designers may use social welfare functions that map aggregated individual preferences to outcomes, and optimize resource allocation through *mechanism design* (usually through auctions) (Shoham and Leyton-Brown 2008; Wooldridge 2009).

2.3 Virtual Enterprises (VEs) as an Application Area of the MAS Paradigm

As aggregates of self-interested individuals that work together for common goals, organizations easily lend themselves to being modeled as MASs. An established example of meta-model for artificial organization in MASs is Aalaadin (Ferber and Gutknecht 1998). It is based on three main concepts: agent, group, and role. An agent can belong to one or more groups and handle one or more roles contained by groups.

Example models of architectures for artificial organizations that invite agent-based solutions are (Cunha and Putnik 2006):

- Supply Chain Management (SCM);
- Extended Enterprise;
- Agile Enterprise Manufacturing;
- Virtual Enterprise (VE)/Virtual Organization (VO);

- BM_VEARM Agile/Virtual Enterprise (A/VE);
- One-Product Integrated Manufacturing (OPIM).

In this chapter, our focus is on the VE and A/VE models, which are very illustrative of the applicability of the MAS paradigm and have generated considerable research interest (Camarinha-Matos and Afsarmanesh 2001; Cunha and Putnik 2006; Protogeros 2007).

2.3.1 VEs: Definitions, Properties, Integration in the MAS Framework

VEs are temporary alliances (i.e., dynamic networks) of organizations and/or individuals that share resources, risks, and markets in a virtual environment with the goal of exploiting a business opportunity (Protogeros 2007). Virtuality in this context refers to overcoming geographical and legal barriers—even if a VE has no physical correspondent, it appears so to the outside world by simulating a single entity. That is, the infrastructure (physical, legal, hardware, etc.) is hidden. Typically, VEs are not vertically integrated and have no physical headquarters. Such cooperation between enterprises is made possible by the advancements in information and communication technologies (ICT) (Cunha and Putnik 2006). A more general concept is the virtual organization (VO), i.e., an alliance of organizations not necessarily driven by commercial purposes (Camarinha-Matos and Afsarmanesh 2001). However, research questions regarding VOs are essentially the same as for VEs. As a particular case, the latter enjoyed the most attention, and we will refer to VE research throughout the paper, with the mention that much of the review (though not all, as we will argue in the Conclusion and Perspectives section) applies to VOs as well.

Through VEs, independent players on the market join their knowledge and competencies to supply a specified good or service (Protogeros 2007). As opportunistic alliances, VEs join competitors with the aim of generating added value: partners join the alliance and retreat depending upon whether or not they add value to the VE. Each VE partner attempts to simultaneously maximize its own profit and the one of the VE (Cunha and Putnik 2006). Such partnerships are especially opportune when the product they attempt to deliver neither has hitherto been demanded, nor is it guaranteed to be demanded henceforth. Essential characteristics of VEs include (Protogeros 2007):

- Temporary;
- Geographically distributed;
- Goal-oriented;
- Commitment-based;
- Based on communication and information flow;
- Sharing skills, costs, and profits;
- Decentralized.

These features make **Agent-Oriented Software Engineering (AOSE)** a very suitable approach to developing VE platforms. Moreover, VEs regularly form through auctions and, due to their self-interested nature, negotiate throughout the life cycle of the VE in order to reach agreements. Such a desired behavior naturally leads to the adoption of agent-based solutions (Protogeros 2007).

2.3.2 VEs and Web Services

The Service-Oriented Architecture (SOA) is regarded by many authors as appropriate for modeling VEs (Protogeros 2007; Wooldridge 2009). Web services are web-based software systems meant to provide specific functionalities upon request by other software systems over the network. To this end, they are discoverable by other software systems and capable of interacting with them. The key characteristic of a web service is interoperability, i.e., platform- and implementation-independence. Therefore, the analogy between web services and VE is quite straightforward: a VE is composed of "service agents" that produce and consume services from one another in an electronic market. Mapping web service concepts to the VE terminology, the service requester becomes the *VE Initiator* (i.e., the agent that announces its requirements on the market), service providers become *interested partners*, and *partner selection* (i.e., the process through which the VE is configured) is a particularization of service matchmaking (Protogeros 2007).

2.3.3 The Holonic Perspective

In many cases, VE designers implicitly or explicitly adopt a holonic perspective (Mella 2009; Ulieru et al. 2002). In fact, in the Japanese literature, VEs/VOs are referred to as **holonic networks**, or **enterprises**, or **organizations**, with the mention that holonic networks also include related concepts such as Holonic Manufacturing Systems or Agile Manufacturing Systems (Mella 2009).

A *holon* is an entity that simultaneously represents a whole for its parts and a part of a larger whole. When it observes its interior, it perceives itself an independent whole; when it observes its exterior, it perceives itself as a dependent part. Holons are simultaneously independent head-holons with respect to their parts and dependent member-holons with respect to whole they belong to. They are emergent, self-transcendent entities that facilitate creative change. Holons are vertically integrated into arborescent recursive structures called holarchies (Mella 2009).

Holons are autonomous interactive entities, which makes them suitable candidates for integration in the MAS framework (Ulieru et al. 2002). Applying the holonic perspective to an individual organization, the organization may be thought of as modular holarchy. Individual members of the organizations may be viewed as base holons (i.e., leaves in the holarchy), integrated into larger holons such as departments or divisions, which in turn form an even larger holon—the organization (Mella 2009).

VEs may be conceptualized as interconnected holarchies. VE design reflects the principle of inter-firm specialization: each firm engages in highly specialized, narrow activities, meant to be integrated in complex chains of input/output. While VE partners collaborate in order to achieve a common goal, they also seek to preserve their competitive advantages, to keep their know-how regarding products and/or processes private. The concept of holarchy alone is not enough to model both cooperation and competition: while MASs can be purely competitive, interactions in a holarchy are necessarily governed by cooperation. In order for competition to also be modeled, horizontal relationships between holons may be conceptualized as forming holonic networks, in which case the notion of co-opetition comes into play (Ulieru et al. 2002).

Collaboration in a holonic network can be achieved using a mediator design pattern. A mediator holon is a holon that acts as a broker to its sub-holons and as a facilitator (i.e., an interface, a representative) for the exterior. An example behavior of the mediator may be as follows: it sets a goal (to be achieved by the holon it represents), then decomposes the goal into sub-tasks, and finally clusters resources offered by the global electronic market to which it distributes the sub-tasks (Ulieru et al. 2002).

2.3.4   Major Concerns in Applied VE Research

VEs are basically meant to decrease production costs and time to market, concomitantly with increasing the quality (robustness, applicability, etc.) of the products (Cunha and Putnik 2006). VEs emerged in response to trends in economics aimed at addressing these goals, such as outsourcing and strategic alliances, and understanding organizations through a combined technological and socio-organizational perspective. It is worth mentioning, however, that VEs are not meant to totally replace human decisions, but to complement and support these decisions (Protogeros 2007).

Mainstream research in the VE field is primarily concerned with two main problems:

- Designing and implementing architectures for **VE platforms**, usually electronic markets. Note that a VE is, by definition, a temporary and dynamic entity. Hence, what VE researchers focus on developing are not particular VEs, but software environments that allow the creation and storage of reusable modules (implementing VE concepts such as clients, brokers, partners, etc.) that may constitute the building blocks of as many VEs as possible. The goal of VE platforms is to support the ad-hoc creation (in response to a specific market opportunity), and sometimes the operation and dynamic reconfiguration of VEs.
- Proposing **methods and algorithms** for solving problems characteristic to the phases of a VE life cycle. For example, VE formation is usually framed as a multi-objective (Pareto) optimization problem and most commonly solved using genetic algorithms (see, e.g., Ding et al. 2006; Wang et al. 2009; Zhang et al. 2013).

Simply put, VE research is concerned with designing and implementing the functionalities associated with each stage of evolution of a VE, as well as infrastructures to support these functionalities. VE researchers have originally identified four major phases in the life cycle of a VE (Camarinha-Matos and Afsarmanesh 2001):

- **Creation** (configuration, formation), wherein essential functionalities include partner selection and contract negotiation;
- **Operation**, where communication and coordination issues are of primary interest;
- **Evolution**, or reconfiguration (with inherently the same functionalities required in the creation phase);
- **Dissolution**, where functionalities of interest include negotiating liability clauses and assessing the performance of partners (with the generated information to be included in partner selection for future VEs).

Initially, VE creation garnered the most attention among researchers. However, contemporary market dynamics require enterprises to be more agile, more versatile, more reactive (Protogeros 2007). Product redesign or simply evolving in the product life cycle, partners leaving the VE because either party (the VE or the partner) ceased to produce added value—all these events count as reasons for reconfiguration (Cunha and Putnik 2006).

The response for high reconfiguration dynamics is **agility**: the capacity of an enterprise to maintain its competitive advantages (lower prices, superior quality, faster delivery, etc.) by rapidly adapting to changes in the market. Agility is about surviving and thriving in a competitive environment characterized by continuous unpredictable change and by client-designed products (Cunha and Putnik 2006). To this end, some researchers (e.g., Cunha and Putnik 2006) called for a shift in research focus from creation to reconfiguration. Though creation and reconfiguration functionalities are basically the same, high reconfiguration dynamics imposes a redefinition of performance criteria for these functionalities. Requirements for an agile enterprise include quasi-instantaneous flexible access to optimal resources, independence of spatial barriers, and minimizing reconfiguration time (Cunha and Putnik 2006; Protogeros 2007).

An established generic architecture for electronic markets that supports agility is BM_VEARM (BM_Virtual Enterprise Architecture Reference Model; Cunha and Putnik 2006). At the core of the BM_VEARM stand the concepts of *resource* and *resource brokerage*. The Market of Resources offers support for a basic BM_VEARM-compliant scenario for VE formation: one enterprise (the Client, which is the A/VE Owner) contracts other enterprises (Resource Providers) through a Broker with the aim of obtaining a specific good or a service. The Market of Resources is an infrastructure that supports the creation and reconfiguration of VEs according to the extended VE life cycle proposed by Cunha and Putnik (2006), where contractualization with the market and VE redesign can take place many times during VE operation (see Fig. 1).

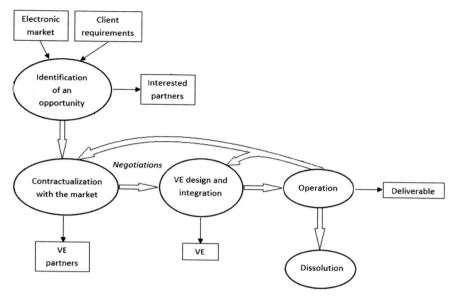

Fig. 1 VE lifecycle (adapted from Cunha and Putnik 2006; Protogeros 2007)

3 Results

In this section, we present an example MAS platform that provides a suitable infrastructure for VE applications. Of the multitude of extant MAS frameworks, we have chosen to describe the **JANUS platform** (Cossentino et al. 2007; Gaud et al. 2008). The reason for our choice is this platform's organizational approach to MAS design. Moreover, JANUS offers native management of holons.

JANUS is intended to bridge the gap between holonic MAS (HMAS) design and implementation. It supports nested hierarchies through holon composition: agents are atomic (non-composed) holons nested under groups, which in turn are nested under super-holons. This design is based on the CRIO meta-model, wherein the concepts of organization and role are first-class entities.

The key concepts in JANUS are the following (see Fig. 2):

- *Organization*—implemented as a singleton class, an organization is composed of roles and instantiated in the form of groups;
- *Group*—each group implements an organization and possesses a subset of roles of the organization it belongs to;
- *Role*—two types of role are modeled: roles that represent compositions of tasks, requiring a set of capacities; holonic roles assigned to holons in their super-holons;
- *Holon*—each holon possesses a set of capacities, which allow it to adhere to those groups where it can play roles according to its capacities; agents are atomic (non-composed) holons;

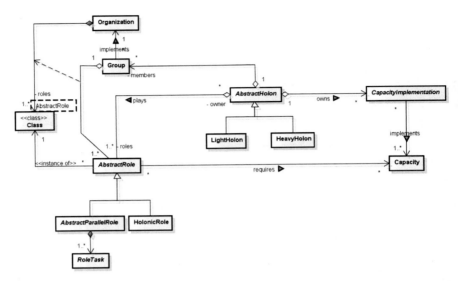

Fig. 2 Class diagram illustrating the metamodel JANUS is based on (adapted from Gaud et al. 2008)

- *Capacity*—models the skills or competences of a holon, as well as the requirements of a role; a holon's access to a role is conditioned upon implementing the required capacities.

Inside each composed holon, each sub-holon may play one of the following holonic roles, which dictate its interactions with other holons:

- *Head*—the decision maker;
- *Representative*—the holon's interface with the outside world;
- *Part*—member of this holon only;
- *Multi-part*—also part of other holons.

Cossentino et al. (2010) also offered an instantiation of this design for an existing automotive plant in Eastern France. The authors implemented a HMAS for traffic simulation meant to assist transportation-related decision-making. According to the holonic decomposition principle, the *plant* is an **organization** that can be decomposed into *zones*, which constitute smaller organizations and may in turn be decomposed into even smaller *zones*, with *road segments* as examples of lowest level *zones*. *Zone simulations* are organizations that can be recursively decomposed into smaller *Zone simulations*, with *Traffic simulation* organizations at the lowest level of the holarchy. A *Traffic simulation* organization is composed of three **roles**: *crossroad*, *road segment*, and *road user*. **Holons** playing such roles form **groups** that implement the *Traffic simulation* organization. These roles may be played by *connection holons*, *zone holons* (which are both **light holons**), and *vehicles* (**heavy holons**), respectively. The **holonic role** of *Head* is automatically assigned to holons playing the *road segment* role, which then choose a *Representative* among them. The role of *road user* is composed of *motion parameter computation* and *crossing decision* **role tasks**, requiring the **capacity** to *choose the route*, which is owned by *vehicles*.

JANUS's meta-model provides increased modularity through the holonic approach, and reusability through the separation of roles from agents and of organizations from groups. Modern design and implementation of electronic markets supporting VEs as holonic networks is thus facilitated.

4 Conclusion and Perspectives

Although MAS research has reached a relatively high level of maturity, there is still room for improvement even at a fundamental level. For example, advances in both MAS theory and applied research could be stimulated by the wider adoption of organizational and holonic perspectives. Such models extend the notion of agency to hierarchical structures and provide adequate frameworks for modeling increasingly complex interactions between agents.

Second, as suggested by Protogeros (2007), the future of VEs seems indissolubly linked to developments in the web service technology, SOA, and ontologies. The more interoperable web services become, the easier the collaboration of multiple development teams; the greater the advances in the comprehensiveness and integrity of ontologies, the easier the formation and reconfiguration of VEs for an ever increasing set of problem domains. Another potential fruitful direction for future research would be adjusting the level of autonomy in VE formation and reconfiguration. Further efforts should also be devoted to developing an integrated electronic market architecture based on both SOA and holonic principles.

Last but not least, in our opinion, more research attention should be directed towards the more general VOs, as the number of beneficiaries would dramatically increase if public services (such as public administration, health, or education) would also capitalize on the developments in the VE field, but the inherently different goals of activities not related to production may require a twist in the perspectives underlying the design of VE applications.

Overall, improving the quality and availability of VE/VO applications could have a substantial positive impact on all economic sectors by creating a more flexible, more dynamic market able to quickly adapt to ever changing client demands.

References

Camarinha-Matos, L. M., Afsarmanesh, H.: Virtual enterprise modeling and support infrastructures: applying multi-agent system approaches. In: Luck, M., Marik, V., Stpankova, O., Trappl, R. (eds.). LNAI, vol. 2086, pp. 335–364. Springer (2001)

Carley, K.M., Gasser, L.: Computational organization theory. In: Weiss, G. (ed.) Multiagent Systems: A Modern Approach to Distributed Modern Approach to Artificial Intelligence. The MIT Press, Cambridge, Massachusetts (1999)

Cossentino, M., Gaud, N., Galland, S., Hilaire, V., & Koukam, A.: A holonic metamodel for agent-oriented analysis and design. In: Marik, V., Vyatkin, V., Colombo, A. W. (eds.) Holonic and Multi-Agent Systems for Manufacturing: Third International Conference on Industrial Applications of Holonic and Multi-Agent Systems, HoloMAS 2007, vol. 4659, pp. 237–246. Springer Science & Business Media (2007)

Cossentino, M., Gaud, N., Hilaire, V., Galland, S., Koukam, A.: ASPECS: an agent-oriented software process for engineering complex systems. Auton. Agents Multi-Agent Syst. **20**(2), 260–304 (2010)

Cunha, M.M., Putnik, G.: Agile Virtual Enterprises: Implementation and Management Support. IGI Global, Hershey, New York (2006)

Ding, H., Benyoucef, L., Xie, X.: A simulation-based multi-objective genetic algorithm approach for networked enterprises optimization. Eng. Appl. Artif. Intell. **19**(6), 609–623 (2006)

Doran, J., Palmer, M., Meliars, P.: The EOS project: modelling upper palaeolithic social change. In: Doran, J., Gilbert, N. (eds.) Simulating Societies: The Computer Simulation of Social Phenomena, pp. 195–221. UCL Press, London (1994)

Ferber, J., Gutknecht, O.: A meta-model for the analysis and design of organizations in multi-agent systems. In: Proceedings of the 3rd International Conference on Multi Agent Systems ICMAS'98, pp. 128–135. IEEE Computer Society Washington, DC, USA (1998)

Fisher, M., Hepple, A.: Executing logical agent specifications. In: Bordini, R.H., et al. (eds.) Multi-Agent Programming: Languages, Tools and Applications, pp. 3–30. Springer, Heidelberg (2009)

Gaud, N., Galland, S., Hilaire, V., Koukam, A.: An Organisational Platform for Holonic and Multi-agent Systems. In: Hindriks, K.V., Pokahr, A., Sardina, S. (eds.) ProMAS 2008. LNAI, vol. 5442, pp. 104–119. Springer-Verlag, Berlin, Heidelberg (2009)

Hindriks, K.V.: Programming rational agents in GOAL. In: Bordini, R.H., et al. (eds.) Multi-Agent Programming: Languages, Tools and Applications, pp. 119–158. Springer, Heidelberg (2009)

Huhns, M.N., Stephens, L.M.: Multiagent systems and societies of agents. In: Weiss, G. (ed.) Multiagent Systems: A Modern Approach to Distributed Modern Approach to Artificial Intelligence. The MIT Press, Cambridge, Massachusetts (1999)

Mella, P.: The holonic perspective in management and manufacturing. Int. Manag. Rev. **5**(2), 19–30 (2009)

Pokahr, A., Braubach, L.: A survey of agent-oriented development tools. In: Bordini, R.H., et al. (eds.) Multi-Agent Programming: Languages, Tools and Applications, pp. 289–332. Springer, Heidelberg (2009)

Poutakidis, D., Winikoff, M., Padgham, L., Zhang, Z.: Debugging and testing of multi-agent systems using design artefacts. In: Bordini, R.H., et al. (eds.) Multi-Agent Programming: Languages, Tools and Applications, pp. 215–258. Springer, Heidelberg (2009)

Protogeros, N.: Agent and Web Service Technologies in Virtual Enterprises. IGI Global, Hershey, New York (2007)

Sandholm, T.W.: Distributed rational decision making. In: Weiss, G. (ed.) Multiagent Systems: A Modern Approach to Distributed Modern Approach to Artificial Intelligence. The MIT Press, Cambridge, Massachusetts (1999)

Shoham Y., Leyton-Brown K.: Multiagent Systems: Algorithmic, Game-Theoretic, and Logical Foundations. Cambridge University Press (2008)

Sierhuis, M., Clancey, W.J., van Hoof, R.J.J.: Brahms. In: Bordini, R.H., et al. (eds.) Multi-Agent Programming: Languages, Tools and Applications, pp. 73–118. Springer, Heidelberg (2009)

Ulieru, M., Brennan, R.W., Walker, S.S.: The holonic enterprise: a model for Internet-enabled global manufacturing supply chain and workflow management. Integr. Manuf. Syst. **13**, 538–550 (2002)

Wang, Z.J., Xu, X.F., Zhan, D.C.: Genetic algorithm for collaboration cost optimization-oriented partner selection in virtual enterprises. Int. J. Prod. Res **47**(4), 859–881 (2009)

Weiss, G.: Multiagent Systems: A Modern Approach to Distributed Modern Approach to Artificial Intelligence. The MIT Press, Cambridge, Massachusetts (1999)

Wooldridge, M.: An Introduction to Multiagent Systems, 2nd edn. Wiley (2009)

Zhang, Y., Tao, F., Laili, Y., Hou, B., Lv, L., Zhang, L.: Green partner selection in virtual enterprise based on Pareto genetic algorithms. Int. J. Adv. Manuf. Technol. 1–17 (2013)

On Consistency and Incoherence in Analytical Hierarchy Process and Intertemporal Choices Models

Fabrizio Maturo, Viviana Ventre and Angelarosa Longo

Abstract A rational choice is based on conditions of coherence, that are universally accepted rules to which a decision-maker must comply in expressing his/her preferences. In this paper, we focus on two different approaches in decision-making processes, i.e. the Analytical Hierarchy Process and Intertemporal Choices models, highlighting the consistency conditions usually adopted. After a general discussion on consistence and incoherence in the framework of these two different approaches, we show that sometimes it is preferable to weaken or reinforce coherence conditions according to the specific context.

Keywords AHP models · Intertemporal choice models · Coherence
Allowing inconsistency

1 Introduction

A decision model is defined as *"rational"* or *"coherent"* when the decision maker, while operating according to his own convictions, respects some universally accepted rules called *"consistency (or coherence) criteria"*.

The views expressed in the studies are those of the author Angelarosa Longo and do not involve the responsibility of the institution to which she belong.

F. Maturo (✉)
Department of Management and Business Administration, "G. D'Annunzio" University of
Chieti-Pescara, Viale Pindaro, 65100 Pescara, Italy
e-mail: f.maturo@unich.it

V. Ventre
Department of Mathematics and Physics, "Luigi Vanvitelli" University of Campania, Viale
Lincoln, 5, 81100 Caserta, Italy
e-mail: viviana.ventre@unicampania.it

A. Longo
Servizio Gestione Circolazione Monetaria, Banca d'Italia, Via Nazionale, 91, 00184 Rome, Italy
e-mail: angelarosa.longo@bancaditalia.it

C. Flaut et al. (eds.), *Models and Theories in Social Systems*, Studies in Systems,
Decision and Control 179, https://doi.org/10.1007/978-3-030-00084-4_18

Consider some examples:

(a) In *subjective probability*, a decision maker D can subjectively assign probabil-
     ities to a set of n events E_i, that are a partition of the certain event, but must
     follow the coherent conditions that characterize the subjective probability:

$$(P) \quad p_i \geq 0, \quad \sum p_i = 1.$$

(b) In the *theory of decisions*, in determining the preferences among the objects of
     a set S, a decision maker D can follow his opinions. For each X, Y belonging
     to S, we set $(X > Y)$ or $(Y < X)$ if D prefers X to Y, we also place $(X \sim Y)$ if X
     and Y are equally preferred. Moreover $(X > \sim Y)$ or $(Y < \sim X)$ mean that $(X < Y$
     or $X \sim Y)$. The conditions of consistency universally accepted in the theory of
     decisions are:

     (D1) $\forall X, Y \in S$ only one of the following properties applies: $X > Y, X < Y, X \sim Y$;
     (unicity)
     (D2) $\forall X, Y, Z \in S, (X > \sim Y, Y > \sim Z) \Rightarrow X > \sim Z$. (transitivity)

(c) In financial mathematics, we indicate with (C, t) the possession of capital C at
     a time t. The universally accepted consistency conditions are as follows:

     (MF1) $\forall C_1, C_2, t, C_1 < C_2 \Rightarrow (C_1, t) < (C_2, t)$;
     (MF2) $\forall t_1, t_2, C, t_1 < t_2 \Rightarrow (C, t_1) > (C, t_2)$;
     (MF3) $\forall t_1, t_2, C_1, C_2, h > 0, (C_1, t_1) < (C_2, t_2) \Leftrightarrow (C_1, t_1 + h) < (C_2, t_2 + h)$.

(d) In inferential statistics, the consistency of a null hypothesis with respect to the
     observed facts is measured based on the value assumed by a certain "statistic".

     In some contexts, the need arises to weaken or reinforce the conditions of coher-
ence; for example, some studies consider non-additive measures (Maturo et al. 2006a,
2006b, 2010).

     Decision making in neuroscience (an emerging area of research whose goal is
to integrate research in neuroscience and behavioral decision-making) has call into
question the theories of choice which assume that decisions derive from an assessment of the future outcomes of various options and alternatives through some type
of cost-benefit analyses, which ignore influence of emotions on decision-making.
This field of study explores the neural "*road map*" for the physiological processes
intervening between knowledge and behavior, and the potential interruptions that
lead to a disconnection between what one knows and what one decides to do.

     The studies of decision-making in neurological patients who can no longer process
emotional information normally suggest that people make judgments not only by
evaluating the consequences and their probability of occurring, but also and even
sometimes primarily at a gut or emotional level (Bechara 2004). In a series of studies
(see, e.g., Bechara et al. 1997; Damasio 1994; Nussbaum 2001) using a gambling
task, it emerges that individuals with emotional dysfunction tend to perform poorly
compared with those who are endowed with intact emotional processes. Bechara
et al. (1997) has demonstrated that normal people possess anticipatory SCRs (Skin

Conductance Response) - indices of somatic states - which represent unconscious biases that are linked to prior experiences with reward or punishment and produce inconsistent preferences. These biases alarm the normal subject about selecting a disadvantageous course of action, even before the subject becomes aware of the goodness or badness of the choice he is about to make (Longo and Ventre 2015).

On the other side, in some contexts, individuals that are deprived of normal emotional reactions might make better decisions than normal individuals because of the loss of self-control, as found by Damasio in studying the behavior of people with ventromedial prefrontal damage (Damasio 1994). Temptations are manifestations of loss of self-control and in many cases induce disadvantageous behavior. Indeed, as far as temptation increases the best long run interest of the problem solver conflicts with his short run desires, moreover impulsive behavior may fail to evaluate the consequences of his behavior appropriately (Ventre and Ventre 2012, Urban and Hoskova-Mayerova 2017). Other evidences suggest that even relatively mild negative emotions, that do not result in a loss of self-control, can play a counterproductive role among normal individuals in some situations. When gambles that involve some possible loss are presented one at a time, most people display extreme levels of risk aversion toward the gambles, a condition known as *myopic loss aversion*. Shiv et al. (2005) have shown that individuals that are deprived of normal emotional reactions might, in certain situations, make more advantageous decisions than those not deprived of such reactions; thus, the lack of emotional reactions may lead to more advantageous decisions. Inconsistent preference is the greatest contradiction of rational theory in decision problems. This behavior can be typically seen in psychiatric disorders (alcoholism, drug abuse), but also in more ordinary phenomena (overeating, credit card debt) (Ventre and Ventre 2012).

The problem of decision making arises in several different contexts (e.g. Hošková-Mayerová et al. 2013; Maturo and Maturo 2014; Longo and Ventre 2016; Maturo and Hošková-Mayerová 2017; Hošková- Mayerová and Maturo 2018; Cruz Rambaud et al. 2015, 2017; Maturo and Di Battista 2018); however, this research limits its attention to the problem of consistence and incoherence in the framework of two approaches, i.e. the Analytical Hierarchy Process and Intertemporal Choices models. Specifically, we show that sometimes it is preferable to weaken or reinforce coherence conditions according to the specific context. The remainder of this paper is structured as follows: Sects. 2 and 3 introduces the Analytical Hierarchy Process method and its inconsistency issues. Sections 4 and 5 focus on Intertemporal Choice models. Finally, a discussion on the role of information and mental models in decision processes is presented in Sect. 6.

2 Coherence in Analytical Hierarchy Process (AHP)

Analytical Hierarchy Process (AHP) is a procedure to support a decision process. It allows us synthesizing intangible and tangible, objective and subjective, rational and emotional components in decision-making situations. Here the decision-making process is carried out by a scale of reasoning that captures the perceived reality. AHP involves decomposing a complex decision into a hierarchy of objectives, criteria, sub-criteria, and alternatives by comparing the properties of each possible pair of elements, at each level, as a matrix, and synthesizing the priorities (Saaty 1988; Szczypińska and Piotrowski 2009; Marcarelli et al. 2013; Maturo and Ventre 2018). This decomposition into a hierarchy makes the problem easier to understand, given that it is possible to analyze each sub-problem separately. *"Once the hierarchy is built, the decision-makers systematically evaluate its various elements by comparing them to one another two at a time, with respect to their impact on an element above them in the hierarchy"* (Arabameri 2014). Saaty (1988, 2008) has exposed that, in specifying the preferences in AHP method, it is necessary the measurement of intangible factors that establishes priorities between them in a consistent way. It allows processing and converting comparisons in numerical values that are part of a model. However, these comparisons are not always consistent. According to Amirabadi (2011), AHP method requires that decision-maker remains consistent in making pairwise comparisons among numerous decision criteria, and it becomes more difficult when alternatives increase. In fact, the ability of humans, for accurately expressing their knowledge, decreases with increasing problem complexity. Thus, as the number of criteria in AHP increases, decision-makers are likely to make inconsistent judgments during pairwise comparisons.

Let us analyze the idea of consistency in the AHP method of pairwise comparison. With reference to a given decision criterion, a finite set S of alternatives are compared. If (a, b) is an ordered pair of alternatives of S, if a is preferable to b, the decision maker D assigns to the pair (a, b) a score $s \in \{2, 3,\ldots, 9\}$ which indicates the extent to which a is preferable to b. If a is indifferent to b, D gives to (a, b) the score 1, and finally, if b is preferable to a, the reciprocal of the score assigned to (b, a) is attributed to (a, b).

Let us generalize the notation $a > \sim b$. For every $s \in \{1, 2, \ldots, 9\}$ let us write $a >_s$ b to denote that the pair (a, b) has the score s. For $s = 1$ we have $a \sim b$, and, for $s > 1$, we say that a is preferable to b at level s.

Then we introduce the *"strong transitivity"* as follows:

(D2S) $\forall a, b, c \in S, (a >_x b, b >_y c) \Rightarrow a >_z c, z \geq \max\{x, y\}.$ (strong transitivity)

In general, if a decision maker performs the pairwise comparison on three elements of s, a matrix of the following type is obtained:

The pairwise comparison table is consistent if, for every row, the scores are in the same order. If $a >_x b$, $b >_y c$ with $x \geq 1$ and $y \geq 1$, in the second row there is the order $a \geq b \geq c$. To be able to have the same order in the first and third rows must then be $z \geq x$ and $z \geq y$, respectively. Then $z \geq \max\{x, y\}$. In conclusion, the strong transitivity that we have introduced is equivalent to the fact that, for each subset of three elements of S, the consistency of the comparison in pairs is worth.

If $|S| = 2$, each evaluation is automatically consistent, for $|S| = 3$ it is easy enough to get coherent evaluations, but for $|S| = n > 3$, as n increases, the conditions for having a perfect coherence increase enormously as n increases, for which it is necessary to

Table 1 Example of pairwise comparison table in AHP

	a	b	c
a	1	x	z
b	1/x	1	y
c	1/z	1/y	1

define an index of incoherence evaluation and to be satisfied if this index does not exceed a pre-established threshold.

The condition of strong transitivity can be imposed even if the evaluations belong to a totally ordered set S. It can also be extended to the case where S is a lattice by replacing max{x, y} with sup{x, y}. From a numerical point of view the perfect coherence is obtained only if the rows of the matrix of Table 1 are proportional. This is equivalent to the more restrictive condition of (D2S):

(D2N) $\forall a, b, c \in S, \left(a >_x b, b >_y c\right) \Rightarrow a >_z c, z = xy.$ (numerical transitivity)

If $|S| = n > 3$ then it is practically impossible to obtain the numerical transitivity, for which Saaty has proposed a global index of consistency CI that measures the discrepancy between the matrix M of the pairwise comparison and an ideal matrix with the proportional lines (see also Bozóki and Rapcsák 2008).

Saaty's formula is:

$$CI = (\lambda_{max} - n)/(n - 1)$$

where λ_{max} is the maximum eigenvalue of the matrix M.

Saaty suggests that the matrix M should be considered consistent if CI<0.1. However, in Fig. 1, with reference to the data considered in Table 1, the values of CI are reported with the variation of x and y fixed the value of z.

The Consistency Index is a widely accepted rule to measure the degree of consistency of pairwise comparisons in AHP methodology. Once the inconsistency is located, the decision maker would be advised to revise his preferences to eradicate the measurement errors. Once the consistency is established, the model would be considered valid, unless the decision maker deems it unsatisfactory for reasons outside the model (Dadkhah and Zahedi 1993). AHP method is a very useful way of helping decision makers to make complex decisions, to make decisions in the school environment, especially in the school of the first cycle of education, where teaching is also linked to social problems (Delli Rocili and Maturo 2013; 2017). The possibility to identify and reduce inconsistence to an acceptable level may improve the quality of decisions and the reliability of the analysis.

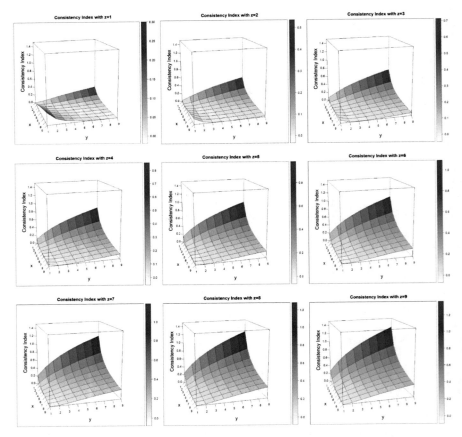

Fig. 1 Values of CI depending on the variation of x and y, fixed the value of z (see also Table 1)

3 Allowing Inconsistency in AHP

Inconsistency may arise due to the use of inadequate information, inadequate time to concentrate on the problem at hand, inexperience in preference elicitation, or the ambiguity in the dimensions of the problem attributes. The analysis and treatment of inconsistency in input data depends on which assumption better characterizes the problem at hand. If we assume that the inconsistency is because of a measurement error, then a move towards consistency is desirable (Dadkhah and Zahedi 1993). However, in the presence of complex problems, for example in the social sphere, the search for excessive coherence can lead to simplistic procedures and conclusions. Various experimental research done in some works related to teaching in the school of the first cycle and social issues lead to believe that for $n > 3$ it is appropriate to accept even a weaker condition of consistency $0.1 \leq CI < 0.2$ (Forcini et al. 2013; Maturo et al. 2014; Maturo and Zappacosta 2017).

In many situations it has been well proved that AHP intransitivity has a clear foundation and is not the consequence of errors or lack of knowledge. Intransitive decisions may not be necessarily irrational and, in fact, may even represent better the preferences of a decision maker. Some authors, like Linares (2009), defend that the best way to handle the AHP methods is to simply accept intransitivity and explore its implications. Allowing for inconsistency, in the input matrix of pairwise comparisons, has made the AHP an interesting decision analysis approach, but it also has been the source of a great deal of confusion. If we assume that the input matrix is always consistent, the computation of local relative weights reduces to a simple normalization of one of the columns of the input matrix, and all computation methods would produce an identical result (Dadkhah and Zahedi 1993).

In real-life decision problems, pairwise comparison matrices are rarely consistent; in effect, to explain the real world, inconsistency is necessary (example: A beats B, B beats C, and C beats A). In any case, decision makers are interested in the level of consistency of the judgments, which somehow expresses the goodness or "*harmony*" of pairwise comparisons totally, because inconsistent judgments may lead to senseless decisions (Bozóki and Rapcsák 2008).

Traditionally, some theories against the inconsistency have axioms that prohibit the intransitivity. However, Karapetrovic and Rosenbloom (1999) expressed that the posterior revision of pairwise comparisons is generally not very successful, improving consistency is not really improving significantly the validity of the model compared to the "*true*" preferences of the decision makers. Often a decision maker has not made a mistake. Usually, decision makers are quite conscientious in evaluating the pairwise comparisons. Therefore, the major problem in solving multiattribute decision problems is to find weights which order the attributes and reflect the recorded judgments in situations where the judgments are not transitive, that is, the judgment matrix is inconsistent.

4 Coherence in Intertemporal Choice Models for Decision-Making Process

In the context of Intertemporal Choice Models, the decision problem is constituted by an intertemporal choice where the subject must choose between several alternatives with different amounts of money and maturities. In this framework, traditionally, to choose the best option, the discounted utility (DU) is widely adopted (Loewenstein and Prelec 1992). This model is used to describe how people make intertemporal choices, it is based on the utility of some future event as perceived at the present time as opposed to at the time of its occurrence. It states that economic agents make intertemporal choices maximizing intertemporal utility function of their consumption profiles $(C_t, C_{t+1}, \ldots, C_{t+n})$, which can be described by the following form:

$$(DU) \quad U^t(C_t, C_{t+1}, \ldots, C_{t+n}) = \sum_{k=0}^{n} D(k)u(C_{t+k})$$

where $u(C_{t+k})$ is the utility of the capital (or non-monetary asset) C_{t+k}, available at time $t+k$ and $D(k)$ is the discount factor for C_{t+k}. The DU model assumes $D(k) = (1+r)^{-k}$, with r a constant discount rate. Because of the discount factor formula, we obtain the classical consistency conditions of financial mathematics apply (MF1), (MF2), and (MF3). So, with the same temporal options and the same information, later preferences confirm earlier preferences (Longo et al. 2015; Olivieri et al. 2016). In the DU method, a priori, it is expected that subjects care more about the present than the future and that decisions taken by employing this model will be consistent in the future.

5   Allowing Inconsistency in Intertemporal Choice Models

One of the most common anomalies in intertemporal choices is inconsistency (Loewenstein and Prelec 1992) in the preference order caused by the temporal distance that is when the discount rate used is not constant across situations and over time. Specifically, the most common case of non-constant time discount is when there is a negative relation between the discount rate and the delay time of the reward (Loewenstein and Thaler 1989). Traditionally, the inconsistency in intertemporal choice is described as a change in the preference between two (monetary or not) rewards when they are postponed the same period. For example, an individual may prefer an apple today (0) to two apples in one week (1 w). But, if both rewards are postponed for one year (1 y), it is possible that the subject chooses two apples available in one year and one week (1 y + 1 w) over one apple available in one year.

On the other hand, it is well known that the indifference curves in intertemporal choices are analytically given by the concept of a discount function F(t). In this context, the negative slope of an indifference line in the interval $[t_1, t_2]$ can be calculated through the so-called unitary discount defined as

$$D(t_1, t_2) = 1 - f(t_1, t_2), \tag{1}$$

where $f(t_1, t_2) = \frac{F(t_2)}{F(t_1)}$ is the discount fraction corresponding to the discount function F(t) in the interval $[t_1, t_2]$. This ratio represents the patience in the interval $[t_1, t_2]$ whilst $D(t_1, t_2)$ represents impatience. Therefore, the inconsistency in intertemporal choice means an increase or a decrease of the patience in intervals of the same length.

However, considering that

$$f(t_1, t_2) = exp\left\{-\int_{t_1}^{t_2} \delta(x)\,dx\right\} \tag{2}$$

where $\delta(x) = -\frac{d\ln F(z)}{dz}\Big|_{z=x}$ is the instantaneous discount rate at time x, we can state that a necessary and sufficient condition for inconsistency is that the instantaneous discount rate increases or decreases. Observe that (x) does not represent patience but impatience. Thus, when referring to inconsistency, we can distinguish between increasing and decreasing impatience,

The standard model supposes, by simplicity that the preferences are consistent over time. To explain the inconsistency two theories have been employed (Calderon Güémez et al. 2004):

- Conductivity theories. It relies in the attempt to parameterize the preferences in function of time by introducing an additional discount rate to the standard model of discounted utility that increases the significance ascribed to the present in every moment.
- Axiomatic theories. The traditional framework of intertemporal choice is conserved, and it is proposed to aggregate axioms that admit the possibility of inconsistency.

The model of discount utility assumes that the discount in the utility between two periods is given by a constant, it means that a subject discount equal to advance/delay a payment a day regardless what day is. Experiments show a trend to reduce the "*premium*" as the delay maturity increases (common difference effect). To model this type of situations it is necessary that the discount rate value would be non-constant in function of time. It would have to decrease when the period between the present moment and the delay moment increases. The standard model uses a single discount factor in the utility function. To correct the possible inconsistencies in the utility function it is necessary to develop a model able to show the preferences by the present (by giving a greater weight to the utility in the closer present than in the future). Therefore, there is considerable agreement among psychologists and economists that the notion of exponential discounting should be replaced by some form of hyperbolic discounting, which can point out the delay effect (or present bias), that is the tendency of the individuals to increasingly choose a smaller-sooner reward over a larger-later reward as the delay occurs sooner in time (Longo and Ventre 2015).

A hyperbolic utility function that represents the preferences that contain a bias for the present, could be given by:

$$U^t(u_t, u_{t+1} \ldots u_T) \equiv \delta^t u_t + \beta \sum_{\tau=t+1}^{T} \delta^\tau u_\tau \tag{3}$$

for all value of t, where: $0 < \beta \leq 1$, and $0 < \delta \leq 1$.

As we can see, it differs from the traditional model of discounted utility in the aggregation of a constant parameter β that decreases the utility in the future (preferences skewed by the present moment). Another function that represents a similar behavior to Eq. 3 is:

$$V^t(u_t, u_{t+1} \ldots u_T) \equiv \alpha \delta^t u_t + \sum_{\tau=t+1}^{T} \delta^\tau u_\tau \tag{4}$$

where: $\alpha \geq 1$ and $0 < \delta \leq 1$.

In Eq. 4, it is easier to observe how the instantaneous utility in the moment t generates a higher utility that the future instantaneous utilities, given that u_t is multiplied by a constant factor $\alpha \geq 1$.

In intertemporal choices, the preference for more immediate rewards per se is not always irrational or inconsistent; addicts' behavior is clinically problematic, but economically rational when their choices are time-consistent (if they have large discount rates with an exponential discount function) (Ventre 2014; Longo and Ventre 2015). The hyperbolic model allows to avoid inconsistency, but it not eliminate it; in effect, it is able to represent the inconsistency against the exponential model. Hyperbolic discounting (Green and Myerson 1996) allows us to explain several inconsistent behaviors.

Let's take an example. Suppose we have a capital C at time t and a capital $C+k$ at time $t+h$. We indicate with D(t) the discount factor at time t.

If $D(t) = (1 + r)^{-t}$, with r constant, we have:

$$U^t(C)/U^{t+h}(C + k) = \left[C(1 + r)^{-t}\right]/\left[(C + k)(1 + r)^{-t-h}\right] = C(1 + r)^h/(C + k).$$

This ratio does not depend on t and therefore the coherence condition (MF3) is respected.

Instead if $D(t) = (1 + rt)^{-1}$, we have:

$$U^t(C)/U^{t+h}(C + k) = [C/(1 + rt)]/[(C + k)/(1 + rt + rh)].$$

The ratio between the two utilities depends on t. Let $G(t) = U^t(C)/U^{t+h}(C + k)$. Moreover, let $C + k = (1 + \rho)C$, where $\rho > 0$.

With simple calculations we get

$$G(t) < 1, \text{ and so } U^t(C) < U^{t+h}((1 + \rho)C) \text{ iff } t > (rh - \rho)/(\rho r).$$

Then:

(a) If $rh < \rho$ then for every t we have $U^t(C) < U^{t+h}(C+k)$;
(b) If $rh = \rho$ then $U^t(C) < U^{t+h}(C+k)$ for $t > 0$ and $U^0(C) = U^h(C+k)$;
(c) If $rh > \rho$ there exists a time $t_0 > 0$ in which there is a change in preferences: for $t < t_0$, (C, t) is preferable to $(C+k, t+h)$, for $t = t_0$ they are equally preferred, and for $t > t_0$, $(C+k, t+h)$ is preferable to (C, t).

So we can check the consistency. If we want to maintain the condition of intertemporal consistency (MF3) we just have to impose $rh < \rho$; on the contrary, if $rh > \rho$ there exists a time $t_0 > 0$ in which there is a change in preferences.

An inconsistency source may be the subadditivity of the discount financial law or, in the case of stationary laws, those that represent a rapid discount. Both situations are exemplified by the hyperbolic discount. A discount financial law $A(t, p)$ is subadditive if, in every p, s, t, being $p < s < t$, it is verified that (Muñoz 2004):

$$A(t, p) > A(t, s) A(s, p).$$

In the case of a discount financial law to p, the discount speed in the interval (s, t) to an increment of Δs it is represented by $v(s, t, \Delta s)$ is:

$$v(s, t, \Delta s) = \frac{\delta(s, s + \Delta s, p)}{\delta(t, t + \Delta t, p)}$$

being, $\delta(s, s + \Delta s, p)$ the discount rate in the interval $(s, s + \Delta s)$ and $\delta(t, t + \Delta t, p)$ the discount rate in $(t, t + \Delta t)$ of the financial law $A(t, p)$. Thus, the hyperbolic discount (that is stationary and describes the inconsistent individual behavior) is defined by a discount law subadditive and of rapid discount (Muñoz 2004).

Behavioral neuroeconomics and econophysical studies have proposed two discount models to better describe the neural and behavioral correlates of impulsivity and inconsistency in intertemporal choice.

Q-exponential discount model. This function has been proposed and examined for subjective value V(D) of delayed reward:

$$V(D) = \frac{A}{\exp_q(k_q D)} = A / \left[1 + (1 - q)k_q D \right]^{\frac{1}{1-q}}$$

where D denotes a delay until receipt of a reward, A the value of a reward at $D = 0$, and kq a parameter of impulsivity at delay $D = 0$ (q-exponential discount rate) and the q-exponential function is defined as:

$$\exp_q(x) = (1 + (1 - q))^{\frac{1}{1-q}}$$

The function can distinctly parametrize impulsivity and inconsistency (Longo and Ventre 2015).

Quasi-hyperbolic discount model. Behavioral economists have proposed that the inconsistency in intertemporal choice is attributable to an internal conflict between "multiple selves" within a decision maker. As a consequence, there are (at least) two exponential discounting selves (with two exponential discount rates) in a single human individual; and when delayed rewards are at the distant future (> 1 year), the self with a smaller discount rate wins, while delayed rewards approach to the near future (within a year), the self with a larger discount rate wins, resulting in preference reversal over time. This intertemporal choice behavior can be parametrized in a quasi-hyperbolic discount model (also as a β-δ model). For discrete time τ (the unit assumed is one year) it is defined as:

$$F(\tau) = \beta \delta^t \quad (\text{for } \tau = 1, 2, 3, \ldots) \quad \text{and} \quad F(0) = 1 \quad (0 < \beta < \delta < 1).$$

A discount factor between the present and one-time period later (β) is smaller than that between two future time-periods (δ). In the continuous time, the proposed model is equivalent to the linearly-weighted two-exponential functions (generalized quasi-hyperbolic discounting):

$$V(D) = A[w \exp(-k_1 D) + (1 - w) \exp(-k_2 D)]$$

where w, $0 < w < 1$, is a weighting parameter and k1 and k2 are two exponential discount rates (k1 < k2). Note that the larger exponential discount rate of the two k2, corresponds to an impulsive self, while the smaller discount rate k1 corresponds to a patient self (Longo and Ventre 2015).

6 A Discussion on the Role of Information and Mental Models in Decision Processes

In decision processes, the information known and the way in which it is represented play an important role, this is called *"framing effect"*: different levels of availability or quality of information about a problem influence the dynamics of the problem-solving process, and can produce different reaction, so it can lead to different solution of the same case (Longo and Ventre 2015). Human thinking is intransitive, because e.g. new knowledge requires that we change our minds. Thus, nowadays decision makers must take into consideration more and more alternatives. In such situations, the conservation of consistency of the judgment matrix in AHP is impossible (Szczypińska, Piotrowski 2009). Consistency and inconsistency must coexist given that: consistency allows capturing the knowledge about the world, meanwhile inconsistent behaviors make able to change subject's mentality in the face of new information. In this way, Consistency Index must have a value of 10%, not too much high, but enough to allow the change (Saaty 1988).

The previous AHP example assumes that the decision maker has sufficient and correct information to correctly perform pairwise comparisons between the alternatives and criteria. Thus, if the decision maker does not have the right information, he can make a choice completely away from his real preferences. In the same way, the information effects on the intertemporal choices. The way the phenomenon is described and how the questions are phrased, and graphs used influence the behavior and preferences of the agents, so their utility function and their ability of self-control and finally their choice.

We can affirm that the failure of rational models is first linked with the way in which the problem is presented to the agents. The first aspect that influence the final choice in a decision process is the model of the problem. Saaty (2008), basing on the study of the cognitive psychologist claims that judgment involves the comparison between the stimulus of the immediate information and the impressions in memory of similar stimuli, that is some information held in short-term memory, information about some former comparison stimuli or about some previously experienced measurement scale. Hence, also impressions in memory have a key role, the so called *"priming effect"*. They influence all positive and negative biases in an individual choice. They are another important part in explanation of why emotions do not have always positive or negative effects on decision process. They impact on the mental models which are informal models, quickly constructed by problem solvers, which go on constantly during problem solving. Mental models help us to relate cause and effect, but often in a highly simplified and incomplete way. They are always influenced by our preferences and our personal experiences. Thus, they can be extremely limiting (Longo and Ventre 2015). This explains why emotions do not have always positive or negative effects on decision process and why impulsivity generates sometimes positive and sometimes negative effects (Longo and Ventre 2016). In this situation, in AHP, the personal experience of a decision maker can lead to different results in terms of pairwise comparisons among the criteria. In an intertemporal choice, even if rewards have place in different maturities, in the same way impressions in memory produce inconsistence. Hyperbolic discounting predicted many mechanisms of self-control to model the effects derived from mental models. However, the hyperbolic model, as well as the exponential one, is only a special case of interpreting reality. Common sense highlights how people, when are in front of identical short-term opportunities, perform only sometimes self-control (Longo and Ventre 2015). An interesting research direction for future studies could be focusing on decision making problems using hyperstructures (e.g. see Vougiouklis et al. 1997; Massouros 1999; Hedayati and Ameri 2005; Hedayati 2009; Massouros and Massouros 2011; Vougiouklis 2011; Chvalina and Hoskova-Mayerova 2012; Novak 2012; Bakhshi and Borzooei 2013; Nikolaidou and Vougiouklis 2014; Lygeros and Vougiouklis (2013); Rezaei et al. 2015; Vougiouklis and Vougiouklis 2016; Hoskova-Mayerova, 2017a; 2017b; Maturo and Maturo 2017; Antampoufis and Hoskova-Mayerova 2017; Al Tahan et al. 2018).

References

Amirabadi S. (2011): A mathematical method for preventing inconsistency in decision maker's comparisons. In: XI Symposium of Analytic Hierarchy/Network Process. Sorrento, Naples (Italy)

Antampoufis, N., Hoskova-Mayerova, S.: A brief survey on the two different approaches of fundamental equivalence relations on hyperstructures. Ratio Math. **33**, 47–60 (2017). doi:http://dx.doi.org/10.23755/rm.v33i0.388

Al Tahan, M., Hoskova-Mayerova, S., Davvaz, B.: An overview of topological hypergroupoids. J. Intell. Fuzzy Syst. **34**(3), 1907–1916 (2018)

Arabameri, A.: Application of the analytic hierarchy process (AHP) for locating fire stations: case study Maku City. Merit Res. J. Art Soc. Sci. Humanities **2**(1) (2014)

Bakhshi, M., Borzooei, R.: Ordered Polygroups. Ratio Math. **24**(1), 31–40 (2013)

Bechara, A.: The role of emotion in decision-making: evidence from neurological patients with orbitofrontal damage. Brain Cogn. **55**, 30–40 (2004)

Bechara, A., Damasio, H., Tranel, D., Damasio, A.R.: Deciding advantageously before knowing the advantageous strategy. Science **275**, 1293–1295 (1997)

Bozóki, S., Rapcsák, T.: On Saaty's and Koczkodaj's inconsistencies of pairwise comparison matrices. J. Global Optim. **42**(2), 157–175 (2008)

Calderon Güémez, G., Elbittar, A.A., Lever Guzmán, C.: Inconsistencias en la teoría de la elección intertemporal: un enfoque económico. In Santoyo Velasco, C., Vázquez Pineda (Coord.). Teoría Conductual de la Elección: Decisiones Que Se Revierten, pp. 207–229 (2004)

Chvalina, J., Hoskova-Mayerova, S.: General ω-hyperstructures and certain applications of those. Ratio Math. **23**(1), 3–20 (2012)

Cruz Rambaud, S., Maturo, F., Sánchez Pérez, A.M.: Expected present and final value of an annuity when some non-central moments of the capitalization factor are unknown: theory and an application using R. Stud. Syst. Decis. Control **104**, 233–248 (2017). https://doi.org/10.1007/978-3-3 19-54819-7_16

Cruz Rambaud, S., Maturo, F., Sánchez, A.: Approach of the value of an annuity when non-central moments of the capitalization factor are known: An R application with interest rates following normal and beta distributions. Ratio Math. **28**(1), 15–30 (2015). https://doi.org/10.23755/rm.v2 8i1.25

Dadkhah, K.M., Zahedi, F.: A mathematical treatment of inconsistency in the analytic hierarchy process. Math. Comput. Model. **17**(4), 111–122 (1993)

Damasio, A.R.: Descartes' Error: Emotion, Reason, and the Human Brain. Grosset/Putnam, New York (1994)

Delli Rocili, L., Maturo., A.: Teaching mathematics to children: social aspects, psychological problems and decision-making models. Interdisciplinary Approaches in Social Sciences. Editura Universitatii A.I. Cuza, Iasi, Romania (2013)

Delli, Rocili L., Maturo, A.: Social problems and decision making for teaching approaches and relationship management in an elementary school. Stud. Syst. Decis. Control **104**, 81–94 (2017). https://doi.org/10.1007/978-3-319-54819-7_7

Forcini, S., Maturo A., Ventre, A.G.S.: The role of folk dance in the processes of individual and social wellbeing: a comparison with other popular recreational activities through models of decision theory and game theory. Proc.: Soc. Behav. Sci. **84**, 1750–1756 (2013)

Green, L., Myerson, J.: Exponential versus hyperbolic discounting of delayed outcomes: risk and waiting time. Am. Zool. **36**(4), 496–505 (1996)

Hedayati, H., Ameri, R.: Construction of k-Hyperideals by P-Hyperoperations. Ratio Math. **15**, 75–89 (2005)

Hedayati, H.: On properties of fuzzy subspaces of vectorspaces. Ratio Math. **19**(1), 1–10 (2009)

Hoskova-Mayerova, S.: An overview of topological and fuzzy topological hypergroupoids. Ratio Math. **33**, 21–38 (2017). https://doi.org/10.23755/rm.v33i0.389

Hošková-Mayerová, Š.: Quasi-order hypergroups and T-hypergroups. Ratio Math. **32**, 37–44 (2017). https://doi.org/10.23755/rm.v32i0.333

Hošková-Mayerová, Š., Maturo, A.: Decision-making process using hyperstructures and fuzzy structures in social sciences. Stud. Fuzziness Soft Comput. **357**, 103–111 (2018). https://doi.or g/10.1007/978-3-319-60207-3_7

Hošková-Mayerová, Š., Talhofer, V., Hofmann, A.: Decision-making process with respect to the reliability of geo-database. Stud. Fuzziness Soft Comput. **357**, 179–194 (2013). https://doi.org/1 0.1007/978-3-642-35635-3_15

Karapetrovic, S., Rosenbloom, E.S.: A quality control approach to consistency paradoxes in AHP. Eur. J. Oper. Res. **119**(3), 704–718 (1999)

Linares, P.: Are inconsistent decisions better? An experiment with pairwise comparisons. Eur. J. Oper. Res. **193**(2), 492–498 (2009)

Loewenstein, G., Thaler, R.H.: Anomalies: intertemporal choice. J. Econ. Persp. **3**(4), 181–193 (1989)

Loewenstein, G., Prelec, D.: Anomalies in intertemporal choice: evidence and an interpretation. Q. J. Econ. **107**(2), 573–597 (1992)

Longo, A., Ventre, V.: Influence of information on behavioral effects in decision processes. Ratio Math. **28**, 31–43 (2015). https://doi.org/10.23755/rm.v28i1.26

Longo, A., Ventre, V.: The level of information held by a problem solver influences decision processes. J. Math. Econ. Finan. **2**(2) (2016)

Longo, A., Squillante, M., Ventre, A.G.S., Ventre, V.: The intertemporal choice behavior: the role of emotions in a multi-agent decision problem. Atti Accad. Pelorit. Pericol. Cl. Sci. Fis. Mat. Nat. **93**(2) (2015)

Lygeros, N., Vougiouklis, T.: The LV-hyperstructures. Ratio Math. **25**, 59–66 (2013)

Marcarelli, G., Simonetti, B., Ventre, V.: Analyzing AHP Matrix by Robust Regression. Studies in Computational Intelligence, pp. 223–231 (2013). https://doi.org/10.1007/978-3-642-32903-6_16

Massouros, C.G., Massouros, G.G.: The transposition axiom in hypercompositional structures. Ratio Math. **21**, 75–90 (2011)

Massouros, G.G.: Hypercompositional structures from the computer theory. Ratio Math. **13**, 37–42 (1999)

Maturo, A., Zappacosta, M.G.: Mathematical models for the comparison of teaching strategies in primary school. Sci. Philos. **5**(2), 25–38 (2017). https://doi.org/10.23756/sp.v5i2.392

Maturo, A., D'Orazio, A., De Crescenzo, A.: A decision model for the sustainable protection of human rights in Italian Prison system. Sci. Philos. **2**(2), 91–100 (2014)

Maturo, A., Squillante, M., Ventre, A.G.S.: Consistency for assessments of uncertainty evaluations in non-additive setting. In: Metodi, Modelli e Tecnologie dell'Informazione a Supporto delle Decisioni, Franco Angeli, Milano, pp. 75–88 (2006b)

Maturo, A., Squillante, M., Ventre, A.G.S.: Consistency for Non Additive Measures: Analytical and Algebraic Methods in Computational Intelligence, Theory and Applications, pp. 29–40. Springer, Berlin (2006)

Maturo, A., Squillante, M., Ventre, A.G.S.: Coherence for Fuzzy Measures and Applications to Decision Making. Studies in Fuzziness and Soft Computing, vol. 257, pp. 291–304. Springer, Berlin (2010). https://doi.org/10.1007/978-3-642-15976-3_17

Maturo, F., Di Battista, T.: A functional approach to Hill's numbers for assessing changes in species variety of ecological communities over time. Ecol. Ind. **84**, 70–81 (2018). https://doi.org/10.1016/j.ecolind.2017.08.016

Maturo, A., Maturo, F.: On Some Applications of the Vougiouklis Hyperstructures to Probability Theory. Ratio Math. **33**, 5–20 (2017). https://doi.org/10.23755/rm.v33i0.372

Maturo, A., Maturo, F.: Finite geometric spaces, steiner systems and cooperative games. Analele Universitatii "Ovidius" Constanta - Seria Matematica, **22**(1), (2014) https://doi.org/10.2478/auom-2014-0015

Maturo, F., Ventre, V.: Consensus in Multiperson Decision Making Using Fuzzy Coalitions. Studies in Fuzziness and Soft Computing, vol. 357, pp. 451–464 (2018). https://doi.org/10.1007/978-3-319-60207-3_26

Maturo, F., Hošková-Mayerová, Š.:: Fuzzy Regression Models and Alternative Operations for Economic and Social Sciences. Studies in Systems, Decision and Control, vol. 66, pp. 235–247 (2017). https://doi.org/10.1007/978-3-319-40585-8_21

Muñoz Torrecillas, M.J.: Anomalías en la elección intertemporal: obtención de la tasa social de descuento. Universidad de Almería (PhD Thesis) (2004)

Nikolaidou, P., Vougiouklis, T.: The Lie-Santilli admissible hyperalgebras of type An. Ratio Math. **26**(1), 113–128 (2014)

Novak, M.: EL-hyperstructures: an overview. Ratio Math. **23**(1), 65–80 (2012)

Nussbaum, M.C.: Upheavals of Thought. The Intelligence of Emotions. Cambridge University Press, Cambridge (2001)

Olivieri, M., Squillante, M., Ventre, V.: Information and intertemporal choices in multi-agent decision problems. Ratio Math. **31**(1), 3–24 (2016). https://doi.org/10.23755/rm.v31i0.316

Rezaei, A., Saeid, A.Borumand, Smarandache, F.: Neutrosophic filters in BE-algebras. Ratio Math. **29**(1), 65–79 (2015). https://doi.org/10.23755/rm.v29i1.23

Saaty, T.L.: What is the analytic hierarchy process? Mathematical Models for Decision Support, pp. 109–121. Springer, Berlin Heidelberg (1988)

Saaty, T.L.: Decision making with the analytic hierarchy process. Int. J. Serv. Sci. **1**(1), 83–98 (2008)

Shiv, B., Loewenstein, G., Bechara, A., Damasio, H., Damasio, A.R.: Investment behavior and the negative side of emotion. Psychol. Sci. **16**, 435–439 (2005)

Szczypińska, A., Piotrowski, E.W.: Inconsistency of the judgment matrix in the AHP method and the decision maker's knowledge. Physica A **388**(6), 907–915 (2009)

F. Maturo et al.

Urban, R., Hošková-Mayerová, Š.: Threat life cycle and its dynamics. Deturope **9**(2), 93–109 (2017)

Ventre, A.G.S., Ventre, V.: The intertemporal choice behavior: classical and alternative delay discounting models and control techniques. Atti Accad. Pelorit. Pericol. Cl. Sci. Fis. Mat. Nat. **90**, Suppl. No. 1, C3 (2012)

Ventre, V.: The intertemporal choice behavior: the role of emotions in a multiagent decision problem. Ratio Math. **27**(1), 91–110 (2014). https://doi.org/10.23755/rm.v27i1.36

Vougiouklis, T.: Bar and theta hyperoperations. Ratio Math. **21**(1), 27–42 (2011)

Vougiouklis, T., Vougiouklis, S.: Helix-hopes on finite hyperfields. Ratio Math. **31**(1), 65–78 (2016). https://doi.org/10.23755/rm.v31i0.321

Vougiouklis, T., Spartalis, S., Kessoglides, M.: Weak hyperstructures on small sets. Ratio Math. **12**(1), 90–96 (1997)

On Some Applications of Fuzzy Sets for the Management of Teaching and Relationships in Schools

Šárka Hošková-Mayerová and Antonio Maturo

Abstract Often in the scholastic context the efficiency of the teaching depends to a large extent on the system of relationships established within the classroom. In this paper we show how the relatively recent mathematical theories on fuzzy sets and algebraic hyperstructures can make a significant contribution to the understanding of the system of relationships within the class. Furthermore, it is possible to evaluate the impact of interventions aimed at improving the system of relationships and therefore to establish a serene participation of the students in the learning process.

Keywords Relations and directed graph · Fuzzy relations
Algebraic hyperstructures · Strategies for teaching and managing relationships

1 Introduction

In recent decades, research on teaching efficiency has attracted the interest of world-wide scholars (e.g. Ceccatelli et al. 2013a, b, Fortuna and Maturo 2018; Maturo et al. 2018).

Various experiences have shown that, in school, especially in primary school, the teacher's preparation and teaching skills can not achieve acceptable results if there is no serene and collaborative relationship within the school (see e.g. Delli Rocili and Maturo 2013, 2017, 2018). The study of existing relationships between students is then a necessary prerequisite for to be able to plan interventions aimed at obtaining adequate teaching efficiency.

Š. Hošková-Mayerová (✉)
Department of Mathematics and Physics, University of Defence Brno,
Kounicova 65, 66210 Brno, Czech Republic
e-mail: sarka.mayerova@seznam.cz

A. Maturo
Department of Architecture, University "G. d'Annunzio" of Chieti-Pescara,
Viale Pindaro 42, 65127 Pescara, Italy
e-mail: antomato75@gmail.com

© Springer Nature Switzerland AG 2019
C. Flaut et al. (eds.), *Models and Theories in Social Systems*, Studies in Systems,
Decision and Control 179, https://doi.org/10.1007/978-3-030-00084-4_19

Let S be the set of the students of a given scholastic class K. A scientific study of relationships between students in the class leads to identify a finite set R of relations, and, for each $\rho \in R$, a directed graph (S, ρ) with vertices the students of the class and arcs the pairs $(x, y) \in S \times S$ such that $x \rho y$.

Social Relations in a school environment are described with a set of binary crisp relations by many authors (e.g. Moreno 1951, 1953; Sciarra 2007). The study of the most efficient teaching methodologies in relation to the system of relationships in a scholastic class was also examined in some recent works by Hošková-Mayerová (2011, 2014, 2016), Rosická and Hošková-Mayerová (2014), Svatoňová and Hošková-Mayerová (2017), Vougiouklis (2011) and Vougiouklis and Voougiouklis (2016).

In some of our papers (Hošková-Mayerová and Maturo 2013, 2016, 2017) it has been observed that a more in-depth knowledge of relationships in the scholastic environment can be obtained through fuzzy relations, which take into account the semantic uncertainty and the degree of intensity of relationships. In this work we resume and deepen the ideas introduced in our previous works, showing various points of view. We also show that an additional tool to represent and evaluate uncertainty is given by algebraic hyperstructures, which, considering operations with more results, are more flexible than ordinary operations to represent the possible results of agreements between individuals.

The work takes place according to the following itinerary. In Sect. 2 a formalization of the system of relationships in a school class is given. In Sect. 3, starting from the Moreno approach to social relations in a school class, some indicators are presented to evaluate the social group of students. An extension of Moreno's approach with three-valued logic is studied in Sect. 4.

Section 5 studies a fuzzy extension of Moreno's theory and, finally, Sect. 6 presents some conclusions and research perspectives.

2   A Formalization of the System of Relationships in a School Class

Let $S = \{s_1, s_2, \ldots, s_n\}$ be the set of students in a school class K. The system of relations within the class can be represented by a set $R = \{\rho_1, \rho_2, \ldots, \rho_m\}$ of relations that take into account at least three criteria of association: *play, work, leisure* (Delli Rocili and Maturo 2018). We will assume that every relation $\rho \in R$ is reflexive, i.e. $x \rho x$, $\forall x \in S$. Moreover we assume that, for every relation $\rho \in R$, also the inverse $\rho^{-1} \in R$.

For every relation ρ in S and for every $x, y \in S$, we write $x \rho y = 1$ if $x \rho y$ and $x \rho y = 0$ otherwise. In particular, the 0-relation (or null relation) is the relation ω such that $x \omega y = 0$, $\forall x, y \in S$ and the 1-relation (or total relation) υ satisfies the condition $x \upsilon y = 1$, $\forall x, y \in S$.

For each $\rho_j \in R$, we have:

- the *directed graph* (S, ρ_j) *associated* with ρ_j, having as vertices the students of the class K and arcs the pairs $(x, y) \in S \times S$ such that $x \rho_j y$;
- the *binary matrix* $M_j = (m^j_{rs})$ *associated* with ρ_j, with $m^j_{rs} = 1$ if $x \rho_j y$ and $m^j_{rs} = 0$ otherwise;
- the *symmetrical closure* of ρ_j, defined as the union $\rho_j \cup \rho_j^{-1}$, i.e. the relation σ_j such that $x \sigma_j y \Leftrightarrow (x \rho_j y$ or $y \rho_j x)$. The graph (S, σ_j) is the undirected graph associated with (S, ρ_j);
- the *binary matrix* $N_j = (n^j_{rs})$ *associated* with σ_j.

For every ordered pair (x, y) of students, if there exists a path of length k that goes from x to y, but not a path of length $k - 1$, we say that the relation ρ_j (resp. σ_j) from x to y is at level k and we write $x \rho_j^k y$ (resp. $x \sigma_j^k y$). Obviously, since there are n elements, the maximum level of a relation is less than or equal to n-1.

A first study is to verify if the class is divided into groups without relations between them. This leads to the consideration of the connected components of the graphs (S, ρ_j) and (S, σ_j). Let $\rho_j^0 = \cup \{\rho_j^k, k = 1, 2, ..., n - 1\}$, $\sigma_j^0 = \cup \{\sigma_j^k, k = 1, 2, ..., n - 1\}$. They are the *transitive closure* of ρ_j and σ_j, respectively. Then the quotient sets S/ρ_j^0 is the set of the strongly connected components of (S, ρ_j) and the elements of S/σ_j^0 are the connected component of (S, σ_j) (and of (S, ρ_j)).

If (S, ρ_j) is a strongly connected digraph then in some way the school class can be considered balanced with respect to the relation ρ_j; if instead there are more strongly connected components, the school class is divided and it is necessary to intervene to improve the relationship between the students. Moreover, if (S, σ_j) is a connected graph then there is some connection with a path of (S, σ_j) for each pair of students; however this connection can have a complex interpretation with respect to the relation ρ_j, in particular it can be one-way. The number of connected components and the number of strongly connected components indicate, from two different points of view, the degree of fragmentation of the class with respect to the relations ρ_j.

From the matrix point of view, it is necessary to introduce into the set $\Im^n$ of binary matrices, of the same order n, the following operations. For any $A = (a_{rs})$, $B = (b_{rs})$ belonging to $\Im^n$:

- *bounded addition* $+_b$. $A +_b B = (\min\{1, a_{rs} + b_{rs}\})$;
- *bounded multiplication.* $A \cdot_b B = (\min\{1, c_{rs}\})$, where c_{rs} is the generic element of the usual matrix product $A \cdot_b B$;
- *bounded subtraction* $-_b$. $A -_b B = (\max\{0, a_{rs} - b_{rs}\})$.

Moreover, the bounded power of exponent n of A is the bounded product $A^{n,b}$ of n matrices equal to A.

If $M_j = (m^j_{rs})$ is the binary matrix associated with ρ_j, then $M_j^{(2)} = M_j^{2,b} -_b M_j$ is the binary matrix associated to ρ_j^2, and, by recurrence, $M_j^{(k)} = M_j^{k,b} -_b (M_j +_b M_j^{2,b} +_b ... +_b M_j^{k-1,b})$ is the binary matrix associated with ρ_j^k.

For every relation ρ in S, the contrary of ρ is the relation $-\rho$ such that, $\forall x, y \in S$, $x(-\rho)y = 1 - x \rho y$. For every pair (ρ_r, ρ_s) of relations in S we have:

- the *union* $\rho_r \cup \rho_s$ defined as the relation such that $\forall x, y \in S$, $x\ \rho_r \cup \rho_s\ y = \max\{x\ \rho_r\ y, x\ \rho_s\ y\}$;
- the *intersection* $\rho_r \cap \rho_s$ defined as the relation such that $\forall x, y \in S$, $x\ \rho_r \cap \rho_s\ y = \min\{x\ \rho_r\ y, x\ \rho_s\ y\}$;
- the *implication*, $\rho_r \rightarrow \rho_s$ defined as the the relation $\rho_r \rightarrow \rho_s = (-\rho_r) \cup \rho_s$.

Starting from the set R of relations introduced in the scholastic class, we can define:

- the *atoms associated* with R as the non-null relations $\alpha_1 \cap \alpha_2 \cap \ldots \cap \alpha_m$, with $\alpha_j \in \{\rho_j, -\rho_j\}$;
- the *Boolean algebra generated* by R as the set having as elements the null relation, the atoms and all the relations obtained as unions of atoms.

3   The Moreno Approach to the Social Relations in a School Class

In the Moreno approach the system of relationships considered in a school class is identified starting from a set of choices made by the students of a class, each of which belongs to a different sphere of interests, called *association criterion*: for example in the fields of school work, activities recreational, trust, esteem, feeling of friendship (Moreno 1951, 1953; Sciarra 2007; Hošková-Mayerová and Maturo 2013; Hošková-Mayerová and Maturo 2016, 2017; Delli Rocili and Maturo 2013, 2017, 2018). Each boy is subjected to a set of questions that tends to identify his preferences. Consider, for example, the following questions:

(Q1) (school work) With which classmates would you like to work to build a geographical map?
(Q2) (recreational activities) Which kids do you prefer to play during recreation?
(Q3) (trust) Which guys do you think can help you in case of difficulty?
(Q4) (estimate) Which friends do you think are worthy of being imitated for their skills?
(Q5) (friendship) Who would you like to invite for your birthday?

A positive integer $h < n$ is set, which represents the maximum number of choices a student can make. In particular, for $h = 1$, a boy can indicate only one classmate, for $h = n - 1$ he can indicate all. Experiences conducted in the classroom have shown that $h = 3$ can be a good choice (Delli Rocili and Maturo 2018).

After examining the answers given by the students to the questionnaire, with each question (QJ), a relationship ρ_j can be associated, defined as follows:

$$\forall x, y \in S,\ x\ \rho_j\ y \Leftrightarrow (x = y \text{ or } x \text{ has indicated } y);$$
$$\forall x, y \in S,\ x\ \rho_{j+5}\ y \Leftrightarrow (x = y \text{ or } y \text{ has indicated } x).$$

$R^+ = \{\rho_j, j = 1, 2, ..., 5\}$ is said to be the set of *active relations*, and $R^- = \{\rho_j, j = 6, 7, ..., 10\}$ is the set of *passive relations*. $R = R^+ \cup R^-$ is the set of relations that represent the system of relations in the school class K. In particular, $\rho_{j+5} = \rho_j^{-1}, j = 1, 2, ..., 5$.

From the Moreno approach we can calculate some indices (Moreno 1951, 1953; Sciarra 2007; Hošková-Mayerová and Maturo 2016). For each $x \in S$ we indicate with $c_j^1(x)$ the number of elements y such that $x \rho_j y, j = 1, 2, ..., 10$.

For $j = 1, 2, ..., 5$, the numbers $c_j^1(x)$ and $c_{j+5}^1(x)$ are called *degree of integration* and *degree of prestige* of x, respectively, with respect to the relation ρ_j. So, with respect to the relation ρ_j, every boy x is represented by the point of the plane $(c_j^1(x), c_{j+5}^1(x))$ and, as x varies in S we obtain a cloud of n points of the plane, with non-negative coordinates. By introducing a metric in R^2, we can evaluate many social aspects within the classroom, in particular, by activating cluster analysis methods we identify the groups in which students are divided (Delli Rocili and Maturo 2018).

For a discussion independent of the number of students in the class, it is better to replace the numbers $c_j^1(x)$ and $c_{j+5}^1(x)$ with with the quotients $\gamma_j^1 = c_j^1(x)/n$, $\gamma_{j+5}^1 = c_{j+5}^1(x)/n$, called *integration index* and *prestige index* of x, respectively.

Further integration and prestige measures can be obtained by considering, for each $j \in \{1, 2, ..., 5\}$, the relations of level k, ρ_j^k, $k = 1, 2, ..., n - 1$. For each $x \in S$ we indicate with $c_j^k(x)$ the number of elements y such that $x \rho_j^k y$, $j = 1, 2, ..., 10$; i.e. $c_j^k(x)$ is the number of elements y of the digraph (S, ρ_j) such that there exists a path of length k that goes from x to y, but not a path of length k-1. The numbers $c_j^k(x)$ and $c_{j+5}^k(x)$ are called *degree of integration at level* k and *degree of prestige at level* k of x, respectively, with respect to the relation ρ_j.

The *integration index* and *prestige index at level* k, are the numbers $\gamma_j^k(x) = c_j^k(x)/n$, $\gamma_{j+5}^k(x) = c_{j+5}^k(x)/n$, respectively.

If $\alpha_1, \alpha_2, ..., \alpha_{n-1}$ are positive real numbers with the conditions:

(1) $\alpha_1 > \alpha_2 > ... > \alpha_{n-1}$;
(2) $\alpha_1 + \alpha_2 + ... + \alpha_{n-1} = 1$;
(3) α_k it is the degree of importance attributed to the k level of the relation ρ_j;

then we can define the *global integration index* and *prestige index* associated to the relation ρ_j as the weighted averages:

$$\gamma_j^0 = \alpha_1 \gamma_j^1 + \alpha_2 \gamma_j^2 + ... + \alpha_{n-1}\gamma_j^{n-1}; \quad \gamma_{j+5}^0 = \alpha_1 \gamma_{j+5}^1 + \alpha_2 \gamma_{j+5}^2 + ... + \alpha_{n-1}\gamma_{j+5}^{n-1}$$

$$(3.1)$$

respectively.

4 An Extension of Moreno's Approach with Three-Valued Logic

Taking up the considerations made in the works of Reichenbach (1944), Fadini (1979), de Finetti (1970), Gentilhomme (1968), Maturo (1993) and others, we can consider in the set S of students a set R of three-valued relations. For every x, y∈S and for every $\rho \in R$, we assume that x ρ y can assume three values of truth: T = "true", F = "false" and a I = "third truth value", intermediate between true and false that, using the notation of Reichenbach, we will call "indeterminate". We will also place T = 1, F = 0, I = i, where i can be any real number greater than zero or less than 1. More generally, i can be an element of a lattice other than 0 and 1. In any case we assume F < I < T.

A simple application in the scholastic setting is obtained if each student of the school class is asked to answer each of the questions (Q1),..., (Q5) indicating both the students he wants to choose and those he wants to refuse. Then, for every relation ρ_j, j = 1, 2, ..., 5 and for any pair (x, y) of students, we put x ρ_j y = 1, 0, i, if x choose y, refuse y, or does not choose and does not refuse y, respectively.

Then, for each x∈S, and for any j ∈{1, 2, ..., 5}, we have the numbers:

- $c_j^1(x)$ = the number of elements y such that x ρ_j y = 1, called *degree of integration* if j ≤ 5, and *degree of prestige* if j > 5;
- $c_j^{1*}(x)$ = the number of elements y such that x ρ_j y = i, called *degree of weak integration* if j ≤ 5, and *degree of weak prestige* if j > 5;
- $c_j^{1**}(x)$ = the number of elements y such that x ρ_j y = 0, *degree of anti-integration* if j ≤ 5, and *degree of anti-prestige* if j > 5;

satisfying the conditions of non-negativity and such that their sum is equal to n.

In correspondence, for any j∈{1, 2, ..., 5}, we obtain the following indices:

- $\gamma_j^1(x) = c_j^1(x)/n$, $\gamma_{j+5}^1 = c_{j+5}^1(x)/n$, called *integration index* and *prestige index* of x, respectively;
- $\gamma_j^{1*}(x) = c_j^{1*}(x)/n$, $\gamma_{j+5}^{1*} = c_{j+5}^{1*}(x)/n$, called *weak integration index* and *weak prestige index* of x, respectively;
- $\gamma_j^{1**}(x) = c_j^{1**}(x)/n$, $\gamma_{j+5}^{1**} = c_{j+5}^{1**}(x)/n$, called *anti-integration index* and *anti-prestige index* of x, respectively;

satisfying the conditions of non-negativity and such that their sum is equal to 1.

In order to define degrees and indexes of level k > 1 it is necessary to extend the logical operations between relations to three-value relations. For every three-valued relation ρ in S, the contrary of ρ is the relation $-\rho$ such that, ∀x, y∈S, x(−ρ)y = 1, i, 0, if x ρ y = 0, i, 1, respectively. For every pair (ρ_r, ρ_s) of three-valued relations in S we define:

- the *union* $\rho_r \cup \rho_s$, as the relation such that ∀x, y∈S, x $\rho_r \cup \rho_s$ y = max{x ρ_r y, x ρ_s y};
- the *intersection* $\rho_r \cap \rho_s$, as the relation such that ∀x, y∈S, x $\rho_r \cap \rho_s$ y = min{x ρ_r y, x ρ_s y}.

As far as implication is concerned, the following conditions are generally accepted:

(extension of the binary implication)$F \rightarrow F = T$; $F \rightarrow T = T$; $T \rightarrow F = 0$; $T \rightarrow T = 1$;

(monotonicity) $\forall (H, K) \in \{T, I, F\}^2$, $(H \leq K) \Rightarrow (H \rightarrow K = T)$.

The majority of the authors define the "standard implication" $\rightarrow_s$ assuming the conditions of extension and monotonicity and the:

(one step rule) $I \rightarrow_s F = I$; $T \rightarrow_s I = I$.

Some authors define an "alternative implication" $\rightarrow_a$ by assuming the conditions of extension and monotonicity, but not the "one step rule" that is replaced by the conditions:

(circular rule) $I \rightarrow_a F = T$; $T \rightarrow_a I = F$.

From a different point of view the "quasi-implication" $\rightarrow_q$ is defined assuming:

(indetermination rule) $\forall (H, K) \in \{T, I, F\}^2$, $(H \neq T) \Rightarrow (H \rightarrow_q K = I)$.
(second - term rule) $\forall K \in \{T, I, F\}$, $T \rightarrow_q K = K$.

The quasi-implication is used in Subjective Probability (see e.g. de Finetti 1970; Maturo 1993), in which a conditional event E|H is defined as a three-valued proposition that assumes the values T, F, I, if E∩H, (−E)∩H or (−H) is true, respectively. If for E and H the value I is also allowed, then it results $E \mid H = H \rightarrow_q E$.

Let us be given the students x, y $\in$S and the three-value relation $\rho \in$R. If x ρ y is not known, we can consider the events:

$$H = \text{"x} \rho \text{y} \in \{0, 1\}\text{"}, E = \text{"x} \rho \text{y} = 1\text{"}.$$

In this case, the unknown value assumed by x ρ y can be identified with the conditional event E | H.

5   A Fuzzy Extension of Moreno's Approach

If we modify the questions (Q1),..., (Q5), substituting "With which classmate ..." the phrase "To what extent for each of the classmates ...", we can formalize the answers as fuzzy relations in S, i.e. functions defined in S × S and with values in [0, 1]. For more information on concepts related to fuzzy relationships, see, for example, Zadeh (1965, 1975), Klir and Yuan (1995), Ross (1995). Experiences of fuzzy logic in teaching and in the measure of uncertainty are also in Delli Rocili and Maturo (2015); Maturo and Porreca (2016). Applications of fuzzy sets and fuzzy relations in Social Sciences are in Ragin (2000), Maturo et al. (2008), Maturo and

Ventre (2017), Hoskova-Mayerova and Maturo (2016, 2017), Maturo and Hošková-Mayerová (2017), Maturo and Fortuna (2016), Maturo (2016).

We can therefore ask the following questions:

(Q1 ~) (fuzzy school work) To what extent for each of the classmates you like to work to build a geographical map?

(Q2 ~) (fuzzy recreational activities) To what extent for each of the classmates do you prefer to play during recreation?

(Q3 ~) (fuzzy trust) To what extent for each of the classmates do you think can help you in case of difficulty?

(Q4 ~) (fuzzy estimate) To what extent for each of the classmates do you think are worthy of being imitated for their skills?

(Q5 ~) (fuzzy friendship) To what extent for each of the classmates you like to invite for your birthday?

We must impose that the answers of the students belong to an ordered set, e.g. the scores from 0 to 10, or some qualitative values expressed by adjectives like "for nothing", "a little", "sufficiently", "very", completely". Subsequently we record the answers to each question (QJ ~) as values of a fuzzy relation $\rho_j\tilde{\ }$, dividing by 10 in the case of scores from 1 to 10, or by assigning numerical values from 0 to 1 to the qualitative values, setting, for example:

"for nothing $= 0$", "a little $= 0.25$", "sufficiently $= 0.50$", "very $= 0.75$", completely $= 1$".

We assume, for every $x \in S$, $x \, \rho_j\tilde{\ } \, x = 1$ (reflexivity). Moreover, for every $j \in \{1, 2, ..., 5\}$ we consider the inverse fuzzy relation $\rho_{j+5}\tilde{\ }$ defined as $y \, \rho_{j+5}\tilde{\ } \, x = x \, \rho_j\tilde{\ } \, y$.

$R^+ = \{\rho_j\tilde{\ }, j = 1, 2, ..., 5\}$ is said to be the set of *active fuzzy relations*, and $R^- = \{\rho_j\tilde{\ }, j = 6, 7, ..., 10\}$ is the set of *passive fuzzy relations*. $R = R^+ \cup R^-$ is the set of fuzzy relations that represent the system of relations in the school class K.

For every $x \in S$, we indicate with $c_j^{1\tilde{\ }}(x)$ the sum of the numbers $x \, \rho_j\tilde{\ } \, y$, $j = 1, 2, ..., 10$. For $j = 1, 2, ..., 5$, $c_j^{1\tilde{\ }}(x)$ and $c_{j+5}^{1\tilde{\ }}(x)$ are the *fuzzy degree of integration* and the *fuzzy degree of prestige* of x, respectively, associated with $\rho_j\tilde{\ }$. The *fuzzy index of integration* and the *fuzzy index of prestige* of x are the numbers $\gamma_j^{1\tilde{\ }}(x) = c_j^{1\tilde{\ }}(x)/n$, $\gamma_{j+5}^{1\tilde{\ }}(x) = c_{j+5}^{1\tilde{\ }}(x)/n$.

The construction of partial hypergroups and hyperpergroupoids associated with assigned fuzzy relationships can be done starting from a generalization of the Definition 2.1 from the paper (Hošková-Mayerová and Maturo 2018).

Definition 6.1 Let $\otimes$ be an operation in [0, 1], and ρ a reflexive fuzzy relation on S. For every $k \in (0, 1]$, we define:

- *attive hyperoperation* (eventually partial hyperoperation) at level k, associated with $(\otimes, \rho)$ the function $\otimes_\rho^{a,k}$: $(x, y) \in S \times S \to x \otimes_\rho^{a,k} y = \{z \in S: (x \, \rho \, z) \otimes (y \, \rho \, z) \geq k\}$;
- *passive hyperoperation* (eventually partial hyperoperation) at level k, associated with $(\otimes, \rho)$ the function $\otimes_\rho^{p,k}$: $(x, y) \in S \times S \to x \otimes_\rho^{p,k} y = \{z \in S: (z \, \rho \, x) \otimes (z \, \rho \, y) \geq k\}$;

- *circular hyperoperation* (eventually partial hyperoperation) at level k, associated with $(\otimes, \rho)$ the function $\otimes_\rho^{c,k}: (x, y) \in S \times S \to x \otimes_\rho^{c,k} y = \{z \in S: (x \rho z) \otimes (z \rho y) \geq k\}$;
- *inverse circular hyperoperation* (eventually partial hyperoperation) at level k, associated with $(\otimes, \rho)$ the function $\otimes_\rho^{i,k}: (x, y) \in S \times S \to x \otimes_\rho^{i,k} y = \{z \in S: (z \rho x) \otimes (y \rho z) \geq k\}$.

Particular cases:

Let $\otimes$ be the union $\cup$ in [0, 1], defined as $a \cup b = \max\{a, b\}$. From reflexivity, for all the above functions $\otimes_\rho^{s,k}$, $s \in \{a, p, c, i\}$, $\otimes_\rho^{s,k}(x, y) \supseteq \{x, y\}$. Then $(S, \cup_\rho^{s,k})$ is a closed quasi-hypergroup for every $k \in (0, 1]$. As $\cup$ is commutative, the associate active hyperoperation is commutative.

Let $\otimes$ be the intersection $\cap$ in [0, 1], defined as $a \cap b = \min\{a, b\}$. Then $(x \rho z) \otimes (y \rho z) = \min\{(x \rho z), (y \rho z)\}$. The function $\cap_\rho^{a,k}$ is a commutative partial hyperoperations and is an hyperoperations if and only if the following condition holds: $\forall x, y \in S, \exists t \in S:$ $x \rho t \geq k$ and $y \rho t \geq k$. The student t can be defined as "a *passive mediator at level* k" between x and y. So each pair of students must have at least one passive mediator at level k. The smaller the k the more the possibility of having a hypergroup is increased. By examining the answers to the questionnaires, one can try to identify the maximum value of k that allows to have a hypergroup. Similar considerations can be made for the functions $\cap_\rho^{s,k}$, $s \in \{p, c, i\}$.

There are many possible implication definitions in [0, 1], each of which satisfies the condition of being an extension of the classical implication in {0, 1} of the binary logic. This implies that, $\forall k \in (0, 1]$, for every implication $\to$, $\forall x, y \in S, x \to_\rho^{s,k} y \supseteq \{y\}$. Then every implication is associated with a hypergroupoid.

6 Conclusions and Research Perspectives

In some of our papers (see e.g., Hoskova-Mayerova and Maturo 2013, 2016, 2017) many types of hyperstructures, from points of view different from those considered in this work, have also been examined. From Sect. 5 we can see the use of fuzzy relations in place of binary relations seems to give much more precise and detailed results. The embarrassment of giving scores to all the companions can be overcome by giving qualitative judgments.

References

Ceccatelli, C., Di Battista, T., Fortuna, F., Maturo, F.: L'item response theory come strumento di valutazione delle eccellenze nella scuola. Sci. Philos. 1(1), 143–156 (2013)

Ceccatelli, C., Di Battista, T., Fortuna, F., Maturo, F.: Best Practices to Improve the Learning of Statistics: The Case of the National Olympics of Statistic in Italy. Procedia - Social and Behavioral Sciences, vol. 93, pp. 2194–2199 (2013b). https://doi.org/10.1016/j.sbspro.2013.186

de Finetti, B.: Teoria delle Probabilità, vol. I e II, Einaudi, Torino (1970)

Delli, R.L., Maturo, A.: Teaching mathematics to children: social aspects, psychological problems and decision-making models. In: Soitu, D., Gavriluta, C., Maturo, A. (eds.) Interdisciplinary Approaches in Social Sciences. Editura Universitatii A.I. Cuza, Iasi, Romania (2013)

Delli, R.L., Maturo, A.: Interdisciplinarity, logic of uncertainty and fuzzy logic in primary school. Sci. Philos. 3(2), 11–26 (2015)

Delli, R.L., Maturo, A.: Social problems and decision making for teaching approaches and relationship management in an elementary school. (2017). https://doi.org/10.1007/978-3-319-5481 9-7_7

Delli, R.L., Maturo, A.: Problems and Decision-Making Models in the First Cycle of Education, in press, to appear (2018)

Fadini, A.: Introduzione alla teoria degli insiemi sfocati. Liguori, Napoli (1979)

Fortuna, F., Maturo, F.: K-means clustering of item characteristic curves and item information curves via functional principal component analysis. Qual. Quant. (2018). https://doi.org/10.100 7/s11135-018-0724-7

Gentilhomme, M.Y.: Les ensembles flous en linguistiques. Cahiers de linguistique theorique et appliquée, Bucarest 5(47), 47–65 (1968)

Hošková-Mayerová, Š.: Operational program"Education for competitive advantage", preparation of study materials for teaching in English. Proc. – Soc. Behav. Sci. 15, 3800–3804 (2011). https://doi.org/10.1016/j.sbspro.2011.04.376

Hošková-Mayerová, Š.: The effect of language preparation on communication skills and growth of students' self - confidence, Proc. – Soc. Behav. Sci. 114, 644–648 (2014). https://doi.org/10.101 6/j.sbspro.2013.12.761

Hošková-Mayerová, Š.: Education and Training in Crisis Management. In: ICEEPSY 2016 - 7th International Conference on Education and Educational Conference, pp. 849–856. Future Academy (2016). ISSN 2357-1330

Hošková-Mayerová, Š.: Quasi-order hypergroups determinated by T-hypergroups. Ratio Math. 32(2017), 37–44 (2017). https://doi.org/10.23755/rm.v32i0.333

Hošková-Mayerová, Š., Maturo, A.: Hyperstructures in social sciences. In: AWER Procedia Information Technology and Computer Science, vol. 3, 547—552. Barcelona, Spain (2013)

Hošková-Mayerová, Š., Maturo, A.: An analysis of Social Relations and Social Group behaviors with fuzzy sets and hyperstructures. Int. J. Algebraic Hyperstruct. Appl. 2(1), 91–99 (2016). ISSN 2383-2851

Hošková-Mayerová, Š., Maturo, A.: Fuzzy sets and algebraic hyperoperations to model interpersonal relations. In: Recent Trends in Social Systems: Quantitative Theories and Quantitative Models. Studies in Systems, Studies in Systems, Decision and Control, 66 (2016). https://doi.org/10.100 7/978-3-319-40583-8_19

Hošková-Mayerová, Š., Maturo, A.: On some applications of algebraic hyperstructures for the management of teaching and relationships in schools (2018, submitted)

Klir, G., Yuan, B.: Fuzzy Sets and Fuzzy Logic: Theory and Applications. Prentice Hall, Upper Saddle River (1995)

Maturo, A.: Struttura algebrica degli eventi generalizzati. Periodico di Matematiche 4(1993), 18–26 (1993)

Maturo, A., Porreca, A.: Algebraic hyperstructures and fuzzy logic in the treatment of uncertainty. Sci. Philos. 4(1), 31–42 (2016)

Maturo, A., Sciarra, E., Tofan, I.: A formalization of some aspects of the social organization by means of the fuzzy set theory. Ratio Sociol. 1(2008), 5–20 (2008)

Maturo., F.: Dealing with randomness and vagueness in business and management sciences: the fuzzy-probabilistic approach as a tool for the study of statistical relationships between imprecise variables. Ratio Math. 30(1), 45–58 (2016). doi:http://dx.doi.org/10.23755/rm.v30i1.8

Maturo, F., Fortuna, F.: Bell-shaped fuzzy numbers associated with the normal curve. In: Topics on Methodological and Applied Statistical Inference, pp. 131–144 (2016). https://doi.org/10.1007/ 978-3-319-44093-4_13

Maturo, F., Fortuna, F., Di Battista, T.: Testing equality of functions across multiple experimental conditions for different ability levels in the IRT context: the case of the IPRASE TLT 2016 Survey. Soc. Indic. Res. (2018). https://doi.org/10.1007/s11205-018-1893-4

Maturo, F., Hošková-Mayerová, Š.: Fuzzy regression models and alternative operations for economic and social sciences. In: Recent Trends in Social Systems: Quantitative Theories and Quantitative Models. Series Studies in Systems, Decision and Control **66**, 235–247 (2017). https://doi.org/10.1007/978-3-319-40585-8_21

Maturo, F., Ventre, V.: Consensus in multiperson decision making using fuzzy coalitions. Stud. Fuzziness Soft Comp. **357**, 451–464 (2017). https://doi.org/10.1007/978-3-319-60207-3_26

Moreno, J.L.: Who Shall Survive? Beacon Press, New York (1953)

Moreno, J.L.: Sociometry. Experimental Methods and the Science of Society. Beacon Press, New York (1951)

Ragin, C.C.: Fuzzy-Set Social Science. University Chicago Press, Chicago (2000)

Reichenbach, H.: Philosophic Foundations of Quantum Mechanics. University of California Press, Berkeley (1944)

Ross, T.J.: Fuzzy logic with engineering applications. McGraw-Hill, New York (1995)

Rosická, Z., Hošková-Mayerová, Š., Motivation to study and work with talented students, Proc. – Soc. Behav. Sci. **114**, 234–238 (2014). https://doi.org/10.1016/j.sbspro.2013.12.691

Sciarra, E.: Paradigmi e metodi di ricerca sulla socializzazione autorganizzante. Sigraf Edizioni Scientifiche, Pescara (2007)

Svatoňová, H., Hošková-Mayerová, Š.: Social aspects of teaching: Subjective preconditions and objective evaluation of interpretation of image data. (2017). https://doi.org/10.1007/978-3-319-54819-7_13

Vougiouklis, T.: Hyperstructures as models in social sciences. Ratio Math. **21**(2011), 27–42 (2011)

Vougiouklis, T., Vougiouklis, S.: Helix-hopes on finite hyperfields. Ratio Math. **31**(2016), 65–78 (2016)

Zadeh, L.A.: Fuzzy Sets. Inf. Control **8**, 338–358 (1965)

Zadeh, L.A.: The concept of a linguistic variable and its application to approximate reasoning. Inf. Sci. (1975);8 Part I:199—249, Part II 301—357, Part III. Inf Sci 1975;9: 43—80

Resources and Capabilities for Academic Spin-Offs' Development. An Empirical Analysis of the Italian Context

Stefania Migliori and Francesco De Luca

Abstract The transfer of scientific knowledge from the academic context to the market can generate positive effects on the economic system. Universities have mainly carried out these activities by supporting the creation of spin-off firms. Drawing on a resource-based view theory, the aim of this study is to investigate which resources can affect more than others the creation and successful development of university spin-offs (USOs). In order to analyze the relevance of these drivers, we use a sample of 100 Italian USOs and focus our attention on the following factors: the availability of resources for the spin-off at the earlier stage such as know-how, financial assets, and managerial skills; the number of patents that have been successfully issued by the spin-off; the cultural background of the founders; the frequency of interactions with stakeholders outside the spin-off; and the propensitity and the ability to innovation. Our analysis shows that Italian spin-offs appears to be quite innovative, but they generally need time and probably more funding to protect their innovation through patents issuing. In this sense, established spin-offs suffer for the difficulties in raising funds, the high costs of developing ideas (together with the difficulties in monitoring them), and the lack of governmental support.

Keywords University spin-off · Technology transfer
Resource-based view theory · Resources and capabilities

S. Migliori (✉) · F. De Luca
Department of Management and Business Administration, "G. D' Annunzio" University of Chieti-Pescara, V.le Pindaro, n. 42, Pescara, Italy
e-mail: stefania.migliori@unich.it

F. De Luca
e-mail: francesco.deluca@unich.it

© Springer Nature Switzerland AG 2019
C. Flaut et al. (eds.), *Models and Theories in Social Systems*, Studies in Systems, Decision and Control 179, https://doi.org/10.1007/978-3-030-00084-4_20

1 Introduction

University spin-offs (USOs) are one of the main mechanisms through which scientific knowledge which is produced by universities, can be enhanced, applied and put on the market (Fontes 2005; Kirwan et al. 2006). The literature and the empirical studies highlight that the passage of scientific knowledge from the academic world to the market can produce positive effects in terms of employment and improvement of competitiveness and growth within the economic context (Fini and Lacetera 2010). However, some research highlights the difficulties of growth and development of university spin-offs compared to other spin-off or start-up companies (Ensley and Hemieleski 2005; Zahara et al. 2007; Wennberg et al. 2011; Zhang 2009) and that some universities produce a greater number of spin-offs.

In consequence, today the debate about the economic impact produced by university spin-offs is still open and many aspects have yet to be investigated. In particular, researches are trying to answer to the following research questions: Why are some universities more likely to create spin-offs? What factors and/or resources can influence the chances of success of university spin-offs?

According to these arguments, this study aims to contribute to a better understanding of USOs by analyzing the factors that can influence their emergence and development.

In order to pursue our objective, we adopt the resource-based view perspective. This theory describes companies as bundle of resources (Penrose 1959). Specifically, the resources and capabilities possessed by the organization at a particular time, affecting their future performance (Penrose 1959; Wernerfelt 1995). Then, resources and capabilities determine firm behaviour and performance (Lockett and Wright 2005) distinguishing an organization from one another. Following this perspective, we consider that the processes of creation and development of the spin-offs are determined by the joint action of the resources and competences of the university with those owned by the spin-off and by the subjects that are part of it. Then, in our theoretical framework we consider the following aspects:

- the policies and strategies of universities;
- the university's internal resources for the production of spin-offs;
- the initial resources of USOs;
- the network capabilities of USOs.

We use a sample of 100 Italian university spin-offs and particularly focus on the resources and capability of USOs. Our research emphasizes that Italian spin-offs have a high level of know how, appears to be quite innovative, but they generally need time and probably more funding to protect their innovation through patents issuing. In this sense, established spin-offs suffer for the difficulties in raising funds, the lack of managerial skills, the high costs of developing ideas (together with the difficulties in monitoring them), and the lack of governmental support.

Our analysis contributes to the literature on university spin-offs in different ways. First, it provides a clear representation of the initial resources that characterize the

Italian USOs. This makes it possible to better understand any differences in the growth and development processes of the USO in different countries. Secondly, our study can help the scholars and the policy makers to identify the strengths and weaknesses of the Italian USOs during the start-up phase, in order to define even more effective actions aimed at supporting the creation and development of USOs.

The remainder of this study is organized as follow: Section 2 proposes a review of the theoretical and empirical literature regarding factors that may affect the development of spin-off. Section 3 describes the methodological approach adopted, the dataset and the descriptive statistics and the outcomes. Finally, Sect. 4 presents our conclusions.

2 Literature Background

2.1 The Policies and Strategies of Universities

The policies and strategies adopted by the university can represent a key element in the development processes of university spin-offs. In particular, universities can implement actions that can ease or hamper companies' spinning-out process and the chance to have successful spin-offs. In fact, the more or less favorable orientation of the university towards spin-off operations can strongly influence the context in which individuals interested in starting up spin-off companies operate. In particular, the policies and strategies adopted by the university can strongly affect both the chance of starting up spin-offs operations and the probability of their success or failure. In consequence, the direction taken by the university is a strategic resource for the spin-off in its starting phase likewise other resources (e.g. charter members' *skills*, abilities, available know-how).

The study of Locket et al. (2003) on a sample of 57 UK spin-offs highlights that the universities considered most successful based on their ability to produce righteous spin-offs are those that have developed and adopted a clear-cut and proactive strategy concerning aspects such as:

- the inclination to implement spinning-out processes;
- the clear preference between license agreements and spin-offs operations;
- the role of academic entrepreneurs and external consultants in the management positions of USOs.

With reference to the first aspect, it is possible to identify two distinct kinds of policies adopted by universities and which lead to the identification of two distinct kinds of university spin-offs. The first type of spin-off is linked to individuals' initiative who have identified a market opportunity to commercially exploit the outcomes of the research activity. The literature defines such spin-offs as "spontaneously emerged spin-offs" (Steffensen et al. 2000), or "pull spin-offs" (Pirnay 1998), since it is the market itself that draws academic entrepreneurs towards itself. The second type, however, is related to spin-offs produced after promotion and support activities carried out by the university towards academic entrepreneurship. In this case the literature calls such spin-offs as "planned spin-offs" (Steffensen et al. 2000), or "push-spin-off" (Pirnay 1998), because it is the university that drives academics to put on the market the researches' outcomes through spin-off operations.

With reference to the second aspect, it should be noted that an increasing focus has been placed on the *spinning-out companies* process of universities considering the limits highlighted by applying the *technology licensing* model (Siegel et al. 1999) which has represented for a long time the most employed method for passing on scientific knowledge to the private sector.

The University can adopt policies and strategies aimed at encouraging or discouraging academics who have produced the new scientific knowledge suitable for marketing to taking on entrepreneurial roles.

Within the *parent organization*, the academic can choose between different possible degrees of involvement in the management and administration of the spin-off: from complete involvement as an entrepreneur-founder up to minimal types of involvement and which contemplate the intervention of external subjects as entrepreneurs or managers of the spin-off. The involvement of these individuals can be directed at finding business skills or even financial resources which generally academics do not have.

Each of the two cases shows advantages and disadvantages. In fact, the direct involvement of the academic in the business management can bring advantages arising from his complete knowledge of the features of the new technology to be put on the market. From another perspective, however, the academic may not have proper knowledge of the target market and adequate managerial skills. Furthermore, the University can interpret the academic's entrepreneurial involvement as a time taken away from research and/or teaching activities. The literature suggests that the greater the involvement of academics in the management of spin-offs, the greater the possibility of developing conflicts of interest between academic activities and entrepreneurial and commercial ones (Lockett et al. 2003).

With reference to the hypothesis of external subjects' involvement in the academic spin-off, the empirical research emphasizes that the spin-off companies set up with external subjects and with forms of "academic" support, reach larger dimensions than those set up and managed only by academics (Chrisman et al. 1995). The most successful spin-offs turn out to be those towards which the membership university adopts a clear strategy about the involvement of external entrepreneurs and managers to whom to entrust the business development (Lockett 2003).

2.2   University's Internal Resources for the Creation of Spin-Offs

The existing literature identifies three main kinds of resources owned by universities and which can effectively influence the spin-off creation processes (Lockett et al. 2005; Pazos et al. 2012):

- the stock of technological resources;
- the capabilities and routines possessed and developed;
- the level of experience in producing spin-offs.

Moreover, the most recent literature attaches a key role to the presence of staff units devoted to technology transfer. In fact, these organizational units can carry out a relevant supporting action, both in identifying market opportunities and in boosting the entrepreneurial activity of academics. An additional factor which can affect the spin-off creation processes within the university is represented by the availability of financial resources. In fact, it often happens that the assessment of the potential of commercial exploitation integrated in a research output requires specialized *skills* and expertise which the university structure does not own and therefore must be obtained outside.

Furthermore, the number of patents and the level of financial resources allocated to the research are other factors revealing the level of stock of potentially transferable technological knowledge.

In addition to the stock of resources, a relevant role is assigned to the existing skills and routines developed by the University. In particular, the skills linked to the various issues which characterize the development of the spin-off business, the practices to protect the rights of the business idea and its management and marketing. Furthermore, the level of experience which the University has gained in producing spin-offs over time is also particularly significant (Lockett et al. 2005). The University's long involvement in activities aimed at technology transfer allows it to develop a specific experience base (Lockett et al. 2004; O'Shea et al. 2005; Powers and McDougall 2005). This enables the University to have a greater ability in starting *spinning-out* processes and a greater efficiency in their management. These skills, being linked to each specific University structure (Teece et al. 1997), can create strong differences among universities in their ability to produce spin-offs.

The joint action of stocks of resources, skills and experience developed by the University will bring about the success of the *spinning-out* operations. In fact, the consequent outcome of the availability of a resource also depends on the effects of synergy arising from the relations of this resource with the others existing in the company (Teece 1986). These are the resource structures that managers establish by using the so-called *dynamic capabilities* (Eisenhardt and Martin 2000) which bring about the long-term success of the company and, in this particular case, the spin-off. In this regard, the existing empirical researches clearly show that the effect of synergy among the resources is responsible for driving the growth of spin-offs of the research (Brush et al. 2001; Chandler and Hanks 1998; Roberts 1991). Besides, the failure to take into account the effects of synergy among the resources is one of the feasible explanations of conflicting results about the effect of initial resources affecting the success of the company (Hirman and Clarisse 2004).

2.3 The Resources of USOs in Their Start-Up Phase

The processes of starting up university spin-offs are strongly affected not only, as above mentioned, by policies and strategies adopted by the University and the existing resources within it, but also by the resources owned by academic-inventors involved in these processes. These resources take on particular importance since they are able to affect performances and therefore the success of the different USOs.

In this regard, the literature identifies a plurality of types of resources which if owned by the academics involved in starting up the spin-offs can strongly rebound on the success of such initiatives, including the cognitive, technological, financial, organizational resources and human capital (Landry et al. 2006).

Some scholars point out that other factors are also relevant for the purpose of starting up and implementing academic entrepreneurial initiatives. These factors are related to the individual sphere of the academic-entrepreneurship such as: risk propensity, fear of losing academic status and the level of cohesion of the research team.

In particular the research carried out by Compagno et al. (2008) suggests an integrated model of tangible and intangible resources related to the sphere of the individual academic-researcher and which may influence the spin-offs' starting up initiatives. The research performed on 249 researchers from Italian Universities has highlighted that the critical resources appear to be the following ones: relational capital, experience, research funds obtained outside the university, easily applicable research outcomes and the dimensions of the laboratory. Moreover, successful initiatives appear to be linked to the presence of researchers who have a greater inclination towards economic risk and greater cohesion within the research group.

The entrepreneurial resources, both human and financial, thus take on particular importance. However, start-ups can hardly rely on appropriate financial resources and/or specialized skills and knowledge regarding the business. With reference to the availability of adequate financial resources, the literature has pointed out that in the phase of creation of the innovative idea, the presence of asymmetric information and the absence of sufficient self-financing conditions lead to resort to the entrepreneur's own capital (or to people to whom he is linked by family ties) or investors such as *business angels*. Even in the immediately following phases (*early stage*), the persistence of high information asymmetries on the innovation project and on the company's future behavior and the absence of an established reputation of the company among the possible sponsors, make it too difficult the access to loan capital. Accordingly, the increase in the financial requirements connected with this life course phase is achieved by using the venture capital offered by institutional investors capable of providing, in addition to financial resources, also knowledge and skills. The difficulty in obtaining these resources emphasizes the significance of other factors such as the entrepreneur's relational capital (linked to personal relationships, networks or strategic alliances).

2.4 The Network Capabilities of Spin-Offs

For USOs, the relationships that the founding academics establish with other companies, different organizations and research centers as well as with institutes for technology transfer and public organizations play a crucial role. Relations with external agents can be a way to obtain resources and skills which do not exist within the spin-off, especially in the first start-up phase. In this regard, this can include the recurrent case in which the academic-founder lacks all the managerial skills necessary for managing and developing the spin-off as well as the case of the possible shortage of scientific and technological knowledge necessary for the full commercial exploitation of innovation. Indeed, the literature points out the key role of the ability and regularity of interactions that the company establishes with external subjects for the success of innovation processes in the different market sectors (Galbraith 1973). The development of relationships with external subjects is therefore a way by which to obtain resources held by others in the environment in which the company operates (Hoang and Antonic 2003; Nicolau and Birley 2003; Shan and Stuart 2002). The importance of these relationships has been emphasized in several studies which have shown how an increasing portion of the value produced by the company is linked to its ability to establish relationships within a network and develop lasting alliances (Kale et al. 2002). Moreover, the ability and regularity of interaction with external subjects must be taken into account since they allow the spin-off to obtain and exchange information aimed at better understanding the demands of the market which affect the market attractiveness/market potential of the business idea. This aspect plays a significant role in the start-up phase of academic spin offs when the technological knowledge of the new business idea to be put into market is still in the process of being finalized. Interactions with other subjects make it possible to gather information and knowledge useful to allow the innovation embedded in the business idea to be attractive on the market (Walter et al. 2006).

The literature has adopted various terminologies in order to describe the company's ability to develop relationships and interact with other organizations, such as "alliance capability" or "relational capital" (Kale et al. 2002) or "network capability" (Walter et al. 2006) and has employed different theoretical approaches for understanding the capabilities which enable the company to grow and develop within a backdrop of relations. All these approaches highlight the nature of "capability" of the company's inclination for relations which should be understood as a special kind of resource which is implied, non-transferable within the organization and is likely to enhance the effectiveness and efficiency of other resources held by the company (Eisenhardt and Martin 2000). Walter et al. (2006) in this perspective introduced the concept of network capabilities of the spin-off which must be seen "*as its abilities to initiate, maintain, and utilize relationship with various external partners. The term network therefore expresses that managing relationship goes beyond coping with single relationship and alliance".* The authors, in particular, in their study carried out on one hundred and forty-nine Italian USOs, emphasize the joint impact of network capabilities and entrepreneurial direction on performances and the moderating role played by the network capabilities in this relationship.

3 Method and Empirical Analysis

The research was performed in 2017 on a sample of Italian USOs. The survey was carried out through spreading a questionnaire via e-mail in addition to an introductory and explanatory letter about the research project and the pursued objectives. The empirical analysis has been carried out with the aim of collecting information and data useful for understanding the possible determinants of the development process of university spin-offs. In order to identify the Italian USOs, various sources have been consulted (books, articles in specialized magazines, university websites and spin-offs networks, and personal contacts). In this phase, 468 USOs target have been identified. After the survey questionnaire that was sent via e-mail, 100 questionnaires correctly filled have been collected.

The questionnaire, developed as part of a broader research project, was divided into fifteen questions related to the following six macro-areas: (1) general information about the USOs; (2) availability of initial resources and capabilities of USOs; (3) entrepreneurial, competitive and market orientation of the USOs; (4) conditions of external environment; (5) innovation capability of USOs; (6) USOs' performance.

In the empirical analysis of this study, we refer to the aspects indicated by the macro-areas 2 and 5. Table 1 shows the specific aspects which have been analyzed.

3.1 *Empirical Analysis and Descriptive Statistics*

The empirical anaysis has been carried out on the basis of the main variables that can either ease or hinder the establishment and the development of academic spin-offs.

Therefore, five main aspects (Lei-Yu Wu 2007) have been considered as follows:

- the availability of resources for the spin-off at the earlier stage such as know-how, financial assets, and managerial skills;
- the number of patents that have been successfully issued by the spin-off;
- the cultural background of the founders (i.e.: competences, studies, degree);
- the propensity and the ability to innovation;
- the frequency of interactions with stakeholders outside the spin-off.

Table 1 Map of the variables considered in our study

Macro-areas	Variable	Question	Reference
USOs' resources and capabilities in the starting-up phase	Level of know expertise Financial Resources Managerial skills	1. In the early stage, your own specialized know how was 2. In the early stage, your own capital was 3. In the early stage, your own managerial skills was Likert scale: 1 (minimum) to 7 (maximum)	Lei-Yu, W. (2007). 'Entrepreneurial resources, dynamic capabilities and start-up performance of Taiwan's high-tech firms', Journal of Business Research, 60, pp. 549–555.
	Scientific area of the management degree	sciences medicine architecture engineering mathematics economics and statistics law other	
	Frequency of interaction with external subjects	The variable refers to the frequency of interaction that academic founders, before starting up the new venture, used to have with: 1. Research institution 2. Institution for technological transfer of research results 3. Public organization 4. Financial Institutions 5. Industry association 6. Other agents (Likert scale 1 low frequency – 7 very high frequency)	Grandi, A. and Grimaldi, R. (2005). 'Academics' organizational characteristics and the generation of successful business ideas', Journal of Business Venturing, 20, 821-845.

(continued)

Table 1 (continued)

Macro-areas	Variable	Question	Reference
USOs' innovation capability	Innovativeness capability	1. Our company frequently tries out new ideas 2. Our company seeks out new ways to do things 3. Our company is creative in its methods of operation 4. Our company is often the first to market with new products and services 5. Innovation in our company is perceived as to risky and is resisted 6. Our new product introduction as increased over the last five years (Likert scale 1 very much in disagreement – 7 very high agreement)	Calantone et al. (2002). 'Learning orientation, firm innovation capability, and firm performance'. Industrial Marketing Management, 31, 515–524.
	Ability to develop patents	Number of issued patents	

In the following Figs. 1 and 2, we propose some indicators of initial resources availability for spin-offs, according to a scale from 1 (lower amount) and 7 (highest amount). We easily observe that know how is always available at higher level, whilst financial assets are often lacking.

With respect to the know-how variable, Fig. 2 highlights that 48% of respondents has evaluated its level of know-how as very high and innovative. On the other hand, only 12% of respondents believes that its level of know-how is very low and not innovative enough.

On the contrary, when we observe in the same Fig. 1 the response about the perception of financial assets and capital available in the early stage, most part of spin-offs has evaluated it as very low. See Fig. 3 for the frequency distribution of answers for the financial assets availability. It results that 75% of respondent spin-offs considers that its capital at the early stage is low, while only 20% believes it is sufficient for its entrepreneurial purposes. In this sense, we consider capital and the possibility to access financial funding sources as a fundamental variable that impacts considerably on the probability of success of a spin-off.

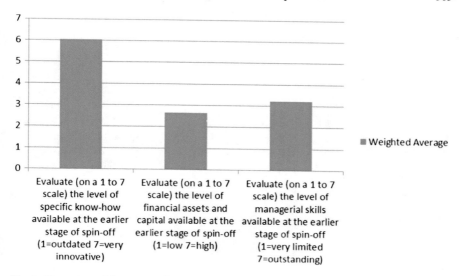

Fig. 1 Entrepreneurial rens at earlier stage of spin-offs

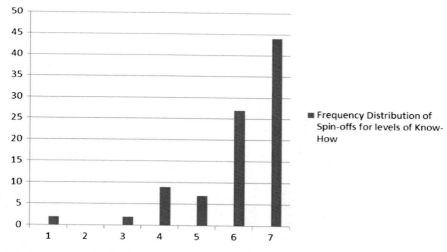

Fig. 2 Frequency distribution of respondent spin-offs on the evaluation scale of Know-How

If we then consider the third variable of available resources for spin-offs, namely the managerial skills available at the early stage of development, we find a not homogeneus trend in the responses. In fact, as it is shown in Fig. 1 the average of respondent spin-offs has evaluated its managerial skills at the medium level of the evaluation scale. Moreover, Fig. 4 shows that 65% of respondent spin-offs has evaluated its managerial skills as limited or very limited.

S. Migliori and F. De Luca

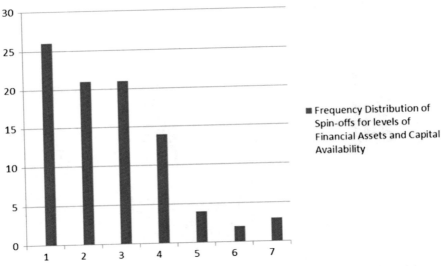

Fig. 3 Frequency distribution of respondent spin-offs on the evaluation scale of financial assets and capital availability

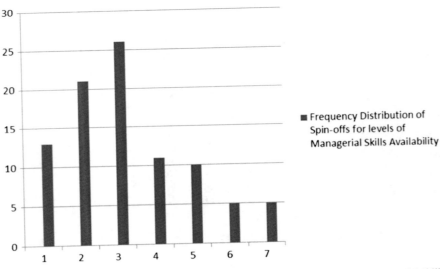

Fig. 4 Frequency distribution of respondent spin-offs on the evaluation scale of managerial skills availability

At the same time there is a 22% of respondents that consider their managerial skills as satisfying for the entrepreneurial development of the spin-off.

The number of issued patents from respondent spin-offs (97 valid responses of a total amount of 100 respondents spin-offs) is represented in Fig. 5. It appears that

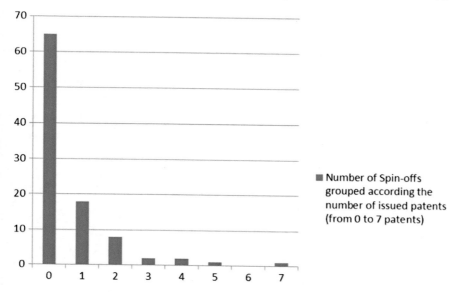

Fig. 5 Number of issued patents from respondent spin-offs

67% of the sample has not issued any patent yet. On the other hand, only 19% has issued one patent and 8% has issued two patents. (Fig. 5)

Another variable that impacts on the probability of development of a spin-off is the cultural background of the founder/s. The analyzed sample shows that more than 41% of respondents (96 valid answers on a total of 100 respondents spin-offs) has obtained an engineer master degree and 30% has a natural science master degree (such as medicine, biology, chemistry, geology). Only 3% has an economics/statisics studies background and this could represent a weakness in terms of managerial capability of running a business. Figure 6 shows the articulation of cultural background in respect to the number of respondent spin-offs.

If we consider the innovativeness capabilities of spin-offs, within the 89 valid responses of 100 total respondents, we observe that the highest values are for the propensity to find new ways in doing things (5,26 average value). The items about propensity to creativity and to new ideas developing follow with values higher than 5. Rignt under this threshold, we find the item about the capability of being first mover in introducing new product and services to the markets (with 4,38 average value). Innovation perceived as high risky is the item with the lowest score (2,76 average value). In a middle position, lies the item about the rate of new product/services introduced along the last five years (3,81 average value).

This analisys confirms the expected scenario where spin-offs normally present a strong propensity to innovate and to risk given that their probability of success strictly depends on their capability to create something new both in terms of product/process innovation and of new solutions to previous demands.

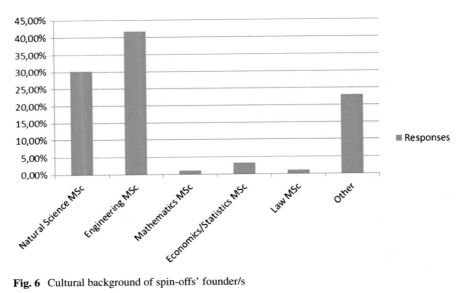

Fig. 6 Cultural background of spin-offs' founder/s

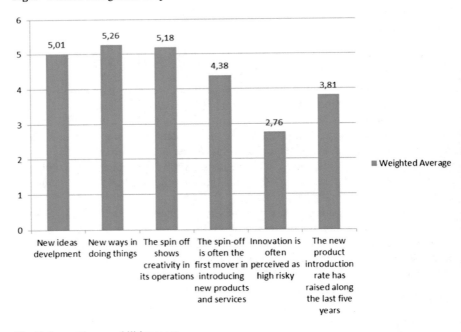

Fig. 7 Innovation capabilities score

Figure 7 shows the results of the innovativeness capabilities analysis.

The last variable that has been considered for this analysis, is the intensity of interaction between spin-offs and external stakeholders and operators, specifically

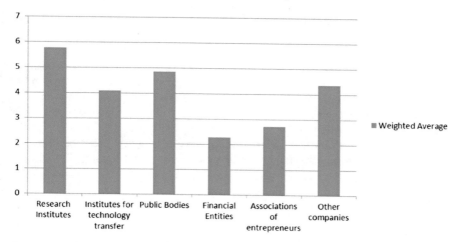

Fig. 8 Intensity and frequency of interaction with external stakeholders

within the period right before the foundation of the spin-off. Figure 8 shows the results of the survey with respect to the above item on a scale from 0 (absent or low frequency) to 7 (very high frequency of interaction).

It appears clearly that 80% of respondent spin-offs (for this item we collected 91 valid responses on a total of 100 respondent spin-offs) has already had a high level of interaction with research institutes. After that, the public bodies are the second group of operator with high frequency interaction with spin-offs (42%). On the other hand, 78% of respondent spin-offs show a low frequency of interaction with financial entities. This is quite intuitive but could represent a weak point of newborn spin-offs as a consequence, especially because access to funding sources is a critical item for spin-off success and growth.

4   Concluding Remarks

The main findings of the empirical research contribute to the present debate about University spin-off development and probability of success, and could provide some suggestions both for existing spin off and for policy makers.

Italian academic spin-offs are often undersized firms mainly because of their early stage; they mainly belong to technology and environmental industry and their founder/s largely present/s an engineering degree. Italian academic spin-offs are weak in terms of financial assets and capital availability at their early stage of development. This clearly represents a threat to the ability of the spin-off to overcome the start-up phase. The reason of this may lie on the low level of managerial skills of the people involved into the spin-off and this could bring to the low reliability of developed project for financial entities and investors. Therefore, the Italian spin-off environment could benefit of a strongest support from both Universities and policy makers because the success of established spin-off is affected by managerial support (i.e.: universities could act as incubator to support to spin-offs), and specific way of easy credit terms.

Moreover, Italian USOs appears to be propense to innovation and to risk of introducing new ideas and ways of doing things, but at the same time the average number of issued patents is quite low. Once again Italian USOs appears to be quite innovative but they generally need more time and probably more funding to protect their innovation through patents issuing. In this sense, established USOs suffer for the difficulties in raising funds, the high costs of developing ideas (together with the difficulties in monitoring them), and the lack of governmental support.

References

Brush, C.G., Green, P.G., Hart, M.M.: From initial idea to unique advantage: the entrepreneurial challenge of constructing a resource base. Acad. Manag. Exec. **15**, 64–78 (2001)

Calantone, R.J., Cavusgil, S.T., Zhao, Y.: Learning orientation, firm innovation capability, and firm performance. Ind. Mark. Manage. **31**, 515–524 (2002)

Chandler, G., Hanks, S.H.: Market attractiveness, Resource- based capabilities, Venture strategies and Venture performance. J. Bus. **9**, 331–349 (1998)

Chrisman, J.J., Hynes, T., Fraser, S.: Faculty entrepreneurship and economic development: the case of the university of Calgary. Bus. Ventur. **10**, 267–281 (1995)

Compagno, C., Lauto, G., Fornasier, E.: La genesi degli spin-off accademici di successo, Paper presented at IX Workshop dei Docenti e dei Ricercatori di Organizzazione Aziendale - L'organizzazione fa la differenza?. Università Ca' Foscari –Venezia. (2008).

Eisenhardt, K.M., Martin, J.A.: Dynamic capabilities: what are they? Strateg. Manag. J. **21**, 1105–1121 (2000)

Ensley, M., Hmieleski, K.: A comparative study of new venture top management team composition, dynamics and performance between university-based and independent start-ups. Res. Policy **34**, 1091–1105 (2005)

Fini, R., Lacetera, N.: Different yokes for different folks: individual preferences, institutional logics, and the commercialization of academic research. In: Advance in the study of entrepreneurship, innovation and Economic Growth, emerald group publishing, vol. 21, 1–25 (2010)

Fontes, M.: The process of transformation of scientific and technological knowledge into economic value conducted by biotechnology spin-offs. Technovation **25**, 339–347 (2005)

Galbraith, J.R.: Designing Complex Organization. Addison-Wesley, Reading, MA (1973)

Grandi, A., Grimaldi, R.: Academics' organizational characteristics and the generation of successful business ideas. J. Bus. Ventur. **20**, 821–845 (2005)

Hirman, A., Clarisse, B.: The initial resources and market strategy to create high growth firms. Working Paper Steunpunt OOI, October (2004)

Hoang, H., Antonic, B.: Network-based research in entrepreneurship: a critical review. J. Bus. Ventur. **18**, 165–187 (2003)

Kale, P., Dyer, J.H., Singh, H.: Alliance capability, stock market response, and long-term alliance resources: the role of the alliance function. Strateg. Manag. J. **23**, 747–767 (2002)

Kirwan, P., Sijde van der, P., Groen, A.: Assessing the needs of new technology based firms (NTBFs): An investigation among spin-off companies from six European Universities. Int. Entrepreneurship Manag. J. **2**, 173–187 (2006)

Landry, R., Amara, N., Rherrad, I.: Why are some university researchers more likely to create spin-offs than others? Evidence from Canadian universities. Res. Policy **35**, 1599–1615 (2006)

Lei-Yu, W.: Entrepreneurial resources, dynamic capabilities and start-up performance of Taiwan's high-tech firms. J. Bus. Res. **60**, 549–555 (2007)

Lockett, A., Wright, M.: Resources, capabilities, risk capital and the creation of university spin-out companies. Res. Policy **34**, 891–1122 (2005)

Lockett, A., Siegel, D., Wright, M., Ensley, M.: The creation of spin-off firms at public research institutions: managerial and policy implications. Res. Policy **34**, 981–993 (2005)

Lockett, A., Wright, M., Franklin, S.: Technology transfer and universities' spin-out strategies. Small Bus. Econ. **20**, 185–200 (2003)

Lockett, A., Wright, M., Vohora, A.: Resources, Capabilities, Risk Capital and the Creation of University Spin-Out Companies. University of Sussex, SPRU-Science and Technology Policy Research (2004)

Nicolau, N., Birley, S.: Academic network in a tricothomous categorisation of university spinout phenomenon. J. Bus. Ventur. **18**, 333–359 (2003)

O'Shea, R.P., Chugh, H., Allen, J.T.: Determinants and consequences of university spinoff activity: A conceptual framework. Int. J. Technol. Transf. **33**, 653–666 (2008)

O'Shea, R.P., Allen, T.J., Chevalier A., Roche F.: Entrepreneurial orientation, technology transfer and spinoff performance of U.S. universities. Res. Policy **34**, 994–1009 (2005)

Pazos, D., López, S., González, L., Sandiás, A.: A resource-based view of university spin-off activity: New evidence from the Spanish caseUna aplicación de la teoría de los recursos a la creación de spin-offs universitarias: nuevas evidencias desde el caso español. Revista Europea de Dirección y Economía de la Empresa **21**, 255–265 (2012)

Penrose, E.T.: The Theory of the Growth of the Firm. Wiley, New York (1959)

Pirnay, F.: Spin-off et essaimage de quoi s'agit-il? Une revue de la literature, In: 4ème Colloque International Francophone sur la PME, Metz-Nancy, 22–24 October (1998)

Powers, J.B., McDougall, P.P.: University start-up formation and technology licensing with firms that go public: A resource based view of academic entrepreneurship. J. Bus. Ventur. **20**, 291–311 (2005)

Roberts, E.B.: The technological base of the new enterprise. Res. Policy **20**, 283–298 (1991)

Shan, S., Stuart, T.E.: Organizational endowments and the performance of university start-ups. Manag. Sci. **48**, 154–170 (2002)

Siegel, D., Waldman, D., Link, A.: Assessing the impact of organizational practices on the productivity of university technology transfer offices. An exploratory study. In: NBER Working Paper n. 7252, July (1999)

Steffensen, M., Everet, M.R., Speakman, K.: Spin-offs from research centers at a research university. J. Bus. Ventur. **15**, 93–111 (2000)

Teece, D.: Profiting from technological innovation: implication for integration, collaboration, licensing and public policy. Res. Policy **15**, 285–305 (1986)

Teece, D.J., Pisano, G., Shane, A.: Dynamic capabilities and strategic management. Strateg. Manag. J. **18**, 509–534 (1997)

Walter, A., Auer, M., Ritter, T.: The impact of network capabilities and entrepreneurial orientation on university spin-off performance. J. Bus. Ventur. **21**, 541–567 (2006)

Wennberg, K.J., Wiklund, J., Wrigth, M.: The effectiveness of university knowledge spillovers: performance differences between university spinoff and corporate spinoffs. Res. Policy **40**, 1128–1143 (2011)

Wernerfelt, B.: The resource-based view of the firm: Ten years after. Strateg. Manag. J. **16**, 171–174 (1995)

Zahra, S., Van de Velde, E., Larraneta, B.: Knowledge conversion capabilities and the performance of corporate and university spin-offs. Ind. Corp. Change **16**, 569–608 (2007)

Zhang, J.: The performance of university spin-offs: An exploratory analysis using venture capital data. J. Technol. Transfer **34**, 255–285 (2009)

Part IV
Mathematical Methods in Social Sciences

A Fixed Point Result on the Interesting Abstract Space: Partial Metric Spaces

Erdal Karapınar

Abstract In this chapter, we shall investigate the existence of fixed point of certain mappings via simulation functions in the framework of an interesting abstract space, namely, partial metric spaces. The main results of this manuscript not only extend, but also generalize, improve and unify several existing results on the literature of metric fixed point theory.

1 Introduction and Preliminaries

The notion of distance is as old as the mathematics due to its nature. On the other hand, it was axiomatically formulated for the first time by Frechét (1906) as a "L-function" and it was called "metric" by Felix Hausdorff. Clearly, the concept of metric is not only the cornerstone for analysis and many other branches of mathematics, but also for quantitative sciences (see e.g. the report of Matthews (1994) and references therein).

It is an indispensable fact that metric notion can be considered as a milestone of our modern daily life. Indeed, using this notion, from social to quantitative sciences, we are able to invent, discover and create many new notions and tools that helps to understand the rules of the real world and make our lives easier.

One of the basic example for the solution of the real life problem via metric notion can be found in all smart cellphones, in particular in navigation programs. Indeed, any navigation program is based on the taxi-cab metric

$$d((x_1, x_2), (y_1, y_2)) = |x_1 - y_1| + |x_2 - y_2|,$$
$$\text{for every } (x_1, x_2), (y_1, y_2) \in \mathbb{R}^2 := \mathbb{R} \times \mathbb{R},$$

where $\mathbb{R}$ stands for the set of real numbers. Moreover, this metric and the traffic density of the road can be combined in a new metric, to reach from one point to another with maximum benefits that can be time, money, distance. Further, according to these

E. Karapınar (✉)
Atilim University, 06836 Incek, Ankara, Turkey
e-mail: erdalkarapinar@yahoo.com

© Springer Nature Switzerland AG 2019
C. Flaut et al. (eds.), *Models and Theories in Social Systems*, Studies in Systems, Decision and Control 179, https://doi.org/10.1007/978-3-030-00084-4_21

parameters, computer program, which appears as a smart cellphone application, may offer different options.

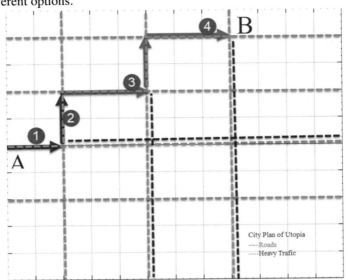

In the above example, the time and distance are considered as the main parameters. In the near future, it is possible to consider "*space-taxi-cab*" metric $d(A, B) = \sum_{i=1}^{3} |a_i - b_i|$ for the new air-vehicle that can move in both horizontally and vertically, where $A = (a_1, a_2, a_3)$, $B = (b_1, b_2, b_3) \in \mathbb{R}^3$.

Throughout this chapter, $\mathbb{N}$ and $\mathbb{N}_0$ denote the set of positive integers and the set of nonnegative integers, respectively. Furthermore, $\mathbb{R}^+$ and $\mathbb{R}_0^+$ represent the set of positive and nonnegative real numbers, respectively.

First, we recall the definition of partial metric space that was introduced by Matthews (1994) in 1992 to handle the problems of the research fields of computer science, particularly, "domain theory" and "semantics":

Definition 1 A partial metric on a nonempty set X is a function $p : X \times X \to \mathbb{R}_0^+$ such that for all $x, y, z \in X$,

(p_1) $x = y$ if and only if $p(x, x) = p(x, y) = p(y, y)$;
(p_2) $p(x, x) \leq p(x, y)$;
(p_3) $p(x, y) = p(y, x)$;
(p_4) $p(x, y) \leq p(x, z) + p(z, y) - p(z, z)$.

A pair (X, p) is called a partial metric space.

At the first glance, the concept of partial metric and the notion of standard metric are very close the each other. On the other hand, the nature of these space are very different due to the main difference of partial metric: non-zero self-distance. More precisely, in partial metric space, self-distance, $p(x, x)$, is not necessarily equal to zero. Furthermore, the topologies of these spaces are also quite different from each other. In particular, in partial metric space the limit is not unique.

When it is heard first that $p(x, x) \neq 0$, one can think that it is ridiculous/irrational /unnatural axiom. Actually, in the frame of computer scientist, it is very natural and working with partial metric space is more "economical" than studying with the standard metric. To clarify why it is a very natural of the axiom, it is enough to consider the operator of "maximum" on the positive half-line with zero, that is $(\mathbb{R}_0^+, \max\{\cdot, \cdot\})$. Indeed, $\max\{a, a\} = a > 0$ for any $a \in \mathbb{R}^+$ and $\max\{a, a\} = 0$ only if $a = 0$. Accordingly, $(\mathbb{R}_0^+, \max\{\cdot, \cdot\})$ satisfies (p_1). Regarding that the others conditions are standard, we deduce that $(\mathbb{R}_0^+, \max\{\cdot, \cdot\})$ fulfills all axioms of partial metrics.

To understand why the partial metric space is more economical for the computer scientist, we shall examine the distance function d, on the set of all infinite sequences ω, defined as follows:

$$d : \omega \times \omega : [0, \infty), \tag{1}$$

such that

$$d(x, y) = 2^{-\sup\{n | \forall i < n \text{ such that } x_i = y_i\}}. \tag{2}$$

It is clear that (ω, d) provides all axioms of a standard metric and it is known as the Baire metric in the literature. At this point, we shall include the aspects of researchers in computer science. Although, in mathematics, investigation of the infinite sequences is interesting, finite sequences have been preferred in computer science, in particular, in software programming. More precisely, the "good" and the "correct" software program is a program which terminates after finite number of step. It is clear that it is not "good" and not "economic", if the program does not terminate after it runs due to infinite sequences, or it processes in infinite loops. Regarding this fact, Matthews (1994) modified the Baire metric space (ω, d) by replacing the domain ω with its extension, $\omega \cup \omega_f$, where ω_f is the set of all finite sequence, that is,

$$d : (\omega \cup \omega_f) \times (\omega \cup \omega_f) : [0, \infty), \tag{3}$$

such that

$$p(x, y) = 2^{-\sup\{n | \forall i < n \text{ such that } x_i = y_i\}}. \tag{4}$$

It is clear that the pair $(\omega \cup \omega_f, p)$ provides all axioms of a partial metric space. Notice that, here $p(x, x)$ is not necessarily zero. Indeed, for $x \in \omega_f$ such that $x = (x_1, x_2, x_3, \ldots, x_n)$ for some $n \in \mathbb{N}$, we have $p(x, x) = \frac{1}{2^n} \neq 0$.

Inspired from such examples, for getting "good" and "economic" software programming Matthews (1994) improved the notion of standard metric to partial metric. Furthermore, he discussed the topology of the new abstract space to be able to transfer the techniques and knowledge of the standard topology of metric space to corresponding topology of partial metric space. By the help of the new structure, it is possible to terminate the software program by a fixed point of an operator that is obtained by a proper transformation of the software programming problem to a fixed point problem. This direction was also initiated by Matthews (1994) who obtained the mimic of famous Banach Contraction Mapping Principle in the context of partial metric spaces. Inspired from this pioneer result, a number of papers has released on the existence and uniqueness of distinct operators in the setting of the partial metric spaces (see e.g. Karapınar et al. 2018; Abdeljawad et al. 2012; Agarwal et al. 2012; Ali et al. 2014; Alsulami et al. 2014; Altun and Erduran 2011; Altun and Simsek 2008; Altun et al. 2010; Aydi 2011a, b, c, d, 2012; Aydi and Karapınar 2012; Aydi et al. 2011, 2012a, b, 2015a, b, c; Banach 1922; Bukatin et al. 2009; Chen and Karapınar 2013; Chi et al. 2012; Ćirić et al. 2011; Ćirić 1974a, b; Escardo 1996; Gulyaz and Karapınar 2013; Heckmann 1999; Hitzler and Seda 2011; Hošková-Mayerová 2016, 2017; Ilić et al. 2011, 2012; Jleli et al. 2013a, b, c; Kannan 1968; Karapınar 2011a, c, 2012a, b; Karapınar and Erhan 2011, 2012; Karapınar and Romaguera 2013; Karapınar and Samet 2012; Karapınar and Yuksel 2011; Karapınar et al. 2012; Karapınar et al. 2013a; b; Karapınar et al. 2015; 2018; Kopperman et al. 2004; Kramosil and Michalek 1975; Künzi et al. 2006; Matthews 1992; Mohammadi et al. 2012; Oltra and Valero 2004; Paesano and Vetro 2012; Popa 1997; Reich 1971; Roldan et al. 2013; Romaguera 2010, 2011, 2012; Romaguera and Schellekens 2002, 2005; Romaguera and Valero 2009; Samet et al. 2011, 2013; Schellekens 2003, 2004; Shatanawi et al. 2012; Shobkolaei et al. 2011; Stoy 1981; Valero 2005; Vetro and Vetro 2013; Vetro and Radenović 2012; Waszkiewicz 2003, 2006).

The details of the following example can be found in the recent paper of Karapınar et al. (2018).

Example 1 Let (X, d) be standard metric space and (X, p) be a partial metric space. Consider the mappings $q_i : X \times X \longrightarrow \mathbb{R}_0^+$ $(i \in \{1, 2, 3\})$ defined by

$$q_1(x, y) = d(x, y) + p(x, y)$$
$$q_2(x, y) = d(x, y) + \max\{\omega(x), \omega(y)\}$$
$$q_3(x, y) = d(x, y) + a$$

It is clear that the functions q_1, q_2, q_3 form partial metrics on X, where $\omega : X \longrightarrow \mathbb{R}_0^+$ is an arbitrary function and $a \geq 0$.

Notice that if we take the metric d as taxi-cab metric or space-taxi-cab metric in Example 1, we get distinct partial metric space. For example, one can take $\omega(x)$ as a traffic density of road, or the traffic lights on the road, or some other criteria that makes the journey more comfortable.

The details of the following two interesting examples can be found in the paper of Matthews (1994).

Example 2 Let $X = \{[a, b] : a, b, \in \mathbb{R}, \ a \leq b\}$ and define

$$p([a, b], [c, d]) = \max\{b, d\} - \min\{a, c\}.$$

Then, the pair (X, p) forms a partial metric space.

Example 3 Let $X := [0, 1] \cup [2, 3]$ and define $p : X \times X \to [0, \infty)$ by

$$p(x, y) = \begin{cases} \max\{x, y\} & \text{if } \{x, y\} \cap [2, 3] \neq \emptyset, \\ |x - y| & \text{if } \{x, y\} \subset [0, 1]. \end{cases}$$

Then, the pair (X, p) forms a partial metric space.

Remark 1 If $p(x, y) = 0$, then from (p_1) and (p_2), we have $x = y$. The converse may not hold as it is discussed above.

Each partial metric p on X generates a T_0 topology τ_p on X which has as a base the family of open p-balls $\{B_p(x, \gamma) : x \in X, \gamma > 0\}$, where $B_p(x, \gamma) = \{y \in X : p(x, y) < p(x, x) + \gamma\}$ for all $x \in X$ and $\gamma > 0$. If p is a partial metric on X, then the function $d_p : X \times X \to \mathbb{R}_0^+$ given by

$$d_p(x, y) = 2p(x, y) - p(x, x) - p(y, y)$$

form a metric on X. Moreover, the functions d_m^p and d_{p_0} defined on $X \times X$ by

$$\begin{aligned} d_m^p(x, y) &= \max\{p(x, y) - p(x, x), p(x, y) - p(y, y)\} \\ &= p(x, y) - \min\{p(x, x), p(y, y)\}, \end{aligned} \tag{5}$$

and

$$d_{p_0}(x, x) = \begin{cases} p(x, y), & \text{for all } x \neq y, \\ 0, & \text{otherwise,} \end{cases} \tag{6}$$

form metrics on X (see Altun and Acar (2012) and Sehgal (1974), respectively). Furthermore, we observe that

$$\tau_p \subseteq \tau_{d_p} = \tau_{d_p^m} \subseteq \tau_{d_{p_0}}.$$

Pay attention to the fact that in the partial metric space (X, p) mentioned in Remark 1 both d_p and d_p^m are the Euclidean metric on X.

For the partial metric space, Matthews (1994) suggested the following topological notions which are the analogs of the corresponding concepts in the setting of the standard metric space.

Definition 2 Let (X, p) be a partial metric space. Then

(1) a sequence $\{x_n\}$ in (X, p) converges to $x \in X$ if $p(x, x) = \lim_{n \to \infty} p(x, x_n)$;
(2) a sequence $\{x_n\}$ in (X, p) is called a Cauchy sequence if $\lim_{m,n \to \infty} p(x_m, x_n)$ exists (and is finite);
(3) (X, p) is said to be complete if every Cauchy sequence $\{x_n\}$ in X converges, with respect to τ_p, to a point $x \in X$ such that $p(x, x) = \lim_{m,n \to \infty} p(x_m, x_n)$;
(4) a subset A of a partial metric space (X, p) is closed in (X, p) if it contains its limit points, that is, if a sequence $\{x_n\}$ in A converges to some $x \in X$, then $x \in A$.

(5) a subset A of a partial metric space (X, p) is bounded in (X, p) if there exist $x_0 \in X$ and $M \in \mathbb{R}$ such that for all $a \in A$, we have $a \in B_p(x_0, M)$, that is, $p(x_0, a) < p(a, a) + M$.

Definition 3 Let (X, p) be a partial metric space. A self-mapping T on X is called continuous, if for each sequence $\{x_n\}$ in X which converges to $u \in X$, that is,

$$\lim_{n \to \infty} p(x_n, u) = \lim_{n \to \infty} p(x_n, x_{n+k}) = p(u, u) \tag{7}$$

provides

$$\lim_{n \to \infty} p(Tx_n, Tu) = \lim_{n \to \infty} p(Tx_n, Tx_{n+k}) = p(Tu, Tu). \tag{8}$$

Notice that the equality (8) can be expressed as

$$\lim_{n \to \infty} p(Tx_n, Tu) = \lim_{n \to \infty} p(x_{n+1}, Tu) = \lim_{n \to \infty} p(x_{n+1}, x_{n+k+1}) = p(u, u)$$
$$= \lim_{n \to \infty} p(Tx_n, Tx_{n+k}) = P(Tu, Tu). \tag{9}$$

Remark 2 The limit in a partial metric space may not be unique. For example, consider the sequence $\{\frac{1}{n^2+n}\}_{n \in \mathbb{N}}$ in the partial metric space (X, p) where $p(x, y) = \max\{x, y\}$. Note

$$p(1, 1) = \lim_{n \to \infty} p\left(1, \frac{1}{n^2 + n}\right) \quad \text{and} \quad p(2, 2) = \lim_{n \to \infty} p\left(2, \frac{1}{n^2 + n}\right).$$

Matthews (1994) proved the following lemma.

Lemma 1 *(i) $\{x_n\}$ is a Cauchy sequence in a partial metric space (X, p) if and only if it is a Cauchy sequence in the metric space (X, d_p);*

(ii) A partial metric space (X, p) is complete if and only if the metric space (X, d_p) is complete. Furthermore, $\lim_{n \to \infty} d_p(x_n, x) = 0$ if and only if $p(x, x) = \lim_{n \to \infty} p(x_n, x) = \lim_{n \to \infty} p(x_n, x_m)$.

(iii) If $\{x_n\}$ is a convergent sequence in (X, d_p), then it is a convergent sequence in the partial metric space (X, p).

Abedeljawad et al. (2011) proved the following lemma.

Lemma 2 *Let $\{x_n\}$ and $\{y_n\}$ be two sequences in a partial metric space X such that*

$$\lim_{n \to \infty} p(x_n, x) = \lim_{n \to \infty} p(x_n, x_n) = p(x, x), \tag{10}$$

and

$$\lim_{n \to \infty} p(y_n, y) = \lim_{n \to \infty} p(y_n, y_n) = p(y, y), \tag{11}$$

then $\lim_{n \to \infty} p(x_n, y_n) = p(x, y)$. In particular, $\lim_{n \to \infty} p(x_n, z) = p(x, z)$ for every $z \in X$.

For the sake of completeness, we recall following lemma of Karapınar (2011b).

Lemma 3 *Let* (X, p) *be a partial metric space. Then*

(A) If $p(x, y) = 0$ *then* $x = y$,
(B) If $x \neq y$, *then* $p(x, y) > 0$.

Existence and uniqueness of different operators in the frame of partial metric spaces have been investigated and reported by a number of researchers (see e.g., Karapınar et al. 2018 and references therein). In 2013, Haghi et al. (2013) showed that some fixed point results in the setting of partial metric spaces are equivalent to the corresponding ones in the frame of standard metric. Although some of the announced results lack of novelty, as it is shown in the paper of Haghi et al. (2013), the obtained results still have a worth due to several aspects:

1. It is economical in the frame of computer scientist.
2. Since the topology of partial metric has some weakness, the proofs need to special painstaking.
3. Since the natures of the standard metric and partial metric are quite distinct, it needs an additional efforts to get the existence results in the context of partial metric.

For the sake of completeness, we also recall the approaches of Haghi et al. (2013) to avoid getting worthless results: For a mapping $T : X \to X$ where $X \neq \emptyset$ it was observed, in Haghi et al. (2013), that

$$M_d^T(x, y) = M_p^T(x, y), \text{ with}$$

$$M_p^T(x, y) = \max\{\rho(x, y), \rho(x, Tx), \rho(Ty, y), \rho(Tx, y), \rho(x, Ty)\}$$

where $\rho = d$, p are metric, and partial metric, respectively. We underline that this approach is not applicable for our results.

Let Φ be the family of nondecreasing functions $\phi : [0, \infty) \to [0, \infty)$ satisfying the following conditions:

(Φ_1) $\phi(t) < t$, for any $t \in \mathbb{R}^+$;
(Φ_2) the series $\sum_{k=1}^{\infty} \phi^k(t)$ converges for any $t \in \mathbb{R}^+$.

The following interesting lemma proved by Rus (2001).

Lemma 4 *If* $\phi \in \Phi$, *then the following hold:*

(i) $(\phi^n(t))_{n \in \mathbb{N}}$ *converges to* 0 *as* $n \to \infty$ *for all* $t \in \mathbb{R}^+$;
(ii) ϕ *is continuous at* 0;

In what follows, we recall the concept of α-orbital admissiblity introduced by Popescu (2014), a refinement of the α-admissiblity notion defined in Samet et al. (2012).

Definition 4 Let $T : X \to X$ be a mapping and $\alpha : X \times X \to [0, \infty)$ be a function. We say that T is an α-orbital admissible if

$$\alpha(x, Tx) \geq 1 \Rightarrow \alpha(Tx, T^2x) \geq 1.$$

Further, T is called a triangular α-orbital admissible if T is α-orbital admissible and

$$\alpha(x, y) \geq 1 \text{ and } \alpha(y, Ty) \geq 1 \Rightarrow \alpha(x, Ty) \geq 1.$$

Notice that each α-admissible mapping is an α-orbital admissible.

In what follows we recall the notion of *simulation function* that was suggested by Khojasteh et al. (2015) to combine certain existing fixed point results on the topic in the literature.

Definition 5 A *simulation function* is a mapping $\zeta : [0, \infty) \times [0, \infty) \to \mathbb{R}$ satisfying the following conditions:

(ζ_1) $\zeta(0, 0) = 0$;
(ζ_2) $\zeta(t, s) < s - t$ for all $t, s > 0$;
(ζ_3) if $\{t_n\}, \{s_n\}$ are sequences in $(0, \infty)$ such that $\lim_{n \to \infty} t_n = \lim_{n \to \infty} s_n > 0$, then

$$\limsup_{n \to \infty} \zeta(t_n, s_n) < 0. \tag{12}$$

Let $\mathscr{Z}$ denote the family of all simulation functions $\zeta : [0, \infty) \times [0, \infty) \to \mathbb{R}$. Due to the axiom ($\zeta_2$), we have

$$\zeta(t, t) < 0 \text{ for all } t > 0. \tag{13}$$

We recollect the examples from the recent results of Khojasteh et al. (2015) and Roldán-López-de-Hierro et al. (2015).

Example 4 Let $\mu_i : [0, \infty) \to [0, \infty)$ be continuous functions with $\mu_i(t) = 0$ if, and only if, $t = 0$. For $i = 1, 2, 3, 4, 5, 6$, we define the mappings $\zeta_i : [0, \infty) \times [0, \infty) \to \mathbb{R}$, as follows

(i) $\zeta_1(t, s) = \mu_1(s) - \mu_2(t)$ for all $t, s \in [0, \infty)$, where $\mu_1(t) < t \leq \mu_2(t)$ for all $t > 0$.

(ii) $\zeta_2(t, s) = s - \dfrac{f(t, s)}{g(t, s)} t$ for all $t, s \in [0, \infty)$, where $f, g : [0, \infty)^2 \to (0, \infty)$ are two continuous functions with respect to each variable such that $f(t, s) > g(t, s)$ for all $t, s > 0$.

(iii) $\zeta_3(t, s) = s - \mu_3(s) - t$ for all $t, s \in [0, \infty)$.

(iv) If $\varphi : [0, \infty) \to [0, 1)$ is a function such that $\limsup_{t \to r^+} \varphi(t) < 1$ for all $r > 0$, and we define

$$\zeta_4(t, s) = s\varphi(s) - t \qquad \text{for all } s, t \in [0, \infty).$$

(v) If $\eta : [0, \infty) \rightarrow [0, \infty)$ is an upper semi-continuous mapping such that $\eta(t) < t$ for all $t > 0$ and $\eta(0) = 0$, and we define

$$\zeta_5(t, s) = \eta(s) - t \qquad \text{for all } s, t \in [0, \infty).$$

(vi) If $\mu : [0, \infty) \rightarrow [0, \infty)$ is a function such that $\int_0^\varepsilon \mu(u)du$ exists and $\int_0^\varepsilon \mu(u)du > \varepsilon$, for each $\varepsilon > 0$, and we define

$$\zeta_6(t, s) = s - \int_0^t \mu(u)du \qquad \text{for all } s, t \in [0, \infty).$$

It is clear that each function ζ_i $(i = 1, 2, 3, 4, 5, 6)$ forms a simulation function.

Suppose (X, d) is a metric space, T is a self-mapping on X and $\zeta \in \mathscr{Z}$. We say that T is a $\mathscr{Z}$-contraction with respect to ζ, if

$$\zeta(d(Tx, Ty), d(x, y)) \geq 0 \qquad \text{for all } x, y \in X. \tag{14}$$

Again by (ζ_2), we have the following inequality

$$q(Tx, Ty) \neq q(x, y) \text{ for all distinct } x, y \in X. \tag{15}$$

Thus, we conclude that T cannot be an isometry whenever T is a $\mathscr{Z}$-contraction. In other words, if a $\mathscr{Z}$-contraction T in a metric space has a fixed point, then it is necessarily unique.

Theorem 1 *Every $\mathscr{Z}$-contraction on a complete metric space has a unique fixed point. In fact, each Picard sequence converges to its unique fixed point.*

In this chapter, we investigate the existence and uniqueness of a fixed point of certain mapping satisfying a special contraction that is a combination of $\alpha - \phi$-contraction and $\mathscr{Z}$-contraction. The presented results extend and improve several existing results in the literature, in particular the results of Karapınar et al. (2018).

2 Main Results

In this section, we shall state and prove a fixed point theorem that cover and unify several existing results in the literature.

Definition 6 Let T be a self-mapping defined on a partial metric space (X, p) and $\zeta \in \mathscr{Z}$. We say that T is an $(\alpha - \phi)$-type Z_K-contraction if there exist mappings $\alpha : X \times X \rightarrow [0, \infty)$ and $\phi \in \Phi$ such that

$$\zeta(\alpha(x, y)p(Tx, Ty), \phi(K(x, y))) \geq 0, \text{ for all } x, y \in X, \tag{16}$$

where

$$K(x, y) := a_1 p(x, y) + a_2 p(x, Tx) + a_3 p(y, Ty) + a_4[p(x, Ty) + p(y, Tx)], \tag{17}$$

with $0 \le a_i \le 1$, $i = 1, 2, 3, 4$, and $a_1 + a_2 + a_3 + 2a_4 \le 1$.

Theorem 2 *Let (X, p) be a complete partial metric space and let $T : X \to X$ be an $(\alpha - \phi)$-type Z_K-contraction. Suppose that*

(i) *T is α-orbital admissible;*
(ii) *there exists $x_0 \in X$ such that $\alpha(x_0, Tx_0) \ge 1$;*
(iii) *T is continuous.*

Then, there exists $u \in X$ such that $Tu = u$ and $p(u, u) = 0$.

Proof The condition (ii) guarantee that there exists $x_0 \in X$ such that $\alpha(x_0, Tx_0) \ge 1$. Starting from this point, we shall built a recursive sequence $\{x_n\}$ in X by

$$x_{n+1} = Tx_n \text{ for all } n \ge 0. \tag{18}$$

Without loss of generality, throughout the proof, we suppose that

$$x_n \ne x_{n+1} \text{ for all } n.$$

Indeed, if there exists $n_0 \in \mathbb{N}_0$ such that $x_{n_0} = x_{n_0+1}$, then $u = x_{n_0}$ forms a fixed point of T and the proof is over.

On account of the fact that T is α-admissible, we derive

$$\alpha(x_0, x_1) = \alpha(x_0, Tx_0) \ge 1 \Rightarrow \alpha(Tx_0, Tx_1) = \alpha(x_1, x_2) \ge 1.$$

Recursively, we obtain that

$$\alpha(x_n, x_{n+1}) \ge 1, \text{ for all } n = 0, 1, \ldots \tag{19}$$

From (16) and (18), we find that

$$\begin{aligned}
0 &\le \zeta(\alpha(x_n, x_{n-1})p(Tx_n, Tx_{n-1}), \phi(K(x_n, x_{n-1}))) \\
&= \zeta(\alpha(x_n, x_{n-1})p(x_{n+1}, x_n), \phi(K(x_n, x_{n-1}))) \\
&< \phi(K(x_n, x_{n-1})) - \alpha(x_n, x_{n-1})p(x_{n+1}, x_n).
\end{aligned} \tag{20}$$

Combining (19) and (20), we get that

$$p(x_{n+1}, x_n) = p(Tx_n, Tx_{n-1}) \le \alpha(x_n, x_{n-1})p(Tx_n, Tx_{n-1}) \le \phi(K(x_n, x_{n-1})), \tag{21}$$

for all $n \ge 1$, where

$$K(x_n, x_{n-1}) = a_1 p(x_n, x_{n-1}) + a_2 p(x_n, Tx_n) + a_3 p(x_{n-1}, Tx_{n-1})$$
$$+ a_4[p(x_n, Tx_{n-1}) + p(x_{n-1}, Tx_n)]$$
$$= a_1 p(x_n, x_{n-1}) + a_2 p(x_n, x_{n+1}) + a_3 p(x_{n-1}, x_n)$$
$$+ a_4[p(x_n, x_n) + p(x_{n-1}, x_{n+1})].$$

On account of the axiom (p_4) of partial metric, we find that

$$p(x_{n-1}, x_{n+1}) + p(x_n, x_n) \leq p(x_n, x_{n-1}) + p(x_n, x_{n+1}).$$

So, we deduce that

$$K(x_n, x_{n-1}) = (a_1 + a_3 + a_4)p(x_n, x_{n-1}) + (a_2 + a_4)p(x_n, x_{n+1}).$$

If the case $p(x_n, x_{n-1}) \leq p(x_n, x_{n+1})$ occurs for some $n \in \mathbb{N}$, then, since $a_1 + a_2 + a_3 + a_4 \leq 1$, we find that

$$K(x_n, x_{n-1}) \leq (a_1 + a_2 + a_3 + 2a_4)p(x_n, x_{n+1}) \leq p(x_n, x_{n+1}), \quad (22)$$

Regarding that ϕ is a nondecreasing function and keeping (22) in the mind, we derive from the inequality (21) that

$$p(x_{n+1}, x_n) \leq \phi(K(x_n, x_{n-1})) \leq \phi(p(x_n, x_{n+1})) < p(x_n, x_{n+1}),$$

a contradiction. Accordingly, for all $n \geq 1$, we deduce that

$$p(x_n, x_{n+1}) \leq p(x_n, x_{n-1}). \quad (23)$$

Employing (21) and (23), we find that

$$p(x_{n+1}, x_n) \leq \phi(p(x_n, x_{n-1})), \quad (24)$$

for all $n \geq 1$. Iteratively, we obtain that

$$p(x_{n+1}, x_n) \leq \phi^n(p(x_1, x_0)), \text{ for all } n \geq 1. \quad (25)$$

Owing to Lemma (25) (i), we get that

$$\lim_{n \to \infty} p(x_{n+1}, x_n) = 0. \quad (26)$$

Again by taking (25) into consideration and employing the triangular inequality (p_4), for all $k \geq 1$, we have

$$p(x_n, x_{n+k}) \leq p(x_n, x_{n+1}) + \cdots + p(x_{n+k-1}, x_{n+k}) - \sum_{j=1}^{k-1}(p(x_{n+j}, x_{n+j}))$$

$$\leq \sum_{j=n}^{n+k-1} \phi^j(p(x_1, x_0))$$

$$\leq \sum_{j=n}^{+\infty} \phi^j(p(x_1, x_0)) \to 0 \text{ as } n \to \infty.$$

It yields that

$$\lim_{n \to \infty} p(x_n, x_{n+k}) = 0,$$

and thus $\{x_n\}$ is a Cauchy sequence in (X, d). On account of the fact that (X, p) is complete, there is $u \in X$ such that

$$\lim_{n \to \infty} p(x_n, u) = 0 = \lim_{n \to \infty} p(x_n, x_{n+k}) = p(u, u). \tag{27}$$

Regarding the continuity of the mapping T, and keeping Definition 3 in mind, we deduce from (27) that

$$\lim_{n \to \infty} p(x_{n+1}, Tu) = \lim_{n \to \infty} p(Tx_n, Tu) = 0. \tag{28}$$

Due to Lemma 2 together with (27) and (28), we get that u is a fixed point of T, that is, $Tu = u$.

Definition 7 A non-empty set X is called regular if each recursive sequence $\{x_n\}$ in X fulfills the following condition:

(R) if $\alpha(x_n, x_{n+1}) \geq 1$ for all n and $x_n \to x \in X$ as $n \to \infty$, then there exists a subsequence $\{x_{n(k)}\}$ of $\{x_n\}$ such that $\alpha(x_{n(k)}, x) \geq 1$ for all k.

Theorem 3 *Let (X, p) be a complete partial metric space and let $T : X \to X$ be an $(\alpha - \phi)$-type Z_K-contraction. Suppose that*

(i) *T is α-orbital admissible;*
(ii) *there exists $x_0 \in X$ such that $\alpha(x_0, Tx_0) \geq 1$;*
(iii) *X is regular.*

Then there exists $u \in X$ such that $Tu = u$ and $p(u, u) = 0$.

Proof It is clear that the distinction between Theorems 2 and 3 is the condition (iii). By mimic of the lines in the proof of Theorem 2, we deduce that there is a recursive sequence $\{x_n\}$ defined by $x_{n+1} = Tx_n$ for all $n \in \mathbb{N}_0$ which converges to some $u \in X$. Taking (19) into account together with the condition (iii) of Theorem 3, we get a subsequence $\{x_{n(k)}\}$ of $\{x_n\}$ such that $\alpha(x_{n(k)}, u) \geq 1$ for all k. Employing (16), for all k, we find that

$$0 \leq \zeta(\alpha(x_{n(k)}, u)p(Tx_{n(k)}, Tu), \phi(K(x_{n(k)}, u)))$$
$$= \phi(K(x_{n(k)}, u)) - \alpha(x_{n(k)}, u)p(Tx_{n(k)}, Tu) \tag{29}$$

Accordingly, we have

$$p(x_{n(k)+1}, Tu) = p(Tx_{n(k)}, Tu) \leq \alpha(x_{n(k)}, u)p(Tx_{n(k)}, Tu)$$
$$\leq \phi(K(x_{n(k)}, u)) < K(x_{n(k)}, u), \tag{30}$$

where,

$$K(x_{n(k)}, u) = a_1 p(x_{n(k)}, u) + a_2 p(x_{n(k)}, x_{n(k)+1}) + a_3 p(u, Tu)$$
$$+ a_4[p(x_{n(k)}, Tu) + p(u, x_{n(k)+1})].$$

Taking $k \to \infty$ in the equality (30), we obtain that

$$p(u, Tu) \leq (a_3 + a_4)p(u, Tu), \tag{31}$$

which is a contradiction. Thus we have $p(u, Tu) = 0$, that is, $u = Tu$.

For the uniqueness of a fixed point derived in Theorems 2 and 3, we shall need an extra hypothesis. Let T be a self-mapping on a non-empty set X and $\alpha : X \times X \to \mathbb{R}_0^+$. Suppose that $Fix(T)$ denotes the set of all fixed points of the self-mapping T. Then, we say that T provides the (U) condition, if

(U) For all $u, v \in Fix(T)$, then $\alpha(u, v) \geq 1$,

Theorem 4 *Putting the condition (U) to the statements of Theorem 2 (resp. Theorem 3), we find that u is the unique fixed point of T.*

Proof Let u, v be two distinct fixed point of T and $p(u, v) > 0$. Note that in case $p(u, v) = 0$, there is nothing to prove. Taking the property of ϕ into account, we derive that $\phi(p(u, v)) > 0$.

On account of the condition (U) and the assumption of the Theorem 2 (resp. Theorem 3)

$$0 \leq \zeta(\alpha(u, v)p(Tu, Tv), \phi(K(u, v)))$$
$$= \phi(K(u, v)) - \alpha(u, v)p(Tu, Tv) \tag{32}$$

and hence

$$p(u, v) \leq \alpha(u, v)p(Tu, Tv)$$
$$\leq \phi(K(u, v)) = \phi(p(u, v))$$
$$< p(u, v),$$

which is a contradiction. Thus, $u = v$.

Definition 8 Let T be a self-mapping defined on a partial metric space (X, p) and $\zeta \in \mathscr{Z}$. We say that T is an $(\alpha - \phi)$-type Z_N-contraction if there exist mappings $\alpha : X \times X \to [0, \infty)$ and $\phi \in \Phi$ such that

$$\zeta(\alpha(x, y)p(Tx, Ty), \phi(N(x, y))) \geq 0, \text{ for all } x, y \in X, \tag{33}$$

where

$$N(x, y) := a_1 p(x, y) + a_2[p(x, Tx) + p(y, Ty)] + a_3[p(x, Ty) + p(y, Tx)], \tag{34}$$

where $0 \leq a_i \leq 1$, $i = 1, 2, 3$, and $a_1 + 2a_2 + 2a_3 \leq 1$.

Theorem 5 *Let (X, p) be a complete partial metric space and let $T : X \rightarrow X$ be an $(\alpha - \phi)$-type Z_N-contraction. Suppose that*

(i) *T is α-orbital admissible;*
(ii) *there exists $x_0 \in X$ such that $\alpha(x_0, Tx_0) \geq 1$;*
(iii) *either T is continuous, or, X is regular.*

Then there exists $u \in X$ such that $Tu = u$ and $p(u, u) = 0$.

Since $N(x, y) \leq K(x, y)$ for all $x, y \in X$, the proof of Theorem 5 can be derived as a consequence of Theorems 2 and 3.

On the other hand, for the uniqueness of a fixed point of the operator, defined in Theorem 5, we shall propose a weaker condition:

(H) For all $x, y \in \text{Fix}(T)$, there exists $z \in X$ such that $\alpha(x, z) \geq 1$ and $\alpha(y, z) \geq 1$.

Theorem 6 *Adding condition (H) to the hypotheses of Theorem 5, we obtain that u is the unique fixed point of T.*

Proof Suppose that v is another fixed point of T. From (H), there exists $z \in X$ such that

$$\alpha(u, z) \geq 1 \text{ and } \alpha(v, z) \geq 1. \tag{35}$$

Since T is α-admissible and u, v are the fixed point of T, the inequalities in (35) yield that

$$\alpha(u, T^n z) \geq 1 \text{ and } \alpha(v, T^n z) \geq 1, \text{ for all } n. \tag{36}$$

We construct an iterative sequence $\{z_n\}$ in X by $z_{n+1} = Tz_n$ for all $n \geq 0$ and $z_0 = z$. From (36), for all n, we have

$$0 \leq \zeta(\alpha(u, z_n)p(Tu, Tz_n), \phi(K(u, z_n))) = \phi(N(u, z_n) - \alpha(u, z_n)p(Tu, Tz_n), \tag{37}$$

and hence

$$p(u, z_{n+1}) = p(Tu, Tz_n) \leq \alpha(u, z_n)p(Tu, Tz_n) \leq \phi(N(u, z_n)), \tag{38}$$

where

$$
\begin{aligned}
N(u, z_n) &= a_1 p(u, z_n) + a_2 p(u, Tu) + a_3 p(z_n, Tz_n) \\
&= a_1 p(u, z_n) + a_2 p(u, u) + a_3 p(z_n, z_{n+1}) \\
&\leq a_1 p(u, z_n) + a_2[p(u, z_{n+1}) + p(u, z_{n+1}) - p(z_{n+1}, z_{n+1})] \\
&\quad + a_3[p(u, z_{n+1}) + p(u, z_n) - p(u, u)] \\
&\leq (a_1 + a_3) p(u, z_n) + (2a_2 + a_3) d(u, z_{n+1}).
\end{aligned}
$$

Without lost of the generality, we can suppose that $p(u, z_n) > 0$ for all n. If we have $p(u, z_n) \leq p(u, z_{n+1})$, then due to the monotone property of ϕ, and the inequality (38), we find that

$$
p(u, z_{n+1}) \leq \phi((a_1 + 2a_2 + 2a_3 + 2a_4) p(u, z_{n+1})). \tag{39}
$$

By keeping, (38) and the monotone property of ϕ, in the mind, if $\max\{p(u, z_n), p(u, z_{n+1})\} = p(u, z_{n+1})$, we get that, for all n,

$$
p(u, z_{n+1}) \leq \phi((a_1 + 2a_2 + 2a_3) p(u, z_{n+1})) \leq \phi(p(u, z_{n+1})) < p(u, z_{n+1}),
$$

which is a contradiction. Thus we have $\max\{p(u, z_n), p(u, z_{n+1})\} = p(u, z_n)$, and

$$
p(u, z_{n+1}) \leq \phi(p(u, z_n)),
$$

for all n. This implies that

$$
p(u, z_n) \leq \phi^n(p(u, z_0)), \quad \text{for all } n \geq 1.
$$

Letting $n \to \infty$ in the above inequality, we obtain

$$
\lim_{n \to \infty} p(z_n, u) = 0. \tag{40}
$$

Similarly, one can show that

$$
\lim_{n \to \infty} p(z_n, v) = 0. \tag{41}
$$

From (40) and (41), it follows that $u = v$. Thus we proved that u is the unique fixed point of T.

2.1 Some Immediate Consequences Due to Choice of Coefficients a_1, a_2, a_3, a_4

In what follows, we shall list some consequences of Theorems 2–6 by choosing various combinations of coefficients a_1, a_2, a_3, a_4.

Theorem 7 *Let (X, p) be a complete partial metric space, $\zeta \in \mathscr{Z}$, and $\alpha : X \times X \rightarrow [0, \infty)$ and $\phi \in \Phi$. Suppose that a mapping $T : X \rightarrow X$ fulfills*

$$\zeta(\alpha(x, y)p(Tx, Ty), \phi(a_1 p(x, y) + a_2 p(x, Tx) + a_3 p(y, Ty))) \geq 0, \quad (42)$$

for all $x, y \in X$, where $0 \leq a_i \leq 1$, $i = 1, 2, 3$, and $a_1 + a_2 + a_3 \leq 1$. Assume also the assumptions (i)–(iii) in the statement of Theorem 5 fulfill. Then there exists $u \in X$ such that $Tu = u$ and $p(u, u) = 0$.

Proof It is evident that

$$a_1 p(x, y) + a_2 p(x, Tx) + a_3 p(y, Ty) \leq K(x, y) \text{ for all } x, y \in X.$$

On account of the property (Φ_1), the desired results is observed from Theorem 4.

Theorem 8 *Let (X, p) be a complete partial metric space, $\zeta \in \mathscr{Z}$, and $\alpha : X \times X \rightarrow [0, \infty)$ and $\phi \in \Phi$. Suppose that a mapping $T : X \rightarrow X$ fulfills*

$$\zeta(\alpha(x, y)p(Tx, Ty), \phi(a_1 p(x, y) + a_2[p(x, Tx) + p(y, Ty)])) \geq 0, \quad (43)$$

for all $x, y \in X$, where $0 \leq a_i \leq 1$, $i = 1, 2$, and $a_1 + 2a_2 \leq 1$. Suppose also that the assumptions (i)–(iii) in the statement of Theorem 5 satisfy. Then there exists $u \in X$ such that $Tu = u$ and $p(u, u) = 0$.

Proof It is clear that

$$a_1 p(x, y) + a_2[p(x, Tx) + p(y, Ty)] \leq N(x, y) \text{ for all } x, y \in X.$$

Taking the property (Φ_1) into account, the desired results is derived from Theorem 6.

Theorem 9 *Let (X, p) be a complete partial metric space, $\zeta \in \mathscr{Z}$, and $\alpha : X \times X \rightarrow [0, \infty)$ and $\phi \in \Phi$. Suppose that a mapping $T : X \rightarrow X$ fulfills*

$$\zeta(\alpha(x, y)p(Tx, Ty), \phi(b_1 p(x, y) + b_2[p(x, Ty) + p(y, Tx)])) \geq 0, \quad (44)$$

for all $x, y \in X$, where $0 \leq b_i \leq 1$, $i = 1, 2$, and $b_1 + 2b_2 \leq 1$. Suppose that the assumptions (i)–(iii) in statement of Theorem 5 hold. Then there exists $u \in X$ such that $Tu = u$ and $p(u, u) = 0$.

Proof Keeping the property (Φ_1) in mind, together with the inequality below,

$$b_1 p(x, y) + b_2[p(x, Ty) + p(y, Tx)] \leq N(x, y) \text{ for all } x, y \in X,$$

we conclude the desired result from Theorem 6.

Theorem 10 *Let (X, p) be a complete partial metric space, $\zeta \in \mathscr{Z}$, and $\alpha : X \times X \to [0, \infty)$ and $\phi \in \Phi$. Suppose that a mapping $T : X \to X$ fulfills*

$$\zeta\left(\alpha(x, y)p(Tx, Ty), \phi\left(\frac{[p(x, Tx) + p(y, Ty)]}{2}\right)\right) \geq 0, \qquad (45)$$

for all $x, y \in X$. Suppose that the assumptions (i)–(iii) in the statement of Theorem 5 satisfy. Then there exists $u \in X$ such that $Tu = u$ and $p(u, u) = 0$. Additionally, if the condition (H) is satisfied, then u is the unique fixed point of T.

Proof The result follows from Theorem 6 due to the property (Φ_1) together with the inequality below:

$$\frac{[p(x, Tx) + p(y, Ty)]}{2} \leq N(x, y) \text{ for all } x, y \in X.$$

Theorem 11 *Let (X, p) be a complete partial metric space, $\zeta \in \mathscr{Z}$, and $\alpha : X \times X \to [0, \infty)$ and $\phi \in \Phi$. Suppose that a mapping $T : X \to X$ fulfills*

$$\zeta\left(\alpha(x, y)p(Tx, Ty), \phi\left(\frac{p(x, Ty) + p(y, Tx)}{2}\right)\right) \geq 0, \qquad (46)$$

for all $x, y \in X$. Suppose also that the assumptions (i)–(iii) in the statement of Theorem 5 fulfill. Then there exists $u \in X$ such that $Tu = u$ and $p(u, u) = 0$.

Proof The result is derived from Theorem 6 due to the property (Φ_1) and the inequality below:

$$\frac{[p(x, Ty) + p(y, Tx)]}{2} \leq N(x, y) \text{ for all } x, y \in X.$$

Theorem 12 *Let (X, p) be a complete partial metric space, $\zeta \in \mathscr{Z}$, and $\alpha : X \times X \to [0, \infty)$ and $\phi \in \Phi$. Suppose that a mapping $T : X \to X$ fulfills*

$$\zeta(\alpha(x, y)p(Tx, Ty), \phi(p(x, y))) \geq 0, \qquad (47)$$

for all $x, y \in X$. Suppose also that the assumptions (i)–(iii) in the statement of Theorem 5 hold. Then there exists $u \in X$ such that $Tu = u$ and $p(u, u) = 0$.

Proof Keeping Theorem 6 in mind, the inequality

$$p(x, y) \leq N(x, y) \text{ for all } x, y \in X,$$

and the property (Φ_1) yields the result.

Remark 3 Additionally, if the condition (U) is satisfied, in Theorems 7–12 then, we guarantee that u is the unique fixed point of T.

Notice that, in this section, for the uniqueness of the fixed point, we use property (U) only in Theorem 7. For the other theorems, we use property (H) instead of (U). It is clear that the condition (U) is stronger than the condition (H).

2.2   Some Consequences Due to Choice of φ

In this subsection, by choosing $\phi(t) = kt$ for $k \in [0, 1)$, we derive several consequences of Theorems 2–12. Here, we just list the results but do not give a detailed proof. Indeed, choosing $\phi(t) = kt$ for $k \in [0, 1)$ and setting $\alpha_i = ka_i$ for $i = 1, 2, 3, 4$ yields $\alpha_1 + \alpha_2 + \alpha_3 + 2\alpha_4 < 1$. Consequently, under this setting, Theorems 2–12 turns in to the following results.

Theorem 13 *Let (X, p) be a complete partial metric space, $\zeta \in \mathscr{Z}$, and $\alpha : X \times X \to [0, \infty)$. Suppose that a mapping $T : X \to X$ satisfies*

$$\zeta(\alpha(x, y)p(Tx, Ty), K_0(x, y)) \geq 0, \tag{48}$$

for all $x, y \in X$, where

$$K_0(x, y) = \alpha_1 p(x, y) + \alpha_2 p(x, Tx) + \alpha_3 p(y, Ty) + \alpha_4[p(x, Ty) + p(y, Tx)]$$

$0 \leq \alpha_i$, $i = 1, 2, 3, 4$, and $\alpha_1 + \alpha_2 + \alpha_3 + 2\alpha_4 < 1$. Suppose also that the assumptions (i)–(iii) in the statement of Theorem 5 hold. Then there exists $u \in X$ such that $Tu = u$ and $p(u, u) = 0$.

Theorem 14 *Let (X, p) be a complete partial metric space, $\zeta \in \mathscr{Z}$, and $\alpha : X \times X \to [0, \infty)$. Suppose that a mapping $T : X \to X$ fulfills*

$$\zeta(\alpha(x, y)p(Tx, Ty), N_0(x, y)) \geq 0, \tag{49}$$

for all $x, y \in X$, where

$$N_0(x, y) = \alpha_1 p(x, y) + \alpha_2[p(x, Tx) + p(y, Ty)] + \alpha_3[p(x, Ty) + p(y, Tx)],$$

$0 \leq \alpha_i$, $i = 1, 2, 3$, and $\alpha_1 + 2\alpha_2 + 2\alpha_3 < 1$. Suppose also that the assumptions (i)–(iii) in the statement of Theorem 5 satisfy. Then there exists $u \in X$ such that $Tu = u$ and $p(u, u) = 0$.

Theorem 15 *Let (X, p) be a complete partial metric space, $\zeta \in \mathscr{Z}$, and $\alpha : X \times X \to [0, \infty)$. Suppose that a mapping $T : X \to X$ provides*

$$\zeta(\alpha(x, y)p(Tx, Ty), \alpha_1 p(x, y) + \alpha_2 p(x, Tx) + \alpha_3 p(y, Ty)) \geq 0, \tag{50}$$

for all $x, y \in X$, *where* $0 \le \alpha_i$, $i = 1, 2, 3$, *and* $\alpha_1 + \alpha_2 + \alpha_3 < 1$. *Suppose also that the assumptions* (i)–(iii) *in the statement of Theorem 5 fulfill. Then there exists* $u \in X$ *such that* $Tu = u$ *and* $p(u, u) = 0$.

Theorem 16 *Let* (X, p) *be a complete partial metric space,* $\zeta \in \mathscr{Z}$, *and* $\alpha : X \times X \to [0, \infty)$. *Suppose that a mapping* $T : X \to X$ *satisfies*

$$\zeta(\alpha(x, y)p(Tx, Ty), \alpha_1 p(x, y) + \alpha_2[p(x, Tx) + p(y, Ty)]) \ge 0, \qquad (51)$$

for all $x, y \in X$, *where* $0 \le \alpha_i$, $i = 1, 2$, *and* $\alpha_1 + 2\alpha_2 < 1$. *Suppose also that the assumptions* (i)–(iii) *in the statement of Theorem 5 hold. Then there exists* $u \in X$ *such that* $Tu = u$ *and* $p(u, u) = 0$.

Theorem 17 *Let* (X, p) *be a complete partial metric space,* $\zeta \in \mathscr{Z}$, *and* $\alpha : X \times X \to [0, \infty)$. *Suppose that a mapping* $T : X \to X$ *fulfills*

$$\zeta(\alpha(x, y)p(Tx, Ty), \alpha_1 p(x, y) + \alpha_3[p(x, Ty) + p(y, Tx)]) \ge 0, \qquad (52)$$

for all $x, y \in X$, *where* $0 \le \alpha_i$, $i = 1, 3$, *and* $\alpha_1 + 2\alpha_3 < 1$. *Suppose also that the assumptions* (i)–(iii) *in the statement of Theorem 5 hold. Then there exists* $u \in X$ *such that* $Tu = u$ *and* $p(u, u) = 0$.

Theorem 18 *Let* (X, p) *be a complete partial metric space,* $\zeta \in \mathscr{Z}$, *and* $\alpha : X \times X \to [0, \infty)$. *Suppose that a mapping* $T : X \to X$ *fulfills*

$$\zeta(\alpha(x, y)p(Tx, Ty), \alpha_2[p(x, Tx) + p(y, Ty)]) \ge 0, \qquad (53)$$

for all $x, y \in X$, *where* $0 \le 2\alpha_2 < 1$. *Suppose also that the assumptions* (i)–(iii) *in the statement of Theorem 5 satisfy. Then there exists* $u \in X$ *such that* $Tu = u$ *and* $p(u, u) = 0$.

Theorem 19 *Let* (X, p) *be a complete partial metric space,* $\zeta \in \mathscr{Z}$, *and* $\alpha : X \times X \to [0, \infty)$. *Suppose that a mapping* $T : X \to X$ *fulfills*

$$\zeta(\alpha(x, y)p(Tx, Ty), \alpha_3[p(x, Ty) + p(y, Tx)]) \ge 0, \qquad (54)$$

for all $x, y \in X$, *where* $0 \le 2\alpha_3 < 1$. *Suppose also that the assumptions* (i)–(iii) *in the statement of Theorem 5 hold. Then there exists* $u \in X$ *such that* $Tu = u$ *and* $p(u, u) = 0$.

Theorem 20 *Let* (X, p) *be a complete partial metric space,* $\zeta \in \mathscr{Z}$, *and* $\alpha : X \times X \to [0, \infty)$. *Suppose that a mapping* $T : X \to X$ *satisfies*

$$\zeta(\alpha(x, y)p(Tx, Ty), \alpha_1 p(x, y)) \ge 0, \qquad (55)$$

for all $x, y \in X$, *where* $0 \le \alpha_1 < 1$. *Suppose also that the assumptions* (i)–(iii) *in the statement of Theorem 5 fulfill. Then there exists* $u \in X$ *such that* $Tu = u$ *and* $p(u, u) = 0$.

Remark 4 Additionally, if the condition (U) is satisfied, in Theorems 13–20 then, we guarantee that u is the unique fixed point of T.

Notice that the given results in this subsection are not the whole lists of the consequence of Theorems 2–6, since the class of Φ is too rich. Accordingly, one can observe further consequences.

2.3 Consequences in the Frame of Partial Metric Spaces with a Partial Order

Ran and Reurings (2003) reported fixed point results in on metric spaces endowed with partial orders and considered their application to matrix equations. After this initial result, studying metric fixed point on metric spaces endowed with partial orders turns to be a trend and a number of paper has appeared. In this section, we shall show that how such theorems can be derived easily from our results by choosing α in a proper way.

Definition 9 Let $(X, \preceq)$ be a partially ordered set, and $T : X \to X$ be a given mapping. We say that T is nondecreasing with respect to $\preceq$ if

$$x, y \in X, \ x \preceq y \Longrightarrow Tx \preceq Ty.$$

Definition 10 Let $(X, \preceq)$ be a partially ordered set. A sequence $\{x_n\} \subset X$ is said to be nondecreasing with respect to $\preceq$ if $x_n \preceq x_{n+1}$ for all n.

Definition 11 Let $(X, \preceq)$ be a partially ordered set and p be a metric on X. We say that $(X, \preceq, p)$ is regular if for every nondecreasing sequence $\{x_n\} \subset X$ such that $x_n \to x \in X$ as $n \to \infty$, there exists a subsequence $\{x_{n(k)}\}$ of $\{x_n\}$ such that $x_{n(k)} \preceq x$ for all k.

We have the following result.

Theorem 21 *Let $(X, \preceq)$ be a partially ordered set, $\zeta \in \mathscr{Z}$, and p be a partial metric on X such that (X, p) is complete. Let $T : X \to X$ be a nondecreasing mapping with respect to $\preceq$. Suppose that there exists a function $\phi \in \Phi$ such that*

$$\zeta(p(Tx, Ty), \phi(K(x, y))) \geq 0,$$

for all $x, y \in X$ with $x \succeq y$, where $K(x, y)$ is defined as in (17). Suppose also that the following conditions hold:

(i) there exists $x_0 \in X$ such that $x_0 \preceq Tx_0$;
(ii) T is continuous or $(X, \preceq, p)$ is regular.

Then T has a fixed point.

Proof Set $\alpha : X \times X \to [0, \infty)$ as

$$\alpha(x, y) = \begin{cases} 1 \text{ if } x \preceq y \text{ or } x \succeq y, \\ 0 \text{ otherwise.} \end{cases}$$

Obviously, T is a $(\alpha - \phi)$-type Z_K-contraction, that is,

$$\zeta(\alpha(x, y)p(Tx, Ty), \phi(K(x, y))) \geq 0,$$

for all $x, y \in X$. On account of the condition (i), we have $\alpha(x_0, Tx_0) \geq 1$. Additionally, for all $x, y \in X$, from the monotone property of T, we have

$$\alpha(x, y) \geq 1 \implies x \succeq y \text{ or } x \preceq y \implies Tx \succeq Ty \text{ or } Tx \preceq Ty \implies \alpha(Tx, Ty) \geq 1.$$

Hence, we find that T is α-admissible.

For the case, when T is continuous, the existence of a fixed point can be derived from Theorem 2.

Let us consider the other case. Assume that $(X, \preceq, p)$ is regular. Let $\{x_n\}$ be a sequence in X such that $\alpha(x_n, x_{n+1}) \geq 1$ for all n and $x_n \to x \in X$ as $n \to \infty$. From the regularity hypothesis, there exists a subsequence $\{x_{n(k)}\}$ of $\{x_n\}$ such that $x_{n(k)} \preceq x$ for all k. Keeping the definition of α in mind, we find that $\alpha(x_{n(k)}, x) \geq 1$ for all k. Consequently, the existence of a fixed point follows from Theorem 3.

Theorem 22 *Let $(X, \preceq)$ be a partially ordered set, $\zeta \in \mathcal{Z}$, and p be a partial metric on X such that (X, p) is complete. Let $T : X \to X$ be a nondecreasing mapping with respect to $\preceq$. Suppose that there exists a function $\phi \in \Phi$ such that*

$$\zeta(p(Tx, Ty), \phi(N(x, y))) \geq 0,$$

for all $x, y \in X$ with $x \succeq y$, where $K(x, y)$ is defined as in (34). Suppose also that the conditions (i) and (ii) of Theorem 21 hold. Then T has a fixed point. Moreover, if for all $x, y \in X$ there exists $z \in X$ such that $x \preceq z$ and $y \preceq z$, we have uniqueness of the fixed point.

Proof By mimicking the proof of Theorem 21, we guarantee the existence of a fixed point of T. It is sufficient to indicate that it is unique. Let $x, y \in X$ be two fixed points of T. By hypothesis, there exists $z \in X$ such that $x \preceq z$ and $y \preceq z$, which implies from the definition of α that $\alpha(x, z) \geq 1$ and $\alpha(y, z) \geq 1$. Thus we deduce the uniqueness of the fixed point by Theorem 6.

Remark 5 Note that the techniques, used in this section, can be applied to all other results, Theorems 7–20. Regarding the analogy, we skip the statement of these result here.

2.4 Consequences in the Frame of Cyclic Contractive Mappings

In the last decade, another trend in the research of metric fixed point theory was initiated by Kirk et al. (2003) who investigated the existence and uniqueness of cyclic mappings. In this subsection, we shall show that how such results can be concluded from our results by choosing the α function in a suitable way.

Corollary 1 *Let* $\{A_i\}_{i=1}^2$ *be nonempty closed subsets of a complete partial metric space* (X, p), $\zeta \in \mathcal{Z}$, *and* $T : Y \to Y$ *be a given mapping, where* $Y = A_1 \cup A_2$. *Suppose that the following conditions hold:*

(I) $T(A_1) \subseteq A_2$ *and* $T(A_2) \subseteq A_1$;
(II) *there exists a function* $\phi \in \Phi$ *such that*

$$\zeta(p(Tx, Ty), \phi(K(x, y))) \geq 0,$$

for all $(x, y) \in A_1 \times A_2$, *where* $K(x, y)$ *is defined as in (17).*

Then T *has a fixed point that belongs to* $A_1 \cap A_2$.

Proof It is clear that (Y, p) is complete since A_1 and A_2 are closed subsets of the complete partial metric space (X, p). We set the function $\alpha : Y \times Y \to [0, \infty)$ as

$$\alpha(x, y) = \begin{cases} 1 \text{ if } (x, y) \in (A_1 \times A_2) \cup (A_2 \times A_1), \\ 0 \text{ otherwise.} \end{cases}$$

On account of (II) and the definition of α, we derive that

$$\zeta(\alpha(x, y)p(Tx, Ty), \phi(K(x, y))) \geq 0,$$

for all $x, y \in Y$. Accordingly, T is a $(\alpha - \phi)$-type Z_K-contraction.

Let $(x, y) \in Y \times Y$ such that $\alpha(x, y) \geq 1$. If $(x, y) \in A_1 \times A_2$, from (I), $(Tx, Ty) \in A_2 \times A_1$, which yields that $\alpha(Tx, Ty) \geq 1$. If $(x, y) \in A_2 \times A_1$, from (I), $(Tx, Ty) \in A_1 \times A_2$, which implies that $\alpha(Tx, Ty) \geq 1$. As a result, we have $\alpha(Tx, Ty) \geq 1$. Hence, we get that T is α-admissible.

Further, due to (I), for any $a \in A_1$, we have $(a, Ta) \in A_1 \times A_2$, which implies that $\alpha(a, Ta) \geq 1$.

Now, let $\{x_n\}$ be a sequence in X such that $\alpha(x_n, x_{n+1}) \geq 1$ for all n and $x_n \to x \in X$ as $n \to \infty$. On account of the definition of α that

$$(x_n, x_{n+1}) \in (A_1 \times A_2) \cup (A_2 \times A_1), \text{ for all } n.$$

Since $(A_1 \times A_2) \cup (A_2 \times A_1)$ is a closed set with respect to the Euclidean metric, we get that

$$(x, x) \in (A_1 \times A_2) \cup (A_2 \times A_1),$$

which implies that $x \in A_1 \cap A_2$. Thus we get immediately from the definition of α that $\alpha(x_n, x) \geq 1$ for all n.

Corollary 2 *Let* $\{A_i\}_{i=1}^2$ *be nonempty closed subsets of a complete partial metric space* (X, d), $\zeta \in \mathscr{Z}$, *and* $T : Y \to Y$ *be a given mapping, where* $Y = A_1 \cup A_2$. *Suppose that the following conditions hold:*

(I) $T(A_1) \subseteq A_2$ *and* $T(A_2) \subseteq A_1$;
(II) *there exists a function* $\phi \in \Phi$ *such that*

$$\zeta(p(Tx, Ty), \phi(N(x, y))), \text{ for all } (x, y) \in A_1 \times A_2,$$

where $K(x, y)$ *is defined as in (34).*

Then T *has a unique fixed point that belongs to* $A_1 \cap A_2$.

Proof By mimicking the proof of Theorem 1, we guarantee the existence of a fixed point of T. In what follows, we indicate the uniqueness of it. Let x, y be distinct fixed point of T. On account of (I), we derive that $x, y \in A_1 \cap A_2$. Accordingly, for any $z \in Y$, we have $\alpha(x, z) \geq 1$ and $\alpha(y, z) \geq 1$. As a result, the condition (H) is fulfilled. Consequently, all the hypotheses of Theorem 6 hold. Thus, T has a unique fixed point that belongs to $A_1 \cap A_2$ (from (I)).

Remark 6 As it is mentioned in the previous sections, by employing this techniques to Theorems 7–20, we obtain more consequences.

2.5 *More Consequences in the Frame of Partial Metric Spaces*

By setting $\alpha(x, y) = 1$ for all $x, y \in X$, in Theorems 2–6, the following consequences are obtained immediately.

Theorem 23 *Let* (X, p) *be a complete partial metric space,* $\zeta \in \mathscr{Z}$, *and* $T : X \to X$ *be a continuous mapping. Suppose that there exists a function* $\phi \in \Phi$ *such that*

$$\zeta(p(Tx, Ty), \phi(K(x, y))) \geq 0,$$

for all $x, y \in X$, *where* $K(x, y)$ *is defined as in (17). Then* T *has a unique fixed point.*

Theorem 24 *Let* (X, p) *be a complete partial metric space,* $\zeta \in \mathscr{Z}$, *and* $T : X \to X$ *be a continuous mapping. Suppose that there exists a function* $\phi \in \Phi$ *such that*

$$\zeta(p(Tx, Ty), \phi(N(x, y))) \geq 0,$$

for all $x, y \in X$, *where* $K(x, y)$ *is defined as in (34). Then* T *has a unique fixed point.*

Theorem 25 *Let (X, p) be a complete partial metric space, $\zeta \in \mathscr{Z}$, and $T : X \to X$ be a continuous mapping. Suppose that there exists a function $\phi \in \Phi$ such that*

$$\zeta(p(Tx, Ty), \phi(a_1 p(x, y) + a_2 p(x, Tx) + a_3 p(y, Ty))) \geq 0,$$

for all $x, y \in X$, where $0 \leq a_i$, $i = 1, 2, 3$, and $a_1 + a_2 + a_3 < 1$. Then T has a unique fixed point.

Theorem 26 *Let (X, p) be a complete partial metric space, $\zeta \in \mathscr{Z}$, and $T : X \to X$ be a continuous mapping. Suppose that there exists a function $\phi \in \Phi$ such that*

$$\zeta(p(Tx, Ty), \phi(a_1 p(x, y) + a_4[p(x, Ty) + p(y, Tx)])) \geq 0,$$

for all $x, y \in X$, where $0 \leq a_i$, $i = 1, 4$, and $a_1 + 2a_4 < 1$. Then T has a unique fixed point.

Theorem 27 *Let (X, p) be a complete partial metric space, $\zeta \in \mathscr{Z}$, and $T : X \to X$ be a continuous mapping. Suppose that there exists a function $\phi \in \Phi$ such that*

$$\zeta(p(Tx, Ty), \phi(b_1 p(x, y) + b_2[p(x, Tx) + p(y, Ty)])) \geq 0,$$

for all $x, y \in X$, where $0 \leq b_i$, $i = 1, 2$, and $b_1 + 2b_2 < 1$. Then T has a unique fixed point.

Theorem 28 *Let (X, p) be a complete partial metric space, $\zeta \in \mathscr{Z}$, and $T : X \to X$ be a continuous mapping. Suppose that there exists a function $\phi \in \Phi$ such that*

$$\zeta\left(p(Tx, Ty) \leq \phi\left(\frac{[p(x, Tx) + p(y, Ty)]}{2}\right)\right) \geq 0,$$

for all $x, y \in X$. Then T has a unique fixed point.

Theorem 29 *Let (X, p) be a complete partial metric space, $\zeta \in \mathscr{Z}$, and $T : X \to X$ be a continuous mapping. Suppose that there exists a function $\phi \in \Phi$ such that*

$$\zeta\left(p(Tx, Ty), \phi\left(\frac{[p(x, Ty) + p(y, Tx)]}{2}\right)\right) \geq 0,$$

for all $x, y \in X$. Then T has a unique fixed point.

Notice that in the following theorem the continuity of the mapping T is not required. In fact, the contraction criteria already yields the continuity of T.

Theorem 30 *Let (X, p) be a complete partial metric space, $\zeta \in \mathscr{Z}$, and $T : X \to X$ be a given mapping. Suppose that there exists a function $\phi \in \Phi$ such that*

$$\zeta(p(Tx, Ty), \phi(p(x, y))) \geq 0,$$

for all $x, y \in X$. *Then* T *has a unique fixed point.*

Taking $\phi(t) = kt$ with $k \in [0, 1)$, we get the analog of the initial result, appeared in Khojasteh et al. (2015), in this direction

Theorem 31 *Let* (X, p) *be a complete partial metric space and* $T : X \to X$ *be a given mapping. Suppose that there exists a* $k \in [0, 1)$ *such that*

$$\zeta(p(Tx, Ty), kp(x, y)) \geq 0,$$

for all $x, y \in X$. *Then* T *has a unique fixed point.*

References

Abedeljawad, T., Karapınar, E., Taş, K.: Existence and uniqueness of common fixed point on partial metric spaces. Appl. Math. Lett. **24**, 1894–1899 (2011)

Abdeljawad, T., Karapınar, E., Tas, K.: A generalized contraction principle with control functions on partial metric spaces. Comput. Math. Appl. **63**(3), 716–719 (2012)

Agarwal, R.P., Alghamdi, M.A., Shahzad, N.: Fixed point theory for cyclic generalized contractions in partial metric spaces. Fixed Point Theory Appl. **2012**, 40 (2012)

Ali, M.U., Kamran, T., Karapınar, E.: On (α, ϕ, η)-contractive multivalued mappings. Fixed Point Theory Appl. **2014**, 7 (2014)

Alsulami, H.H., Karapınar, E., Khojasteh, F., Roldán-López-de-Hierro, A.F.: A proposal to the study of contractions in quasi-metric spaces. Discret. Dyn. Nat. Soc. (2014), Article ID 269286

Altun, I., Acar, O.: Fixed point theorems for weak contractions in the sense of Berinde on partial metric spaces. Topol. Appl. **159**, 2642–2648 (2012)

Altun, I., Erduran, A.: Fixed point theorems for monotone mappings on partial metric spaces. Fixed Point Theory Appl. **2011** (2011) Article ID 508730

Altun, I., Simsek, H.: Some fixed point theorems on dualistic partial metric spaces. J. Adv. Math. Stud. **1**, 1–8 (2008)

Altun, I., Sola, F., Simsek, H.: Generalized contractions on partial metric spaces. Topol. Appl. **157**, 2778–2785 (2010)

Aydi, H.: Some coupled fixed point results on partial metric spaces. Int. J. Math. Math. Sci. (2011a), Article ID 647091

Aydi, H.: Some fixed point results in ordered partial metric spaces. J. Nonlinear Sci. Appl. **4**(2), 210–217 (2011b)

Aydi, H.: Fixed point results for weakly contractive mappings in ordered partial metric spaces. J. Adv. Math. Stud. **4**(2), 1–12 (2011c)

Aydi, H.: Fixed point theorems for generalized weakly contractive condition in ordered partial metric spaces. J. Nonlinear Anal. Optim.: Theory Appl. **2**(2), 33–48 (2011d)

Aydi, H.: Common fixed point results for mappings satisfying (ϕ, ϕ)-weak contractions in ordered partial metric space. Int. J. Math. Stat. **12**(2), 53–64 (2012)

Aydi, H., Karapınar, E.: A Meir-Keeler common type fixed point theorem on partial metric spaces. Fixed Point Theory Appl. **2012**, 26 (2012)

Aydi, H., Karapınar, E., Shatanawi, W.: Coupled fixed point results for (ϕ, φ)- weakly contractive condition in ordered partial metric spaces. Comput. Math. Appl. **62**(12), 4449–4460 (2011)

Aydi, H., Abbas, M., Vetro, C.: Partial Hausdorff metric and Nadler's fixed point theorem on partial metric spaces. Topol. Appl. **159**, 3234–3242 (2012a)

Aydi, H., Vetro, C., Sintunavarat, W., Kumam, P.: Coincidence and fixed points for contractions and cyclical contractions in partial metric spaces. Fixed Point Theory Appl. **2012**, 124 (2012b)

Aydi, H., Vetro, C., Karapınar, E.: On Ekeland's variational principle in partial metric spaces. Appl. Math. Inf. Sci. **9**(1), 257–262 (2015a)

Aydi, H., Bilgili, N., Karapınar, E.: Common fixed point results from quasi-metric spaces to G-metric spaces. J. Egypt. Math. Soc. **23**(2), 356–361 (2015b)

Aydi, H., Jellali, M., Karapınar, E.: Common fixed points for generalized α-implicit contractions in partial metric spaces: consequences and application. RACSAM **109**(2), 367–384 (2015c)

Banach, S.: Sur les opérations dans les ensembles abstraits et leur application aux équations intégrales. Fundam. Math. **3**, 133–181 (1922)

Bukatin, M., Kopperman, R., Matthews, S., Pajoohesh, H.: Partial metric spaces. Am. Math. Mon. **116**(8), 708–718 (2009)

Chen, C.-M., Karapınar, E.: Fixed point results for the α-Meir-Keeler contraction on partial Hausdorff metric spaces. J. Inequal. Appl. **2013**, 410 (2013)

Chi, K.P., Karapınar, E., Thanh, T.D.: A generalized contraction principle in partial metric spaces. Math. Comput. Model. **55**, 1673–1681 (2012). https://doi.org/10.1016/j.mcm.2011.11.005

Ćirić, L.j., Samet, B., Aydi, H., Vetro, C.: Common fixed points of generalized contractions on partial metric spaces and an application. Appl. Math. Comput. **218**, 2398–2406 (2011)

Ćirić, L.B.: On some maps with a nonunique fixed point. Publications de L'Institut Mathématique **17**, 52–58 (1974a)

Ćirić, L.B.: A generalization of Banach's contraction principle. Proc. Am. Math. Soc. **45**(2), 267–273 (1974b)

Escardo, M.H.: PCF extended with real numbers. Theor. Comput. Sci. **162**, 79–115 (1996)

Frechét, M.R.: Sur quelques points du calcul fonctionnel. Rend. Circ. Mat. Palermo **22**, 174 (1906)

Gulyaz, S., Karapınar, E.: Coupled fixed point result in partially ordered partial metric spaces through implicit function. Hacet. J. Math. Stat. **42**(4), 347–357 (2013)

Haghi, R.H., Rezapour, Sh, Shahzad, N.: Be careful on partial metric fixed point results. Topol. Appl. **160**(3), 450–454 (2013)

Heckmann, R.: Approximation of metric spaces by partial metric spaces. Appl. Categ. Struct. **7**, 71–83 (1999)

Hitzler, P., Seda, A.: Mathematical Aspects of Logic Programming Semantics. CRC Press, Taylor and Francis Group, Boca Raton, Studies in Informatics Series. Chapman and Hall (2011)

Hošková-Mayerová, Š., Maturo, F., Kacprzyk, J.: Recent Trends in Social Systems: Quantitative Theories and Quantitative Models Edition: Studies in System, Decision and Control 66. Springer International Publishing AG, Switzerland (2016), 426 p. ISSN 2198-4182. ISBN 978-3-319-40583-4

Hošková-Mayerová, Š., Maturo, F., Kacprzyk, J.: Mathematical-Statistical Models and Qualitative Theories for Economic and Social Sciences. Springer International Publishing, New York (2017), 437 p. ISBN 978-3-319-54819-7

Ilić, D., Pavlović, V.: Rakočević, V.: Some new extensions of Banachs contraction principle to partial metric space. Appl. Math. Lett. **24**(8), 1326–1330 (2011)

Ilić, D., Pavlović, V.: Rakočević, V.: Extensions of the Zamfirescu theorem to partial metric spaces. Original Research Article. Math. Comput. Model. **55**(34), 801–809 (2012)

Jleli, M., Karapınar, E., Samet, B.: Best proximity points for generalized $\alpha - \phi$-proximal contractive type mappings, J. Appl. Math. (2013a) Article ID 534127

Jleli, M., Karapınar, E., Samet, B.: Fixed point results for $\alpha - \phi_\lambda$ contractions on gauge spaces and applications. Abstract Appl. Anal. (2013b) Article ID 730825

Jleli, M., Karapınar, E., Samet, B.: Further remarks on fixed point theorems in the context of partial metric spaces. Abstract Appl. Anal. (2013c) Article Id: 715456

Kannan, R.: Some results on fixed points. Bull. Calcutta Math. Soc. **60**, 71–76 (1968)

Karapınar, E.: Generalizations of Caristi Kirk's theorem on partial metric spaces. Fixed Point Theory Appl. **2011**, 4 (2011a)

Karapınar, E.: A note on common fixed point theorems in partial metric spaces. Miskolc Math. Notes **12**(2), 185–191 (2011b)

Karapınar, E.: Some fixed point theorems on the class of comparable partial metric spaces on comparable partial metric spaces. Appl. General Topol. **12**(2), 187–192 (2011c)

Karapınar, E.: Weak ϕ-contraction on partial metric spaces. J. Comput. Anal. Appl. **14**(2), 206–210 (2012a)

Karapınar, E.: Ćirić types nonunique fixed point theorems on partial metric spaces. J. Nonlinear Sci. Appl. **5**, 74–83 (2012b)

Karapınar, E., Erhan, I.M.: Fixed point theorems for operators on partial metric spaces. Appl. Math. Lett. **24**, 1900–1904 (2011)

Karapınar, E., Erhan, I.M.: Fixed point theorem for cyclic maps on partial metric spaces. Appl. Math. Inf. Sci. **6**, 239–244 (2012)

Karapınar, E., Romaguera, S.: Nonunique fixed point theorems in partial metric spaces. Filomat **27**(7), 1305–1314 (2013)

Karapınar, E., Samet, B.: Generalized $(\alpha - \phi)$-contractive type mappings and related fixed point theorems with applications. Abstract Appl. Anal. (2012), Article ID 793486

Karapınar, E., Yuksel, U.: Some common fixed point theorems in partial metric spaces. J. Appl. Math. (2011) Article ID 263621

Karapınar, E., Shobkolaei, N., Sedghi, S., Vaezpour, S.M.: A common fixed point theorem for cyclic operators on partial metric spaces. Filomat **26**(2), 407–414 (2012)

Karapınar, E., Erhan, I., Ozturk, A.: Fixed point theorems on quasi-partial metric spaces. Math. Comput. Model. **57**(9–10), 2442–2448 (2013a)

Karapınar, E., Kuman, P., Salimi, P.: On $\alpha - \phi$-Meri-Keeler contractive mappings. Fixed Point Theory Appl. **2013**, 94 (2013b)

Karapınar, E., Alsulami, H.H., Noorwali, M.: Some extensions for Geragthy type contractive mappings. J. Inequal. Appl. **2015**, 303 (2015)

Karapınar, E., Taş, K., Rakočević, V.: Advances on fixed point results on partial metric spaces. In: Tas, K., Tenreiro, J.A., Baleanu, D. (eds.) Mathematical Methods in Engineering: Theory, pp. 1–59. Springer, Berlin (2018)

Khojasteh, F., Shukla, S., Radenović, S.: A new approach to the study of fixed point theorems via simulation functions. Filomat **29**(6), 1189–1194 (2015)

Kirk, W.A., Srinivasan, P.S., Veeramani, P.: Fixed points for mappings satisfying cyclical contractive conditions. Fixed Point Theory **4**(1), 79–89 (2003)

Kopperman, R.D., Matthews, S.G., Pajoohesh, H.: What do partial metrics represent? Notes distributed at the 19th Summer Conference on Topology and Its Applications, University of CapeTown (2004)

Kramosil, O., Michalek, J.: Fuzzy metric and statistical metric spaces. Kybernetika **11**, 326–334 (1975)

Künzi, H.P.A., Pajoohesh, H., Schellekens, M.P.: Partial quasi-metrics. Theor. Comput. Sci. **365**(3), 237–246 (2006)

Matthews, S.G.: Partial metric topology. Research report 212. Department of Computer Science. University of Warwick (1992)

Matthews, S.G.: Partial metric topology. In: Proceedings of the 8th Summer of Conference on General Topology and Applications (Ann. N.Y. Acad. Sci. **728**, 183–197) (1994)

Mohammadi, B., Rezapour, Sh, Shahzad, N.: Some results on fixed points of α-ϕ-Ciric generalized multifunctions. Fixed Point Theory Appl. **2013**, 24 (2013)

Oltra, S., Valero, O.: Banach's fixed point theorem for partial metric spaces. Rend. Istid. Math. Univ. Trieste **36**, 17–26 (2004)

Paesano, D., Vetro, P.: Suzuki's type characterizations of completeness for partial metric spaces and fixed points for partially ordered metric spaces. Topol. Appl. **159**(3), 911–920 (2012)

Popa, V.: Fixed point theorems for implicit contractive mappings. Stud. Cerc. St. Ser. Mat. Univ. Bacau **7**, 129–133 (1997)

Popescu, O.: Some new fixed point theorems for α-Geraghty contractive type maps in metric spaces. Fixed Point Theory Appl. **2014**, 190 (2014)

Ran, A.C.M., Reurings, M.C.B.: A fixed point theorem in partially ordered sets and some applications to matrix equations. Proc. Am. Math. Soc. **132**, 1435–1443 (2003)

Reich, S.: Kannans fixed point theorem. Boll. Un. Mat. Ital. **4**(4), 111 (1971)

Roldan, A., Martinez-Moreno, J., Roldan, C., Karapınar, E.: Multidimensional fixed point theorems in partially ordered complete partial metric spaces under (ψ, φ)-contractivity conditions. Abstract Appl. Anal. (2013) Article Id: 634371

Roldán-López-de-Hierro, A.F., Karapınar, E., Roldán-López-de-Hierro, C., Martínez-Moreno, J.: Coincidence point theorems on metric spaces via simulation functions. J. Comput. Appl. Math. **275**, 345–355 (2015)

Romaguera, S.: A Kirk type characterization of completeness for partial metric spaces. Fixed Point Theory Appl. (2010), Article ID 493298

Romaguera, S.: Matkowskis type theorems for generalized contractions on (ordered) partial metric spaces. Appl. General Topol. **12**(2), 213–220 (2011)

Romaguera, S.: Fixed point theorems for generalized contractions on partial metric spaces. Topol. Appl. **159**, 194–199 (2012)

Romaguera, S., Schellekens, M.: Duality and quasi-normability for complexity spaces. Appl. General Topol. **3**, 91–112 (2002)

Romaguera, S., Schellekens, M.: Partial metric monoids and semivaluation spaces. Topol. Appl. **153**(5–6), 948–962 (2005)

Romaguera, S., Valero, O.: A quantitative computational model for complete partial metric spaces via formal balls. Math. Struct. Comput. Sci. **19**(3), 541–563 (2009)

Rus, I.A.: Generalized Contractions and Applications. Cluj University Press, Cluj-Napoca (2001)

Samet, B., Rajović, M., Lazović, R., Stoiljković, R.: Common fixed point results for nonlinear contractions in ordered partial metric spaces. Fixed Point Theory Appl. **2011**, 71 (2011)

Samet, B., Vetro, C., Vetro, P.: Fixed point theorem for $\alpha - \phi$ contractive type mappings. Nonlinear Anal. **75**, 2154–2165 (2012)

Samet, B., Vetro, C., Vetro, F.: From metric spaces to partial metric spaces. Fixed Point Theory Appl. **2013**, 5 (2013)

Schellekens, M.P.: A characterization of partial metrizability: domains are quantifiable. Theor. Comput. Sci. **305**(13), 409–432 (2003)

Schellekens, M.P.: The correspondence between partial metrics and semivaluations. Theor. Comput. Sci. **315**(1), 135–149 (2004)

Sehgal, V.M.: Some fixed and common fixed point theorems in metric spaces. Can. Math. Bull. **17**(2), 257–259 (1974)

Shatanawi, W., Samet, B., Abbas, M.: Coupled fixed point theorems for mixed monotone mappings in ordered partial metric spaces. Math. Comput. Model. **55**(3–4), 680–687 (2012)

Shobkolaei, N., Vaezpour, S.M., Sedghi, S.: A common fixed point theorem on ordered partial metric spaces. J. Basic Appl. Sci. Res. **1**(12), 3433–3439 (2011)

Stoy, J.E.: Denotational Semantics: The Scott-Strachey Approach to Programming Language Theory. MIT Press, Cambridge (1981)

Valero, O.: On Banach fixed point theorems for partial metric spaces. Appl. General Topol. **6**(2), 229–240 (2005)

Vetro, C., Vetro, F.: Common fixed points of mappings satisfying implicit relations in partial metric spaces. J. Nonlinear Sci. Appl. **6**(3), 152–161 (2013)

Vetro, C., Vetro, F.: Metric or partial metric spaces endowed with a finite number of graphs: a tool to obtain fixed point results. Topol. Appl. **164**, 125–137 (2014)

Vetro, F., Radenović, S.: Nonlinear ϕ-quasi-contractions of Ćirić-type in partial metric spaces. Appl. Math. Comput. **219**(4), 1594–1600 (2012)

Waszkiewicz, P.: Quantitative continuous domains. Appl. Categ. Struct. **11**, 4167 (2003)

Waszkiewicz, P.: Partial metrisability of continuous posets. Math. Struct. Comput. Sci. **16**(2), 359–372 (2006)

Geometric Properties of Mittag-Leffler Functions

Dorina Răducanu

Keywords Convex · Starlike · Close-to-convex · Mittag-leffler function

In recent decades the attention towards Mittag-Leffler type functions has deepened due to their direct involvement in problems of physics, biology, chemistry, engineering and other applied sciences. More precisely, applications of Mittag-Leffler functions appear in stochastic systems (Polito and Scalas 2016), statistical distribution with results obtained by Pillai (1990), dynamical models investigated by An et al. (2012) etc. Special emphasis should be placed on the applications of Mittag-Leffler type functions in fractional calculus (Kilbas et al. 2004; Srivastava and Tomovski 2009) and also fractional differential and integral equations such as: diffusion equation with results obtained by Langlands (2006) and Yu and Zhang (2006), telegraph equation (Camargo et al. 2012), kinetic equation (Metzler and Klafter 2000), Abel type integral equations investigated by Kilbas and Saigo (1995) just to mention a few.

Starlikeness, convexity, close-to-convexity (see Bansal and Prajapat 2016), univalency (Srivastava et al. 2017) and differential subordination results obtained by Attiya (2016), Răducanu (2017) for Mittag-Leffler type functions have been recently investigated.

In this chapter certain geometric properties for two-parametric Mittag-Leffler function are presented.

D. Răducanu (✉)
Faculty of Mathematics and Computer Science, Transilvania University of Brașov,
Iuliu Maniu, 50, 500091 Brașov, Romania
e-mail: draducanu@unitbv.ro

© Springer Nature Switzerland AG 2019
C. Flaut et al. (eds.), *Models and Theories in Social Systems*, Studies in Systems,
Decision and Control 179, https://doi.org/10.1007/978-3-030-00084-4_22

1 Mittag-Leffler Functions

The function $E_\alpha(z)$ defined by

$$E_\alpha(z) = \sum_{n=0}^{\infty} \frac{z^n}{\Gamma(\alpha n + 1)}, \quad \alpha \in \mathbb{C}, \ \Re\alpha > 0 \tag{1}$$

was introduced by Mittag-Leffler (1903) and it is, therefore, known as Mittag-Leffler function. Note that the series (1) is convergent in the whole complex plane for $\Re\alpha > 0$.

A first generalization of the function $E_\alpha(z)$ is a two-parametric function, defined by

$$E_{\alpha,\beta}(z) = \sum_{n=0}^{\infty} \frac{z^n}{\Gamma(\alpha n + \beta)}, \quad z \in \mathbb{C} \ \alpha, \beta \in \mathbb{C}, \ \Re\alpha > 0, \ \Re\beta > 0. \tag{2}$$

This function was introduced by Wiman (1905). Note that

$$E_{\alpha,1}(z) = E_\alpha(z).$$

For some special values of parameters α and β, the function $E_{\alpha,\beta}(z)$ coincides with some elementary and special functions:

$$E_{1,2}(z) = \frac{e^z - 1}{z}, \quad E_{2,1}(z) = \cosh(\sqrt{z}), \quad E_{2,2}(z) = \frac{\sinh(\sqrt{z})}{\sqrt{z}}. \tag{3}$$

Another generalization of function (1) was considered by Prabhakar (1997). This three-parametric function is defined by the series

$$E_{\alpha,\beta}^{\gamma}(z) = \sum_{n=0}^{\infty} \frac{(\gamma)_n}{\Gamma(\alpha n + \beta)} \frac{z^n}{n!}, \quad z \in \mathbb{C}, \alpha, \beta, \gamma \in \mathbb{C}, \Re\alpha > 0, \Re\beta > 0, \Re\gamma > 0, \tag{4}$$

where

$$(\gamma)_n = \frac{\Gamma(n+\gamma)}{\Gamma(\gamma)} = \begin{cases} 1 & , n = 0 \\ \gamma(\gamma+1)\ldots(\gamma+n-1) & , n \in \mathbb{N} = \{1, 2, \ldots\} \end{cases}$$

is the well-known Pochhammer symbol. Note that

$$E_{\alpha,\beta}^1(z) = E_{\alpha,\beta}(z) \quad \text{and} \quad E_{\alpha,1}^1(z) = E_\alpha(z).$$

Further, an extension of Mittag-Leffler function to four parameters was defined by Salim (2009) and is given by the series

$$E_{\alpha,\beta}^{\gamma,\delta}(z) = \sum_{n=0}^{\infty} \frac{(\gamma)_n}{\Gamma(\alpha n + \beta)(\delta)_n} z^n, \ z \in \mathbb{C} \tag{5}$$

where $\alpha, \beta, \gamma, \delta \in \mathbb{C}$ and $\Re\alpha > 0, \Re\beta > 0, \Re\gamma > 0, \Re\delta > 0$.

Multi-parametric analogous of the Mittag-Leffler function $E_\alpha(z)$ were introduced and studied, among others, by Kilbas et al. (2013), Kiryakova (2010), Salim (2012) etc.

A detailed account of various generalizations, properties and applications of Mittag-Leffler type functions can be found in the works of Gorenflo et al. (1998), Gorenflo et al. (2014), Haubold et al. (2011) and Lavault (2017).

2 Geometric Properties

Numerous studies have been recently devoted to the geometric properties of special functions. Starlikeness, convexity, close-to-convexity and univalency have been usually investigated. In this context, there are many results available in the literature regarding generalized hypergeometric functions (Ponnusamy 1997; Ponnusamy and Vuorinen 1998), Bessel functions (Baricz 2008; Mondal and Swaminathan 2012), Struve functions (Yagmur and Orhan 2013) and Wright functions (Prajapat 2014; Raza et al. 2016).

In this section geometric properties for two-parametric normalized Mittag-Leffler function are studied.

2.1 *Definitions and Notations*

Let $\mathcal{H}$ denote the class of all analytic functions in the open unit disk

$$\mathbb{U} = \{z \in \mathbb{C} : |z| < 1\}.$$

The class of all functions $p \in \mathcal{H}$ with $p(0) = 1$ satisfying the condition

$$\Re p(z) > \gamma, \ z \in \mathbb{U}, \ \gamma \in [0, 1) \tag{6}$$

will be denoted by $\mathcal{P}(\gamma)$. In particular, $\mathcal{P}(0) = \mathcal{P}$ is the well-known Caratheódory class of functions with positive real part in $\mathbb{U}$ (see Goodman 1983).

Let $\mathcal{A}$ be the class of functions $f \in \mathcal{H}$ of the form

$$f(z) = z + \sum_{n=2}^{\infty} a_n z^n. \tag{7}$$

Denote by $\mathcal{S}$ the subclass of $\mathcal{A}$ which consists of univalent functions. Further, denote by $\mathcal{S}^*(\gamma)$, $C(\gamma)$ and $\mathcal{K}(\gamma)$ the classes of starlike, convex and close-to-convex functions of order γ, respectively. They are defined by (see Goodman 1983):

$$\mathcal{S}^*(\gamma) = \left\{ f \in \mathcal{A} : \Re \frac{zf'(z)}{f(z)} > \gamma, z \in \mathbb{U}, \ \gamma \in [0, 1) \right\} \tag{8}$$

$$C(\gamma) = \left\{ f \in \mathcal{A} : \Re \left(1 + \frac{zf''(z)}{f'(z)} \right) > \gamma, z \in \mathbb{U}, \ \gamma \in [0, 1) \right\} \tag{9}$$

$$\mathcal{K}(\gamma) = \left\{ f \in \mathcal{A} : \Re \frac{f'(z)}{g'(z)} > \gamma, z \in \mathbb{U}, \ \gamma \in [0, 1), \ g \in C \right\}. \tag{10}$$

In particular, $\mathcal{S}^*(0) = \mathcal{S}^*$, $C(0) = C$ and $\mathcal{K}(0) = \mathcal{K}$ are the classes of starlike, convex and close-to-convex functions, respectively.

Let $k - \mathcal{UCV}$ and $k - \mathcal{ST}$ be the subclasses of $\mathcal{S}$ consisting of functions which are k-uniformly convex and k-starlike, respectively (see Kanas and Wisniowska 1999 and Kanas and Wisniowska 2000). They are given by:

$$k - \mathcal{UCV} = \left\{ f \in \mathcal{S} : \Re \left(1 + \frac{zf''(z)}{f'(z)} \right) > k \left| \frac{zf''(z)}{f'(z)} \right|, z \in \mathbb{U}, \ k \geq 0 \right\} \tag{11}$$

$$k - \mathcal{ST} = \left\{ f \in \mathcal{S} : \Re \frac{zf'(z)}{f(z)} > k \left| \frac{zf'(z)}{f(z)} - 1 \right|, z \in \mathbb{U}, \ k \geq 0 \right\}. \tag{12}$$

Consider the function $E_{\alpha,\beta}(z)$ defined by (2). This two-parametric Mittag-Leffler function does not belong to the class $\mathcal{A}$. Therefore, we consider the next normalization of the function $E_{\alpha,\beta}(z)$:

$$\mathbb{E}_{\alpha,\beta}(z) = \Gamma(\beta)zE_{\alpha,\beta}(z) = z + \sum_{n=1}^{\infty} \frac{\Gamma(\beta)}{\Gamma(\alpha n + \beta)} z^{n+1}, \ z \in \mathbb{U}. \tag{13}$$

The above formula holds for complex-valued α, β and $z \in \mathbb{C}$. However, throughout this chapter, we restrict our attention to the case of real-valued α, β with $\alpha > 0$, $\beta > 0$ and $z \in \mathbb{U}$. Several particular cases of $\mathbb{E}_{\alpha,\beta}(z)$ are:

$$\begin{cases} \mathbb{E}_{1,1}(z) = ze^z & \mathbb{E}_{1,2}(z) = e^z - 1 \\ \mathbb{E}_{2,1}(z) = z \cosh(\sqrt{z}) & \mathbb{E}_{2,2}(z) = \sqrt{z} \sinh(\sqrt{z}). \end{cases} \tag{14}$$

2.2 Starlikeness, Convexity and Close-to-Convexity

Several geometric properties for normalized Mittag-Leffler function $\mathbb{E}_{\alpha,\beta}(z)$ given by (13) were obtained by Bansal and Prajapat (2016). They proved the following two results.

Theorem 2.1 *If $\alpha \geq 1$ and $\beta \geq (3 + \sqrt{17})/2$ then, the function $\mathbb{E}_{\alpha,\beta}(z)$ is starlike, i.e $\mathbb{E}_{\alpha,\beta} \in \mathcal{S}^*$.*

Theorem 2.2 *Let $\alpha \geq 1$ and $\gamma \in [0, 1)$. If*

$$\beta \geq \frac{(3 - \gamma) + \sqrt{5\gamma^2 - 19\gamma + 12}}{2(1 - \gamma)}$$

then, the function $\mathbb{E}_{\alpha,\beta}(z)$ is starlike of order γ, i.e $\mathbb{E}_{\alpha,\beta} \in \mathcal{S}^*(\gamma)$.

In the sequence, convexity of order γ, close-to-convexity of order $(1 + \gamma)/2$ for normalized Mittag-Leffler function $\mathbb{E}_{\alpha,\beta}(z)$ are investigated. Certain sufficient conditions for $\mathbb{E}_{\alpha,\beta}(z)$ to be in the classes $\mathcal{P}(\gamma)$, $\mathcal{S}^*(\gamma)$, $C(\gamma)$, $k - \mathcal{UCV}$ and $k - \mathcal{ST}$ are also given.

The following lemmas will be required to prove the next results.

Lemma 2.1 (Owa et al. 2002) *If $f \in \mathcal{A}$ satisfies the inequality*

$$|zf''(z)| < \frac{1 - \gamma}{4}, \quad z \in \mathbb{U}, \ \gamma \in [0, 1) \tag{15}$$

then

$$\Re f'(z) > \frac{1 + \gamma}{2}, \quad z \in \mathbb{U}, \ \gamma \in [0, 1). \tag{16}$$

Lemma 2.2 (Silverman 1975) *Let $\gamma \in [0, 1)$. A sufficient condition for $f \in \mathcal{A}$, given by (7), to be in $\mathcal{S}^*(\gamma)$ and $C(\gamma)$, respectively, is*

$$\sum_{n=2}^{\infty} (n - \gamma)|a_n| \leq 1 - \gamma \tag{17}$$

and

$$\sum_{n=2}^{\infty} n(n - \gamma)|a_n| \leq 1 - \gamma, \tag{18}$$

respectively.

Lemma 2.3 (Kanas and Wisniowska 1999, 2000) *Let $f \in \mathcal{A}$ be given by (7). If for some $k \geq 0$*

$$\sum_{n=2}^{\infty} n(n - 1)|a_n| \leq \frac{1}{k + 2} \tag{19}$$

and

$$\sum_{n=2}^{\infty}[n + k(n - 1)]|a_n| \le 1 \tag{20}$$

then $f \in k - \mathcal{UCV}$ and $f \in k - \mathcal{ST}$, respectively.

Lemma 2.4 (Ozaki 1935) *Let $f(z) = z + \sum_{n=1}^{\infty} b_{2n+1}z^{2n+1}$ be an odd function. If*

$$1 \ge 3b_3 \ge \ldots \ge (2n + 1)b_{2n+1} \ge \ldots \ge 0 \tag{21}$$

or

$$1 \le 3b_3 \le \ldots \le (2n + 1)b_{2n+1} \le \ldots \le 2 \tag{22}$$

then the function f is univalent in $\mathbb{U}$.

Following the proof of Ozaki (1935) it can be proved that if an odd function f satisfies (21) or (22) then, f is close-to-convex with respect to the convex function $2^{-1}\log((1 + z)/(1 - z))$.

Convexity and close-to-convexity properties for $\mathbb{E}_{\alpha,\beta}(z)$ are obtained in the next result.

Theorem 2.3 *Let $\alpha, \beta, \gamma \in \mathbb{R}$ with $\gamma \in [0, 1)$. The following assertions are true:*

(i) *If $\alpha \ge 1$ and $\beta > \left((8 - 3\gamma) + \sqrt{17\gamma^2 - 68\gamma + 76}\right)/2(1 - \gamma)$ then $\mathbb{E}_{\alpha,\beta} \in C(\gamma)$.*

(ii) *If $\alpha \ge 1$ and $\beta > (10 + 2\sqrt{26 - \gamma})/(1 - \gamma)$ then $\mathbb{E}_{\alpha,\beta} \in \mathcal{K}(\frac{1+\gamma}{2})$.*

(iii) *If $\alpha \ge 1$ and $\beta > (1 + \sqrt{5 - 4\gamma})/2(1 - \gamma)$ then $\mathbb{E}_{\alpha,\beta}(z)/z \in \mathcal{P}(\gamma)$.*

Proof To prove the first assertion, we need to show that

$$\left|\frac{z\mathbb{E}''_{\alpha,\beta}(z)}{\mathbb{E}'_{\alpha,\beta}(z)}\right| < 1 - \gamma.$$

Under the hypothesis, the inequality $\Gamma(n + \beta) \le \Gamma(\alpha n + \beta)$ holds and it is equivalent to

$$\frac{\Gamma(\beta)}{\Gamma(\alpha n + \beta)} \le \frac{1}{\beta(\beta + 1)\ldots(\beta + n - 1)} = \frac{1}{(\beta)_n}, \quad n \in \mathbb{N}. \tag{23}$$

Moreover, under the given condition

$$\frac{n}{\beta(\beta + 1)\ldots(\beta + n - 1)} \le \frac{1}{\beta(\beta + 1)_{n-2}}, \quad n \in \mathbb{N}\setminus\{1\} \tag{24}$$

and

$$\frac{n(n-1)}{\beta(\beta+1)\dots(\beta+n-1)} \le \frac{1}{\beta(\beta+1)_{n-3}}, \quad n \in \mathbb{N} \setminus \{1, 2\}. \tag{25}$$

If $z \in \mathbb{U}$ then, by using (23), (24) and relations

$$(\beta)_n = \beta(\beta+1)_{n-1}, \quad (\beta)_n \ge \beta^n, \quad n \in \mathbb{N} \tag{26}$$

we get

$$|\mathbb{E}'_{\alpha,\beta}(z)| = \left|1 + \sum_{n=1}^{\infty} \frac{\Gamma(\beta)(n+1)}{\Gamma(\alpha n + \beta)} z^n\right| < 1 + \sum_{n=1}^{\infty} \frac{\Gamma(\beta)(n+1)}{\Gamma(\alpha n + \beta)} \le 1 + \sum_{n=1}^{\infty} \frac{n+1}{(\beta)_n}$$

$$= 1 + \sum_{n=1}^{\infty} \frac{n}{(\beta)_n} + \sum_{n=1}^{\infty} \frac{1}{(\beta)_n} \le 1 + \frac{1}{\beta} + \frac{2}{\beta} \sum_{n=0}^{\infty} \left(\frac{1}{\beta+1}\right)^n = \frac{\beta^2 + 3\beta + 2}{\beta^2}. \tag{27}$$

For the reverse inequality, we have

$$|\mathbb{E}'_{\alpha,\beta}(z)| \ge 1 - \left|\sum_{n=1}^{\infty} \frac{\Gamma(\beta)(n+1)}{\Gamma(\alpha n + \beta)} z^n\right| > 1 - \sum_{n=1}^{\infty} \frac{\Gamma(\beta)(n+1)}{\Gamma(\alpha n + \beta)}$$

$$\ge 1 - \frac{1}{\beta} - \frac{2}{\beta} \sum_{n=0}^{\infty} \left(\frac{1}{\beta+1}\right)^n = \frac{\beta^2 - 3\beta - 2}{\beta^2}. \tag{28}$$

Combining (27) and (28), we obtain

$$\frac{\beta^2 - 3\beta - 2}{\beta^2} \le |\mathbb{E}'_{\alpha,\beta}(z)| \le \frac{\beta^2 + 3\beta + 2}{\beta^2}, \quad z \in \mathbb{U}. \tag{29}$$

Further, by using (23), (25) and (26), a simple computation gives

$$|z\mathbb{E}''_{\alpha,\beta}(z)| = \left|\sum_{n=1}^{\infty} \frac{\Gamma(\beta)n(n+1)}{\Gamma(\alpha n + \beta)} z^n\right| < \sum_{n=1}^{\infty} \frac{\Gamma(\beta)n(n+1)}{\Gamma(\alpha n + \beta)} \le \sum_{n=1}^{\infty} \frac{n(n+1)}{(\beta)_n}$$

$$\le \frac{4}{\beta} + \frac{1}{\beta} \sum_{n=0}^{\infty} \left(\frac{1}{\beta+1}\right)^n \le \frac{5\beta+1}{\beta^2}.$$

Hence

$$|z\mathbb{E}''_{\alpha,\beta}(z)| \le \frac{5\beta+1}{\beta^2}, \quad z \in \mathbb{U}. \tag{30}$$

In view of left-hand side of (29) and (30), we have

$$\left| \frac{z\mathbb{E}''_{\alpha,\beta}(z)}{\mathbb{E}'_{\alpha,\beta}(z)} \right| \leq \frac{5\beta + 1}{\beta^2 - 3\beta - 2}, \quad z \in \mathbb{U}.$$

It follows that $\mathbb{E}_{\alpha,\beta}(z)$ is convex of order γ in $\mathbb{U}$ if $(5\beta + 1)/(\beta^2 - 3\beta - 2) < 1 - \gamma$ which is a consequence of hypothesis $\beta > \left((8 - 3\gamma) + \sqrt{17\gamma^2 - 68\gamma + 76} \right) / 2(1 - \gamma)$.

(ii) Making use of inequality (30) and Lemma 2.1, we have

$$|z\mathbb{E}''_{\alpha,\beta}(z)| \leq \frac{5\beta + 1}{\beta^2} < \frac{1 - \gamma}{4}, \quad z \in \mathbb{U},$$

where $0 \leq \gamma < 1 - 4(5\beta + 1)/\beta^2$ or $\beta > \left(10 + 2\sqrt{26 - \gamma} \right)/(1 - \gamma)$. This proves that

$$\Re(\mathbb{E}'_{\alpha,\beta}(z)) > \frac{1 + \gamma}{2}$$

and thus $\mathbb{E}_{\alpha,\beta} \in \mathcal{K}(\frac{1+\gamma}{2})$.

(iii) Let $p(z)$ be the function defined by $p(z) = (\mathbb{E}_{\alpha,\beta}(z)/z - \gamma)/(1 - \gamma)$. The function $p(z)$ is analytic in $\mathbb{U}$ and $p(0) = 1$. To prove the result we have to show that $|p(z) - 1| < 1$. If $z \in \mathbb{U}$, from (23) and (26), we obtain

$$|p(z) - 1| = \left| \frac{1}{1 - \gamma} \sum_{n=1}^{\infty} \frac{\Gamma(\beta)}{\Gamma(\alpha n + \beta)} z^n \right| < \frac{1}{1 - \gamma} \sum_{n=1}^{\infty} \frac{\Gamma(\beta)}{\Gamma(\alpha n + \beta)}$$

$$\leq \frac{1}{1 - \gamma} \frac{1}{\beta} \sum_{n=0}^{\infty} \left(\frac{1}{\beta + 1} \right)^n = \frac{\beta + 1}{\beta^2 (1 - \gamma)}.$$

Under the hypothesis, the inequality $(\beta + 1)/(\beta^2(1 - \gamma)) < 1$ holds and therefore $\mathbb{E}_{\alpha,\beta}(z)/z \in \mathcal{P}(\gamma)$.   □

If we take $\gamma = 0$ in Theorem 2.3 then, we have the following result.

Corollary 2.1 *Let $\alpha, \beta \in \mathbb{R}$. The following assertions are true:*

(i) If $\alpha \geq 1$ and $\beta > 4 + \sqrt{19}$ then $\mathbb{E}_{\alpha,\beta} \in C$.
(ii) If $\alpha \geq 1$ and $\beta > 10 + 2\sqrt{26}$ then $\mathbb{E}_{\alpha,\beta} \in \mathcal{K}(1/2)$.
(iii) If $\alpha \geq 1$ and $\beta > (1 + \sqrt{5})/2$ then $\mathbb{E}_{\alpha,\beta}(z)/z \in \mathcal{P}$.

Exercise 2.1 Consider $\alpha = 1$ and $\beta = 2$ in Corollary 2.1 (iii). It follows that the function $\mathbb{E}_{1,2}(z)/z = (e^z - 1)/z$ is in the class $\mathcal{P}$ and therefore $\Re(e^z - 1)/z > 0, z \in \mathbb{U}$.

A result in which sufficient conditions for $\mathbb{E}_{\alpha,\beta}$ to be in the classes $\mathcal{S}^*(\gamma)$ and $C(\gamma)$ is given in the following theorem.

Theorem 2.4 *Let $\alpha \geq 1$, $\beta \geq 1$ and $\gamma \in [0, 1)$.*
(i) If inequality

$$\mathbb{E}'_{\alpha,\beta}(1) - \gamma \mathbb{E}_{\alpha,\beta}(1) \leq 2(1-\gamma) \tag{31}$$

holds, then $\mathbb{E}_{\alpha,\beta} \in \mathcal{S}^*(\gamma)$.
(ii) If inequality

$$\mathbb{E}''_{\alpha,\beta}(1) + (1-\gamma)\mathbb{E}'_{\alpha,\beta}(1) \leq 2(1-\gamma) \tag{32}$$

holds, then $\mathbb{E}_{\alpha,\beta} \in C(\gamma)$.

Proof (i) Since $\mathbb{E}_{\alpha,\beta}(z) = z + \sum\limits_{n=2}^{\infty} b_{n-1} z^n$, where $b_{n-1} = \dfrac{\Gamma(\beta)}{\Gamma(\alpha(n-1)+\beta)}$, in view of Lemma 2.2, we only need to show that

$$\sum_{n=2}^{\infty} (n-\gamma)|b_{n-1}| \leq 1-\gamma.$$

We have

$$\sum_{n=2}^{\infty} (n-\gamma)|b_{n-1}| = \sum_{n=2}^{\infty} nb_{n-1} - \gamma \sum_{n=2}^{\infty} b_{n-1}$$

$$= (\mathbb{E}'_{\alpha,\beta}(1) - 1) - \gamma(\mathbb{E}_{\alpha,\beta}(1) - 1) = \mathbb{E}'_{\alpha,\beta}(1) - \gamma \mathbb{E}_{\alpha,\beta}(1) - 1 + \gamma.$$

Under hypothesis, the last sum is bounded above by $1 - \gamma$ if inequality (31) holds.
(ii) From Lemma 2.2, it follows that $\mathbb{E}_{\alpha,\beta} \in C(\gamma)$ if

$$\sum_{n=2}^{\infty} n(n-\gamma)|b_{n-1}| \leq 1-\gamma.$$

A simple computation gives

$$\sum_{n=2}^{\infty} n(n-\gamma)|b_{n-1}| = \sum_{n=2}^{\infty} n(n-1+1-\gamma)b_{n-1} = \sum_{n=2}^{\infty} n(n-1)b_{n-1} + (1-\gamma)\sum_{n=2}^{\infty} nb_{n-1}$$

$$= \mathbb{E}''_{\alpha,\beta}(1) + (1-\gamma)(\mathbb{E}'_{\alpha,\beta}(1) - 1) = \mathbb{E}''_{\alpha,\beta}(1) + (1-\gamma)\mathbb{E}'_{\alpha,\beta}(1) - (1-\gamma).$$

This sum is bounded above by $1 - \gamma$ if (32) holds. $\qquad\square$

Next, conditions for $\mathbb{E}_{\alpha,\beta}$ to be in the class $k - \mathcal{UCV}$ and $k - \mathcal{ST}$, respectively, are given.

Theorem 2.5 *Let $\alpha \geq 1$, $\beta \geq 1$ and $k \geq 0$.*
(i) A sufficient condition for $\mathbb{E}_{\alpha,\beta}$ to be in $k - \mathcal{ST}$ is

$$\mathbb{E}'_{\alpha,\beta}(1) - \frac{k}{k+1}\mathbb{E}_{\alpha,\beta}(1) \le \frac{2}{k+1}. \tag{33}$$

(ii) A sufficient condition for $\mathbb{E}_{\alpha,\beta}$ to be in $k - \mathcal{UCV}$ is

$$\mathbb{E}''_{\alpha,\beta}(1) \le \frac{1}{k+2}. \tag{34}$$

Proof The proof is similar to the proof of Theorem 2.4 and therefore, it is omitted. □

The odd function $\mathbb{E}_{\alpha,\beta}(z^2)/z$ is close-to-convex. This property is proved in the next theorem.

Theorem 2.6 *If $\alpha \ge 1$ and $\beta \ge 1$ then, $\mathbb{E}_{\alpha,\beta}(z^2)/z$ is close-to-convex with respect to the convex function $2^{-1}\log(\frac{1+z}{1-z})$.*

Proof We have

$$\frac{\mathbb{E}_{\alpha,\beta}(z^2)}{z} = z + \sum_{n=2}^{\infty} A_{2n-1}z^{2n-1},$$

where

$$A_{2n-1} = \frac{\Gamma(\beta)}{\Gamma(\alpha(n-1)+\beta)}, \quad n \ge 2.$$

Note that $A_1 = 1$ and $A_{2n-1} > 0$ for all $n \ge 2$. In view of Lemma 2.4, we have to prove that $\{(2n-1)A_{2n-1}\}_{n\ge2}$ is a decreasing sequence. A simple computation gives

$$(2n-1)A_{2n-1} - (2n+1)A_{2n+1} = \Gamma(\beta)\left[\frac{2n-1}{\Gamma(\alpha(n-1)+\beta)} - \frac{2n+1}{\Gamma(\alpha n+\beta)}\right]$$

$$= \Gamma(\beta)\frac{(2n-1)\Gamma(\alpha n+\beta) - (2n+1)\Gamma(\alpha(n-1)+\beta)}{\Gamma(\alpha(n-1)+\beta)\Gamma(\alpha n+\beta)}, \quad n \ge 2.$$

Since

$$(2n-1)\Gamma(\alpha n+\beta) \ge (2n-1)[\alpha(n-1)+\beta]\Gamma(\alpha(n-1)+\beta) \ge (2n+1)\Gamma(\alpha(n-1)+\beta)$$

for all $n \ge 2$, it follows that $(2n-1)A_{2n-1} - (2n+1)A_{2n+1} > 0$ and thus, the assertion is proved. □

For a function $f \in \mathcal{A}$ given by (7), the Alexander transform $A(f) : \mathbb{U} \to \mathbb{C}$ is defined by (see Alexander 1915)

$$A(f)(z) = \int_0^z \frac{f(u)}{u}du = z + \sum_{n=2}^{\infty} \frac{a_n}{n}z^n.$$

A condition for $A(\mathbb{E}_{\alpha,\beta})$ to be in the class $\mathcal{S}^*$ is given below.

Theorem 2.7 *Let $\alpha \geq 1$ and $\beta \geq 1$. If $\mathbb{E}_{\alpha,\beta}(1) \leq 2$ then $A(\mathbb{E}_{\alpha,\beta}) \in \mathcal{S}^*$.*

Proof Since

$$\frac{\mathbb{E}_{\alpha,\beta}(z)}{z} = 1 + \sum_{n=2}^{\infty} \frac{\Gamma(\beta)}{\Gamma(\alpha(n-1)+\beta)} z^{n-1} = 1 + \sum_{n=2}^{\infty} b_{n-1} z^{n-1}$$

it follows that

$$A(\mathbb{E}_{\alpha,\beta})(z) = \int_0^z \frac{\mathbb{E}_{\alpha,\beta}(u)}{u} du = \sum_{n=1}^{\infty} a_n z^n,$$

where $a_1 = 1$ and $a_n = \frac{b_{n-1}}{n}$, $n \geq 2$. From (17), we have that $A(\mathbb{E}_{\alpha,\beta}) \in \mathcal{S}^*(0) = \mathcal{S}^*$ if inequality

$$\sum_{n=2}^{\infty} n|a_n| \leq 1$$

holds. Note that

$$\sum_{n=2}^{\infty} n|a_n| = \sum_{n=2}^{\infty} n \frac{b_{n-1}}{n} = \sum_{n=2}^{\infty} b_{n-1} = \mathbb{E}_{\alpha,\beta}(1) - 1 \leq 1$$

only if $\mathbb{E}_{\alpha,\beta}(1) \leq 2$. $\qquad\qquad\qquad\square$

3 Conclusions

In the section "Geometric properties" we have considered the normalized two-parametric Mittag-Leffler function. For this function, we have obtained new properties including convexity of order γ and close-to-convexity of order $(1 + \gamma)/2$. Sufficient conditions for normalized Mittag-Leffler function to be in certain classes of univalent functions have been also given. Moreover, we have proved that the Alexander transform of Mittag-Leffler function is starlike.

References

Alexander, J.W.: Functions which map the interior of the unit circle upon simple regions. Ann. Math. **17**, 12–22 (1915)

An, J., Van Hese, E., Baes, M.: Phase-space consistency of stellar dynamical models determined by separable augmented densities. Mon. Nat. R. Astron. Soc. **422**(1), 652–664 (2012)

Attiya, A.A.: Some applications of Mittag-Leffler function in the unit disk. Filomat **30**, 2075–2081 (2016)

Bansal, D., Prajapat, J.K.: Certain geometric properties of the Mittag-Leffler functions. Complex Var. Elliptic Equ. **61**(3), 338–350 (2016)

Baricz, Á.: Geometric properties of generalized Bessel functions. Publ. Math. **73**, 155–178 (2008). Debrecen

Camargo, R.F., Oliveira, A.C., Vaz Jr., J.: On the generalized Mittag-Leffler function and its application in a fractional telegraph equation. Math. Phys. Anal. Geom. **15**, 1–16 (2012)

Goodman, A.E.: Univalent Functions, vol. 1–2. Mariner Publishing Company, Inc., Florida (1983)

Gorenflo, R., Mainardi, F., Rogosin, S.V.: On the generalized Mittag-Leffler functions. Integral Transform. Spec. Funct. **7**, 215–224 (1998)

Gorenflo, R., Kilbas, A.A., Mainardi, F., Rogosin, S.V.: Mittag-Leffler Functions. Related Topics and Applications. Springer, Heildeberg (2014)

Haubold, H.J., Mathai, A.M., Saxena, R.K.: Mittag-Leffler functions and their applications. J. Appl. Math. 51 (2011). Art.ID 298628

Kanas, S., Wisniowska, A.: Conic regions and k-uniform convexity. J. Comput. Appl. Math. **105**, 327–336 (1999)

Kanas, S., Wisniowska, A.: Conic regions and k-starlike functions. Rev. Roumaine Math. Pures Appl. **45**(4), 647–657 (2000)

Kilbas, A.A., Saigo, M.: Solution of Abel type integral equations of the second kind and of differential equations of fractional order. Integral Transform. Spec. Funct. **95**(5), 29–34 (1995)

Kilbas, A.A., Saigo, M., Saxena, R.K.: Generalized Mittag-Leffler function and generalized fractional calculus operators. Integral Transform. Spec. Funct. **15**, 31–49 (2004)

Kilbas, A.A., Koroleva, A.A., Rogosin, S.V.: Multi-parametric Mittag-Leffler functions and their extension. Frac. Calc. Appl. Anal. **16**(2), 378–404 (2013)

Kiryakova, V.: The multi-index Mittag-Leffler functions as an important class of special functions in fractional calculus. Comput. Math. Appl. **59**(5), 1885–1895 (2010)

Langlands, T.A.M.: Solution of modified fractional diffusion equation. Physica A **367**, 136–144 (2006)

Lavault C.: Fractional calculus and generalized Mittag-Leffler functions, arXiv:1703.01912v2, 2017

Metzler, Klafter J.: The random walk's guides to anomalous diffusion: a fractional kinetic equation. Phys. Rep. **339**, 1–77 (2000)

Mittag-Leffler, G.M.: Sur la nouvelle fonction $E_\alpha(x)$. C. R. Acad. Sci. Paris **137**, 554–558 (1903)

Mondal, S.R., Swaminathan, A.: Geometric properties of Bessel functions. Bull. Malays. Math. Sci. Soc. **35**(1), 179–194 (2012)

Owa, S., Nunokawa, M., Saitoh, H., Srivastava, H.M.: Close-to-convexity, starlikeness and convexity of certain analytic functions. Appl. Math. Lett. **15**, 63–69 (2002)

Ozaki, S.: On the theory of multivalent functions. Sci. Rep. **2**, 167–188 (1935). Tokyo Bunrika Daigaku

Pillai, R.N.: On Mittag-Leffler functions and related distributions. Ann. Inst. Stat. Math. **42**(1), 157–161 (1990)

Polito, F., Scalas, E.: A generalization of the space fractional Poisson process and its connection to some Lévy process. Electron. Commun. Prob. **21**(20) (2016)

Ponnusamy, S.: Close-to-convexity properties of Gaussian hypergeometric functions. J. Comput. Appl. Math. **88**, 328–337 (1997)

Ponnusamy, S., Vuorinen, M.: Univalence and convexity properties of confluent hypergeometric functions. Complex. Var. Elliptic Equ. **36**, 73–97 (1998)

Prabhakar, T.R.: A singular integral equation with a generalized Mittag-Leffler function in the kernel. Yokohama Math. J. **19**, 7–15 (1997)

Prajapat, J.K.: Certain geometric properties of the Wright function. Integral Transform. Spec. Funct. **26**(3), 203–312 (2014)

Răducanu, D.: Third-order differential subordinations for analytic functions associated with generalized Mittag-Leffler functions. Mediterr. J. Math. **14**(167), 18 (2017)

Raza, M., Din, M.U., Malik, S.N.: Certain geometric properties of normalized Wright functions. J. Funct. Space, 8 (2016). Art.ID 1896154

Salim, T.: Some properties relating to the generalized Mittag-Leffler function. Adv. Appl. Math. Anal. **4**, 21–30 (2009)

Salim, T., Faraj, A.: A generalization of Mittag-Leffler function and integral operator associated with fractional calculus. J. Frac. Calc. Appl. **3**(5), 1–13 (2012)

Silverman, H.: Univalent functions with negative coefficients. Proc. Am. Math. Soc. **51**, 109–116 (1975)

Srivastava, H.M., Tomovski, Z.: Fractional calculus with an integral operator containing a generalized Mittag-Leffler function in the kernel. Appl. Math. Comput. **211**(1), 198–210 (2009)

Srivastava, H.M., Frasin, B.A., Pescar, V.: Univalence of integral operators involving Mittag-Leffler functions. Appl. Math. Inf. Sci. **11**(3), 635–461 (2017)

Wiman, A.: Über den Fundamental satz in der Theorie der Funcktionen $E_\alpha(x)$. Acta Math. **29**, 191–201 (1905)

Yagmur, N., Orhan, H.: Starlikeness and convexity of generalized Struve functions. Abst. Appl. Anal. 6 (2013). Art.ID 954513

Yu, R., Zhang, H.: New function of Mittag-Leffler type and its applications in the fractional diffusion-wave equation. Chaos Solitons Fractals **30**, 946–955 (2006)

Special Numbers, Special Quaternions and Special Symbol Elements

Diana Savin

Abstract Most mathematical notions have connections with real life. Although the theory of rings and algebras is abstract, however, this theory has many applications, some indirect, in real life. Many sets of real-life objects, taken together with one or more laws of composition, form algebraic structures with interesting properties. Quaternion algebras and of symbol algebras have applications in various branches of mathematics, but also in computer science, physics, signal theory. In this paper we define and we study properties of $(l, 1, p + 2q, q \cdot l) -$ numbers, $(l, 1, p + 2q, q \cdot l) -$ quaternions, $(l, 1, p + 2q, q \cdot l) -$ symbol elements. Finally, we obtain an algebraic structure with these elements.

Keywords Quaternion algebras; symbol algebras · Fibonacci numbers · Lucas numbers · Fibonacci–Lucas quaternions · Pell- Fibonacci–Lucas quaternions · $(l, 1, p + 2q, q \cdot l) -$ quaternions · $(l, 1, p + 2q, q \cdot l) -$ symbol elements

2000 AMS Subject Classification 15A24 · 15A06 · 16G30 · 11R52 · 11B39 · 11R54

1 Introduction

Most mathematical notions have connections with real life. Although the theory of rings and algebras is abstract, however, this theory has many applications, some indirect, in real life. Many sets of real-life objects, taken together with one or more laws of composition, form algebraic structures with interesting properties. In this book chapter we study some properties of quaternion algebras and symbol algebras.

Quaternion algebras and of symbol algebras have applications in various branches of mathematics, but also in computer science, physics, signal theory.

D. Savin (✉)
Faculty of Mathematics and Computer Science, Ovidius University,
Bd. Mamaia 124, 900527 Constanta, Romania
e-mail: savin.diana@univ-ovidius.ro; dianet72@yahoo.comaaaa
URL: http://www.univ-ovidius.ro/math/

© Springer Nature Switzerland AG 2019
C. Flaut et al. (eds.), *Models and Theories in Social Systems*, Studies in Systems, Decision and Control 179, https://doi.org/10.1007/978-3-030-00084-4_23

In this chapter we introduce special numbers, special quaternions, special symbol elements, and we present some of their properties and their applications in combinatorics, number theory and associative algebra theory.

Let K be a field with $char(K) \neq 2$ and let $\alpha, \beta \in K \backslash \{0\}$. We recall that *the generalized quaternion algebra* $H_K(\alpha, \beta)$ is an algebra over the field K with a basis $\{e_1, e_2, e_3, e_4\}$ (where $e_1 = 1$) and the following multiplication:

$\cdot$	1	e_2	e_3	e_4
1	1	e_2	e_3	e_4
e_2	e_2	α	e_4	αe_3
e_3	e_3	$-e_4$	β	$-\beta e_2$
e_4	e_4	$-\alpha e_3$	βe_2	$-\alpha\beta$

Let x be element from $H_K(\alpha, \beta)$, $x = x_1 \cdot 1 + x_2 e_2 + x_3 e_3 + x_4 e_4$, where $x_i \in K$, $(\forall) i \in \{1, 2, 3, 4\}$ and let $\overline{x}$ be the conjugate of x, $\overline{x} = x_1 \cdot 1 - x_2 e_2 - x_3 e_3 - x_4 e_4$. The trace of x is $t(x) = x + \overline{x} = 2x_1$. The norm of x is $n(x) = x \cdot \overline{x} = x_1^2 - \alpha x_2^2 - \beta x_3^2 + \alpha\beta x_4^2$.

If $K = \mathbb{R}$ and $\alpha = \beta = -1$, we obtain Hamilton quaternion algebra $\mathbb{H}_{\mathbb{R}}(-1, -1)$, with the basis $\{1, i, j, k\}$.

The generalization of a quaternion algebra is a symbol algebra.

Let n be an arbitrary positive integer, $n \geq 3$ and let K be a field with $char(K)$, which does not divide n, containing ξ, where ξ is a primitive nth root of unity. Let $a, b \in K \backslash \{0\}$. The algebra A over K generated by elements x and y where

$$x^n = a, \ y^n = b, \ yx = \xi xy$$

is called a *symbol algebra* and it is denoted by $\left(\frac{a, b}{K, \xi}\right)$. Symbol algebras are also known as *power norm residue algebras*. For $n = 2$, we obtain the quaternion algebra over the field K. Quaternion algebras and symbol algebras are associative but non-commutative algebras, of dimension n^2 over K. Also, they are central simple algebras over the field K (this means they are simple algebras and their centers are equal to K). Theoretical aspects about these algebras can be found in the books: Pierce (1982), Lam (2004), Milne (1997); Gille and Szamuely (2006), Ledet (2005), Alsina and Bayer (2004), Vigneras (1980), Voight (2010), Kohel; Milnor (1971). Several properties of these algebras and their applications in number theory, combinatorics, associative algebra, geometry, coding theory, mechanics can be found in the articles Akyigit et al. (2014), Flatley (2012), Flaut et al. (2013), Flaut and Savin (2014a, b), Flaut and Savin (2015a, b), Flaut and Savin (2017), Flaut and Savin (2018), Flaut and Shpakivskyi (2013a, b), Halici (2012), Horadam (1963), Jafari and Yayli (2013), Karatas and Halici (2017), Linowitz (2012), Ramirez (2015), Savin et al. (2009), Savin (2014b), Savin (2016a, b), Savin (2017a, b), Swamy (1973), Tarnauceanu (2013). Another interesting view to applications of algebras to coding theory can be found in Saeid et al. (2018).

Many mathematicians studied the Fibonacci numbers, Lucas numbers, Pell numbers, Pell–Lucas numbers, generalized Fibonacci- Lucas numbers, the generalized Pell- Fibonacci–Lucas numbers, Fibonacci polynomials, Jacobsthal–Lucas polynomials, Fibonacci quaternions, the generalized Fibonacci–Lucas quaternions, the generalized Pell–Fibonacci–Lucas quaternions (see Horadam 1963; Catarino 2015; Catarino and Morgado 2016; Catarino 2016; Flaut and Savin 2015a, 2018; Flaut and Shpakivskyi 2013a; Halici 2012; Savin 2014a; Swamy 1973; Yilmaz and Yazlik 2017, etc.).

In book chapter we define $(l, 1, p + 2q, q \cdot l) -$ numbers, $(l, 1, p + 2q, q \cdot l) -$ quaternions, $(l, 1, p + 2q, q \cdot l) -$ symbol elements. We also study properties and applications of these elements.

This book chapter is organized as follow: Section 2 is a preliminary section, containing theoretical notions which we will then use in our results. In Sect. 3 we introduce two special number sequences (namely $(a_n)_{n \geq 0}$, $(b_n)_{n \geq 0}$), we obtain some interesting properties of these sequences and we also obtain some quaternion algebras which split or some division quaternion algebras. After these, in the same section, we introduce $(l, 1, p + 2q, q \cdot l) -$ numbers, $(l, 1, p + 2q, q \cdot l) -$ quaternions, $(l, 1, p + 2q, q \cdot l) -$ symbol elements and we obtain interesting properties and applications of them.

2 Preliminaries

First of all, we recall some results about prime integers, about diophantine equations or about the Fibonacci numbers, properties which will be necessary (in the next section) for to study some quaternion algebras.

Proposition 2.1 (Cucurezeanu 2006) *Let m be a fixed positive integer. The diophantine equation $x^2 + my^2 = z^2$ has an infinity of solutions:*

$$x = a^2 - mb^2, y = 2ab, z = a^2 + mb^2, a, b \in \mathbb{Z}.$$

Theorem 2.2 (Alexandru and Gosoniu 1999) *Let n be a positive integer. Then, there exist integers x, y such that $n = x^2 + y^2$ if and only if the exponent of any prime $p \equiv 3 \pmod 4$ that divides n is even.*

Proposition 2.3 (Savin 2014a) *For each positive integer n, $n \equiv 7 \pmod{16}$, there exist integer numbers x, y so that, the Fibonacci number f_n can be written as $f_n = x^2 + 9y^2$.*

Let K be a field with $char(K) \neq 2$, let $\alpha, \beta \in K \backslash \{0\}$ and let the generalized quaternion algebra $H_K(\alpha, \beta)$. $H_K(\alpha, \beta)$ is a division algebra if and only if for $x \in H_K(\alpha, \beta)$ we have $n(x) = 0$ only for $x = 0$.

We recall that $H_K(\alpha, \beta)$ is called split by K if it is isomorphic with a matrix algebra over K (see Pierce 1982; Lam 2004, Milne; Gille and Szamuely 2006). It is known the following remark about the quaternion algebras.

Remark 2.4 (Lam 2004; Ledet 2005) Let K be a field with char$K \neq 2$ and let $\alpha, \beta \in K \backslash \{0\}$. Then, the quaternion algebra $H_K (\alpha, \beta)$ is either split or a division algebra. In the book Gille and Szamuely (2006) appears the following criterion to decide if a quaternion algebra splits.

Proposition 2.5 (Gille and Szamuely 2006) *Let K be a field with char$K \neq 2$ and let $\alpha, \beta \in K \backslash \{0\}$. The quaternion algebra $\mathbb{H}_K(\alpha, \beta)$ splits if and only if the conic C $(\alpha, \beta): \alpha x^2 + \beta y^2 = z^2$ has a rational point over K (i.e. if there are $x_0, y_0, z_0 \in K$, not all zero such that $\alpha x_0^2 + \beta y_0^2 = z_0^2$).*

3 Some Properties of Special Quaternions and Special Symbol Elements

Let l be a nonzero natural number. We consider the sequence $(a_n)_{n \geq 0}$

$$a_n = l \cdot a_{n-1} + a_{n-2}, \ n \geq 2, a_0 = 0, a_1 = 1$$

and let the sequence $(b_n)_{n \geq 0}$

$$b_n = l \cdot b_{n-1} + b_{n-2}, \ n \geq 2, b_0 = 2, b_1 = l.$$

Let $\alpha = \frac{l + \sqrt{l^2 + 4}}{2}$ and $\beta = \frac{l - \sqrt{l^2 + 4}}{2}$. It results immediately the following relations:

Binet's Formula for the Sequence $(a_n)_{n \geq 0}$.

$$a_n = \frac{\alpha^n - \beta^n}{\alpha - \beta} = \frac{\alpha^n - \beta^n}{\sqrt{l^2 + 4}}, \quad (\forall) n \in \mathbb{N}.$$

Binet's Formula for the Sequence $(b_n)_{n \geq 0}$.

$$b_n = \alpha^n + \beta^n, \quad (\forall) n \in \mathbb{N}.$$

In the following, we show that the product of two elements belonging to the sequences $(a_n)_{n \geq 0}$, $(b_n)_{n \geq 0}$ are transformed into sums of elements belonging to the same sequences. Also, we find another properties of these sequences.

Proposition 3.1 *Let $(a_n)_{n \geq 0}$, $(b_n)_{n \geq 0}$ be the sequences previously defined. Then, the following equalities are true:*
(i)

$$b_n b_{n+m} = b_{2n+m} + (-1)^n b_m, \quad (\forall) n, m \in \mathbb{N};$$

(ii)

$$a_n b_{n+m} = a_{2n+m} + (-1)^{n+1} a_m, \quad (\forall) n, m \in \mathbb{N};$$

(iii)

$$a_{n+m}b_n = a_{2n+m} + (-1)^n a_m, \ (\forall) \, n, m \in \mathbb{N};$$

(iv)

$$a_n a_{n+m} = \frac{1}{l^2 + 4} \left[b_{2n+m} + (-1)^{n+1} b_m \right], \ (\forall) \, n, m \in \mathbb{N};$$

(v)

$$b_n + b_{n+2} = \left(l^2 + 4 \right) \cdot a_{n+1}, \ (\forall) \, n \in \mathbb{N};$$

(vi)

$$a_n^2 + a_{n+1}^2 = a_{2n+1}, \ (\forall) \, n \in \mathbb{N};$$

(vii)

$$b_n^2 + b_{n+1}^2 = \left(l^2 + 4 \right) \cdot a_{2n+1}, \ (\forall) \, n \in \mathbb{N};$$

Proof Let n, m be two positive integers. Applying Binet's formulae, we have:
(i)

$$b_n b_{n+m} = \left(\alpha^n + \beta^n \right) \cdot \left(\alpha^{n+m} + \beta^{n+m} \right) =$$

$$= \alpha^{2n+m} + \beta^{2n+m} + \alpha^n \beta^n \left(\alpha^m + \beta^m \right) = b_{2n+m} + (-1)^n b_m.$$

(ii)

$$a_n b_{n+m} = \frac{\alpha^n - \beta^n}{\alpha - \beta} \left(\alpha^{n+m} + \beta^{n+m} \right) =$$

$$= \frac{\alpha^{2n+m} - \beta^{2n+m}}{\alpha - \beta} - \frac{\alpha^n \beta^n \left(\alpha^m - \beta^m \right)}{\alpha - \beta} = a_{2n+m} + (-1)^{n+1} a_m.$$

(iii)

$$a_{n+m} b_n = \frac{\alpha^{n+m} - \beta^{n+m}}{\alpha - \beta} \left(\alpha^n + \beta^n \right) =$$

$$= \frac{\alpha^{2n+m} - \beta^{2n+m}}{\alpha - \beta} + \frac{\alpha^n \beta^n \left(\alpha^m - \beta^m \right)}{\alpha - \beta} = a_{2n+m} + (-1)^n a_m.$$

(iv)

$$a_n a_{n+m} = \frac{\alpha^n - \beta^n}{\alpha - \beta} \cdot \frac{\alpha^{n+m} - \beta^{n+m}}{\alpha - \beta} =$$

$$= \frac{1}{(\alpha - \beta)^2} \cdot \left[\alpha^{2n+m} + \beta^{2n+m} - \alpha^n \beta^n \left(\alpha^m + \beta^m \right) \right] =$$

$$= \frac{1}{l^2 + 4} \left[b_{2n+m} + (-1)^{n+1} b_m \right].$$

(v)

$$b_n + b_{n+2} = \alpha^n + \beta^n + \alpha^{n+2} + \beta^{n+2} =$$

$$= \alpha^{n+1} \cdot \left(\alpha + \frac{1}{\alpha} \right) + \beta^{n+1} \cdot \left(\beta + \frac{1}{\beta} \right) =$$

$$= \alpha^{n+1} \cdot \sqrt{l^2 + 4} - \beta^{n+1} \cdot \sqrt{l^2 + 4} = \left(l^2 + 4 \right) \cdot a_{n+1}.$$

(vi) Applying (iv) for $m = 0$, we have:

$$a_n^2 + a_{n+1}^2 = \frac{1}{l^2 + 4} \left[b_{2n} + (-1)^{n+1} b_0 + b_{2n+2} + (-1)^{n+2} b_0 \right] =$$

$$= \frac{1}{l^2 + 4} \left[b_{2n} + b_{2n+2} \right].$$

Applying (v) we obtain:

$$a_n^2 + a_{n+1}^2 = a_{2n+1}.$$

(vii) Applying (v) we have:

$$b_n^2 + b_{n+1}^2 = \left(\alpha^n + \beta^n \right)^2 + \left(\alpha^{n+1} + \beta^{n+1} \right)^2 =$$

$$= \alpha^{2n} + \beta^{2n} + 2 (-1)^n + \alpha^{2n+2} + \beta^{2n+2} + 2 (-1)^{n+1} =$$

$$= b_{2n} + b_{2n+2} = \left(l^2 + 4 \right) \cdot a_{2n+1}. \qquad \square$$

Let $(f_n)_{n \geq 0}$ be the Fibonacci sequence and let $(l_n)_{n \geq 0}$ be the Lucas sequence. There are well known the Cassini's identities for Fibonacci and Lucas numbers:

$$f_{n+1} f_{n-1} - f_n^2 = (-1)^n, \quad (\forall) \ n \in \mathbb{N}^*,$$

and

$$l_{n+1} l_{n-1} - l_n^2 = 5 \cdot (-1)^{n-1}, \quad (\forall) \ n \in \mathbb{N}^*,$$

Now, we obtain similarly results for the the sequences $(a_n)_{n \geq 0}$, $(b_n)_{n \geq 0}$.

Proposition 3.2 Let $(a_n)_{n \geq 0}$, $(b_n)_{n \geq 0}$ be the sequences previously defined. Then, the following identities are true:
(i)

$$a_{n+1} a_{n-1} - a_n^2 = (-1)^n, \quad (\forall) \ n \in \mathbb{N}^*;$$

(ii)

$$b_{n+1} b_{n-1} - b_n^2 = (-1)^{n-1} \cdot \left(l^2 + 4 \right), \quad (\forall) \ n \in \mathbb{N}^*.$$

Proof (i)

$$a_{n+1}a_{n-1} - a_n^2 = \frac{\alpha^{n+1} - \beta^{n+1}}{\sqrt{l^2 + 4}} \cdot \frac{\alpha^{n-1} - \beta^{n-1}}{\sqrt{l^2 + 4}} - \frac{(\alpha^n - \beta^n)^2}{l^2 + 4} =$$

$$= \frac{-(-1)^{n-1} \cdot (\alpha^2 + \beta^2) + 2(-1)^n}{l^2 + 4} = \frac{(-1)^n \cdot (b_2 + 2)}{l^2 + 4} = (-1)^n.$$

(ii)

$$b_{n+1}b_{n-1} - b_n^2 = \left(\alpha^{n+1} + \beta^{n+1}\right) \cdot \left(\alpha^{n-1} + \beta^{n-1}\right) - \left(\alpha^n + \beta^n\right)^2 =$$

$$= (-1)^{n-1} \cdot (\alpha^2 + \beta^2) - 2(-1)^n = (-1)^{n-1} \cdot (b_2 + 2) = (-1)^{n-1} \cdot \left(l^2 + 4\right).$$

□

Proposition 3.3 *Let* $(b_n)_{n \geq 0}$ *be the sequence previously defined. Then, the followings are true:*
(i) if l is even, then b_n *is even* $(\forall) \ n \in \mathbb{N}$;
(ii) if l is odd, then b_n *is even if and only if* $n \equiv 0 \ (mod \ 3)$;
(iii) if l is odd and $n \equiv 0 \ (mod \ 6)$, *then* $b_{n-1} \cdot b_{n+1} \equiv 3 \ (mod \ 4)$;
(iv) if l is odd and $n \equiv 3 \ (mod \ 6)$, *then* $b_{n-1} \cdot b_{n+1} \equiv 1 \ (mod \ 4)$.

Proof For (i), (ii), (iii) and (iv) the proof is immediate, using the principle of mathematics induction (after $n \in \mathbb{N}$). □

Proposition 3.4 *Let* $(a_n)_{n \geq 0}$, $(b_n)_{n \geq 0}$ *be the sequences previously defined. Then, the followings are true:*
(i) The quaternion algebra $\mathbb{H}_{\mathbb{Q}}(-1, f_{2n+1})$ *splits,* $(\forall) \ n \in \mathbb{N}^*$;
(ii) The quaternion algebra $\mathbb{H}_{\mathbb{Q}}(-1, 5f_{2n+1})$ *splits,* $(\forall) \ n \in \mathbb{N}^*$;
(iii) The quaternion algebra $\mathbb{H}_{\mathbb{Q}}(-1, a_{2n+1})$ *splits,* $(\forall) \ n \in \mathbb{N}^*$;
(iv) The quaternion algebra $\mathbb{H}_{\mathbb{Q}}\left(-1, (l^2 + 4) \cdot a_{2n+1}\right)$ *splits,* $(\forall) \ n \in \mathbb{N}^*$;
(v) The quaternion algebra $\mathbb{H}_{\mathbb{Q}}(-1, f_{2n+1}f_{2n-1})$ *splits,* $(\forall) \ n \in \mathbb{N}^*$;
(vi) The quaternion algebra $\mathbb{H}_{\mathbb{Q}}(-1, a_{2n+1}a_{2n-1})$ *splits,* $(\forall) \ n \in \mathbb{N}^*$;
(vii)The quaternion algebra $\mathbb{H}_{\mathbb{Q}}(-1, -b_{n+1}b_{n-1})$ *is a division algebra,* $(\forall) \ n \in \mathbb{N}^*$;
(viii) The quaternion algebra $\mathbb{H}_{\mathbb{Q}}(1, b_{n+1}b_{n-1})$ *splits,* $(\forall) \ n \in \mathbb{N}^*$;
(ix) If l is odd and $n \equiv 0(mod \ 6)$, *then the quaternion algebra* $\mathbb{H}_{\mathbb{Q}}(-1, b_{n+1}b_{n-1})$ *is a division algebra,* $(\forall) \ n \in \mathbb{N}^*$;
(x) If $6 \nmid n$ *and the exponent of any prime* $p \equiv 3(mod \ 4)$ *that divides* $b_{n+1}b_{n-1}$ *is even, then the quaternion algebra* $\mathbb{H}_{\mathbb{Q}}(-1, b_{n+1}b_{n-1})$ *splits,* $(\forall) \ n \in \mathbb{N}^*$;
(xi) The quaternion algebra $\mathbb{H}_{\mathbb{Q}}(-9, f_n)$ *splits,* $(\forall) \ n \in \mathbb{N}^*$, $n \equiv 7 \ (mod \ 16)$.

Proof Since (iii) is a generalization of (i), we are proving directly (iii).
(iii) If we consider the equation $-x^2 + a_{2n+1} \cdot y^2 = z^2$, we apply Proposition 3.1 (vi) and we obtain that it has the following solution in $\mathbb{Q} \times \mathbb{Q} \times \mathbb{Q} : (x_0, y_0, z_0) = (a_n, 1, a_{n+1})$. According to Proposition 2.5, it results that the quaternion algebra $\mathbb{H}_{\mathbb{Q}}(-1, a_{2n+1})$ splits, $(\forall) \ n \in \mathbb{N}^*$.

(iv) Using Proposition 3.1 (vii), it results that the equation $-x^2 + \left(l^2 + 4\right) \cdot a_{2n+1} \cdot y^2 = z^2$ has a solution in $\mathbb{Q} \times \mathbb{Q} \times \mathbb{Q}$, namely $(x_0, y_0, z_0) = (b_n, 1, b_{n+1})$. Applying Proposition 2.5, it results that the quaternion algebra $\mathbb{H}_{\mathbb{Q}} \left(-1, \left(l^2 + 4\right) \cdot a_{2n+1}\right)$ splits, $(\forall)\ n \in \mathbb{N}^*$.

(ii) This is a particular case of (ii) (for $l = 1$).

(vi) This is a generalization of (v), so we are proving only (vi).

Using Proposition 3.2 (i), we find the following solution in $\mathbb{Q} \times \mathbb{Q} \times \mathbb{Q}$ for the equation $-x^2 + a_{2n+1} \cdot a_{2n-1} \cdot y^2 = z^2$: $(x_0, y_0, z_0) = \left((-1)^n, 1, a_{2n}\right)$. Applying Proposition 2.5, we obtain that the quaternion algebra $\mathbb{H}_{\mathbb{Q}} (-1, a_{2n+1}a_{2n-1})$ splits, $(\forall)\ n \in \mathbb{N}^*$.

(vii) Let the quaternion algebra $\mathbb{H}_{\mathbb{Q}} (-1, -b_{n+1}b_{n-1})$ and let $\{1, e_2, e_3, e_4\}$ a basis in this algebra. Let $x = x_1 \cdot 1 + x_2 \cdot e_2 + x_3 \cdot e_3 + x_4 \cdot e_4 \in \mathbb{H}_{\mathbb{Q}} (-1, -b_{2n+1}b_{2n-1})$. The norm of x is $n(x) = x \cdot \overline{x} = x_1^2 + x_2^2 + b_{n-1}b_{n+1}x_3^2 + b_{n-1}b_{n+1}x_4^2$. Since $b_n \in \mathbb{N}^*$, for $(\forall)\ n \in \mathbb{N}^*$, it results that $n(x) = 0$ if and only if $x = 0$. So, $\mathbb{H}_{\mathbb{Q}} (-1, -b_{n+1}b_{n-1})$ is a division algebra for $(\forall)\ n \in \mathbb{N}^*$.

Similarly, it results immediately that $\mathbb{H}_{\mathbb{Q}} (-1, -f_{n+1}f_{n-1})$, $\mathbb{H}_{\mathbb{Q}} (-1, -a_{n+1}a_{n-1})$, $\mathbb{H}_{\mathbb{Q}} (-1, -l_{n+1}l_{n-1})$ are division algebras for $(\forall)\ n \in \mathbb{N}^*$.

(viii) We study if the equation $x^2 + b_{n+1}b_{n-1} \cdot y^2 = z^2$ has rational solutions. Applying Proposition 2.1, it results that the equation $x^2 + b_{n+1}b_{n-1} \cdot y^2 = z^2$ has solutions in integer numbers, so it has solutions in the set of rational numbers. Using Proposition 2.5, we obtain that the quaternion algebra $\mathbb{H}_{\mathbb{Q}} (1, b_{n+1}b_{n-1})$ splits.

(ix) If $n \equiv 0 \pmod 6$, according to Proposition 3.3, $b_{n-1} \cdot b_{n+1} \equiv 3 \pmod 4$. We study if the equation $-x^2 + b_{n+1}b_{n-1} \cdot y^2 = z^2$ has integer solutions. We suppose that this equation has a solution $(x_0, y_0, z_0) \in \mathbb{Z} \times \mathbb{Z} \times \mathbb{Z} \setminus \{(0, 0, 0)\}$, g.c.d.$(x_0, y_0) =$ g.c.d.$(y_0, z_0) =$ g.c.d.$(y_0, z_0) = 1$. We have: $b_{n+1}b_{n-1} \cdot y_0^2 \equiv 0$ or $3 \pmod 4$, but $x_0^2 + z_0^2 \equiv 1$ or $2 \pmod 4$, so we cannot have $b_{n+1}b_{n-1} \cdot y_0^2 = x_0^2 + z_0^2$. It results that the equation $-x^2 + b_{n+1}b_{n-1} \cdot y^2 = z^2$ does not have integer solutions. We obtain immediately that the equation $-x^2 + b_{n+1}b_{n-1} \cdot y^2 = z^2$ does not have solutions in the set of rational numbers, so the quaternion algebra $\mathbb{H}_{\mathbb{Q}} (-1, b_{n+1}b_{n-1})$ does not split.

Applying Remark 2.4, we obtain that the quaternion algebra $\mathbb{H}_{\mathbb{Q}} (-1, b_{n+1}b_{n-1})$ is a division algebra, $(\forall)\ n \in \mathbb{N}^*$.

(x) Case 1 : l is odd.

If $n \equiv 3 \pmod 6$, according to Proposition 3.3 (iv) $b_{n-1} \cdot b_{n+1} \equiv 1 \pmod 4$. If $n \equiv 1$ or 2 or 4 or $5 \pmod 6$, according to Proposition 3.3 (ii) $b_{n-1} \cdot b_{n+1}$ is even.

Case 2 : l is even, according to Proposition 3.3 (i) $b_{n-1} \cdot b_{n+1}$ is even.

In all these cases, it is possible to exist a prime $p \equiv 3 \pmod 4$ that divides $b_{n+1}b_{n-1}$. If the exponent of any prime $p \equiv 3 \pmod 4$ that divides $b_{n+1}b_{n-1}$ is even, according to Theorem 2.2 there exist integers $x_0; z_0$ such that $b_{n+1}b_{n-1} = x_0^2 + z_0^2$. This implies that $(x_0, 1, z_0)$ is a solution in integer numbers for the equation $-x^2 + b_{n+1}b_{n-1} \cdot y^2 = z^2$, so, according Proposition 2.5, the quaternion algebra $\mathbb{H}_{\mathbb{Q}} (-1, b_{n+1}b_{n-1})$ splits, $(\forall)\ n \in \mathbb{N}^*$.

(xi) Let n be a positive integer number, $n \equiv 7 \pmod{16}$. Using Proposition 2.3 we obtain that there are $x_0, z_0 \in \mathbb{Z}$ such that $(x_0, 1, z_0)$ is a solution of the equation $-9x^2 + f_n \cdot y^2 = z^2$. Applying Proposition 2.5, we obtain that the quaternion algebra $\mathbb{H}_{\mathbb{Q}} (-9, f_n)$ splits, $(\forall)\ n \in \mathbb{N}^*, n \equiv 7 \pmod{16}$. ∎

Let p, q be two arbitrary integers and $(a_n)_{n \geq 0}$, $(b_n)_{n \geq 0}$ are the sequences previously defined. If $n \in N^*$, $a_{-n} = (-1)^{n+1} \cdot a_n$.
Let the sequence $(u_n)_{n \geq 0}$,

$$u_{n+1} = p a_n + q b_{n+1}, \ n \geq 0.$$

To avoid confusion, we will use the notation $u_n^{p,q}$ for u_n.
We remark that $u_n = l u_{n-1} + u_{n-2}$, $(\forall) n \in \mathbb{N}, n \geq 2$,
We calculate $u_0 = p a_{-1} + q b_0 = p + 2q$, $u_1 = p a_0 + q b_1 = q \cdot l$. We call the elements of the sequence $(u_n)_{n \geq 0}$ the $(l, 1, p + 2q, q \cdot l)$ −numbers.

Remark 3.5 Let p, q be two arbitrary integers, and let $\left(u_n^{p,q}\right)_{n \geq 0}$ the sequence previously defined. Then, we have:

$$p a_{n+1} + q b_n = u_n^{p,q} + u_{n+1}^{pl,o}, \ \forall n \in \mathbb{N} - \{0\}.$$

Proof We compute

$$p a_{n+1} + q b_n = p l a_n + p a_{n-1} + q b_n = u_n^{p,q} + u_{n+1}^{pl,o}. \qquad \blacksquare$$

Let $\alpha, \beta \in \mathbb{Q}^*$. We consider the generalized quaternion algebra $\mathbb{H}_\mathbb{Q} (\alpha, \beta)$ with basis $\{1, e_2, e_3, e_4\}$. We define the nth $(l, 1, p + 2q, q \cdot l) -$ quaternion to be the element of the form

$$U_n^{p,q} = u_n^{p,q} \cdot 1 + u_{n+1}^{p,q} \cdot e_2 + u_{n+2}^{p,q} \cdot e_3 + u_{n+3}^{p,q} \cdot e_4.$$

Remark 3.6 Let $U_n^{p,q}$ be the nth $(l, 1, p + 2q, q \cdot l) -$ quaternion. Then, we have:

$$U_n^{p,q} = 0 \ if \ and \ only \ if \ p = q = 0.$$

Proof "$\Leftarrow$" It is trivial.
"$\Rightarrow$" If $U_n^{p,q} = 0$, using the fact that $\{1, e_2, e_3, e_4\}$ is a basis in quaternion algebra $\mathbb{H}_\mathbb{Q} (\alpha, \beta)$, we obtain that $u_n^{p,q} = 0, u_{n+1}^{p,q} = 0, u_{n+2}^{p,q} = 0, u_{n+3}^{p,q} = 0$.
From the recurrence relation of the sequence $\left(u_n^{p,q}\right)_{n \geq 1}$, it results that $u_{n-1}^{p,q} = 0$, $u_{n-2}^{p,q} = 0, ..., s_1^{p,q} = 0, u_0^{p,q} = 0$. So, $q = 0$ and $p = 0$. $\qquad \blacksquare$

About the generalized Fibonacci–Lucas quaternions $\left(G_n^{p,q}\right)_{n \geq 0}$, in the paper Flaut and Savin (2015a) (Theorem 3.5), we proved that:
(i) *The set*

$$M = \left\{ \sum_{i=1}^{n} 5 G_{n_i}^{p_i, q_i} | n \in \mathbb{N}^*, p_i, q_i \in \mathbb{Z}, (\forall) i = \overline{1, n} \right\} \cup \{1\}$$

has a ring structure with quaternion addition and multiplication.

(ii) *The set M is an order of the quaternion algebra* $\mathbb{H}_{\mathbb{Q}}(\alpha, \beta)$.

We generalized these results for $(1, a, p + 2q, q)$ quaternions $\left(S_n^{p,q}\right)_{n \geq 0}$ in the paper Flaut and Savin (2017) (Proposition 5.4), namely:
Let a be a nonzero natural number and let O be the set

$$O = \left\{ \sum_{i=1}^{n} (1 + 4a) \, S_{n_i}^{p_i, q_i} \mid n \in \mathbb{N}^*, \, p_i, q_i \in \mathbb{Z}, \, (\forall)i = \overline{1, n} \right\} \cup \{1\}.$$

Then O is an order of the quaternion algebra $\mathbb{H}_{\mathbb{Q}}(\alpha, \beta)$.

Similarly, in the paper Flaut and Savin (2018), we introduced the generalized Pell- Fibonacci–Lucas numbers $\left(r_n^{p,q}\right)_{n \geq 0}$, the generalized Pell- Fibonacci–Lucas quaternions $\left(R_n^{p,q}\right)_{n \geq 0}$, and we proved that (Proposition 3.7. from the paper Flaut and Savin 2018) the set

$$O = \left\{ \sum_{i=1}^{n} 8 R_{n_i}^{p_i, q_i} \mid n \in \mathbb{N}^*, \, p_i, q_i \in \mathbb{Z}, \, (\forall)i = \overline{1, n} \right\} \cup \{1\}$$

is an order of the quaternion algebra $\mathbb{H}_{\mathbb{Q}}(\alpha, \beta)$.

Here, we generalized these numbers and these quaternions: the sequence $(a_n)_{n \geq 0}$ is the generalization for the Pell sequence $(P_n)_{n \geq 0}$ and the sequence $(b_n)_{n \geq 0}$ is the generalization for the Pell–Lucas sequence $(Q_n)_{n \geq 0}$. Also, the sequence $\left(u_n^{p,q}\right)_{n \geq 0}$ is the generalization for the sequence $\left(r_n^{p,q}\right)_{n \geq 0}$ of Pell-Fibonacci–Lucas numbers and the sequence of the $(l, 1, p + 2q, q \cdot l) -$ quaternions $\left(U_n^{p,q}\right)_{n \geq 0}$ is the generalization for the sequence of the generalized Pell- Fibonacci–Lucas quaternions $\left(R_n^{p,q}\right)_{n \geq 0}$.

Let ϵ be a primitive root of the unity of order 3 and let K be a field with the property $\epsilon \in K$. Let $\alpha_1, \alpha_2 \in K \backslash \{0\}$ and let $A = \left(\frac{\alpha_1, \alpha_2}{K, \epsilon}\right)$ be the symbol algebra of degree 3. A has a K-basis $\left\{x^{j_1} y^{j_2} \mid 0 \leq j_1, j_2 < 3\right\}$, with $x^3 = \alpha_1$, $y^3 = \alpha_2$, $yx = \epsilon xy$.

In the paper Flaut and Savin (2014a), we defined the nth Fibonacci symbol element

$$F_n = f_n \cdot 1 + f_{n+1} \cdot x + f_{n+2} \cdot x^2 + f_{n+3} \cdot y + f_{n+4} \cdot xy +$$

$$+ f_{n+5} \cdot x^2 y + f_{n+6} \cdot y^2 + f_{n+7} \cdot xy^2 + f_{n+8} \cdot x^2 y^2.$$

In the paper Flaut et al. (2013) we defined the nth Lucas symbol element

$$L_n = l_n \cdot 1 + l_{n+1} \cdot x + l_{n+2} \cdot x^2 + l_{n+3} \cdot y + l_{n+4} \cdot xy +$$

$$+ l_{n+5} \cdot x^2 y + l_{n+6} \cdot y^2 + l_{n+7} \cdot xy^2 + l_{n+8} \cdot x^2 y^2.$$

Now, we define the nth $(l, 1, p + 2q, q \cdot l) -$ symbol element to be the element of the form

$$\mathbb{U}_n^{p,q} = u_n^{p,q} \cdot 1 + u_{n+1}^{p,q} \cdot x + u_{n+2}^{p,q} \cdot x^2 + u_{n+3}^{p,q} \cdot y +$$

$$+u_{n+4}^{p,q} \cdot xy + u_{n+5}^{p,q} \cdot x^2 y + u_{n+6}^{p,q} \cdot y^2 + u_{n+7}^{p,q} \cdot xy^2 + u_{n+8}^{p,q} \cdot x^2 y^2.$$

Remark 3.7 Let $\mathbb{U}_n^{p,q}$ be the nth $(l, 1, p + 2q, q \cdot l)$ − symbol element. Then, we have:

$$\mathbb{U}_n^{p,q} = 0 \; if \; and \; only \; if \, p = q = 0.$$

The proof of this remark is similar to the proof of Remark 3.6.

With proof ideas similar to those in the Theorem 3.5 from the paper Flaut and Savin (2015a), Proposition 5.4 from the paper Flaut and Savin (2017), Proposition 3.7. from the paper Flaut and Savin (2018), we obtain the following results:

Proposition 3.8 *Let l be a nonzero natural number and let M_1 be the set*

$$M_1 = \left\{ \sum_{i=1}^{n} (l^2 + 4) \, U_{n_i}^{p_i, q_i} \, | n \in \mathbb{N}^*, \, p_i, q_i \in \mathbb{Z}, \, (\forall) i = \overline{1, n} \right\} \cup \{1\}.$$

Then M_1 is an order of the quaternion algebra $\mathbb{H}_{\mathbb{Q}} (\alpha, \beta)$.

Proposition 3.9 *Let l be a nonzero natural number and let M_2 be the set*

$$M_2 = \left\{ \sum_{i=1}^{n} (l^2 + 4) \, \mathbb{U}_{n_i}^{p_i, q_i} \, | n \in \mathbb{N}^*, \, p_i, q_i \in \mathbb{Z}, \, (\forall) i = \overline{1, n} \right\} \cup \{1\}.$$

Then M_2 is an order of the symbol algebra $A = \left(\frac{\alpha_1, \alpha_2}{K, \epsilon} \right)$.

Since the proofs of Proposition 3.8 and Proposition 3.9 are similar, we only prove one of them - Proposition 3.9.

Proof We prove that M_2 is a free $\mathbb{Z}-$ submodule of rank 9 of the symbol algebra $A = \left(\frac{\alpha_1, \alpha_2}{K, \epsilon} \right)$.

According to Remark 3.7, $\mathbb{U}_n^{0,0} = 0 \in O$.

Let $n, m \in \mathbb{N}^*, \, p, q, p', q', c, d \in \mathbb{Z}$. We have:

$$cu_n^{p,q} + du_m^{p',q'} = u_n^{cp,cq} + u_m^{dp',dq'}.$$

This implies that

$$c\mathbb{U}_n^{p,q} + d\mathbb{U}_m^{p',q'} = \mathbb{U}_n^{cp,cq} + \mathbb{U}_m^{dp',dq'} \tag{1}$$

So, M_2 is a free $\mathbb{Z}-$ submodule of rank 9 of the symbol algebra A.

We consider the set $M_3 = \left\{ \sum_{i=1}^{n} (l^2 + 4) \, u_{n_i}^{p_i, q_i} \, | n \in \mathbb{N}^*, \, p_i, q_i \in \mathbb{Z}, \, (\forall) i = \overline{1, n} \right\}$.

We are proving that M_2 is a subring of the symbol algebra A. Taking into account the relation (1), it is enough to prove that $(l^2 + 4) \, \mathbb{U}_n^{p,q} \, (l^2 + 4) \, \mathbb{U}_m^{p',q'} \in M_2$. For this, it is enough to prove that $(l^2 + 4) u_n^{p,q} (l^2 + 4) u_m^{p',q'} \in M_3$.

Let m, n be two integers, $n < m$. We calculate:

$$(l^2 + 4) u_n^{p,q} (l^2 + 4) u_m^{p',q'} =$$

$$= (l^2 + 4) (p a_{n-1} + q b_n) (l^2 + 4) \left(p' a_{m-1} + q' b_m \right) =$$

$$= (l^2 + 4)^2 pp' a_{n-1} a_{m-1} + (l^2 + 4)^2 pq' a_{n-1} b_m +$$

$$+ (l^2 + 4)^2 p' q a_{m-1} b_n + (l^2 + 4)^2 qq' b_n b_m.$$

Using Proposition 3.1, we have:

$$(l^2 + 4) u_n^{p,q} \cdot (l^2 + 4) u_m^{p',q'} =$$

$$= (l^2 + 4)^2 pp' \frac{1}{l^2 + 4} [b_{n+m-2} + (-1)^n b_{m-n}] + (l^2 + 4)^2 pq' [a_{m+n-1} + (-1)^n a_{m-n+1}] +$$

$$+ (l^2 + 4)^2 p' q [a_{n+m-1} + (-1)^n a_{m-n-1}] + (l^2 + 4)^2 qq' [b_{n+m} + (-1)^n b_{m-n}] =$$

$$= (l^2 + 4)^2 \left[pq' a_{n+m-1} + qq' b_{m+n} \right] + (l^2 + 4)^2 \left[(-1)^n p' q a_{m-n-1} + (-1)^n qq' b_{m-n} \right] +$$

$$+ (l^2 + 4) \left[(-1)^{n-1} (l^2 + 4) pq' a_{m-n+1} + (-1)^n pp' b_{m-n} \right] +$$

$$+ (l^2 + 4) \left[(l^2 + 4) p' q a_{n+m-1} + pp' b_{n+m-2} \right]$$

Applying the definition of the sequence $(u_n)_{n \geq 0}$ and Remark 3.5, we obtain:

$$(l^2 + 4) u_n^{p,q} \cdot (l^2 + 4) u_m^{p',q'} =$$

$$= (l^2 + 4) u_{m+n}^{(l^2+4)pq',(l^2+4)qq'} + (l^2 + 4) u_{m-n}^{(-1)^n (l^2+4)p'q,(-1)^n (l^2+4)qq'} +$$

$$+ (l^2 + 4) u_{m-n}^{(-1)^{n-1}(l^2+4)pq',(-1)^n pp'} + (l^2 + 4) u_{m-n+1}^{(-1)^{n-1}l(l^2+4)pq',0} +$$

$$+ (l^2 + 4) u_{m+n-2}^{(l^2+4)p'q,pp'} + (l^2 + 4) u_{m+n-1}^{l(l^2+4)p'q,0}.$$

So, $(l^2 + 4) u_n^{p,q} \cdot (l^2 + 4) u_m^{p',q'} \in M_3$.

It results that M_2 is an order of the symbol algebra $A = \left(\frac{\alpha_1, \alpha_2}{K, \epsilon} \right)$. ∎

4 Conclusions

In this paper we found some split quaternion algebras or division quaternion algebras involving Fibonacci numbers or another sequence numbers obtained from a difference equation of degree two. Also, we found certain sets of $(l, 1, p + 2q, q \cdot l) -$ quaternions, respectively of $(l, 1, p + 2q, q \cdot l) -$ symbol elements which are orders in the sense of ring theory in a quaternion algebra, respectively in a symbol algebra.

The results obtained in this paper can constitute the start for a further research in which we intend to find another properties and applications of $(l, 1, p + 2q, q \cdot l) -$ numbers, $(l, 1, p + 2q, q \cdot l) -$ quaternions, $(l, 1, p + 2q, q \cdot l) -$ symbol elements.

Acknowledgements The author dedicates this book chapter to her mother, Prof. Elena Savin.

References

Akyigit, M., Kosal, H.H., Tosun, M.: Fibonacci generalized quaternions. Adv. Appl. Clifford Algebras **24**(3), 631–641 (2014)

Alexandru, V., Gosoniu, N.M.: Elements of Number Theory (in Romanian). Bucharest University, Romania (1999)

Alsina, M., Bayer, P.: Quaternion Orders, Quadratic Forms and Shimura Curves. CRM Monograph Series, vol. 22. American Mathematical Society, Providence (2004)

Catarino, P.: A note on $h(x)$ Fibonacci quaternion polynomials. Chaos, Solitons Fractals **77**, 1–5 (2015)

Catarino, P., Morgado, M.L.: On generalized Jacobsthal and Jacobsthal-Lucas polynomials. An. St. Univ. Ovidius Constanta Mat. Ser. **24**(3), 61–78 (2016)

Catarino, P.: The modified Pell and the modified K-Pell quaternions and octonions. Adv. Appl. Clifford Algebra **26**(2), 577–590 (2016)

Cucurezeanu, I.: Equations in Integer Numbers (in Romanian). Aramis, Romania (2006)

Flatley, R.: Trace forms of symbol algebras. Algebra Colloq. **19**, 1117–1124 (2012)

Flaut, C., Savin, D., Iorgulescu, G.: Some properties of Fibonacci and Lucas symbol elements. J. Math. Sci. Adv. Appl. **20**, 37–43 (2013)

Flaut, C., Savin, D.: Some properties of symbol algebras of degree 3. Math. Rep. **16**(66)(3), 443–463 (2014a)

Flaut, C., Savin, D.: About quaternion algebras and symbol algebras. Bull. Univ. Transilv. Brasov Seria III **7**(56)(2), 59–64 (2014b)

Flaut, C., Savin, D.: Quaternion algebras and generalized Fibonacci-Lucas quaternions. Adv. Appl. Clifford Algebra **25**(4), 853–862 (2015a)

Flaut, C., Savin, D.: Some examples of division symbol algebras of degree 3 and 5. Carpathian J. Math. **31**(2), 197–204 (2015b)

Flaut, C., Savin D.: Some remarks regarding a, b, x_0, x_1 numbers and a, b, x_0, x_1 quaternions, submitted (2017)

Flaut, C., Savin, D.: Some special number sequences obtained from a difference equation of degree three. Chaos, Solitons Fractals **106**, 67–71 (2018)

Flaut, C., Shpakivskyi, V.: On generalized Fibonacci quaternions and Fibonacci-Narayana quaternions. Adv. Appl. Clifford Algebra **23**(3), 673–688 (2013a)

Flaut, C., Shpakivskyi, V.: Real matrix representations for the complex quaternions. Adv. Appl. Clifford Algebra **23**(3), 657–671 (2013b)

Gille, P., Szamuely, T.: Central Simple Algebras and Galois Cohomology. Cambridge University Press, Cambridge (2006)

Halici, S.: On Fibonacci quaternions. Adv. Appl. Clifford Algebras **22**(2), 321–327 (2012)

Horadam, A.F.: Complex Fibonacci numbers and Fibonacci quaternions. Am. Math. Mon. **70**, 289–291 (1963)

Jafari, M., Yayli, Y.: Rotation in four dimensions via generalized Hamilton operators. Kuwait J. Sci. **40**(1), 67–79 (2013)

Karatas, A., Halici, S.: Horadam octonions. An. St. Univ. Ovidius Constanta Mat. Ser. **25**(3), 97–106 (2017)

Lam, T.Y.: Introduction to Quadratic Forms over Fields. AMS, Providence (2004)

Ledet, A.: Brauer Type Embedding Problems. American Mathematical Society, Providence (2005)

Linowitz, B.: Selectivity in quaternion algebras. J. Number Theory **132**, 1425–1437 (2012)

Milne J.S.: Class Field Theory. https://www.jmilne.org/math/CourseNotes/CFT310.pdf

Milnor, J.: Introduction to Algebraic K-Theory. Annals of Mathematics Studies. Princeton University Press, Princeton (1971)

Pierce, R.S.: Associative Algebras. Springer, Berlin (1982)

Ramirez, J.L.: Some combinatorial properties of the k-Fibonacci and the k-Lucas quaternions. An. St. Univ. Ovidius Constanta Mat. Ser. **23**(2), 201–212 (2015)

Saeid, A.B., Flaut, C., Hoskova-Mayerova, S., Afshar, M., Rafsanjani, M.K.: Some connections between BCKalgebras and nary block codes. Soft Comput. Fusion Found. Methodol. Appl. **22**(1), 41–46 (2018)

Savin, D., Flaut, C., Ciobanu, C.: Some properties of the symbol algebras. Carpathian J. Math. **25**(2), 239–245 (2009)

Savin, D.: Fibonacci primes of special forms. Notes Number Theory Discret. Math. **20**(2), 10–19 (2014a)

Savin, D.: About some split central simple algebras. An. St. Univ. Ovidius Constanta Mat. Ser. **22**(1), 263–272 (2014b)

Savin, D.: About division quaternion algebras and division symbol algebras. Carpathian J. Math. **32**(2), 233–240 (2016a)

Savin, D.: Quaternion algebras and symbol algebras over algebraic number field K, with the degree $[K : \mathbb{Q}]$ even. Gulf J. Math. **4**(4), 16–21 (2016b)

Savin, D.: About special elements in quaternion algebras over finite fields. Adv. Appl. Clifford Algebras **27**(2), 1801–1813 (2017a)

Savin D.: About split quaternion algebras over quadratic fields and symbol algebras of degree n. Bull. Math. Soc. Sci. Math. Roum. Tome **60**(108)(3), 307–312 (2017b)

Swamy, M.N.S.: On generalized Fibonacci quaternions. Fibonacci Quaterly **11**(5), 547–549 (1973)

Tarnauceanu, M.: A characterization of the quaternion group. An. St. Univ. Ovidius Constanta **21**(1), 209–214 (2013)

Vigneras, M.F.: Arithmetique des Algebres de Quaternions. Lecture Notes in Mathematics, vol. 800. Springer, Berlin (1980)

Voight J.: The arithmetic of quaternion algebras (2010). http://www.math.dartmouth.edu/jvoight/crmquat/book/quat-modforms-041310.pdf

Yilmaz, N., Yazlik, Y., Taskara, N.: On the bi-periodic Lucas octonions. Adv. Appl. Clifford Algebras **27**(2), 1927–1937 (2017)

An Algebraic Model for Real Matrix Representations. Remarks Regarding Quaternions and Octonions

Cristina Flaut

Abstract In this chapter, we present some applications of quaternions and octonions. We present the real matrix representations for complex octonions and some of their properties which can be used in computations where these elements are involved. Moreover, we give a set of invertible elements in split quaternion algebras and in split octonion algebras.

Keywords Quaternion algebras · Octonion algebras · Matrix representation

2000 AMS Subject Classification 17A35 · 15A06 · 15A24 · 16G30

1 Introduction

The mathematical objects have, directly or indirectly, many applications in our lives. It is very interesting to remark how abstract notions and abstract theories can have such an influence, leading to a significant development in various domains.

As examples in this direction, Quaternions and Octonions are some of such objects which influenced, in a good sense, our lives over the time.

Quaternions were discovered in October 1843 by Sir William Rowan Hamilton, when he introduced an algebraic system formed by one real part and three imaginary parts. This system is nothing else than the well known real quaternion division algebra, (Hanson 2006, p. 5).

Since quaternions generalize complex numbers, after quaternions were discovered, the question arose whether this structure can be generalized. In December 1843, John T. Graves, Hamilton's friend, generalized quaternions to octonions and gave them the name "octaves", obtaining an 8-dimensional algebra. Independently

C. Flaut (✉)

Faculty of Mathematics and Computer Science, Ovidius University, Bd. Mamaia 124, 900527 Constanta, Romania

e-mail: cflaut@univ-ovidius.ro; cristina_flaut@yahoo.comaaaa

URL: http://www.cristinaflaut.wikispaces.com/; http://www.univ-ovidius.ro/math/

© Springer Nature Switzerland AG 2019

C. Flaut et al. (eds.), *Models and Theories in Social Systems*, Studies in Systems, Decision and Control 179, https://doi.org/10.1007/978-3-030-00084-4_24

431

by Graves, Arthur Cayley discovered octonions and published his result in 1845. Octonions are also called *Cayley numbers,* (Hanson 2006, p. 9).

A generalized real quaternion algebra, $\mathbb{H}(\beta_1, \beta_2)$, is an algebra with the elements of the form $a = a_0 + a_1 e_1 + a_2 e_2 + a_3 e_3$, where $a_i \in \mathbb{R}$, $i \in \{0, 1, 2, 3\}$, and the basis $\{1, e_1, e_2, e_3\}$, with the multiplication given in the following Table:

$\cdot$	1	e_1	e_2	e_3
1	1	e_1	e_2	e_3
e_1	e_1	$-\beta_1$	e_3	$-\beta_1 e_2$
e_2	e_2	$-e_3$	$-\beta_2$	$\beta_2 e_1$
e_3	e_3	$\beta_1 e_2$	$-\beta_2 e_1$	$-\beta_1 \beta_2$

If $a \in \mathbb{H}(\beta_1, \beta_2)$, $a = a_0 + a_1 e_1 + a_2 e_2 + a_3 e_3$, then $\bar{a} = a_0 - a_1 e_1 - a_2 e_2 - a_3 e_3$ is called the *conjugate* of the element a. We denote by

$$t(a) = a + \bar{a} \in \mathbb{R}$$

and

$$n(a) = a\bar{a} \in \mathbb{R},$$

the trace and *the norm* of a real quaternion a. The norm of a generalized quaternion has the following expression

$$n(a) = a_1^2 + \beta_1 a_2^2 + \beta_2 a_3^2 + \beta_1 \beta_2 a_4^2.$$

A generalized octonion algebra $\mathbb{O}(\alpha, \beta, \gamma)$ is an algebra with the elements of the form $a = a_0 + a_1 e_1 + a_2 e_2 + a_3 e_3 + a_4 e_4 + a_5 e_5 + a_6 e_6 + a_7 e_7$, where $a_i \in \mathbb{R}$, $i \in \{0, 1, 2, 3, 4, 5, 6, 7\}$, and the basis $\{1, e_1, ..., e_7\}$, with multiplication given in the following Table:

$\cdot$	1	e_1	e_2	e_3	e_4	e_5	e_6	e_7
1	1	e_1	e_2	e_3	e_4	e_5	e_6	e_7
e_1	e_1	$-\alpha$	e_3	$-\alpha e_2$	e_5	$-\alpha e_4$	$-e_7$	αe_6
e_2	e_2	$-e_3$	$-\beta$	βe_1	e_6	e_7	$-\beta e_4$	$-\beta e_5$
e_3	e_3	αe_2	$-\beta e_1$	$-\alpha\beta$	e_7	$-\alpha e_6$	βe_5	$-\alpha\beta e_4$
e_4	e_4	$-e_5$	$-e_6$	$-e_7$	$-\gamma$	γe_1	γe_2	γe_3
e_5	e_5	αe_4	$-e_7$	αe_6	$-\gamma e_1$	$-\alpha\gamma$	$-\gamma e_3$	$\alpha\gamma e_2$
e_6	e_6	e_7	βe_4	$-\beta e_5$	$-\gamma e_2$	γe_3	$-\beta\gamma$	$-\beta\gamma e_1$
e_7	e_7	$-\alpha e_6$	βe_5	$\alpha\beta e_4$	$-\gamma e_3$	$-\alpha\gamma e_2$	$\beta\gamma e_1$	$-\alpha\beta\gamma$

The algebra $\mathbb{O}(\alpha, \beta, \gamma)$ is non-commutative, non-associative but it is *alternative* (i.e. $x^2 y = x (xy)$ and $yx^2 = (yx) x$, $\forall x, y \in \mathbb{O}(\alpha, \beta, \gamma)$), *flexible* (i.e. $x (yx) = (xy) x$, $\forall x, y \in \mathbb{O}(\alpha, \beta, \gamma)$) and *power-associative* (i.e. for each $x \in \mathbb{O}(\alpha, \beta, \gamma)$, the subalgebra generated by x is an associative algebra). For other details regarding quaternions and octonions, the reader is referred to (Schafer 1966).

If $a \in \mathbb{O}(\alpha, \beta, \gamma)$, $a = a_0 + a_1 e_1 + a_2 e_2 + a_3 e_3 + a_4 e_4 + a_5 e_5 + a_6 e_6 + a_7 e_7$, then $\bar{a} = a_0 - a_1 e_1 - a_2 e_2 - a_3 e_3 - a_4 e_4 - a_5 e_5 - a_6 e_6 - a_7 e_7$ is called the *conjugate* of the element a. Let $a \in \mathbb{O}(\alpha, \beta, \gamma)$. We have the *trace*, respectively, the *norm* of the element a given by the relations

$$\mathbf{t}(a) = a + \bar{a} \in \mathbb{R}$$

and

$$\mathbf{n}(a) = a\bar{a} = a_0^2 + \alpha a_1^2 + \beta a_2^2 + \alpha\beta a_3^2 + \gamma a_4^2 + \alpha\gamma a_5^2 + \beta\gamma a_6^2 + \alpha\beta\gamma a_7^2 \in \mathbb{R}.$$

We remark that the following relation holds

$$a^2 - \mathbf{t}(a)\, a + \mathbf{n}(a) = 0,$$

for each $a \in A$, where $A \in \{\mathbb{H}(\beta_1, \beta_2), \mathbb{O}(\alpha, \beta, \gamma)\}$.

We know that a finite-dimensional algebra A is *a division* algebra if and only if A does not contain zero divisors, (Schafer 1966). With the above notations, for $\beta_1 = \beta_2 = 1$, we obtain the real division algebra $\mathbb{H}$ and for $\alpha = \beta = \gamma = 1$, we obtain the real division octonion algebra $\mathbb{O}$.

If a quaternion algebra and an octonion algebra are not division algebras, we call them *a split quaternion algebra*, respectively, *a split octonion algebra*.

From the above properties, we can see that an algebra A, with $A \in \{\mathbb{H}(\beta_1, \beta_2), \mathbb{O}(\alpha, \beta, \gamma)\}$, is a division algebra if and only if $\mathbf{n}(x) \neq 0$, for each $x \in A$, $x \neq 0$, (Schafer 1966, p. 27). Therefore, depending on the values of the numbers $\beta_1, \beta_2, \alpha, \beta, \gamma$, we can obtain division algebras or split algebras.

A real quaternion algebra is isomorphic to $\mathbb{H}$, when we have a division algebra, or it is isomorphic to the algebra $\mathcal{M}_2(R)$, the algebra of 2×2 real matrices, when we have a split algebra, (Schafer 1966, p. 25).

We consider the following real algebra

$$\mathbb{A} = \left\{ \begin{pmatrix} a & u \\ v & b \end{pmatrix}, a, b \in \mathbb{R}, u, v \in \mathbb{R}^3 \right\},$$

with the usual addition and scalar multiplication of matrices. We consider the following multiplication

$$\begin{pmatrix} a & u \\ v & b \end{pmatrix} \begin{pmatrix} c & z \\ w & d \end{pmatrix} = \begin{pmatrix} ac + (u, w) & az + du - v \times w \\ cv + bw + u \times z & bd + \langle v, z \rangle \end{pmatrix}, \quad (1.1)$$

with $\langle, \rangle$ the usual dot product and $\times$ the cross product of vectors from $\mathbb{R}^3$. With matrices addition, scalar multiplication of matrices and the multiplication given in relation (1.1) , we obtain a unitary non-associative algebra of dimension 8, called the *Zorn's vector-matrix algebra*. Therefore, a real octonion algebra is isomorphic to $\mathbb{O}$, when we have a division algebra, or it is isomorphic to the algebra $\mathbb{A}$, when we have a split algebra, see (Kostrikin and Shafarevich 1995). A famous Hurwitz's theorem states that $\mathbb{R}$, $\mathbb{C}$, $\mathbb{H}$ and $\mathbb{O}$ are the only real alternative division algebras, see (Baez 2002).

Division algebras are used to built space-time block codes and division quaternion algebras are also used for this purpose. Quaternions are used in Coding Theory and in digital signal processing, see (Alfsmann et al. 2007). There are a lot of examples of codes based on quaternion algebras, first of them, Alamouti code, appeared in 1998, see (Alamouti 1998) and (Unger and Markin 2011).

Another application of quaternions is in representing rotations in $\mathbb{R}^3$. Usually, a rotation in $\mathbb{R}^3$ around an axis is given by a square orthogonal matrix of order three, with its determinant equal with 1. When we compute two rotations, a lot of computations are involved. In this situations, quaternions give a much better representation for a rotation. In data registration, when we want to find a transformation for a set of data points such that these points fit better to a shape model, quaternions are used for solving this problem, since it is necessary to find a rotation and a translation with some good properties, see (Jia 2017).

Octonions have many applications in processing of color images, as for example in color image edge detection, see (Chen and Tu 2013), in remote sensing images, (Li 2011), in 2D and 3D signal processing, (Snopek 2015), in artificial neural networks, where Octonionic neural networks are used as a computational models, (Klco et al. 2017), in electrodynamics, see (Chanyal et al. 2011) and (Chen and Tu 2013), in wireless data communication (Jouget 2013) and examples can continue.

Since the above applications used quaternion and octonion matrices or matrix representations of these algebras, in this chapter, we present real matrix representations for complex octonions and some of their properties and we provide examples of invertible elements in split quaternion algebras and in split octonion algebras.

2 Real Matrix Representations for Complex Octonions

We know that in an alternative algebra A, the following identities, called *Moufang identities*, are true:

$$(xzx)\, y = x[z\,(xy)], \tag{2.1}$$

$$y\,(xzx) = [(yx)\,z]x, \tag{2.2}$$

$$(xy)\,(zx) = x\,(yz)\,x, \tag{2.3}$$

for all x, y, $z \in A$. Since octonions form an alternative algebra, we can use these relations in computations.

Let $\mathbb{O}$ be the real division octonion algebra, the algebra of the elements of the form $a = a_0 + a_1 i + a_2 j + a_3 ij + a_4 k + a_5 ik + a_6 jk + a_7 (ij)k$, where

$$a_i \in \mathbb{R}, i^2 = j^2 = k^2 = -1,$$

and

$$ij = -ji, jk = -kj, ki = -ik, (ij)k = -i(jk).$$

The set $\{1, i, j, ij, k, ik, jk, (ij)k\}$ is a basis in $\mathbb{O}$.

We consider K to be the field $\{\begin{pmatrix} a & -b \\ b & a \end{pmatrix} \mid a, b \in \mathbb{R}\}$ and the map

$$\varphi : \mathbb{C} \to K, \varphi(a + bi) = \begin{pmatrix} a & -b \\ b & a \end{pmatrix},$$

where $i^2 = -1$. The map φ is a fields morphism and the element $\varphi(z) = \begin{pmatrix} a & -b \\ b & a \end{pmatrix}$ is called the matrix representation of the complex element $z = a + bi \in \mathbb{C}$.

A complex octonion is an element of the form $A = a_0 + a_1 e_1 + a_2 e_2 + a_3 e_3 + a_4 e_4 + a_5 e_5 + a_6 e_6 + a_7 e_7$, where $a_m \in \mathbb{C}, m \in \{0, 1, 2, ..., 7\}$,

$$e_m^2 = -1, \quad m \in \{0, 1, 2, ..., 7\}$$

and

$$e_m e_n = -e_n e_m = \gamma_{mn} e_t, \gamma_{mn} \in \{-1, 1\}, m \neq n, m, n \in \{0, 1, 2, ..., 7\},$$

γ_{mn} and e_t are uniquely determined by e_m and e_n. We denote by $\mathbb{O}_C$ the algebra of the complex octonions, called *the complex octonion algebra*. This algebra is an algebra over the field $\mathbb{C}$ and the set $\{1, e_1, e_2, e_3, e_4, e_5, e_6, e_7\}$ is a basis in $\mathbb{O}_C$.

The map $\delta : \mathbb{R} \to \mathbb{C}, \delta(a) = a$ is the inclusion morphism between $\mathbb{R}$ and $\mathbb{C}$ as $\mathbb{R}$-algebras. We denote by $\mathbb{G}$ the following $\mathbb{C}$-subalgebra of the algebra $\mathbb{O}_C$,

$$\mathbb{G} = \{A \in \mathbb{O}_C \mid A = \sum_{m=0}^{7} a_m e_m, a_m \in \mathbb{R}, m \in \{0, 1, 2, 3, ..., 7\}\}.$$

It is clear that $\mathbb{G}$ becomes an algebra over $\mathbb{R}$, with the following multiplication "$\cdot$"

$$a \cdot A = \delta(a) A = aA, a \in \mathbb{R}, A \in \mathbb{G}.$$

We denote this algebra by $\mathbb{O}_\mathbb{R}$ and the map

$$\phi : \quad \mathbb{O} \to \mathbb{O}_{\mathbb{R}}, \phi \, (a_0 + a_1 i + a_2 j + a_3 ij + a_4 k + a_5 ik + a_6 jk + a_7(ij)k) =$$
$$= \quad a_0 + a_1 e_1 + a_2 e_2 + a_3 e_3 + a_4 e_4 + a_5 e_5 + a_6 e_6 + a_7 e_7,$$

where $a_m \in \mathbb{R}, m \in \{0, 1, 2, 3, ..., 7\}$ is an algebra isomorphism. Due to this isomorphism, we have that $\phi \, (1) = 1, \phi \, (i) = e_1, \phi \, (j) = e_2, \phi \, (ij) = e_3,$
$\phi \, (k) = e_4, \phi \, (ik) = e_5, \phi \, (jk) = e_6, \phi \, ((ij)k) = e_7$ and the algebras $\mathbb{O}_{\mathbb{R}}$ and $\mathbb{O}_C$ have the same basis $\{1, e_1, e_2, e_3, ..., e_7\}$. With these notations, in the rest of the paper, we denote the real octonion $a_0 + a_1 i + a_2 j + a_3 ij + a_4 k + a_5 ik + a_6 jk + a_7(ij)k$ with the octonion $a_0 + a_1 e_1 + a_2 e_2 + a_3 e_3 + a_4 e_4 + a_5 e_5 + a_6 e_6 + a_7 e_7,$ and viceversa, where $a_m \in \mathbb{R}, m \in \{0, 1, 2, 3, ..., 7\}$ and we use the notation $\mathbb{O}$ instead of $\mathbb{O}_{\mathbb{R}}$.

We consider the complex octonion
$A = a_0 + a_1 e_1 + a_2 e_2 + a_3 e_3 + a_4 e_4 + a_5 e_5 + a_6 e_6 + a_7 e_7,$
$a_m \in \mathbb{C}, m \in \{0, 1, 2, 3, ..., 7\}$. This octonion can be written under the following form
$A = (x_0 + iy_0) + (x_1 + iy_1)e_1 + (x_2 + iy_2)e_2 + (x_3 + iy_3)e_3 +$
$+(x_4 + iy_4)e_4 + (x_5 + iy_5)e_5 + (x_6 + iy_6)e_6 + (x_7 + iy_7)e_7$, where $x_m, y_m \in \mathbb{R},$
$m \in \{0, 1, 2, 3, ..., 7\}$ and $i^2 = -1$.

It results that

$$A = x + iy, \tag{2.4}$$

with $x, y \in \mathbb{O}, x = x_0 + x_1 e_1 + x_2 e_2 + x_3 e_3 + x_4 e_4 + x_5 e_5 + x_6 e_6 + x_7 e_7$ and $y = y_0 + y_1 e_1 + y_2 e_2 + y_3 e_3 + y_4 e_4 + y_5 e_5 + y_6 e_6 + y_7 e_7.$

The conjugate of this octonion is the element $\overline{A} = a_0 - a_1 e_1 - a_2 e_2 - a_3 e_3 - a_4 e_4 - a_5 e_5 - a_6 e_6 - a_7 e_7$, therefore, with the above notations, we have

$$\overline{A} = \overline{x} + i\overline{y}. \tag{2.5}$$

To an octonion $a = a_0 + a_1 e_1 + a_2 e_2 + a_3 e_3 + a_4 e_4 + a_5 e_5 + a_6 e_6 + a_7 e_7 \in \mathbb{O},$ we associate the following element

$$a^* = a_0 + a_1 e_1 - a_2 e_2 - a_3 e_3 - a_4 e_4 - a_5 e_5 - a_6 e_6 - a_7 e_7. \tag{2.6}$$

We remark that

$$(a^*)^* = a \tag{2.7}$$

and

$$(a + b)^* = a^* + b^*, \tag{2.8}$$

The identity $(ab)^* = a^* b^*$, for all $a, b \in \mathbb{O}$, in general is not true. Since the real octonion $a_0 + a_1 i + a_2 j + a_3 ij + a_4 k + a_5 ik + a_6 jk + a_7(ij)k$ can be written under the form $a = q_1 + q_2 k$, where $q_1 = a_0 + a_1 i + a_2 j + a_3 ij$ and $q_2 = a_4 + a_5 ei + a_6 j + a_7 ij$ are two real quaternions, it results

$$a^* = q_1^* - q_2 k, \tag{2.9}$$

where $q_1^* = a_0 + a_1 i - a_2 j - a_3 i j$.

With the above notations, we define

$$\tilde{a} = q_1 - q_2 k, \tag{2.10}$$

$$a_+ = q_1^* + q_2^* k, \ a^+ = \overline{q}_1 + q_2 k. \tag{2.11}$$

Considering the real division quaternion algebra $\mathbb{H}$, in (Tian 2000) were defined the following maps

$$\lambda : \mathbb{H} \to \mathcal{M}_4 (\mathbb{R}), \lambda (a) = \begin{pmatrix} a_0 & -a_1 & -a_2 & -a_3 \\ a_1 & a_0 & -a_3 & a_2 \\ a_2 & a_3 & a_0 & -a_1 \\ a_3 & -a_2 & a_1 & a_0 \end{pmatrix} \tag{2.12}$$

and

$$\rho : \mathbb{H} \to \mathcal{M}_4 (\mathbb{R}), \ \rho (a) = \begin{pmatrix} a_0 & -a_1 & -a_2 & -a_3 \\ a_1 & a_0 & a_3 & -a_2 \\ a_2 & -a_3 & a_0 & a_1 \\ a_3 & a_2 & -a_1 & a_0 \end{pmatrix}, \tag{2.13}$$

where $a = a_0 + a_1 i + a_2 j + a_3 i j \in \mathbb{H}$.

We remark that λ is an isomorphism between $\mathbb{H}$ and the algebra of the matrices

$$\left\{ \begin{pmatrix} a_0 & -a_1 & -a_2 & -a_3 \\ a_1 & a_0 & -a_3 & a_2 \\ a_2 & a_3 & a_0 & -a_1 \\ a_3 & -a_2 & a_1 & a_0 \end{pmatrix}, a_0, a_1, a_2, a_3 \in \mathbb{R} \right\}.$$

The columns of the matrix $\lambda (a) \in \mathcal{M}_4 (\mathbb{R})$ are represented by the coefficients in $\mathbb{R}$ of the elements $\{a, ai, aj, a(ij)\}$, considered in respect to the basis $\{1, i, j, ij\}$.

The matrix $\lambda (a)$ is called *the left matrix representation* of the element $a \in \mathbb{H}$.

The map ρ is an isomorphism between $\mathbb{H}$ and the algebra of the matrices

$$\left\{ \begin{pmatrix} a_0 & -a_1 & -a_2 & -a_3 \\ a_1 & a_0 & a_3 & -a_2 \\ a_2 & -a_3 & a_0 & a_1 \\ a_3 & a_2 & -a_1 & a_0 \end{pmatrix}, a_0, a_1, a_2, a_3 \in \mathbb{R} \right\}.$$

In a similar way, we remark that the columns of the matrix $\rho (a) \in \mathcal{M}_4 (\mathbb{R})$ are the coefficients in $\mathbb{R}$ of the elements $\{a, ia, ja, (ij)a\}$, considered in respect to the basis $\{1, i, j, ij\}$. The matrix $\rho (a)$ is called *the right matrix representation* of the quaternion $a \in \mathbb{H}$.

With these notations, in (Tian 2000) were defined the left and right real representations of the octonion $a = q_1 + q_2 k$, namely

$$\Lambda(a) = \begin{pmatrix} \lambda(q_1) & -\rho(q_2) M_1 \\ \lambda(q_2) M_1 & \rho(q_1) \end{pmatrix} \in \mathcal{M}_8(\mathbb{R}) \tag{2.14}$$

and

$$\Delta(a) = \begin{pmatrix} \rho(q_1) & -\lambda(\overline{q}_2) \\ \lambda(q_2) & \rho(\overline{q}_1) \end{pmatrix} \in \mathcal{M}_8(\mathbb{R}), \tag{2.15}$$

where $q_1 = a_0 + a_1 i + a_2 j + a_3 ij$, $q_2 = a_4 + a_5 ei + a_6 j + a_7 ij$ are two real

quaternions and $M_1 = \begin{pmatrix} 1 & 0 & 0 & 0 \\ 0 & -1 & 0 & 0 \\ 0 & 0 & -1 & 0 \\ 0 & 0 & 0 & -1 \end{pmatrix}$. We remark that $L_1 = \lambda(i) = \begin{pmatrix} 0 & -1 & 0 & 0 \\ 1 & 0 & 0 & 0 \\ 0 & 0 & 0 & -1 \\ 0 & 0 & 1 & 0 \end{pmatrix}$

and $R_1 = \rho(i) = \begin{pmatrix} 0 & -1 & 0 & 0 \\ 1 & 0 & 0 & 0 \\ 0 & 0 & 0 & 1 \\ 0 & 0 & -1 & 0 \end{pmatrix}$.

In (Flaut and Shpakivskyi 2013), using the above notations, were obtained the following matrices

$$\Gamma(Q) = \begin{pmatrix} \lambda(a) & -\lambda(b^*) \\ \lambda(b) & \lambda(a^*) \end{pmatrix} \tag{2.16}$$

and

$$\Theta(Q) = \begin{pmatrix} \rho(a) & -\rho(b) \\ \rho(b^*) & \rho(a^*) \end{pmatrix}, \tag{2.17}$$

where $Q = a + ib$ is a complex quaternion, with $a = a_0 + a_1 e_1 + a_2 e_2 + a_3 e_3 \in \mathbb{H}$, $b = b_0 + b_1 e_1 + b_2 e_2 + b_3 e_3 \in \mathbb{H}$, $a^* = a_0 + a_1 e_1 - a_2 e_2 - a_3 e_3 \in \mathbb{H}$, $b^* = b_0 + b_1 e_1 - b_2 e_2 - b_3 e_3 \in \mathbb{H}$ and $i^2 = -1$. The matrix $\Gamma(Q) \in \mathcal{M}_8(\mathbb{R})$ is called *the left real matrix representation for the complex quaternion* Q and $\Theta(Q) \in \mathcal{M}_8(\mathbb{R})$ is called the *right real matrix representation* for the complex quaternion Q.

Using some ideas developed above, in the following, we define the left and the right real matrix representations for the complex octonions and we investigate some of their properties.

We consider the complex octonion $A = x + iy$, with $x, y \in \mathbb{O}$, $x = x_0 + x_1 e_1 + x_2 e_2 + x_3 e_3 + x_4 e_4 + x_5 e_5 + x_6 e_6 + x_7 e_7$ and $y = y_0 + y_1 e_1 + y_2 e_2 + y_3 e_3 + y_4 e_4 + y_5 e_5 + y_6 e_6 + y_7 e_7$.

The matrices

$$\Phi(A) = \begin{pmatrix} \Lambda(x) & -\Lambda(y) \\ \Lambda(y^*) & \Lambda(x^*) \end{pmatrix} \in \mathcal{M}_{16}(\mathbb{R}), \tag{2.18}$$

and

$$\Psi(A) = \begin{pmatrix} \Delta(x) & -\Delta(y) \\ \Delta(y^*) & \Delta(x^*) \end{pmatrix} \in \mathcal{M}_{16}(\mathbb{R}) \tag{2.19}$$

are called *the left real matrix representation* and *the right real matrix representation* for the complex octonion A.

Definition 2.1 Let $A \in \mathbb{O}_{\mathbb{C}}$, $A = x + iy$. We consider the following matrix

$$\overrightarrow{A} = (\overrightarrow{x}^t, \overrightarrow{y}^t)^t = \begin{pmatrix} \overrightarrow{x} \\ \overrightarrow{y} \end{pmatrix} \in \mathcal{M}_{16 \times 1}(\mathbb{R}),$$

called *the vector representation* of the element X, where $x, y \in \mathbb{O}$,
$x = x_0 + x_1 e_1 + x_2 e_2 + x_3 e_3 + x_4 e_4 + x_5 e_5 + x_6 e_6 + x_7 e_7$,
$y = y_0 + y_1 e_1 + y_2 e_2 + y_3 e_3 + y_4 e_4 + y_5 e_5 + y_6 e_6 + y_7 e_7$,
$\overrightarrow{x} = (x_0, x_1, x_2, x_3, x_4, x_5, x_6, x_7)^t \in \mathcal{M}_{8 \times 1}(\mathbb{R})$,
$\overrightarrow{y} = (y_0, y_1, y_2, y_3, y_4, y_5, y_6, y_7)^t \in \mathcal{M}_{8 \times 1}(\mathbb{R})$ are the vector representations for the real octonions x and y, as were defined in (Tian 2000). If $\overrightarrow{x} = (x_0, x_1, x_2, x_3, x_4, x_5, x_6, x_7)^t \in \mathcal{M}_{8 \times 1}(\mathbb{R})$ and
$\overrightarrow{y} = (y_0, y_1, y_2, y_3, y_4, y_5, y_6, y_7)^t \in \mathcal{M}_{8 \times 1}(\mathbb{R})$, we have

$$\overrightarrow{xy} = \Lambda(x) \overrightarrow{y} \tag{2.20}$$

and

$$\overrightarrow{yx} = \Delta(x) \overrightarrow{y}. \tag{2.21}$$

From the same paper, Tian (2000), Theorems 2.1, 2.3 and 2.9, we have that

$$\overrightarrow{ax} = \Lambda(a) \overrightarrow{x}, \overrightarrow{xa} = \Delta(a) \overrightarrow{x}, \tag{2.22}$$

$$\Lambda(a^2) = \Lambda(a) \Lambda(a), \Delta(a^2) = \Delta(a) \Delta(a), \tag{2.23}$$

$$\Lambda(a) \Delta(a) = \Delta(a) \Lambda(a), \tag{2.24}$$

where a, x are real octonions.

Remark 2.2 With the above notations, we have

$$\varepsilon \overrightarrow{x} = \overrightarrow{\overline{x}}$$

and

$$\tau \overrightarrow{x} = \overrightarrow{x_+},$$

where $\varepsilon = diag(1, 1, 1, 1, -1, -1, -1, -1) \in \mathcal{M}_8(\mathbb{R})$,
$\tau = diag(1, 1, -1, -1, 1, 1, -1, -1) \in \mathcal{M}_8(\mathbb{R})$. Indeed,

$$\varepsilon \, \vec{x} \; = \; \begin{pmatrix} 1 & 0 & 0 & 0 & 0 & 0 & 0 & 0 \\ 0 & 1 & 0 & 0 & 0 & 0 & 0 & 0 \\ 0 & 0 & 1 & 0 & 0 & 0 & 0 & 0 \\ 0 & 0 & 0 & 1 & 0 & 0 & 0 & 0 \\ 0 & 0 & 0 & 0 & -1 & 0 & 0 & 0 \\ 0 & 0 & 0 & 0 & 0 & -1 & 0 & 0 \\ 0 & 0 & 0 & 0 & 0 & 0 & -1 & 0 \\ 0 & 0 & 0 & 0 & 0 & 0 & 0 & -1 \end{pmatrix} \begin{pmatrix} y_0 \\ y_1 \\ y_2 \\ y_3 \\ y_4 \\ y_5 \\ y_6 \\ y_7 \end{pmatrix} = \begin{pmatrix} y_0 \\ y_1 \\ y_2 \\ y_3 \\ -y_4 \\ -y_5 \\ -y_6 \\ -y_7 \end{pmatrix} = \vec{x} \, . \text{ In the same way we obtain}$$

$$\tau \, \vec{x} = \begin{pmatrix} 1 & 0 & 0 & 0 & 0 & 0 & 0 & 0 \\ 0 & 1 & 0 & 0 & 0 & 0 & 0 & 0 \\ 0 & 0 & -1 & 0 & 0 & 0 & 0 & 0 \\ 0 & 0 & 0 & -1 & 0 & 0 & 0 & 0 \\ 0 & 0 & 0 & 0 & 1 & 0 & 0 & 0 \\ 0 & 0 & 0 & 0 & 0 & 1 & 0 & 0 \\ 0 & 0 & 0 & 0 & 0 & 0 & -1 & 0 \\ 0 & 0 & 0 & 0 & 0 & 0 & 0 & -1 \end{pmatrix} \begin{pmatrix} y_0 \\ y_1 \\ y_2 \\ y_3 \\ y_4 \\ y_5 \\ y_6 \\ y_7 \end{pmatrix} = \begin{pmatrix} y_0 \\ y_1 \\ -y_2 \\ -y_3 \\ y_4 \\ y_5 \\ -y_6 \\ -y_7 \end{pmatrix} = \vec{x}_+ \, .$$

Proposition 2.3 *Using the above notations and definitions, we have*

$$\sigma \lambda (q_1) \sigma = \lambda \left(q_1^* \right)$$

and

$$\sigma \rho (q_1) \sigma = \rho \left(q_1^* \right),$$

where $\sigma = diag(1, 1, -1, -1)$.

Proof Indeed, $\sigma \lambda (q_1) \sigma =$

$$= \begin{pmatrix} 1 & 0 & 0 & 0 \\ 0 & 1 & 0 & 0 \\ 0 & 0 & -1 & 0 \\ 0 & 0 & 0 & -1 \end{pmatrix} \begin{pmatrix} a_0 & -a_1 & -a_2 & -a_3 \\ a_1 & a_0 & -a_3 & a_2 \\ a_2 & a_3 & a_0 & -a_1 \\ a_3 & -a_2 & a_1 & a_0 \end{pmatrix} \begin{pmatrix} 1 & 0 & 0 & 0 \\ 0 & 1 & 0 & 0 \\ 0 & 0 & -1 & 0 \\ 0 & 0 & 0 & -1 \end{pmatrix} =$$

$$= \begin{pmatrix} a_0 & -a_1 & a_2 & a_3 \\ a_1 & a_0 & a_3 & -a_2 \\ -a_2 & -a_3 & a_0 & -a_1 \\ -a_3 & a_2 & a_1 & a_0 \end{pmatrix} = \lambda \left(q_1^* \right).$$

In the same way, we obtain $\sigma \rho (q_1) \sigma =$

$$= \begin{pmatrix} 1 & 0 & 0 & 0 \\ 0 & 1 & 0 & 0 \\ 0 & 0 & -1 & 0 \\ 0 & 0 & 0 & -1 \end{pmatrix} \begin{pmatrix} a_0 & -a_1 & -a_2 & -a_3 \\ a_1 & a_0 & a_3 & -a_2 \\ a_2 & -a_3 & a_0 & a_1 \\ a_3 & a_2 & -a_1 & a_0 \end{pmatrix} \begin{pmatrix} 1 & 0 & 0 & 0 \\ 0 & 1 & 0 & 0 \\ 0 & 0 & -1 & 0 \\ 0 & 0 & 0 & -1 \end{pmatrix} =$$

$$= \begin{pmatrix} a_0 & -a_1 & a_2 & a_3 \\ a_1 & a_0 & -a_3 & a_2 \\ -a_2 & a_3 & a_0 & a_1 \\ -a_3 & -a_2 & -a_1 & a_0 \end{pmatrix} = \rho \left(q_1^* \right).$$

Proposition 2.4 *Let $x, y \in \mathbb{O}$, be two real octonions. The following relations hold:*
(1) $iy = y^*i$;
(2) $(iy) x = i (y^*x^*)^*$;
(3) $x (iy) = i (x^*y)$;
(4) $(iy) (ix) = - (yx^*)^*$.

Proof (1) We have $iy = e_1 (y_0 + y_1e_1 + y_2e_2 + y_3e_3 + y_4e_4 + y_5e_5 + y_6e_6 + y_7e_7)$
$= -y_1 + y_0e_1 - y_3e_2 + y_2e_3 - y_5e_4 + y_4e_5 + y_7e_6 - y_6e_7$ and
$y^*i = (y_0 + y_1e_1 - y_2e_2 - y_3e_3 - y_4e_4 - y_5e_5 - y_6e_6 - y_7e_7) e_1 =$
$= -y_1 + y_0e_1 - y_3e_2 + y_2e_3 - y_5e_4 + y_4e_5 + y_7e_6 - y_6e_7$.
(2) From the above, it results $(iy) x = (y^*i) x$. We have $((iy) x) i = ((y^*i) x) i$.
We apply relation (2.2) and we obtain $((iy) x) i = y^* (ixi) = - y^*x^*$. Therefore $(((iy) x) i) i = (-y^*x^*) i$, then, from alternativity, we get $(iy) x = (y^*x^*) i =$
$i (y^*x^*)^*$.

(3) From relation (2.1), the following relation holds

$$i (x (iy)) = (ixi) y = -x^*y.$$

Using again alternativity, it results $i (i(x (iy))) = -i(x^*y)$, that means $x (iy) = i (x^*y)$.

(4) We apply relation (2.3) and we have $(iy) (ix) = (iy) (x^*i) = i (yx^*) i =$
$- (yx^*)^*$.
$\qquad\qquad\qquad\qquad\qquad\qquad\qquad\qquad\qquad\qquad\qquad\qquad\qquad\qquad\square$

Proposition 2.5 *For real octonion $a = q_1 + q_2k$, with q_1, q_2 two real quaternions, we have the following relations:*

$$\varepsilon \Lambda (a) \varepsilon = \Lambda (\tilde{a}) , \varepsilon \Delta (a) \varepsilon = \Delta (\tilde{a})$$

and

$$\tau \Lambda (a) \tau = \Lambda (a_+) , \tau \Delta (a) \tau = \Delta (a_+).$$

Proof (1) Since $\varepsilon = diag(1, 1, 1, 1, -1, -1, -1, -1) \in \mathcal{M}_8 (\mathbb{R})$, we have $\varepsilon = \begin{pmatrix} I_4 & O_4 \\ O_4 & -I_4 \end{pmatrix}$, where $I_4 \in \mathcal{M}_4 (\mathbb{R})$ is the unit matrix and $O_4 \in \mathcal{M}_4 (\mathbb{R})$ is the zero matrix.

$$\varepsilon \Lambda (a) \varepsilon = \begin{pmatrix} I_4 & 0 \\ 0 & -I_4 \end{pmatrix} \begin{pmatrix} \lambda (q_1) & -\rho (q_2) M_1 \\ \lambda (q_2) M_1 & \rho (q_1) \end{pmatrix} \begin{pmatrix} I_4 & 0 \\ 0 & -I_4 \end{pmatrix} =$$
$$= \begin{pmatrix} \lambda (q_1) & -\rho (q_2) M_1 \\ -\lambda (q_2) M_1 & -\rho (q_1) \end{pmatrix} \begin{pmatrix} I_4 & 0 \\ 0 & -I_4 \end{pmatrix} = \begin{pmatrix} \lambda (q_1) & \rho (q_2) M_1 \\ -\lambda (q_2) M_1 & \rho (q_1) \end{pmatrix} =$$
$$= \Lambda (\tilde{a}).$$

We have $\varepsilon \Delta (a) \varepsilon = \begin{pmatrix} I_4 & 0 \\ 0 & -I_4 \end{pmatrix} \begin{pmatrix} \rho (q_1) & -\lambda (\overline{q}_2) \\ \lambda (q_2) & \rho (\overline{q}_1) \end{pmatrix} \begin{pmatrix} I_4 & 0 \\ 0 & -I_4 \end{pmatrix} =$
$$= \begin{pmatrix} \rho (q_1) & -\lambda (\overline{q}_2) \\ -\lambda (q_2) & -\rho (\overline{q}_1) \end{pmatrix} \begin{pmatrix} I_4 & 0 \\ 0 & -I_4 \end{pmatrix} = \begin{pmatrix} \rho (q_1) & \lambda (\overline{q}_2) \\ -\lambda (q_2) & \rho (\overline{q}_1) \end{pmatrix} = \Delta (\tilde{a}).$$

(2) Since $\tau = diag(1, 1, -1, -1, 1, 1, -1, -1)$, and $\sigma = diag(1, 1, -1, -1)$, we have

$$\tau \Lambda (a) \tau = \begin{pmatrix} \sigma & 0 \\ 0 & \sigma \end{pmatrix} \begin{pmatrix} \lambda (q_1) & -\rho (q_2) M_1 \\ \lambda (q_2) M_1 & \rho (q_1) \end{pmatrix} \begin{pmatrix} \sigma & 0 \\ 0 & \sigma \end{pmatrix} =$$

$$= \begin{pmatrix} \sigma \lambda (q_1) & -\sigma \rho (q_2) M_1 \\ \sigma \lambda (q_2) M_1 & \sigma \rho (q_1) \end{pmatrix} \begin{pmatrix} \sigma & 0 \\ 0 & \sigma \end{pmatrix} =$$

$$= \begin{pmatrix} \sigma \lambda (q_1) \sigma & -\sigma \rho (q_2) M_1 \sigma \\ \sigma \lambda (q_2) M_1 \sigma & \sigma \rho (q_1) \sigma \end{pmatrix} = \begin{pmatrix} \lambda (q_1^*) & -\rho (q_2^*) M_1 \\ \lambda (q_2^*) M_1 & \rho (q_1^*) \end{pmatrix} = \Lambda (a_+),$$

since $M_1 \sigma = \sigma M_1$.

$$\tau \Delta (a) \tau = \begin{pmatrix} \sigma & 0 \\ 0 & \sigma \end{pmatrix} \begin{pmatrix} \rho (q_1) & -\lambda (\overline{q}_2) \\ \lambda (q_2) & \rho (\overline{q}_1) \end{pmatrix} \begin{pmatrix} \sigma & 0 \\ 0 & \sigma \end{pmatrix} =$$

$$= \begin{pmatrix} \sigma \rho (q_1) & -\sigma \lambda (\overline{q}_2) \\ \sigma \lambda (q_2) & \sigma \rho (\overline{q}_1) \end{pmatrix} \begin{pmatrix} \sigma & 0 \\ 0 & \sigma \end{pmatrix} = \begin{pmatrix} \sigma \rho (q_1) \sigma & -\sigma \lambda (\overline{q}_2) \sigma \\ \sigma \lambda (q_2) \sigma & \sigma \rho (\overline{q}_1) \sigma \end{pmatrix} =$$

$$= \begin{pmatrix} \rho (q_1^*) & -\lambda (\overline{q}_2^*) \\ \lambda (q_2^*) & \rho (\overline{q}_1^*) \end{pmatrix} = \Delta (a_+). \qquad \square$$

Proposition 2.6 Let $A, X \in \mathbb{O}_\mathbb{C}$, $A = x + iy$, $A' = x + iy^*$, $X = v + iw$, $x, y, v, w \in \mathbb{O}$, then:

$$\overrightarrow{AX} = \Phi (A') \overrightarrow{X}.$$

Proof We have

$$\Phi (A') \overrightarrow{X} = \begin{pmatrix} \Lambda (x) & -\Lambda (y^*) \\ \Lambda (y) & \Lambda (x^*) \end{pmatrix} \begin{pmatrix} \overrightarrow{v} \\ \overrightarrow{w} \end{pmatrix} =$$

$$= \begin{pmatrix} \Lambda (x) \overrightarrow{v} - \Lambda (y^*) \overrightarrow{w} \\ \Lambda (y) \overrightarrow{v} + \Lambda (x^*) \overrightarrow{w} \end{pmatrix}.$$

We have $AX = (x + iy)(v + iw) = xv + x(iw) + (iy)v + (iy)(iw) = xv - (yw^*)^* + i(x^*w + (y^*v^*)^*)$. From Proposition 2.4, (i) and relations (2.22), (2.23) and (2.24), it follows that

$$\overrightarrow{(yw^*)^*} = \Lambda (y^*) \overrightarrow{w}.$$

Indeed,

$$\overrightarrow{(yw^*)^*} = -i \overrightarrow{(yw^*)} i = -\Lambda (i) \Delta (i) \overrightarrow{yw^*} =$$

$$= -\Lambda (i) \Delta (i) \Lambda (y) \overrightarrow{w^*} = \Lambda (i) \Delta (i) \Lambda (y) \overrightarrow{iwi} =$$

$$= \Lambda (i) \Delta (i) \Lambda (y) \Lambda (i) \Delta (i) \overrightarrow{w} =$$

$$= -\Delta (i) \Lambda (iyi) \Delta^{-1} (i) \overrightarrow{w} =$$

$$= -\Lambda\,(iy)\,\Lambda\,(i)\,\overrightarrow{w} = \Lambda\,(-1)\,\Lambda\,(iy)\,\Lambda\,(i)\,\overrightarrow{w} = \Lambda\,(i)\,\Lambda\,(i)\,\Lambda\,(iy)\,\Lambda\,(i)\,\overrightarrow{w} =$$

$$= \Lambda\,(i)\,\Lambda\,(i(iy)i)\,\overrightarrow{w} = -\Lambda\,(i)\,\Lambda\,(yi)\,\overrightarrow{w} = -\overrightarrow{i((yi)\,w)} = -\overrightarrow{(iyi)w} =$$

$$= \overrightarrow{y^*w} = \Lambda\,(y^*)\,\overrightarrow{w}.$$

We also used relations (2.13) and (2.20) from (Tian 2000), namely

$$\Lambda\,(xyx) = \Lambda\,(x)\,\Lambda\,(y)\,\Lambda\,(x)\,,\ \Delta\,(xyx) = \Delta\,(x)\,\Delta\,(y)\,\Delta\,(x)\,,$$

where x, y are real octonions, and

$$\Lambda\,(xy) = \Delta\,(x)\,\Lambda\,(x)\,\Lambda\,(y)\,\Delta^{-1}\,(x)\,,\ \Delta\,(xy) = \Lambda\,(y)\,\Delta\,(y)\,\Delta\,(x)\,\Lambda^{-1}\,(y)\,,$$

where $x \neq 0$, $y \neq 0$ are real octonions. Therefore, we have

$$\overrightarrow{AX} = \begin{pmatrix} \overrightarrow{x\overrightarrow{v}} - \overrightarrow{(yw^*)^*} \\ \overrightarrow{x^*w} + \overrightarrow{(y^*v^*)^*} \end{pmatrix} = \begin{pmatrix} \overrightarrow{x\overrightarrow{v}} - \Lambda\,(y^*)\,\overrightarrow{w} \\ \overrightarrow{x^*w} + \Lambda\,(y)\,\overrightarrow{v} \end{pmatrix} =$$

$$= \begin{pmatrix} \Lambda\,(x)\,\overrightarrow{v} - \Lambda\,(y^*)\,\overrightarrow{w} \\ \Lambda\,(y)\,\overrightarrow{v} + \Lambda\,(x^*)\,\overrightarrow{w} \end{pmatrix} =$$

$$= \begin{pmatrix} \Lambda\,(x) & -\Lambda\,(y^*) \\ \Lambda\,(y) & \Lambda\,(x^*) \end{pmatrix} \begin{pmatrix} \overrightarrow{v} \\ \overrightarrow{w} \end{pmatrix} = \Phi\,(A')\,\overrightarrow{X}.\ \square$$

Let M, N be the matrices

$$N = (1, e_1, e_2, e_3, e_4, e_5, e_6, e_7)^t\,,$$

$$M = (1, -e_1, -e_2, -e_3, -e_4, -e_5, -e_6, -e_7)^t\,.$$

We remark that $N^t M = 8$.

Proposition 2.7 If $A \in \mathbb{O}$, $a = a_0 + a_1e_1 + a_2e_2 + a_3e_3 + a_4e_4 + a_5e_5 + a_6e_6 + a_7e_7$ we have:
(i) $\Lambda\,(a)\,M = Ma$.
(ii) $\theta M = Me_1$.
(iii) $\Lambda\,(a)\,N = \bar{a}N$.

Proof (i) We have $\Lambda\,(a)\,M =$

$$
= \begin{pmatrix}
a_0 & -a_1 & -a_2 & -a_3 & -a_4 & -a_5 & -a_6 & -a_7 \\
a_1 & a_0 & -a_3 & a_2 & -a_5 & a_4 & a_7 & -a_6 \\
a_2 & a_3 & a_0 & -a_1 & -a_6 & -a_7 & a_4 & a_5 \\
a_3 & -a_2 & a_1 & a_0 & -a_7 & a_6 & -a_5 & a_4 \\
a_4 & a_5 & a_6 & a_7 & a_0 & -a_1 & -a_2 & -a_3 \\
a_5 & -a_4 & a_7 & -a_6 & a_1 & a_0 & a_3 & -a_2 \\
a_6 & -a_7 & -a_4 & a_5 & a_2 & -a_3 & a_0 & a_1 \\
a_7 & a_6 & -a_5 & -a_4 & a_3 & a_2 & -a_1 & a_0
\end{pmatrix}
\begin{pmatrix}
1 \\ -e_1 \\ -e_2 \\ -e_3 \\ -e_4 \\ -e_5 \\ -e_6 \\ -e_7
\end{pmatrix} =
$$

$$
= \begin{pmatrix}
a_0 + a_1 e_1 + a_2 e_2 + a_3 e_3 + a_4 e_4 + a_5 e_5 + a_6 e_6 + a_7 e_7 \\
a_1 - a_0 e_1 + a_3 e_2 - a_2 e_3 + a_5 e_4 - a_4 e_5 - a_7 e_6 + a_6 e_7 \\
a_2 - a_3 e_1 - a_0 e_2 + a_1 e_3 + a_6 e_4 + a_7 e_5 - a_4 e_6 - a_5 e_7 \\
a_3 + a_2 e_1 - a_1 e_2 - a_0 e_3 + a_7 e_4 - a_6 e_5 + a_5 e_6 - a_4 e_7 \\
a_4 - a_5 e_1 - a_6 e_2 - a_7 e_3 - a_0 e_4 + a_1 e_5 + a_2 e_6 + a_3 e_7 \\
a_5 + a_4 e_1 - a_7 e_2 + a_6 e_3 - a_1 e_4 - a_0 e_5 - a_3 e_6 + a_2 e_7 \\
a_6 + a_7 e_1 + a_4 e_2 - a_5 e_3 - a_2 e_4 + a_3 e_5 - a_0 e_6 - a_1 e_7 \\
a_7 - a_6 e_1 + a_5 e_2 + a_4 e_3 - a_3 e_4 - a_2 e_5 + a_1 e_6 - a_0 e_7
\end{pmatrix} =
$$

$$
= \begin{pmatrix}
a_0 + a_1 e_1 + a_2 e_2 + a_3 e_3 + a_4 e_4 + a_5 e_5 + a_6 e_6 + a_7 e_7 \\
-e_1 (a_0 + a_1 e_1 + a_2 e_2 + a_3 e_3 + a_4 e_4 + a_5 e_5 + a_6 e_6 + a_7 e_7) \\
-e_2 (a_0 + a_1 e_1 + a_2 e_2 + a_3 e_3 + a_4 e_4 + a_5 e_5 + a_6 e_6 + a_7 e_7) \\
-e_3 (a_0 + a_1 e_1 + a_2 e_2 + a_3 e_3 + a_4 e_4 + a_5 e_5 + a_6 e_6 + a_7 e_7) \\
-e_4 (a_0 + a_1 e_1 + a_2 e_2 + a_3 e_3 + a_4 e_4 + a_5 e_5 + a_6 e_6 + a_7 e_7) \\
-e_5 (a_0 + a_1 e_1 + a_2 e_2 + a_3 e_3 + a_4 e_4 + a_5 e_5 + a_6 e_6 + a_7 e_7) \\
-e_6 (a_0 + a_1 e_1 + a_2 e_2 + a_3 e_3 + a_4 e_4 + a_5 e_5 + a_6 e_6 + a_7 e_7) \\
-e_7 (a_0 + a_1 e_1 + a_2 e_2 + a_3 e_3 + a_4 e_4 + a_5 e_5 + a_6 e_6 + a_7 e_7)
\end{pmatrix} =
$$

$$= Ma.$$

(ii)

$$
\theta M = \begin{pmatrix}
0 & -1 & 0 & 0 & 0 & 0 & 0 & 0 \\
1 & 0 & 0 & 0 & 0 & 0 & 0 & 0 \\
0 & 0 & 0 & -1 & 0 & 0 & 0 & 0 \\
0 & 0 & 1 & 0 & 0 & 0 & 0 & 0 \\
0 & 0 & 0 & 0 & 0 & -1 & 0 & 0 \\
0 & 0 & 0 & 0 & 1 & 0 & 0 & 0 \\
0 & 0 & 0 & 0 & 0 & 0 & 0 & 1 \\
0 & 0 & 0 & 0 & 0 & 0 & -1 & 0
\end{pmatrix}
\begin{pmatrix}
1 \\ -e_1 \\ -e_2 \\ -e_3 \\ -e_4 \\ -e_5 \\ -e_6 \\ -e_7
\end{pmatrix} =
$$

$$= \begin{pmatrix} e_1 \\ 1 \\ e_3 \\ -e_2 \\ e_5 \\ -e_4 \\ -e_7 \\ e_6 \end{pmatrix} = M e_1.$$

(iii)

$$\Lambda\,(a)\,N = \begin{pmatrix} a_0 & -a_1 & -a_2 & -a_3 & -a_4 & -a_5 & -a_6 & -a_7 \\ a_1 & a_0 & -a_3 & a_2 & -a_5 & a_4 & a_7 & -a_6 \\ a_2 & a_3 & a_0 & -a_1 & -a_6 & -a_7 & a_4 & a_5 \\ a_3 & -a_2 & a_1 & a_0 & -a_7 & a_6 & -a_5 & a_4 \\ a_4 & a_5 & a_6 & a_7 & a_0 & -a_1 & -a_2 & -a_3 \\ a_5 & -a_4 & a_7 & -a_6 & a_1 & a_0 & a_3 & -a_2 \\ a_6 & -a_7 & -a_4 & a_5 & a_2 & -a_3 & a_0 & a_1 \\ a_7 & a_6 & -a_5 & -a_4 & a_3 & a_2 & -a_1 & a_0 \end{pmatrix} \begin{pmatrix} 1 \\ e_1 \\ e_2 \\ e_3 \\ e_4 \\ e_5 \\ e_6 \\ e_7 \end{pmatrix} =$$

$$= \begin{pmatrix} a_0 - a_1 e_1 - a_2 e_2 - a_3 e_3 - a_4 e_4 - a_5 e_5 - a_6 e_6 - a_7 e_7 \\ a_1 + a_0 e_1 - a_3 e_2 + a_2 e_3 - a_5 e_4 + a_4 e_5 + a_7 e_6 - a_6 e_7 \\ a_2 + a_3 e_1 + a_0 e_2 - a_1 e_3 - a_6 e_4 - a_7 e_5 + a_4 e_6 + a_5 e_7 \\ a_3 - a_2 e_1 + a_1 e_2 + a_0 e_3 - a_7 e_4 + a_6 e_5 - a_5 e_6 + a_4 e_7 \\ a_4 + a_5 e_1 + a_6 e_2 + a_7 e_3 + a_0 e_4 - a_1 e_5 - a_2 e_6 - a_3 e_7 \\ a_5 - a_4 e_1 + a_7 e_2 - a_6 e_3 + a_1 e_4 + a_0 e_5 + a_3 e_6 - a_2 e_7 \\ a_6 - a_7 e_1 - a_4 e_2 + a_5 e_3 + a_2 e_4 - a_3 e_5 + a_0 e_6 + a_1 e_7 \\ a_7 + a_6 e_1 - a_5 e_2 - a_4 e_3 + a_3 e_4 + a_2 e_5 - a_1 e_6 + a_0 e_7 \end{pmatrix} =$$

$$= \begin{pmatrix} a_0 - a_1 e_1 - a_2 e_2 - a_3 e_3 - a_4 e_4 - a_5 e_5 - a_6 e_6 - a_7 e_7 \\ (a_0 - a_1 e_1 - a_2 e_2 - a_3 e_3 - a_4 e_4 - a_5 e_5 - a_6 e_6 - a_7 e_7)\, e_1 \\ (a_0 - a_1 e_1 - a_2 e_2 - a_3 e_3 - a_4 e_4 - a_5 e_5 - a_6 e_6 - a_7 e_7)\, e_2 \\ (a_0 - a_1 e_1 - a_2 e_2 - a_3 e_3 - a_4 e_4 - a_5 e_5 - a_6 e_6 - a_7 e_7)\, e_3 \\ (a_0 - a_1 e_1 - a_2 e_2 - a_3 e_3 - a_4 e_4 - a_5 e_5 - a_6 e_6 - a_7 e_7)\, e_4 \\ (a_0 - a_1 e_1 - a_2 e_2 - a_3 e_3 - a_4 e_4 - a_5 e_5 - a_6 e_6 - a_7 e_7)\, e_5 \\ (a_0 - a_1 e_1 - a_2 e_2 - a_3 e_3 - a_4 e_4 - a_5 e_5 - a_6 e_6 - a_7 e_7)\, e_6 \\ (a_0 - a_1 e_1 - a_2 e_2 - a_3 e_3 - a_4 e_4 - a_5 e_5 - a_6 e_6 - a_7 e_7)\, e_7 \end{pmatrix} =$$

$$= \bar{a} N. \qquad \square$$

Proposition 2.8 *Let* $A = x + iy$ *be a complex octonion with* x, y *two real octonions and* $a = q_1 + q_2 k$ *be a real octonion, with* q_1, q_2 *two real quaternions. The following relations are true.*

(i) $T \Lambda (a) T = \Delta (a^+)$, where $T = \begin{pmatrix} M_1 & O_4 \\ O_4 & I_4 \end{pmatrix} \in \mathcal{M}_8 (\mathbb{R})$ and $O_4 \in \mathcal{M}_8 (\mathbb{R})$ is zero matrix.

(ii) $S \Phi (A) S = \Psi (A^+)$, where $A^+ = x^+ + iy^+$, $S = \begin{pmatrix} T & O_8 \\ O_8 & T \end{pmatrix} \in \mathcal{M}_{16} (\mathbb{R})$ and $O_8 \in \mathcal{M}_8 (\mathbb{R})$ is zero matrix.

Proof (i) From (Tian 2000), relations (1.17) and (1.18), we know that $\rho (q) = M_1 \lambda^t (q) M_1 = M_1 \lambda (\bar{q}) M_1$, with q a real quaternion. It results

$$\begin{pmatrix} M_1 & O_4 \\ O_4 & I_4 \end{pmatrix} \Lambda (A) \begin{pmatrix} M_1 & O_4 \\ O_4 & I_4 \end{pmatrix} =$$

$$= \begin{pmatrix} M_1 & O_4 \\ O_4 & I_4 \end{pmatrix} \begin{pmatrix} \lambda (q_1) & -\rho (q_2) M_1 \\ \lambda (q_2) M_1 & \rho (q_1) \end{pmatrix} \begin{pmatrix} M_1 & O_4 \\ O_4 & I_4 \end{pmatrix} =$$

$$= \begin{pmatrix} M_1 \lambda (q_1) & -M_1 \rho (q_2) M_1 \\ \lambda (q_2) M_1 & \rho (q_1) \end{pmatrix} \begin{pmatrix} M_1 & O_4 \\ O_4 & I_4 \end{pmatrix} =$$

$$= \begin{pmatrix} M_1 \lambda (q_1) M_1 & -M_1 \rho (q_2) M_1 \\ \lambda (q_2) & \rho (q_1) \end{pmatrix} = \begin{pmatrix} \rho (\bar{q}_1) & -M_1 \rho (q_2) M_1 \\ \lambda (q_2) & \rho (q_1) \end{pmatrix} =$$

$$= \begin{pmatrix} \rho (\bar{q}_1) & -\lambda (\bar{q}_2) \\ \lambda (q_2) & \rho (q_1) \end{pmatrix} = \Delta (a^+).$$

(ii) We have

$$\begin{pmatrix} T & O_8 \\ O_8 & T \end{pmatrix} \Phi (A) \begin{pmatrix} T & O_8 \\ O_8 & T \end{pmatrix} =$$

$$= \begin{pmatrix} T & O_8 \\ O_8 & T \end{pmatrix} \begin{pmatrix} \Lambda (x) & -\Lambda (y) \\ \Lambda (y^*) & \Lambda (x^*) \end{pmatrix} \begin{pmatrix} T & O_8 \\ O_8 & T \end{pmatrix} =$$

$$= \begin{pmatrix} T \Lambda (x) & -T \Lambda (y) \\ T \Lambda (y^*) & T \Lambda (x^*) \end{pmatrix} \begin{pmatrix} T & O_8 \\ O_8 & T \end{pmatrix} = \begin{pmatrix} T \Lambda (x) T & -T \Lambda (y) T \\ T \Lambda (y^*) T & T \Lambda (x^*) T \end{pmatrix} =$$

$$= \begin{pmatrix} \Delta (x^+) & -\Delta (y^+) \\ \Delta (y^{*+}) & \Delta (x^{*+}) \end{pmatrix} = \Psi (A^+). \qquad \square$$

Proposition 2.9 *With the above notations, we have*

$$\Psi (A^+) \overrightarrow{X} = S \overrightarrow{A'X^+},$$

where $X^+ = v^+ + iw^+$.

Proof Since $T\vec{v} = \vec{v^+}$, it results

$$\Psi\left(A^+\right)\vec{X} = S\Phi\left(A\right)S\vec{X} = S\Phi\left(A\right)\begin{pmatrix} T & O_8 \\ O_8 & T \end{pmatrix}\begin{pmatrix} \vec{v} \\ \vec{w} \end{pmatrix} =$$

$$= S\Phi\left(A\right)\begin{pmatrix} T\vec{v} \\ T\vec{w} \end{pmatrix} = S\Phi\left(A\right)\begin{pmatrix} \vec{v^+} \\ \vec{w^+} \end{pmatrix} = S\Phi\left(A\right)\vec{X^+} = S\overrightarrow{A'X^+}.$$

$\square$

3 A Set of Invertible Elements in Split Quaternion and Octonion Algebras

In a split quaternion algebra and in a split octonion algebra there are nonzero elements such that their norms are zero. In such algebras, it is very good to know sets of invertible elements, that means nonzero elements with their norms nonzero. In the following, we give a method to find such a sets.

Let n be an arbitrary positive integer and let a, b, c, x_0, x_1, x_2 be arbitrary integers. We consider the following difference equation of degree three

$$X_n = aX_{n-1} + bX_{n-2} + cX_{n-3}, X_0 = x_0, X_1 = x_1, X_2 = x_2. \quad (3.1)$$

We consider the following degree three equation

$$x^3 - ax^2 - bx - c = 0. \quad (3.2)$$

We consider that this equation has three real solutions $\sigma_1 > \sigma_2 > \sigma_3$, with $\sigma_1 > 1$. For this case, we have the following Binet's formula:

$$X_n = A\sigma_1^n + B\sigma_2^n + C\sigma_3^n, \quad (3.3)$$

where A, B, C are solutions of the following linear system:

$$\begin{cases} A + B + C = x_0 \\ A\sigma_1 + B\sigma_2 + C\sigma_3 = x_1 \\ A\sigma_1^2 + B\sigma_2^2 + C\sigma_3^2 = x_2 \end{cases}.$$

Since we obtain a Vandermonde determinant, the system has a unique solution.

We consider the real generalized quaternion algebra $\mathbb{H}\left(\beta_1, \beta_2\right)$ and we define the quaternions

$$W_n = X_n + X_{n+1}e_2 + X_{n+2}e_3 + X_{n+3}e_4,$$

where X_n is the nth number given by the relation (3.3).

From the above, we can compute the following limit

$$\lim_{n\to\infty} n(W_n) = \lim_{n\to\infty} (X_n^2 + \beta_1 X_{n+1}^2 + \beta_2 X_{n+2}^2 + \beta_1\beta_2 X_{n+3}^2) =$$

$$= \lim_{n \to \infty} ((A\sigma_1^n + B\sigma_2^n + C\sigma_3^n)^2 + \beta_1 \left(A\sigma_1^{n+1} + B\sigma_2^{n+1} + C\sigma_3^{n+1}\right)^2 +$$
$$\beta_2 \left(A\sigma_1^{n+2} + B\sigma_2^{n+2} + C\sigma_3^{n+2}\right)^2 + \beta_1\beta_2 \left(A\sigma_1^{n+3} + B\sigma_2^{n+3} + C\sigma_3^{n+3}\right)^2).$$

Let $f(\beta_1, \beta_2) = A^2(1 + \beta_1\sigma_1^2 + \beta_2\sigma_1^4 + \beta_1\beta_2\sigma_1^6)$. If $f(\beta_1, \beta_2) \neq 0$, it results that

$$\lim_{n \to \infty} n(W_n) = signf(\beta_1, \beta_2) \cdot \infty.$$

Therefore, for all $\beta_1, \beta_2 \in \mathbb{R}$ with $f(\beta_1, \beta_2) \neq 0$, in the algebra $\mathbb{H}(\beta_1, \beta_2)$ there is a natural number n_0 such that $n(W_n) \neq 0$. From here, we have that W_n is an invertible element for all $n \geq n_0$.

Now, we consider the real octonion algebra $\mathbb{O}(\alpha, \beta, \gamma)$. We define the octonions

$$Z_n = X_n + X_{n+1}e_2 + X_{n+2}e_3 + X_{n+3}e_4 + X_{n+4}e_5 + X_{n+5}e_6 + X_{n+6}e_7,$$

where X_n is the nth number given by the relation (3.3).

From the above, we can compute the following limit

$$\lim_{n \to \infty} n(Z_n) =$$

$$= \lim_{n \to \infty} (X_n^2 + \alpha X_{n+1}^2 + \beta X_{n+2}^2 + \alpha\beta X_{n+3}^2 + \gamma X_{n+4}^2 + \alpha\gamma X_{n+5}^2 + \beta\gamma X_{n+6}^2 + \alpha\beta\gamma X_{n+7}^2) =$$

$$= \lim_{n \to \infty} ((A\sigma_1^n + B\sigma_2^n + C\sigma_3^n)^2 + \alpha \left(A\sigma_1^{n+1} + B\sigma_2^{n+1} + C\sigma_3^{n+1}\right)^2 +$$

$$+ \quad \beta \left(A\sigma_1^{n+2} + B\sigma_2^{n+2} + C\sigma_3^{n+2}\right)^2 + \alpha\beta \left(A\sigma_1^{n+3} + B\sigma_2^{n+3} + C\sigma_3^{n+3}\right)^2 +$$

$$+\gamma \left(A\sigma_1^{n+4} + B\sigma_2^{n+4} + C\sigma_3^{n+4}\right)^2 + \alpha\gamma \left(A\sigma_1^{n+5} + B\sigma_2^{n+5} + C\sigma_3^{n+5}\right)^2 +$$

$$+\beta\gamma \left(A\sigma_1^{n+6} + B\sigma_2^{n+6} + C\sigma_3^{n+6}\right)^2 + \alpha\beta\gamma \left(A\sigma_1^{n+7} + B\sigma_2^{n+7} + C\sigma_3^{n+7}\right)^2.$$

Let $g(\alpha, \beta, \gamma) = A^2(1 + \alpha\sigma_1^2 + \beta\sigma_1^4 + \alpha\beta\sigma_1^6 + \gamma\sigma_1^8 + \alpha\gamma\sigma_1^{10} + \beta\gamma\sigma_1^{12} + \alpha\beta\gamma\sigma_1^{14})$.
If $g(\alpha, \beta, \gamma) \neq 0$, it results that

$$\lim_{n \to \infty} n(Z_n) = signg(\alpha, \beta, \gamma) \cdot \infty.$$

Therefore, for all $\alpha, \beta, \gamma \in \mathbb{R}$ with $g(\alpha, \beta, \gamma) \neq 0$, in the algebra $\mathbb{O}(\alpha, \beta, \gamma)$ there is a natural number n_0 such that $n(Z_n) \neq 0$. From here, we have that z_n is an invertible element for all $n \geq n_0$.

Since algebras $\mathbb{H}(\beta_1, \beta_2)$ and $\mathbb{O}(\alpha, \beta, \gamma)$ are not always division algebras, finding examples of invertible elements in such algebras can be a difficult problem. The above elements, W_n and Z_n, provide us an infinite set of invertible elements in a split quaternion algebra and in a split octonion algebra.

Conclusions In this chapter, we gave some properties of the real matrix representations for complex octonions and we provided sets of invertible elements in a split quaternion algebra and in a split octonion algebra. Due their applications, the study of these representations and the study of these elements can give us other properties and applications.

References

Alamouti, S.M.: A simple transmit diversity technique for wireless communications. IEEE J. Sel. Areas Commun. **16**(8), 1451–1458 (1998)

Alfsmann, D., Göckler, H.G., Sangwine, S.J., Ell, T., A.: Hypercomplex algebras in digital signal processing: benefits and drawbacks. In: 15th European Signal Processing Conference (EUSIPCO, 2007), Poznan, Poland, pp. 1322–1326 (2007)

Baez, J.C.: The octonions. B. Am. Math. Soc. **39**(2), 145–205 (2002)

Chanyal, B. C.: Octonion massive electrodynamics. Gen. Relat. Gravit. **46**, article ID: 1646 (2014)

Chanyal, B.C., Bisht, P.S., Negi, O.P.S.: Generalized split-octonion electrodynamics. Int. J. Theor. Phys. **50**(6), 1919–1926 (2011)

Chen, J., Tu, A.: Fabric image edge detection based on octonion and echo state networks. Appl. Mech. Mater. **263–266**, 2483–2487 (2013)

Flaut, C., Shpakivskyi, V.: Real matrix representations for the complex quaternions. Adv. Appl. Clifford Algebras **23**(3), 657–671 (2013)

Hanson, A.J.: Visualizing Quaternions. Elsevier Morgan Kaufmann Publishers, Burlington (2006)

Jia, Y.B.: Quaternion and Rotation, Com S 477/577 Notes, (2017)

Jouget, P.: Sécurité et performance de dispositifs de distribution quantique de clés à variables continues. Ph.D Thesis, TELECOM ParisTech (2013)

Klco, P., Smetana, M., Kollarik, M., Tatar, M.: Application of octonions in the cough sounds classification. Adv. Appl. Sci. Res. **8**(2), 30–37 (2017)

Kostrikin, A.I., Shafarevich, I.R. (eds.): Algebra VI. Springer, Berlin (1995)

Li, X.M.: Hyper-Complex Numbers and its Applications in Digital Image Processing. Seminars and Distinguished Lectures (2011)

Schafer, R.D.: An Introduction to Nonassociative Algebras. Academic, New York (1966)

Snopek, K., M.: Quaternions and Octonions in Signal Processing - Fundamentals and Some New Results, Przeglad Telekomunikacyjny - Wiadomoś ci Telekomunikacyjne, SIGMA NOT, **134**(6), 619–622 (2015)

Tian, Y.: Matrix representations of octonions and their applications. Adv. in Appl. Clifford Algebras **10**(1), 61–90 (2000)

Unger, T., Markin, N.: Quadratic forms and space-time block codes from generalized quaternion and biquaternion algebras. IEEE Trans. Inf. Theory **57**(9), 6148–6156 (2011)

A Theory of Quaternionic G-Monogenic Mappings in E_3

T. S. Kuzmenko and V. S. Shpakivskyi

Abstract We consider a class of so-called quaternionic G-monogenic (differentiable in the sense of Gâteaux) mappings and propose a description of all mappings in this class by using four analytic functions of complex variable. For G-monogenic mappings we generalize some analogues of classical integral theorems of the holomorphic function theory of one complex variable (the surface and the curvilinear Cauchy integral theorems, the Morera theorem), and Taylor and Laurent expansions. Moreover, we introduce a new class of quaternionic H-monogenic (differentiable in the sense of Hausdorff) mappings and establish the relation between G-monogenic and H-monogenic mappings. In addition, we prove the theorem of equivalence of different definitions of a G-monogenic mapping.

Keywords Algebra of complex quaternions · G-monogenic mappings
Constructive description · Integral theorems · Taylor and Laurent expansions
Singular points · H-monogenic mappings

1 Introduction

The quaternionic analysis arose long ago. It is now extensively developed as a separate direction of mathematics due to its numerous applications in various fields, in particular in mathematical physics and differential equations (see, e.g., Gürlebeck and Sprössig 1997; Kravchenko and Shapiro 1996; Kravchenko 2003).

T. S. Kuzmenko · V. S. Shpakivskyi (✉)
Department of Complex Analysis and Potential Theory, Institute of Mathematics
of the National Academy of Science of Ukraine, 3, Tereshchenkivs'ka st.,
Kyiv 01004, Ukraine
e-mail: shpakivskyi86@gmail.com; shpakivskyi@imath.kiev.ua

T. S. Kuzmenko
e-mail: kuzmenko.ts15@gmail.com; kuzmenko@imath.kiev.ua

© Springer Nature Switzerland AG 2019
C. Flaut et al. (eds.), *Models and Theories in Social Systems*, Studies in Systems,
Decision and Control 179, https://doi.org/10.1007/978-3-030-00084-4_25

451

A realization of this approach requires the introduction of special classes of quaternionic "holomorphic" functions which satisfy certain Cauchy–Riemann type operators with quaternionic coefficients.

The quaternionic analysis in $\mathbb{R}^3$ was originated by Moisil and Theodoresco Moisil and Theodoresco (1931) who proposed a three-dimensional analog of the Cauchy–Riemann system of equations. They introduced the notion of *holomorphic vector* as a quaternion-valued function of three real variables

$$U(xi + yi + zk) := u_0(x, y, z) + iu_1(x, y, z) + ju(x, y, z) + ku(x, y, z),$$

whose components $u_n(x, y, z)$, $n = 0, 1, 2, 3$, are continuously differentiable and satisfy the above-mentioned system, which has been called the Moisil–Theodoresco system (where i, j and k are quaternionic imaginary units). The mentioned Moisil–Theodoresco system is equivalent to the following quaternionic equation

$$\frac{\partial U}{\partial x} i + \frac{\partial U}{\partial y} j + \frac{\partial U}{\partial z} k = 0. \tag{1}$$

In the scientific literature Eq. (1) is called sometimes Dirac equation and sometimes Cauchy–Riemann equation.

In the same paper Moisil and Theodoresco (1931), the authors proved an analog of the Morera theorem and analogs of the integral Cauchy formula. The investigations originated in Moisil and Theodoresco (1931) were continued in Bitsadze (1966), where the notion of Cauchy-type integral was introduced, the existence of its boundary values was investigated, and the applications of this integral to systems of singular integral equations were discussed.

In Fueter (1935) constructed a four-dimensional generalization of the Moisil–Theodoresco system which is equivalent to the following equation

$$\frac{\partial V}{\partial t} + \frac{\partial V}{\partial x} i + \frac{\partial V}{\partial y} j + \frac{\partial V}{\partial z} k = 0. \tag{2}$$

Fueter proved analogs of the classical results of complex analysis for functions satisfying Eq. (2). These results were generalized in Sudbery (1979) and, together with the applications to some models of mathematical physics, presented in the monograph Kravchenko and Shapiro (1996). It is also worth noting that the so-called α-*holomorphic functions* f investigated in Kravchenko and Shapiro (1996) satisfy the three-dimensional Helmholtz equation

$$(\Delta_3 + \alpha)f := \frac{\partial^2 f}{\partial x^2} + \frac{\partial^2 f}{\partial y^2} + \frac{\partial^2 f}{\partial z^2} + \alpha f = 0,$$

where α is a complex quaternion.

The last investigations in this field (see, e.g., Herus 2011; Gerus and Shapiro 2003; Schneider 2006) can be regarded as various generalizations of the results obtained in Kravchenko and Shapiro (1996).

Another (relatively new) direction of quaternionic analysis in $\mathbb{R}^3$ and $\mathbb{R}^4$ is represented by the so-called modified quaternionic analysis originated by Leutwiler in the early 1990s (see, e.g., Leutwiler 1992; Hempfling and Leutwiler 1998; Eriksson-Bique 2001). In the space $\mathbb{R}^3$, Leutwiler investigates solutions of the following generalization of the Cauchy–Riemann system:

$$\begin{cases} t\left(\dfrac{\partial u}{\partial x} - \dfrac{\partial v}{\partial y} - \dfrac{\partial w}{\partial t}\right) + w = 0, \\ \dfrac{\partial u}{\partial y} = -\dfrac{\partial v}{\partial x}, \quad \dfrac{\partial u}{\partial t} = -\dfrac{\partial w}{\partial x}, \quad \dfrac{\partial v}{\partial t} = \dfrac{\partial w}{\partial y}. \end{cases}$$

Note that the first two components of his *hyperholomorphic functions* $f(x + iy + jt) = u(x, y, t) + iv(x, y, t) + jw(x, y, t)$ satisfy the Laplace–Beltrami equation

$$t\Delta_3 u - \frac{\partial u}{\partial t} = 0,$$

and the third component w satisfies the equation

$$t^2\Delta_3 w - t\frac{\partial w}{\partial t} + w = 0.$$

In Leutwiler (1992) one can find the expansion of a hyperholomorphic function in a series of quaternionic polynomials. For more information see Leutwiler and Zeilinger (2004), Eriksson and Leutwiler (2009).

In $\mathbb{R}^4$ (see Hempfling and Leutwiler 1998), Leutwiler finds solutions of the generalized Cauchy–Riemann system

$$\begin{cases} s\left(\dfrac{\partial u}{\partial x} - \dfrac{\partial v}{\partial y} - \dfrac{\partial w}{\partial t} - \dfrac{\partial r}{\partial s}\right) + 2r = 0, \\ \dfrac{\partial u}{\partial y} = -\dfrac{\partial v}{\partial x}, \quad \dfrac{\partial u}{\partial t} = -\dfrac{\partial w}{\partial x}, \quad \dfrac{\partial u}{\partial s} = -\dfrac{\partial r}{\partial x}, \\ \dfrac{\partial v}{\partial t} = \dfrac{\partial w}{\partial y}, \quad \dfrac{\partial v}{\partial s} = \dfrac{\partial r}{\partial y}, \quad \dfrac{\partial w}{\partial s} = \dfrac{\partial r}{\partial t} \end{cases}$$

in the form of a quaternion-valued function $f(x + iy + jt + ks) = u(x, y, t, s) + iv(x, y, t, s) + jw(x, y, t, s) + kr(x, y, t, s)$.

Unlike (Kravchenko and Shapiro 1996; Moisil and Theodoresco 1931; Fueter 1935; Sudbery 1979), in the Leutwiler approach a power function is hyperholomorphic and the partial derivatives of a hyperholomorphic function are also hyperholomorphic. At the same time, there exists a relationship between both directions

described above (see Eriksson-Bique 2001). Today, the ideas of Leutwiler successfully develop and find new applications (see Bryukhov and Kähler 2017).

We can also mention a new theory in the quaternionic analysis, namely, the theory
of so-called *s-regular functions* introduced by Gentili and Struppa (2006) on the
basis of development of Cullen's idea (Cullen 1965). This idea can be formulated as
follows: Let

$$x = x_0 + x_1 i + x_2 j + x_3 k =: x_0 + \operatorname{Im} x,$$

where x_0, x_1, x_2, x_3 are real numbers and i, j, k are quaternionic imaginary units.
Every quaternion $x = x_0 + \operatorname{Im} x$ with $x \neq x_0$ can be represented in the form of a
"complex number" with a new imaginary unit I: $x = x_0 + I \mid \operatorname{Im} x \mid$, where $I :=
\frac{\operatorname{Im} x}{|\operatorname{Im} x|}$ and $|\cdot|$ is the modulus of a quaternion. It is clear that $I^2 = -1$. In the same
form, one can also represent a quaternion-valued function: $f(x) = U(x_0, \mid \operatorname{Im} x \mid) +
I\, V(x_0, \mid \operatorname{Im} x \mid)$.

Then the function f is called an *s*-regular function (see Gentili and Struppa 2006)
if the "complex-valued" function $f = U + IV$ is a holomorphic function of the
"complex" variable $x = x_0 + I \mid \operatorname{Im} x \mid$. It is proved that all quaternionic polynomials
and special power series are *s*-regular. At present, the theory of *s*-regular functions
is widely developed (see Colombo et al. 2011; Gentili et al. 2013; Alpay et al. 2016;
Colombo et al. 2016).

The mentioned variety of different approaches poses a natural question of classification of generalized analytic function theories (Kisil 1998). Such a classification
can be derived from the symmetry group of respective theory. Moreover, it is possible to build new theories from a given group representation following the scheme in
Kisil (1999, 2012).

In our paper Shpakivskyi and Kuzmenko (2016), we introduced a special class of
mappings in the algebra of complex quaternions, which is not covered by the above-
mentioned theories. Note that the commutative algebra of bicomplex numbers (or of
Segre commutative quaternions Segre 1892; Luna-Elizarrarás et al. 2015) is a subalgebra of the algebra of complex quaternions $\mathbb{H}(\mathbb{C})$. In this subalgebra, we selected a
three-dimensional real subspace, E_3, and consider mappings Φ defined in a domain
Ω of this subspace E_3 and taking values in the entire algebra of complex quaternions. These mappings are continuous and Gâteaux differentiable. They are called
G-monogenic and represent the main object of our investigations. It is shown that not
only quaternionic polynomials but also quaternionic power series are G-monogenic.
Moreover, in the paper Shpakivskyi and Kuzmenko (2016), we proposed a construc-
tive description of all G-monogenic mappings of the form $\Phi : E_3 \supset \Omega \to \mathbb{H}(\mathbb{C})$
based on the use of four analytic functions of complex variable. As a consequence,
the Gâteaux derivative of a G-monogenic mapping is, in turn, a G-monogenic mapping. In addition, we study the relationship between G-monogenic mappings and
three-dimensional partial differential equations. In particular, we discuss several
applications of monogenic mappings to the construction of solutions of the three-
dimensional Laplace equation.

In the paper Shpakivskyi and Kuzmenko (2016), we proved analogs of classical integral theorems of the holomorphic function theory: the Cauchy integral the-

orems for surface and curvilinear integrals, and the Cauchy integral formula for G-monogenic mappings of the form $\Phi : E_3 \supset \Omega \to \mathbb{H}(\mathbb{C})$. Furthermore, in Kuzmenko (2016) a curvilinear Cauchy integral theorem for G-monogenic mappings in the case where a curve of integration lies on the boundary of a domain of G-monogeneity has been proved.

The analogs of the Cauchy integral theorems (see Shpakivskyi and Kuzmenko 2016) are of the form

$$\int_\Gamma \widehat{\Phi}\, \sigma = 0, \qquad \int_\Gamma \sigma\, \Phi = 0,$$

where Γ is a closed surface (or a closed curve), σ is a special differential form, and $\widehat{\Phi}$, Φ are left-G-monogenic mapping and right-G-monogenic mapping, respectively.

In the paper Kuzmenko and Shpakivskyi (2017) we generalized analogs of the surface and curvilinear Cauchy integral theorems for G-monogenic mappings to "two-sided" integrals. Namely, under some assumptions we proved the equality

$$\int_\Gamma \widehat{\Phi}\, \sigma\, \Phi = 0. \tag{3}$$

Taylor and Laurent expansions of G-monogenic mappings of the form $\Phi : E_3 \supset \Omega \to \mathbb{H}(\mathbb{C})$ are obtained and singularities of these mappings are classified in the paper Kuzmenko (2015).

In Shpakivskyi and Kuzmenko (2017), we introduce quaternionic H-monogenic (differentiable in the sense of Hausdorff) mappings and establish a relation between G- and H-monogenic mappings which are defined in a domain of the space E_3. The equivalence of different definitions of a G-monogenic mapping is proved.

We note that the research methods in this Chapter are similar to the methods of paper Flaut (2017), where some connections between Hilbert algebras and binary block-codes are studied.

In the present Chapter, we give from a unified point of view the results of the papers Shpakivskyi and Kuzmenko (2016), Shpakivskyi and Kuzmenko (2016), Kuzmenko (2016), Kuzmenko and Shpakivskyi (2017), Kuzmenko (2015), Shpakivskyi and Kuzmenko (2017); we present some new theorems, simplify some proofs, and give new examples. The purpose of this publication is to collect in one work the latest achievements in the theory of quaternionic G-monogenic mappings that defined in a domain of E_3.

2   The Algebra of Complex Quaternions

Let us consider the algebra of quaternions $\mathbb{H}(\mathbb{C})$ over the field of complex numbers $\mathbb{C}$ with the basis $\{1, I, J, K\}$, whose elements satisfy the following multiplication rules:

$$I^2 = J^2 = K^2 = -1,$$

$$IJ = -JI = K, \qquad JK = -KJ = I, \qquad KI = -IK = J.$$

In the algebra $\mathbb{H}(\mathbb{C})$ there exists another basis $\{e_1, e_2, e_3, e_4\}$:

$$e_1 = \frac{1}{2}(1 + iI), \quad e_2 = \frac{1}{2}(1 - iI), \quad e_3 = \frac{1}{2}(iJ - K), \quad e_4 = \frac{1}{2}(iJ + K),$$

where i is the complex imaginary unit. The multiplication table in the new basis has the form (see Cartan 1898)

$$
\begin{array}{c||c|c|c|c}
\cdot & e_1 & e_2 & e_3 & e_4 \\
\hline\hline
e_1 & e_1 & 0 & e_3 & 0 \\
\hline
e_2 & 0 & e_2 & 0 & e_4 \\
\hline
e_3 & 0 & e_3 & 0 & e_1 \\
\hline
e_4 & e_4 & 0 & e_2 & 0
\end{array}
\tag{4}
$$

where the unit of the algebra is decomposed as $1 = e_1 + e_2$.

The norm of the complex quaternion

$$a = \sum_{k=1}^{4} a_k e_k, \ a_k \in \mathbb{C}$$

is given by the formula

$$\|a\| := \sqrt{\sum_{k=1}^{4} |a_k|^2}. \tag{5}$$

It is easily seen, that the basis vectors $\{e_1, e_2\}$ are idempotents, which generate a semi-simple algebra. Note also that this subalgebra is the algebra of bicomplex numbers or the Segre algebra of commutative quaternions (Segre 1892; Luna-Elizarrarás et al. 2015).

Recall that (see, e.g., Van der Waerden 1991), a subset $\mathcal{I} \subset \mathbb{H}(\mathbb{C})$ is called *the right ideal* if the condition $x \in \mathcal{I}$ implies that $xy \in \mathcal{I}$, and a subset $\mathcal{I}$ is called *the left ideal* if the condition $x \in \mathcal{I}$ implies that $yx \in \mathcal{I}$ for any $y \in \mathbb{H}(\mathbb{C})$.

The algebra $\mathbb{H}(\mathbb{C})$ contains two right maximal ideals

$$\mathcal{I}_1 := \{\lambda_2 e_2 + \lambda_4 e_4 : \lambda_2, \lambda_4 \in \mathbb{C}\}, \qquad \mathcal{I}_2 := \{\lambda_1 e_1 + \lambda_3 e_3 : \lambda_1, \lambda_3 \in \mathbb{C}\}$$

and two left maximal ideals

$$\widehat{\mathcal{I}}_1 := \{\lambda_2 e_2 + \lambda_3 e_3 : \lambda_2, \lambda_3 \in \mathbb{C}\}, \qquad \widehat{\mathcal{I}}_2 := \{\lambda_1 e_1 + \lambda_4 e_4 : \lambda_1, \lambda_4 \in \mathbb{C}\}.$$

Since the radical consists of the zero element only the algebra $\mathbb{H}(\mathbb{C})$ is semi-simple (see, e.g. Hille and Phillips 1948).

The obvious equalities

$$\mathcal{I}_1 \cap \mathcal{I}_2 = \widehat{\mathcal{I}}_1 \cap \widehat{\mathcal{I}}_2 = 0, \qquad \mathcal{I}_1 \cup \mathcal{I}_2 = \widehat{\mathcal{I}}_1 \cup \widehat{\mathcal{I}}_2 = \mathbb{H}(\mathbb{C})$$

yield the following decomposition into the direct sum:

$$\mathbb{H}(\mathbb{C}) = \mathcal{I}_1 \oplus \mathcal{I}_2 = \widehat{\mathcal{I}}_1 \oplus \widehat{\mathcal{I}}_2.$$

We introduce linear functionals $f_1 : \mathbb{H}(\mathbb{C}) \to \mathbb{C}$ and $f_2 : \mathbb{H}(\mathbb{C}) \to \mathbb{C}$ by setting

$$f_1(e_1) = f_1(e_3) = 1, \qquad f_1(e_2) = f_1(e_4) = 0,$$

$$f_2(e_2) = f_2(e_4) = 1, \qquad f_2(e_1) = f_2(e_3) = 0,$$

where maximal ideals $\mathcal{I}_1$, $\mathcal{I}_2$ are the kernels of the functionals f_1, f_2, i.e., $f_1(\mathcal{I}_1) = f_2(\mathcal{I}_2) = 0$. We also define linear functionals $\widehat{f}_1 : \mathbb{H}(\mathbb{C}) \to \mathbb{C}$ and $\widehat{f}_2 : \mathbb{H}(\mathbb{C}) \to \mathbb{C}$ by the equalities

$$\widehat{f}_1(e_1) = \widehat{f}_1(e_4) = 1, \qquad \widehat{f}_1(e_2) = \widehat{f}_1(e_3) = 0,$$

$$\widehat{f}_2(e_2) = \widehat{f}_2(e_3) = 1, \qquad \widehat{f}_2(e_1) = \widehat{f}_2(e_4) = 0.$$

It is clear that $\widehat{f}_1(\widehat{\mathcal{I}}_1) = \widehat{f}_2(\widehat{\mathcal{I}}_2) = 0$.

We say that a functional $f : \mathbb{H}(\mathbb{C}) \to \mathbb{C}$ is *right-multiplicative* (or *left-multiplicative*) if, for any $x \in \mathbb{H}(\mathbb{C})$ and $y \in E_3$, the following equality is true: $f(yx) = f(y)f(x)$ (or $f(xy) = f(x)f(y)$).

Lemma 1 *The functionals $f_1 : \mathbb{H}(\mathbb{C}) \to \mathbb{C}$ and $f_2 : \mathbb{H}(\mathbb{C}) \to \mathbb{C}$ are continuous and right-multiplicative and the functionals $\widehat{f}_1 : \mathbb{H}(\mathbb{C}) \to \mathbb{C}$ and $\widehat{f}_2 : \mathbb{H}(\mathbb{C}) \to \mathbb{C}$ are continuous and left-multiplicative.*

Proof The corresponding multiplicativity of all functionals is directly verified. The continuity of the functionals follows from their boundedness, namely, if

$$a = \sum_{k=1}^{4} a_k e_k \in \mathbb{H}(\mathbb{C}),$$

then, e.g., for f_1, we have

$$\frac{|f_1(a)|}{\|a\|} \leq \frac{|a_1| + |a_3|}{\sqrt{|a_1|^2 + |a_2|^2 + |a_3|^2 + |a_4|^2}} \leq 2.$$

The continuity of the other functionals is proved in a similar way. $\qquad\square$

3 *G*-Monogenic Mappings

Let

$$i_1 = 1, \qquad i_2 = a_1 e_1 + a_2 e_2, \qquad i_3 = b_1 e_1 + b_2 e_2 \tag{6}$$

with $a_k, b_k \in \mathbb{C}$, $k = 1, 2$, be a triple of linearly independent vectors over the field of real numbers $\mathbb{R}$ (see, e.g., Plaksa and Pukhtaievych 2014). This means that the equality

$$\alpha_1 i_1 + \alpha_2 i_2 + \alpha_3 i_3 = 0, \qquad \alpha_1, \alpha_2, \alpha_3 \in \mathbb{R}$$

holds if and only if $\alpha_1 = \alpha_2 = \alpha_3 = 0$.

In the algebra $\mathbb{H}(\mathbb{C})$ we select a linear span

$$E_3 := \{\zeta = x i_1 + y i_2 + z i_3 : x, y, z \in \mathbb{R}\}$$

over the field $\mathbb{R}$ generated by the vectors i_1, i_2, and i_3.

With a set $S \subset \mathbb{R}^3$ we associate the set $S_\zeta := \left\{\zeta = x i_1 + y i_2 + z i_3 : (x, y, z) \in S\right\}$ in E_3. Note that topological properties of the set S_ζ in E_3 are understood as the corresponding topological properties of the set S in $\mathbb{R}^3$.

We introduce the notation:

$$\xi_1 := f_1(\zeta) = \widehat{f_1}(\zeta) = x + y a_1 + z b_1,$$

$$\xi_2 := f_2(\zeta) = \widehat{f_2}(\zeta) = x + y a_2 + z b_2.$$

For an element $\zeta \in E_3$ in accordance to (5) the norm is

$$\|\zeta\| = \sqrt{|\xi_1|^2 + |\xi_2|^2}. \tag{7}$$

Note that, in what follows, a significant role is played by the assumption $f_1(E_3) = f_2(E_3) = \mathbb{C}$, where $f_k(E_3)$ for $k = 1, 2$ is the image of the set E_3 under the mapping f_k. It is obvious that this assumption is true if and only if one of the components of each pair (a_1, b_1) and (a_2, b_2) belongs to $\mathbb{C} \setminus \mathbb{R}$.

Let Ω_ζ be a domain in E_3.

A continuous mapping $\Phi : \Omega_\zeta \to \mathbb{H}(\mathbb{C})$ (or $\widehat{\Phi} : \Omega_\zeta \to \mathbb{H}(\mathbb{C})$) is called *right-G-monogenic* (or *left-G-monogenic*) in the domain $\Omega_\zeta \subset E_3$ if Φ (or $\widehat{\Phi}$) is differentiable in the sense of Gâteaux at every point of Ω_ζ, i.e. for every $\zeta \in \Omega_\zeta$ there exists the element $\Phi'(\zeta) \in \mathbb{H}(\mathbb{C})$ $\left(\text{or } \widehat{\Phi}'(\zeta) \in \mathbb{H}(\mathbb{C})\right)$ such that

$$\lim_{\varepsilon \to 0+0} \frac{\Phi(\zeta + \varepsilon h) - \Phi(\zeta)}{\varepsilon} = h \Phi'(\zeta) \quad \forall h \in E_3 \tag{8}$$

$$\left(\text{or} \quad \lim_{\varepsilon \to 0+0} \frac{\widehat{\Phi}(\zeta + \varepsilon h) - \widehat{\Phi}(\zeta)}{\varepsilon} = \widehat{\Phi}'(\zeta)h \quad \forall h \in E_3\right),$$

where $\Phi'(\zeta)$ is *the right Gâteaux derivative* of the mapping Φ and $\widehat{\Phi}'(\zeta)$ is *the left Gâteaux derivative* of the mapping $\widehat{\Phi}$ at the point ζ.

Theorem 3.1 *A mapping* $\Phi : \Omega_\zeta \to \mathbb{H}(\mathbb{C})$ *of the form*

$$\Phi(\zeta) = \sum_{k=1}^{4} U_k(x, y, z)e_k, \quad x, y, z \in \mathbb{R}, \tag{9}$$

where $U_k : \Omega \to \mathbb{C}$ *are differentiable functions in the domain* Ω, *is left-G-monogenic or right-G-monogenic in the domain* $\Omega_\zeta \subset E_3$ *if and only if the conditions*

$$\begin{aligned}
&\frac{\partial U_1}{\partial y} = a_1 \frac{\partial U_1}{\partial x}, \quad \frac{\partial U_2}{\partial y} = a_2 \frac{\partial U_2}{\partial x}, \quad \frac{\partial U_3}{\partial y} = a_2 \frac{\partial U_3}{\partial x}, \quad \frac{\partial U_4}{\partial y} = a_1 \frac{\partial U_4}{\partial x}, \\
&\frac{\partial U_1}{\partial z} = b_1 \frac{\partial U_1}{\partial x}, \quad \frac{\partial U_2}{\partial z} = b_2 \frac{\partial U_2}{\partial x}, \quad \frac{\partial U_3}{\partial z} = b_2 \frac{\partial U_3}{\partial x}, \quad \frac{\partial U_4}{\partial z} = b_1 \frac{\partial U_4}{\partial x}.
\end{aligned} \tag{10}$$

or

$$\begin{aligned}
&\frac{\partial U_1}{\partial y} = a_1 \frac{\partial U_1}{\partial x}, \quad \frac{\partial U_2}{\partial y} = a_2 \frac{\partial U_2}{\partial x}, \quad \frac{\partial U_3}{\partial y} = a_1 \frac{\partial U_3}{\partial x}, \quad \frac{\partial U_4}{\partial y} = a_2 \frac{\partial U_4}{\partial x}, \\
&\frac{\partial U_1}{\partial z} = b_1 \frac{\partial U_1}{\partial x}, \quad \frac{\partial U_2}{\partial z} = b_2 \frac{\partial U_2}{\partial x}, \quad \frac{\partial U_3}{\partial z} = b_1 \frac{\partial U_3}{\partial x}, \quad \frac{\partial U_4}{\partial z} = b_2 \frac{\partial U_4}{\partial x},
\end{aligned} \tag{11}$$

respectively, are satisfied.

Proof Necessity. If mapping (9) is right-G-monogenic in the domain Ω_ζ, then, for $h = i_1$, equality (8) has the form

$$\Phi'(\zeta) = \sum_{k=1}^{4} \frac{\partial U_k(x, y, z)}{\partial x} e_k, \quad \zeta = xi_1 + yi_2 + zi_3 \in \Omega_\zeta.$$

Setting in equality (8) first $h = i_2$ and then $h = i_3$ and using the rules of multiplication for basis elements, we get conditions (10) for the components of the right-G-monogenic mapping (9).

Sufficiency. Let $\zeta = xi_1 + yi_2 + zi_3 \in \Omega_\zeta$, $h := h_1 i_1 + h_2 i_2 + h_3 i_3$, where h_1, $h_2, h_3 \in \mathbb{R}$, and let ε be a positive number such that $\zeta + \varepsilon h \in \Omega_\zeta$. In view of conditions (10), we get

$$\left(\Phi(\zeta + \varepsilon h) - \Phi(\zeta)\right)\varepsilon^{-1} - h \sum_{k=1}^{4} \frac{\partial U_k(x, y, z)}{\partial x} e_k =$$

$$= \varepsilon^{-1} \sum_{k=1}^{4} \left(U_k(x + \varepsilon h_1, y + \varepsilon h_2, z + \varepsilon h_3) - U_k(x, y, z)\right)e_k -$$

$$- \left(\frac{\partial U_1}{\partial x} h_1 + a_1 \frac{\partial U_1}{\partial x} h_2 + b_1 \frac{\partial U_1}{\partial x} h_3\right) e_1 - \left(\frac{\partial U_2}{\partial x} h_1 + a_2 \frac{\partial U_2}{\partial x} h_2 + b_2 \frac{\partial U_2}{\partial x} h_3\right) e_2 -$$

$$- \left(\frac{\partial U_3}{\partial x} h_1 + a_1 \frac{\partial U_3}{\partial x} h_2 + b_1 \frac{\partial U_3}{\partial x} h_3\right) e_3 - \left(\frac{\partial U_4}{\partial x} h_1 + a_2 \frac{\partial U_4}{\partial x} h_2 + b_2 \frac{\partial U_4}{\partial x} h_3\right) e_4 =$$

$$= \varepsilon^{-1} \sum_{k=1}^{4} \left(U_k(x + \varepsilon h_1, y + \varepsilon h_2, z + \varepsilon h_3) - U_k(x, y, z) - \right.$$

$$\left. - \frac{\partial U_k(x, y, z)}{\partial x} \varepsilon h_1 - \frac{\partial U_k(x, y, z)}{\partial y} \varepsilon h_2 - \frac{\partial U_k(x, y, z)}{\partial z} \varepsilon h_3\right)e_k. \qquad (12)$$

By virtue of the differentiability of the functions U_k in the domain Ω, the relations

$$U_k(x + \varepsilon h_1, y + \varepsilon h_2, z + \varepsilon h_3) - U_k(x, y, z) - \frac{\partial U_k(x, y, z)}{\partial x} \varepsilon h_1 -$$

$$- \frac{\partial U_k(x, y, z)}{\partial y} \varepsilon h_2 - \frac{\partial U_k(x, y, z)}{\partial z} \varepsilon h_3 = o(\varepsilon), \quad \varepsilon \to 0, \quad k = \overline{1, 4}$$

are true. Passing to the limit as $\varepsilon \to 0$, in equality (12), we obtain equality (8). For the left-G-monogenic mapping, the proof is carried out in a similar way. □

Note that conditions (10) and (11) are analogs of the Cauchy–Riemann conditions. In a compact form, these conditions can be written as

$$\frac{\partial \Phi}{\partial y} = i_2 \frac{\partial \Phi}{\partial x}, \qquad \frac{\partial \Phi}{\partial z} = i_3 \frac{\partial \Phi}{\partial x} \qquad (13)$$

for a right-G-monogenic mapping and

$$\frac{\partial \widehat{\Phi}}{\partial y} = \frac{\partial \widehat{\Phi}}{\partial x} i_2, \qquad \frac{\partial \widehat{\Phi}}{\partial z} = \frac{\partial \widehat{\Phi}}{\partial x} i_3 \qquad (14)$$

for a left-G-monogenic mapping.

We now consider examples of right- and left-G-monogenic mappings. In view of the representation $\zeta = \xi_1 e_1 + \xi_2 e_2$ for an element ζ and the table of multiplication for the algebra $\mathbb{H}(\mathbb{C})$, we obtain $\zeta^n = \xi_1^n e_1 + \xi_2^n e_2$.

By using conditions (13) and (14), we readily verify that the mapping $\Phi(\zeta) = \zeta^n$ is simultaneously right- and left-G-monogenic in the entire space E_3. Similarly, we check that the mapping

$$\Phi(\zeta) = \sum_{k=0}^{n} \zeta^k c_k, \quad c_k \in \mathbb{H}(\mathbb{C}) \tag{15}$$

is right-G-monogenic in E_3 and the mapping

$$\widehat{\Phi}(\zeta) = \sum_{k=0}^{n} c_k \zeta^k, \quad c_k \in \mathbb{H}(\mathbb{C})$$

is left-G-monogenic in E_3.

4   A Constructive Description of G-Monogenic Mappings

In the next lemma we obtain an expansion of the resolvent $(t - \zeta)^{-1}$.

Lemma 2 *An expansion of the resolvent is of the form*

$$(t - \zeta)^{-1} = \frac{1}{t - \xi_1} e_1 + \frac{1}{t - \xi_2} e_2, \quad \forall\, t \in \mathbb{C} : t \neq \xi_1, \, t \neq \xi_2. \tag{16}$$

Proof We now find for which $t \in \mathbb{C}$ the element $(t - \zeta)^{-1}$ exist and determine the coefficients A_k of its decomposition in the basis:

$$(t - \zeta)^{-1} = \sum_{k=1}^{4} A_k e_k.$$

By using representation (6) for the elements i_1, i_2, and i_3 in the basis $\{e_1, e_2, e_3, e_4\}$ and the table of multiplication for the algebra $\mathbb{H}(\mathbb{C})$, we get

$$1 = (t - \zeta)(t - \zeta)^{-1} = \left((t - \xi_1)e_1 + (t - \xi_2)e_2 \right) \sum_{k=1}^{4} A_k e_k =$$

$$= (t - \xi_1)A_1 e_1 + (t - \xi_1)A_3 e_3 + (t - \xi_2)A_2 e_2 + (t - \xi_2)A_4 e_4 = e_1 + e_2.$$

Equating the coefficients of the corresponding basis units, we arrive at the decomposition (16). □

It follows from equality (16) that the points $(x, y, z) \in \mathbb{R}^3$, corresponding to the noninvertible elements

$$\zeta = xi_1 + yi_2 + zi_3 \in E_3$$

lie on the straight lines

$$L^1 : x + y \operatorname{Re} a_1 + z \operatorname{Re} b_1 = 0, \qquad y \operatorname{Im} a_1 + z \operatorname{Im} b_1 = 0,$$

$$L^2 : x + y \operatorname{Re} a_2 + z \operatorname{Re} b_2 = 0, \qquad y \operatorname{Im} a_2 + z \operatorname{Im} b_2 = 0$$

in $\mathbb{R}^3$.

A domain $\Omega_\zeta \subset E_3$ is called convex in the direction of the straight line L if it contains every segment parallel to the straight line L and which connect two points of this domain.

Similar to the proof of Lemma 1 Shpakivskyi and Kuzmenko (2016) the following statements can be proved.

Lemma 3 *Let a domain $\Omega \subset \mathbb{R}^3$ be convex in the direction of the straight lines L^1 and L^2, let $f_1(E_3) = f_2(E_3) = \mathbb{C}$, and let the mapping $\Phi : \Omega_\zeta \to \mathbb{H}(\mathbb{C})$ be right-G-monogenic in the domain Ω_ζ. If the points $\zeta_1, \zeta_2 \in \Omega_\zeta$ are such that*

$$\zeta_1 - \zeta_2 \in \{\zeta = xi_1 + yi_2 + zi_3 : (x, y, z) \in L^1\},$$

then

$$\Phi(\zeta_1) - \Phi(\zeta_2) \in \mathcal{I}_1 . \tag{17}$$

If points $\zeta_1, \zeta_2 \in \Omega_\zeta$ are such that

$$\zeta_1 - \zeta_2 \in \{\zeta = xi_1 + yi_2 + zi_3 : (x, y, z) \in L^2\},$$

then

$$\Phi(\zeta_1) - \Phi(\zeta_2) \in \mathcal{I}_2 . \tag{18}$$

Proof Let us prove relation (17). Let (x_1, y_1, z_1) and (x_2, y_2, z_2) be points of the domain Ω such that the segment that connects them is parallel to the straight line L^1.

In the domain Ω we construct two surfaces with common edge. Namely a surface Q that contains the point (x_1, y_1, z_1) and a surface Σ that contains the point (x_2, y_2, z_2), such that the restrictions of the functional f_1 to the corresponding subsets Q_ζ and Σ_ζ of the domain Ω_ζ are bijections of these subsets to the same domain G of the complex plane. Moreover, at every point $\zeta_0 \in Q_\zeta$ (or $\zeta_0 \in \Sigma_\zeta$) one has

$$\lim_{\varepsilon \to 0+0} (\Phi(\zeta_0 + \varepsilon(\zeta - \zeta_0)) - \Phi(\zeta_0)) \varepsilon^{-1} = \Phi'(\zeta_0)(\zeta - \zeta_0) \tag{19}$$

for all $\zeta \in Q_\zeta$ such that $\zeta_0 + \varepsilon(\zeta - \zeta_0) \in Q_\zeta$ for any $\varepsilon \in (0, 1)$ (or, respectively, for all $\zeta \in \Sigma_\zeta$ such that $\zeta_0 + \varepsilon(\zeta - \zeta_0) \in \Sigma_\zeta$ for any $\varepsilon \in (0, 1)$).

As the surface Q in the domain Ω we take a fixed equilateral triangle with vertices A_1, A_2, and A_3 centered at the point (x_1, y_1, z_1), the plane of which is perpendicular to the straight line L^1. We now continue the construction of the surface Σ.

Consider the triangle with vertices A'_1, A'_2, and A'_3 centered at the point (x_2, y_2, z_2), lying in the domain Ω, and such that its sides $A'_1 A'_2$, $A'_2 A'_3$, and $A'_1 A'_3$ are parallel to the segments $A_1 A_2$, $A_2 A_3$, and $A_1 A_3$ respectively, and have smaller lengths than the sides of the triangle $A_1 A_2 A_3$. Since the domain Ω is convex in the direction of the straight line L^1, we conclude that the prism with vertices A'_1, A'_2, A'_3, A''_1, A''_2, and A''_3 such that the points A''_1, A''_2, and A''_3 lie in the plane of the triangle $A_1 A_2 A_3$ and its edges $A'_m A''_m$, $m = \overline{1, 3}$ are parallel to the straight line L^1 is completely contained in Ω.

We now fix a triangle with vertices B_1, B_2, and B_3 such that the point B_m lies on the segment $A'_m A''_m$, $m = \overline{1, 3}$ and the truncated pyramid with vertices A_1, A_2, A_3, B_1, B_2, and B_3 and lateral edges $A_m B_m$, $m = \overline{1, 3}$, is completely contained in the domain Ω.

Finally, in the plane of the triangle $A'_1 A'_2 A'_3$, we fix a triangle T with vertices C_1, C_2, and C_3 such that its sides $C_1 C_2$, $C_2 C_3$, and $C_1 C_3$ are parallel to the segments $A'_1 A'_2$, $A'_2 A'_3$, and $A'_1 A'_3$ respectively, and have smaller lengths than the sides of the triangle $A'_1 A'_2 A'_3$. By construction, the truncated pyramid with vertices B_1, B_2, B_3, C_1, C_2, and C_3 and lateral edges $B_m C_m$, $m = \overline{1, 3}$, is completely contained in the domain Ω.

Let Σ denote the surface formed by the triangle T and the lateral surfaces of the truncated pyramids $A_1 A_2 A_3 B_1 B_2 B_3$ and $B_1 B_2 B_3 C_1 C_2 C_3$.

Since the surfaces Q and Σ have a common edge, the sets Q_ζ and Σ_ζ are mapped by the functional f_1 onto the same domain G of the complex plane. In the domain G we define two complex-valued functions H_1 and H_2 such that, for every $\xi_1 \in G$, one has

$$H_1(\xi_1) := f_1(\Phi(\zeta)), \quad \text{where } \xi_1 = f_1(\zeta) \text{ and } \zeta \in Q_\zeta,$$

$$H_2(\xi_1) := f_1(\Phi(\zeta)), \quad \text{where } \xi_1 = f_1(\zeta) \text{ and } \zeta \in \Sigma_\zeta.$$

Let us show that H_1 and H_2 are functions of the complex variable ξ_1 monogenic in G. Note that, acting by the functional f_1 on equality (19) and using the linearity, continuity, and right-multiplicativity of the functional, we get

$$\lim_{\varepsilon \to 0+0} \left(f_1(\Phi(\zeta_0 + \varepsilon(\zeta - \zeta_0))) - f_1(\Phi(\xi_1))\right) \varepsilon^{-1} = f_1(\Phi'(\zeta_0))(f_1(\zeta) - f_1(\zeta_0)).$$

This implies that the functions H_1 and H_2 have derivatives at the point $f_1(\zeta_0) \in G$ in all directions, and, furthermore, these derivatives are equal for each of the functions H_1 and H_2. Therefore, according to Theorem 21 in Trohimchuk (1964), the functions H_1, and H_2 are monogenic in the domain G.

According to the definition of the functions H_1 and H_2, we have $H_1(\xi_1) \equiv H_2(\xi_1)$ on the boundary of the domain G. By virtue of the monogeneity of the functions H_1 and H_2 in the domain G the identity $H_1(\xi) \equiv H_2(\xi)$ holds everywhere in G. Consequently, for $\zeta_1 := x_1 e_1 + y_1 e_2 + z_1 e_3$ and $\zeta_2 := x_2 e_1 + y_2 e_2 + z_2 e_3$ one has

$$f_1(\Phi(\zeta_2) - \Phi(\zeta_1)) = f_1(\Phi(\zeta_2)) - f_1(\Phi(\zeta_1)) = 0,$$

i.e. $\Phi(\zeta_2) - \Phi(\zeta_1)$ belongs to the kernel $\mathcal{I}_1$ of the functional f_1. $\square$

The proof of the next lemma is similar.

Lemma 4 *Let a domain $\Omega \subset \mathbb{R}^3$ be convex in the direction of the straight lines L^1 and L^2, let $f_1(E_3) = f_2(E_3) = \mathbb{C}$, and let the mapping $\widehat{\Phi} : \Omega_\zeta \to \mathbb{H}(\mathbb{C})$ be left-G-monogenic in the domain Ω_ζ. If points $\zeta_1, \zeta_2 \in \Omega_\zeta$ are such that*

$$\zeta_1 - \zeta_2 \in \{\zeta = x i_1 + y i_2 + z i_3 : (x, y, z) \in L^1\},$$

then

$$\widehat{\Phi}(\zeta_1) - \widehat{\Phi}(\zeta_2) \in \widehat{\mathcal{I}}_1 .$$

If points $\zeta_1, \zeta_2 \in \Omega_\zeta$ are such that

$$\zeta_1 - \zeta_2 \in \{\zeta = x i_1 + y i_2 + z i_3 : (x, y, z) \in L^2\},$$

then

$$\widehat{\Phi}(\zeta_1) - \widehat{\Phi}(\zeta_2) \in \widehat{\mathcal{I}}_2 .$$

Theorem 4.1 *Every right-G-monogenic mapping $\Phi : \Omega_\zeta \to \mathbb{H}(\mathbb{C})$ in the domain Ω_ζ can be expressed in the form*

$$\Phi(\zeta) = \Phi_1(\zeta) + \Phi_2(\zeta),$$

where $\Phi_1 : \Omega_\zeta \to \mathcal{I}_1$, $\Phi_2 : \Omega_\zeta \to \mathcal{I}_2$ are the some right-G-monogenic in the domain Ω_ζ mappings taking values in the right maximal ideals $\mathcal{I}_1$, $\mathcal{I}_2$.

Proof It follows from the decomposition of the unit $1 = e_1 + e_2$ that any mapping $\Phi : \Omega_\zeta \to \mathbb{H}(\mathbb{C})$ is expressed in the form

$$\Phi = e_1 \Phi + e_2 \Phi,$$

where $e_1 \Phi \in \mathcal{I}_2$ and $e_2 \Phi \in \mathcal{I}_1$.

We introduce the notation $\Phi_1 := e_2 \Phi$, $\Phi_2 := e_1 \Phi$ and show that the mappings Φ_1, Φ_2 are right-G-monogenic in the domain Ω_ζ. To this end, we multiply from the left equality (8) by e_1:

$$\lim_{\varepsilon \to 0+0} e_1 \frac{\Phi(\zeta + \varepsilon h) - \Phi(\zeta)}{\varepsilon} = e_1 h \Phi'(\zeta) \quad \forall h \in E_3. \tag{20}$$

Since the elements e_1 and h belong to the commutative subalgebra with the basis $\{e_1, e_2\}$, we have $e_1 h = h e_1$. Equality (20) yields the equality

$$\lim_{\varepsilon \to 0+0} \frac{e_1 \Phi(\zeta + \varepsilon h) - e_1 \Phi(\zeta)}{\varepsilon} = h e_1 \Phi'(\zeta),$$

which proves that the mapping Φ_2 is right-G-monogenic in the domain Ω_ζ. Similarly we prove that the mapping Φ_1 is also right-G-monogenic. $\qquad\square$

Theorem 4.2 *Every left-G-monogenic mapping $\widehat{\Phi} : \Omega_\zeta \to \mathbb{H}(\mathbb{C})$ in the domain Ω_ζ can be expressed in the form*

$$\widehat{\Phi}(\zeta) = \widehat{\Phi}_1(\zeta) + \widehat{\Phi}_2(\zeta), \tag{21}$$

where $\widehat{\Phi}_1 : \Omega_\zeta \to \widehat{\mathcal{I}}_1$, $\widehat{\Phi}_2 : \Omega_\zeta \to \widehat{\mathcal{I}}_2$ are some left-G-monogenic in the domain Ω_ζ mappings taking values in the left maximal ideals $\widehat{\mathcal{I}}_1$, $\widehat{\mathcal{I}}_2$.

The next theorem describes all right-G-monogenic mappings taking values in the ideals $\mathcal{I}_1$ and $\mathcal{I}_2$ using holomorphic functions of the corresponding complex variable.

By D_k we denote a domain of the complex plane $\mathbb{C}$ onto which the domain Ω_ζ is mapped by the functional f_k, $k = 1, 2$.

Theorem 4.3 *Suppose that a domain Ω is convex in the direction of the straight line L^2 and that $f_1(E_3) = f_2(E_3) = \mathbb{C}$. Then each right-$G$-monogenic mapping $\Phi_1 : \Omega_\zeta \to \mathcal{I}_1$ has the form*

$$\Phi_1(\zeta) = F_2(\xi_2)e_2 + F_4(\xi_2)e_4, \tag{22}$$

where F_2, F_4 are some holomorphic in the domain D_2 functions of the variable ξ_2, and every right-G-monogenic mapping $\Phi_2 : \Omega_\zeta \to \mathcal{I}_2$ taking values in the ideal $\mathcal{I}_2$ can be expressed in the form

$$\Phi_2(\zeta) = F_1(\xi_1)e_1 + F_3(\xi_1)e_3, \tag{23}$$

where F_1, F_3 are some holomorphic in the domain D_1 functions of the variable ξ_1.

Proof Since Φ_1 takes values in the ideal $\mathcal{I}_1$, the following equality is true:

$$\Phi_1(\zeta) = V_2(x, y, z)e_2 + V_4(x, y, z)e_4, \tag{24}$$

where $V_2 : \Omega \to \mathbb{C}$ and $V_4 : \Omega \to \mathbb{C}$.

The mapping Φ_1 satisfies the conditions of right-G-monogeneity (13) for $\Phi = \Phi_1$. Substituting relations (6) and (24) in these conditions, in view of the uniqueness of the decomposition of the elements of the algebra $\mathbb{H}(\mathbb{C})$ in the basis $\{e_1, e_2, e_3, e_4\}$, we arrive at the following system for the functions V_2 and V_4:

$$\frac{\partial V_2}{\partial y} = a_2 \frac{\partial V_2}{\partial x},$$

$$\frac{\partial V_4}{\partial y} = a_2 \frac{\partial V_4}{\partial x},$$

$$\frac{\partial V_2}{\partial z} = b_2 \frac{\partial V_2}{\partial x},$$ (25)

$$\frac{\partial V_4}{\partial z} = b_2 \frac{\partial V_4}{\partial x}.$$

We determine the function V_2 from the first and third equations of system (25). To this end, we separate the real and imaginary parts of the variable ξ_2 :

$$\xi_2 = (x + y \operatorname{Re} a_2 + z \operatorname{Re} b_2) + i(y \operatorname{Im} a_2 + z \operatorname{Im} b_2) := \tau_2 + i\eta_2$$

and note that these equations lead to the equalities

$$\frac{\partial V_2}{\partial \eta_2} \operatorname{Im} a_2 = i\frac{\partial V_2}{\partial \tau_2} \operatorname{Im} a_2, \quad \frac{\partial V_2}{\partial \eta_2} \operatorname{Im} b_2 = i\frac{\partial V_2}{\partial \tau_2} \operatorname{Im} b_2.$$ (26)

Since the equality $f_1(E_3) = f_2(E_3) = \mathbb{C}$ implies that at least one of the numbers $\operatorname{Im} a_2$ or $\operatorname{Im} b_2$ is nonzero, in view of (26), we get the relation

$$\frac{\partial V_2}{\partial \eta_2} = i\frac{\partial V_2}{\partial \tau_2}.$$

As in the proof of Theorem 2 in Plaksa and Shpakovskii (2011), by using Lemma 3 and Theorem 6 in Tolstov (1950), we prove the equality

$$V_2(x_1, y_1, z_1) = V_2(x_2, y_2, z_2)$$

for the points (x_1, y_1, z_1), $(x_2, y_2, z_2) \in \Omega$ such that the segment connecting these points is parallel to the straight line L^2. This implies that a function V_2 of the form

$$V_2(x, y, z) := F_1(\xi_2),$$

where F_1 is a certain analytic function in the domain D_2, is a general solution of the system consisting of the first and third equations of system (25).

By using the second and fourth equations of system (25), we establish, in a similar way, that the function V_4 has the form

$$V_4(x, y, z) := F_4(\xi_2),$$

where F_4 is a certain analytic function in the domain D_2. Equality (22) is proved in a similar way.   □

The following theorem, which is proved similarly to the proof of Theorem 4.3, describes all left-G-monogenic mappings taking values in the ideals $\widehat{\mathcal{I}}_1$ and $\widehat{\mathcal{I}}_2$ by means of holomorphic functions of the corresponding complex variable.

Theorem 4.4 *Assume that a domain Ω is convex in the direction of the straight line L^2 and that $f_1(E_3) = f_2(E_3) = \mathbb{C}$. Then every left-$G$-monogenic in the domain Ω_ζ mapping $\widehat{\Phi}_1 : \Omega_\zeta \to \widehat{\mathcal{I}}_1$ taking values in the ideal $\widehat{\mathcal{I}}_1$ can be expressed in the form*

$$\widehat{\Phi}_1(\zeta) = \widehat{F}_2(\xi_2)e_2 + \widehat{F}_3(\xi_2)e_3, \tag{27}$$

where $\widehat{F}_2, \widehat{F}_3$ are some holomorphic in the domain D_2 functions of the variable ξ_2, and every left-G-monogenic $\widehat{\Phi}_2 : \Omega_\zeta \to \widehat{\mathcal{I}}_2$ taking values in the ideal $\widehat{\mathcal{I}}_2$ can be expressed in the form

$$\widehat{\Phi}_2(\zeta) = \widehat{F}_1(\xi_1)e_1 + \widehat{F}_4(\xi_1)e_4, \tag{28}$$

where $\widehat{F}_1, \widehat{F}_4$ are some holomorphic in the domain D_1 functions of the variable ξ_1.

Using Theorems 4.1 and 4.3, we have the following statement.

Theorem 4.5 *Let a domain Ω be convex in the direction of the straight lines L^1 and L^2 and let $f_1(E_3) = f_2(E_3) = \mathbb{C}$. Then each right-$G$-monogenic mapping $\Phi : \Omega_\zeta \to \mathbb{H}(\mathbb{C})$ has the form*

$$\Phi(\zeta) = F_1(\xi_1)e_1 + F_2(\xi_2)e_2 + F_3(\xi_1)e_3 + F_4(\xi_2)e_4 \tag{29}$$

where F_1, F_3 are some holomorphic functions of the variable ξ_1 in the domain D_1 and F_2, F_4 are some holomorphic functions of the variable ξ_2 in the domain D_2.

It is clear that mapping (15) is right-G-monogenic in E_3 because, for this mapping, the functions F_1, F_2, F_3, and F_4 are polynomials. Moreover, in the corresponding domain, not only a polynomial of the form (15) is a right-G-monogenic mapping but also the series

$$\Phi(\zeta) = \sum_{k=0}^{\infty} \zeta^k c_k, \qquad c_k \in \mathbb{H}(\mathbb{C}), \tag{30}$$

for which complex power series playing the role of analytic functions F_1, F_2, F_3, and F_4 are convergent.

Similarly, using Theorems 4.2 and 4.4, we obtain the following statement, which describes all left-G-monogenic mappings.

Theorem 4.6 *Let a domain Ω be convex in the direction of the straight lines L^1 and L^2 and let $f_1(E_3) = f_2(E_3) = \mathbb{C}$. Then every left-$G$-monogenic mapping $\widehat{\Phi} : \Omega_\zeta \to \mathbb{H}(\mathbb{C})$ can be expressed in the form*

$$\widehat{\Phi}(\zeta) = \widehat{F}_1(\xi_1)e_1 + \widehat{F}_2(\xi_2)e_2 + \widehat{F}_3(\xi_2)e_3 + \widehat{F}_4(\xi_1)e_4, \tag{31}$$

where $\widehat{F}_1$, $\widehat{F}_4$ *are some holomorphic functions of the variable* ξ_1 *in the domain* D_1 *and* $\widehat{F}_2$, $\widehat{F}_3$ *are some holomorphic functions of the variable* ξ_2 *in the domain* D_2.

By analogy with (30), the mapping

$$\widehat{\Phi}(\zeta) = \sum_{k=0}^{\infty} c_k \, \zeta^k, \qquad c_k \in \mathbb{H}(\mathbb{C}), \tag{32}$$

is left-G-monogenic.

Obviously, formula (29) makes it possible to explicitly construct all right-G-monogenic mappings and formula (31) indicates the way to construct any left-G-monogenic mapping by means of four holomorphic functions of a corresponding complex variable. Note that, in Flaut and Shpakivskyi (2015), the analytic functions of a complex variable were used to construct the so-called A_t-*hyperholomorphic functions* in any Cayley–Dickson algebra A_t over the field $\mathbb{R}$. Also we note, that in paper Shpakivskyi (2016) a constructive description of all G-monogenic functions taking values in an arbitrary commutative associative algebra is obtained.

Equating the right-hand sides of equalities (31) and (29), we arrive at the conclusion that the mapping $\Psi(\zeta)$ is simultaneously right- and left-G-monogenic if and only if it has the form

$$\Psi(\zeta) = F_1(\xi_1)e_1 + F_2(\xi_2)e_2 + c_3e_3 + c_4e_4 \,,$$

where c_3, $c_4 \in \mathbb{C}$. In the case where $c_3 = c_4 = 0$, we have a constructive description of all bicomplex holomorphic ($\mathbb{BC} - holomorphic$) functions (Luna-Elizarrarás et al. 2015). It is now obvious that the mapping $\Psi(\zeta) = \zeta^n = \xi_1^n \, e_1 + \xi_2^n \, e_2$ is simultaneously right- and left-G-monogenic in E_3.

Now using the decomposition (16) and the multiplication rules (4), we obtain the following integral representation of the right-G-monogenic mapping

$$\Phi(\zeta) = \frac{1}{2\pi i} \int_{\Gamma_1} (t - \zeta)^{-1}\Big(F_1(t)e_1 + F_3(t)e_3\Big)dt+$$

$$+ \frac{1}{2\pi i} \int_{\Gamma_2} (t - \zeta)^{-1}\Big(F_2(t)e_2 + F_4(t)e_4\Big)dt, \tag{33}$$

and the left-G-monogenic mapping

$$\widehat{\Phi}(\zeta) = \frac{1}{2\pi i} \int_{\Gamma_1} \Big(F_1(t)e_1 + F_4(t)e_4\Big)(t - \zeta)^{-1}dt+$$

$$+ \frac{1}{2\pi i} \int_{\Gamma_2} \Big(F_2(t)e_2 + F_3(t)e_3\Big)(t - \zeta)^{-1}dt, \tag{34}$$

where Γ_k is a closed Jordan rectifiable curve in D_k, which surrounds point ξ_k and does not contain point ξ_q, $k, q = 1, 2$, $k \neq q$.

Note also that the right Gâteaux derivative expressed by formula

$$\Phi'(\zeta) = F_1'(\xi_1)e_1 + F_2'(\xi_2)e_2 + F_3'(\xi_1)e_3 + F_4'(\xi_2)e_4 \tag{35}$$

and the left Gâteaux derivative expressed by formula

$$\widehat{\Phi}'(\zeta) = F_1'(\xi_1)e_1 + F_2'(\xi_2)e_2 + F_3'(\xi_2)e_3 + F_4'(\xi_1)e_4 .$$

The next statement follows directly from the equalities (29) and (31) whose right-hand sides are, respectively, right- and left-G-monogenic mappings in the domain

$$\Pi_\zeta := \{ \zeta \in E_3 : f_1(\zeta) \in D_1, \ f_2(\zeta) \in D_2 \}.$$

Theorem 4.7 *Let a domain Ω be convex in the direction of the straight lines L^1 and L^2 and let $f_1(E_3) = f_2(E_3) = \mathbb{C}$, let the mapping $\Phi : \Omega_\zeta \to \mathbb{H}(\mathbb{C})$ be right-G-monogenic, and let the mapping $\widehat{\Phi} : \Omega_\zeta \to \mathbb{H}(\mathbb{C})$ be left-G-monogenic in the domain Ω_ζ. Then Φ and $\widehat{\Phi}$ can be extended to right- and left-G-monogenic mappings in the domain Π_ζ, respectively.*

The following statement is a fundamental consequence of equalities (29) and (31), which is true for an arbitrary domain Ω_ζ.

Theorem 4.8 *Let $f_1(E_3) = f_2(E_3) = \mathbb{C}$, let a mapping $\Phi : \Omega_\zeta \to \mathbb{H}(\mathbb{C})$ be right-G-monogenic, and let a mapping $\widehat{\Phi} : \Omega_\zeta \to \mathbb{H}(\mathbb{C})$ be left-G-monogenic in the domain Ω_ζ. Then the Gâteaux s-th derivative $\Phi^{(s)}$ is right-G-monogenic and $\widehat{\Phi}^{(s)}$ is left-G-monogenic mapping in the domain Ω_ζ for all s.*

Proof Since the ball $\Theta \subset \Omega$ with the center at the point $(x_0, y_0, z_0) \in \Omega$ is a convex set in the direction of the straight lines L^1 and L^2, in the neighborhood $\Theta_\zeta := \{\zeta = xi_1 + yi_2 + zi_3 : (x, y, z) \in \Theta\}$ of the point $\zeta_0 = x_0 i_1 + y_0 i_2 + z_0 i_3$ the equalities (29) and (35) are true. In the same time the components of the decomposition (35) are holomorphic functions of the corresponding complex variable, it means that the expression for $\Phi'(\zeta)$ has the form (29) and $\Phi'(\zeta)$ is right-G-monogenic mapping.

The statement for the left-G-monogenic mappings is proved completely analogously. $\qquad\square$

Using the integral expression (33) of the right-G-monogenic mapping $\Phi : \Omega_\zeta \to \mathbb{H}(\mathbb{C})$ in the case where the domain Ω_ζ is a convex set in the direction of the straight lines L^1 and L^2, we obtain the following expression for the right Gâteaux s-th derivative $\Phi^{(s)}$:

$$\Phi^{(s)}(\zeta) = \frac{s!}{2\pi i} \int_{\Gamma_1} \left((t-\zeta)^{-1}\right)^{s+1} \left(F_1(t)e_1 + F_3(t)e_3\right) dt +$$

$$+ \frac{s!}{2\pi i} \int_{\Gamma_2} \left((t-\zeta)^{-1}\right)^{s+1} \left(F_2(t)e_2 + F_4(t)e_4\right) dt.$$

In the same way we obtain the left Gâteaux s-th derivative $\widehat{\Phi}^{(s)}$ of the left-G-monogenic mapping $\widehat{\Phi} : \Omega_\zeta \to \mathbb{H}(\mathbb{C})$:

$$\widehat{\Phi}^{(s)}(\zeta) = \frac{s!}{2\pi i} \int_{\Gamma_1} \left(F_1(t)e_1 + F_4(t)e_4\right) \left((t-\zeta)^{-1}\right)^{s+1} dt +$$

$$+ \frac{s!}{2\pi i} \int_{\Gamma_2} \left(F_2(t)e_2 + F_3(t)e_3\right) \left((t-\zeta)^{-1}\right)^{s+1} dt.$$

5 Relations Between G-Monogenic Mappings and Partial Differential Equations

Consider a linear partial differential equation with constant coefficients:

$$\mathcal{L}_n U(x, y, z) := \sum_{\alpha+\beta+\gamma=n} C_{\alpha,\beta,\gamma} \frac{\partial^n U}{\partial x^\alpha \partial y^\beta \partial z^\gamma} = 0, \qquad C_{\alpha,\beta,\gamma} \in \mathbb{R}. \tag{36}$$

If the mapping Φ is n-times Gâteaux right-differentiable and the mapping $\widehat{\Phi}$ is n-times Gâteaux left-differentiable at every point of Ω_ζ, then

$$\frac{\partial^{\alpha+\beta+\gamma}\Phi}{\partial x^\alpha \partial y^\beta \partial z^\gamma} = i_1^\alpha i_2^\beta i_3^\gamma \, \Phi^{(\alpha+\beta+\gamma)}(\zeta) = i_2^\beta i_3^\gamma \, \Phi^{(n)}(\zeta)$$

and

$$\frac{\partial^{\alpha+\beta+\gamma}\widehat{\Phi}}{\partial x^\alpha \partial y^\beta \partial z^\gamma} = \widehat{\Phi}^{(\alpha+\beta+\gamma)}(\zeta) \, i_1^\alpha i_2^\beta i_3^\gamma = \widehat{\Phi}^{(n)}(\zeta) \, i_2^\beta i_3^\gamma.$$

Therefore, due to the equality

$$\mathcal{L}_n \Phi(\zeta) = \sum_{\alpha+\beta+\gamma=n} C_{\alpha,\beta,\gamma} \, i_2^\beta i_3^\gamma \, \Phi^{(n)}(\zeta) \tag{37}$$

every n-times Gâteaux right-differentiable mapping Φ, under the condition $\Phi^{(n)}(\zeta) \neq 0$ and

$$\sum_{\alpha+\beta+\gamma=n} C_{\alpha,\beta,\gamma}\, i_2^\beta\, i_3^\gamma = 0 \tag{38}$$

satisfies the equation

$$\mathcal{L}_n \Phi(\zeta) = 0.$$

Similarly, by virtue of the equality

$$\mathcal{L}_n \widehat{\Phi}(\zeta) = \widehat{\Phi}^{(n)}(\zeta) \sum_{\alpha+\beta+\gamma=n} C_{\alpha,\beta,\gamma}\, i_2^\beta\, i_3^\gamma \tag{39}$$

every n-times Gâteaux left-differentiable mapping $\widehat{\Phi}$, under the condition $\Phi^{(n)}(\zeta) \neq 0$ and the equality (38), satisfies the equation

$$\mathcal{L}_n \widehat{\Phi}(\zeta) = 0.$$

Accordingly, if the condition (38) is satisfied, then all real-valued components in the decompositions of the mappings Φ and $\widehat{\Phi}$ in the basis

$$\{e_1, e_2, e_3, e_4, ie_1, ie_2, ie_3, ie_4\}$$

are solutions of Eq. (36).

In the case where

$$f_1(E_3) = f_2(E_3) = \mathbb{C}, \tag{40}$$

it follows from Theorem 4.8 that the equalities (37) and (39) hold for every right-G-monogenic mapping $\Phi : \Omega_\zeta \to \mathbb{H}(\mathbb{C})$ and left-G-monogenic mapping $\widehat{\Phi} : \Omega_\zeta \to \mathbb{H}(\mathbb{C})$, respectively.

Thus, to construct solutions of Eq. (36) in the form of components of the right- or the left-G-monogenic mapping, we must find three linearly independent vectors (6) over $\mathbb{R}$ satisfying the characteristic Eq. (38) and verifying condition (40).

In the next theorem we assign a special class of Eq. (36) for which condition (40) is true. Let us introduce the polynomial

$$P(a, b) := \sum_{\alpha+\beta+\gamma=n} C_{\alpha,\beta,\gamma}\, a^\beta\, b^\gamma. \tag{41}$$

Theorem 5.1 *Suppose that the algebra $\mathbb{H}(\mathbb{C})$ contains a triple of linearly independent vectors over $\mathbb{R}$ of form (6) satisfying equality (38). If $P(a, b) \neq 0$ for all real a, b, then the relation (40) is true.*

Proof Using the multiplication table of $\mathbb{H}(\mathbb{C})$ we obtain the equalities

$$i_2^\beta = a_1^\beta e_1 + a_2^\beta e_2, \qquad i_3^\gamma = b_1^\gamma e_1 + b_2^\gamma e_2.$$

Now equality (38) takes the form

$$\sum_{\alpha+\beta+\gamma=n} C_{\alpha,\beta,\gamma} \left(a_1^\beta \, b_1^\gamma \, e_1 + a_2^\beta \, b_2^\gamma \, e_2\right) = 0.$$

Or, equivalently,

$$\sum_{\alpha+\beta+\gamma=n} C_{\alpha,\beta,\gamma} \, a_k^\beta \, b_k^\gamma = 0, \qquad k = 1, 2. \tag{42}$$

Since the solution of system (42) exists by the condition of the theorem and $P(a, b) \neq 0$ for all real a and b, equalities (42) can be true only in the case where at least one element in each pair (a_1, b_1) and (a_2, b_2) belongs to $\mathbb{C} \setminus \mathbb{R}$. $\square$

Note that if $P(a, b) \neq 0$, then $C_{n,0,0} \neq 0$, because otherwise $P(a, b) = 0$ for $a = b = 0$. In addition, since the function $P(a, b)$ is continuous on $\mathbb{R}^2$, the condition $P(a, b) \neq 0$ has the following meaning: either $P(a, b) > 0$ or $P(a, b) < 0$ for all $a, b \in \mathbb{R}$.

It is clear also that the elliptic equation of the form (36) always satisfies the condition $P(a, b) \neq 0$ for all $a, b \in \mathbb{R}$. At the same time, there exist equations of the form (36) for which $P(a, b) > 0$ but these equations are not elliptic. For example, this is true for the equation

$$\frac{\partial^5 u}{\partial x^5} + \frac{\partial^5 u}{\partial x \partial y^2 \partial z^2} + \frac{\partial^5 u}{\partial x \partial z^4} = 0.$$

5.1 Example

We show now a relation between the G-monogenic mappings and the three-dimensional Laplace equation

$$\Delta_3 U(x, y, z) := \frac{\partial^2 U}{\partial x^2} + \frac{\partial^2 U}{\partial y^2} + \frac{\partial^2 U}{\partial z^2} = 0. \tag{43}$$

The characteristic Eq. (38) for Eq. (43) has the form

$$1 + i_2^2 + i_3^2 = 0. \tag{44}$$

A triple of linearly independent vectors i_1, i_2, i_3 over $\mathbb{R}$ is called *a harmonic triple*, if the equality (44) is true and the conditions $i_2^2 \neq 0$, $i_3^2 \neq 0$ are satisfied (see, e.g., Ketchum 1928).

Substituting equalities (6) into conditions (44), we obtain the following

Proposition 5.1 *A harmonic triple in the algebra* $\mathbb{H}(\mathbb{C})$ *are vectors, which are decomposed with respect to the basis* $\{e_1, e_2\}$ *in the form (6) and complex numbers satisfy the system of the equations*

$$1 + a_1^2 + a_2^2 = 0, \qquad 1 + b_1^2 + b_2^2 = 0. \tag{45}$$

In particular, the system (45) is satisfied by the expressions

$$a_1 = i \sin t, \quad a_2 = i \cos t, \quad b_1 = i \sin \tau, \quad b_2 = i \cos \tau$$

corresponding to the variables

$$\xi_1 = x + iy \sin t + iz \cos t, \quad \xi_2 = x + iy \sin \tau + iz \cos \tau, \qquad t, \tau \in \mathbb{C}. \tag{46}$$

Since for the Laplace equation

$$P(a, b) = 1 + a^2 + b^2 > 0,$$

it follows that the conditions of Theorem 5.1 are satisfied. It means that every G-monogenic mapping satisfies the Eq. (43). Mappings (29) and (31) for which ξ_1 and ξ_2 are given by equalities (46), define G-monogenic mappings in $\mathbb{H}(\mathbb{C})$ associated with Eq. (43). Hence, solutions of the Eq. (43) are the real and imaginary parts of the function

$$U(x, y, z) = F(x + iy \sin t + iz \cos t),$$

where $t \in \mathbb{C}$ and F is an arbitrary holomorphic function.

6 Cauchy Integral Theorem for a Surface Integral

Let Ω_ζ be a bounded domain in E_3. For a continuous mapping $\varphi : \Omega_\zeta \to \mathbb{H}(\mathbb{C})$ of the form

$$\varphi(\zeta) = \sum_{k=1}^{4} U_k(x, y, z)e_k + i \sum_{k=1}^{4} V_k(x, y, z)e_k,$$

where $(x, y, z) \in \Omega$ and $U_k : \Omega \to \mathbb{R}$, $V_k : \Omega \to \mathbb{R}$, we define *a volume integral* by the equality

$$\int_{\Omega_\zeta} \varphi(\zeta) dx dy dz :=$$

$$\sum_{k=1}^{4} e_k \int_{\Omega} U_k(x, y, z) dx dy dz + i \sum_{k=1}^{4} e_k \int_{\Omega} V_k(x, y, z) dx dy dz.$$

Let Σ_ζ be a piece-wise smooth surface in E_3. For a continuous mappings $\varphi : \Omega_\zeta \to \mathbb{H}(\mathbb{C})$ and $\psi : \Omega_\zeta \to \mathbb{H}(\mathbb{C})$ of the forms

$$\varphi(\zeta) = \sum_{k=1}^{4} U_k(x, y, z)e_k + i \sum_{k=1}^{4} V_k(x, y, z)e_k, \qquad (47)$$

$$\psi(\zeta) = \sum_{m=1}^{4} P_m(x, y, z)e_m + i \sum_{m=1}^{4} Q_m(x, y, z)e_m, \qquad (48)$$

where $(x, y, z) \in \Sigma$, $U_k : \Sigma \to \mathbb{R}$, $V_k : \Sigma \to \mathbb{R}$ and $P_m : \Sigma \to \mathbb{R}$, $Q_m : \Sigma \to \mathbb{R}$, we define *a surface integral* along a piece-wise smooth surface Σ_ζ with the differential form

$$\sigma := dydz + dzdxi_2 + dxdyi_3$$

by the equality

$$\int_{\Sigma_\zeta} \varphi(\zeta)\, \sigma\, \psi(\zeta) := \sum_{k,m=1}^{4} e_k e_m \int_{\Sigma} (U_k\, P_m - V_k\, Q_m)dydz$$

$$+ \sum_{k,m=1}^{4} e_k i_2 e_m \int_{\Sigma} (U_k\, P_m - V_k\, Q_m)dzdx$$

$$+ \sum_{k,m=1}^{4} e_k i_3 e_m \int_{\Sigma} (U_k\, P_m - V_k\, Q_m)dxdy$$

$$+ i \sum_{k,m=1}^{4} e_k e_m \int_{\Sigma} (V_k\, P_m + U_k\, Q_m)dydz$$

$$+ i \sum_{k,m=1}^{4} e_k i_2 e_m \int_{\Sigma} (V_k\, P_m + U_k\, Q_m)dzdx$$

$$+ i \sum_{k,m=1}^{4} e_k i_3 e_m \int_{\Sigma} (V_k\, P_m + U_k\, Q_m)dxdy.$$

If a domain $\Omega \subset \mathbb{R}^3$ has a closed piece-wise smooth boundary $\partial\Omega$ and a mappings $\varphi : \Omega_\zeta \to \mathbb{H}(\mathbb{C})$ and $\psi : \Omega_\zeta \to \mathbb{H}(\mathbb{C})$ are continuous together with the partial derivatives of the first order up to the boundary $\partial\Omega_\zeta$, then the following analog of *the Gauss—Ostrogradsky formula* is true:

$$\int_{\partial\Omega_\zeta} \varphi(\zeta)\,\sigma\,\psi(\zeta) =$$

$$\int_{\Omega_\zeta} \left(\frac{\partial\varphi}{\partial x}\psi + \varphi\frac{\partial\psi}{\partial x} + \frac{\partial\varphi}{\partial y}i_2\psi + \varphi i_3\frac{\partial\psi}{\partial y} + \frac{\partial\varphi}{\partial z}i_3\psi + \varphi i_3\frac{\partial\psi}{\partial z} \right) dxdydz. \quad (49)$$

Using equality (49) and conditions (13), (14), we obtain the following theorem.

Theorem 6.1 *Suppose that a domain Ω_ζ has a closed piece-wise smooth boundary $\partial\Omega_\zeta$. Suppose also that $\Phi : \Omega_\zeta \to \mathbb{H}(\mathbb{C})$ is a right-G-monogenic mapping, $\widehat{\Phi} : \Omega_\zeta \to \mathbb{H}(\mathbb{C})$ is a left-G-monogenic in the domain Ω_ζ and continuous together with partial derivatives of the first order up to the boundary $\partial\Omega_\zeta$. Then*

$$\int_{\partial\Omega_\zeta} \widehat{\Phi}(\zeta)\,\sigma\,\Phi(\zeta) =$$

$$\int_{\Omega_\zeta} \left[\widehat{\Phi}'(\zeta)(1 + i_2^2 + i_3^2)\Phi(\zeta) + \widehat{\Phi}(\zeta)(1 + i_2^2 + i_3^2)\Phi'(\zeta) \right] dxdydz. \quad (50)$$

Proof Using conditions (13) and (14) we have

$$\int_{\partial\Omega_\zeta} \widehat{\Phi}(\zeta)\,\sigma\,\Phi(\zeta) =$$

$$\int_{\Omega_\zeta} \left(\widehat{\Phi}'\,\Phi + \widehat{\Phi}\,\Phi' + \widehat{\Phi}'\,i_2^2\,\Phi + \widehat{\Phi}\,i_2^2\,\Phi' + \widehat{\Phi}'\,i_3^2\,\Phi + \widehat{\Phi}\,i_3^2\,\Phi' \right) dxdydz =$$

$$\int_{\Omega_\zeta} \left[(\widehat{\Phi}' + \widehat{\Phi}'\,i_2^2 + \widehat{\Phi}'\,i_3^2)\,\Phi + \widehat{\Phi}(\Phi' + i_2^2\,\Phi' + i_3^2\,\Phi') \right] dxdydz =$$

$$\int_{\Omega_\zeta} \left[\widehat{\Phi}'\,(1 + i_2^2 + i_3^2)\,\Phi + \widehat{\Phi}(1 + i_2^2 + i_3^2)\,\Phi' \right] dxdydz. \qquad \square$$

The following statement is a consequence of Theorem 6.1.

Theorem 6.2 *Under the conditions of Theorem 6.1 with the additional assumption $1 + i_2^2 + i_3^2 = 0$, (i.e., the mappings Φ and $\widehat{\Phi}$ are solutions of the three-dimensional Laplace equation), the equality (50) can be rewritten in the form*

$$\int_{\partial \Omega_\zeta} \widehat{\Phi}(\zeta)\, \sigma\, \Phi(\zeta) = 0.$$

7 Cauchy Integral Theorem for a Curvilinear Integral

Let γ_ζ be a Jordan rectifiable curve in E_3. For a continuous mappings $\varphi : \gamma_\zeta \to \mathbb{H}(\mathbb{C})$ and $\psi : \gamma_\zeta \to \mathbb{H}(\mathbb{C})$ of forms (47) and (48), respectively, where $(x, y, z) \in \Sigma$, $U_k : \Sigma \to \mathbb{R}$, $V_k : \Sigma \to \mathbb{R}$ and $P_m : \Sigma \to \mathbb{R}$, $Q_m : \Sigma \to \mathbb{R}$, we define *a curvilinear integral* along a Jordan rectifiable curve γ_ζ by the equality:

$$\int_{\gamma_\zeta} \varphi(\zeta)\, d\zeta\, \psi(\zeta) := \sum_{k,m=1}^{4} e_k e_m \int_\gamma (U_k\, P_m - V_k\, Q_m)\, dx$$

$$+ \sum_{k,m=1}^{4} e_k i_2 e_m \int_\gamma (U_k\, P_m - V_k\, Q_m)\, dy$$

$$+ \sum_{k,m=1}^{4} e_k i_3 e_m \int_\gamma (U_k\, P_m - V_k\, Q_m)\, dz$$

$$+ i \sum_{k,m=1}^{4} e_k e_m \int_\gamma (V_k\, P_m - U_k\, Q_m)\, dx$$

$$+ i \sum_{k,m=1}^{4} e_k i_2 e_m \int_\gamma (V_k\, P_m - U_k\, Q_m)\, dy$$

$$+ i \sum_{k,m=1}^{4} e_k i_3 e_m \int_\gamma (V_k\, P_m - U_k\, Q_m)\, dz,$$

where $d\zeta := dx + i_2 dy + i_3 dz$.

If mappings $\varphi : \Omega_\zeta \to \mathbb{H}(\mathbb{C})$ and $\psi : \Omega_\zeta \to \mathbb{H}(\mathbb{C})$ are continuous together with partial derivatives of the first order in Ω_ζ and Σ_ζ is an arbitrary piece-wise smooth surface in Ω_ζ with a rectifiable Jordan edge γ_ζ, then the following analogue of *the Stokes formula* is true:

$$\int_{\gamma_\zeta} \varphi(\zeta)\,d\zeta\,\psi(\zeta) = \int_{\Sigma_\zeta} \left(\frac{\partial\varphi}{\partial x} i_2 \psi + \varphi i_2 \frac{\partial\psi}{\partial x} - \frac{\partial\varphi}{\partial y}\psi - \varphi\frac{\partial\psi}{\partial y} \right) dx\,dy +$$

$$\left(\frac{\partial\varphi}{\partial y} i_3 \psi + \varphi i_3 \frac{\partial\psi}{\partial y} - \frac{\partial\varphi}{\partial z} i_2 \psi - \varphi i_2 \frac{\partial\psi}{\partial z} \right) dy\,dz +$$

$$\left(\frac{\partial\varphi}{\partial z} \psi + \varphi\frac{\partial\psi}{\partial z} - \frac{\partial\varphi}{\partial x} i_3 \psi - \varphi i_3 \frac{\partial\psi}{\partial x} \right) dz\,dx. \tag{51}$$

In the next theorem we show that the right-hand side of equality (51) equals zero for a right-G-monogenic mapping $\Phi : \Omega_\zeta \to \mathbb{H}(\mathbb{C})$ and a left-G-monogenic mapping $\widehat{\Phi} : \Omega_\zeta \to \mathbb{H}(\mathbb{C})$. Note that this theorem is a generalization of Theorem 1 of Shpakivskyi and Kuzmenko (2016).

Theorem 7.1 *Suppose that $\Phi : \Omega_\zeta \to \mathbb{H}(\mathbb{C})$ is a right-G-monogenic mapping and $\widehat{\Phi} : \Omega_\zeta \to \mathbb{H}(\mathbb{C})$ is a left-G-monogenic mapping in Ω_ζ, and γ_ζ is a rectifiable Jordan edge of some piece-wise smooth surface in Ω_ζ. Then*

$$\int_{\gamma_\zeta} \widehat{\Phi}(\zeta)\,d\zeta\,\Phi(\zeta) = 0. \tag{52}$$

Proof Using formula (51) and conditions (13) and (14) we obtain

$$\int_{\gamma_\zeta} \widehat{\Phi}(\zeta)\,d\zeta\,\Phi(\zeta) = \int_{\Sigma_\zeta} \left(\frac{\partial\widehat{\Phi}}{\partial x} i_2 \Phi + \widehat{\Phi} i_2 \frac{\partial\Phi}{\partial x} - \frac{\partial\widehat{\Phi}}{\partial y}\Phi - \widehat{\Phi}\frac{\partial\Phi}{\partial y} \right) dx\,dy +$$

$$\left(\frac{\partial\widehat{\Phi}}{\partial y} i_3 \Phi + \widehat{\Phi} i_3 \frac{\partial\Phi}{\partial y} - \frac{\partial\widehat{\Phi}}{\partial z} i_2 \Phi - \widehat{\Phi} i_2 \frac{\partial\Phi}{\partial z} \right) dy\,dz +$$

$$\left(\frac{\partial\widehat{\Phi}}{\partial z} \Phi + \widehat{\Phi}\frac{\partial\Phi}{\partial z} - \frac{\partial\widehat{\Phi}}{\partial x} i_3 \Phi - \widehat{\Phi} i_3 \frac{\partial\Phi}{\partial x} \right) dz\,dx =$$

$$\int_{\Sigma_\zeta} \left(\widehat{\Phi}'(\zeta) i_2 \Phi(\zeta) + \widehat{\Phi}(\zeta) i_2 \Phi'(\zeta) - \widehat{\Phi}'(\zeta) i_2 \Phi(\zeta) - \widehat{\Phi}(\zeta) i_2 \Phi'(\zeta) \right) dx\,dy +$$

$$\left(\widehat{\Phi}'(\zeta) i_2 i_3 \Phi(\zeta) + \widehat{\Phi}(\zeta) i_3 i_2 \Phi'(\zeta) - \widehat{\Phi}'(\zeta) i_3 i_2 \Phi(\zeta) - \widehat{\Phi}(\zeta) i_2 i_3 \Phi'(\zeta) \right) dy\,dz +$$

$$\left(\widehat{\Phi}'(\zeta) i_3 \Phi(\zeta) + \widehat{\Phi}(\zeta) i_3 \Phi'(\zeta) - \widehat{\Phi}'(\zeta) i_3 \Phi(\zeta) - \widehat{\Phi}(\zeta) i_3 \Phi'(\zeta) \right) dz\,dx = 0.$$

$\square$

Denote by $\partial \triangle_\zeta$ the boundary of the triangle $\triangle_\zeta$ in relative topology of its plane.

Since every triangle $\triangle_\zeta \subset \Omega_\zeta$ can be included into a convex subset of a domain Ω_ζ, the following statement is a consequence of Theorem 7.1.

Corollary 1 *If $\Omega_\zeta \subset E_3$ is a convex domain, a mapping $\Phi : \Omega_\zeta \to \mathbb{H}(\mathbb{C})$ is right-G-monogenic and a mapping $\widehat{\Phi} : \Omega_\zeta \to \mathbb{H}(\mathbb{C})$ is left-G-monogenic, then for an arbitrary triangle $\triangle_\zeta$ such that $\overline{\triangle_\zeta} \subset \Omega_\zeta$, the following equality is true:*

$$\int_{\triangle_\zeta} \widehat{\Phi}(\zeta) \, d\zeta \, \Phi(\zeta) = 0. \tag{53}$$

Let us consider the algebra $\widetilde{\mathbb{H}}(\mathbb{R})$ with the basis $\{e_k, ie_k\}_{k=1}^4$ over $\mathbb{R}$. In the algebra $\widetilde{\mathbb{H}}(\mathbb{R})$ there exist another basis $\{i_k\}_{k=1}^8$, where the vectors i_1, i_2, i_3 are the same as in equalities (6).

For an element $a := \sum_{k=1}^8 a_k i_k$, $a_k \in \mathbb{R}$, we define the Euclidean norm

$$\|a\| := \sqrt{\sum_{k=1}^8 a_k^2}\,.$$

Accordingly, $\|\zeta\| = \sqrt{x^2 + y^2 + z^2}$ and $\|i_1\| = \|i_2\| = \|i_3\| = 1$.

By virtue of the theorem of equivalence of norms, the inequalities

$$|b_{1k} + ib_{2k}| \le \sqrt{\sum_{k=1}^4 (b_{1k}^2 + b_{2k}^2)} \le c\|b\|, \tag{54}$$

where c is a positive constant independent of b, hold for an arbitrary element $b := \sum_{k=1}^4 (b_{1k} + ib_{2k})e_k$, $b_{1k}, b_{2k} \in \mathbb{R}$.

Lemma 5 *If γ is a closed Jordan rectifiable curve in $\mathbb{R}^3$ and mappings $\phi, \psi : \gamma_\zeta \to \mathbb{H}(\mathbb{C})$ are continuous, then*

$$\left\| \int_{\gamma_\zeta} \varphi(\zeta) \, d\zeta \, \psi(\zeta) \right\| \le c \int_{\gamma_\zeta} \|\varphi(\zeta)\| \|d\zeta\| \|\psi(\zeta)\|, \tag{55}$$

where c is a positive absolute constant.

Proof Using the representation of function φ and ψ in forms (47) and (48) for $(x, y, z) \in \gamma$, we obtain

$$\left\| \int_{\gamma_\zeta} \varphi(\zeta)\, d\zeta\, \psi(\zeta) \right\| \le \sum_{q,r=1}^{4} \| e_q e_r \| \int_{\gamma} |U_q + i V_q| \cdot |P_r + i Q_r|\, dx$$

$$+ \sum_{q,r=1}^{4} \| e_q\, i_2\, e_r \| \int_{\gamma} |U_q + i V_q| \cdot |P_r + i Q_r|\, dy +$$

$$+ \sum_{q,r=1}^{4} \| e_q\, i_3\, e_r \| \int_{\gamma} |U_q + i V_q| \cdot |P_r + i Q_r|\, dz.$$

Now, taking into account inequality (54) and the inequalities $\| e_q\, i_k\, e_r \| \le c_u$, $k = 1, 2, 3$, where c_k are positive absolute constants, we obtain relation (55). □

Now we apply a scheme of the proof of the corresponding lemma for a function given in the complex plane (see, e.g., Privalov 1977) to the proof of the following statement.

Lemma 6 *Suppose that $\varphi : \Omega_\zeta \to \mathbb{H}(\mathbb{C})$ and $\psi : \Omega_\zeta \to \mathbb{H}(\mathbb{C})$ are continuous mappings in a simply connected domain Ω_ζ, and γ_ζ is a rectifiable curve in Ω_ζ. Then for an arbitrary $\varepsilon > 0$ there exists a broken line $\Lambda_\zeta \subset \Omega_\zeta$, vertices of which lie on the curve γ_ζ, such that*

$$\left\| \int_{\gamma_\zeta} \varphi(\zeta)\, d\zeta\, \psi(\zeta) - \int_{\Lambda_\zeta} \varphi(\zeta)\, d\zeta\, \psi(\zeta) \right\| < \varepsilon. \tag{56}$$

Proof Let us consider a closed domain $\overline{\Delta}_\zeta \subset \Omega_\zeta$, containing a curve γ_ζ in its interior. Since φ and ψ are continuous in $\overline{\Delta}_\zeta$, then it is uniformly continuous in this domain. It means that the product of these mappings is uniformly continuous too. Thus, for an arbitrary $\varepsilon_1 > 0$ there exists a number $\delta(\varepsilon) > 0$ such that

$$\| \varphi(\zeta')\, \psi(\zeta') - \varphi(\zeta'')\, \psi(\zeta'') \| < \varepsilon_1, \tag{57}$$

if $|\zeta' - \zeta''| < \delta(\varepsilon)$, where ζ', ζ'' are any points of the domain $\overline{\Delta}_\zeta$.

In addition, under the same assumptions, the following inequalities are true:

$$\| \varphi(\zeta')\, i_2\, \psi(\zeta') - \varphi(\zeta'')\, i_2\, \psi(\zeta'') \| < \varepsilon_2, \tag{58}$$

$$\| \varphi(\zeta')\, i_3\, \psi(\zeta') - \varphi(\zeta'')\, i_3\, \psi(\zeta'') \| < \varepsilon_3. \tag{59}$$

Let us divide the curve γ_ζ into n arcs $Q_\zeta^0, Q_\zeta^1, \ldots, Q_\zeta^{n-1}$ so that the length of each of them where less than δ and inscribe the broken curve Λ_ζ in the curve γ_ζ so that their broken links $L_\zeta^0, L_\zeta^1, \ldots, L_\zeta^{n-1}$ connect these arcs. The vertices of the broken curve Λ_ζ are denoted by $\zeta_0, \zeta_1, \ldots, \zeta_{n-1}, \zeta_n$. Since the length of every arc Q_ζ^k is less than δ the distance between any two points on the same arc less than δ. The same is true for links L_ζ^k.

We compare the value of the integral along the curve γ_ζ with the value of the integral along the broken curve Λ_ζ. For do this we consider an integral sum, which is an approximate value of the integral $\int_{\gamma_\zeta} \varphi(\zeta)\,d\zeta\,\psi(\zeta)$:

$$S := \varphi(\zeta_0)\,\Delta\zeta_0\,\psi(\zeta_0) + \varphi(\zeta_1)\,\Delta\zeta_1\,\psi(\zeta_1) + \cdots + \varphi(\zeta_{n-1})\,\Delta\,\zeta_{n-1}\psi(\zeta_{n-1}). \quad (60)$$

Since $\Delta\zeta_k = \int_{\varrho_\zeta^k} d\zeta$ then equality (60) can be represented in the form

$$S := \int_{\varrho_\zeta^0} \varphi(\zeta_0)\,d\zeta\,\psi(\zeta_0) + \int_{\varrho_\zeta^1} \varphi(\zeta_1)\,d\zeta\,\psi(\zeta_1) + \cdots + \int_{\varrho_\zeta^{n-1}} \varphi(\zeta_{n-1})\,d\zeta\,\psi(\zeta_{n-1}).$$
$$(61)$$

On the other hand, the integral $\int_{\gamma_\zeta} \varphi(\zeta)\,d\zeta\,\Psi(\zeta)$ can be represented in the form of the sum of the integrals along the arcs ϱ_ζ^k:

$$\int_{\gamma_\zeta} \varphi(\zeta)\,d\zeta\,\psi(\zeta) = \int_{\varrho_\zeta^0} \varphi(\zeta)\,d\zeta\,\psi(\zeta)+$$

$$\int_{\varrho_\zeta^1} \varphi(\zeta)\,d\zeta\,\psi(\zeta) + \cdots + \int_{\varrho_\zeta^{n-1}} \varphi(\zeta)\,d\zeta\,\psi(\zeta). \quad (62)$$

Subtracting Eqs. (62) and (61) we get:

$$\int_{\gamma_\zeta} \varphi(\zeta)\,d\zeta\,\psi(\zeta) - S = \int_{\varrho_\zeta^0} \Big(\varphi(\zeta)\,d\zeta\,\psi(\zeta) - \varphi(\zeta_0)\,d\zeta\,\psi(\zeta_0)\Big)+$$

$$\int_{\varrho_\zeta^1} \Big(\varphi(\zeta)\,d\zeta\,\psi(\zeta) - \varphi(\zeta_1)\,d\zeta\,\psi(\zeta_1)\Big) + \cdots +$$

$$\int_{\varrho_\zeta^{n-1}} \Big(\varphi(\zeta)\,d\zeta\,\psi(\zeta) - \varphi(\zeta_{n-1})\,d\zeta\,\psi(\zeta_{n-1})\Big) =$$

$$\int_{\varrho^0} \Big(\varphi(\zeta)\,\psi(\zeta) - \varphi(\zeta_0)\,\psi(\zeta_0)\Big)dx + \int_{\varrho^0} \Big(\varphi(\zeta)\,i_2\,\psi(\zeta) - \varphi(\zeta_0)\,i_2\,\psi(\zeta_0)\Big)dy+$$

$$\int_{Q^0} \Big(\varphi(\zeta)\, i_3\, \psi(\zeta) - \varphi(\zeta_0)\, i_3\, \psi(\zeta_0)\Big) dz + \cdots + \int_{Q^{n-1}} \Big(\varphi(\zeta)\, \psi(\zeta) - \varphi(\zeta_0)\, \psi(\zeta_0)\Big) dx +$$

$$\int_{Q^{n-1}} \Big(\varphi(\zeta)\, i_2\, \psi(\zeta) - \varphi(\zeta_0)\, i_2\, \psi(\zeta_0)\Big) dy + \int_{Q^{n-1}} \Big(\varphi(\zeta)\, i_3\, \psi(\zeta) - \varphi(\zeta_0)\, i_3\, \psi(\zeta_0)\Big) dz.$$

Since on every arc Q^k_ζ inequalities (57)–(59) are true we obtain

$$\left\| \int_{\gamma_\zeta} \varphi(\zeta)\, d\zeta\, \psi(\zeta) - S \right\| < \big(\varepsilon_1\, Q^0_x + \varepsilon_2\, Q^0_y + \varepsilon_3\, Q^0_z\big) + \ldots$$

$$\ldots + \big(\varepsilon_1 \cdot Q^{n-1}_x + \varepsilon_2\, Q^{n-1}_y + \varepsilon_3\, Q^{n-1}_z\big) < \varepsilon\, Q^0 + \cdots + \varepsilon\, Q^{n-1} < \varepsilon\, L, \qquad (63)$$

where Q^j_x, Q^j_y, Q^j_z are the lengths of the projections of the arc Q^j onto the axes Ox, Oy, Oz respectively, $\varepsilon := \max\{\varepsilon_1, \varepsilon_2, \varepsilon_3\}$ and L is the length of the curve γ_ζ.
In the same way we estimate the difference $\int_{\Lambda_\zeta} \varphi(\zeta)\, d\zeta\, \psi(\zeta) - S$. Since $\Delta\zeta_k = \int_{L^k_\zeta} d\zeta$, equality (60) is represented in the form

$$S := \int_{L^0_\zeta} \varphi(\zeta_0)\, d\zeta\, \psi(\zeta_0) + \int_{L^1_\zeta} \varphi(\zeta_1)\, d\zeta\, \psi(\zeta_1) + \cdots + \int_{L^{n-1}_\zeta} \varphi(\zeta_{n-1})\, d\zeta\, \psi(\zeta_{n-1}).$$

$$(64)$$

On the other hand, the integral $\int_{\Lambda_\zeta} \varphi(\zeta)\, d\zeta\, \psi(\zeta)$ can be represented in the form of the sum of integrals along the links L^k_ζ:

$$\int_{\Lambda_\zeta} \varphi(\zeta)\, d\zeta\, \psi(\zeta) =$$

$$\int_{L^0_\zeta} \varphi(\zeta)\, d\zeta\, \psi(\zeta) + \int_{L^1_\zeta} \varphi(\zeta)\, d\zeta\, \psi(\zeta) + \cdots + \int_{L^{n-1}_\zeta} \varphi(\zeta)\, d\zeta\, \psi(\zeta). \qquad (65)$$

Subtracting Eqs. (65) and (64) we get:

$$\int_{\Lambda_\zeta} \varphi(\zeta)\, d\zeta\, \psi(\zeta) - S = \int_{L_\zeta^0} \Big(\varphi(\zeta)\, d\zeta\, \psi(\zeta) - \varphi(\zeta_0)\, d\zeta\, \psi(\zeta_0)\Big)+$$

$$\int_{L_\zeta^1} \Big(\varphi(\zeta)\, d\zeta\, \psi(\zeta) - \varphi(\zeta_1)\, d\zeta\, \psi(\zeta_1)\Big) + \cdots +$$

$$\int_{L_\zeta^{n-1}} \Big(\varphi(\zeta)\, d\zeta\, \psi(\zeta) - \varphi(\zeta_{n-1})\, d\zeta\, \psi(\zeta_{n-1})\Big) =$$

$$\int_{L^0} \Big(\varphi(\zeta)\, \psi(\zeta) - \varphi(\zeta_0)\, \psi(\zeta_0)\Big) dx + \int_{L^0} \Big(\varphi(\zeta)\, i_2\, \psi(\zeta) - \varphi(\zeta_0)\, i_2\, \psi(\zeta_0)\Big) dy+$$

$$\int_{L^0} \Big(\varphi(\zeta)\, i_3\, \psi(\zeta) - \varphi(\zeta_0)\, i_3\, \psi(\zeta_0)\Big) dz + \cdots + \int_{L^{n-1}} \Big(\varphi(\zeta)\, \psi(\zeta) - \varphi(\zeta_0)\, \psi(\zeta_0)\Big) dx+$$

$$\int_{L^{n-1}} \Big(\varphi(\zeta)\, i_2\, \psi(\zeta) - \varphi(\zeta_0)\, i_2\, \psi(\zeta_0)\Big) dy + \int_{L^{n-1}} \Big(\varphi(\zeta)\, i_3\, \psi(\zeta) - \varphi(\zeta_0)\, i_3\, \psi(\zeta_0)\Big) dz.$$

Since inequalities (57)–(59) are true on every link L_ζ^k, we have

$$\left\| \int_{\gamma_\zeta} \varphi(\zeta)\, d\zeta\, \psi(\zeta) - S \right\| < \left(\varepsilon_1\, L_x^0 + \varepsilon_2\, L_y^0 + \varepsilon_3\, L_z^0\right) + \ldots$$

$$\left(\varepsilon_1 \cdot L_x^{n-1} + \varepsilon_2\, L_y^{n-1} + \varepsilon_3\, L_z^{n-1}\right) < \varepsilon\, L^0 + \cdots + \varepsilon\, L^{n-1} < \varepsilon L, \qquad (66)$$

where L_x^j, L_y^j, L_z^j are lengths of the projections of the link L^j onto the axes Ox, Oy, Oz respectively, $\varepsilon := \max\{\varepsilon_1, \varepsilon_2, \varepsilon_3\}$ and L is the length of the curve γ_ζ.

Taking into account inequalities (63) and (66), we have

$$\left\| \int_{\gamma_\zeta} \varphi(\zeta)\, d\zeta\, \psi(\zeta) - \int_{\Lambda_\zeta} \varphi(\zeta)\, d\zeta\, \psi(\zeta) \right\| \le \left\| \int_{\gamma_\zeta} \varphi(\zeta)\, d\zeta\, \psi(\zeta) - S \right\|+$$

$$\left\| S - \int_{\Lambda_\zeta} \varphi(\zeta)\, d\zeta\, \psi(\zeta) \right\| < 2\varepsilon L. \qquad \square$$

Now, using Corollary 1 and Lemma 6, we prove the following analogue of the Cauchy theorem for an arbitrary rectifiable curve in a convex domain.

Theorem 7.2 *Suppose that* $\Phi : \Omega_\zeta \to \mathbb{H}(\mathbb{C})$ *is right-G-monogenic and* $\widehat{\Phi} : \Omega_\zeta \to \mathbb{H}(\mathbb{C})$ *is left-G-monogenic mappings in the convex domain* Ω_ζ. *Then for any closed rectifiable Jordan curve* $\gamma_\zeta \subset \Omega_\zeta$ *equality (52) is true.*

Proof Using Lemma 6 we inscribe the broken curve Λ_ζ into the curve γ_ζ such that inequality (56) hold. Then we divide the broken curve Λ_ζ by the diagonals into triangles. Since the domain Ω_ζ is convex, all obtained triangles are in Ω_ζ. By Corollary 1 the integral along every triangle equals zero. Then the integral along the broken curve equals zero too:

$$\int_{\Lambda_\zeta} \varphi(\zeta) \, d\zeta \, \psi(\zeta) = 0. \tag{67}$$

Now the consequence of equalities (56) and (67) is equality (52). $\qquad\square$

In the case where Ω_ζ is an arbitrary domain, similarly to the proof of Theorem 3.2 Blum (1955), we can prove the following statements.

Theorem 7.3 *Let* $\Phi : \Omega_\zeta \to \mathbb{H}(\mathbb{C})$ *be a right-G-monogenic mapping and* $\widehat{\Phi} : \Omega_\zeta \to \mathbb{H}(\mathbb{C})$ *be a left-G-monogenic mapping in* Ω_ζ. *Then for every closed Jordan rectifiable curve* γ_ζ *homotopic to zero in* Ω_ζ *equality (52) is true.*

Proof Let a curve γ_ζ be defined by the equality $\zeta = \phi(t)$, $0 \le t \le 1$, where $\phi(0) = \phi(1) = \zeta_0$, and let γ_ζ be homotopic to the point ζ_0. Then there exists a continuous on the square $Q := [0, 1] \times [0, 1]$ mapping $H(s, t)$ of two real variables s and t taking values in the domain Ω_ζ, such that

$$H(0, t) = \phi(t), \qquad H(1, t) \equiv \zeta_0 \quad \forall t \in [0, 1],$$

$$H(s, 0) = H(s, 1) = \zeta_0 \quad \forall s \in [0, 1].$$

Since the mapping H is continuous on a compact set Q, its image $K := \{H(s, t) : (s, t) \in Q\}$ is a compact set in Ω_ζ.

Let $\rho := \min_{\zeta' \in K, \, \zeta'' \in \partial\Omega_\zeta} ||\zeta' - \zeta''||$.

The mapping H is also uniformly continuous on the set Q. It means that there exists $\delta > 0$ such that

$$\forall (s, t), (s', t') : |s' - s| < \delta, \ |t' - t| < \delta \Rightarrow ||H(s', t') - H(s, t)|| < \frac{\rho}{2}. \tag{68}$$

Let us choose a set of numbers $0 = t_0 < t_1 < \ldots < t_n = 1$, which are satisfying the inequalities $t_j - t_{j-1} < \delta$, $j = 1, 2, \ldots, n$, and put $s_1 = t_1$. Let $\zeta_{0,j} := H(0, t_j)$, $\zeta_{1,j} := H(s_1, t_j)$ for $j = 1, 2, \ldots, n - 1$ and denote by L_ζ^j a segment, beginning at the point $\zeta_{0,j}$ and ending at the point $\zeta_{1,j}$. Also consider a curve $\gamma_\zeta^{[1]} := \{H(s_1, t) : 0 \le t \le 1\}$.

Denote by $\gamma_\zeta[\zeta_1, \zeta_2]$ the arc of a Jordan oriented curve γ_ζ, beginning at the point ζ_1 and ending at the point ζ_2.

Because of inequality (68) the arcs $\gamma_\zeta[\zeta_0, \zeta_{01}]$, $\gamma_\zeta^{[1]}[\zeta_0, \zeta_{11}]$ and the segment L_ζ^1 are contained in the ball $S(\zeta_0) := \{\zeta \in E_3 : ||\zeta - \zeta_0|| < \rho\}$. Since $S(\zeta_0)$ is a convex set and is contained in the domain Ω_ζ, the following equality is a consequence of Theorem 7.2:

$$\int_{\gamma_\zeta[\zeta_0, \zeta_{01}]} \widehat{\Phi}(\zeta)\, d\zeta\, \Phi(\zeta) + \int_{L_\zeta^1} \widehat{\Phi}(\zeta)\, d\zeta\, \Phi(\zeta) = \int_{\gamma_\zeta^{[1]}[\zeta_0, \zeta_{11}]} \widehat{\Phi}(\zeta)\, d\zeta\, \Phi(\zeta). \qquad (69)$$

Inequalities (68) imply

$$||\zeta - \zeta_{0,j}|| < \frac{\rho}{2} \qquad \forall \zeta \in \gamma_\zeta[\zeta_{0,j}, \zeta_{0,j+1}],$$

$$||\zeta - \zeta_{1,j}|| < \frac{\rho}{2} \qquad \forall \zeta \in \gamma_\zeta^{[1]}[\zeta_{1,j}, \zeta_{1,j+1}], \qquad ||\zeta_{1,j} - \zeta_{0,j}|| < \frac{\rho}{2}$$

for $j = 1, 2, \ldots, n - 2$. Then the arcs $\gamma_\zeta[\zeta_{0,j}, \zeta_{0,j+1}]$, $\gamma_\zeta^{[1]}[\zeta_{1,j}, \zeta_{1,j+1}]$ and the segments L_ζ^j, L_ζ^{j+1} are contained in the ball $S(\zeta_{0,j}) := \{\zeta \in E_3 : ||\zeta - \zeta_{0,j}|| < \rho\}$ for $j = 1, 2, \ldots, n - 2$. Since $S(\zeta_{0,j})$ is a convex set and is contained in Ω_ζ, Theorem 7.2 imply equalities

$$-\int_{L_\zeta^j} \widehat{\Phi}(\zeta)\, d\zeta\, \Phi(\zeta) + \int_{\gamma_\zeta[\zeta_{0,j}, \zeta_{0,j+1}]} \widehat{\Phi}(\zeta)\, d\zeta\, \Phi(\zeta) +$$

$$+ \int_{L_\zeta^{j+1}} \widehat{\Phi}(\zeta)\, d\zeta\, \Phi(\zeta) = \int_{\gamma_\zeta^{[1]}[\zeta_{1,j}, \zeta_{1,j+1}]} \widehat{\Phi}(\zeta)\, d\zeta\, \Phi(\zeta) \qquad (70)$$

for $j = 1, 2, \ldots, n - 2$.

Finally, similarly to equality (69) we obtain equality

$$-\int_{L_\zeta^{n-1}} \widehat{\Phi}(\zeta)\, d\zeta\, \Phi(\zeta) + \int_{\gamma_\zeta[\zeta_{0,n-1}, \zeta_0]} \widehat{\Phi}(\zeta)\, d\zeta\, \Phi(\zeta) = \int_{\gamma_\zeta^{[1]}[\zeta_{1,n-1}, \zeta_0]} \widehat{\Phi}(\zeta)\, d\zeta\, \Phi(\zeta). \qquad (71)$$

Adding all the equalities (69)–(71), we obtain the equality

$$\int_{\gamma_\zeta} \widehat{\Phi}(\zeta)\, d\zeta\, \Phi(\zeta) = \int_{\gamma_\zeta^{[1]}} \widehat{\Phi}(\zeta)\, d\zeta\, \Phi(\zeta). \qquad (72)$$

Then we put $s_j = t_j$ and consider a curve $\gamma_\zeta^{[j]} := \{H(s_j, t) : 0 \leq t \leq 1\}$ for $j = 1, 2, \ldots, n$. Similarly to equality (72), we obtain the equalities

$$\int_{\gamma_\zeta^{[1]}} \widehat{\Phi}(\zeta)\, d\zeta\, \Phi(\zeta) = \int_{\gamma_\zeta^{[2]}} \widehat{\Phi}(\zeta)\, d\zeta\, \Phi(\zeta) = \cdots = \int_{\gamma_\zeta^{[n]}} \widehat{\Phi}(\zeta)\, d\zeta\, \Phi(\zeta).$$

Hence, we have

$$\int_{\gamma_\zeta} \widehat{\Phi}(\zeta)\, d\zeta\, \Phi(\zeta) = \int_{\gamma_\zeta^{[n]}} \widehat{\Phi}(\zeta)\, d\zeta\, \Phi(\zeta),$$

where the curve $\gamma_\zeta^{[n]}$ degenerates to the point, since $H(1, t) \equiv \zeta_0$. Now, taking into account the equality

$$\int_{\gamma_\zeta^{[n]}} \widehat{\Phi}(\zeta)\, d\zeta\, \Phi(\zeta) = 0$$

we complete the proof of the theorem. $\qquad\square$

Now, similarly to the proof of Theorem 2 of Kuzmenko (2016) we can prove the curvilinear Cauchy integral theorem for G-monogenic mappings in the case where a curve of integration lies on the boundary of a domain of G-monogeneity.

Let on the boundary $\partial\Omega_\zeta$ of a domain Ω_ζ their be given a closed Jordan rectifiable curve $\gamma_\zeta \equiv \gamma_\zeta(t)$, where $0 \leq t \leq 1$, homotopic to an interior point $\zeta_0 \in \Omega_\zeta$. It means that there exists the mapping $H(s, t)$ continuous on the square $[0, 1] \times [0, 1]$, such that $H(0, t) = \gamma_\zeta(t)$, $H(1, t) \equiv \zeta_0$, and all curves $\gamma_\zeta^s \equiv \gamma_\zeta^s(t) := \{\zeta = H(s, t) : 0 \leq t \leq 1\}$ for $0 < s < 1$ are contained in the domain Ω_ζ.

Consider also the curves $\Gamma_\zeta^t \equiv \Gamma_\zeta^t(s) := \{\zeta = H(s, t) : 0 \leq s \leq 1\}$. Denote by $\Gamma[\zeta_1, \zeta_2]$ the arc of the Jordan oriented rectifiable curve, beginning at the point ζ_1 and ending at the point ζ_2, and denote by the mes a linear Lebesgue measure of a rectifiable curve.

Theorem 7.4 *Suppose that $\Phi : \overline{\Omega}_\zeta \to \mathbb{H}(\mathbb{C})$ and $\widehat{\Phi} : \overline{\Omega}_\zeta \to \mathbb{H}(\mathbb{C})$ are continuous mappings in the closure $\overline{\Omega}_\zeta$ of a domain Ω_ζ, Φ is right-G-monogenic and $\widehat{\Phi}$ is left-G-monogenic mapping in Ω_ζ. Suppose also that $\gamma_\zeta \subset \partial\Omega_\zeta$ is any closed Jordan rectifiable curve homotopic to a point $\zeta_0 \in \Omega_\zeta$ such that the curves of the family $\{\Gamma_\zeta^t : 0 \leq t \leq 1\}$ are rectifiable and the set $\{mes\, \gamma_\zeta^s : 0 \leq s \leq 1\}$ is bounded. Then equality (52) is true.*

8 The Morera Theorem

Denote by $s[\zeta_1, \zeta_2]$ the segment beginning at the point ζ_1 and ending at the point ζ_2.

Theorem 8.1 *Let* $f_1(E_3) = f_2(E_3) = \mathbb{C}$. *If a mapping* $\Phi : \Omega \to \mathbb{H}(\mathbb{C})$ *is continuous in a domain* Ω_ζ *and satisfies the equality*

$$\int_{\partial \Delta_\zeta} d\zeta \Phi(\zeta) = 0 \qquad (73)$$

for every triangle $\Delta_\zeta \subset \Omega_\zeta$, *such that the closure* $\overline{\Delta}_\zeta \subset \Omega_\zeta$, *then the mapping* Φ *is right-G-monogenic in the domain* Ω_ζ.

Proof Let us fix a point a in the domain Ω_ζ. Consider the mapping

$$\Psi(\zeta) := \int_{s[a,\zeta]} d\tau \, \Phi(\tau)$$

and show that it is right-G-monogenic in Ω_ζ, moreover

$$\Psi'(\zeta) = \Phi(\zeta). \qquad (74)$$

Let $h \in E_3$ and $\varepsilon > 0$ such that a triangle Δ_ζ with the vertices a, ζ, $\zeta + \varepsilon h$ is contained in the domain Ω_ζ.

Consider the difference

$$\Psi(\zeta + \varepsilon h) - \Psi(\zeta) = \int_{s[a,\zeta+\varepsilon h]} d\tau \, \Phi(\tau) - \int_{s[a,\zeta]} d\tau \, \Phi(\tau) =$$

$$= \int_{s[a,\zeta+\varepsilon h]} d\tau \, \Phi(\tau) + \int_{s[\zeta,a]} d\tau \, \Phi(\tau) + \int_{s[\zeta+\varepsilon h,\zeta]} d\tau \, \Phi(\tau) - \int_{s[\zeta+\varepsilon h,\zeta]} d\tau \, \Phi(\tau) =$$

$$= \int_{\partial \Delta_\zeta} d\tau \, \Phi(\tau) + \int_{s[\zeta,\zeta+\varepsilon h]} d\tau \, \Phi(\tau) = \int_{s[\zeta,\zeta+\varepsilon h]} d\tau \, \Phi(\tau). \qquad (75)$$

Now, using equality (75), inequality (55) and continuity of the mapping Φ, we obtain:

$$\left\| \frac{\Psi(\zeta + \varepsilon h) - \Psi(\zeta)}{\varepsilon} - h\Phi(\zeta) \right\| = \left\| \frac{\int_{s[\zeta,\zeta+\varepsilon h]} d\tau \, \Phi(\tau)}{\varepsilon} - h\Phi(\zeta) \right\| =$$

$$= \frac{1}{\varepsilon} \left\| \int_{s[\zeta,\zeta+\varepsilon h]} d\tau \Big(\Phi(\tau) - \Phi(\zeta) \Big) \right\| \leq \frac{c}{\varepsilon} \int_{s[\zeta,\zeta+\varepsilon h]} \| \Phi(\tau) - \Phi(\zeta) \| \, \| d\tau \| \leq$$

$$\leq \frac{c}{\varepsilon} \sup_{\tau,\zeta \in \Omega_\zeta,\, \|\tau-\zeta\|\leq\varepsilon} \|\Phi(\tau) - \Phi(\zeta)\| \int_{s[\zeta,\zeta+\varepsilon h]} \|d\tau\| \leq$$

$$\leq c\,\|h\| \sup_{\tau,\zeta \in \Omega_\zeta,\, \|\tau-\zeta\|\leq\varepsilon} \|\Phi(\tau) - \Phi(\zeta)\| \to 0, \qquad \varepsilon \to 0. \tag{76}$$

The relation (76) imply the equality

$$\lim_{\varepsilon\to 0+0} \frac{\Psi(\zeta + \varepsilon h) - \Psi(\zeta)}{\varepsilon} = h\Phi(\zeta),$$

the consequence of which is the equality (74).

Since in an arbitrary neighborhood of the point ζ the mapping Φ is the Gâteaux derivative of the right-G-monogenic mapping $\Psi : \Omega_\zeta \to \mathbb{H}(\mathbb{C})$, then using Theorem 4.8 the mapping Φ is right-G-monogenic in the domain Ω_ζ. $\qquad\square$

Theorem 8.2 *Let $f_1(E_3) = f_2(E_3) = \mathbb{C}$. If a mapping $\widehat{\Phi} : \Omega \to \mathbb{H}(\mathbb{C})$ is continuous in a domain Ω_ζ and satisfies the equality*

$$\int_{\partial\triangle_\zeta} \widehat{\Phi}(\zeta)d\zeta = 0 \tag{77}$$

for every triangle $\triangle_\zeta \subset \Omega_\zeta$, such that the closure $\overline{\triangle}_\zeta \subset \Omega_\zeta$, then the mapping $\widehat{\Phi}$ is left-G-monogenic in the domain Ω_ζ.

9 Cauchy Integral Formula for a Curvilinear Integral

To establish the Cauchy integral formula for a curvilinear integral, consider the following auxiliary statement.

Lemma 7 *Suppose that a domain $\Omega \subset \mathbb{R}^3$ is convex in the direction of the straight lines L^1, L^2 and $f_1(E_3) = f_2(E_3) = \mathbb{C}$. Suppose also that $\Phi : \Omega_\zeta \to \mathbb{H}(\mathbb{C})$ is a right-G-monogenic mapping in Ω_ζ, $\widehat{\Phi} : \Omega_\zeta \to \mathbb{H}(\mathbb{C})$ is a left-G-monogenic mapping in Ω_ζ, and γ_ζ is an arbitrary rectifiable curve in Ω_ζ. Then*

$$\int_{\gamma_\zeta} \widehat{\Phi}(\zeta)d\zeta\, \Phi(\zeta) =$$

$$= e_1 \int_{\gamma_1} \widehat{F}_1(\xi_1)F_1(\xi_1)d\xi_1 + \widehat{F}_3(\xi_2)F_4(\xi_2)d\xi_2$$

$$+e_2 \int_{\gamma_1} \widehat{F}_2(\xi_2) F_2(\xi_2) d\xi_2 + \widehat{F}_2(\xi_1) F_3(\xi_1) d\xi_1$$

$$+e_3 \int_{\gamma_1} \widehat{F}_1(\xi_1) F_3(\xi_1) d\xi_1 + \widehat{F}_3(\xi_2) F_2(\xi_2) d\xi_2$$

$$+e_4 \int_{\gamma_1} \widehat{F}_2(\xi_2) F_2(\xi_2) d\xi_2 + \widehat{F}_2(\xi_1) F_1(\xi_1) d\xi_1 . \tag{78}$$

Proof The equality (78) follows immediately from representations (29), (31), the equality $d\zeta = d\xi_1 e_1 + d\xi_2 e_2$ and multiplication rules (4).   □

Let $\zeta \in E_3$. By formula (16) the inverse element ζ^{-1} is of the following form:

$$\zeta^{-1} = \frac{1}{\xi_1} e_1 + \frac{1}{\xi_2} e_2 \tag{79}$$

and it exists if and only if $\xi_1 \neq 0$ and $\xi_2 \neq 0$.

Let $\zeta_0 = \xi_1^{(0)} e_1 + \xi_2^{(0)} e_2$ be a point in a domain $\Omega_\zeta \subset E_3$. In a neighborhood of ζ_0 contained in Ω_ζ let us take a circle $C(\zeta_0)$ centered in ζ_0. By $C_k \subset \mathbb{C}$ we denote the image of $C(\zeta_0)$ under the mapping f_k, $k = 1, 2$. We assume that the circle $C(\zeta_0)$ *embraces the set* $\{\zeta - \zeta_0 : \zeta \in L_\zeta^1 \cup L_\zeta^2\}$. It means that C_k bounds some domain D_k' and $\xi_k^{(0)} \in D_k'$, $k = 1, 2$.

We say that the curve $\gamma_\zeta \subset \Omega_\zeta$ *embraces one time the set* $\{\zeta - \zeta_0 : \zeta \in L_\zeta^1 \cup L_\zeta^2\}$, if there exists a circle $C(\zeta_0)$ which embraces the mentioned set and is homotopic to γ_ζ in the domain $\Omega_\zeta \setminus \{\zeta - \zeta_0 : \zeta \in L_\zeta^1 \cup L_\zeta^2\}$.

The following theorem is an analogue of the Cauchy integral formula for G-monogenic mappings.

Theorem 9.1 *Suppose that a domain* $\Omega \subset \mathbb{R}^3$ *is convex in the direction of the straight lines* L^1, L^2 *and* $f_1(E_3) = f_2(E_3) = \mathbb{C}$. *Suppose also that* $\Phi : \Omega_\zeta \to \mathbb{H}(\mathbb{C})$ *is a right-G-monogenic mapping in* Ω_ζ *and* $\widehat{\Phi} : \Omega_\zeta \to \mathbb{H}(\mathbb{C})$ *is a left-G-monogenic mapping in* Ω_ζ. *Then for every point* $\zeta_0 \in \Omega_\zeta$ *the following equality is true:*

$$\widehat{\Phi}(\zeta_0) \cdot \Phi(\zeta_0) = \frac{1}{2\pi i} \int_{\gamma_\zeta} \widehat{\Phi}(\zeta) \, (\zeta - \zeta_0)^{-1} d\zeta \, \Phi(\zeta), \tag{80}$$

where γ_ζ *is an arbitrary closed Jordan rectifiable curve in* Ω_ζ, *that embraces one time the set* $\{\zeta - \zeta_0 : \zeta \in L_\zeta^1 \cup L_\zeta^2\}$.

Proof Inasmuch as γ_ζ is homotopic to $C(\zeta_0)$ in the domain

$$\Omega_\zeta \setminus \{\zeta - \zeta_0 : \zeta \in L_\zeta^1 \cup L_\zeta^2\},$$

then it follows from Theorem 7.3 that

$$\frac{1}{2\pi i}\int_{\gamma_\zeta}\widehat{\Phi}(\zeta)\,(\zeta-\zeta_0)^{-1}d\zeta\,\Phi(\zeta)=\frac{1}{2\pi i}\int_{C(\zeta_0)}\widehat{\Phi}(\zeta)\,(\zeta-\zeta_0)^{-1}d\zeta\,\Phi(\zeta).$$

Now, using representations (79), (78) and the Cauchy integral formula for holomorphic functions F_n, we obtain the following equalities:

$$\frac{1}{2\pi i}\int_{C(\zeta_0)}\widehat{\Phi}(\zeta)\,(\zeta-\zeta_0)^{-1}d\zeta\,\Phi(\zeta)=$$

$$=e_1\left(\frac{1}{2\pi i}\int_{C_1}\frac{\widehat{F}_1(\xi_1)\,F_1(\xi_1)}{\xi_1-\xi_{10}}\,d\xi_1+\frac{1}{2\pi i}\int_{C_2}\frac{\widehat{F}_3(\xi_2)\,F_4(\xi_2)}{\xi_2-\xi_{20}}\,d\xi_2\right)$$

$$+e_2\left(\frac{1}{2\pi i}\int_{C_2}\frac{\widehat{F}_2(\xi_2)\,F_2(\xi_2)}{\xi_2-\xi_{20}}\,d\xi_2+\frac{1}{2\pi i}\int_{C_1}\frac{\widehat{F}_4(\xi_1)\,F_3(\xi_1)}{\xi_1-\xi_{10}}\,d\xi_1\right)$$

$$+e_3\left(\frac{1}{2\pi i}\int_{C_1}\frac{\widehat{F}_1(\xi_1)\,F_3(\xi_1)}{\xi_1-\xi_{10}}\,d\xi_1+\frac{1}{2\pi i}\int_{C_2}\frac{\widehat{F}_3(\xi_2)\,F_2(\xi_2)}{\xi_2-\xi_{20}}\,d\xi_2\right)$$

$$+e_4\left(\frac{1}{2\pi i}\int_{C_2}\frac{\widehat{F}_2(\xi_2)\,F_4(\xi_2)}{\xi_2-\xi_{20}}\,d\xi_2+\frac{1}{2\pi i}\int_{C_1}\frac{\widehat{F}_4(\xi_1)\,F_1(\xi_1)}{\xi_1-\xi_{10}}\,d\xi_1\right)=$$

$$=e_1\left(\widehat{F}_1(\xi_{10})\,F_1(\xi_{10})+\widehat{F}_3(\xi_{20})\,F_4(\xi_{20})\right)$$

$$+e_2\left(\widehat{F}_2(\xi_{20})\,F_2(\xi_{20})+\widehat{F}_4(\xi_{10})\,F_3(\xi_{10})\right)$$

$$+e_3\left(\widehat{F}_1(\xi_{10})\,F_3(\xi_{10})+\widehat{F}_3(\xi_{20})\,F_2(\xi_{20})\right)$$

$$+e_4\left(\widehat{F}_1(\xi_{10})\,F_1(\xi_{10})+\widehat{F}_3(\xi_{20})\,F_4(\xi_{20})\right)$$

$$=\widehat{\Phi}(\zeta_0)\cdot\Phi(\zeta),$$

where $\zeta_0=\xi_1^{(0)}e_1+\xi_2^{(0)}e_2$.   □

10 The Taylor Expansion

Considering the problem on an expansion of the G-monogenic mapping into the Taylor power series, we assume that Ω_ζ is a bounded domain.

Let $\zeta_0 := x_0 i_1 + y_0 i_2 + z_0 i_3$ be an arbitrary fixed point in the domain Ω_ζ, $\xi_{10} := x_0 + a_1 y_0 + b_1 z_0$, $\xi_{20} := x_0 + a_2 y_0 + b_2 z_0$ be points of the complex plane corresponding to the point ζ_0 by formulas $\xi_{10} = f_1(\zeta_0)$, $\xi_{20} = f_2(\zeta_0)$, where a_k, b_k are coefficients from decomposition (6).

Let $R_0 := \min\limits_{\zeta \in \partial \Omega_\zeta} \|\zeta - \zeta_0\|$, where $\partial \Omega_\zeta$ is the boundary of the domain Ω_ζ in E_3.

Consider the ball

$$\Theta(\zeta_0, R_0) := \{\zeta \in E_3 : \|\zeta - \zeta_0\| < R_0\}$$

in E_3 with the radius R_0 and the center at the point ζ_0. Also denote by $\widetilde{D}_k$ the domain in the complex plane $\mathbb{C}$, onto which the ball $\Theta(\zeta_0, R_0)$ is mapped by the functional f_k for $k = 1, 2$.

Let

$$R := \min\left\{R_0 \, , \, \min\limits_{\tau_k \in \partial \widetilde{D}_k} |\tau_k - \xi_{k0}| \right\},$$

where $\partial \widetilde{D}_k$ is the boundary of the domain $\widetilde{D}_k$. By

$$U(\xi_{k0}, R) := \{\xi_k \in \mathbb{C} : |\xi_k - \xi_{k0}| < R\}$$

we denote the disk in the complex plane with the radius R and with the center at the point ξ_{k0} for $k = 1, 2$.

Applying to a G-monogenic mapping a method similar to the method for expanding holomorphic functions, which is based on the expansion of the Cauchy kernel into a power series (see, e.g., Shabat 1976, p. 107), we obtain immediately the following expansion of the right-G-monogenic mapping Φ into the power series

$$\Phi(\zeta) = \sum_{n=0}^{\infty} (\zeta - \zeta_0)^n p_n \tag{81}$$

and of the left-G-monogenic mapping $\widehat{\Phi}$ into the power series

$$\widehat{\Phi}(\zeta) = \sum_{n=0}^{\infty} \widehat{p}_n (\zeta - \zeta_0)^n \tag{82}$$

in the ball with the center at the fixed point $\zeta_0 \in E_3$ and with the radius which is less than a distance between ζ_0 and the boundary of the domain Ω_ζ. Here

$$p_n = \frac{\Phi^{(n)}(\zeta_0)}{n!} = \frac{1}{2\pi i} \int_{\gamma_\zeta} \left((\tau - \zeta_0)^{-1} \right)^{n+1} d\tau \, \Phi(\tau);$$

$$\widehat{p}_n = \frac{\widehat{\Phi}^{(n)}(\zeta_0)}{n!} = \frac{1}{2\pi i} \int_{\gamma_\zeta} \widehat{\Phi}(\tau) \left((\tau - \zeta_0)^{-1} \right)^{n+1} d\tau,$$

where γ_ζ is an arbitrary closed Jordan rectifiable curve in Ω_ζ such that its embraces one time the set $\{\zeta - \zeta_0 : \zeta \in L_\zeta^1 \cup L_\zeta^2\}$ and lies in a ball, which is contained in the domain Ω_ζ. This is due to the fact that in the inequality $\|ab\| \le c \|a\| \|b\|$ the constant c cannot be replaced by the unit 1.

Further we show that representation (29) provides a possibility to obtain an expansion of the right-G-monogenic mapping Φ into power series (81) and representation (31) provides a possibility to obtain an expansion of the left-G-monogenic mapping $\widehat{\Phi}$ into power series (82) in the domain

$$B(\zeta_0, R) := \{\zeta \in E_3 : f_k(\zeta) \in U(\xi_{k0}, R)\}, \quad k = 1, 2.$$

Since by the construction the domain $B(\zeta_0, R)$ is convex with respect to the straight lines L_ζ^1, L_ζ^2, it follows that the right-G-monogenic mapping Φ is expressed in form (29) and the left-G-monogenic mapping $\widehat{\Phi}$ is expressed in form (31) in the domain $B(\zeta_0, R)$.

Theorem 10.1 *Let $f_1(E_3) = f_2(E_3) = \mathbb{C}$. If a mapping $\Phi : \Omega_\zeta \to \mathbb{H}(\mathbb{C})$ is right-G-monogenic in $\Omega_\zeta \subset E_3$ and $\zeta_0 \in \Omega_\zeta$, then the mapping Φ is expressed as the sum of the convergent power series (81) in the domain $B(\zeta_0, R)$. In this case*

$$p_n = a_n e_1 + b_n e_2 + c_n e_3 + d_n e_4, \tag{83}$$

where a_n, b_n, c_n, d_n are coefficients of the Taylor series:

$$F_1(\xi_1) = \sum_{n=0}^{\infty} a_n (\xi_1 - \xi_{10})^n, \qquad F_2(\xi_2) = \sum_{n=0}^{\infty} b_n (\xi_2 - \xi_{20})^n,$$
$$F_3(\xi_1) = \sum_{n=0}^{\infty} c_n (\xi_1 - \xi_{10})^n, \qquad F_4(\xi_2) = \sum_{n=0}^{\infty} d_n (\xi_2 - \xi_{20})^n, \tag{84}$$

where F_1, F_2, F_3, F_4 are functions included in equality (29) for $\zeta \in B(\zeta_0, R)$.

Proof Inasmuch as in equality (29) the functions F_1, F_3 are holomorphic in the disk $U(\xi_{10}, R)$ and the functions F_2, F_4 are holomorphic in the disk $U(\xi_{20}, R)$, series (84) are absolutely convergent in the corresponding disks. Then we rewrite equality (29) in the form

$$\Phi(\zeta) = \sum_{n=0}^{\infty} a_n (\xi_1 - \xi_{10})^n e_1 + \sum_{n=0}^{\infty} b_n (\xi_2 - \xi_{20})^n e_2 +$$

$$+ \sum_{n=0}^{\infty} c_n (\xi_1 - \xi_{10})^n e_3 + \sum_{n=0}^{\infty} d_n (\xi_2 - \xi_{20})^n e_4.$$

Now, using the relations

$$(\zeta - \zeta_0)^n e_1 = (\xi_1 - \xi_{10})^n e_1, \quad (\zeta - \zeta_0)^n e_2 = (\xi_2 - \xi_{20})^n e_2, \\ (\zeta - \zeta_0)^n e_3 = (\xi_1 - \xi_{10})^n e_3, \quad (\zeta - \zeta_0)^n e_4 = (\xi_2 - \xi_{20})^n e_4 \tag{85}$$

for all $\zeta \in E_3$ and $n = 0, 1, \ldots$, we obtain expression (81), where coefficients are defined by equality (83) and series (81) is absolutely convergent in the domain $B(\zeta_0, R)$. $\qquad \square$

A similar statement is true for left-G-monogenic mappings.

Theorem 10.2 *Let $f_1(E_3) = f_2(E_3) = \mathbb{C}$. If a mapping $\widehat{\Phi} : \Omega_\zeta \to \mathbb{H}(\mathbb{C})$ is left-G-monogenic in $\Omega_\zeta \subset E_3$ and $\zeta_0 \in \Omega_\zeta$, then the mapping $\widehat{\Phi}$ is expressed as the sum of the convergent power series (82), where*

$$\widehat{p}_n = \widehat{a}_n e_1 + \widehat{b}_n e_2 + \widehat{c}_n e_3 + \widehat{d}_n e_4 \tag{86}$$

and $\widehat{a}_n, \widehat{b}_n, \widehat{c}_n, \widehat{d}_n$ are coefficients of the Taylor series:

$$\widehat{F}_1(\xi_1) = \sum_{n=0}^{\infty} \widehat{a}_n (\xi_1 - \xi_{10})^n, \quad \widehat{F}_2(\xi_2) = \sum_{n=0}^{\infty} \widehat{b}_n (\xi_2 - \xi_{20})^n,$$
$$\widehat{F}_3(\xi_2) = \sum_{n=0}^{\infty} \widehat{c}_n (\xi_2 - \xi_{20})^n, \quad \widehat{F}_4(\xi_1) = \sum_{n=0}^{\infty} \widehat{d}_n (\xi_1 - \xi_{10})^n, \tag{87}$$

where $\widehat{F}_1, \widehat{F}_2, \widehat{F}_3, \widehat{F}_4$ are functions included in equality (31) for $\zeta \in B(\zeta_0, R)$.

The following theorem is an analogue of the uniqueness theorem for the right-G-monogenic mappings taking values in the algebra $\mathbb{H}(\mathbb{C})$.

Theorem 10.3 *Let $f_1(E_3) = f_2(E_3) = \mathbb{C}$. If two right-$G$-monogenic mappings $\Phi_1 : \Omega_\zeta \to \mathbb{H}(\mathbb{C})$, $\Phi_2 : \Omega_\zeta \to \mathbb{H}(\mathbb{C})$ in $\Omega_\zeta \subset E_3$ coincide in a neighborhood of an interior point in the domain Ω_ζ, then they are identically equal everywhere in the domain Ω_ζ.*

Proof Let in the neighborhood

$$\omega(\zeta_0, R) := \{\zeta \in E_3 : \|\zeta - \zeta_0\| < R\}$$

of a point $\zeta_0 \in \Omega_\zeta$ the following equality is true:

$$\Phi_1(\zeta) \equiv \Phi_2(\zeta). \tag{88}$$

Since the ball $\omega(\zeta_0, R)$ is a convex set, the mappings Φ_1, Φ_2 can be represented in form (29):

$$\Phi_1(\zeta) = F_1(\xi_1)e_1 + F_2(\xi_2)e_2 + F_3(\xi_1)e_3 + F_4(\xi_2)e_4,$$

$$\Phi_2(\zeta) = H_1(\xi_1)e_1 + H_2(\xi_2)e_2 + H_3(\xi_1)e_3 + H_4(\xi_2)e_4.$$

Now the equalities

$$F_1 \equiv H_1, \quad F_3 \equiv H_3 \quad \text{in the domain} \quad f_1(\omega(\zeta_0, R)), \tag{89}$$

$$F_2 \equiv H_2, \quad F_4 \equiv H_4 \quad \text{in the domain} \quad f_2(\omega(\zeta_0, R)) \tag{90}$$

follow from equality (88). Using the uniqueness theorem for holomorphic functions of a complex variable (see, e.g., Shabat 1976, p. 118), equalities (89) are true everywhere in the domain $f_1(\Omega_\zeta)$ and equalities (90) are true everywhere in the domain $f_2(\Omega_\zeta)$. Now using the uniqueness of decomposition with respect to a basis, we have that equality (88) holds everywhere in the domain Ω_ζ. □

The same statement is true for the left-G-monogenic mappings taking values in the algebra $\mathbb{H}(\mathbb{C})$.

Theorem 10.4 *Let $f_1(E_3) = f_2(E_3) = \mathbb{C}$. If two left-$G$-monogenic mappings $\widehat{\Phi}_1 : \Omega_\zeta \to \mathbb{H}(\mathbb{C})$, $\widehat{\Phi}_2 : \Omega_\zeta \to \mathbb{H}(\mathbb{C})$ in domain $\Omega_\zeta \subset E_3$ coincide in a neighborhood of an interior point in the domain Ω_ζ, then they are identically equal everywhere in the domain Ω_ζ.*

Note, that the coincidence of mappings $\Phi_1 : \Omega_\zeta \to \mathbb{H}(\mathbb{C})$ and $\Phi_2 : \Omega_\zeta \to \mathbb{H}(\mathbb{C})$ on the set of the points that contains at least one limit point of the domain Ω_ζ is not sufficient to be identically equality of these mappings in the whole domain Ω_ζ. For example, the values of the G-monogenic mappings $\Phi_1(\zeta) = \zeta^2 \cdot e_1$ and $\Phi_2(\zeta) = \sin \zeta \cdot e_3$ coincide for all $\zeta \in L_\zeta^1$, but does not coincide identically.

11 The Laurent Expansion

Consider the problem on an expansion of the right-G-monogenic mapping $\Phi : \mathcal{K}_\zeta \to \mathbb{H}(\mathbb{C})$ and the left-G-monogenic mapping $\widehat{\Phi} : \mathcal{K}_\zeta \to \mathbb{H}(\mathbb{C})$ into the Laurent series about the point $\zeta_0 = x_0 i_1 + y_0 i_2 + z_0 i_3$ in the unbounded domain

$$\mathcal{K}_\zeta := \{\zeta \in E_3 : 0 \le r < |\xi_k - \xi_{k0}| < R \le \infty\}, \quad k = 1, 2.$$

Theorem 11.1 *Let $f_1(E_3) = f_2(E_3) = \mathbb{C}$. Then every right-$G$-monogenic mapping $\Phi : \mathcal{K}_\zeta \to \mathbb{H}(\mathbb{C})$ is expressed in the domain $\mathcal{K}_\zeta$ as the sum of the convergent series*

$$\Phi(\zeta) = \sum_{n=-\infty}^{\infty} (\zeta - \zeta_0)^n \, p_n \,, \tag{91}$$

where $(\zeta - \zeta_0)^n := \left((\zeta - \zeta_0)^{-1}\right)^{-n}$ for $n = -1, -2, \ldots$ and coefficients p_n are the same as in the equality (83), in which a_n, b_n, c_n, d_n are coefficients of the Laurent series

$$F_1(\xi_1) = \sum_{n=-\infty}^{\infty} a_n(\xi_1 - \xi_{10})^n, \qquad F_2(\xi_2) = \sum_{n=-\infty}^{\infty} b_n(\xi_2 - \xi_{20})^n,$$

$$F_3(\xi_1) = \sum_{n=-\infty}^{\infty} c_n(\xi_1 - \xi_{10})^n, \qquad F_4(\xi_2) = \sum_{n=-\infty}^{\infty} d_n(\xi_2 - \xi_{20})^n, \tag{92}$$

where $\widehat{F}_1, \widehat{F}_2, \widehat{F}_3, \widehat{F}_4$ are functions included in equality (31) for $\zeta \in \mathcal{K}_\zeta$.

Proof Since in equality (29) the functions F_1, F_3 are holomorphic in the ring $\{\xi_1 \in \mathbb{C}: r < |\xi_1 - \xi_{10}| < R\}$ with the center at the point $\xi_{10} = x_0 + a_1 y_0 + b_1 z_0$ and the functions F_2, F_4 are holomorphic in the ring $\{\xi_2 \in \mathbb{C}: r < |\xi_2 - \xi_{20}| < R\}$ with the center at the point $\xi_{20} = x_0 + a_2 y_0 + b_2 z_0$, they extend into Laurent series (92), which are absolutely convergent in the corresponding rings. Then we rewrite equality (29) in the form

$$\Phi(\zeta) = \sum_{n=-\infty}^{\infty} a_n(\xi_1 - \xi_{10})^n e_1 + \sum_{n=-\infty}^{\infty} b_n(\xi_2 - \xi_{20})^n e_2$$

$$+ \sum_{n=-\infty}^{\infty} c_n(\xi_1 - \xi_{10})^n e_3 + \sum_{n=-\infty}^{\infty} d_n(\xi_2 - \xi_{20})^n e_4 \,.$$

Further, using equalities (85) for all $\zeta \in \mathcal{K}_\zeta$ and integer values n, we obtain the expression of the mapping Φ into series (91), where coefficients are defined by equalities (83). Moreover, series (91) is absolutely convergent in the domain $\zeta \in \mathcal{K}_\zeta$. □

In the same way we can prove the following theorem, which is true for the left-G-monogenic mappings.

Theorem 11.2 *Let $f_1(E_3) = f_2(E_3) = \mathbb{C}$. Then every left-$G$-monogenic mapping $\widehat{\Phi}: \mathcal{K}_\zeta \to \mathbb{H}(\mathbb{C})$ is expressed in the domain $\mathcal{K}_\zeta$ as the sum of the convergent series*

$$\widehat{\Phi}(\zeta) = \sum_{n=-\infty}^{\infty} \widehat{p}_n(\zeta - \zeta_0)^n \,, \tag{93}$$

where $(\zeta - \zeta_0)^n := \left((\zeta - \zeta_0)^{-1}\right)^{-n}$ for $n = -1, -2, \ldots$ and coefficients $\widehat{p}_n$ are the same as in equality (86), in which $\widehat{a}_n$, $\widehat{b}_n$, $\widehat{c}_n$, $\widehat{d}_n$ are coefficients of the Laurent series

$$\widehat{F}_1(\xi_1) = \sum_{n=-\infty}^{\infty} \widehat{a}_n(\xi_1 - \xi_{10})^n, \qquad \widehat{F}_2(\xi_2) = \sum_{n=-\infty}^{\infty} \widehat{b}_n(\xi_2 - \xi_{20})^n,$$

$$\widehat{F}_3(\xi_2) = \sum_{n=-\infty}^{\infty} \widehat{c}_n(\xi_2 - \xi_{20})^n, \qquad \widehat{F}_4(\xi_1) = \sum_{n=-\infty}^{\infty} \widehat{d}_n(\xi_1 - \xi_{10})^n, \tag{94}$$

where $\widehat{F}_1, \widehat{F}_2, \widehat{F}_3, \widehat{F}_4$ are functions included in equality (31) for $\zeta \in \mathcal{K}_\zeta$.

12 A Classification of Isolated Singular Points of G-Monogenic Mappings

The terms of Laurent series (91) and (93) with nonnegative powers form *a regular part*, and the terms with negative powers form *a principal part* of the series (91) and (93).

Let us compactify the algebra $\mathbb{H}(\mathbb{C})$ by means of addition of an infinite point. Let us say that every sequence $w_n : = \tau_{1,n}e_1 + \tau_{2,n}e_2 + \tau_{3,n}e_3 + \tau_{4,n}e_4$ with $\tau_{1,n}$, $\tau_{2,n}$, $\tau_{3,n}$, $\tau_{4,n} \in \mathbb{C}$ converges to the infinite point in the case, where at least one of the sequences $\tau_{1,n}, \tau_{2,n}, \tau_{3,n}, \tau_{4,n}$ converges to the infinity in the extended complex plane.

Now suppose that the right-G-monogenic mapping $\Phi : \mathcal{K}_\zeta^0 \to \mathbb{H}(\mathbb{C})$ and the left-G-monogenic mapping $\widehat{\Phi} : \mathcal{K}_\zeta^0 \to \mathbb{H}(\mathbb{C})$ defined in the domain

$$\mathcal{K}_\zeta^0 := \{\zeta \in E_3 : 0 < |\xi_k - \xi_{k0}| < R \leq \infty\}, \qquad k = 1, 2.$$

Let

$$\widetilde{\mathcal{K}}_\zeta^0 := \{\zeta \in E_3 : |\xi_k - \xi_{k0}| < R\}.$$

The following theorem is true.

Theorem 12.1 *Let* $f_1(E_3) = f_2(E_3) = \mathbb{C}$. *If expansion (91) of the mapping* $\Phi : \mathcal{K}_\zeta^0 \to \mathbb{H}(\mathbb{C})$:

(1) *does not contain the principal part, then the mapping* Φ *has finite limit*

$$\lim_{\substack{\zeta \to \zeta_0 + \zeta^*, \\ \zeta \notin \{\zeta_0 + \zeta^* : \zeta^* \in L_\zeta^1 \cup L_\zeta^2\}}} \Phi(\zeta) \tag{95}$$

(2) *contains only finite numbers of terms in the principal part, then at least for one value* $k = 1, 2$ *the mapping* Φ *has infinite limit*

$$\lim_{\substack{\zeta \to \zeta_0 + \zeta_k^*, \\ \zeta \notin \{\zeta_0 + \zeta_k^* : \zeta_k^* \in L_\zeta^k\}}} \Phi(\zeta) \tag{96}$$

at all points

$$\zeta_0 + \zeta_k^* \in \widetilde{\mathcal{K}}_\zeta^0 \cap \{\zeta_0 + \zeta_k^* : \ \zeta_k^* \in L_\zeta^k\};$$

(3) *contains infinite numbers of terms in the principal part, then at least for one value* $k = 1, 2$ *the mapping* Φ *either has an infinite limit, or has neither finite nor infinite limit at all points*

$$\zeta_0 + \zeta_k^* \in \widetilde{\mathcal{K}}_\zeta^0 \cap \{\zeta_0 + \zeta_k^* : \ \zeta_k^* \in L_\zeta^k\}.$$

Proof The mapping Φ in the domain $\mathcal{K}_\zeta^0$ is expressed in form (29), where the functions F_1, F_3 are holomorphic in the deleted neighborhood $U(\xi_{10}, R) \setminus \{\xi_{10}\}$ of the point ξ_{10}, and the functions F_2, F_4 are holomorphic in the deleted neighborhood $U(\xi_{20}, R) \setminus \{\xi_{20}\}$ of the point ξ_{20}.

Let us consider the case where decomposition (91) does not contain the principal part, namely, it is expressed in form (81). In this case coefficients of Laurent series (92) are related with coefficients of the series (81) by the equalities (83), then due to the equalities $p_n = 0$ for $n = -1, -2, \ldots$, the equalities $a_n = b_n = c_n = d_n = 0$ hold for all negative n. Hence, Laurent series (92) in the neighborhood of the corresponding points ξ_{10}, ξ_{20} are the Taylor series of their sums, and the functions F_1, F_2, F_3, F_4 from equality (29) are holomorphic in the corresponding domains $U(\xi_{10}, R)$, $U(\xi_{20}, R)$. It means that mapping (29) has the finite limits (95) at all points

$$\zeta_0 + \zeta^* \in \widetilde{\mathcal{K}}_\zeta^0 \cap \{\zeta_0 + \zeta^* : \zeta^* \in L_\zeta^1 \cup L_\zeta^2\}.$$

Now consider the case where the principal part of decomposition (91) contains only finite number of terms. Then from relations (83), which associate coefficients of Laurent series (92) with the coefficients of series (91) it follows, that all principal parts of series (92) do not contain infinite number of terms, and the principal part at least one of them does not equal zero. It means that the point ξ_{10} is not an essential singular point for the functions F_1, F_3 and the point ξ_{20} is not an essential singular point for the functions F_2, F_4, but at least one of the functions F_1, F_2, F_3, F_4 has a pole at a corresponding point. It follows, that at least one of the functions F_1, F_2, F_3, F_4 has an infinite limit as $\xi_1 \to \xi_{10}$ or as $\xi_2 \to \xi_{20}$, so limit (96) is also infinite for $k = 1$ or $k = 2$.

Finally, consider the case where the principal part of decomposition (91) contains an infinite number of nonzero members, so there exists an infinite number of nonzero coefficients p_n for negative n. Then from relations (83) it follows that the principal part of at least one of series (92) contains an infinite number of the terms and it means, that either the point ξ_{10} is an essential singularity for the functions F_1, F_3, or the point ξ_{20} is an essential singularity for at least one of the functions F_2 or F_4. Therefore, the mapping Φ can not have a finite limit at all points of the set

$$\widetilde{\mathcal{K}}_\zeta^0 \cap \{\zeta_0 + \zeta^* : \zeta^* \in L_\zeta^1 \cup L_\zeta^2\},$$

but it can have an infinity limit at these points.   □

For example, if ξ_{10} is a pole of the function F_1 and an essential singular point of the function F_3, the point ξ_{20} is an essential singular point of the functions F_2, F_4, then the function F_1 has an infinite limit at the point ξ_{10}. Thus, limit (96) is an infinity at all points

$$\zeta_0 + \zeta_1^* \in \widetilde{\mathcal{K}}_\zeta^0 \cap \{\zeta_0 + \zeta_1^* : \zeta_1^* \in L_\zeta^1\}.$$

In the case where, for example, $F_2 \equiv 0$, $F_3 \equiv 0$, $F_4 \equiv 0$ and the point ξ_{10} is an essential singular point of the function F_1 the mapping Φ has neither the finite, nor the infinite limit (96) at all points

$$\zeta_0 + \zeta_1^* \in \widetilde{\mathcal{K}}_\zeta^0 \cap \{\zeta_0 + \zeta_1^* : \zeta_1^* \in L_\zeta^1\}.$$

Now, for a removable singular point, a pole and a essential singular point of the G-monogenic mapping Φ in a deleted neighborhood of the point $\zeta_0 \in E_3$, one can give the same definitions as for appropriate notions in the complex plane (see, e.g., Shabat 1976, p. 135). Namely, the point ζ_0 is called:

(1) *a removable singular point* of the mapping Φ, if there exists a finite limit

$$\lim_{\substack{\zeta \to \zeta_0, \\ \zeta \notin \{\zeta_0 + \zeta^* : \zeta^* \in L_\zeta^1 \cup L_\zeta^2\}}} \Phi(\zeta) = A;$$

(2) *a pole* of the mapping Φ, if there exists an infinite limit

$$\lim_{\substack{\zeta \to \zeta_0, \\ \zeta \notin \{\zeta_0 + \zeta^* : \zeta^* \in L_\zeta^1 \cup L_\zeta^2\}}} \Phi(\zeta) = \infty;$$

(3) *an essential singular point* of the mapping Φ, if the mapping Φ has neither finite, nor infinite limits as $\zeta \to \zeta_0$ and $\zeta \notin \{\zeta_0 + \zeta^* : \zeta^* \in L_\zeta^1 \cup L_\zeta^2\}$.

It follows from Theorem 12.1, that the isolated singular point of the G-monogenic mapping can be only removable singular point. In the case where the mapping has unremovable singularity at the point ζ_0, the singular points are all the points of the set $\widetilde{\mathcal{K}}_\zeta^0 \cap \{\zeta_0 + \zeta_k^* : \zeta_k^* \in L_\zeta^k\}$ for at least one $k = \{1, 2\}$.

13 H-Monogenic Mappings

F. Hausdorff (1900) proposed the definition of an analytic function in any associative (commutative or non–commutative) algebra $\mathbb{A}$ over $\mathbb{C}$ with unit. It can be formulated as follows. A hypercomplex function

$$f(\eta) = \sum_{k=1}^{n} f_k(\eta_1, \ldots, \eta_n) e_k, \tag{97}$$

where e_k are basis elements of the algebra $\mathbb{A}$, is called an $H-$*analytic function* of the variable $\eta := \sum_{k=1}^{n} \eta_k e_k$, if the components f_k from expansion (97) are analytic functions of the complex variables $\eta_1, \ldots, \eta_n$, and the differential

$$df := \sum_{k=1}^{n} df_k(\eta_1, \ldots, \eta_n)e_k = \sum_{j,k=1}^{n} \frac{\partial f_k}{\partial \eta_j} d\eta_j e_k \qquad (98)$$

is a linear homogeneous polynomial of the differential $d\eta := \sum_{k=1}^{n} d\eta_k e_k$, i.e.,

$$df = \sum_{s=1}^{n^2} A_s \, d\eta \, B_s , \qquad (99)$$

where A_s and B_s are some $\mathbb{A}$–valued functions.

In this case, the values $f'_H(\eta) := \sum_{s=1}^{n^2} A_s B_s$ is called the *Hausdorff derivative* of the function $f(\eta)$.

Note that the analyticity of real–valued components f_k from expansion (97) was assumed in the paper Ringleb (1933) at the definition of an $H-$analytic function in an associative algebra over $\mathbb{R}$. As for the associative algebras over $\mathbb{R}$ or $\mathbb{C}$, they were considered in the paper Rinehart and Wilson (1962), where only the existence of partial derivatives $\frac{\partial f_k}{\partial \eta_j}$ for all $j, k = 1, 2, \ldots, n$, was assumed.

We emphasize that the property of $H-$analyticity of a function is independent of a choice of the basis of an algebra. In addition, if the functions $f(\eta)$ and $g(\eta)$ are $H-$analytic, then the functions $f(\eta) + g(\eta)$ and $f(\eta) \cdot g(\eta)$ are also $H-$analytic. In this case, (see Ringleb 1933; Portman 1959),

$$d(f + g) = df + dg \qquad (100)$$

and

$$d(f \cdot g) = df \cdot g + f \cdot dg. \qquad (101)$$

Now, we realize the Hausdorff approach to quaternionic mappings of the variable $\zeta = xi_1 + yi_2 + zi_3$.

A continuous mapping $\Phi : \Omega_\zeta \to \mathbb{H}(\mathbb{C})$ of form (9) is called *H-monogenic* in a domain $\Omega_\zeta \subset E_3$ if Φ is differentiable in the sense of Hausdorff at every point $\zeta \in \Omega_\zeta$, i.e. the components of the mapping have partial derivatives of the first order with respect to the variables x, y, z, and a formal differential of the mapping

$$d\Phi := \sum_{k=1}^{4} \left(\frac{\partial U_k}{\partial x} dx + \frac{\partial U_k}{\partial y} dy + \frac{\partial U_k}{\partial z} dz \right) e_k \qquad (102)$$

is a linear homogeneous function of the differential $d\zeta = dx + i_2 dy + i_3 dz$, i.e.

$$d\Phi = \sum_{s=1}^{16} A_s \, d\zeta \, B_s \,, \tag{103}$$

where A_s, B_s are certain $\mathbb{H}(\mathbb{C})$ – valued functions.

Note, if partial derivatives of the first order of functions U_q for $r = 1, 2, 3, 4$ exist and continuous, then the formal differential (102) will be the total differential of the mapping Φ, i.e. it will be the principal part of the increment of this mapping.

The value $\Phi'_H(\zeta) := \sum_{s=1}^{16} A_s B_s$ is called *the Hausdorff derivative* of the mapping Φ at the point ζ.

Moreover, the following theorem is true:

Theorem 13.1 *If a mapping* $\Phi : \Omega_\zeta \to \mathbb{H}(\mathbb{C})$ *is H-monogenic in a domain Ω_ζ, then its derivative Φ'_H exists, does not depend on the choice of the functions A_s, B_s in equality (103) and*

$$\Phi'_H(\zeta) = \frac{\partial \Phi}{\partial x_1} \,. \tag{104}$$

Proof The consequence of the H-monogeneity of the mapping Φ is the equality

$$\sum_{s=1}^{16} A_s d\zeta B_s = \sum_{k=1}^{4} \left(\frac{\partial U_k}{\partial x} dx + \frac{\partial U_k}{\partial y} dy + \frac{\partial U_k}{\partial z} dz \right) e_k \,. \tag{105}$$

Let

$$A_s = a_{s1}e_1 + a_{s2}e_2 + a_{s3}e_3 + a_{s4}e_4 \,,$$
$$B_s = b_{s1}e_1 + b_{s2}e_2 + b_{s3}e_3 + b_{s4}e_4 \tag{106}$$

for $s = 1, 2, \ldots, 16$. Using the equalities

$$d\zeta = (dx + a_1 dy + b_1 dz)e_1 + (dx + a_2 dy + b_2 dz)e_2$$

and (106) we obtain:

$$A_s d\zeta B_s = (a_{s1}e_1 + a_{s2}e_2 + a_{s3}e_3 + a_{s4}e_4)\Big((dx + a_1 dy + b_1 dz)e_1 +$$

$$+ (dx + a_2 dy + b_2 dz)e_2 \Big)(b_{s1}e_1 + b_{s2}e_2 + b_{s3}e_3 + b_{s4}e_4) =$$

$$= \Big(a_{s1}b_{s1}(dx + a_1 dy + b_1 dz) + a_{s3}b_{s4}(dx + a_2 dy + b_2 dz) \Big)e_1 +$$

$$+ \Big(a_{s2}b_{s2}(dx + a_2 dy + b_2 dz) + a_{s4}b_{s3}(dx + a_1 dy + b_1 dz) \Big)e_2 +$$

$$+\Big(a_{s1}b_{s3}(dx + a_1dy + b_1dz) + a_{s3}b_{s2}(dx + a_2dy + b_2dz)\Big)e_3+$$

$$+\Big(a_{s2}b_{s4}(dx + a_2dy + b_2dz) + a_{s4}b_{s1}(dx + a_1dy + b_1dz)\Big)e_4. \qquad (107)$$

The relations

$$\frac{\partial U_1}{\partial x} = \sum_{s=1}^{16} a_{s1}b_{s1} + a_{s3}b_{s4}, \qquad \frac{\partial U_2}{\partial x} = \sum_{s=1}^{16} a_{s2}b_{s2} + a_{s4}b_{s3},$$

$$\frac{\partial U_3}{\partial x} = \sum_{s=1}^{16} a_{s1}b_{s3} + a_{s3}b_{s2}, \qquad \frac{\partial U_4}{\partial x} = \sum_{s=1}^{16} a_{s2}b_{s4} + a_{s4}b_{s1} \qquad (108)$$

follows from equalities (105) and (107).

Due to equality (106), we have

$$\Phi'_H(\zeta) := \sum_{s=1}^{16} A_s d\zeta B_s = \sum_{s=1}^{16} \Big((a_{s1}b_{s1} + a_{s3}b_{s4})e_1+$$

$$+(a_{s2}b_{s2} + a_{s4}b_{s3})e_2 + (a_{s1}b_{s3} + a_{s3}b_{s2})e_3 + (a_{s2}b_{s4} + a_{s4}b_{s1})e_4\Big),$$

Then, using relation (108), we obtain

$$\Phi'_H(\zeta) = \frac{\partial U_1}{\partial x} e_1 + \frac{\partial U_2}{\partial x} e_2 + \frac{\partial U_3}{\partial x} e_3 + \frac{\partial U_4}{\partial x} e_4 = \frac{\partial \Phi}{\partial x}. \qquad \square$$

Theorem 13.2 *If mappings* $\Phi : \Omega_\zeta \to \mathbb{H}(\mathbb{C})$ *and* $\Psi : \Omega_\zeta \to \mathbb{H}(\mathbb{C})$ *are H-monogenic in a domain* Ω_ζ, *then the product* $\Phi \cdot \Psi$ *is also H-monogenic mapping in* Ω_ζ *and*

$$d(\Phi \cdot \Psi) = d\Phi \cdot \Psi + \Phi \cdot d\Psi.$$

Proof Let

$$\Phi(\zeta) = \sum_{k=1}^{4} U_k(x, y, z)e_k, \qquad \Psi(\zeta) = \sum_{k=1}^{4} V_k(x, y, z)e_k.$$

Then

$$d\Phi = \sum_{k=1}^{4} \left(\frac{\partial U_k}{\partial x} dx + \frac{\partial U_k}{\partial y} dy + \frac{\partial U_k}{\partial z} dz \right) e_k,$$

$$d\Psi = \sum_{k=1}^{4} \left(\frac{\partial V_k}{\partial x} dx + \frac{\partial V_k}{\partial y} dy + \frac{\partial V_k}{\partial z} dz \right) e_k$$

and

$$d(\Phi \cdot \Psi) = d\left(U_1 V_1 + U_3 V_4 \right) e_1 + d\left(U_2 V_2 + U_4 V_3 \right) e_2$$

$$+ d\left(U_1 V_3 + U_3 V_2 \right) e_3 + d\left(U_2 V_4 + U_4 V_1 \right) e_4$$

$$= \left[\left(\frac{\partial U_1}{\partial x} V_1 + \frac{\partial V_1}{\partial x} U_1 + \frac{\partial U_3}{\partial x} V_4 + \frac{\partial V_4}{\partial x} U_3 \right) dx \right.$$

$$+ \left(\frac{\partial U_1}{\partial y} V_1 + \frac{\partial V_1}{\partial y} U_1 + \frac{\partial U_3}{\partial y} V_4 + \frac{\partial V_4}{\partial y} U_3 \right) dy$$

$$\left. + \left(\frac{\partial U_1}{\partial z} V_1 + \frac{\partial V_1}{\partial z} U_1 + \frac{\partial U_3}{\partial z} V_4 + \frac{\partial V_4}{\partial z} U_3 \right) dz \right] e_1$$

$$+ \left[\left(\frac{\partial U_2}{\partial x} V_2 + \frac{\partial V_2}{\partial x} U_2 + \frac{\partial U_4}{\partial x} V_3 + \frac{\partial V_3}{\partial x} U_4 \right) dx \right.$$

$$+ \left(\frac{\partial U_2}{\partial y} V_2 + \frac{\partial V_2}{\partial y} U_2 + \frac{\partial U_4}{\partial y} V_3 + \frac{\partial V_3}{\partial y} U_4 \right) dy$$

$$\left. + \left(\frac{\partial U_2}{\partial z} V_2 + \frac{\partial V_2}{\partial z} U_2 + \frac{\partial U_4}{\partial z} V_3 + \frac{\partial V_3}{\partial z} U_4 \right) dz \right] e_2$$

$$+ \left[\left(\frac{\partial U_1}{\partial x} V_3 + \frac{\partial V_3}{\partial x} U_1 + \frac{\partial U_3}{\partial x} V_2 + \frac{\partial V_2}{\partial x} U_3 \right) dx \right.$$

$$+ \left(\frac{\partial U_1}{\partial y} V_3 + \frac{\partial V_3}{\partial y} U_1 + \frac{\partial U_3}{\partial y} V_2 + \frac{\partial V_2}{\partial y} U_3 \right) dy$$

$$\left. + \left(\frac{\partial U_1}{\partial z} V_3 + \frac{\partial V_3}{\partial z} U_1 + \frac{\partial U_3}{\partial z} V_2 + \frac{\partial V_2}{\partial z} U_3 \right) dz \right] e_3$$

$$+ \left[\left(\frac{\partial U_2}{\partial x} V_4 + \frac{\partial V_4}{\partial x} U_2 + \frac{\partial U_4}{\partial x} V_1 + \frac{\partial V_1}{\partial x} U_4 \right) dx \right.$$

$$+ \left(\frac{\partial U_2}{\partial y} V_4 + \frac{\partial V_4}{\partial y} U_2 + \frac{\partial U_4}{\partial y} V_1 + \frac{\partial V_1}{\partial y} U_4 \right) dy$$

$$+ \left(\frac{\partial U_2}{\partial z} V_4 + \frac{\partial V_4}{\partial z} U_2 + \frac{\partial U_4}{\partial z} V_1 + \frac{\partial V_1}{\partial z} U_4 \right) dz \Big] e_4 .$$

Let us transform the obtained expression to the following form:

$$\left(V_1 \frac{\partial U_1}{\partial x} dx + V_1 \frac{\partial U_1}{\partial y} dy + V_1 \frac{\partial U_1}{\partial z} dz + V_4 \frac{\partial U_3}{\partial x} dx + V_4 \frac{\partial U_3}{\partial y} dy + V_4 \frac{\partial U_3}{\partial z} dz \right) e_1$$

$$+ \left(V_2 \frac{\partial U_2}{\partial x} dx + V_2 \frac{\partial U_2}{\partial y} dy + V_2 \frac{\partial U_2}{\partial z} dz + V_3 \frac{\partial U_4}{\partial x} dx + V_3 \frac{\partial U_4}{\partial y} dy + V_3 \frac{\partial U_4}{\partial z} dz \right) e_2$$

$$+ \left(V_3 \frac{\partial U_1}{\partial x} dx + V_3 \frac{\partial U_1}{\partial y} dy + V_3 \frac{\partial U_1}{\partial z} dz + V_2 \frac{\partial U_3}{\partial x} dx + V_2 \frac{\partial U_3}{\partial y} dy + V_2 \frac{\partial U_3}{\partial z} dz \right) e_3$$

$$+ \left(V_4 \frac{\partial U_2}{\partial x} dx + V_4 \frac{\partial U_2}{\partial y} dy + V_4 \frac{\partial U_2}{\partial z} dz + V_1 \frac{\partial U_4}{\partial x} dx + V_1 \frac{\partial U_4}{\partial y} dy + V_1 \frac{\partial U_4}{\partial z} dz \right) e_4$$

$$+ \left(U_1 \frac{\partial V_1}{\partial x} dx + U_1 \frac{\partial V_1}{\partial y} dy + U_1 \frac{\partial V_1}{\partial z} dz + U_4 \frac{\partial V_3}{\partial x} dx + U_4 \frac{\partial V_3}{\partial y} dy + U_4 \frac{\partial V_3}{\partial z} dz \right) e_1$$

$$+ \left(U_2 \frac{\partial V_2}{\partial x} dx + U_2 \frac{\partial V_2}{\partial y} dy + U_2 \frac{\partial V_2}{\partial z} dz + U_3 \frac{\partial V_4}{\partial x} dx + U_3 \frac{\partial V_4}{\partial y} dy + U_3 \frac{\partial V_4}{\partial z} dz \right) e_2$$

$$+ \left(U_3 \frac{\partial V_1}{\partial x} dx + U_3 \frac{\partial V_1}{\partial y} dy + U_3 \frac{\partial V_1}{\partial z} dz + U_2 \frac{\partial V_3}{\partial x} dx + U_2 \frac{\partial V_3}{\partial y} dy + U_2 \frac{\partial V_3}{\partial z} dz \right) e_3$$

$$+ \left(U_4 \frac{\partial V_2}{\partial x} dx + U_4 \frac{\partial V_2}{\partial y} dy + U_4 \frac{\partial V_2}{\partial z} dz + U_1 \frac{\partial V_4}{\partial x} dx + U_1 \frac{\partial V_4}{\partial y} dy + U_1 \frac{\partial V_4}{\partial z} dz \right) e_4 ,$$

where we have

$$\left(V_1dU_1 + V_4dU_3\right)e_1 + \left(V_2dU_2 + V_3dU_4\right)e_2 + \left(V_3dU_1 + V_2dU_3\right)e_3+$$

$$+\left(V_4dU_2 + V_1dU_4\right)e_4 + \left(U_1dV_1 + U_3dV_4\right)e_1 + \left(U_2dV_2 + U_4dV_3\right)e_2+$$

$$+\left(U_1dV_3 + U_3dV_2\right)e_3 + \left(U_2dV_4 + U_4dV_1\right)e_4 = d\Phi \cdot \Psi + \Phi \cdot d\Psi. \qquad \square$$

By Theorem 13.2 the set of H-monogenic mappings taking values in the algebra $\mathbb{H}(\mathbb{C})$ forms the functional algebra, since the product of two H-monogenic mappings is H-monogenic mapping too.

In the next theorem we establish a relation between G-monogenic and H-monogenic mappings.

Theorem 13.3 *Every right-G-monogenic mapping $\Phi : \Omega_\zeta \to \mathbb{H}(\mathbb{C})$ and every left-G-monogenic mapping $\widehat{\Phi} : \Omega_\zeta \to \mathbb{H}(\mathbb{C})$ in the domain Ω_ζ is H-monogenic mapping in this domain.*

Proof Let $\Phi : \Omega_\zeta \to \mathbb{H}(\mathbb{C})$ be a right-G-monogenic mapping. Then the existence of the partial derivatives of the first order of the components of the mapping Φ follows from the existence of the Gâteaux derivative (equality (8)). Let us show that the differential

$$d\Phi = \frac{\partial\Phi}{\partial x} dx + \frac{\partial\Phi}{\partial y} dy + \frac{\partial\Phi}{\partial z} dz \qquad (109)$$

can be represented in form (103).

For this we note, that due to equality (109) and conditions (13) the equality

$$d\Phi = \left(dx + i_2dy + i_3dz\right)\frac{\partial\Phi}{\partial x} = d\zeta\, \Phi'(\zeta),$$

is true, so the differential (109) is represented in the form (103), where $A_1 = 1$, $B_1 = \Phi'(\zeta)$.

In a similar way we establish that due to equality (109) for $\Phi = \widehat{\Phi}$ and conditions (14) we obtain the equality

$$d\widehat{\Phi} = \widehat{\Phi}'(\zeta)\, d\zeta,$$

so the differential of the mapping $\widehat{\Phi}$ is represented in the form (103), where $A_1 = \widehat{\Phi}'(\zeta)$, $B_1 = 1$. $\qquad \square$

Since the right- and left-G-monogenic mappings are H-monogenic, then their products are also H-monogenic mappings. Therefore, Theorems 13.2 and 13.3 and representations (29) and (31) yield the following proposition.

Corollary 2 *Let the domain $\Omega \subset \mathbb{R}^3$ be convex in the direction of the straight lines L^1 and L^2, and let $f_1(E_3) = f_2(E_3) = \mathbb{C}$. Then the mappings*

$$\Phi(\zeta) \cdot \widehat{\Phi}(\zeta) =$$

$$= \Big(F_1(\xi_1)\widehat{F}_1(\xi_1) + F_3(\xi_1)\widehat{F}_4(\xi_1)\Big)e_1 + \Big(F_2(\xi_2)\widehat{F}_2(\xi_2) + F_4(\xi_2)\widehat{F}_3(\xi_2)\Big)e_2$$

$$+\Big(F_1(\xi_1)\widehat{F}_3(\xi_2) + F_3(\xi_1)\widehat{F}_2(\xi_2)\Big)e_3 + \Big(F_2(\xi_2)\widehat{F}_4(\xi_1) + F_4(\xi_2)\widehat{F}_1(\xi_1)\Big)e_4 ,$$

$$\widehat{\Phi}(\zeta) \cdot \Phi(\zeta) =$$

$$= \Big(\widehat{F}_1(\xi_1)F_1(\xi_1) + \widehat{F}_3(\xi_2)F_4(\xi_2)\Big)e_1 + \Big(\widehat{F}_2(\xi_2)F_2(\xi_2) + \widehat{F}_4(\xi_1)F_3(\xi_1)\Big)e_2$$

$$+\Big(\widehat{F}_1(\xi_1)F_3(\xi_1) + \widehat{F}_3(\xi_2)F_2(\xi_2)\Big)e_3 + \Big(\widehat{F}_2(\xi_2)F_4(\xi_2) + \widehat{F}_4(\xi_1)F_1(\xi_1)\Big)e_4 ,$$

where the holomorphic functions F_k, $\widehat{F}_k$ are defined in equalities (29) and (31), are H-monogenic in the domain Ω_ζ.

At the same time, there exist H-monogenic mappings that are neither right-G-monogenic, nor left-G-monogenic.

Example The mapping

$$h(\zeta) = (e^{\xi_1} + \xi_2^2)\, e_1 + \xi_1 \sin \xi_2\, e_2 + \xi_2^2\, e_3 + e^{\xi_1}\, e_4$$

is H-monogenic in the space E_3, but it is not left-G-monogenic or right-G-monogenic. Indeed, the differential of this mapping can be represented in form (103):

$$dh = e^{\xi_1} e_1 d\zeta e_1 + \xi_1 \cos \xi_2\, e_2\, d\zeta e_2 + 2\xi_2\, e_3\, d\zeta e_2 + e^{\xi_1}\, e_4\, d\zeta e_1$$

$$+2\xi_2\, e_3\, d\zeta\, e_4 + \sin \xi_2\, e_4\, d\zeta\, e_3 .$$

However, the mapping H is represented neither in form (29), nor in form (31). The H-monogenic mapping Φ, whose differential is represented as

$$d\Phi = d\zeta\, \Phi'_H(\zeta) \tag{110}$$

is called *right-H-monogenic*, and H-monogenic mapping $\widehat{\Phi}$, whose differential is represented as

$$d\widehat{\Phi} = \widehat{\Phi}'_H(\zeta)\, d\zeta \tag{111}$$

is called *left-H-monogenic* in a domain Ω_ζ.

We now obtain the necessary and sufficient conditions of G-monogenicity of a mapping.

Theorem 13.4 *Suppose that the components $U_k : \Omega \to \mathbb{C}$ of the mapping (9) are $\mathbb{R}$-differentiable in a domain Ω. A mapping $\Phi : \Omega_\zeta \to \mathbb{H}(\mathbb{C})$ is right-G-monogenic* *if and only if it is right-H-monogenic, and a mapping $\widehat{\Phi} : \Omega_\zeta \to \mathbb{H}(\mathbb{C})$ is left-G-monogenic if and only if it is left-H-monogenic.*

Proof The necessity is proved in the proof of Theorem 13.3. Let us prove the sufficiency. Let a mapping Φ be right-H-monogenic, so equality (110) hold. The consequence of equalities (109) and (110) is the equality

$$\frac{\partial\Phi}{\partial x}\,dx + \frac{\partial\Phi}{\partial y}\,dy + \frac{\partial\Phi}{\partial z}\,dz = d\zeta\Phi'_H(\zeta).$$

Using equality (104) and the expression $d\zeta = dx + i_2 dy + i_3 dz$ we have the equality

$$\frac{\partial\Phi}{\partial x}\,dx + \frac{\partial\Phi}{\partial y}\,dy + \frac{\partial\Phi}{\partial z}\,dz = \frac{\partial\Phi}{\partial x}\,dx + i_2\frac{\partial\Phi}{\partial x}\,dy + i_3\frac{\partial\Phi}{\partial x}\,dz,$$

from which the Cauchy – Riemann condition (13) follows. Then the mapping Φ is right-G-monogenic. The case of a left-H-monogenic mapping is considered analogously. $\square$

Theorem 13.4 and representations (29), (31) yield the following proposition.

Corollary 3 *If the domain $\Omega \subset \mathbb{R}^3$ is convex in the direction of the straight lines L^1 and L^2, and if $f_1(E_3) = f_2(E_3) = \mathbb{C}$, then every right-$H$-monogenic mapping $\Phi : \Omega_\zeta \to \mathbb{H}(\mathbb{C})$ can be represented in form (29), and every left-H-monogenic mapping $\widehat{\Phi} : \Omega_\zeta \to \mathbb{H}(\mathbb{C})$ can be represented in form (31).*

14 Different Equivalent Definitions of G-Monogenic Mappings

Here we obtain the following theorem which gives different equivalent definitions of G-monogenic mappings in a domain Ω_ζ.

Theorem 14.1 *A mapping $\Phi : \Omega_\zeta \to \mathbb{H}(\mathbb{C})$ $\left(or\ \widehat{\Phi} : \Omega_\zeta \to \mathbb{H}(\mathbb{C})\right)$ is right-G-monogenic $\left(or\ left\text{-}G\text{-}monogenic\right)$ in the domain $\Omega_\zeta \subset E_3$ if and only if one of the following conditions is satisfied:*
   (I) the components $U_k : \Omega \to \mathbb{C}$ of mapping (9) are $\mathbb{R}$-differentiable in the domain Ω and conditions (13) $\left(or\ (14)\right)$ are satisfied in the domain Ω_ζ;
   (II) the components $U_k : \Omega \to \mathbb{C}$ of mapping (9) are $\mathbb{R}$-differentiable in the domain Ω and the mapping Φ $\left(or\ \widehat{\Phi}\right)$ is right-H-monogenic $\left(or\ left\text{-}H\text{-}monogenic\right)$ in the domain Ω_ζ.

Corollary 4 *If $f_1(E_3) = f_2(E_3) = \mathbb{C}$, then the mapping Φ is right-G-monogenic $\left(or\ \widehat{\Phi}\ is\ left\text{-}G\text{-}monogenic\right)$ if and only if one of the following conditions is satisfied:*
   (III) for every point $\zeta_0 \in \Omega_\zeta$ there exists a neighborhood, in which the mapping Φ $\left(or\ \widehat{\Phi}\right)$ is the sum of power series (81) $\left(or\ (82)\right)$;

(IV) *the mapping Φ (or $\widehat{\Phi}$) is continuous in Ω_ζ and satisfies equality (73) (or (77)) for every triangle $\triangle_\zeta$ such that $\overline{\triangle_\zeta} \subset \Omega_\zeta$.*

Corollary 5 *If $f_1(E_3) = f_2(E_3) = \mathbb{C}$ for $k = 1, 2$ and in addition the domain Ω is convex in the direction of the straight lines L^1 and L^2, then the mapping Φ is right-G-monogenic (or $\widehat{\Phi}$ is left-G-monogenic) if and only if*
(V) there exist unique pair holomorphic in the domain D_1 functions F_1, F_3 (or $\widehat{F}_1$, $\widehat{F}_4$) of the variable ξ_1 and unique pair holomorphic in the domain D_2 functions F_2, F_4 (or $\widehat{F}_2$, $\widehat{F}_3$) of the variable ξ_2 such that the mapping Φ (or $\widehat{\Phi}$) is expressed in form (29) (or (31)) in the domain Ω_ζ.

Proof We start with a proof of the Theorem. The equivalence of condition (I) and the property of right-G-monogenicity was established by Theorem 3.1. The equivalence of condition (II) and the notion of right-G-monogenic mapping is established in Theorem 13.4.

Now we continue proofs or corollaries. That condition (III) and the notion of right-G-monogenic mapping are equivalent is a consequence of Theorem 10.1 and the property of convergent series (81) to define a mapping right-G-monogenic in a domain of convergence. The equivalence of condition (IV) and the notion of right-G-monogenic mapping follows from Theorem 8.1 and Theorem 7.3.

Finally, in order to proof the equivalence of condition (V) and the notion of right-G-monogenic mapping Φ, it is sufficient to note that the uniqueness of the functions F_1, F_2, F_3, F_4 in (31) follows from the uniqueness of the decomposition of an element with respect to the basis $\{e_1, e_2, e_3, e_4\}$ of the algebra $\mathbb{H}(\mathbb{C})$, and mapping (31) is right-G-monogenic in Ω_ζ because it satisfies condition (13).

Theorem is proved for the left-G-monogenic mappings in the same way. □

Acknowledgements This research is partially supported by Grant of Ministry of Education and Science of Ukraine (Project No. 0116U001528).

References

Alpay, D., Colombo, F., Sabadini, I.: Slice hyperholomorphic Schur Analysis. Birkhäuser (2016)

Bitsadze, A.V.: Boundary-value Problems for Elliptic Equations of the Second Order. Nauka, Moscow (1966). (in Russian)

Blum, E.K.: A theory of analytic functions in Banach algebras. Trans. Amer. Math. Soc. **78**, 343–370 (1955)

Bryukhov, D., Kähler, U.: The static Maxwell system in three dimensional axially symmetric inhomogeneous media and axially symmetric generalization of the Cauchy-Riemann system. Adv. Appl. Clifford Algebra **27**(2), 993–1005 (2017)

Cartan, E.: Les groupes bilinéares et les systèmes de nombres complexes. Annales de la faculté des sciences de Toulouse **12**(1), 1–64 (1898)

Colombo, F., Sabadini, I., Struppa, D.C.: Entire Slice Regular Functions. Springer (2016)

Colombo, F., Sabadini, I., Struppa, D.C.: Noncommutative functional calculus: theory and applications of slice hyperholomorphic functions. In: Progress in Mathematics, vol. 289 (2011)

Cullen, C.G.: An integral theorem for analytic intrinsic functions on quaternions. Duke Math. J. **32**, 139–148 (1965)

Eriksson, S.-L., Leutwiler, H.: An improved Cauchy formula for hypermonogenic functions. Adv. Appl. Clifford Alg. **19**(2), 269–282 (2009)

Eriksson-Bique, S.-L.: A correspondence of hyperholomorphic and monogenic functions in $\mathbb{R}^4$. Clifford Analysis and Its Applications, NATO Science Series **25**, 71–80 (2001)

Flaut, C., Shpakivskyi, V.: Holomorphic functions in generalized Cayley-Dickson algebras. Adv. Appl. Clifford Algebra **25**(1), 95–112 (2015)

Flaut, C.: Some connections between binary block codes and Hilbert algebras. In: Maturo, A., et al. (ed.) Recent Trends in Social Systems: Quantitative Theories and Quantitative Models, Studies in Systems, Decision and Control, vol. 66, pp. 249–256 (2017)

Fueter, R.: Die Funktionentheorie der Differentialgleichungen $\Delta u = 0$ und $\Delta \Delta u = 0$ mit vier reellen Variablen. Comment. math. helv. **7**, 307–330 (1935)

Gentili, G., Stoppato, C., Struppa, D.: Regular Functions of a Quaternionic Variable. Springer Monographs in Mathematics (2013)

Gentili, G., Struppa, D.C.: A new approach to Cullen-regular functions of a quaternionic variable. Comptes Rendus Mathematique **342**(10), 741–744 (2006)

Gerus, O.F., Shapiro, M.: On the boundary values of a quaternionic generalization of the Cauchy-type integral in $\mathbb{R}^2$ for rectifiable curves. J. Nat. Geom. **24**(1–2), 121–136 (2003)

Gürlebeck, K., Sprössig, W.: Quaternionic and Clifford Calculus for Physicists and Engineers. Wiley (1997)

Hausdorff, F.: Zur Theorie der Systeme complexer Zahlen. Leipziger Berichte **52**, 43–61 (1900)

Hempfling, Th, Leutwiler, H.: Modified quaternionic analysis in $\mathbb{R}^4$, Clifford algebras and their appl. in math. physics, pp. 227–238. Kluwer, Aachen, Dordrecht (1998)

Herus, O.F.: On hyperholomorphic functions of the space variable. Ukr. Math. J. **63**(4), 530–537 (2011)

Hille, E., Phillips, R.S.: Functional analysis and semi-groups. Am. Math, Soc (1948)

Ju, J.: Trohimchuk, Continuous Mappings and Conditions of Monogenity, Israel Program for Scientific Translations. Jerusalem; Daniel Davey and Co., Inc, New York (1964)

Ketchum, P.W.: Analytic functions of hypercomplex variables. Trans. Amer. Math. Soc. **30**(4), 641–667 (1928)

Kisil, V.V.: Analysis in $\mathbf{R}^{1,1}$ or the principal function theory. Complex Var. Theory Appl. **40**(2), 93–118 (1999)

Kisil, V.V.: How Many Essentially Different Function Theories Exist? In: Clifford Algebras and Their Application in Mathematical Physics ,vol. 94, pp. 175–184. Kluwer Academic Publishers, Dordrecht (1998)

Kisil, V.V.: Erlangen programme at large: an overview. In: Rogosin, S.V., Koroleva, A.A. (eds.) Advances in Applied Analysis, vol. 94, pp. 1–94. Birkhäuser Verlag, Basel (2012)

Kravchenko, V.V., Shapiro, M.V.: Integral Representations for Spatial Models of Mathematical Physics. Addison Wesley Longman Inc, Pitman Research Notes in Mathematics (1996)

Kravchenko, V.V.: Applied Quaternionic Analysis. Heldermann Verlag (2003)

Kuzmenko, T.S.: Curvilinear integral theorem for G-monogenic mappings in the algebra of complex quaternion. Int. J. Adv. Res. Math. **6**, 21–25 (2016)

Kuzmenko, T.S.: Power and Laurent series in the algebra of complex quaternion. In: Proceedings of the Institute Mathematics of the National Academy of Sciences of Ukraine, vol. 12, iss. 3, pp. 164–174 (2015). (in Ukrainian)

Kuzmenko, T.S., Shpakivskyi, V.S.: Generalized integral theorems for the quaternionic G-monogenic mappings. J. Math. Sci. **224**(4), 530–540 (2017)

Leutwiler, H., Zeilinger, P.: On quaternionic analysis and its modifications. Comput. Methods Funct. Theory **4**(1), 159–182 (2004)

Leutwiler, H.: Modified quaternionic analysis in $\mathbb{R}^3$. Complex variables theory appl. **20**, 19–51 (1992)

Luna-Elizarrarás, M.E., Shapiro, M., Struppa, D.C., Vajiac, A.: Bicomplex holomorphic functions: the algebra, geometry and analysis of bicomplex numbers. Birkhäuser (2015)

Moisil, G.C., Theodoresco, N.: Functions holomorphes dans l'espace. Mathematica (Cluj) **5**, 142–159 (1931)

Plaksa, S.A., Pukhtaievych, R.P.: Monogenic functions in a finite-dimensional semi-simple commutative algebra. An. Şt. Univ. Ovidius Constanţa **22**(1), 221–235 (2014)

Plaksa, S.A., Shpakovskii, V.S.: Constructive description of monogenic functions in a harmonic algebra of the third rank. Ukr. Math. J. **62**(8), 1251–1266 (2011)

Portman, W.O.: A derivative for Hausdorff-analytic functions. Proc. Amer. Math. Soc. **V**(10), 101–105 (1959)

Privalov, I.I.: Introduction to the Theory of Functions of a Complex Variable. GITTL, Moscow (1977). (in Russian)

Rinehart, R.F., Wilson, J.C.: Two types of differentiability of functions on algebras. Rend. Circ. Matem. Palermo, **II**(11), 204–216 (1962)

Ringleb, F.: Beiträge zur funktionentheorie in hyperkomplexen systemen. I. Rend. Circ. Mat. Palermo **57**(1), 311–340 (1933)

Schneider, B.: Some properties of a Cauchy-type integral for the Moisil-Theodoresco system of partial differential equations. Ukr. Math. J. **58**(1), 105–112 (2006)

Segre, C.: The real representations of complex elements and extension to bicomplex systems. Math. Ann. **40**, 413–467 (1892)

Shabat, B.V.: Introduction to Complex Analysis, Part 1. Nauka, Moskow (1976). (in Russian)

Shpakivskyi, V.S., Kuzmenko, T.S.: Integral theorems for the quaternionic G-monogenic mappings. An. Şt. Univ. Ovidius Constanţa **24**(2), 271–281 (2016)

Shpakivskyi, V.S.: Constructive description of monogenic functions in a finite-dimensional commutative associative algebra. Adv. Pure Appl. Math. **7**(1), 63–76 (2016)

Shpakivskyi, V.S., Kuzmenko, T.S.: On monogenic mappings of a quaternionic variable. J. Math. Sci. **221**(5), 712–726 (2017)

Shpakivskyi, V.S., Kuzmenko, T.S.: On one class of quaternionic mappings. Ukr. Math. J. **68**(1), 127–143 (2016)

Sudbery, A.: Quaternionic analysis. Math. Proc. Camb. Phil. Soc. **85**, 199–225 (1979)

Tolstov, G.P.: On the curvilinear and iterated integral, Trudy Mat. Inst. Steklov. Acad. Sci. USSR **35**, 3–101 (1950). (in Russian)

Van der Waerden, B.L.: Algebra, vol. 1. Springer (1991)

On Bicomplex Fibonacci Numbers and Their Generalization

Serpil Halici

Abstract In this chapter, we consider bicomplex numbers with coefficients from Fibonacci sequence and give some identities. Moreover, we demonstrate the accuracy of such identities by taking advantage of idempotent representations of the bicomplex numbers. And then by this representation, we give some identities containing these numbers. We then make a generalization that includes these new numbers and we call them Horadam bicomplex numbers. Moreover, we obtain the Binet formula and generating function of Horadam bicomplex numbers for the first time. We also obtain two important identities that relate the matrix theory to the second order recurrence relations.

1 Introduction and Preliminaries

Hamilton (1844) tried to define three dimensional algebra and then he recognized the idea behind the multiplication of imaginary units i, j, k and defined real quaternions. Cockle et al. (1848) defined the split quaternions. There are many studies on quaternion algebra and some special quaternions. For some of these, one can look at the references Anastassiu et al. (2003), Cockle et al. (1848), Flaut and Savin (2015), Halici (2012, 2013), Horadam (1963, 1967), Savin (2016), Ward (1997). Segre (1892) defined bicomplex numbers and gave some fundamental properties. Bicomplex numbers are neglected because of their zero divisors but now they are attracting attention of researchers.

For some important papers on bicomplex variables, the readers may refer to Anastassiu et al. (2003), Kabadayi and Yayli (2011), Luna et al. (2012, 2015). Also, there are some studies on bicomplex quantum mechanics Bagchi and Banarjee (2015), Lavoie et al. (2010), Mathieu et al. (2013), Rochon et al. (2004, 2006). Since the mathematical structure of quantum mechanics is studied on the complex number field, there are many authors studying on bicomplex numbers in this area. This

S. Halici (✉)
Department of Mathematics, Faculty of Arts and Sciences, Pamukkale University,
20007 Denizli, Turkey
e-mail: shalici@pau.edu.tr

© Springer Nature Switzerland AG 2019
C. Flaut et al. (eds.), *Models and Theories in Social Systems*, Studies in Systems, Decision and Control 179, https://doi.org/10.1007/978-3-030-00084-4_26

new area, which is an approximation of the standard quantum mechanics, is known as bicomplex quantum mechanics. In recent years, some researchers have studied algebraic, geometric, topological and dynamic properties of bicomplex numbers (see, Anastassiu et al. 2003; Bagchi and Banarjee 2015; Kabadayi and Yayli 2011; Mathieu et al. 2013, Nurkan and Guven 2015). In Nurkan and Guven (2015), the authors described bicomplex Fibonacci and Lucas numbers and gave many equality and identity with respect to these numbers.

In this study, paying attention to the study in Nurkan and Guven (2015), we introduce idempotent representations of these new numbers. With the help of this representation, we show that many algebraic processes can actually be easily played on this newly introduced set.

In the second section of this study, we first examine some properties of this new numbers with idempotent representation by examining the equivalence and norms of bicomplex numbers. Later, we consider bicomplex numbers with coefficient from a special number sequence, and give some fundamental properties of the sequence containing these numbers. We also give three important identities that the elements of this sequence provided. Then, we do a generalization that includes these new numbers and we call them Horadam bicomplex numbers. Moreover, we obtain the Binet formula and generating function of these numbers for the first time. We also obtain two important identities that relate the matrix theory to the second order recurrence relations.

In the third part of this study, using the bicomplex numbers with coefficients from the Fibonacci sequences we give many equations that hold an important place in the literature on bicomplex Fibonacci numbers. We then use the Horadam numbers to define a new set of numbers that generalizes all these types of numbers, and we call it bicomplex Horadam numbers. We also prove some important identities for these numbers, which are first defined by us.

Now, let us give definition and some fundamental properties of bicomplex numbers. If $z_1 = a_1 + ia_2$ and $z_2 = a_3 + ia_4$, then $b \in \mathbb{BC}$ can be given in hypercomplex number form $b = a_1 + ia_2 + ja_3 + ija_4$ where $ij = ji = k$. Then the set of bicomplex numbers is

$$\mathbb{BC} = \{b = z_1 + z_2 j \mid z_1, \ z_2 \in \mathbb{C}, \ j^2 = -1\}.$$

Addition of bicomplex numbers b_1 and b_2 is component wise. And the multiplication of their basis elements can be done according to the following Table 1.

It should be noted that the multiplication of bicomplex numbers is similar to multiplication of real quaternions. If we summarize some differences between these two sets of numbers, we can list them as bicomplex numbers are commutative, have zero divisors and non-trivial idempotent elements, but real quaternions are non-commutative and don't have zero divisors and non-trivial idempotent elements. That is, the following equations are satisfied for elements of the set $\mathbb{BC}$.

$$(i + j)(i - j) = i^2 - ij + ji - j^2 = -1 - k + k + 1 = 0$$

Table 1 Multiplication of imaginary units

.		i	j	ij
i		-1	ij	$-j$
j		ij	-1	$-i$
ij		$-j$	$-i$	1

and

$$\left(\frac{1+ij}{2}\right)^2 = \frac{1+ij}{2}.$$

Furthermore, bicomplex numbers have three different involutions and norms. For any bicomplex number b these involutions are as follows;

$$b = a_1 + a_2 i + a_3 j + a_4 ij,$$

$$\overline{b}_i = a_1 - a_2 i + a_3 j - a_4 ij,$$

$$\overline{b}_j = a_1 + a_2 i - a_3 j - a_4 ij,$$

$$\overline{b}_{ij} = a_1 - a_2 i - a_3 j + a_4 ij.$$

Norms arising from the definitions of involutions are as follows;

$$N_i(b) = b\overline{b}_i = (a_1 + a_2 i + a_3 j + a_4 ij)(a_1 - a_2 i + a_3 j - a_4 ij),$$

$$N_j(b) = b\overline{b}_j = (a_1 + a_2 i + a_3 j + a_4 ij)(a_1 + a_2 i - a_3 j - a_4 ij),$$

$$N_{ij}(b) = b\overline{b}_{ij} = (a_1 + a_2 i + a_3 j + a_4 ij)(a_1 - a_2 i - a_3 j + a_4 ij).$$

We should note all of these norms are isotropic. And for $N_i( )$ we can take $(1 + ij)$ and calculate the norm;

$$N_i(1 + ij) = (1 + ij)(1 - ij) = 1^2 - ij + ij - (ij)^2 = 0.$$

There are two other idempotent elements in $\mathbb{BC}$ apart from the obvious idempotent elements (Luna et al. 2015):

$$e_1 = (1 + ij)/2, \quad e_2 = (1 - ij)/2 \tag{1}$$

and these elements provide the following equations.

$$e_1 e_2 = 0, \quad e_1 + e_2 = 1, \quad e_1 - e_2 = ij. \tag{2}$$

Hence, any b element in $\mathbb{BC}$ can be expressed in terms of the elements e_1 and e_2;

$$b = z_1 + z_2 j = \beta_1 e_1 + \beta_2 e_2 = (z_1 - iz_2)e_1 + (z_1 + iz_2)e_2. \tag{3}$$

The Eq. (3) is called as the idempotent representation of element b. Also, β_1 and β_2 are called as idempotent coefficients (Luna et al. 2015). Then, for the elements b_1, b_2 are written as

$$b_1 = \beta_1 e_1 + \beta_2 e_2, \quad b_2 = \delta_1 e_1 + \delta_2 e_2 \tag{4}$$

and the algebraic operations in this set are done with the help of this representation as follows;

$$b_1 + b_2 = (\beta_1 + \delta_1)e_1 + (\beta_2 + \delta_2)e_2, \tag{5}$$

$$b_1.b_2 = (\beta_1.\delta_1)e_1 + (\beta_2.\delta_2)e_2, \tag{6}$$

$$b_1{}^n = \beta_1^n e_1 + \beta_2^n e_2. \tag{7}$$

Where $\beta_1 = z_1 - iz_2$ and $\beta_2 = z_1 + iz_2$. Also, the conjugates $\overline{b}_i, \overline{b}_j, \overline{b}_{ij}$ are

$$\overline{b}_i = \overline{\beta}_2 e_1 + \overline{\beta}_1 e_2, \tag{8}$$

$$\overline{b}_j = \beta_2 e_1 + \beta_1 e_2, \tag{9}$$

$$\overline{b}_{ij} = \overline{\beta}_1 e_1 + \overline{\beta}_2 e_2. \tag{10}$$

Hence three different norm formulas can be derived as

$$b\overline{b}_i = \beta_1 \overline{\beta}_2 e_1 + \overline{\beta_1 \overline{\beta}_2} e_2, \tag{11}$$

$$b\overline{b}_j = \beta_1 \beta_2, \tag{12}$$

$$b\overline{b}_{ij} = |\beta_1|^2 e_1 + |\beta_2|^2 e_2. \tag{13}$$

Up to now we presented some fundamental properties of bicomplex numbers. In the following sections firstly we recall the definitions of bicomplex Fibonacci and Lucas numbers which is defined in the reference Nurkan and Guven (2015). Secondly, we give some important additional identities for the recurrence relations involving these numbers. And then, we define the bicomplex Horadam numbers which generalizes all Fibonacci-like recurrence relations on bicomplex numbers. Also, we give some fundamental identities for bicomplex Horadam numbers.

2 Bicomplex Fibonacci and Lucas Numbers

It is known that Fibonacci and Lucas sequences are given by the following recurrence relations (see, Atanassov et al. 2002; Koshy 2001),

$$F_{n+2} = F_{n+1} + F_n \ and \ L_{n+2} = L_{n+1} + L_n$$

where $F_0 = 0$, $F_1 = 1$ and $L_0 = 2$, $L_1 = 1$ are the initial values of these sequences, respectively.

Now let's define two bicomplex sequences with coefficients are from Fibonacci and Lucas sequences. And then let us demonstrate them with the symbols $\mathbb{BC}_F$ and $\mathbb{BC}_L$. For $n \geq 0$, these sequences are

$$\mathbb{BC}_F = \{f_n + f_{n+2}j| \ f_n = F_n + iF_{n+1}, \ j^2 = -1\} \tag{14}$$

and

$$\mathbb{BC}_L = \{l_n + l_{n+2}j| \ l_n = L_n + iL_{n+1}, \ j^2 = -1\}. \tag{15}$$

Briefly, let's write these sequences as follows

$$\mathbb{BC}_F = \{Q_0, Q_1, Q_2, \ldots, Q_n \ldots\}, \ \mathbb{BC}_L = \{K_0, K_1, K_2, \ldots, K_n \ldots\}. \tag{16}$$

Then, we can express any element Q_n in $\mathbb{BC}_F$ as

$$Q_n = \alpha_n e_1 + \beta_n e_2 \tag{17}$$

where

$$\alpha_n = f_n - if_{n+2}, \ \beta_n = f_n + if_{n+2}. \tag{18}$$

The idempotent representation (17) of the element Q_n is unique in $\mathbb{BC}_F$. We would like to state the algebraic operations to be performed on $\mathbb{BC}_F$ are easily done with the help of idempotent representation. Moreover, it is also easier to examine the properties of $\mathbb{BC}_F$ with the help of this notation. It is important to note that in this point, in Halici (2012), the author defined the quaternions with Fibonacci coefficient for the first time, and obtained the Binet formula for these quaternions. Since both the bicomplex Fibonacci numbers and the Fibonacci quaternions have the same recurrence relation, their Binet formulas are expected to be the same (see, Halici 2012; Nurkan and Guven 2015).

Notice that using the idempotent representations of Q_n and Q_m, and the algebraic properties of e_1, e_2, it is easy to see the following equations are satisfied.

$$Q_n + Q_m = (\alpha_n + \alpha_m)e_1 + (\beta_n + \beta_m)e_2, \tag{19}$$

$$Q_n Q_m = (\alpha_n \alpha_m)e_1 + (\beta_n \beta_m)e_2, \tag{20}$$

$$Q_n^m = \alpha_n^m e_1 + \beta_n^m e_2. \tag{21}$$

Corollary 1 *For $n \geq 2$, the following recursive relation among the elements of sequence $\mathbb{BC}_F$ is valid.*

$$Q_{n+2} = Q_{n+1} + Q_n, \tag{22}$$

where

$$Q_0 = \alpha_0 e_1 + \beta_0 e_2 , \quad Q_1 = \alpha_1 e_1 + \beta_1 e_2. \tag{23}$$

Proof Using the Eqs. (17) and (18) for the element Q_n, we have

$$Q_n = f_n + f_{n+2} j = (f_n - if_{n+2})e_1 + (f_n + if_{n+2})e_2.$$

Since the coefficients α_n and β_n in the equation

$$Q_n = \alpha_n e_1 + \beta_n e_2$$

also provide recurrence relations are

$$\alpha_{n+2} = \alpha_{n+1} + \alpha_n, \quad \beta_{n+2} = \beta_{n+1} + \beta_n$$

for $n \geq 0$, among the elements of $\mathbb{BC}_F$ we obtain

$$Q_{n+2} = Q_{n+1} + Q_n.$$

Thus, the proof is completed.

Remark 1 For the elements Q_n, Q_m in $\mathbb{BC}_F$, the following equations are satisfied

$$(i) \ (\overline{Q_n})_i = \overline{\beta}_n e_1 + \overline{\alpha}_n e_2, \tag{24}$$

$$(ii) \ (\overline{Q_n})_j = \beta_n e_1 + \alpha_n e_2, \tag{25}$$

$$(iii) \ (\overline{Q_n})_{ij} = \overline{\alpha}_n e_1 + \overline{\beta}_n e_2. \tag{26}$$

Remark 2 In $\mathbb{BC}_F$, for the elements Q_n, Q_m, the following equations are satisfied

$$i) \ |Q_n|_i^2 = Q_n(\overline{Q_n})_j = \alpha_n \overline{\beta}_n e_1 + \overline{(\alpha_n \overline{\beta}_n)}e_2, \tag{27}$$

$$ii) \ |Q_n|_j^2 = Q_n(\overline{Q_n})_i = \alpha_n \beta_n, \tag{28}$$

$$iii) \ |Q_n|_{ij}^2 = Q_n(\overline{Q_n})_{ij} = |\alpha_n|^2 e_1 + |\beta_n|^2 e_2. \tag{29}$$

In the following theorem, we give the formula known as the Binet formula with the aid of the second order linear homogeneous recurrence relation provided by elements of the sequence.

Theorem 1 (Binet Formula) *For the elements Q_n in sequence $\mathbb{BC}_F$, the identity* (30)

$$Q_n = \frac{1}{\sqrt{5}}[(A_1\alpha^n + B_1\beta^n)e_1 + (A_2\alpha^n + B_2\beta^n)e_2] \tag{30}$$

is satisfied. Here, the values α and β are the roots of the characteristic equation in (22). Where, $A_1 = \alpha_1 - \alpha_0\beta$, $B_1 = -\alpha_1 + \alpha_0\alpha$, $A_2 = \beta_1 - \beta_0\beta$, and $B_2 = \beta_1 - \beta_0\alpha$.

Proof By using the facts

$$\alpha_{n+2} = \alpha_{n+1} + \alpha_n, \quad \beta_{n+2} = \beta_{n+1} + \beta_n$$

which have the same relation as Q_{n+2}

$$Q_{n+2} = \alpha_{n+2}e_1 + \beta_{n+2}e_2$$

we get the Binet formulas for α_n and β_n as follows;

$$\alpha_n = \frac{1}{\sqrt{5}}(A_1\alpha^n + B_1\beta^n), \tag{31}$$

where $\alpha_0 = 2$ and $\alpha_1 = 4 - i$ are the initial values of (31) and similarly,

$$\beta_n = \frac{1}{\sqrt{5}}(A_2\alpha^n + B_2\beta^n), \tag{32}$$

where $\beta_0 = 2i - 2$ and $\beta_1 = 3i - 2$ are the initial values of (32). Thus, by using the equalities of α_n and β_n we obtain

$$Q_n = \frac{1}{\sqrt{5}}[(A_1\alpha^n + B_1\beta^n)e_1 + (A_2\alpha^n + B_2\beta^n)e_2]$$

which is the desired result. If we examine this formula given here it is the same with the formula is given in (Theorem 3.1, Halici 2012).

Theorem 2 (Cassini Identity) *For the elements in set $\mathbb{BC}_F$, we have*

$$Q_{n-1}Q_{n+1} - Q_n^2 = 3j(-1)^n(2+i), \quad n \geq 1. \tag{33}$$

Proof The Binet formula, direct proof and induction method can be used to prove. If we use the Binet formula and substitute Q_n, Q_{n-1} and Q_{n+1} in the formula (33), then we can get

$$Q_{n-1}Q_{n+1} - Q_n^2 = \frac{1}{5}A_1B_1(\alpha^{n-1}\beta^{n+1} + \alpha^{n+1}\beta^{n-1} - 2(-1)^n)e_1$$

$$+\frac{1}{5}A_2B_2(\alpha^{n-1}\beta^{n+1} + \alpha^{n+1}\beta^{n-1} - 2(-1)^n)e_2.$$

Considering the relations between α and β and in particular $\alpha = -\beta^{-1}$, we can rearrange the last equation as follow;

$$Q_{n-1}Q_{n+1} - Q_n^2 = A_1B_1(-1)^n e_1 + A_2B_2(-1)^n e_2.$$

After calculating the values A_1B_1 and A_2B_2 we have

$$Q_{n-1}Q_{n+1} - Q_n^2 = (-1)^n\{(-3 + 6i)e_1 + (3 - 6i)e_2\} = 3j(-1)^n(2 + i).$$

Thus, the claim is verified.

The Cassini identity has two important generalizations; one is Catalan and the other is d'Ocagne identities. These identities actually represent that there is a close relationship between the recurrence relations and matrix theory. In some books, a few sections are specified for this identities (for example, see Koshy 2001).

Now let's give Catalan identity for this new sequence.

Theorem 3 (Catalan) *When m, n are integers and $n \geq m$, then we have*

$$Q_{n+m}Q_{n-m} - Q_n^2 = 3j(-1)^{n-m+1}F_m^2(2 + i). \tag{34}$$

Proof Utilizing the Binet formula of Q_n, let us write the following equations

$$Q_{n+m} = \frac{1}{\sqrt{5}}[(A_1\alpha^{n+m} + B_1\beta^{n+m})e_1 + (A_2\alpha^{n+m} + B_2\beta^{n+m})e_2]$$

and

$$Q_{n-m} = \frac{1}{\sqrt{5}}[(A_1\alpha^{n-m} + B_1\beta^{n-m})e_1 + (A_2\alpha^{n-m} + B_2\beta^{n-m})e_2].$$

So, the left side of equation of (34) is

$$Q_{n+m}Q_{n-m} - Q_n^2 = \frac{1}{5}\{(A_1B_1(\alpha^{n+m}\beta^{n-m} + \alpha^{n-m}\beta^{n+m} - 2(-1)^n)e_1)$$
$$+ (A_2B_2(\alpha^{n+m}\beta^{n-m} + \alpha^{n-m}\beta^{n+m} - 2(-1)^n)e_2)\},$$

$$Q_{n+m}Q_{n-m} - Q_n^2 = \frac{1}{5}A_1B_1\{(-1)^n(-1)^{-m}(\alpha^m - \beta^m)^2\}e_1$$
$$+ A_2B_2\{(-1)^n(-1)^{-m}(\alpha^m - \beta^m)^2\}e_2,$$

$$Q_{n+m}Q_{n-m} - Q_n^2 = (-1)^{n-m+1}A_1B_1\{F_m^2 e_1 + A_2B_2F_m^2\}e_2,$$

$$Q_{n+m}Q_{n-m} - Q_n^2 = (-1)^{n-m+1}F_m^2(A_1B_1e_1 + A_2B_2e_2).$$

If we calculate the values A_1B_1 and A_2B_2 and write these in the last equation, then we obtain

$$Q_{n+m}Q_{n-m} - Q_n^2 = 3j(-1)^{n-m+1}F_m^2(2+i).$$

Thus, the claim is true.

Now let's give also an important identity related to the Cassini identity.

Theorem 4 (d'Ocagne) *For integers n and m which are positive and different from each other, the following equation is true.*

$$Q_m Q_{n+1} - Q_n Q_{m+1} = 3j(-1)^m Q_{n-m}(2+i), \quad n \geq m. \tag{35}$$

Proof Let's first write the Binet formula for Q_n and Q_m and then calculate $Q_m Q_{n+1} - Q_n Q_{m+1}$. Then we have

$$Q_m Q_{n+1} - Q_n Q_{m+1} = \frac{1}{5}\{(A_1\alpha^m + B_1\beta^m)e_1 + (A_2\alpha^m + B_2\beta^m)e_2\}.$$

$$\{(A_1\alpha^{n+1} + B_1\beta^{n+1})e_1 + (A_2\alpha^{n+1} + B_2\beta^{n+1})e_2\}$$

$$- \frac{1}{5}\{(A_1\alpha^n + B_1\beta^n)e_1 + (A_2\alpha^n + B_2\beta^n)e_2\}.$$

$$\{(A_1\alpha^{m+1} + B_1\beta^{m+1})e_1 + (A_2\alpha^{m+1} + B_2\beta^{m+1})e_2\}.$$

If we make the necessary calculations, then we have

$$Q_m Q_{n+1} - Q_n Q_{m+1} = \frac{1}{5}\{A_1B_1(\alpha^m\beta^{n+1} + \alpha^{n+1}\beta^m - \alpha^n\beta^{m+1}$$

$$- \alpha^{m+1}\beta^n)e_1 + A_2B_2(\alpha^m\beta^{n+1}$$

$$+ \alpha^{n+1}\beta^m - \alpha^n\beta^{m+1} - \alpha^{m+1}\beta^n)e_2\}.$$

Using the relations between the roots of the characteristic equation, we get

$$Q_m Q_{n+1} - Q_n Q_{m+1} = \frac{1}{5}\{A_1B_1(-1)^m(\alpha - \beta)(\alpha^{n-m} - \beta^{n-m})e_1$$

$$+ A_2B_2(-1)^m(\alpha - \beta)(\alpha^{n-m} - \beta^{n-m})e_2\}.$$

Making some necessary simplifications and using the formula $\alpha^{n-m} - \beta^{n-m} = \sqrt{5}Q_{n-m}$, we derive that

$$Q_m Q_{n+1} - Q_n Q_{m+1} = (-1)^m Q_{n-m}(A_1B_1e_1 + A_2B_2e_2) = 3j(-1)^m Q_{n-m}(2+i).$$

Thus, the proof is completed.

We will give the following theorem without proof.

Theorem 5 *For Q_n in $\mathbb{BC}_F$ and m is a positive integer, $m \neq 1$, the following equations are satisfied.*

$$(i) \quad \sum_{n=1}^{m} Q_{2n} = Q_{2m+1} - (1 + i + 2j + 3ij). \tag{36}$$

$$(ii) \quad \sum_{n=1}^{m} Q_n = Q_{m+2} - (1 + 2i + 3j + 5ij). \tag{37}$$

Theorem 6 (Generating Function) *For the elements in the set $\mathbb{BC}_F^*$, the generating function is*

$$G(t) = \frac{t + i + j(1 + t) + ij(2 + t)}{1 - t - t^2}. \tag{38}$$

Proof For Q_n, the generating function $G(t)$ is

$$G(t) = (\alpha_0 e_1 + \beta_0 e_2) + (\alpha_1 e_1 + \beta_1 e_2)t + \cdots + (\alpha_n e_1 + \beta_n e_2)t^n + \cdots . \tag{39}$$

If we multiply the function $G(t)$ with t and t^2, respectively, and use the recurrence $Q_{n+2} - Q_{n+1} - Q_n = 0$, and making the necessary calculations, then we get

$$G(t)(1 - t - t^2) = (\alpha_0 e_1 + \beta_0 e_2) + (\alpha_1 e_1 + \beta_1 e_2 - \alpha_0 e_1 - \beta_0 e_2)t.$$

Making some needed arrangements, we have

$$G(t) = \frac{t + i + j(1 + t) + ij(2 + t)}{1 - t - t^2}.$$

Accordingly, the desired result is obtained.

Theorem 7 *There is a relationship between any element Q_n in $\mathbb{BC}_F$ and its conjugates*

$$Q_n + (\overline{Q_n})_i + (\overline{Q_n})_j + (\overline{Q_n})_{ij} = 4F_n. \tag{40}$$

Proof One can easily see that the following equations are true.

$$Q_n + (\overline{Q_n})_i = 2(F_n + j F_{n+2}),$$

$$Q_n + (\overline{Q_n})_j = 2(F_n + i F_{n+1}),$$

$$Q_n + (\overline{Q_n})_{ij} = 2(F_n + ij F_{n+3}).$$

Hence, with the help of a simple addition process, the correctness of the claim can be seen.

Corollary 2 *The following relations exist between bicomplex Fibonacci Q_n and bicomplex Lucas numbers K_n;*

$$(i) \ Q_n + K_n = 2Q_{n+1}, \tag{41}$$

$$(ii) \ K_{n-1} + K_{n+1} = 5Q_n, \tag{42}$$

$$(iii) \ 5Q_n^2 - K_n^2 = 12(-1)^{n+1}(2j + k). \tag{43}$$

Corollary 3 *For some sums of bicomplex Lucas numbers K_n, we have*

$$(i) \ \sum_{i=1}^{n} K_{2i-1} = K_{2n} - K_0, \tag{44}$$

$$(ii) \ \sum_{i=1}^{n} K_i = K_{n+2} - K_2, \tag{45}$$

$$(iii) \ \sum_{i=1}^{n} K_{2i} = K_{2n+1} - K_1, \tag{46}$$

$$(iv) \ \sum_{i=1}^{n} K_i^2 = K_n K_{n+1} - (5 - 10i - 12j + 9k). \tag{47}$$

Theorem 8 *For bicomplex Lucas numbers K_n, K_m, we have*

$$K_n K_m + K_{n+1} K_{m+1} = 5(2Q_{n+m+1} + 2F_{n+m+4} - F_{n+m+1}$$
$$- 2F_{n+m+6}i - 2F_{n+m+5}j + 2F_{n+m+4}k). \tag{48}$$

Proof We need firstly to use the identity

$$L_n L_m + L_{n+1} L_{m+1} = 5F_{m+n+1}$$

where L_n is the nth Lucas number (Koshy 2001). So, let us explicitly calculate the right hand side of the equation

$$K_n K_m + K_{n+1} K_{m+1} = [(L_n L_m + L_{n+1} L_{m+1}) - (L_{n+1} L_{m+1} + L_{n+2} L_{m+2})$$

$$-(L_{n+2} L_{m+2} + L_{n+3} L_{m+3}) + (L_{n+3} L_{m+3} + L_{n+4} L_{m+4})]$$

$$+[(L_n L_{m+1} + L_{n+1} L_m) - (L_{n+2} L_{m+3} + L_{n+3} L_{m+2})$$

$$+(L_{n+1}L_{m+2} + L_{n+2}L_{m+1}) - (L_{n+3}L_{m+4} + L_{n+4}L_{m+3})]i$$

$$+[(L_n L_{m+2} + L_{n+2}L_m) - (L_{n+1}L_{m+3} + L_{n+3}L_{m+1})$$

$$+(L_{n+1}L_{m+3} + L_{n+3}L_{m+1}) - (L_{n+2}L_{m+4} + L_{n+4}L_{m+2})]j$$

$$+[(L_n L_{m+3} + L_{n+3}L_m) + (L_{n+1}L_{m+2} + L_{n+2}L_{m+1})$$

$$+(L_{n+1}L_{m+4} + L_{n+4}L_{m+1}) + (L_{n+2}L_{m+3} + L_{n+3}L_{m+2})]ij.$$

If we use the identity
$$L_n L_m + L_{n+1}L_{m+1} = 5F_{m+n+1},$$

then we get

$$K_n K_m + K_{n+1}K_{m+1} = 5(F_{m+n+1} - F_{m+n+3} - F_{m+n+5} + F_{m+n+7})$$

$$+5(F_{m+n+2} + F_{m+n+2} - F_{m+n+6} - F_{m+n+6})i$$

$$+5(F_{m+n+3} + F_{m+n+3} - F_{m+n+5} - F_{m+n+5})j$$

$$+5(F_{m+n+4} + F_{m+n+4} + F_{m+n+4} + F_{m+n+4})ij.$$

After the necessary calculations and arrangements we get the desired result.

Up to now, we give the basic properties and some advanced identities including bicomplex Fibonnaci and Lucas numbers. In the next section, we present Horadam numbers which is defined to generalize all Fibonacci-like numbers and define bicomplex Horadam numbers. Then, we deduce from our definition involving bicomplex Horadam numbers generalizing all bicomplex Fibonacci-like numbers.

3   Bicomplex Horadam Numbers

Horadam defined the Horadam numbers as

$$w_n = w_n(a, b; p, q) = pw_{n-1} + qw_{n-2}; \quad n \geq 2, \quad W_0 = a, \quad W_1 = b. \tag{49}$$

Where a, b and $p, q \in \mathbb{Z}$ (Horadam 1963, 1967, 1965, 1961).
Now, let us define nth bicomplex Horadam number by the Horadam numbers as

$$BH_n = w_n + w_{n+1}i + w_{n+2}j + w_{n+3}k. \tag{50}$$

After some simple operations one can easily see that the recurrence relation involving bicomplex Horadam numbers is

$$BH_{n+2} = pBH_{n+1} + qBH_n, \tag{51}$$

and its initial values are

$$BH_0 = a + bi + (pb + qa)j + (p^2b + pqa + qb)k \tag{52}$$

and

$$BH_1 = b + (pb + qa)i + (p^2b + pqa + qb)j + (p^3b + p^2qa + 2pqb + q^2a)k. \tag{53}$$

We should note that if we take $(a, b; p, q) = (0, 1; 1, 1)$ and $(a, b; p, q) = (2, 1; 1, 1)$ in the equations (51)–(53), then we get the bicomplex Fibonacci and Lucas numbers, respectively.

In the following theorem, we give the Binet formula for bicomplex Horadam numbers.

Theorem 9 *For bicomplex Horadam numbers, the Binet formula is*

$$BH_n = \frac{A\underline{\alpha}\alpha^n - B\underline{\beta}\beta^n}{\alpha - \beta}. \tag{54}$$

Where

$$A = b - a\beta, \quad B = b - a\alpha, \quad \underline{\alpha} = 1 + \alpha i + \alpha^2 j + \alpha^3 k \text{ and } \underline{\beta} = 1 + \beta i + \beta^2 j + \beta^3 k. \tag{55}$$

Proof Using the roots of characteristic equation $t^2 - pt - q = 0$ related to Horadam numbers that is,

$$\alpha = \frac{p + \sqrt{p^2 + 4q}}{2}, \quad \beta = \frac{p - \sqrt{p^2 + 4q}}{2} \tag{56}$$

and by doing the necessary calculations we obtain

$$BH_n = \frac{A\underline{\alpha}\alpha^n - B\underline{\beta}\beta^n}{\alpha - \beta} \tag{57}$$

which is desired.

It should be noted that, the formula (57) is a generalization of Binet formula of bicomplex Fibonacci and Lucas numbers. In fact, in the Eqs. (51)–(53), if we choose the values a, b and p, q as 0, 1 and 1, 1 respectively, then we have the Binet formula for bicomplex Fibonacci numbers and then using the values $a = 2, b = 1$ and

$p = 1, q = 1$ we get the Binet formula for bicomplex Lucas numbers (see, Nurkan and Guven 2015, Theorem 4).

Theorem 10 *Generating function for bicomplex Horadam numbers is*

$$g(t) = \frac{BH_0 + (BH_1 - pBH_0)t}{1 - pt - qt^2}. \tag{58}$$

Proof First, let's write the generating function involving these numbers as follows;

$$g(t) = \sum_{n=0}^{\infty} BH_n t^n. \tag{59}$$

Hence, let's calculate the equations $ptg(t)$ and $qt^2 g(t)$;

$$ptg(t) = \sum_{n=0}^{\infty} pBH_n t^{n+1}; \qquad qt^2 g(t) = \sum_{n=0}^{\infty} qBH_n t^{n+2}. \tag{60}$$

Also, if we take advantage of the characteristic equation of these numbers, $t^2 - pt - q = 0$, we find an explicit form for the generating function

$$g(t) = \frac{BH_0 + (BH_1 - pBH_0)t}{1 - pt - qt^2}.$$

Thus, the proof is completed.

In the equation (58), using the initial values of Fibonacci sequence , the formula of generating function for bicomplex Fibonacci numbers is obtained. In particular, it is not difficult to see that the following formula can be obtained

$$\frac{K_0 + (K_1 - K_0)t}{1 - t - t^2} \tag{61}$$

if the initial values of Lucas numbers are written.

In the following theorem, we give the identity which is an important generalization of Cassini identity called as Catalan identity. Then, we will see that we can get the Catalan identity for all bicomplex Fibonacci-like numbers using the Binet formula (54).

Theorem 11 *The Catalan identity for bicomplex Horadam numbers is*

$$BH_n^2 - BH_{n+r} BH_{n-r} = \frac{AB \, \underline{\alpha\beta} \, q^{n-r}}{p^2 + 4q} (\alpha^r - \beta^r)^2. \tag{62}$$

Where $\underline{\alpha}$ and $\underline{\beta}$ are the same as in the equation (55).

Proof Let's use the Binet formula we found for Horadam numbers on the left side of the Catalan equality (62). Then, we can write this.

$$\left(\frac{A\underline{\alpha}\alpha^n - B\underline{\beta}\beta^n}{\alpha - \beta}\right)^2 - \frac{A\underline{\alpha}\alpha^{n+r} - B\underline{\beta}\beta^{n+r}}{\alpha - \beta} \frac{A\underline{\alpha}\alpha^{n-r} - B\underline{\beta}\beta^{n-r}}{\alpha - \beta}. \tag{63}$$

If we make some adjustments and calculations in the Eq. (63), then we have the following formula.

$$\frac{1}{(\alpha - \beta)^2}(AB\,\underline{\alpha\beta}\,q^{n-r})(\alpha^r - \beta^r)^2. \tag{64}$$

With the help of the Eqs. (63) and (64), we obtain the following equality.

$$BH_n^2 - BH_{n+r}\,BH_{n-r} = \frac{AB\,\underline{\alpha\beta}\,q^{n-r}}{p^2 + 4q}(\alpha^r - \beta^r)^2.$$

Thus, the proof is completed.

Now finally, in the following theorem, we will give an important equality which is a special case of the Catalan identity and which relates the matrix theory to the recurrence relations.

Theorem 12 *The Cassini identity for bicomplex Horadam numbers is*

$$BH_n^2 - BH_{n+1}BH_{n-1} = AB\,\underline{\alpha\beta}\,q^{n-1}. \tag{65}$$

Proof The proof of this theorem can easily done by an oversimplification of the Catalan identity. For this purpose, in the Eq. (62), it is enough to write $r = 1$.

4 Conclusion

In this chapter, we first describe bicomplex numbers with coefficients from the Fibonacci and Lucas sequences. We give many identities that takes an important place in the literature on bicomplex numbers. Then, we use the Horadam numbers to define a new set of numbers that generalizes all these types of numbers, and we call as bicomplex Horadam numbers. We also prove some important identities for these numbers.

References

Anastassiu, H.T., Prodromos, E.A., Dimitra, I.K.: Application of bicomplex (quaternion) algebra to fundamental electromagnetics: a lower order alternative to the Helmholtz equation. IEEE Trans. Antennas Propag. **51**(8), 2130–2136 (2003)

Atanassov, K.T., Atanassova, V., Shannon, A.G., Turner, J.C.: New Visual Perspectives on Fibonacci Numbers. World Scientific, Singapore (2002)

Bagchi, B., Banerjee, A.: Bicomplex hamiltonian systems in quantum mechanics. J. Phys. A: Math. Theor. **48**, 50 (2015)

Cockle, J., Davies, T.S.: On certain functions resembling quaternions, and on a new imaginary in algebra. Philos. Mag. Ser. Taylor & Francis **33**(224), 435–439 (1848)

Flaut, C.: Savin, Diana: Quaternion algebras and generalized FibonacciLucas quaternions. Adv. Appl. Clifford Algebras **25**(4), 853–862 (2015)

Halici, S.: On fibonacci quaternions. Adv. Appl. Clifford Algebras **22**(2), 321–327 (2012)

Halici, S.: On complex fibonacci quaternions. Adv. Appl. Clifford Algebras 1–8 (2013)

Hamilton, W.R.: Ii on quaternions; or on a new system of imaginaries in algebra. Philos. Mag. Ser. Taylor & Francis **25**(163), 10–13 (1844)

Horadam, A.F.: A generalized Fibonacci sequence. Am. Math. Mon. **68**(5), 455–459 (1961)

Horadam, A.F.: Complex Fibonacci numbers and Fibonacci quaternions. Am. Math. Mon. **70**(3), 289–291 (1963)

Horadam, A.F.: Basic properties of a certain generalized sequence of numbers. Fibonacci Q. **3**(3), 161–176 (1965)

Horadam, A.F.: Special properties of the sequence W_n (a, b; p, q). Fibonacci Q. **5**(5), 424–434 (1967)

Jrmie, M., Marchildon, L., Rochon, D.: The bicomplex quantum Coulomb potential problem. Canadian J. Phys. **91**(12), 1093–1100 (2013)

Kabadayi, H., Yayli, Y.: Homothetic motions at E^4 with bicomplex numbers. Adv. Appl. Clifford Algebras **21**(3), 541–546 (2011)

Koshy, T.: Fibonacci and Lucas Numbers with Applications, vol. 1. Wiley, New Jersey (2001)

Lavoie, R.G., Louis, M., Rochon, D.: The Bicomplex Quantum Harmonic Oscillator (2010). arXiv:1001.1149

Luna-Elizarraras, M.E., Shapiro, M., Struppa, D.C., Vajiac, A.: Bicomplex numbers and their elementary functions. Cubo (Temuco) **14**(2), 61–80 (2012)

Luna-Elizarraras, M.E., Shapiro, M., Struppa, D.C., Vajiac, A.: Bicomplex Holomorphic Functions: The Algebra. Geometry and Analysis of Bicomplex Numbers, Birkhuser (2015)

Nurkan, S.K., Guven, I.A.: A Note On Bicomplex Fibonacci and Lucas Numbers (2015). arXiv:1508.03972

Rochon, D., Tremblay, S.: Bicomplex quantum mechanics: I. The generalized Schrdinger equation. Adv. Appl. Clifford Algebras **14**(2), 231–248 (2004)

Rochon, D., Tremblay, S.: Bicomplex quantum mechanics: II. The Hilbert space. Adv. Appl. Clifford Algebras **16**(2), 135–157 (2006)

Savin, D.: About division quaternion algebras and division symbol algebras. Carpathian J. Math. 233–240 (2016)

Segre, C.L.: rappresentazioni reali delle forme complesse a gli enti iperalgebrici. Mathematische Annalen **40**, 413–467 (1892)

Ward J.P.: Quaternions and Cayley Numbers: Algebra and Applications. Mathematics and Its Applications, vol. 403. Kluwer, Dordrecht (1997)

Cluster Analysis: An Application to a Real Mixed-Type Data Set

G. Caruso, S. A. Gattone, A. Balzanella and T. Di Battista

Abstract When you dispose of multivariate data it is crucial to summarize them, so as to extract appropriate and useful information, and consequently, to make proper decisions accordingly. Cluster analysis fully meets this requirement; it groups data into meaningful groups such that both the similarity within a cluster and the dissimilarity between groups are maximized. Thanks to its great usefulness, clustering is used in a broad variety of contexts; this explains its huge appeal in many disciplines. Most of the existing clustering approaches are limited to numerical or categorical data only. However, since data sets composed of mixed types of attributes are very common in real life applications, it is absolutely worth to perform clustering on them. In this paper therefore we stress the importance of this approach, by implementing an application on a real world mixed-type data set.

Keywords Clusters analysis · Numeric data · Categorical data · Mixed data
Cluster algorithm

1 Introduction

In order to discover interesting groups, cluster analysis has been applied to a variety of scientific areas (Caruso et al. 2018). In marketing, it can help to identify different customers clusters and to use this knowledge to create targeted campaigns (Valentini et al. 2011). In the field of crime prevention, clustering analysis is used in the search

G. Caruso (✉) · S. A. Gattone · T. Di Battista
University G. d'Annunzio, Pescara, Italy
e-mail: giulia.caruso@unich.it

S. A. Gattone
e-mail: gattone@unich.it

T. Di Battista
e-mail: dibattis@unich.it

A. Balzanella
University of Campania Luigi Vanvitelli, Caserta, Italy
e-mail: antonio.balzanella@unicampania.it

© Springer Nature Switzerland AG 2019
C. Flaut et al. (eds.), *Models and Theories in Social Systems*, Studies in Systems,
Decision and Control 179, https://doi.org/10.1007/978-3-030-00084-4_27

of credit card frauds or in monitoring criminal activities in electronic commerce (Nie et al. 2010; Peng et al. 2005). In the educational sector, cluster analysis can be used to identify specific common patterns among test items (Di Battista and Fortuna 2016). In medicine and in biology the clustering of shapes is routinely applied by practitioners in order to discover different structures in the set of objects (Brignell et al. 2010; Gattone et al. 2017). In environmental studies, clustering is applied in order to classify ecological communities on the basis of their diversity (Di Battista 2002; Di Battista and Gattone 2003; Fortuna and Maturo 2018; Maturo 2018).

To perform such analysis it is necessary to group a data set into homogeneous clusters, and to efficiently interpret them. Traditionally, cluster analysis only focus on purely numerical data. K-means method, due to its huge efficiency in processing large data sets, is the most popular clustering algorithm, especially for data mining (Di Battista and Gattone 2004; MacQueen 1967). Nevertheless, this algorithm has a big disadvantage: it is often limited to numerical attributes, since it is based on the Euclidean distance measure between data points and clusters-means (Everitt 1974). Furthermore, data often contains both numeric and categorical values. A way to overcome this problem is to transform the categorical values into quantitative ones, such as the binary strings, and then apply the numerical-value based clustering methods.

However, this approach would cause a loss of knowledge, ignoring the similarity information enclosed in the categorical attributes (Ahmad and Dey 2007). Hence, it is desirable to overcome this problem by finding a unified similarity metric for both categorical and numerical data, in order to eliminate the metric gap between continuous and categorical data. Some papers have tried to find a unified similarity metric for categorical and quantitative attributes, but a computational efficient similarity measure has yet to be implemented.

Huang (1997) presents a clustering algorithm to solve data partition problems. Whilst it is based on the K-means paradigm, it removes the numeric data only limitation, preserving, at the same time, its efficiency. This algorithm clusters objects with quantitative and categorical attributes, similarly to K-means. It is called K-prototypes algorithm, because objects are clustered against K prototypes instead of K means.

Cheung and Jia (2013) method is based on the concept of object-cluster similarity. They propose a new metric for both quantitative and qualitative attributes, so that the object cluster similarity for both of them has a uniform criterion. In this way, they (Cheung and Jia 2013) eliminate the need of transformation and parameter adjustment between categorical and numerical values.

The rest of the paper is organized as follows. In Sect. 2, two methods for clustering mixed-data are reviewed. In Sect. 3 we present an application on a real mixed-type data set to show the interaction between quantitative and qualitative attributes in the clustering process. Finally, in Sect. 4 we draw some conclusions and discuss some suggestions for future research.

2 Clustering Mixed Data

Let $\mathbf{X} = \{X_1, X_2, \ldots, X_n\}$ indicate a set of n objects and $X_i = [x_{i1}, x_{i2}, \ldots, x_{iM}]$ denote an object constituted by M variables. We consider the case in which the M variables are both continuous and categorical. Let $M = Q + C$ where Q is the number of numeric variables and C is the number of categorical variables. Let us define the two indicator variables subsets which identify the different types of variables as follows: $\mathcal{C} = \{m_1^C, \ldots, m_C^C\}$ denotes the categorical variables and $\mathcal{Q} = \{m_1^Q, \ldots, m_Q^Q\}$ denotes the numeric variables. The aim of clustering is to divide the n objects contained in $\mathbf{X}$ into K separate clusters. Since for a given n the number of possible partitions is significant, it is not advisable to examine each of them to find a better one, but instead try to maximize (or minimize) a suitably chosen objective function (Everitt 1974; Huang 1997). When clustering mixed data sets the main problem is to determine *how close* or *how far apart* objects are from each other. In what follows we consider two approaches that present two different ways to combine in a single cost function distance measures for numeric variables and dissimilarity measures for categorical variables.

2.1 Huang Method

Huang (1997) presented a so-called *K-prototypes* algorithm, which is based on the *K-means* method, but overcomes its quantitative data limitation, preserving, at the same time, its efficiency. The algorithm groups the objects in clusters against K prototypes. The updates occurs in a dynamical manner so to minimize the following objective function:

$$E = \sum_{k=1}^{K} \sum_{i=1}^{n} u_{ik} s(X_i, P_k), \tag{1}$$

where u_{ik} is an element of a *partition matrix* $U_{n \times k}$, and s is a dissimilarity measure between the objects X_i and P_k. $P_k = [p_{k1}, p_{k2}, \ldots, p_{kM}]$ is the *prototype* or *representative vector* for cluster k. U represents a *hard partition* matrix where $u_{ik} \in \{0, 1\}$ and $u_{ik} = 1$ if X_i is allocated to cluster k.

The dissimilarity measure is defined as:

$$s(X_i, P_k) = \sum_{m \in \mathcal{Q}} (x_{im} - p_{km})^2 + \gamma_k \sum_{m \in \mathcal{C}} \delta(x_{im}, p_{km}), \tag{2}$$

where the first term is the squared Euclidean distance, while the second term is defined as $\delta(r, t) = 0$ for $r = t$ and $\delta(r, t) = 1$ for $r \neq t$. γ_k is a weight for categorical attributes in cluster k.

Let define the internal term in Eq. (1) as $E_k = \sum_{i=1}^{n} u_{ik} s(X_i, P_k)$. This term measures the total dissimilarity of objects in cluster k from their prototype P_k, otherwise known as the total cost of allocating $\mathbf{X}$ to cluster k. This term may be rewritten as:

$$E_k = \sum_{i=1}^{n} u_{ik} \sum_{m \in \mathcal{Q}} (x_{im} - p_{km})^2 + \gamma_k \sum_{i=1}^{n} u_{ik} \sum_{m \in \mathcal{C}} \delta(x_{im}, p_{km})$$
$$= E_k^{\mathcal{Q}} + \gamma_k E_k^{\mathcal{C}}, \qquad (3)$$

where $E_k^{\mathcal{Q}}$ and $E_k^{\mathcal{C}}$ represent the dissimilarity of the objects in cluster k, coming from the numerical and the categorical variables, respectively. In order to minimize these two components, let $P_k^{\mathcal{Q}}$ and $P_k^{\mathcal{C}}$ be the prototype for cluster k for the numerical and categorical variables, respectively.

$E_k^{\mathcal{Q}}$ is minimized with the usual update of the K-means algorithm for continuous variables, i.e. the generic component of $P_k^{\mathcal{Q}}$ is calculated by

$$p_{km} = \frac{1}{n_k} \sum_{i=1}^{n} u_{ik} x_{im} \qquad m \in \mathcal{C}, \qquad (4)$$

where n_k is the number of objects in cluster k.

Let $V_m = \{v_{m_1}, v_{m_2}, \dots\}$ be the set enclosing the distinct values of the m-th categorical variable and let $\mathrm{pr}(v_{m_j}|k)$ be the probability that value v_{m_j} is observed in cluster k.

It is possible to rewrite $E_k^{\mathcal{C}}$ in (3) as

$$E_k^{\mathcal{C}} = \sum_{m \in \mathcal{C}} n_k \left[1 - \mathrm{pr}(p_{km} \in V_m|k) \right]. \qquad (5)$$

From (5), $E_k^{\mathcal{C}}$ is minimized by selecting the categorical values of the prototype $P_k^{\mathcal{C}}$, such that

$$\mathrm{pr}(p_{km} \in V_m|k) \geq \mathrm{pr}(v_{m_j} \in V_m|k)$$

for $p_{km} \neq v_{m_j}$ for all categorical attributes.

2.2 Cheung and Jia Method

Cheung and Jia (2013) provide a unified similarity metric which can be used with mixed attributes. For the categorical variables they define the similarity between a categorical attribute value $x_{im}^{\mathcal{C}}$ and cluster k as:

$$s(x_{im}^{\mathcal{C}}, P_l) = \mathrm{pr}(x_{im}^{\mathcal{C}} \in P_k|k). \qquad (6)$$

In the Huang method, the contribution of each categorical attribute is fixed in each cluster k and has to be chosen in a subjective way. Cheung and Jia propose an automatic procedure to compute the importance of each categorical attribute. In particular, the importance of any categorical attribute $V_m (m \in C)$ can be calculated by the average information content of all its possible values:

$$H_{V_m} = -\frac{1}{m_L} \sum_{m_l=1}^{m_L} p(v_{m_l}) \log p(v_{m_l}), \tag{7}$$

where $p(v_{m_l}) = p(v_{m_l} \in V_m)$. The weight of each attribute is then computed as

$$\gamma_m = \frac{H_{V_m}}{\sum_{m \in C} H_{V_m}}. \tag{8}$$

Finally, the similarity for the categorical attributes between the i-th object and cluster k is given by

$$s(X_i^C, P_k) = \sum_{m \in C} \gamma_m s(x_{im}^C, P_k). \tag{9}$$

The similarity metric for numerical attributes is rescaled, so to fall into the interval $[0, 1]$. The normalization is given by

$$s(X_i^Q, P_k) = \frac{exp[-0, 5D(X_i^Q, P_k)]}{\sum_{k=1}^{K} exp[-0.5D(X_i^Q, P_k)]}, \tag{10}$$

where $D(\cdot)$ is any distance function suitable for numerical variables. According to Eqs. (9) and (10), the object-cluster similarity metric for mixed data is defined as

$$s(X_i, P_k) = \frac{C}{C+1} s(X_i^C, P_k) + \frac{1}{C+1} s(X_i^Q, P_k), \tag{11}$$

where $i = 1, 2, \ldots, N$ and $k = 1, 2, \ldots, K$.

3 An Application on a Real Mixed-Type Data Set

Delay is one of the most relevant indicators of any transportation system. Flight delays provoke negative impacts, mainly economic, for airlines, airports and passengers. We used, as data source, the Bureau of Transportation Statistics database.

This dataset shows departure and arrival delays for US domestic flights and it contains both qualitative and quantitative information. For our illustrative example, we analyzed data regarding the 1st of August 2017.

Table 1 Description of the variables

Variable	Type	Description
Airline	CATEGORICAL NOMINAL	The code assigned by IATA to identify a unique airline, also known as carrier
Origin	CATEGORICAL NOMINAL	Geographic area of the origin airport
Destination	CATEGORICAL NOMINAL	Geographic area of the destination airport
Departure Performance, composed by		
DepDelay	CONTINUOUS	The difference (in minutes) between the scheduled and the actual departure time Early departures show negative numbers
Taxi time out	CONTINUOUS	Time between off-block and take off
Arrival Performance, composed by		
ArrDelay	CONTINUOUS	The difference (in minutes) between the scheduled and the actual arrival time Early arrival show negative numbers
Taxi time in	CONTINUOUS	Time between landing and in-block
Flight Summaries, composed by		
Distance	CONTINUOUS	The distance between airports, expressed in miles
Cause of delay	CATEGORICAL NOMINAL	It specifies the reason for cancellation: A: carrier B: weather C: national air system D: security

We considered each commercial flights throughout the United States and Canada from the State of New York. The flights data consists of 1426 instances and 9 features, which are summarized in Table 1.

We run a cluster analysis with a number of clusters equal to $K = 3$ and compared the results of the following three methods:

1. Huang method (described in Sect. 2.1)
2. Cheung and Jia method (described in Sect. 2.2)
3. Standard K-means method (applied on numerical variables only).

Table 2 displays, for each cluster, the mean value of numerical attributes, while Table 3 the distributions of categorical attributes. Figure 1 represents a scatter-plot of Departure Delay vs Arrival Delay and displays the cluster labels recovered by the three methods. Table 2 shows that the patterns between the Huang and the K-means methods are very similar between them.

Interestingly, the Huang method highlights a very strong clustering structure among all numerical attributes whereas, using the K-means method, the cluster means of the variable "Taxi time out" appear to be very similar between them. Furthermore, Fig. 1 illustrates how clusters resulting from the Huang method produce a better separation in the space defined by the Departure and Arrival delays.

Table 2 Variable mean values for each cluster

Method	Cluster	Size	DepDelay	Taxi time out	ArrDelay	Taxi time in	Distance
Huang	1	59	1.45	21.07	1.32	6.93	689
	2	586	0.03	30.62	0.01	10.51	1461
	3	780	0.01	16.36	−0.10	6.95	782
Cheung and Jia	1	472	0.10	22.33	0.06	7.63	883
	2	248	0.02	30.80	−0.02	11.12	2376
	3	705	0.09	16.84	−0.03	7.96	710
K-means	1	54	1.51	21.61	1.38	6.83	640
	2	309	0.02	23.17	−0.01	12.31	2187
	3	1062	0.02	22.24	−0.06	7.36	750

Table 3 Geographic area of the origin and destination airport for each cluster

Method	Cluster	Size	Origin				Destination			
			NE	NW	SE	SW	NE	NW	SE	SW
Huang	1	59	0.70	0.02	0.27	0.02	0.83	0	0.17	0
	2	586	0.68	0.06	0.12	0.14	0.53	0.08	0.20	0.19
	3	780	0.68	0.03	0.24	0.05	0.79	0.01	0.19	0.01
Cheung and Jia	1	472	0.49	0.15	0	0.36	0.51	0.15	0	0.34
	2	248	0.99	0	0.01	0	0.30	0.04	0.58	0.08
	3	705	0.54	0.02	0.39	005	0.99	0	0	0.01
K-means	1	54	0.70	0	0.28	0.02	0.85	0	0.15	0
	2	309	0.50	0.15	0.06	0.29	0.51	0.17	0.02	0.30
	3	1062	0.74	0	0.23	0.03	0.72	0	0.25	0.03

Table 3 shows that, using the Huang method, the 1st cluster has a prevalence of flights with origin and destination NE (North East) and SE (South East). In the 2nd group the relative frequency of flights with origin and destination SW (South West) is higher than in the other groups. The 3rd group is very similar to 1st group, while in Fig. 1 clusters appear very separate. The qualitative variables have a greater impact with the Cheung and Jia method. It is evident, indeed, in Table 3 that the distributions of the qualitative variables are very different in the three clusters. Finally, the Origin and the Destination distributions observed in the clusters recovered by the K-means method look similar to the ones observed in the Huang method.

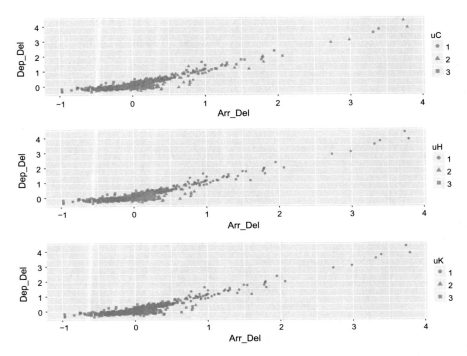

Fig. 1 Scatter-plot of Departure Delay vs Arrival Delay, together with cluster labels: Cheung and Jia (first panel), Huang (second panel) and K-means method (third panel)

4 Conclusions and Future Research

In this work we have shown a real application on clustering real mixed-data. The results have shown how both the Huang and Cheung and Jia algorithms retain the K-means efficiency, whilst removing, simultaneously, its quantitative data only limitation. In the application we have seen that the results obtained by the Huang method are better for the numerical attributes, while the Cheung and Jia results show a higher discrimination for the categorical attributes. Further work has to be implemented to provide a way to automatically choose a balance between numerical and categorical attributes.

References

Ahmad, A., Dey, L.: A k-mean clustering algorithm for mixed numeric and categorical data. Data Knowl. Eng. **63**, 503–527 (2007)

Brignell, C.J., Dryden, I.L., Gattone, S.A., Park, B., Browne, W.J.: Surface shape analysis with an application to brain surface asymmetry in schizophrenia. Biostatistics **11**(4), 1–22 (2010)

Caruso, G., Gattone, S.A., Fortuna, F., Di Battista, T.: Cluster analysis as a decision-making tool: a methodological review. In: Bucciarelli, E., Chen, S., Corchado, J.M., (eds.) Decision Economics: In the Tradition of Herbert A. Simon's Heritage. Advances in Intelligent Systems and Computing, vol. 618, pp. 48–55. Springer International Publishing (2018)

Cheung, Y., Jia, H.: Categorical-and-numerical-attribute data clustering based on a unified similarity metric without knowing cluster number. Pattern Recognit. **46**, 2228–2238 (2013)

Di Battista, T.: Diversity index estimation by adaptive sampling. Environmetrics **13**(2), 209–214 (2002)

Di Battista, T., Fortuna, F.: Clustering dichotomously scored items through functional data analysis. Electron. J. Appl. Stat. Anal. **9**(2), 433–450 (2016)

Di Battista, T., Gattone, S.A.: Multivariate bootstrap confidence regions for abundance vector using data depth. Environ. Ecol. Stat. **11**(4), 355–365 (2004)

Di Battista, T., Gattone, S.A.: Nonparametric tests and confidence regions for intrinsic diversity profiles of ecological populations. Environmetrics **14**(8), 733–741 (2003)

Everitt, B.: Cluster Analysis. Heinemann Educational Books Ltd. (1974)

Fortuna, F., Maturo, F.: K-means clustering of item characteristic curves and item information curves via functional principal component analysis. Qual. Quant. (2018). https://doi.org/10.1007/s11135-018-0724-7

Gattone, S.A., De Sanctis, A., Russo, T., Pulcini, D.: A shape distance based on the Fisher-Rao metric and its application for shapes clustering. Phisica A **487**, 93–102 (2017)

Huang, Z.: Clustering large data sets with mixed numeric and categorical values. In: Proceedings in the First Pacific-Asia Conference on Knowledge Discovery and Data Mining, pp. 21–34 (1997)

MacQueen, J.B.: Some methods for classification and analysis of multivariate observations. In: Proceedings of 5th Berkeley Symposium on Mathematical Statistics and Probability, vol. 1, pp. 281–297. University of California Press, Berkeley (1967)

Maturo, F.: Unsupervised classification of ecological communities ranked according to their biodiversity patterns via a functional principal component decomposition of Hills numbers integral functions. Ecol. Indic. **90**, 305–315 (2018)

Nie, G., Chen, Y., Zhang, L., Guo, Y.: Credit card customer analysis based on panel data clustering. Procedia Comput. Sci. **1**(1), 2489–2497 (2010)

Peng, Y., Kou, G., Shi. Y., Chen, Z.: Improving clustering analysis for credit card accounts classification. In: Proceedings of the 5th International Conference on Computational Science—ICCS 2005, Part III, pp. 548–553. Springer Berlin Heidelberg (2005)

Valentini, P., Di Battista, T., Gattone, S.: Heterogeneity measures in customer satisfaction analysis. J. Classif. **28**, 38–52 (2011)

Ordering in the Algebraic Hyperstructure Theory: Some Examples with a Potential for Applications in Social Sciences

Michal Novák

Abstract In this chapter we include several examples of concepts of the algebraic hyperstructure theory, which are all based on the concept of *ordering*. We also show how these concepts could be linked. The reason why we make this selection, is the fact that, in social sciences, objects are often linked in two different ways, which can be represented by an *operation* (or a *hyperoperation*) and a *relation*. The algebraic *hyperstructure* theory is useful in considerations of social sciences because, in this theory, the result of an interaction of two objects is, generally speaking, a *set* of objects instead of *one particular* object.

1 Introduction

The *algebraic hyperstructure theory* is a generalization of classical algebra. Whereas in classical algebraic structures, such as semigroups, groups, fields or lattices, the rules for assigning elements of a given set assign an *element* of the set to a pair (or to an *n*–tuple, or sometimes – in case of an unary operation – to an element) of the set, in algebraic hyperstructures the assignment results in a *subset*. Thus, if the carrier set is denoted H, we can regard a mapping $H \times H \to H$ or a mapping $H \times H \to \mathcal{P}^*(H)$. If the image of such a mapping is H, we speak of an *operation* (or a *composition*) while if it is $\mathcal{P}^*(H)$, i.e. a set of non-empty subsets of H, we speak of a *hyperoperation* (or a *hypercomposition*). It can be easily proved that the defining properties of a group can be formulated for both cases, or rather in general with a group being a special case of a set endowed with an operation – and a *hypergroup* being another special case.

The algebraic hyperstructure theory was initiated when Marty (1934) studied the issue of quotients of a group by its non-normal subgroups (see also e.g. Ştefănescu (1996) for a brief historical overview). A number of important results or applications have been achieved in this area. Recall e.g. the idea of

M. Novák (✉)
Faculty of Electrical Engineering and Communication, Brno University of Technology, Brno, Czech Republic
e-mail: novakm@feec.vutbr.cz

© Springer Nature Switzerland AG 2019
C. Flaut et al. (eds.), *Models and Theories in Social Systems*, Studies in Systems, Decision and Control 179, https://doi.org/10.1007/978-3-030-00084-4_28

join space introduced as a consequence of hyperstructural approach to geometry studied by Prenowitz and Jantosciak in a sequel of papers climaxing with Prenowitz and Jantosciak (1972, 1979) (or refer to an easy-to-follow introductory overview paper Jantosciak 1985). The applications of the algebraic hyperstructure theory can be found in a variety of contexts: e.g. *graph theory* (where a hyperoperation on the set of vertices of a graph can denote a path), *biology* or *genetics* (where a hyperoperation on the set of individuals can denote a subset of individuals with a given feature), *image processing* (where a hyperoperation on a set of pixels can denote a contour of an object), etc.

In social sciences, Maturo, Ventre and Hošková–Mayerová (see e.g. Maturo and Ventre 2008 or Hošková-Mayerová and Maturo 2016) used elements of hyperstructure theory to describe the processes of decision making. Chvalina and M. Novák (2009) related the topic of this chapter to the issue of consumers' *preference relations*, which is a topic of *microeconomics*. Also, Vougiouklis, the author of the concept of H_v–*structures* (see also Nikolaidou and Vougiouklis 2014; Vougiouklis and Vougiouklis 2016; Vougiouklis et al. 1997; Lygeros 2013; Maturo and Maturo 2017), i.e. hyperstructures in which the usual associativity and commutativity are replaced by its weak variations (equality is replaced by a non-empty intersection), studied the idea of applying hyperstructure theory in social sciences – in his case in preparing and processing of questionnaires. In Vougiouklis and Kambaki-Vougioukli (2005) he advocates the use of a bar instead of the Likert scale while in Vougiouklis (2011) he provides mathematical background to his concepts (recall also Kambaki-Vougioukli and Vougiouklis 2008).

The idea of using a bar instead of Likert scale can be seen as an example of a wider class of *hyperoperations based on geometry*, which can be traced back to the very beginnings of the theory with join spaces being another example. Antampoufis, Dramalidis and Vougiouklis have recently used this approach. In Antampoufis et al. (2011) they show how the hyperstructure theory can be used to describe some urban problems. In order to do this they make use of a *geometrical hyperoperation* and a *constant arc hyperoperation* constructed on the set complex numbers. These are also discussed in Antampoufis (2008) (where, in fact, the *constant arc hyperoperation* was introduced) and in Dramalidis (1996); in Dramalidis (2013) some related hyperstructure properties are visualized for better understanding.

Notice that in many aspects related to social sciences the population, i.e. the elements of the carrier set, on which the operation or a hyperoperation is constructed, are somehow put in relations. A typical example of this are *family relations*, in which the set of individuals are linked in two ways: by *mating operation* (or *hyperoperation*) and in *descendant – ancestor relation*. Mathematically, we speak not only of a pair (H, f), where f is an operation (or a hyperoperation) on f but also of a pair (H, R), where R is a relation. In some cases, f and R are linked by certain rules – such as when $(H, \cdot, \leq)$ is an *ordered semigroup*, i.e. when "$\cdot$", i.e. f, is associative on H, "$\leq$" is a reflexive, antisymmetric and transitive relation and, moreover, for all $a, b, c \in H$ such that $a \leq b$ there is $a \cdot c \leq b \cdot c$ and $c \cdot a \leq c \cdot b$. Ordered semigroups have a counterpart in hyperstructure theory. These are called *ordered hyperstructures* and are based on the idea that the relation is constructed on an already existing semihypergroup, i.e. on a pair (H, f), where f is an associative hyperoperation on H. Ordered hyperstructures were introduced by Heidari and Davvaz (2011) and have been studied by a number of authors since. In this chapter we concentrate on a somewhat similar yet in principle different idea: EL–hyperstructures, i.e. pairs (H, f), where f is a hyperoperation on H constructed from an ordered (or rather, quasi-ordered) semigroup. EL–hyperstructures were introduced by Chvalina (1995) and studied, from the theoretical point of view, by Novák (2012, 2014, 2013, 2015), later together with Křehlík and Novák (2016), Novák and Křehlík (2018) and Novák and Cristea (2018). Notice that yet another approach (out of the infinitely many available) has been studied. Chvalina (1994) classified certain types of hyperstructures which in Chvalina (1995) and later in Chvalina and Chvalinová (1996) were called *quasi-order hypergroups*. These are pairs (H, f), where f is a hyperoperation on H, which is constructed by means of an ordered set (H, R).

For definitions of concepts not defined below cf. e.g. Corsini (1993).

2 EL–Semihypergroups: Definition and Examples

In this section we give a formal definition of the concept of EL–hyperstructures and include some examples. The abbreviation EL in "EL–hyperstructures" stands for "Ends lemma", which is a nickname for the following collection of results.

Lemma 1 (Chvalina 1995) *Let $(S, \cdot, \leq)$ be a partially ordered semigroup. Binary hyperoperation $* : S \times S \rightarrow \mathcal{P}^*(S)$ defined by*

$$a * b = [a \cdot b)_{\leq} = \{x \in S \mid a \cdot b \leq x\} \tag{1}$$

is associative. The semihypergroup $(S, *)$ is commutative if and only if the semigroup $(S, \cdot)$ is commutative.

Lemma 2 (Chvalina 1995) *Let $(S, \cdot, \leq)$ be a partially ordered semigroup. The following conditions are equivalent:*

1^0 *For any pair $a, b \in S$ there exists a pair $c, c' \in S$ such that $b \cdot c \leq a$ and $c' \cdot b \leq a$.*
2^0 *The semi-hypergroup $(S, *)$ defined by (1) is a hypergroup.*

If $(S, \cdot)$ is a group, then we get some more advanced results.

Lemma 3 (Novák 2013, Račková 2009) *Let $(S, \cdot, \leq)$ be a (commutative) partially ordered group. Then $(S, *)$, constructed by means of (1), is a transposition hypergroup (join space).*

However, using the lemma we are not able to obtain *canonical hypergroups* because, in EL–hyperstructures, there do not exist scalar, i.e. one-element, identities.

Theorem 1 (Novák 2013) *Let* $(S, \cdot, \leq)$ *be a non-trivial quasi-ordered semigroup such that* "$\leq$" *is not the identity relation. Then no element of the EL–semihypergroup* $(S, *)$ *is a scalar identity.*

When examining proofs of the above theorems one can easily see that the requirement that $(S, \cdot, \leq)$ is a *partially ordered* semigroup can be in most cases replaced by $(S, \cdot, \leq)$ being *quasi-ordered* semigroup, i.e. "$\leq$" is reflexive and transitive, yet it need not be antisymmetric. In fact the only exception is the implication "$(S, *)$ is commutative $\Rightarrow$ $(S, \cdot)$ is commutative" in Lemma 1. Naturally, the main obstacle with "downgrading" from partial ordering to quasi-ordering is the fact that we loose notions such as the *greatest* or the *lowest element* and encounter rather big difficulties when manipulating *maximal* or *minimal elements*. Notice that, in EL–hyperstructures, this is an important limitation because we work with sets $[a]_{\leq} = \{x \in S \mid a \leq x\}$ and very often we need to prove that $[a]_{\leq} = \{a\}$, or rather that for no $b \neq a$ there is $b \in [a]_{\leq}$ or that if $b \in [a]_{\leq}$, then $b = a$.

Remark 1 We prefer saying "quasi-ordering" to "preorder". Also, we prefer saying "partial ordering" to "partial order" or simply "ordering".

The construction of EL–hyperstructures seems to be both *general* and *simple* enough to be applied in a sufficiently wide range of contexts. At this place we provide several examples so that one can see that this context may be straightforward as well as sophisticated.

One of the earliest occurrences of EL–hyperstructures (of course, not using this name) can be found in 1967 when (Pickett 1967) gives Example 1. We give an exact quote of this example including its parts which are not relevant for our present considerations (the term *multigroup* is equivalent to an *n–ary hypergroup*, here $n = 2$, i.e. we have a hypergroup).

Example 1 Let $(X, \wedge, \vee, \leq)$ be a lattice and define $a \cdot b = \{x \mid x \geq a \wedge b\}, a \circ b = \{x \mid x \leq a \vee b\}$. Both $(X, \cdot)$ and $(X\circ)$ are commutative multigroups and every element is a unit. The only coset decomposition is determined by X, for if Y is determined a coset decomposition for $(X, \cdot)$, say, then if a and b are any two elements of X, $Y \cdot (a \wedge b)$ meets both a and b. Hence they are in the same coset.

The following two examples are natural and obvious.

Example 2 Consider the set $\mathbb{N}$ of all natural numbers (excluding 0). Obviously $(\mathbb{N}, \cdot, \leq)$, where "$\cdot$" is the usual multiplication and "$\leq$" is the natural ordering of natural numbers by size, is a partially ordered semigroup. Thus if we define $a * b = [a \cdot b]_{\leq} = \{x \in \mathbb{N} \mid a \cdot b \leq x\}$, for all $a, b \in \mathbb{N}$, then $(N, *)$ is a commutative semihypergroup.

Example 3 If we regard the divisibility relation "$\mid$" in Example 2, then $(N, *)$, where $a * b = [a \cdot b]_{\mid} = \{x \in \mathbb{N} \mid a \cdot b \mid x\}$, for all $a, b \in \mathbb{N}$, is a semihypergroup.

Notice that in the following example the interval may represent probabilities while the single-valued operation "·" may represent simultaneous probability of independent events. Also, "min" and "max" may represent the event, probability of which is smaller or greater, respectively.

Example 4 The set of all real numbers from the interval $\langle 0, 1 \rangle$ together with the operation of multiplication and the usual ordering of real numbers by size is a partially ordered semigroup. Thus if we define $a * b = [a \cdot b]_{\leq} = \{x \in \langle 0, 1 \rangle \mid a \cdot b \leq x\}$, for all $a, b \in \langle 0, 1 \rangle$, we get that $(\langle 0, 1 \rangle, *)$ is a commutative semihypergroup. The same holds for intervals $(0, 1\rangle$ or $\langle 0, 1)$ or $(0, 1)$.

In their study of *braid groups*, Al Tahan and Davvaz (2016) use the idea of the "Ends lemma" to construct a cyclic hypergroup of an arbitrary braid group B_n of n strands. In this respect it is important to notice that they take an abstract structure and a particular property of their elements – in their case the shortest presentation of the product of elements $\sigma_i^{n_i} \in B_n$. This technique can be generalized: if one can describe properties of elements of a given set by means of natural / whole / rational / real, etc. numbers, one can construct examples such as the following – intentionally simplistic (!) – Example 5.

Example 5 Let S be a set of apple trees and $p(s)$ the average weight of apples collected from a given tree s (in kilos). For arbitrary $r, s \in S$ define $r * s = \{t \in S \mid p(r) + p(s) \leq p(t)\}$. Then $(S, *)$ is a commutative hypergroup.

For a more mathematical example of the previous reasoning, one can consider e.g. an automaton with a given input alphabet I and denote by $l(a)$ the length of a word constructed from the letters of this alphabet. In this case $l(a) + l(b) = l(a\&b)$, where $a\&b$ is a word constructed by the operation of *catenation* of two words a, b. Obviously, $(I^*, \&)$, where I^* is a set of words defined over I, is a monoid, where the neutral element is the empty word. For some results of this approach see Chvalina (1995), Chap. 6.

Example 6 Denote $I = \langle 0, 1 \rangle$ and $C(I)$ the set of all real continuous functions on $C(I)$ and for two functions $f, g \in C(I)$ define that $f \leq g$ if for all $x \in I$ there is $f(x) \leq g(x)$. Since $(C(I), +, \leq)$ is a commutative partially ordered group, we get that $(C(I), *)$, where $f * g = [f + g]_{\leq} = \{h \in C(I) \mid f + g \leq h\}$, for all $f, g \in C(I)$, and "$+$" denotes the usual pointwise addition of functions, is a join space.

Also Phanthawimol and Kemprasit (2010) in fact work in the EL-context. Notice that their hyperoperation $x \circ_N y = xyN$ for all $x, y \in G$, where $(G, \cdot)$ is a group and N is a normal subgroup of G, is included in Corsini (1993) and in fact relates to the original Marty's construction.

3   Relations in Hyperstructure Theory: A Selection of Concepts and Approaches

In the introduction we mentioned several concepts linking sets, operations, hyperoperations and relations. In this section we include their formal definitions. Later on, we will relate EL–hyperoperations to some of these concepts.

The connection between *ordering*, i.e. a special type of a *relation*, and the idea of a *hyperoperation*, i.e. a mapping from H^n to the set of non-empty subsets of H (denoted as $\mathcal{P}^*(H)$), is a very natural one. It was in fact the idea of a line segment being generated by its endpoints, i.e. a hyperoperation

$$a * b = \{x \in \mathbb{R} \mid a \le x \le b\} \tag{2}$$

for all $a, b \in \mathbb{R}$, that was the motivation with which Prenowitz and Jantosciak in 1972 introduced the concept of a *join space* Prenowitz and Jantosciak (1972). Already in 1948, Iwasava (1948) worked with a hyperoperation "$*$" defined on a linearly ordered group G by

$$a * b = \{x \in G \mid \min\{a, b\} \le x \le \max\{a, b\}\}. \tag{3}$$

Corsini (1994, 2003) used a very much similar idea when constructing join spaces from fuzzy sets; this idea was later developped e.g. by Ştefănescu and Cristea (2008). Notice that the very basic idea of fuzzy sets is replacing states $0 - 1$ (or true / false) by a continuum between these extremes, which is also related to Vougiouklis and his idea of using a bar instead of Likert scale. For an example of use of fuzzy sets related to the hyperstructure theory in social sciences cf. e.g. research by Hošková-Mayerová and Maturo (2016), Hošková-Mayerová (2017), Hošková-Mayerová and Maturo (2017, 2018).

Definition 1 Let X be a non-empty set. Any function $\mu : X \to \langle 0, 1 \rangle$ is called a *membership function* of X. The pair (X, μ) is called a *fuzzy set*. For a given $x \in X$, $\mu(x)$ is called the *grade of membership* of x in (X, μ).

Suppose we have a hypergroupoid $(H, *)$. For a given element $u \in H$ denote by $Q(u)$ the set of all pairs $a, b \in H$ such that $u \in a * b$, i.e. $Q(u) = \{a, b \in H \mid u \in a * b\}$ and by $q(u)$ denote the cardinality of $Q(u)$, i.e. $q(u) = |Q(u)|$. Now, for all $u \in H$, define

$$\tilde{\mu}(u) = \frac{\sum\limits_{(x,y) \in Q(u)} \frac{1}{|x*y|}}{q(u)}. \tag{4}$$

If $Q(u) = \emptyset$ or in case of infinite cardinality of $Q(u)$, we define $\tilde{\mu}(u) = 0$. Obviously, $\tilde{\mu}(u) \in \langle 0, 1 \rangle$ for all $u \in H$, i.e. $\tilde{\mu}$ is a membership function of H. Corsini (2003) showed that if we define on H a hyperoperation "$*_{\tilde{\mu}}$" by

$$x *_{\tilde{\mu}} y = \{z \in H \mid \min\{\tilde{\mu}(x), \tilde{\mu}(y)\} \le \tilde{\mu}(z) \le \max\{\tilde{\mu}(x), \tilde{\mu}(y)\}\} \tag{5}$$

for all $x, y \in H$, then $(H, *_{\tilde{\mu}})$ is a join space.

In 1975, Varlet (1975) established a *connection between distributive lattices and join spaces* which had been introduced only 3 years before. Notice that the hyper-operation he uses is defined on a lattice $(L, \wedge, \vee, \leq)$ by

$$a \diamond b = \{x \in L \mid a \wedge b \leq x \leq a \vee b\} \tag{6}$$

for all $a, b \in L$.

The papers that we make reference to above are a tiny selection. However, it may not be so easy to identify elements of hyperstructure theory in many works before 1980s since the terminology had not been codified for long. For example, when *using join spaces for classification of median algebras*, Bandelt and Hedlíková (1983) use the term "operation" when referring to a hyperoperation (see Bandelt and Hedlíková 1983, p. 7). Even though the paper in fact relies on the hyperstructure theory, it never uses the prefix "hyper–" to make this connection obvious.

As has been mentioned in the introduction, the notion of *quasi-order hypergroups* was introduced by Chvalina.

Definition 2 By a *quasi-order hypergroup* we mean a commutative hypergroup $(H, *)$ such that, for all $a \in H$, there is $a * a * a = a * a$, i.e. $a^3 = a^2$.

Later on, Chvalina (1995) approaches the topic from a slightly different perspective than in Chvalina (1994) and defines, on the set of quasi-order hypergroups, a binary relation $R_* \subseteq H * H$ by

$$R_* = \{(a, b) \in H * H \mid a * b * a = a * a\} \tag{7}$$

and shows that R_* is a quasi-ordering on the hypergroup H which is moreover antisymmetric if and only if, for all $a, b \in H$, the fact that $a^2 = b^2$ implies $a = b$. Also, Chvalina showed that if (H, R) is a quasi-ordered set, then $(H, *_R)$, where

$$a *_R b = R(a) \cup R(b) \tag{8}$$

for all $a, b \in H$, is an extensive commutative hypergroup. Corsini and Leoreanu, inspired by Chvalina (1994), include the following definition and proposition.

Definition 3 Let $(H, *)$ be a hypergroupoid. We say that H is a *quasi-order hypergroup*, i.e. a *hypergroup determined by a quasi-order*, if, for all $a, b \in H$, $a \in a^3 \subseteq a^2$ and $a * b = a^2 \cup b^2$. Moreover, if the following implication holds,

$$a^2 = b^2 \Rightarrow a = b, \tag{9}$$

for all $a, b \in H$, then $(H, *)$ is called an *order hypergroup*.

Proposition 1 *A hypergroupoid is a (quasi–) order hypergroup if and only if there exists a (quasi–) order R on the set H such that, for all $a, b \in H$*

$$a * b = R(a) \cup R(b).$$

Notice that, in our $[a]_\leq$ notation, this rewrites to

$$a * b = [a]_\leq \cup [b]_\leq. \tag{10}$$

Starting with Heidari and Davvaz (2011), *ordered hyperstructures*, in which the triple *set – hyperoperation – relation* is used, have been studied. Notice that in the following definition we use the symbol "$\preceq$" not in the sense of a preorder, i.e. a quasi-ordering, but as a symbol for partial ordering on a hyperstructure. This will enable us to easily distinguish between quasi-ordered semigroups $(S, \cdot, \leq)$ and ordered semihypergroups $(S, *, \preceq)$.

Definition 4 An *ordered semihypergroup* $(S, *, \preceq)$ is a semihypergroup $(S, *)$ together with a partial ordering "$\preceq$" which is compatible with the hyperoperation, i.e.

$$x \preceq y \Rightarrow a * x \preceq a * y \text{ and } x * a \preceq y * a \tag{11}$$

for all $a, x, y \in S$. By $a * x \preceq a * y$ we mean that for every $c \in a * x$ there exists $d \in a * y$ such that $c \preceq d$.

Papers written on the topic of ordered hyperstructures are numerous; for a collection of some results see a recently published book (Davvaz 2016).

Finally, one must not forget the issue of *hyper BCK–algebras* which are generalizations of *BCI–* or rather *BCK–algebras* introduced in 1966 by Iséki (1966) and brought to a shape by Iséki and Tanaka in papers such as Iséki (1978). For a collection of results on *BCI–/BCK–*algebras see Huang (2006); notice that the original motivation for introducing such structures lies in combinatory logic and propositional calculus. Also, notice that, in the context of *BCK–*algebras, the standard symbol "$*$" does *not* stand for a hyperoperation.

Definition 5 An algebra $(X; *, 0)$ of type $(2, 0)$ is called a *BCI–algebra* if it, for all $x, y, z \in X$, satisfies the following conditions:

1. $((x * y) * (x * z)) * (z * y) = 0$,
2. $(x * (x * y)) * y = 0$,
3. $x * x = 0$,
4. simultaneous validity of $x * y = 0$ and $y * x = 0$ implies that $x = y$.

In *BCI–*algebras we can define, for all $x, y \in X$, a relation "$\leq$", called a *BCI–ordering*, by setting

$$x \leq y \text{ whenever } x * y = 0. \tag{12}$$

It is easy to show that "$\leq$" is a partial ordering on X. However, when attempting to show that the *BCI–*ordering "$\leq$" is compatible with the operation "$*$" (which is crucial for our further considerations), we run into difficulties as, for all $x, y, z \in X$, the fact that $x \leq y$ implies that $x * z \leq y * z$ yet $z * x \geq z * y$ (given as Huang 2006, Proposition 1.1.4). Thus we see that $(X, *, \leq)$ are not partially ordered semigroups.

Even though there cannot be much common ground between the broadest general case of BCI–algebras and EL–hyperstructures, this is not true for the case of *lower BCK–semilattices*.

Definition 6 A BCI–algebra $(X; *, 0)$ is called a BCK–*algebra* if, for all $x \in X$, there is $0 * x = 0$. A BCK-algebra is called *commutative* if, for all $x, y \in X$ there is $x * (x * y) = y * (y * x)$. A BCK–algebra is called a *lower BCK–semilattice* if $(X, \leq)$, where "$\leq$" is a BCI–ordering, is a lower semilattice.

4 Links Between EL–Semihypergroups and Some Other Concepts

First, we are going to relate EL–semihypergroups and ordered semihypergroups. Recall that *compatibility in the sense of single-valued structures* means that

$$x \leq y \Rightarrow a \cdot x \leq a \cdot y \text{ and } x \cdot a \leq y \cdot a \tag{13}$$

for all $x, y, a \in S$, while *compatibility in the sense of hyperstructures* means that

$$x \preceq y \Rightarrow a * x \preceq a * y \text{ and } x * a \preceq y * a \tag{14}$$

for all $a, x, y \in S$, where by $a * x \preceq a * y$ we mean that for every $c \in a * x$ there exists $d \in a * y$ such that $c \preceq d$.

First of all, suppose that the relations "$\leq$" *and* "$\preceq$" *are the same.* Thus, when rewriting the compatibility condition (14), we get that

$$x \leq y \Rightarrow [a \cdot x)_{\leq} \leq [a \cdot y)_{\leq} \text{ and } [x \cdot a)_{\leq} \leq [y \cdot b)_{\leq}, \tag{15}$$

for all $x, y, a \in S$, i.e. the fact that $x \leq y$ implies that for all $a \in S$ we have that for every $c \in S$ such that $a \cdot x \leq c$ there must exist $d \in S$ such that $a \cdot y \leq d$ and $c \leq d$ and for every $f \in S$ such that $x \cdot a \leq f$ there must exist $g \in S$ such that $y \cdot a \leq g$ and $f \leq g$. The following lemma is obvious.

Lemma 4 *In the EL–semihypergroup $(S, *)$ of a quasi-ordered semigroup $(S, \cdot, \leq)$ there is*

$$x \leq y \Rightarrow a * y \subseteq a * x \text{ and } y * a \subseteq x * a \tag{16}$$

for all $x, y, a \in S$.

Proof Suppose assumptions of the lemma and an arbitrary $k \in a * y = \{k \in S \mid a \cdot y \leq k\}$. Since $(S, \cdot, \leq)$ is a quasi-ordered semigroup, the fact that $x \leq y$ implies $a \cdot x \leq a \cdot y$, for all $a \in S$, and from transitivity of "$\leq$", we have that $a \cdot x \leq k$, i.e. $k \in a * x$. Proving the other inclusion is analogous. □

Now we need to establish, in the context of (1), the relation between $a * x$ and $a * y$ for $x \leq y$ as described by (14). For every element $c \in a * x$ we need to find an element $d \in a * y$ such that $c \leq d$ and for every element $f \in x * a$ we need to find an element $g \in y * a$ such that $f \leq g$. This is an easy and straightforward task in the following two special cases.

Lemma 5 *The EL–semihypergroup $(S, *)$ of a quasi-ordered semigroup $(S, \cdot, \leq)$ is an ordered semihypergroup $(S, *, \leq)$ if:*

1. *$(S, \leq)$ has the greatest element or,*
2. *the relation "$\leq$" is linear ordering.*

Proof We will show the proof for elements c, d (in the sense of the above text) only as reasoning for elements f, g of the other-sided multiplication is analogous.

1. If $(S \leq)$ has the greatest element, then, for an arbitrary $c \in S$, the desired element $d \in a * y$ is exactly this greatest element of $(S, \leq)$. Of course, in such a case, "$\leq$" must be partial ordering.
2. Since every two elements $x, y \in S$ are in relation "$\leq$" and, if $x \leq y$, there is $a \cdot x \leq a \cdot y$ and $a * y \subseteq a * x$, for all $a \in S$, then due to the construction of sets $a * x$ and $a * y$ the statement is obvious. □

Thus we see that, in order to obtain a general answer, we must focus on such cases of $c \in a * x$, where $c \notin a * y$. If $c \leq a \cdot y$, then it is enough to set $d = a \cdot y$ because reflexivity of "$\leq$" provides that $d \in a * y$. Therefore, we must focus on cases of $c \in a * x$, where there is simultaneously $c \notin a * y$ and elements $a \cdot y$ and c are not in relation "$\leq$". Notice that this means that we focus on cases where c is not in relation with any element of $a * y$. In other words, that to such an element $c \in a * x$ with these properties there exists no element $d \in a * y$ such that $c \leq d$.

Given this perspective, the following theorem becomes obvious and we can see that Lemma 5 is in fact its corollary.

Theorem 2 *The EL–semihypergroup $(S, *)$ of a partially ordered semigroup $(S, \cdot, \leq)$ is an ordered semihypergroup $(S, *, \leq)$ if an arbitrary pair of elements $x, y \in S$ has an upper bound.*

Proof Obvious because the existence of an upper bound of an arbitrary two element subset of $(S, \leq)$ prevents the situation described before the theorem. □

If for a pair of elements $x, y \in S$ there is $x \leq y$, then condition (14) must be valid for *all* $a \in S$. This means that if $(S, \cdot)$ is a monoid with a unit u, there must be also $u * x \leq u * y$, i.e. $[u \cdot x]_\leq \leq [u \cdot y]_\leq$ (and also $x * u \leq y * u$), which means $[x]_\leq \leq [y]_\leq$. This justifies the following theorem.

Theorem 3 *The EL–semihypergroup $(S, *)$ of a quasi-ordered monoid $(S, \cdot, \leq)$ is not an ordered semihypergroup $(S, *, \leq)$ if there exists a pair of elements such that it does not have an upper bound yet has a lower bound.*

Fig. 1 To Theorem 3:
$[x]_\le = \{x, c, y\}$ while
$[y]_\le = \{y\}$

Proof The assumptions of the theorem are such that there exists a triple of elements $c, x, y \in S$ such that $x \le c$, $x \le y$ while c and y are not related and do not have an upper bound, i.e. no element from S is simultaneously greater than both c and y (see Fig. 1). Since $(S, \cdot)$ is a monoid and $x \le y$, there must be – should $(S, *, \le)$ be a partially ordered semihypergroup – also $[x]_\le \le [y]_\le$. Yet to our $c \in [x]_\le$ there obviously does not exist any element $d \in [y]_\le$ such that $c \le d$. Therefore, $(S, *, \le)$ is not a partially ordered semigroup. $\qquad\square$

Remark 2 Notice that in Theorem 3 pairs of elements which do not have a lower bound need not be tested for the existence of their upper bound. Indeed, imagine that in Fig. 1 elements c and x are not related. Then $c \notin [x]_\le$ and the problem of finding a suitable element $d \in [y]_\le$ such that $c \le d$ disappears.

Out of the infinitely many ways of defining relation "$\preceq$" by means of "$\le$", one stands out. Let us, for all $x, y \in S$, define that

$$x \preceq y \text{ whenever } y \le x. \tag{17}$$

This turns out to be a universal way of obtaining an ordered semihypergroup from an arbitrary EL–semihypergroup.

Theorem 4 *Let $(S, *)$ be the EL–semihypergroup of a partially ordered semigroup $(S, \cdot, \le)$. For an arbitrary pair of elements $x, y \in S$ define $x \preceq y$ whenever $y \le x$. Then the relation "$\preceq$" is compatible with the hyperoperation "$*$".*

Proof After we rewrite condition (14) and take into account our definition of relation "$\preceq$", we get that, for all $x, y, a \in S$, the fact that $y \le x$ implies that to every element $c \in S$ such that $a \cdot x \le c$ there exists an element $d \in S$ such that $a \cdot y \le d$ and $d \le c$. However, since $(S, \cdot, \le)$ is a partially ordered semigroup, the fact that $y \le x$ implies that $a \cdot y \le a \cdot x$, i.e. we can, for an arbitrary $c \in a * x$ set $d = a \cdot x$ because $a \cdot x \in a * y$. Obviously, the same reasoning can be used for multiplication by an arbitrary $a \in S$ from the right. $\qquad\square$

Remark 3 Naturally, the issue of "$\preceq$" being a quasi-ordering and the properties of "$\preceq$" must be discussed separately. Obviously, the fact that "$\le$" is a partial ordering, means that also "$\preceq$" is a partial ordering. Notice that even though Heidari and Davvaz (2011) originally mention *ordered hyperstructures* only, in e.g. Ghazavi et al. (2015) *quasi-ordered hyperstructures* are discussed as well. The motivation to study *partial ordering* on hyperstructures lies in the fact that including antisymmetry is suitable for description of hyperstructure generalizations of lattices.

Given the definition of *quasi-order hypergroups* and Proposition 1 the quest for relationship between quasi-order hypergroups and EL–hyperstructures is not so straightforward. When constructing EL–hyperstructures, we *start* with a relation "$\leq$" while in order to prove that a hyperstructure is a quasi-order hypergroup, we have to *find* it.

If $(S, \cdot, \leq)$ is an idempotent quasi-ordered semigroup, then $(S, *)$ is commutative and such that, for all $a \in S$, there is $a^3 = a^2$, i.e., by original Chvalina's Definition 2, $(S, *)$ is a quasi-order hypergroup. Now, the condition $a * b = a^2 \cup b^2$ of Definition 3 in this case turns into

$$[a \cdot b)_\leq = [a)_\leq \cup [b)_\leq. \tag{18}$$

If the relation "$\leq$", which we use to construct the EL–semihypergroup $(S, *)$, has this property, then $(S, *)$ can be viewed as a quasi-order hypergroup, or rather as an order hypergroup because the implication (9) holds, for idempotent "$\cdot$", trivially.

Definition 7 A hyperoperation "$*$" on H is called *extensive*[1] if for all $a, b \in H$ there is $\{a, b\} \subseteq a * b$. A hypergroupoid $(H, *)$ with an extensive hyperoperation is called an *extensive hypergroupoid*.

As can be proved easily, the following theorem is a special case of a more general result applicable on *every* extensive semihypergroup.

Theorem 5 *Every extensive EL–semihypergroup is a hypergroup.*

Proof Obvious because extensivity in EL–semihypergroups means that $a \cdot b \leq a$ for all $a, b \in S$. Thus it is sufficient to set $c = c' = a$ and apply Lemma 2. $\square$

Example 7 Consider the EL–semihypergroup $(\mathbb{Z}, *)$ constructed from a partially ordered semigroup $(\mathbb{Z}, \min, \leq)$. Obviously, the operation "min" is idempotent and $[\min\{a, b\})_\leq = [a)_\leq \cup [b)_\leq$ for all $a, b \in \mathbb{Z}$. Thus $(\mathbb{Z}, *)$ is a quasi-order hypergroup. Notice that the same is true when we change $\mathbb{Z}$ to $\mathbb{N}$. There is no ambiguity in terminology because "$*$" is an extensive hyperoperation, i.e. $(\mathbb{Z}, *)$ and $(\mathbb{N}, *)$ are, by Theorem 5, hypergroups.

In *lower BCK–semilattices* $(X, *, 0)$ we denote $x \wedge y$ the greatest lower bound of an arbitrary pair of elements $x, y \in X$. Obviously, in all lower BCK–semilattices,

---

[1] We use the name "extensive" as this can be found in a number of works by Chvalina and / or his students and collaborators. However, some authors, such as Massouros, use a much more suitable name "closed" as this can be easily contrasted with "open" in the geometrical sense. For basic definitions see e.g. Massouros (2016), Massouros and Massouros (2011), Massouros (1999).

i.e. not only in such that X is a commutative BCK–algebra, there is $x \wedge y = y \wedge x$ and $(X, \wedge)$ is a semigroup. If we define[2]

$$x \leq_n y \text{ whenever } x \wedge y = x \tag{19}$$

for all $x, y \in X$, we see that "$\leq_n$" is compatible with the operation "$\wedge$" and we can construct the following.

Example 8 Suppose a lower BCK–semilattice $(X, \leq)$, where "$\wedge$" is the greatest lower bound of elements $x, y \in X$. For all $x, y \in X$ we define "$\leq_n$" by (19). It is easy to show that the relation "$\leq_n$" is a partial ordering. Indeed, since $x \wedge x = x$, the relation is reflexive. Also, if $x \leq_n y$, there is $x \wedge y = x$ and if $y \leq_n x$, there is $y \wedge x = y$, yet since $x \wedge y = y \wedge x$, there is $x = y$. Finally, if $x \leq_n y$ and $y \leq_n z$, then $x \wedge y = x$ and $y \wedge z = y$ and

$$x \wedge z = (x \wedge y) \wedge z = x \wedge (y \wedge z) = x \wedge y = x,$$

i.e. $x \leq_n z$. The compatibility condition also holds because given an arbitrary $a \in X$ and a pair $x, y \in X$ such that $x \leq_n y$ we have

$$(x \wedge a) \wedge (y \wedge a) = (x \wedge y) \wedge (a \wedge a) = x \wedge a,$$

i.e. $x \wedge a \leq_n y \wedge a$. And, thanks to commutativity of "$\wedge$", we also have that $a \wedge x \leq_n a \wedge y$. Altogether, we have that $(X, \wedge, \leq_n)$ is a partially ordered semigroup and as such it can be used to construct an EL–hyperstructure $(X, *)$ by defining the hyperoperation "$*$" by

$$x * y = [x \wedge y)_{\leq_n} = \{z \in X \mid x \wedge y \leq_n z\} \tag{20}$$

for all $x, y \in X$. And we immediately have that $(X, *)$ is a semihypergroup.

In fact, it is irrelevant whether X is a lower BCK–semilattice as it is important that it is a *semilattice*. Example 8 is thus in fact an example supporting the following lemma, which – by a nice loop – brings us back in time to Pickett's Example 1; or to Varlet and (6).

Lemma 6 *Every semilattice* $(X, \wedge, \leq)$ *or* $(X, \vee, \leq)$ *can be used to construct an* EL–*hypergroup* $(X, *)$, *where, for all* $a, b \in X$, *there is*

$$a * b = [a \wedge b)_{\leq} = \{x \in X \mid a \wedge b \leq x\}, \tag{21}$$

or $a * b = [a \vee b)_{\leq} = \{x \in X \mid x \leq a \vee b\}$ *respectively, or an* EL–*semihypergroup* $(X, *)$, *where, for all* $a, b \in X$, *there is*

---

[2]Notice that the lower index "n" in "$\leq_n$" stands for "new" to distinguish (19) from (12), which is defined by means of $x * y = 0$.

$$a * b = [a \vee b)_{\leq} = \{x \in X \mid a \vee b \leq x\}.$$

Proof Given the above reasoning, obvious. The fact that the semihypergroup $(X, *)$ is a hypergroup, follows from Theorem 5, because the hyperoperation is extensive, and from the duality principle. $\square$

Bounded BCI/BCK–algebras are such BCI/BCK–algebras that have the greatest element, which is usually denoted by 1. Yet this fact is exactly what Lemma 5 assumes. Therefore, we easily obtain the following theorem, proof of which immediately follows from Lemma 5.

Theorem 6 *An EL-semihypergroup $(X, *)$ constructed from a bounded lower BCK–semilattice $(X, \wedge, \leq)$ by means of (21) is an ordered semihypergroup.*

5 EL–Semihypergroups Based on Quasi-orderings

When Antampoufis, Vougiouklis and Dramalids discuss their urban applications (Antampoufis et al. 2011) or *geometrical hyperoperations*, they do so in the Gaussian plane, i.e. they work in the complex domain. As has been mentioned above, the relation "$\leq$", which is used to construct EL–hyperstructures need not be antisymmetrical. Therefore, we now include a few examples, taken from Novák (2017), which also make use of the set of complex numbers.

Example 9 On the set $\mathbb{C}$ of all complex numbers regard a binary operation "$\cdot_{|z|}$" defined as multiplication of absolute values, i.e. for all $z_1, z_2 \in \mathbb{C}$ define $z_1 \cdot_{|z|} z_2 = |z_1| \cdot |z_2|$, and a relation "$\leq_{|z|}$" defined as equality of absolute values, i.e. for all $z_1, z_2 \in \mathbb{C}$ put $z_1 \leq_{|z|} z_2$ whenever $|z_1| = |z_2|$. Obviously, $(\mathbb{C}, \cdot_{|z|}, \leq_{|z|})$ is a quasi-ordered semigroup (and "$\leq_{|z|}$" is not antisymmetric, yet it is symmetric). Thus if we define, for all $z_1, z_2 \in \mathbb{C}$, $z_1 * z_2 = \{x \in \mathbb{C} \mid |z_1| \cdot |z_2| = |x|\}$, we get that $(\mathbb{C}, *)$ is an EL–semihypergroup.

Example 10 If we denote by $|\mathbb{C}|_0^1$ the set of all complex numbers such that their absolute value is smaller than or equal to one 1 (i.e. we regard a unit disc of the Gaussian plane) and regard "$\cdot_{|z|}$" multiplication of absolute values and set that $z_1 \leq_{|z|} z_2$ whenever $|z_1| \leq |z_2|$, then we get that $(|\mathbb{C}|_0^1, \cdot_{|z|}, \leq_{|z|})$ is a proper quasi-ordered semigroup. Moreover, "$\leq_{|z|}$" is not antisymmetric. We define a hyperoperation on $|\mathbb{C}|_0^1$ by

$$z_1 * z_2 = [z_1 \cdot_{|z|} z_2)_{\leq_{|z|}} = \{x \in |\mathbb{C}|_0^1 \mid |z_1| \cdot |z_2| \leq |x|\}.$$

And we have that $(\mathbb{C}, *)$ is an EL–semihypergroup.

Example 11 Regard the additive group of complex numbers $(\mathbb{C}, +)$ and define, for all $z_1, z_2 \in \mathbb{C}$, relation "$\leq_{|z|^{-1}}$" by $z_1 \leq_{|z|^{-1}} z_2$ whenever $|z_1| \geq |z_2|$, where $|z|$ stands for the absolute value of $z \in \mathbb{C}$. It is easy to verify that $(\mathbb{C}, +, \leq_{|z|^{-1}})$ is a commutative quasi-ordered group, where "$\leq_{|z|^{-1}}$" is obviously not antisymmetric. Therefore, if we define, for all $z_1, z_2 \in \mathbb{C}$, that $z_1 * z_2 = \{x \in \mathbb{C} \mid |x| \leq |z_1 + z_2|\}$, then $(\mathbb{C}, *)$ is an EL–hypergroup.

6 Modified EL–Hyperstructures and the Issue of Fragmented Carrier Set

In Novák and Křehlík (2018), Novák and Křehlík studied the "Ends lemma" from the point of view of its applicational potential. In order to increase it, we sought ways reducing the assumptions of the construction. We also examined some expected contexts – namely the structure of the carrier set. In this way reached the idea of *modified EL–hyperstructures*. Since there is no need to repeat the results included in Novák and Křehlík (2018), let us concentrate on the main issue, which is connected to social sciences, or to be more precise, *family relations*.

The original "Ends lemma" does not take into account the usual situation of "incompatibility" of elements. One must often take into account that for some a and b the hyperoperation (or the operation) is meaningless as e.g. mating is possible between some individuals only. If the original concept of does not regard this, how can we test associativity?

Example 12 Suppose that $H = T_1^o \cup T_2^o$, where T_1^o is the set of males and T_2^o is the set of females. Let "$\cdot_1$" and "$\cdot_2$" be the operation of mating. For most living organisms, mating is not possible within the same sex as a male and a female are needed. Therefore, we need a "criss-cross" operation "$\cdot$" while speaking about hypergroupoids $(T_1^o, \cdot_1)$ or $(T_2^o, \cdot_2)$ makes no sense in this context. Moreover, the relation "$\leq$" may be the descendent relation valid for all individuals regardless of sex, i.e. $a \leq b$ may mean that a is the offspring (or parent) of b. Obviously, a product of mating between individuals with distinguished sex is an individual of exactly one sex (that is, moreover, related to their parents, i.e. elements of both T_1^o and T_2^o).

In Novák and Křehlík (2018) some results concerning the above explained situation are included. Notice that the nature of the definitions included in Novák and Křehlík (2018) is such that it is easy to apply the "Ends lemma" in contexts where the carrier set is partitioned into more than just two subsets. In this respect more fragmented one-parameter sets can be studied.

References

Al Tahan, M., Davvaz, B.: On a special single-power cyclic hypergroup and its automorphisms. Discret. Math. Algorithm Appl. **8**(4), 1650059 (2016)

Antampoufis, N.: Hypergroups and H_b-groups in complex numbers. J. Basic Sci. **4**(1), 17–25 (2008)

Antampoufis, N., Vougiouklis, T., Dramalidis, A.: Geometrical and circle hyperoperations in urban applications. Ratio Sociologica **4**(2), 53–66 (2011)

Bandelt, J.H., Hedlíková, J.: Median algebras. Discret. Math. **45**, 1–30 (1983)

Davvaz, B.: Semihypergroup Theory. Academic, Cambridge (2016)

Dramalidis, A.: Dual H_v-rings. Riv. Mat. Pura Appl. **17**, 55–62 (1996)

Dramalidis, A.: Visualization of algebraic properties of special H_v-structures. Ratio Mathematica **24**, 41–52 (2013)

Chvalina, J.: Commutative hypergroups in the sense of Marty and ordered sets. In: General Algebra and Ordered Sets, Proceedings of the International Conference Olomouc, pp. 19–30 (1994)

Chvalina, J.: Functional Graphs. Quasi-ordered Sets and Commutative Hypergroups. Masaryk University, Brno (1995). (in Czech)

Chvalina, J., Chvalinová, L.: State hypergroups of automata. Acta Math. et Inform. Univ. Ostraviensis **4**(1), 105–120 (1996)

Chvalina, J., Novák, M.: Hyperstructures of preference relations. In: The 10th International Congress Algebraic Hyperstructures and Applications (AHA 2008): Proceedings, Brno, University of Defence, pp. 131–140 (2009)

Corsini, P.: Prolegomena of Hypergroup Theory. Aviani Editore, Tricesimo (1993)

Corsini, P.: Join spaces, power sets, fuzzy sets. In: Proceedings of the Fifth International Congress on A.H.A., 1993, Iaşi, Romania, pp. 45–52. Hadronic Press, Florida (1994)

Corsini, P.: A new connection between hypergroups and fuzzy sets. Southeast Asian Bull. Math. **27**, 221–229 (2003)

Ghazavi, S.H., Anvariyeh, S.M., Mirvakili, S.: EL^2-hyperstructures derived from (partially) quasi-ordered hyperstructures. Iran. J. Math. Sci. Inform. **10**(2), 99–114 (2015)

Heidari, D., Davvaz, B.: On ordered hyperstructures. U.P.B. Sci. Bull. Ser. A **73**(2), 85–96 (2011)

Hošková-Mayerová, Š, Maturo, A.: Fuzzy Sets and Algebraic Hyperoperations to Model Interpersonal Relations, pp. 211–221. Springer, Switzerland (2016); Recent Trends in Social Systems: Quantitative Theories and Quantitative Models. Studies in Systems, Decision and Control, vol. 66

Hošková-Mayerová, Š., Maturo, F.: Fuzzy Regression Models and Alternative Operations for Economic and Social Sciences, pp. 235–247. Switzerland: Springer, : Recent Trends in Social Systems: Quantitative Theories and Quantitative Models, p. 66. Studies in Systems, Decision and Control. vol (2016)

Hošková-Mayerová, Š., Maturo, A.: Decision-Making Process Using Hyperstructures and Fuzzy Structures in Social Sciences, pp. 103–111. Springer International Publishing AG, Switzerland (2017); Soft Computing Applications for Group Decision-making and Consensus Modeling. Studies in Fuzziness and Soft Computing, vol. 357

Hošková-Mayerová, Š.: An overview of topological and fuzzy topological hypergroupoids. Ratio Mathematica **33**, 21–38 (2017)

Hošková-Mayerová, Š., Maturo, A.: Algebraic hyperstructures and social relations. Ital. J. Pure Appl. Math. **39**, 701–709 (2018)

Huang, Y.: BCI-Algebra. Science Press, Beijing (2006)

Iséki, K.: An algebra related with a propositional calculus. Proc. Jpn. Acad. **42**, 26–29 (1966)

Iséki, K.: An introduction to the theory of BCK-algebras. Math. Japon. **23**, 1–26 (1978)

Iwasava, K.: On linearly ordered groups. J. Math. Soc. **1**(1), 1–9 (1948)

Jantosciak, J.: Classical geometries as hypergroups. In: Corsini, P. (ed.) Convegno su: Ipergruppi, altre strutture multivoche e loro applicazioni, pp. 93–104. Italy, Udine (1985)

Kambaki-Vougioukli, P., Vougiouklis, Th: Bar instead of scale. Ratio Sociologica **3**, 49–56 (2008)

Křehlík, Š., Novák, M.: From lattices to H_v-matrices. An. Şt. Univ. Ovidius Constanţa **24**(3), 209–222 (2016)

Lygeros, N., Vougiouklis, Th: The Lv-hyperstructures. Ratio. Math. **25**, 59–66 (2013)

Marty, F.: Sur une généralisation de la notion de groupe, pp. 45–49. Stockholm, IV Congrès des Mathématiciens Scandinaves (1934)

Massouros, ChG: Hypercompositional structures from the computer theory. Ratio Math. **13**, 37–42 (1999)

Massouros, ChG, Massouros, G.G.: The transposition axiom in hypercompositional structures. Ratio Math. **21**, 75–90 (2011)

Massouros, ChG: On path hypercompositions in graphs and automata. MATEC Web Conf. **41**, 05003 (2016)

Maturo, A., Maturo, F.: On some applications of the Vougiouklis hyperstructures to probability theory. Ratio Math. **33**, 5–20 (2017)

Maturo, A., Ventre, A.G.S.: Multiperson decision making, consensus and associated hyperstructures. In: The 10th International Congress Algebraic Hyperstructures and Applications (AHA 2008): Proceedings, Brno, University of Defence, pp. 241–250 (2009)

Nikolaidou, P., Vougiouklis, Th: The Lie-Santilli admissible hyperalgebras of type An. Ratio Math. **26**(1), 113–128 (2014)

Novák, M.: EL-hyperstructures: an overview. Ratio Math. **23**(1), 65–80 (2012)

Novák, M.: Some basic properties of EL-hyperstructures. Eur. J. Combin. **34**, 446–459 (2013)

Novák, M.: n-ary hyperstructures constructed from binary quasi-ordered semigroups. An. Şt. Univ. Ovidius Constanţa **22**(3), 147–168 (2014)

Novák, M.: On EL–semihypergroups. Eur. J. Combin. **44**(Part B), 274–286 (2015)

Novák, M.: EL-semihypergroups in which the quasi-ordering is not antisymmetric. In: Mathematics, Information Technologies and Applied Sciences 2017: Post-conference Proceedings of Extended Versions of Selected Papers, Brno, University of Defence, pp. 183–192 (2017)

Novák, M., Cristea, I.: Composition in EL–hyperstructures. Hacet. J. Math. Stat. (accepted)

Novák, M., Křehlík, Š.: EL–hyperstructures revisited. Soft Comput. (2017) (in press – online ready)

Phanthawimol, W., Kemprasit, Y.: Homomorphisms and epimorhisms of some hypergroups. Ital. J. Pure Appl. Math **27**, 305–312 (2010)

Pickett, H.E.: Homomorphisms and subalgebras of multialgebras. Pac. J. Math. **21**(2), 327–342 (1967)

Prenowitz, W., Jantosciak, J.: Geometries and join spaces. J. Reine Angew. Math. **257**, 100–128 (1972)

Prenowitz, W., Jantosciak, J.: Join Geometries. UTM. Springer, New York (1979)

Račková, P.: Hypergroups of symmetric matrices. In: 10 th International Congress of Algebraic Hyperstructures and Applications, Proceedings of AHA 2008, University of Defence, Brno, pp. 267–272 (2009)

Ştefănescu, M.M.: Constructions of hypergroups. In: Vougiouklis, Th (ed.) New Frontiers in Hyperstructures, pp. 68–83. Hadronic Press, Florida (1996)

Ştefănescu, M., Cristea, I.: On the fuzzy grade of hypergroups. Fuzzy Sets Syst. **159**(9), 1097–1106 (2008)

Varlet, J.V.: Remarks on distributive lattices. Bull. de l'Acad. Polonnaise des Sciences, Serie des Sciences Math., Astr. et Phys., **23**, 1143–1147 (1975)

Vougiouklis, Th: Bar and theta hyperoperations. Ratio Math. **21**, 27–42 (2011)

Vougiouklis, Th, Kambaki-Vougiouki, P.: On the use of the bar. China-USA Bus. Rev. **10**(6), 197–206 (2005)

Vougiouklis, Th, Vougiouklis, S.: Helix-hopes on finite hyperfields. Ratio Math. **31**(1), 65–78 (2016)

Vougiouklis, Th, Spartalis, S., Kessoglides, M.: Weak hyperstructures on small sets. Ratio Math. **12**(1), 90–96 (1997)

Classical and Weakly Prime
L-Submodules

Razieh Mahjoob and Shaheen Qiami

Abstract Most of problems in biology, economics, ecology, engineering, environmental science, medical science, social science etc. have various uncertainties. Fuzzy set theory, rough set theory, vague set theory, interval mathematics probability, soft set theory are different ways of expressing uncertainty. Let L be a complete lattice. We introduce and characterize classical prime and weakly prime L-submodules of a unitary module over a commutative ring with identity. Also, we topologize Cl.L-Spec(M), the collection of all classical prime L- submodules of M, and investigate the properties of this topological space.

Keywords L-submodule · Weakly prime L-submodule · Classical prime L-submodule · Zariski like-topology

1 Introduction and Preliminaries

Most of problems in biology, economics, ecology, engineering, environmental science, medical science, social science etc. have various uncertainties. Fuzzy set theory, rough set theory, vague set theory, interval mathematics probability, soft set theory are different ways of expressing uncertainty.

As it is well known Zadeh (1965), introduced the notion of a fuzzy subset of a nonempty set X as a function from X to unit real interval $I = [0, 1]$. There have been some developments in the study focusing on a fusian of algebra and theories modelling imprecision. The study of fuzzy algebraic structures, especially of fuzzy groups, dates back to early seventies. Famous mathematicians were involved in it, like Rosenfeld who introduced the notion of fuzzy groups (Rosenfeld 1971). Fuzzy submodules were introduced by Negoita and Ralescu (1975). The scope and

R. Mahjoob (✉) · S. Qiami
Faculty of Mathematics, Statistics and Computer Sciences, Department of Mathematics,
Semnan University, Semnan, Iran
e-mail: mahjoob@semnan.ac.ir

S. Qiami
e-mail: sh.ghiami@semnan.ac.ir

© Springer Nature Switzerland AG 2019
C. Flaut et al. (eds.), *Models and Theories in Social Systems*, Studies in Systems,
Decision and Control 179, https://doi.org/10.1007/978-3-030-00084-4_29

possibilities of this math branch are enormous. Let's just name some examples: Ameri et al. studied multiplicative hyperring of fractions and coprime hyperideals (Ameri et al. 2017). Papers of Maturo and Hoskova-Mayerova show how the relatively recent mathematical theories on fuzzy sets and algebraic hyperstructures can make a significant contribution to the understanding of the system of relationships within the class. see e.g. (Hoskova-Mayerova and Maturo 2017, 2018; Maturo and Hoskova-Mayerova 2017). An overview of topological hypergroupoids can be found in Al Tahan et al. (2018), Hoskova-Mayerova (2017).

Pan (1987) studied fuzzy finitely generated modules and fuzzy quotient modules (also see Sidky 2001).

Let R be a commutative ring with identity and M be a unitary R-module. The prime spectrum, Spec(R), and the topological space obtained by introducing Zarisky topology on the set of prime ideals of a ring R play an important role in the field of commutative algebra, algebraic geometry and lattice theory. Also the notion of prime submodules and Zarisky topology on the Spec(M), the set of all prime submodules M over R studied by many authors (for example see Behboodi and Noori 2009; Hochster 1969; Hoskova-Mayerova 2017; Hoskova-Mayerova and Maturo 2018). For any submodule N of M, we denote the annihilator of $\frac{M}{N}$ by $(N : M)$, i.e. $(N : M) = \{r \in R | rM \subseteq N\}$. In particular $(0 : M)$ is called the annihilator of M that is $Ann(M) = \{r \in R, rM = 0\}$. A prime submodule (or a p-prime submodule) of M is a proper submodule P with $(P : M) = p$, such that $rm \in P$ for $r \in R$ and $m \in M$, either $m \in P$ or $r \in p$.

The set of all prime submodules of M is called the prime spectrum of M or simply the spectrum of M and is denoted by Spec(M). Note that the Spec(M) may be empty for some module M. Such a module is said to be primeless.

For any submodule N of M, $V(N)$ denotes the set of all prime submodule of M containing N, of course $V(M)$ is the empty set and $V(0)$ is Spec(M). For any family of submodule $N_i (i \in I)$ of M, $\bigcap_{i \in I} V(N_i) = V(\sum_{i \in I} N_i)$. Thus if $\zeta(M)$ denotes the collection of all subsets $V(N)$ of Spec(M), then the $\zeta(M)$ contains the empty set and Spec(M) and is closed under arbitrary intersection. If also $\zeta(M)$ is closed under finite union, then $\zeta(M)$ satisfies the axioms of closed subset of topological spaces, which is called Zarisky topology.

In McCasland et al. (1997) a module with Zarisky topology is called top module. Ameri and Mahjoob (2008) introduced and studied the notion of prime L-submodules of a module M over a commutative ring with identity R, where L is a complete lattice. The set of all prime L-submodules of M will be called the prime L-spectrum of M or simply the L-spectrum of M and is denoted by L-Spec(M). They remarked that as ordinary case the L-Spec(M) may be empty for a nonzero module M. Such a module was said to be L-primeless. Then they followed Lu (1999) and topologized L-Spec(M), which is called Zarisky topology and investigated the properties of this topological space. Moreover they studied the relationship between topological spaces L-Spec(M) and and L-Spec($\frac{R}{Ann(M)}$) and finally they found a basis for Zarisky topology for L-Spec(M) (Ameri et al. 2008; Ameri and Mahjoob 2017).

Weakly prime and classical prime submodules are generalizations of prime submodules and they were introduced and studied by Behboodi and Koohy (2004). Weakly prime submodules also have been studied in Ameri et al. (2008), Ameri et al. (2017), Atani and Farzalipoor (2007). In fact a proper submodule N of an R-module M is said to be weakly prime if $0 \neq rx \in N$, for $r \in R$ and $x \in M$, implies that $x \in N$ or $r \in (N : M)$, also N is called classical prime if $abx \in N$ for $a, b \in R$ and $x \in M$, implies that $ax \in N$ or $bx \in N$.

Recently, the notion of classical prime submodules and Zarisky-like topology on Cl-Spec(M), the set of all classical prime submodules of M over R, are studied by Behboodi (2009).

For any submodule N of M, if $\mathbb{V}(N)$ denotes the set of all classical prime submodules of M containing N, of course $\mathbb{V}(M)$ is the empty set and $\mathbb{V}(0)$ is Cl-Spec(M). For any family of submodule $N_i (i \in I)$ of M, $\bigcap_{i \in I} \mathbb{V}(N_i) = \mathbb{V}(\sum_{i \in I} N_i)$. Thus if $\mathbb{C}(M)$ denotes the collection of all subsets $\mathbb{V}(N)$ of Cl-Spec(M), then $\mathbb{C}(M)$ contains the empty set and Cl-Spec(M) and is closed under arbitrary intersection. If also $\mathbb{C}(M)$ is closed under finite union, then M is called a classical Top-module i.e., for every submodules N and L of M, there exists a submodule K of M such that $\mathbb{V}(N) \cap \mathbb{V}(L) = \mathbb{V}(K)$, for in this case, $\mathbb{C}(M)$ satisfies the axioms for the closed subsets of a topological space.

For each submodule N of M, we put $\mathbb{U}(N) = Cl - Spec(M) \setminus \mathbb{V}(N)$ and $\mathbb{B}(M) = \{\mathbb{U}(N) : N \leq M\}$. If $\mathbb{T}(M)$ is the collection of all unions of finite intersections of elements of $\mathbb{B}(M)$, then, $\mathbb{T}(M)$ is a topology on Cl.Spec(M) by the sub-basis $\mathbb{B}(M)$. In this case $\mathbb{T}(M)$ is said Zariski-like topology on M.

In this paper we introduce the notion of weakly prime and classical prime L-submodules of a module M over a commutative ring with identity R, where L is a complete lattice. The set of all classical prime L-submodules M will be called the classical prime L-spectrum of M and is denoted by Cl.L-Spec(M). Also we topologize Cl.L-Spec(M), which is called Zarisky-like topology and investigate its properties. Finally we will show that Zarisky-like topology on Cl.L-Spec(M) is a spectral space.

Throughout of this paper by R we mean a commutative ring with identity, M is a unital R-module and L denotes a complete lattice. By an L-subset μ of a nonempty set X, we mean a function μ from X to L. L^X denotes the set of all L-subsets of X. Let A be a subset of X and $y \in L$. Define $y_A \in L^X$ as follows:

$$y_A(x) = \begin{cases} y & if \ x \in A \\ 0 & otherwise \end{cases}$$

In special case if $A = \{a\}$, we denote $y_{\{a\}}$ by y_a, and it is called an L-point of X.

For $\mu, \nu \in L^X$ we say that μ is contained in ν and we write $\mu \subseteq \nu$ if $\mu(x) \leq \nu(x)$, for all $x \in X$, and the intersection and union, $\mu \cup \nu$, $\mu \cap \nu \in L^X$ are defined by $(\mu \cup \nu)(x) = \mu(x) \vee \nu(x)$ and $(\mu \cap \nu)(x) = \mu(x) \wedge \nu(x)$, for all $x \in X$.

Also for $\mu \in L^X$, $a \in L$, μ_a is defined by,

$$\mu_a = \{x \in M | \mu(x) \geq a\},$$

μ_a is called a-cut or a-level subset of μ.

Let f be a mapping from X into Y and let $\mu \in L^X$, $\nu \in L^Y$. Then $f(\mu) \in L^Y$ and $f^{-1}(\nu) \in L^X$ are defined as follows:
$\forall y \in Y$.

$$f(\mu)(y) = \begin{cases} \bigvee \{\mu(x) | x \in f^{-1}(y)\} & if \ f^{-1}(y) \neq \emptyset \\ 0 & otherwise \end{cases}$$

and $f^{-1}(\nu)(x) = \nu(f(x))$, $\quad \forall x \in X$.

Let M, N be R-modules and $f : M \to N$ be an R-module homomorphism. $\mu \in L^M$ is called f-invariant if $f(x) = f(y)$, implies that $\mu(x) = \mu(y)$ for all $x, y \in M$.

We recall some definitions and theorems from the book (Mordeson and Malik 1998), which we need them for development of our paper.

Definition 1 Let $\mu \in L^R$. Then μ is called L-ideal of R if for every $x, y \in R$ the following conditions are satisfied:

(1) $\mu(x - y) \geq \mu(x) \wedge \mu(y)$;
(2) $\mu(xy) \geq \mu(x) \vee \mu(y)$.

The set of all L-ideals of R is denoted by $LI(R)$.

For $\mu, \nu \in LI(R)$, the product of μ and ν is defined by
$\mu\nu(x) = \bigvee \{\mu(y) \wedge \nu(z) \mid y, z \in R, x = yz\}$, $\forall x \in R$. In Mordeson and Malik (1998), it was proved that $\mu\nu \in LI(R)$

Definition 2 Let $\zeta \in LI(R)$. Then ζ is called prime L-ideal of R if ζ is non-constant and for $\mu, \nu \in LI(R)$,

$$\mu\nu \subseteq \zeta \Longrightarrow \mu \subseteq \zeta \ or \ \nu \subseteq \zeta.$$

By L-Spec(R), we mean the set of all prime L-ideals of R.

Definition 3 Let μ be an L-subset of R. The radical of μ is denoted by $\Re(\mu)$ and is defined by

$$\Re(\mu)(x) = \bigvee_{n \in N} \mu(x^n), \quad \forall x \in R.$$

If $\zeta \in LI(R)$ is prime, then $\Re(\zeta) = \zeta$.

Definition 4 Let $c \in L \setminus \{1\}$. c is called a prime element in L if $a \wedge b \leq c$, implies that $a \leq c$ or $b \leq c$, for all $a, b \in L$.

For $\zeta \in L^R$ and $\mu \in L^M$, define $\zeta \cdot \mu \in L^M$ as follows:
$(\zeta \cdot \mu)(x) = \bigvee \{\zeta(r) \wedge \mu(y) \mid r \in R, y \in M, ry = x\}$ for all $x \in M$.

Definition 5 An L- submodule of M is an L-subset $\mu \in L^M$ such that:

(1) $\mu(0) = 1$;
(2) $\mu(rx) \geq \mu(x)$ for all $r \in R$ and $x \in M$ and
(3) $\mu(x + y) \geq \mu(x) \wedge \mu(y)$ for all $x, y \in M$.

The set of all L-submodules of M is denoted by $L(M)$.

Remark 6 If $M = R$, then it is easy to verify that $\mu \in L^R$ is an L-submodule of M if and only if μ is an L-ideal of R.

Theorem 7 *Let $\mu \in L^M$. Then $\mu \in L(M)$ if and only if each non-empty level subset of μ is a submodule of M. Moreover if $\mu \in L(M)$ then*

$$\mu_* = \{x \in M \mid \mu(x) = 1\}$$

is a submodule of M. $\qquad\square$

Theorem 8 *Let $x \in M$ and $a \in L$. Then an L-submodule of M generated by x_a is the smallest L-submodule of M contains x_a. In Mordeson and Malik (1998) it was proved that*

$$< x_a >= 1_{\{0\}} \cup (\cup\{(rx)_a \mid r \in R\}).$$

$\qquad\square$

Definition 9 For $\mu, \nu \in L^M$ and $\zeta \in L^R$, define $(\mu : \nu) \in L^R$ and $(\mu : \zeta) \in L^M$ as follows:

$$(\mu : \nu) = \bigcup\{\eta \in L^R \mid \eta \cdot \nu \subseteq \mu\};$$

$$(\mu : \zeta) = \bigcup\{\nu \in L^M \mid \zeta \cdot \nu \subseteq \mu\}.$$

In Mordeson and Malik (1998) it was proved that if $\nu \in L^M$, $\mu \in L(M)$ and $\zeta \in LI(R)$ then
$(\mu : \nu) = \bigcup\{\eta \in LI(R) \mid \eta \cdot \nu \subseteq \mu\} \in LI(R)$;
$(\mu : \zeta) = \bigcup\{\nu \in L(M) \mid \zeta \cdot \nu \subseteq \mu\} \in L(M)$.

Theorem 10 *Let N be a submodule of M and $c \in L$. Then*

$$(1_N \cup c_M) : 1_M = 1_{(N:M)} \cup c_R. \square$$

Definition 11 A non-constant L-submodule μ of M is called primary if for $\zeta \in LI(R)$ and $\nu \in L(M)$ such that $\zeta \cdot \nu \subseteq \mu$ then either $\zeta \subseteq \Re(\mu : 1_M)$ or $\nu \subseteq \mu$.

In Mordeson and Malik (1998) it was proved that $\mu \in L(M)$ is primary if and only if $\mu = 1_{\mu_*} \cup c_M$ such that μ_* is a primary submodule of M and c is a prime element in L.

We recall that an L-submodule μ of M is called prime if for $\zeta \in LI(R)$ and $\nu \in L(M)$ such that $\zeta \cdot \nu \subseteq \mu$, then either $\nu \subseteq \mu$ or $\zeta \subseteq (\mu : 1_M)$ (Ameri and Mahjoob 2008).

In Ameri and Mahjoob (2008), it is proved that for $\mu \in L(M)$, μ is prime if and only if $\mu = 1_{\mu_*} \cup c_M$ such that μ_* is a prime submodule of M and c is a prime element in L.

2   Classical Prime L-Submodules

In this section we introduce the notion of classical prime L-submodules and investigate some basic properties of them. We recall that a submodule N of an R-module M is called classical prime, if for any elements $a, b \in R$ and $x \in M$, the condition $abx \in N$ implies that $ax \in N$ or $bx \in N$.

Definition 12 A non-constant L-submodule μ of M is said to be classical prime, if for $\zeta, \eta \in LI(R)$ and $\nu \in L(M)$ such that $\zeta \cdot \eta \cdot \nu \subseteq \mu$, then either $\zeta \cdot \nu \subseteq \mu$ or $\eta \cdot \nu \subseteq \mu$.

Notation: By $Cl.L - Spec(M)$ we mean the set of all classical prime L-submodules of M.

Remark 13 The given definition of classical prime L-submodule is a generalization of the notion of classical prime submodules in module theory.

In the next theorem we give two interesting characterization of classical prime L-submodules. But first we recall this result from Azizi (2008).

Theorem 14 *Let M be an R-module and let N be a proper submodule of M.*
(i) N is a classical prime submodule if and only if for each $x \in M \setminus N$, $(N : x)$ is a prime ideal of R. When this is the case, $\{(N : x)\}_{x \in M \setminus N}$ is a chain of prime ideals of R.
(ii) If N is a classical prime submodule, then $(N : M)$ is prime ideals of R.

Theorem 15 *Let μ be an L-submodule of M. Then μ is classical prime if and only if $\mu = 1_{\mu_*} \cup c_M$ such that μ_* is a classical prime submodule of M and c is a prime element in L.*

Proof First suppose that $\mu \in Cl.L - Spec(M)$. Since μ_* is a submodule of M, then $1 \in \mu(M)$ and $|\mu(M)| \geq 2$. Let $x, y \in M \setminus \mu_*$ and $\mu(x) = c$. Since $\mu(rx) \geq \mu(x) = c$, then, $(rx)_c \subseteq \mu$ for all $r \in R$. Thus,

$$< x_c > = 1_{\{0\}} \cup \{\cup (rx)_c \mid r \in R\} \subseteq \mu.$$

Now let $\zeta = 1_{\{0\}} \cup c_R$, $\eta = 1_R$ and $\upsilon = 1_{<x>} \cup c_M$. $\zeta, \eta \in LI(R)$, $\upsilon \in L(M)$ and

$$1_R \cdot (1_{\{0\}} \cup c_R) = (1_R \cdot 1_{\{0\}}) \cup (1_R \cdot c_R)$$
$$= 1_{\{0\}} \cup c_R,$$

therefore

$$1_R \cdot (1_{\{0\}} \cup c_R) \cdot (1_{<x>} \cup c_M) = (1_{\{0\}} \cup c_R) \cdot (1_{<x>} \cup c_M)$$
$$= (1_{\{0\}} \cdot 1_{<x>}) \cup (1_{\{0\}} \cdot c_M) \cup (c_R \cdot 1_{<x>}) \cup (c_R \cdot c_M)$$
$$= 1_{\{0\}} \cup c_{<x>} \cup c_M = < x_c > \cup c_M \subseteq \mu \cup c_M = \mu.$$

But

$$(1_R \cdot (1_{<x>} \cup c_M))(x) = \bigvee_{x=ry} (1_R(r) \wedge (1_{<x>} \cup c_M)(y))$$
$$= \bigvee_{x=ry} (1_{<x>} \cup c_M)(y) = 1 > \mu(x).$$

Then $\zeta \cdot \upsilon \subseteq \mu$, and so,

$$\mu(y) \geq (\zeta \cdot \upsilon)(y) = 1_{\{0\}} \cup c_{<x>} \cup c_M = c = \mu(x),$$

since $y \neq 0$. Similarly $\mu(x) \geq \mu(y)$.

Hence $\mu(y) = \mu(x)$. Thus $\mid \mu(M) \mid = 2$ and so

$$\mu = 1_{\mu_*} \cup c_M.$$

We prove that μ_* is a classical prime submodule of M. Let $rsx \in \mu_*$, for $r, s \in R$ and $x \in M$. Then $1_{<x>} \in L(M)$, $1_{<r>}, 1_{<s>} \in LI(R)$, and $< r >< s >< x > \subseteq \mu_*$. Therefore

$$1_{<r>} \cdot 1_{<s>} \cdot 1_{<x>} = 1_{<r><s><x>} \subseteq 1_{\mu_*} \subseteq \mu,$$

thus either $1_{<r>} \cdot 1_{<x>} \subseteq \mu$ or $1_{<s>} \cdot 1_{<x>} \subseteq \mu$, and hence $< rx > \subseteq \mu_*$ or $< sx > \subseteq \mu_*$. Then $rx \in \mu_*$ or $sx \in \mu_*$. It means that μ_* is a classical prime submodule of M.

Now suppose that c is not a prime element in L. Then there exist $a, b \in L$ such that $a \nleq c$, $b \nleq c$, and $a \wedge b \leq c$. Thus $a_R \cdot 1_{<x>} \nsubseteq \mu$ and $b_R \cdot 1_{<x>} \nsubseteq \mu$. But $a_R \cdot b_R \cdot 1_{<x>} = (a \wedge b)_R \cdot 1_{<x>} = (a \wedge b)_{<x>} \subseteq \mu$. This contradicts the fact that μ is a classical prime L-submodule of M.

Conversely, suppose that $\mu = 1_{\mu_*} \cup c_M$, where μ_* is a classical prime submodule of M and c is a prime element in L. Let $\zeta, \eta \in LI(R)$, $\nu \in L(M)$ and $\zeta \cdot \eta \cdot \nu \subseteq \mu$. Suppose that $\zeta \cdot \nu \nsubseteq \mu$ and $\eta \cdot \nu \nsubseteq \mu$, then there exist $x, y \in M$ such that $(\zeta \cdot \nu)(x) \nleq \mu(x)$ and $(\eta \cdot \nu)(y) \nleq \mu(y)$. So $x, y \notin \mu_*$, $(\zeta \cdot \nu)(x) \nleq c$ and $(\eta \cdot \nu)(y) \nleq c$. Therefore there exists $r, s \in R$ and $u, v \in M$ such that $x = ru$, $y = sv$ and

$\zeta(r) \wedge \nu(u) \not\leq c, \eta(s) \wedge \nu(v) \not\leq c$. We claim that $rsu \notin \mu_*$ or $rsv \notin \mu_*$. If $rsu \in \mu_*$, then $su \in \mu_*$, and by Theorem 14,

$$(\mu_* : v) \subseteq (\mu_* : u).$$

On the other hand $r \notin (\mu_* : u)$, so $r \notin (\mu_* : v)$, hence $rv \notin \mu_*$, and since $sv \notin \mu_*$ then $rsv \notin \mu_*$. The other case is similar, thus

$$(\zeta(r) \wedge \nu(u)) \wedge (\eta(s) \wedge \nu(v)) \leq \zeta(r) \wedge \eta(s) \wedge \nu(u)$$
$$\leq (\zeta \cdot \eta \cdot \nu)(rsu)$$
$$\leq \mu(rsu) = c.$$

This contradicts the fact that c is a prime element in L. Therefore $\zeta \cdot \nu \subseteq \mu$ or $\eta \cdot \nu \subseteq \mu$. $\qquad \square$

Theorem 16 *Let $\mu \in L(M)$. Then μ is classical prime if and only if μ satisfies the following conditions:*
(i) μ_* is a classical prime submodule of M;
(ii) $(\mu : 1_M)(1)$ is a prime element in L;
(iii) If $r_a \cdot s_b \cdot x_d \subseteq \mu$, $r, s \in R, x \in M$ and $a, b, d \in L$, then either $r_a \cdot x_d \subseteq \mu$ or $s_b \cdot x_d \subseteq \mu$.

Proof First suppose that $\mu \in Cl.L - Spec(M)$. By Theorem 15, $\mu = 1_{\mu_*} \cup c_M$ such that μ_* is a classical prime submodule of M and c is a prime element in L, thus (i) holds.
(ii) $(\mu : 1_M) = (1_{\mu_*:M} \cup c_R)$. By Theorem 14 $(\mu_* : M)$ is a prime ideal of R, then $1 \notin (\mu_* : M)$ and $(\mu : 1_M)(1) = c$ is a prime element in L by Theorem 15
(iii) Suppose that $r_a \cdot s_b \cdot x_d \subseteq \mu$ for $r, s \in R, x \in M$ and $a, b, d \in L$, but $r_a \cdot x_d \not\subseteq \mu$ and $s_b \cdot x_d \not\subseteq \mu$. Then $rx \notin \mu_*$ and $sx \notin \mu_*$. So $rsx \notin \mu_*$, since μ_* is a classical prime submodule of M.
  Also we have

$$(a \wedge d) \wedge (b \wedge d) = a \wedge b \wedge d = r_a(r) \wedge s_b(s) \wedge x_d(x)$$
$$\leq r_a \cdot s_b \cdot x_d(rsx) \leq \mu(rsx) = c.$$

Therefore, $(a \wedge d) \wedge (b \wedge d) \leq c$. Since $r_a \cdot x_d \not\subseteq \mu$ and $s_b \cdot x_d \not\subseteq \mu$, we have

$$a \wedge d = r_a(r) \wedge x_d(x) = (r_a \cdot x_d)(rx) \not\leq \mu(rx) = c$$

and

$$b \wedge d = s_b(s) \wedge x_d(x) = (s_b \cdot x_d)(sx) \not\leq \mu(sx) = c,$$

which is a contradiction. Thus $r_a \cdot x_d \subseteq \mu$ or $s_b \cdot x_d \subseteq \mu$.
  Conversely, suppose that μ satisfies the conditions (i), (ii) and (iii). Let $c = (\mu : 1_M)(1)$ be a prime element in L and $a = 1$. Consider $x \in M \setminus \mu_*$ and $b = \mu(x)$, then

$$(1_b \cdot x_a)(y) = x_b(y) \subseteq \mu(y), \forall y \in M.$$

Thus $1_b \cdot x_a \subseteq \mu$, and so $1_b \subseteq (\mu : 1_M)$. Therefore

$$b = 1_b(1) \subseteq (\mu : 1_M)(1) = \bigvee_{\eta \cdot 1_M \subseteq \mu} \eta(1) = \bigvee_{\eta \cdot 1_M \subseteq \mu} (\eta(1) \wedge 1_M(x))$$

$$\leq \bigvee_{\eta \cdot 1_M \subseteq \mu} (\eta \cdot 1_M)(x) \leq \mu(x) = b.$$

Hence $b = (\mu : 1_M)(1) = \mu(x)$. Then we have $\mu = 1_{\mu^*} \cup c_M$. It means that μ is classical prime.
$\square$

Note that if μ is an L-submodule of M, then $1_R \cdot \mu = \mu$.

Remark 17 Every prime L-submodule is classical prime. But the converse, in general is not true, for example, let R be an integral domain and P a non-zero prime ideal of R. Then for the free R-module $M = R \oplus R$, the submodule $(0 \oplus P)$ is a classical prime submodule, which is not a prime submodule.

For every prime element $t \in L$, define $\mu \in L(M)$ by

$$\mu(x) = \begin{cases} 1 & \text{if } x \in (0 \oplus P) \\ t & \text{otherwise} \end{cases}$$

for all $x \in M$.

Then by Theorem 15 μ is classical prime L-submodule of M, which is not prime.

Theorem 18 *An L-submodule μ of M is classical prime if and only if for every L-submodule ν of M which is not contained in μ, $(\mu : \nu)$ is a prime L-ideal of R.*

Proof Let $\mu \in L(M)$ be classical prime and for $\zeta, \eta \in LI(R)$, $\zeta\eta \subseteq (\mu : \nu)$. Then $\zeta\eta\nu \subseteq \mu$, so $\zeta\nu \subseteq \mu$ or $\eta\nu \subseteq \mu$. Thus $\zeta \subseteq (\mu : \nu)$ or $\eta \subseteq (\mu : \nu)$. Conversely, suppose that for every L-submodule ν of M which is not contained in μ, $(\mu : \nu)$ is prime, and let for $\zeta, \nu \in LI(R)$ and $\nu \in L(M)$, $\zeta \cdot \eta \cdot \nu \subseteq \mu$. Then $\zeta \cdot \eta \subseteq (\mu : \nu)$.
$\square$

Note that in the previous Theorem if ν is contained in μ, then $\zeta \cdot \nu \subseteq \mu$, for all $\zeta \in LI(R)$. Therefore $(\mu : \nu) = 1_R$.

Corollary 19 *If an L-submodule μ of M is classical prime, then $(\mu : 1_M)$ is a prime L-ideal. Also for each $x \in M$ and $a \in L$ provided that $x_a \notin \mu$, we have $(\mu : x_a)$ is a prime L-ideal.*

Recall that an R-module M is called compatible if its classical prime submodules and prime submodules are coincided. For example the field of fractions of a domain R is a compatible R-module. Also all commutative rings, semisimple modules and multiplicative modules are examples of compatible modules. We also recall that an R-module M is called multiplication module if each submodule N of M is of the form $N = IM$, for some ideal I of R.

In a multiplication R-module M every classical prime L-submodule is prime. (By Theorem 15)

Theorem 20 *Let M be an R-module and μ a classical prime L-submodule of M. Then for all L-submodule ξ, ν of M which are not contained in μ, $(\mu : \nu) \subseteq (\mu : \xi)$ or $(\mu : \xi) \subseteq (\mu : \nu)$.*

Proof Let $\mu \in L(M)$ be classical prime and for ν, $\xi \in L(M)$ which are not contained in μ, we have $(\mu : \nu) \nsubseteq (\mu : \xi)$ and $(\mu : \xi) \nsubseteq (\mu : \nu)$. Then there exist $\zeta \subseteq (\mu : \nu)$ and $\eta \subseteq (\mu : \xi)$ such that $\zeta \cdot \xi \nsubseteq \mu$ and $\eta \cdot \nu \nsubseteq \mu$. But we have,

$$\zeta \cdot \eta \cdot (\nu + \xi) = \zeta \cdot \eta \cdot \nu + \zeta \cdot \eta \cdot \xi \subseteq \mu + \mu \subseteq \mu.$$

Thus $\zeta \cdot (\nu + \xi) \subseteq \mu$ or $\eta \cdot (\nu + \xi) \subseteq \mu$. Hence either $\zeta \cdot \nu + \zeta \cdot \xi \subseteq \mu$ or $\eta \cdot \nu + \eta \cdot \xi \subseteq \mu$. Therefore $\zeta \cdot \xi \subseteq \mu$ or $\eta \cdot \nu \subseteq \mu$, which is a contradiction. $\qquad \square$

In the sequel, we will find a relationship between prime L-submodules and classical prime L-submodules.

Theorem 21 *Let M be an R-module and μ a non-constant L-submodule of M. Then μ is prime, if and only if μ is primary and classical prime.*

Proof If μ is prime, the result is obvious. Now let μ be classical prime and primary, $\zeta \in LI(R)$ and $\nu \in L(M)$ such that $\zeta \cdot \nu \subseteq \mu$ and $\nu \nsubseteq \mu$. Then

$$\zeta \subseteq \Re(\mu : 1_M) = (\mu : 1_M),$$

since $(\mu : 1_M)$ is a prime L-ideal of R. Thus μ is prime. $\qquad \square$

Theorem 22 *Let M, N be R-modules and f a homomorphism from M onto N. Then the following statements are satisfied;*
(1) Let $\mu \in Cl.L - Spec(M)$ be f-invariant. Then $f(\mu) \in Cl.L - Spec(N)$.
(2) If $\nu \in Cl.L - Spec(N)$, then $f^{-1}(\nu) \in Cl.L - Spec(M)$.

Proof The proof is similar to the proof of Mordeson and Malik (1998), Theorem 4.6.9. $\qquad \square$

Example 23 (1) Consider the ring of integers $\mathbb{Z}$ as $\mathbb{Z}$-module and let L be an arbitrary lattice. Suppose that $P \in \mathbb{Z}$ is prime. For every prime element $t \in L$, define $P(t) \in L(\mathbb{Z})$ by

$$P(t)(x) = \begin{cases} 1 & \text{if } x \in <P> \\ t & \text{if } x \in \mathbb{Z} \backslash <P> \end{cases}$$

Then by Theorem 15, $P(t)$ is a classical prime L-submodule of M.
(2) Consider $M = \mathbb{R}[x]$ as $\mathbb{R}[x]$-module, where $\mathbb{R}$ is the field of real numbers. For every $P \in \mathbb{R}[x]$ and every $t \in L$, define the fuzzy subset $P(t)$ of $\mathbb{R}[x]$ by

$$P(t)(x) = \begin{cases} 1 & \text{if } x \in <P> \\ t & \text{otherwise} \end{cases}$$

Then by Theorem 15, $P(t)$ is a classical prime L-submodule of M provided that P is irreducible and t is a prime element in L

Example 24 Let $M = R \oplus R$, $N_1 = P_1 \oplus P_2$, and $N_2 = P_1 \oplus R$, where P_1, P_2 are prime ideal of R, and $P_1 \subseteq P_2$. It is easy to verify that N_1 and N_2 are classical prime submodule of M. For every prime element $t \in L$, define ν and $\xi \in L(M)$ by

$$\nu(x) = \begin{cases} 1 & if \ x \in N_1 \\ t & otherwise \end{cases}$$

$$\xi(x) = \begin{cases} 1 & if \ x \in N_2 \\ t & otherwise \end{cases}$$

Then by Theorem 15 ν and ξ are classical prime L-submodules of M. $\qquad\square$

In McCasland (1983), the notion of the envelope of a submodule N of M, $E(N)$, is defined as the set $\{x \in M \mid x = ry, r^n y \in N, \ for \ some \ r \in R, \ y \in M \ and \ n \in \mathbb{N}\}$.

We recall that a module M satisfies the radical formula if for every submodule N of M,

$$< E(N) >= rad(N),$$

where $rad(N)$ is the intersection of all prime submodules of M containing N.

For $\mu \in L(M)$, $Rad(\mu)$ is defined as the intersection of all prime L-submodules of M containing μ, and the envelope of μ is defined by

$$E(\mu)(x) = \vee\{\mu(r^n y)|x = ry, n \in \mathbb{N}\}, \ \forall x \in M.$$

In Mahjoob and Qiami (2018) we investigated some basic properties of $Rad(\mu)$ and $E(\mu)$. Also we say that M satisfies in the L-radical formula if for every $\mu \in L(M)$, $E(\mu) = Rad(\mu)$.

Now we are ready to define the classical Radical of μ, denoted by $cl.Rad(\mu)$, as the intersection of all classical prime L-submodule of M contains μ.

It is clear that, $\mu \subseteq E(\mu) \subseteq cl.Rad(\mu) \subseteq Rad(\mu)$, so if μ is a classical prime L-submodule of M, then $E(\mu) = \mu$.

In Mahjoob and Qiami (2018) also the relationship between $Rad(\mu)_a$ and $Rad(\mu_a)$ is investigated. By a similar argument we can conclude that $cl.Rad(\mu)_a \subseteq cl.Rad(\mu_a)$ and we obtain the following results:

Theorem 25 *Let M be an R-module such that satisfies in the radical formula and L be a dense chain. Then we have*

$$cl.Rad(\mu) = Rad(\mu).$$

Proof Consider $x \in M$. Put $a = Rad(\mu)(x)$, then

$$x \in Rad(\mu)_a \subseteq rad(\mu_a)$$
$$=< E(\mu_a) >$$
$$\subseteq E(\mu)_a$$
$$\subseteq cl.Rad(\mu)_a.$$

So $cl.Rad(\mu)(x) \geq a = Rad(\mu)(x)$, therefore we have $Rad(\mu) \subseteq cl.Rad(\mu)$. $\square$

3 Classical L-Top Module

Let M be a nonzero R-module. For any $\mu \in L(M)$, the L-classical variety of μ is denoted by $\mathbb{V}(\mu)$, and is defined as the set of all classical prime L-submodule containing μ i.e. $\mathbb{V}(\mu) = \{\nu \in Cl.L - Spec(M) \mid \mu \subseteq \nu\}$.

Obviously $\mathbb{V}(1_M)$ is just the empty set and $\mathbb{V}(1_{\{0\}})$ is $Cl.L - Spec(M)$.
Also it is easy to verify that for any family of L-submodules $\{\mu_i\}_{i \in I}$ of M and $\mu, \nu \in L(M)$; $\bigcap_{i \in I} \mathbb{V}(\mu_i) = \mathbb{V}(\sum_{i \in I} \mu_i)$, $\mathbb{V}(\mu) \cup \mathbb{V}(\nu) \subseteq \mathbb{V}(\mu \cap \nu)$.

We remark that if $\mu \in L(M)$, then $Cl.Rad(\mu) = \bigcap_{\nu \in \mathbb{V}(\mu)} \nu$ and equal to 1_M if $\mathbb{V}(\mu) = \emptyset$.

Now we let $\mathbb{C}(M)$ denote the collection of all subsets $\mathbb{V}(\mu)$ of $Cl.L - Spec(M)$. Then $\mathbb{C}(M)$ contains the empty set and $Cl.L - Spec(M)$, and also $\mathbb{C}(M)$ is closed under arbitrary intersections. However, in general, $\mathbb{C}(M)$ is not closed under finite union. Also if $\mathbb{C}(M)$ is closed under finite union, i.e. for every L-submodule μ and ξ of $L(M)$ there exists an L-submodule ν of $L(M)$ such that $\mathbb{V}(\mu) \cup \mathbb{V}(\xi) = \mathbb{V}(\nu)$, for in this case $\mathbb{C}(M)$ satisfies the axioms of the closed subsets of a topological space. A module M for which $\mathbb{C}(M)$ is closed under finite union, is called classical L-top module.

Now let we define classical semiprime and classical extraordinary L-submodules.

An L-submodule $\mu \in L(M)$ is called classical semiprime L-submodule if μ is an intersection of classical prime L-submodules, and μ is called classical extraordinary L-submodule if whenever ν, ξ are classical semiprime L-submodule of M with $\nu \cap \xi \subseteq \mu$ then $\nu \subseteq \mu$ or $\xi \subseteq \mu$.

Theorem 26 *For an R−module M the following statements are equivalent:*
(i) M is a classical L−top module;
(ii) Every classical prime L-submodule of M is classical extraordinary;
(iii) $\mathbb{V}(\mu_1) \cup \mathbb{V}(\mu_2) = \mathbb{V}(\mu_1 \cap \mu_2)$ for every classical semiprime L-submodule $\mu_1, \mu_2 \in L(M)$.

Proof Clearly, if $Cl.L - Spec(M) = \emptyset$, then the result is true. Suppose that $Cl.L - Spec(M) \neq \emptyset$.
$(i) \Rightarrow (ii)$ Consider a classical prime L-submodule ν of M and let $\mu_1, \mu_2 \in L(M)$ be

classical semiprime such that $\mu_1 \cap \mu_2 \subseteq \mu$. Since M is classical L-top module, then there exists an L-submodule μ of M, such that $\mathbb{V}(\mu_1) \cup \mathbb{V}(\mu_2) = \mathbb{V}(\mu)$. Since μ_1 is classical semiprime, then $\mu_1 = \bigcap_{i \in I} \nu_i$, such that ν_i is a classical prime L-submodule of M, for all $i \in I$. Therefore, for each $i \in I$ if $\mu_1 \subseteq \nu_i$, then, $\nu_i \in \mathbb{V}(\mu_1) \subseteq \mathbb{V}(\mu)$. Thus $\mu \subseteq \bigcap_{i \in I} \nu_i = \mu_1$. Similarly $\mu \subseteq \mu_2$, and we obtain that $\mu \subseteq \mu_1 \cap \mu_2$. Also

$$\mathbb{V}(\mu_1) \cup \mathbb{V}(\mu_2) \subseteq \mathbb{V}(\mu_1 \cap \mu_2) \subseteq \mathbb{V}(\mu) = \mathbb{V}(\mu_1) \cup \mathbb{V}(\mu_2).$$

Hence

$$\mathbb{V}(\mu_1) \cup \mathbb{V}(\mu_2) = \mathbb{V}(\mu_1 \cap \mu_2).$$

Now

$$\nu \in \mathbb{V}(\mu_1 \cap \mu_2) \Rightarrow \nu \in \mathbb{V}(\mu_1) \cup \mathbb{V}(\mu_2)$$
$$\Rightarrow \nu \in \mathbb{V}(\mu_1) \text{ or } \nu \in \mathbb{V}(\mu_2)$$
$$\Rightarrow \mu_1 \subseteq \nu \text{ or } \mu_2 \subseteq \nu,$$

it means that ν is classical extraordinary.

$(ii) \Rightarrow (iii)$ Suppose that $\mu_1, \mu_2 \in L(M)$ are classical semiprime. Clearly

$$\mathbb{V}(\mu_1) \cup \mathbb{V}(\mu_2) \subseteq \mathbb{V}(\mu_1 \cap \mu_2). \tag{1}$$

Let $\nu \in \mathbb{V}(\mu_1 \cap \mu_2)$. Then $\mu_1 \cap \mu_2 \subseteq \nu$ and hence by hypothesis $\mu_1 \subseteq \nu$ or $\mu_2 \subseteq \nu$. Thus $\nu \in \mathbb{V}(\mu_1)$ or $\nu \in \mathbb{V}(\mu_2)$. Therefore $\nu \in \mathbb{V}(\mu_1) \cup \mathbb{V}(\mu_2)$ and then

$$\mathbb{V}(\mu_1 \cap \mu_2) \subseteq \mathbb{V}(\mu_1) \cup \mathbb{V}(\mu_2) \tag{2}$$

Now from (1) and (2) it follows that

$$\mathbb{V}(\mu_1) \cup \mathbb{V}(\mu_2) = \mathbb{V}(\mu_1 \cap \mu_2).$$

$(iii) \Rightarrow (i)$ Let $\mu_1, \mu_2 \in L(M)$. If $\mathbb{V}(\mu_1)$ is the empty set, then

$$\mathbb{V}(\mu_1) \cup \mathbb{V}(\mu_2) = \mathbb{V}(\mu_2).$$

So we assume that $\mathbb{V}(\mu_1)$ and $\mathbb{V}(\mu_2)$ are both non-empty. Then $\mathbb{V}(\mu_1) \cup \mathbb{V}(\mu_2) = \mathbb{V}(Cl.Rad(\mu_1)) \cup \mathbb{V}(Cl.Rad(\mu_2))$ and by (iii) since $Cl.Rad(\mu_1)$ and $Cl.Rad(\mu_2)$ are classical semiprime, then
$\mathbb{V}(Cl.Rad(\mu_1)) \cup \mathbb{V}(Cl.Rad(\mu_2)) = \mathbb{V}(Cl.Rad(\mu_1) \cap Cl.Rad(\mu_2))$. By letting $\nu = Cl.Rad(\mu_1) \cap Cl.Rad(\mu_2)$, it concludes that

$$\mathbb{V}(\mu_1) \cup \mathbb{V}(\mu_2) = \mathbb{V}(\nu).$$

This proves (i).   □

In the following, we will show that two concepts of L-top module and classical L-top module are equivalent for certain classes of modules like L-multiplication modules. We recall that an R-module M is called an L-multiplication module if every L-submodule μ of M is of the form $\mu = \zeta.1_M$ for some $\zeta \in LI(R)$. ζ is called a presentation L-ideal for μ. Note that such presentation is not unique. Let μ and ν be L-submodules of an L-multiplication module M where $\mu = \zeta.1_M$ and $\nu = \eta.1_M$ for some L-ideals ζ and η of R. The product of μ and ν is denoted by $\mu\nu$ and is defined by $\mu\nu = \zeta\eta.1_M$. Clearly, $\mu\nu$ is an L-submodule of M and is contained in $\mu \cap \nu$. The product of μ and ν is independent of presentations of μ and ν (Lee and Park 2011).

Lemma 27 *Let M be an L-multiplication module, and $\mu \in L(M)$. Then:*

$$\mu = (\mu : 1_M)1_M.$$

Proof Let $\mu \in L(M)$. So there exists an L-ideal ζ of R such that $\mu = \zeta 1_M$. Then $\zeta \subseteq (\mu : 1_M)$ and

$$\mu = \zeta 1_M \subseteq (\mu : 1_M)1_M \subseteq \mu$$

which implies that $\mu = (\mu : 1_M)1_M$, as required.   □

Lemma 28 *Let M be an L-multiplication module. Then the following statements are equivalent.*
(1) $\mu \in L(M)$ is prime L-submodule;
(2) $(\mu : 1_M)$ is prime L-ideal;
(3) $\mu = \zeta 1_M$ where ζ is a prime L-ideal which is maximal with respect to this property (i.e, $\eta 1_M \subseteq \mu$ implies that $\eta \subseteq \zeta$).

Proposition 29 *Let M be an L-multiplication module. Then every classical prime L-submodule of M is prime.*

Proof Proof of Lemma 28 and Proposition 29 are evident.   □

Now we can conclude that every L-multiplication module is both L-top module and classical L-top module.

Proposition 30 *Let M be a classical L-top module. Then, every homomorphic image of M is a classical L-top module too.*

Proof Let N be a submodule of a classical L-top module M. Let $M' = \frac{M}{N}$. Consider the quotient map $\pi : M \longrightarrow M'$, is defined by $\pi(x) = [x]$, and we denote the image of μ under π by $\pi(\mu)$. In fact, $\pi(\mu)([x]) = \vee\{\mu(z)|z \in [x]\}$. Suppose that $Cl.L - Spec(M') \neq \emptyset$. Clearly, the classical prime L-submodules of M' are precisely the L-submodules $\pi(\mu)$, where μ is a classical prime L-submodule of M. Thus, any classical semiprime L-submodule of M' is of the form $\pi(\nu)$, where ν is a classical semiprime L-submodule of M. Now let $\pi(\mu)$, $\pi(\nu)$ and $\pi(\xi)$ be classical semiprime L-submodule of M' such that $\pi(\mu) \subseteq \pi(\nu) \cap \pi(\xi)$. Then

$$\mu \subseteq \pi^{-1}(\pi(\nu) \cap \pi(\xi)) \subseteq \pi^{-1}(\pi(\nu)) \cap \pi^{-1}(\pi(\xi)),$$

$\pi(\mu) \subseteq \pi(\nu)$ or $\pi(\mu) \subseteq \pi(\xi)$. By Theorem 26, M' is a classical L-top module. $\qquad\square$

4 Zariski-Like Topology as a Spectral Space

Let M be an L-top module. For each L-submodule μ of M, we put $\mathbb{U}(\mu) = Cl.L - Spec(M) \setminus \mathbb{V}(\mu)$ and $\mathbb{B}(M) = \{\mathbb{U}(\mu) : \mu \in L(M)\}$. Then, we define $\mathbb{T}(M)$ to be the collection of all unions of finite intersections of elements of $\mathbb{B}(M)$. In fact, $\mathbb{T}(M)$ is called the topology on $Cl.L - Spec(M)$ by the sub-basis $\mathbb{B}(M)$. $\mathbb{T}(M)$ is Zariski-like topology.

Recall that a topological space (X, τ) is called irreducible if for any decomposition $X = A_1 \cup A_2$, where A_1 and A_2 are closed subsets, we have either $X \subseteq A_1$ or $X \subseteq A_2$. Let Y be a closed subset of a topological space. An element $a \in Y$ is called a generic point of Y if $Y = \overline{\{a\}}$. Note that a generic point of the irreducible closed subset Y of a topological space is unique if the topological space is a T_0-space.

Also we recall that a spectral space is a topological space homeomorphic to the prime spectrum of a commutative ring equipped with Zariski topology. By Hochster's characterization (Hochster 1969), a topology τ on a set X is spectral if and only if the following axioms hold:

(i) X is a T_0-space;
(ii) X is quasi-compact and has a basis of quasi-compact open subsets;
(iii) The family of quasi-compact open subsets of X is closed under finite intersections.
(iv) X is a sober space; i.e., every irreducible closed subset of X has a generic point.

Let M be an R-module. We can easily see that the set, $\{\mathbb{U}(\mu_1) \cap \cdots \cap \mathbb{U}(\mu_k) : \mu_i \in L(M), 1 \le i \le k, \text{ for some } k \in \mathbb{N}\}$ is a basis for Zariski-like topology on $Cl.L - Spec(M)$.

Lemma 31 *Let M be an R-module and Y a nonempty subset of $Cl.L - Spec(M)$. Then,*

$$\overline{Y} = \bigcup_{\mu \in Y} \mathbb{V}(\mu).$$

Proof Clearly, $\overline{Y} \subseteq \bigcup_{\mu \in Y} \mathbb{V}(\mu)$. Suppose that C is any closed subset of X such that $Y \subseteq C$. Then,

$$C = \bigcap_{i \in I}(\bigcup_{j=1}^{n_i} \mathbb{V}(\nu_{ij})), \text{ for some } \nu_{ij} \in L(M), i \in I \text{ and } n_i \in \mathbb{N}. \text{ Consider } \xi \in \bigcup_{\mu \in Y} \mathbb{V}(\mu).$$

Then, there exists $\mu_0 \in Y$ such that $\xi \in \mathbb{V}(\mu_0)$ and so $\mu_0 \subseteq \xi$. Since $\mu_0 \in C$, then for each $i \in I$ there exists $j \in \{1, 2, \ldots, n_i\}$ such that $\nu_{ij} \subseteq \mu_0$, and hence $\nu_{ij} \subseteq \mu_0 \subseteq \xi$. It follows that $\xi \in C$. Therefore, $\bigcup_{\mu \in Y} \mathbb{V}(\mu) \subseteq C$. $\square$

In the next theorem we show that for any R-module M, $Cl.L - Spec(M)$ always is a T_0-space.

Theorem 32 *Let M be an R-module. Then, $Cl.L - Spec(M)$ is a T_0-space*

Proof Let $\mu_1, \mu_2 \in Cl.L - Spec(M)$. Then, by Lemma 31, $\{\overline{\mu_1}\} = \{\overline{\mu_2}\}$ if and only if $\mathbb{V}(\mu_1) = \mathbb{V}(\mu_2)$ if and only if $\mu_1 = \mu_2$. Now, by the fact that a topological space is T_0 if and only if closures of distinct points are distinct, we conclude that for any R-module M, $Cl.L - Spec(M)$ is a T_0-space. $\square$

We say that $L(M)$ is Nothery if any ascending chain of L-submodules of M is terminated, so it is easy to see that M is Notherian R-module if and only if $L(M)$ is Nothery.

Theorem 33 *Let M be a Notherian R-module. Then $Cl.L - Spec(M)$ is a quasi-compact space.*

Proof Let $\mathcal{A}$ be a family of open sets covering $Cl.L - Spec(M)$, and suppose that no finite subfamily of $\mathcal{A}$ covers $Cl.L - Spec(M)$. Since $\mathbb{V}(1_{\{0\}}) = Cl.L - Spec(M)$, then we may use the ACC on L-submodules to choose an L-submodule maximal μ with respect to the property that no finite subfamily of $\mathcal{A}$ covers $\mathbb{V}(\mu)$. We claim that μ is a classical prime L-submodule of M, for if not, there exist $x \in M$, $r, s \in R$ and $a, b, d \in L$, such that $r_a s_b x_d \subseteq \mu$, $s_b x_d \not\subseteq \mu$ and $r_a x_d \not\subseteq \mu$. Thus, $\mu \subseteq \mu + < s_b x_d >$ and $\mu \subseteq \mu + < r_a x_d >$. Hence, without loss of generality, there must exist a finite subfamily $\mathcal{A}_0$ of $\mathcal{A}$ that covers both $\mathbb{V}(\mu + < s_b x_d >)$ and $\mathbb{V}(\mu + < r_a x_d >)$. Let $\nu \in \mathbb{V}(\mu)$. Since $r_a s_b x_d \subseteq \mu$, then $r_a s_b x_d \subseteq \nu$, and since L-submodule ν is classical prime, then $s_b x_d \subseteq \nu$ or $r_a x_d \subseteq \nu$. Thus, either $\nu \subseteq \mathbb{V}(\mu + < s_b x_d >)$ or $\nu \subseteq \mathbb{V}(\mu + < r_a x_d >)$, and therefore,

$$\mathbb{V}(\mu) \subseteq \mathbb{V}(\mu + < s_b x_d >) \cup \mathbb{V}(\mu + < r_a x_d >).$$

So, $\mathbb{V}(\mu)$ is covered by the finite subfamily $\mathcal{A}_0$, which is a contradiction. Thus, μ is a classical prime L-submodule. Now, choose $U \in \mathcal{A}$ such that $\mu \in U$. Then, μ must have a neighborhood $\bigcap_{i=1}^{n} \mathbb{U}(\xi_i)$, for some $\xi_i \in L(M)$ and $n \in \mathbb{N}$, such that $\bigcap_{i=1}^{n} \mathbb{U}(\xi_i) \subseteq U$. We claim that for each i ($1 \leq i \leq n$), $\mu \in \mathbb{U}(\xi_i + \mu) \subseteq \mathbb{U}(\xi_i)$, to see this, assume that $\nu \in \mathbb{U}(\xi_i + \mu)$, i.e., $\xi_i + \mu \not\subseteq \nu$. So, $\xi_i \not\subseteq \nu$, i.e., $\nu \in \mathbb{U}(\xi_i)$. On the other hand, $\mu \in \mathbb{U}(\xi_i)$, i.e., $\xi_i \not\subseteq \mu$. Hence, $\xi_i + \mu \not\subseteq \nu$, i.e., $\nu \in \mathbb{U}(\xi_i + \mu)$. Consequently,

$$\mu \in \bigcap_{i=1}^{n} \mathbb{U}(\xi_i + \mu) \subseteq \bigcap_{i=1}^{n} \mathbb{U}(\xi_i) \subseteq U.$$

Thus $\bigcap\limits_{i=1}^{n} \mathbb{U}(\xi_i')$, where $\xi_i' := \xi_i + \mu$, is a neighborhood of μ such that $\bigcap\limits_{i=1}^{n} \mathbb{U}(\xi_i') \subseteq U$. Since for each $i\,(1 \le i \le n)$, $\mu \not\subseteq \xi_i'$, then $\mathbb{V}(\xi_i')$ can be covered by some finite subfamily $\mathcal{A}_0$ of $\mathcal{A}$. But,

$$\mathbb{V}(\mu) \setminus \left[\bigcup_{i=1}^{n} \mathbb{V}(\xi_i')\right] = \mathbb{V}(\mu) \setminus \left[\bigcap_{i=1}^{n} \mathbb{U}(\xi_i')\right]^c = \left[\bigcap_{i=1}^{n} \mathbb{U}(\xi_i')\right] \cap \mathbb{V}(\mu) \subseteq U,$$

and so $\mathbb{V}(\mu)$ can be covered by $\mathcal{A}_0' \cup \mathcal{A}_0' \cup \cdots \mathcal{A}_0' \cup \{U\}$, contrary to our choice of μ. Then, there must exist a finite subfamily of $\mathcal{A}$ which covers $Cl.Spec(M)$, which means that, $Cl.Spec(M)$ is a quasi-compact space. □

We need to recall the definition of patch topology. Let X be a topological space. By patch topology on X, we mean a topology which has as a sub-basis for its closed sets the closed sets and compact open sets of the original space (or better, which has the quasi-compact open sets and their complements as an open sub-basis). By a patch we mean a set closed in a patch topology.

Definition 34 Let M be an R-module, and let $\mathbb{P}(M)$ be the family of all subsets of $Cl.L - Spec(M)$ of the form $\mathbb{V}(\mu) \cap \mathbb{U}(\nu)$, where $\mu, \nu \in L(M)$. Clearly $\mathbb{P}(M)$ contains $Cl.L - Spec(M)$ and the empty set, since $Cl.L - Spec(M)$ equals $\mathbb{V}(1_{\{0\}}) \cap \mathbb{U}(1_M)$ and the empty set equals $\mathbb{V}(1_M) \cap \mathbb{U}(1_{\{0\}})$. Let $\mathbb{T}_P(M)$ be the collection U of all unions of finite intersections of elements of $\mathbb{P}(M)$. Then, $\mathbb{T}_P(M)$ forms a topology on $Cl.L - Spec(M)$ and is called the patch-like topology (in fact, $\mathbb{T}_P(M)$ is a sub-basis for the patch-like topology on $Cl.L - Spec(M)$).

Any patch-neighborhood of a point $\mu \in Cl.L - Spec(M)$ must contain a neighborhood from U, that is, a neighborhood of the form $\mathbb{V}(\xi) \cap \mathbb{U}(\nu)$, where $\xi \subseteq \mu$ and $\nu \not\subseteq \mu$.

Theorem 35 *Let M be an R-module. Then, $Cl.L - Spec(M)$ with the patch-like topology is a Hausdorff space.*

Proof Consider distinct points $\mu, \nu \in Cl.L - Spec(M)$. Since $\mu \ne \nu$, then either $\mu \not\subseteq \nu$ or $\nu \not\subseteq \mu$. Assume that $\mu \not\subseteq \nu$. By Definition 34, $U_1 := \mathbb{U}(1_M) \cap \mathbb{V}(\mu)$ is a patch-like-neighborhood of μ and $U_2 := \mathbb{U}(\mu) \cap \mathbb{V}(\nu)$ is a patch-like-neighborhood of ν. Clearly, $\mathbb{U}(\mu) \cap \mathbb{V}(\mu) = \emptyset$ and hence $U_1 \cap U_2 = \emptyset$. Thus, $Cl.L - Spec(M)$ is a Hausdorff space. □

Proposition 36 *Let M be a Noetherian R-module. Then, $Cl.L - Spec(M)$ with the patch-like topology is a compact space.*

Proof The proof is similar to the proof of Theorem 33. □

We recall the following evident lemma.

Lemma 37 *Assume τ and τ^* are two topologies on the set X such that $\tau \subseteq \tau^*$. If X is quasi-compact (i.e., any open cover of has a finite subcover) by τ^*, then X is also quasi-compact by τ.*

Theorem 38 *Let M be a Noetherian R-module. Then, for each $n \in \mathbb{N}$ and submodules $\mu_i (1 \leq i \leq n)$ of M, $\mathbb{U}(\mu_1) \cap \mathbb{U}(\mu_2) \cap \cdots \cap \mathbb{U}(\mu_n)$ is a quasi-compact subset of $Cl.L - Spec(M)$ with the Zariski-like topology. Consequently, $Cl.L - Spec(M)$ has a basis of quasi-compact open subsets and the family of Zariski-like quasi-compact open subsets of $Cl.L - Spec(M)$ is closed under finite intersections.*

Proof For any L-submodule $\mu \in L(M)$, $\mathbb{V}(\mu) = \mathbb{U}(\mu) \cap \mathbb{U}(1_M)$ is an open subset of $Cl.L - Spec(M)$ with the patch-like topology (see Definition 34). So, for each L-submodule μ of M, $\mathbb{U}(\mu)$ is a closed subset in $Cl.L - Spec(M)$. Thus, for each $n \in \mathbb{N}$ and $\mu_i \in L(M)(1 \leq i \leq n)$, $\mathbb{U}(\mu_1) \cap \mathbb{U}(\mu_2) \cap \cdots \cap \mathbb{U}(\mu_n)$ is also a closed subset in $Cl.L - Spec(M)$ with the patch-like topology. Since every closed subset of a compact space is compact, then $\mathbb{U}(\mu_1) \cap \mathbb{U}(\mu_2) \cap \cdots \cap \mathbb{U}(\mu_n)$ is compact in $Cl.L - Spec(M)$ with the patch-like topology and so by Lemma 37 it is quasi-compact in $Cl.L - Spec(M)$ with the Zariski-like topology. We Know, $\mathbb{B} = \{\mathbb{U}(\mu_1) \cap \mathbb{U}(\mu_2) \cap \cdots \cap \mathbb{U}(\mu_n) : 1 \leq i \leq n,$ for some $n \in \mathbb{N}\}$ is a basis for the Zariski-like topology of $L(M)$. On the other hand, if U is a Zariski-like quasi-compact open subset of $Cl.L - Spec(M)$, then $U = \bigcup_{i=1}^{m}(\bigcap_{j=1}^{n_i} \mathbb{U}(\mu_j))$. It follows that the family of Zariski-like quasi-compact open subsets of $Cl.L - Spec(M)$ is closed under finite intersections. $\qquad\square$

Proposition 39 *Let M be a Noetherian R-module. Then, every irreducible closed subset of $Cl.L - Spec(M)$ (with the Zariski-like topology) has a generic point.*

Proof Let Y be an irreducible closed subset of $Cl.L - Spec(M)$ (with the Zariski-like topology). First, we show that $Y = \bigcup_{\mu \in Y} \mathbb{V}(\mu)$. Clearly, $Y \subseteq \bigcup_{\mu \in Y} \mathbb{V}(\mu)$. By Lemma 31, for each $\mu \in Y$ we have $\mathbb{V}(\mu) = \overline{\{\mu\}} \subseteq \overline{Y}$, and since $\overline{Y} = Y$, then $\bigcup_{\mu \in Y} \mathbb{V}(\mu) \subseteq Y$. Thus, $Y = \bigcup_{\mu \in Y} \mathbb{V}(\mu)$.

By Definition 34, for each $\mu \in Y$, $\mathbb{V}(\mu)$ is an open subset of $Cl.Spec(M)$ with the patch-like topology. On the other hand, since $Y \subseteq Cl.L - Spec(M)$ is closed with the Zariski-like topology, then the complement of Y is open by this topology. This yields that the complement of Y is open with the patch-like topology, i.e., $Y \subseteq Cl.L - Spec(M)$ is closed with the patch-like topology. By Proposition 36, $Cl.L - Spec(M)$ is compact with the patch-like topology and since $Y \subseteq Cl.L - Spec(M)$ is closed, then Y is also compact. Now, since $Y = \bigcup_{\mu \in Y} \mathbb{V}(\mu)$ and each $\mathbb{V}(\mu)$ is patch-like-open, then there exists a finite subset Y' of Y such that $Y = \bigcup_{\mu \in Y'} \mathbb{V}(\mu)$. Now, since Y is irreducible, then $Y = \mathbb{V}(\mu)$ for some $\mu \in Y'$.

Therefore, we have $Y = \mathbb{V}(\mu) = \overline{\{\mu\}}$ for some $\mu \in Y$, i.e., μ is a generic point for Y. $\square$

Now we are going to present the main result of this section.

Theorem 40 *Let M be a Noetherian R-module. Then, $Cl.L - Spec(M)$ (with the Zariski-like topology) is a spectral space.*

Proof By Theorem 32, Cl.Spec(M) is a T_0-space. Since M is Noetherian, then by Theorem 33, $Cl.L - Spec(M)$ is quasi-compact. By Theorem 38, $Cl.L - Spec(M)$ has a basis of quasi-compact open subsets and the family of quasi-compact open subsets of $Cl.L - Spec(M)$ are closed under finite intersections. Finally, by Proposition 39, each irreducible closed subset of $Cl.L - Spec(M)$ has a generic point. Finally, by Hochster's characterization, $Cl.L - Spec(M)$ is a spectral spaces. $\square$

5 Weakly Prime L-Submodules

This section discusses another main topic of the paper. Our purpose here is to introduce and characterize weakly prime L-submodules in the context that it is based on the definition that is given in Atani and Farzalipoor (2007).

Recall that a proper submodule N of a module M over a commutative ring R with identity is said to be weakly prime if whenever $0 \neq rm \in N$, for $r \in R$ and $m \in M$, then $m \in N$ or $rM \subseteq N$.

Definition 41 Let $\mu \in L(M)$ be non-constant and $\mu \neq 1_{\{0\}}$. μ is called weakly prime if it satisfies the following condition:
if $1_{\{0\}} \neq \zeta \cdot \nu \subseteq \mu$ for $\zeta \in LI(R)$ and $\nu \in L(M)$, then either $\zeta \subseteq (\mu : 1_M)$ or $\nu \subseteq \mu$.

Theorem 42 *Let $1_{\{0\}} \neq \mu \in L(M)$. Then μ is weakly prime if and only if $\mu = 1_{\mu_*} \cup c_M$, where μ_* is a weakly prime submodule of M and c is a prime element of L.*

Proof Let $\mu \in L(M)$ be weakly prime. First μ_* is a submodule of M. Thus $1 \in \mu(M)$ and $|\mu(M)| \geq 2$. Let $x, y \in M \setminus \mu_*$ and $\mu(x) = c$. Since $\mu(rx) \geq \mu(x) = c$, then $(rx)_c \subseteq \mu$ for all $r \in R$. So

$$< x_c > = 1_{\{0\}} \cup \{\cup (rx)_c \mid r \in R\} \subseteq \mu.$$

Note that $c \neq 0$. If $c = 0$, then $\mu = 1_{\{0\}}$, in this case, if L is a chain, then $c = 0$ is a prime element, also $\{0\}$ is a weakly prime submodule.

Now consider $\nu = 1_{<x>}$ and $\zeta = 1_{\{0\}} \cup c_R$. Then $\zeta \in LI(R)$ and $\nu \in L(M)$. Clearly

$$1_{\{0\}} \neq \zeta \cdot \nu = 1_{\{0\}} \cup c_{<x>} = < x_c > \subseteq \mu.$$

Since μ is weakly prime and $1_{<x>} \not\subseteq \mu$, then

$\zeta = 1_{\{0\}} \cup c_R \subseteq (\mu : 1_M)$. Hence

$$\mu(x) = c = \zeta(1) \leq (\mu : 1_M)(1) = \bigvee \{\eta(1) \mid \eta \in LI(R), \eta \cdot 1_M \subseteq \mu\}$$
$$= \bigvee \{(\eta(1) \wedge 1_M)(y) \mid \eta \in LI(R), \eta \cdot 1_M \subseteq \mu\}$$
$$\leq \bigvee \{(\eta \cdot 1_M(y) \mid \eta \in LI(R), \eta \cdot 1_M \subseteq \mu\} \leq \mu(y).$$

Similarly $\mu(y) \leq \mu(x)$. Then $\mu(y) = \mu(x) = c$. Therefore $|\mu(M)| = 2$ and then $\mu = 1_{\mu_*} \cup c_M$. Next we prove that μ_* is a weakly prime submodule of M. Let $r \in R$, $x \in M$ and $0 \neq rx \in \mu_*$. Then $1_{<x>} \in L(M), 1_R \in LI(R), 0 \neq < r >< x >$ and

$$1_{\{0\}} \neq 1_{<r>} \cdot 1_{<x>} = 1_{<r><x>} \subseteq 1_{\mu_*} \subseteq \mu.$$

Thus either $1_{<r>} \subseteq (\mu : 1_M)$ or $1_{<x>} \subseteq \mu$. Note that $1_{<x>} \subseteq \mu$ implies that $< x > \subseteq \mu_*$ and so $x \in \mu_*$. If $1_{<r>} \subseteq (\mu : 1_M) = 1_{(\mu_*:M)} \cup c_R$, then $< r > \subseteq (\mu_* : M)$, and so $r \in (\mu_* : M)$. Therefore, μ_* is a weakly prime submodule of M.

Next we show that c is a prime element in L. Suppose that c is not prime. Then there exists $a, b \in L$ such that $a \not\leq c, b \not\leq c$ and $a \wedge b \leq c$. So $1_{\{0\}} \cup b_M \not\subseteq \mu$ and

$$(1_{\{0\}} \cup a_R)(1) = a \not\leq c = (1_{(\mu_*:M)} \cup c_R)(1) = (\mu : 1_M)(1)$$

and $(1_{\{0\}} \cup a_R) \not\subseteq (\mu : 1_M)$. But $1_{\{0\}} \neq (1_{\{0\}} \cup a_R) \cdot (1_{\{0\}} \cup b_M)$ and

$$(1_{\{0\}} \cup a_R) \cdot (1_{\{0\}} \cup b_M) \subseteq (1_{\{0\}} \cup c_M) \subseteq \mu$$

which is a contradiction. Hence, c is prime.

Conversely, suppose that $\mu = 1_{\mu_*} \cup c_M$, where μ_* is a weakly prime submodule of M and $c \neq 0$ is a prime element in L. Let $\zeta \in LI(R)$ and $\nu \in L(M)$ such that $1_{\{0\}} \neq \zeta \cdot \nu \subseteq \mu, \nu \not\subseteq \mu$ and $\zeta \not\subseteq (\mu : 1_M) = 1_{(\mu_*:M)} \cup c_R$. Then there exist $r \in R$ and $x \in M$ such that $\zeta(r) \not\leq (1_{(\mu_*:M)} \cup c_R)(r)$ and $\nu(x) \not\leq \mu(x)$. Thus $r \notin (\mu_* : M)$, $x \notin \mu_*$. We claim that $rx \neq 0$. If $rx = 0$, then $r \in (0 : M) \subseteq (\mu_* : M)$, because $\{0\}$ is a weakly prime submodule of M and $x \neq 0$. So $0 \neq rx \notin \mu_*$ and so $\mu(rx) = c$. Then we have

$$\zeta(r) \wedge \nu(x) \leq (\zeta \cdot \nu)(rx) = c.$$

This contradicts the fact that c is a prime element in L. Therefore either $\zeta \subseteq (\mu : 1_M)$ or $\nu \subseteq \mu$. $\qquad \square$

Note that if $\mu = 1_{\{0\}}$, then the previous Theorem is valid when L is chain.

Theorem 43 *Let $1_{\{0\}} \neq \mu \in L(M)$. Then μ is weakly prime if and only if it is satisfies the following conditions:*
(1) μ_* is a weakly prime submodule of M;
(2) $(\mu : 1_M)(1)$ is a prime element in L and

(3) $1_{\{0\}} \neq r_a \cdot x_b \subseteq \mu$, for $r \in R$, $x \in M$, and $a, b \in L$, then either $r_a \subseteq (\mu : 1_M)$ or $x_b \subseteq \mu$.

Proof First suppose that μ is a weakly prime L-submodule of M. By Theorem 42, $\mu = 1_{\mu_*} \cup c_M$ which μ_* is a weakly prime submodule of M and c is a prime elemente in L. So μ satisfies (1) and (2). Now suppose that there exist $r \in R$, $x \in M$, and $a, b \in L$ such that $1_{\{0\}} \neq r_a \cdot x_b \subseteq \mu$, but $r_a \not\subseteq (\mu : 1_M) = 1_{\mu_*:M} \cup c_R$ and $x_b \not\subseteq \mu$. Then $r \notin (\mu_* : M)$ and $x \notin \mu_*$. Since μ_* is weakly prime, the same argument as above follows that $0 \neq rx$ and $rx \notin \mu_*$. Thus on the one hand, since $1_{\{0\}} \neq r_a \cdot x_b \subseteq \mu$ and $0 \neq rx \notin \mu_*$, we have $a \wedge b \leq \mu(rx) = c$, and on the other hand, since $r_a \not\subseteq (\mu : 1_M)$ and $x_b \not\subseteq \mu$, we have $a \not\leq (\mu : 1_M)(r) = c$ and $b \not\leq \mu(x) = c$. This contradicts condition (2). Hence μ must satisfies condition (3).

Conversely suppose that μ satisfies the conditions (1), (2) and (3). Let $c = (\mu : 1_M)(1)$ be prime and consider $a = 1$. Let $x \in M \setminus \mu_*$ and $\mu(x) = b$. Then $1_{\{0\}} \neq 1_b \cdot x_a = x_{a \wedge b} \subseteq \mu$. Clearly $x_a \not\subseteq \mu$. Thus it follows from (3) that $1_b \subseteq (\mu : 1_M)$ and so

$$b = 1_b(1) \leq (\mu : 1_M)(1) = c.$$

Hence

$$
\begin{aligned}
c &= \bigvee \{\eta(1) \mid \eta \in LI(R), \eta \cdot 1_M \subseteq \mu\} \\
&= \bigvee \{\eta(1) \wedge 1_M(x) \mid \eta \in LI(R), \eta \cdot 1_M \subseteq \mu\} \\
&\leq \bigvee \{(\eta \cdot 1_M)(x) \mid \eta \in LI(R), \eta \cdot 1_M \subseteq \mu\} \leq \mu(x) = b.
\end{aligned}
$$

Therefore $\mu(x) = (\mu : 1_M)(1) = c$, and so $\mu = 1_{\mu_*} \cup c_M$. Then μ is weakly prime by Theorem 42. $\square$

Corollary 44 *Let μ be a weakly prime L-submodule of M that is not prime. Then*

$$(\mu : 1_M)\mu = 1_{\{0\}} \cup c_M.$$

Proof $(\mu : 1_M)\mu = (1_{(\mu_*:M)} \cup c_M)(1_{\mu_*} \cup c_M) = (1_{(\mu_*:M)M}) \cup c_M = 1_{\{0\}} \cup c_M$ because $(\mu_* : M)M = 0$ by Corollary 2.3 of Atani and Farzalipoor (2007). $\square$

Theorem 45 *Let $\mu \in L(M)$ be weakly prime. Then for each $\xi \in L(M)$ with the proviso that $\xi \not\subseteq \mu$, we have*

$$(\mu : \xi) = (\mu : 1_M) \cup (1_{\{0\}} : \xi).$$

Proof Put $\eta = (\mu : 1_M) \cup (1_{\{0\}} : \xi)$. Clearly $\eta \subseteq (\mu : \xi)$. Let $\zeta \subseteq (\mu : \xi)$. Then $\zeta\xi \subseteq \mu$. If $\zeta\xi \neq 1_{\{0\}}$, then $\zeta \subseteq (\mu : 1_M)$, so $\zeta \subseteq \eta$. If $\zeta\xi = 1_{\{0\}}$, then $\zeta \subseteq (1_{\{0\}} : \xi)$, so $\zeta \subseteq \eta$, and hence we have the equality. $\square$

Lemma 46 *Let M be an L-multiplication module and μ a prime L-submodule of M. Then for L-submodules ν and ξ of M with $\nu\xi \subseteq \mu$, either $\nu \subseteq \mu$ or $\xi \subseteq \mu$.*

Proof Let $\mu = 1_{\mu_*} \cup c_M$ be prime and $\nu\xi \subseteq \mu$, but $\nu \not\subseteq \mu$ and $\xi \not\subseteq \mu$ for L-submodules ν and ξ of M. Suppose that ζ_1 and ζ_2 are presentations L-ideals of ν and ξ, respectively. Then $\xi\nu = \zeta_1\zeta_2 1_M \subseteq \mu$. Thus, there are $x, y \in M$ such that $\nu(x) \not\subseteq \mu(x)$ and $\xi(y) \not\subseteq \mu(y)$. So there are $r, s \in R$ such that

$$a = \zeta_1(r) \not\subseteq c, \text{ for some } u \in M \text{ and } x = ru$$

$$b = \zeta_2(s) \not\subseteq c \text{ for some } v \in M \text{ and } y = sv.$$

But $r_a \cdot s_b \cdot 1_M \subseteq \zeta_1\zeta_2 1_M \subseteq \mu$, $s_b \cdot 1_M \not\subseteq \mu$ and $r_a \not\subseteq (\mu : 1_M)$, which is a contradiction. $\qquad\square$

Theorem 47 *Let M be an L-multiplication module and μ an L-submodule of M. Then μ is weakly prime if and only if for L-submodules ξ and ν of M with $1_{\{0\}} \neq \xi\nu \subseteq \mu$, either $\xi \subseteq \mu$ or $\nu \subseteq \mu$.*

Proof If μ is prime, then the result is clear. So we assume that μ is not prime. Suppose that ν and ξ are L-submodules of M with $1_{\{0\}} \neq \xi\nu \subseteq \mu$, but $\nu \not\subseteq \mu$. Let ζ_1 and ζ_2 be the presentation L-ideals for ξ and ν, respectively. Then $\zeta_1 \cdot \zeta_2 \cdot 1_M \subseteq \mu$. So $\zeta_1 \subseteq (\mu : 1_M)$, therefore $\xi \subseteq \mu$. Conversely, assume that $1_{\{0\}} \neq \xi\nu \subseteq \mu$ implies that $\xi \subseteq \mu$ or $\nu \subseteq \mu$ for $\xi, \nu \in L(M)$. Let $\zeta \in LI(R)$, $\nu \in L(M)$ such that $1_{\{0\}} \neq \zeta \cdot \nu \subseteq \mu$. Set $\xi = \zeta \cdot 1_M$. Then $\xi \cdot \nu = \zeta \cdot (\nu : 1_M)1_M = \zeta \cdot \nu \subseteq \mu$. Thus $1_{\{0\}} \neq \xi \cdot \nu \subseteq \mu$ and hence $\nu \subseteq \mu$ or $\xi \subseteq \mu$. Therefore $\zeta \subseteq (\mu : 1_M)$ or $\nu \subseteq \mu$. $\qquad\square$

Remark 48 We know that if μ is a prime L-submodule of an R-module M, then $(\mu : 1_M)$ is a prime L-ideal of R. Suppose that μ is a weakly prime which is not prime. Contrary to what happens for a prime L-submodule, the L-ideal $(\mu : 1_M)$ is not in general a weakly prime L-ideal of R. For example, let M denotes the cyclic $\mathbb{Z}$-module $\mathbb{Z}/8\mathbb{Z}$ and L be a chain. Take $\mu = 1_{\{0\}}$. Certainly μ is a weakly prime L-submodule of M, but $(\mu : 1_M) = 1_{(0:M)} = 1_{8\mathbb{Z}}$ is not a weakly prime L-ideal of R. But we have the following results:

Theorem 49 *Let R be an integral domain, M a faithful cyclic R-module, L a chain, and μ a weakly prime L-submodule of M. Then $(\mu : 1_M)$ is a weakly prime L-ideal of R.*

Proof We know $(\mu : 1_M) = 1_{(\mu_*:M)} \cup c_R$. By Proposition 2.1 of Atani and Farzalipoor (2007), $(\mu_* : M)$ is a weakly prime ideal of R. Then by Theorem 3.8 of Mahjoob (2016) $(\mu : 1_M)$ is a weakly prime L-ideal of R. $\qquad\square$

Theorem 50 *Let R be an integral domain, M a P-prime R-module, and μ a weakly prime L-submodule of M. Then $(\mu : 1_M)$ is a weakly prime L- ideal of R.*

Proof It is obvious by Proposition 2.2 Atani and Farzalipoor (2007) and Theorem 3.8 of Mahjoob (2016). $\qquad\square$

In the following we investigate the behavior of weakly prime L- submodules under homomorphisms.

Theorem 51 *Let M, N be R-modules and let f be an epimorphism from M into N. Then the following statements are satisfied:*
(i) If μ is a weakly prime f-invariant L-submodule of M, then $f(\mu)$ is a weakly prime L-submodule of N.
(ii) If ν is a weakly prime L-submodule of N, then $f^{-1}(\nu)$ is a weakly prime L-submodule of M.

Proof (i) Let μ be a weakly prime L-submodule of M and f-invariant. By Theorem 42, we have $\mu = 1_{\mu_*} \cup c_M$, where μ_* is a weakly prime submodule of M and c is a prime element in L. It is easy to see that

$$f(\mu) = 1_{f(\mu_*)} \cup c_N.$$

We now show that $f(\mu_*)$ is a weakly prime L-submodule of N. Let $r \in R$, $y \in N$ and

$$0 \neq ry = rf(x) \in f(\mu_*) \text{ for some } x \in M.$$

Then $0 \neq f(rx) = rf(x) \in f(\mu_*)$. Since μ is f-invariant, it follows that $0 \neq rx \in \mu_*$. Also since μ_* is weakly prime, then either $x \in \mu_*$ or $r \in (\mu_* : M)$. Hence either $y = f(x) \in f(\mu_*)$ or $rN = rf(M) = f(rM) \in f(\mu_*)$. Thus $f(\mu_*)$ is a weakly prime submodule of N. Therefore by Theorem 42, $f(\mu)$ is a weakly prime L-submodule of M.

(ii) Let ν be a weakly prime L-submodule of N. By Theorem 42, we have $\nu = 1_{\nu_*} \cup c_N$, where ν_* is a weakly prime submodule of N and c is a prime element in L. it is easy to see that

$$f^{-1}(\nu) = 1_{f^{-1}(\nu_*)} \cup c_M.$$

Now we show that $f^{-1}(\nu_*)$ is a weakly prime submodule of M. Let $r \in R$, $x \in M$ and

$$0 \neq rx \in f^{-1}(\nu_*).$$

Then $0 \neq rf(x) = f(rx) \in \nu_*$. Since ν_* is a weakly prime submodule of N, it follows that either $f(x) \in \nu_*$ or $rN \subseteq \nu_*$. Hence $x \in f^{-1}(\nu_*)$ or $rN = rf(M) = f(rM) \subseteq \nu_*$ and so $rM \subseteq \nu_*$. Thus $f^{-1}(\nu_*)$ is a weakly prime submodule of M. Therefore $f^{-1}(\nu)$ is a weakly prime L-submodule of M. $\qquad\square$

It is easy to check that every weakly prime L-submodule μ of an R-module M is classical prime provided that $\mu \neq 1_{\{0\}}$. $1_{\{0\}}$ is weakly prime L-submodule. But $1_{\{0\}}$ is not classical prime L-submodule, because 0 is not classical prime submodule in general.

References

Al Tahan, M., Hoskova-Mayerova, S., Davvaz, B.: An overview of topological hypergroupoids. J. Intell. Fuzzy Syst. **34**(3), 1907–1916 (2018)

Ameri, R., Mahjoob, R.: Spectrum of prime L-submodules. Fuzzy Sets Syst. **159**, 1107–1115 (2008)

Ameri, R., Mahjoob, R.: Some topological properties of spectrum of fuzzy submodules. Iran. J. Fuzzy Syst. **14**(1), 77–87 (2017)

Ameri, R., Kordi, A., Hoskova-Mayerova, S.: Multiplicative hyperring of fractions and coprime hyperideals. An. St. Univ. Ovidius Constanta **25**(1), 5–23 (2017)

Ameri, R., Mahjoob, R., Motameni, M.: The Zariski topology on the spectrum of prime L-submodules. Soft Comput. **12**(9), 901–908 (2008)

Atani, S.E., Farzalipoor, F.: On weakly prime submodules. Tamkang J. Math. **38**(3), 247–252 (2007)

Azizi, A.: On prime and weakly prime submodules. Vietnam J. Math. **36**(3), 315–325 (2008)

Behboodi, M., Koohy, H.: Weakly prime modules. Vietnam J. Math. **32**(2), 185–195 (2004)

Behboodi, M., Noori, M.J.: Zarisky-like topology on the classical prime spectrumof a modules. Bull. Iran. Math. Soc. **35**(1), 255–271 (2009)

Hochster, M.: Prime ideal structure in commutative rings. Trans. Am. Math. Soc. **137**, 43–60 (1969)

Hoskova-Mayerova, S.: An overview of topological and fuzzy topological hypergroupoids. Ratio Math. **33**, 21–38 (2017)

Hoskova-Mayerova, S., Maturo, A.: Fuzzy sets and algebraic hyperoperations to model interpersonal relations. In: Maturo, A., Hoskova-Mayerova, S., Soitu, D.T., Kacprzyk, J. (eds.) Recent Trends in Social Systems: Quantitative Theories and Quantitative Models. Studies in Systems, Decision and Control, vol. 66, pp. 211–221. Springer, Cham (2017). https://doi.org/10.1007/978-3-319-40585-8_19

Hoskova-Mayerova, S., Maturo, A.: Algebraic hyperstructures and social relations. Ital. J. Pure Appl. Math. **39**, 701–709 (2018)

Lee, D.S., Park, ChH: On fuzzy prime submodules of fuzzy multiplication modules. East Asian Math. J. **27**(1), 75–82 (2011)

Lu, C.: The Zariski topology on the spectrum of a modules. Houst. J. Math. **25**(3), 417–432 (1999)

McCasland, R.: Some commutative ring results generalized to unitary modules, Ph.D. thesis, University of Texas at Arlington (1983)

McCasland, R., Moore, M., Smith, P.: On the spectrum of modules over a commutative ring. Commun. Algebr. **25**(1), 79–103 (1997)

Mahjoob, R.: On weakly prime L-ideal. Ital. J. Pure Appl. Math. **36**, 465–472 (2016)

Mahjoob, R., Qiami, Sh.: Radical formula for L-submodules (2018). (submitted to Iranian journal of science and technology, Transaction A)

Maturo, F., Hoskova-Mayerova, S.: Fuzzy regression models and alternative operations for economic and social sciences. In: Maturo, A., Hoskova-Mayerova, S., Soitu, D.T., Kacprzyk, J. (eds.) Recent Trends in Social Systems: Quantitative Theories and Quantitative Models. Studies in Systems, Decision and Control, vol. 66, pp. 235–247. Springer, Cham, 2017. https://doi.org/10.1007/978-3-319-40585-8_21

Mordeson, J.N., Malik, D.S.: Fuzzy Commutative Algebra. World Scientific Publishing, Singapore (1998)

Negoita, C.V., Ralescu, D.A.: Application of Fuzzy Systems Analysis. Birkhauser, Basel (1975)

Pan, F.Z.: Fuzzy finitely generated modules. Fuzzy Sets Syst. **21**, 105–113 (1987)

Rosenfeld, R.: Fuzzy groups. J. Math. Anal Appl. **35**, 512–517 (1971)

Sidky, F.I.: On radical of fuzzy submodules and primary fuzzy submodules. Fuzzy Sets Syst. **119**, 419–425 (2001)

Zadeh, L.A.: Fuzzy sets. Inf. Control **8**, 338–353 (1965)